Industrial Chemistry

Industrial Chemistry

Editor:

Asad Rahman

Delve Publishing

www.delvepublishing.com

Industrial Chemistry

Editor: Asad Rahman

708 3rd Avenue, 6th Floor
New York, NY 10017
United States of America
www.delvepublishing.com

ISBN: 978-1-68095-893-5

Library of Congress Control Number: 2016951754

About the Editor

Asad Rahman

Asad Rahman obtained MSc Chemical Engineering from University of Nottingham in 2012. Currently He is working towards PhD in Energetic Material Synthesis and its Performance Analysis in Micro Thrusters. His research interests are in Ultrasound based particle treatment, Energetic material synthesis and Micro thruster applications.

Contents

List of Contributors

Danyelle Khadydja F. Santos
Northeast Botechnology Network (RENORBIO), Federal Rural University of Pernambuco, Rua Dom Manoel de Medeiros, s/n, Dois Irmãos, 52171-900 Recife-PE, Brazil
Center of Sciences and Technology, Catholic University of Pernambuco (UNICAP), Rua do Príncipe, 526, Boa Vista, 50050-900 Recife-PE, Brazil

Raquel D. Rufino
Center of Sciences and Technology, Catholic University of Pernambuco (UNICAP), Rua do Príncipe, 526, Boa Vista, 50050-900 Recife-PE, Brazil
Advanced Institute of Technology and Innovation (IATI), Rua Carlos Porto Carreiro, 70, Derby, 50070-090 Recife-PE, Brazil

Juliana M. Luna
Center of Sciences and Technology, Catholic University of Pernambuco (UNICAP), Rua do Príncipe, 526, Boa Vista, 50050-900 Recife-PE, Brazil
Advanced Institute of Technology and Innovation (IATI), Rua Carlos Porto Carreiro, 70, Derby, 50070-090 Recife-PE, Brazil

Valdemir A. Santos
Center of Sciences and Technology, Catholic University of Pernambuco (UNICAP), Rua do Príncipe, 526, Boa Vista, 50050-900 Recife-PE, Brazil
Advanced Institute of Technology and Innovation (IATI), Rua Carlos Porto Carreiro, 70, Derby, 50070-090 Recife-PE, Brazil

Leonie A. Sarubbo
Northeast Botechnology Network (RENORBIO), Federal Rural University of Pernambuco, Rua Dom Manoel de Medeiros, s/n, Dois Irmãos, 52171-900 Recife-PE, Brazil
Center of Sciences and Technology, Catholic University of Pernambuco (UNICAP), Rua do Príncipe, 526, Boa Vista, 50050-900 Recife-PE, Brazil
Advanced Institute of Technology and Innovation (IATI), Rua Carlos Porto Carreiro, 70, Derby, 50070-090 Recife-PE, Brazil

Gcina Mamba
Nanotechnology and Water Sustainability Research Unit, College of Science, Engineering and Technology, University of South Africa, Florida 1709, Johannesburg, South Africa

Ajay Mishra
Nanotechnology and Water Sustainability Research Unit, College of Science, Engineering and Technology, University of South Africa, Florida 1709, Johannesburg, South Africa

B. Gabai
Dept Chemical Engineering, Universidade Federal do Ceará, Brazil

N.A.A. dos Santos
Dept Chemical Engineering, Universidade Federal do Ceará, Brazil

D.C.S. Azevêdo
Dept Chemical Engineering, Universidade Federal do Ceará, Brazil

S. Brandani
Department of Chemical Engineering - 67100, L'Aquila, Italy

C.L. Cavalcante Jr.
Dept Chemical Engineering, Universidade Federal do Ceará, Brazil

Giovanni Neri
Department of Electronic Engineering, Chemistry and Industrial Engineering, University of Messina, Contrada di Dio, Vill. S. Agata, 98166 Messina, Italy

Philip Van den Heede
Magnel Laboratory for Concrete Research, Ghent University, Technologiepark Zwijnaarde 904, Ghent B-9052, Belgium

Niels Ringoot
Magnel Laboratory for Concrete Research, Ghent University, Technologiepark Zwijnaarde 904, Ghent B-9052, Belgium

Arno Beirnaert
Magnel Laboratory for Concrete Research, Ghent University, Technologiepark Zwijnaarde 904, Ghent B-9052, Belgium

Andres Van Brecht
Indaver nv, Dijle 17a, Mechelen B-2800, Belgium

Erwin Van den Brande
Indaver nv, Dijle 17a, Mechelen B-2800, Belgium

Geert De Schutter
Magnel Laboratory for Concrete Research, Ghent University, Technologiepark Zwijnaarde 904, Ghent B-9052, Belgium

Nele De Belie
Magnel Laboratory for Concrete Research, Ghent University, Technologiepark Zwijnaarde 904, Ghent B-9052, Belgium

Xiaolan Song
School of Chemical Engineering and Environment, North University of China, Taiyuan, P.R. China

Yi Wang
School of Materials Science and Engineering, North University of China, Taiyuan, P.R. China

Dan Song
China Ordnance Institutes of Science and Technology, Beijing, P.R. China

Chongwei An
School of Chemical Engineering and Environment, North University of China, Taiyuan, P.R. China

Jingyu Wang
School of Chemical Engineering and Environment, North University of China, Taiyuan, P.R. China

Georg Zocher,
Interfaculty Institute of Biochemistry, Eberhard Karls University Tübingen, Tübingen, Germany

Joel B. Heim
Interfaculty Institute of Biochemistry, Eberhard Karls University Tübingen, Tübingen, Germany

Dominik A. Herbst
Interfaculty Institute of Biochemistry, Eberhard Karls University Tübingen, Tübingen, Germany

Thilo Stehle
Interfaculty Institute of Biochemistry, Eberhard Karls University Tübingen, Tübingen, Germany
Department of Pediatrics, Vanderbilt University School of Medicine, Nashville, Tennessee, United States of America

Orwah Saleh
Pharmaceutical Institute, Eberhard Karls University Tübingen, Tübingen, Germany

Lutz Heide
Pharmaceutical Institute, Eberhard Karls University Tübingen, Tübingen, Germany

Bhaskar Garg
Department of Chemistry, National Tsing Hua University, Hsinchu 30013, Taiwan

Tanuja Bisht
Department of Chemistry, Government Degree College, Champawat 262523, Uttarakhand, India

Yong-Chien Ling
Department of Chemistry, National Tsing Hua University, Hsinchu 30013, Taiwan

Els De Canck
Center for Ordered Materials, Organometallics & Catalysis (COMOC), Department of Inorganic and Physical

Inmaculada Dosuna-Rodríguez
Institute of Condensed Matter and Nanosciences (IMCN)—Division "MOlecules, Solids and reactiviTy—MOST", Université catholique de Louvain, Croix du Sud 2/L7.05.17, Louvain-la-Neuve B-1348, Belgium

Eric M. Gaigneaux
Institute of Condensed Matter and Nanosciences (IMCN)—Division "MOlecules, Solids and reactiviTy—MOST", Université catholique de Louvain, Croix du Sud 2/L7.05.17, Louvain-la-Neuve B-1348, Belgium

Pascal Van Der Voort
Center for Ordered Materials, Organometallics & Catalysis (COMOC), Department of Inorganic and Physical Chemistry, Ghent University, Krijgslaan 281, Building S3, Ghent B-9000, Belgium

Jacques C. Védrine
Laboratoire de Réactivité de Surface, UMR-CNRS 1197, Sorbonne Universités, Université P. & M. Curie, 4 Place Jussieu, 75252 Paris Cedex 05, France

Patrycja Wojciechowska
Faculty of Commodity Science, Poznan' University of Economics, Al. Niepodleglosci 10, 61-875 Poznan', Poland

Zenon Foltynowicz
Faculty of Commodity Science, Poznan' University of Economics, Al. Niepodleglosci 10, 61-875 Poznan', Poland

Marek Nowicki
Faculty of Technical Physics, Poznan' University of Technology, Ul. Nieszawska 13A, 60-965 Poznan', Poland
Received 27 June 2012; Accepted 5 November 2012

Paolo Zucca
Dipartimento di Scienze Biomediche, Università di Cagliari, Cittadella Universitaria, 09042, Monserrato, CA, Italy
Consorzio UNO (University of Oristano Consortium), Oristano, Italy

Antonio Rescigno
Dipartimento di Scienze Biomediche, Università di Cagliari, Cittadella Universitaria, 09042, Monserrato, CA, Italy
Consorzio UNO (University of Oristano Consortium), Oristano, Italy

Manuela Pintus
Dipartimento di Scienze Biomediche, Università di Cagliari, Cittadella Universitaria, 09042, Monserrato, CA, Italy
Consorzio UNO (University of Oristano Consortium), Oristano, Italy

Andrea C Rinaldi
Dipartimento di Scienze Biomediche, Università di Cagliari, Cittadella Universitaria, 09042, Monserrato, CA, Italy
Consorzio UNO (University of Oristano Consortium), Oristano, Italy

Enrico Sanjust
Dipartimento di Scienze Biomediche, Università di Cagliari, Cittadella Universitaria, 09042, Monserrato, CA, Italy
Consorzio UNO (University of Oristano Consortium), Oristano, Italy

Mireia Sala
Institut d'Investigació Tèxtil i Cooperació Industrial de Terrassa (INTEXTER), Universitat Politècnica de Catalunya-BarcelonaTech (UPC), C/Colom 15, 08222 Terrassa, Spain

Víctor López-Grimau
Institut d'Investigació Tèxtil i Cooperació Industrial de Terrassa (INTEXTER), Universitat Politècnica de Catalunya-BarcelonaTech (UPC), C/Colom 15, 08222 Terrassa, Spain

Carmen Gutiérrez-Bouzán
Institut d'Investigació Tèxtil i Cooperació Industrial de Terrassa (INTEXTER), Universitat Politècnica de Catalunya-BarcelonaTech (UPC), C/Colom 15, 08222 Terrassa, Spain

Antonio Botta
Department of Chemistry and Biology, University of Salerno, Via Giovanni Paolo II 132, 84084 Fisciano, Italy

Vincenzo Venditto
Department of Chemistry and Biology, University of Salerno, Via Giovanni Paolo II 132, 84084 Fisciano, Italy

Alfredo Rubino
Department of Industrial Engineering (DIIN), University of Salerno, Via Giovanni Paolo II 132, 84084 Fisciano, Italy

Stefania Pragliola
Department of Chemistry and Biology, University of Salerno, Via Giovanni Paolo II 132, 84084 Fisciano, Italy

Mahesh K. Potdar
Centre for Green Chemistry, School of Chemistry, Monash University, Melbourne, Victoria 3800, Australia

Geoffrey F. Kelso
Centre for Green Chemistry, School of Chemistry, Monash University, Melbourne, Victoria 3800, Australia

Lachlan Schwarz
Centre for Green Chemistry, School of Chemistry, Monash University, Melbourne, Victoria 3800, Australia

Chunfang Zhang
Centre for Green Chemistry, School of Chemistry, Monash University, Melbourne, Victoria 3800, Australia

Milton T. W. Hearn
Centre for Green Chemistry, School of Chemistry, Monash University, Melbourne, Victoria 3800, Australia

Hideki Amii
Division of Molecular Science, Graduate School of Science and Technology, Gunma University, Tenjin-cho, Kiryu, Gunma 376-8515, Japan

Aiichiro Nagaki
Department of Synthetic and Biological Chemistry, Graduate School of Engineering, Kyoto University, Nishikyo-ku, Kyoto 615-8510, Japan

Jun-ichi Yoshida
Department of Synthetic and Biological Chemistry, Graduate School of Engineering, Kyoto University, Nishikyo-ku, Kyoto 615-8510, Japan

Ahmed M. El-Sayed
Department of Chemistry, Faculty of Science, Sohag University, Sohag 82524, Egypt

Omyma A. Abd Allah
Department of Chemistry, Faculty of Science, Sohag University, Sohag 82524, Egypt

Ahmed M. M. El-Saghier
Department of Chemistry, Faculty of Science, Sohag University, Sohag 82524, Egypt

Shaaban K. Mohamed
Chemistry and Environmental Division, Manchester Metropolitan University, Manchester M1 5GD, UK
Chemistry Department, Faculty of Science, Minia University, El-Minia 61519, Egypt

Charalambos M. Andreou
Department of Electrical and Computer Engineering, University of Cyprus, Nicosia 1678, Cyprus

Yiannis Pahitas
Department of Electrical and Computer Engineering, University of Cyprus, Nicosia 1678, Cyprus

Julius Georgiou
Department of Electrical and Computer Engineering, University of Cyprus, Nicosia 1678, Cyprus

Laura Benassi
INSTM and Chemistry for Technologies Laboratory, University of Brescia, via Branze 38, Brescia 25123, Italy

Federica Franchi
INSTM and Chemistry for Technologies Laboratory, University of Brescia, via Branze 38, Brescia 25123, Italy

Daniele Catina
INSTM and Chemistry for Technologies Laboratory, University of Brescia, via Branze 38, Brescia 25123, Italy

Flavio Cioffi
Contento Trade srl, Via Vieris 11, Terenzano (UD) 33050, Italy

Nicola Rodella
INSTM and Chemistry for Technologies Laboratory, University of Brescia, via Branze 38, Brescia 25123, Italy

Laura Borgese
INSTM and Chemistry for Technologies Laboratory, University of Brescia, via Branze 38, Brescia 25123, Italy

Michela Pasquali
INSTM and Chemistry for Technologies Laboratory, University of Brescia, via Branze 38, Brescia 25123, Italy

Laura E. Depero
INSTM and Chemistry for Technologies Laboratory, University of Brescia, via Branze 38, Brescia 25123, Italy

Elza Bontempi
INSTM and Chemistry for Technologies Laboratory, University of Brescia, via Branze 38, Brescia 25123, Italy

Preface

Industrial Chemistry is an extremely diverse field, and it covers almost every single aspect of Chemical industry ranging from Petrochemicals to Pharmaceuticals and biochemical. In simple terms, Industrial Chemistry is the extension of chemistry to chemical engineering and its application at industrial scale. Industrial chemistry not only focuses on the general concepts of chemistry but economics, bulk production, electronics and mechanics.

Recent advances in industry chemistry not only reduce the cost of production, but also provide them at sustainable rate with low environmental hazards. The advances in catalysis increases the reaction yield hence reducing the operating cost, low fuel cost, and mass production. The advances in green technologies produce the already existing products with low environmental effects such as acid recycling. The advances in waste water management and processing reduces the risk to fresh water reservoirs. The single most important breakthrough of industrial chemistry was development of Haber process for synthesis of Ammonia. The Haber process allowed large-scale synthesis of Ammonia gas at reasonable cost, which is an important raw material for several other chemical products such as Fertilizers and Explosives. The use of Ammonia in production of Fertilizer called Ammonium nitrate along with use of pesticides is attributed to increase world production by factor of 4 since 1900 on per acre basis. Therefore, in this book, several articles covering different application areas of industrial chemistry are presented to provide an overview of the importance of industry chemistry.

Editor
Asad Rahman

INTRODUCTION

Industrial chemistry is concerned with using chemical and physical processes to transform raw materials into products that are beneficial to humanity. This includes the manufacture of basic chemicals to produce products for various industries. Industrial chemistry can be thought of as an industry that generates synthetic replacements for natural products. Industrial chemists make use of their broad understanding of chemistry and environmental sustainability at pharmaceutical companies, in polymer manufacturing, petrochemical processing, in the manufacturing industries and in food science. Yet they can also be found in the most unexpected places; like challenging the norm in cutting edge research.

Industrial chemistry is characterized by the very broad nature of the subject, spanning as it does several different disciplines. Apart from chemistry it includes topics such as organization and management of a company, technical economics, chemical engineering and environmental pollution control, and it would not be complete without an in-depth study of several particular sectors of the chemical industry. The latter would be selected as a representative cross-section of the entire industry. Clearly a comprehensive treatment of all these topics would hardly be possible even in a full degree in industrial chemistry, let alone as an option or part of a chemistry degree course.

HISTORY

Chemical industries can be traced back to Middle Eastern artisans, who refined alkali and limestone for the production of glass as early as 7,000 B.C., to the Phoenicians who produced soap in the 6th cent. B.C., and to the Chinese who developed black powder, a primitive explosive around the 10th cent. A.D. In the Middle Ages, alchemists produced small amounts of chemicals and by 1635 the Pilgrims in Massachusetts were producing saltpeter for gunpowder and chemicals for tanning. But, large-scale chemical industries first developed in 19th cent. In 1823, British entrepreneur James Muspratt started mass producing soda ash (needed for soap and glass) using a process developed by Nicolas Leblanc in 1790. Advances in organic chemistry in the last half of the 19th cent. Allowed companies to produce synthetic dyes from coal tar for the textile industry as early as the 1850s.

In the 1890s, German companies began mass producing sulfuric acid and, at about the same time, chemical companies began using the electrolytic method, which required large amounts of electricity and salt, to create caustic soda and chlorine. Man-made fibers changed the textile industry when rayon (made from wood fibers) was introduced in 1914; the introduction of synthetic fertilizers by the American Cyanamid Company in 1909 led to a green revolution in agriculture that dramatically improved crop yields. Advances in the manufacture of plastics led to the invention of celluloid in 1869 and the creation of such products as nylon by Du Pont in 1928. Research in organic chemistry in the 1910s allowed companies in the 1920s and 30s to begin producing chemicals for oil. Today, petrochemicals made from oil are the industry's largest sector. Synthetic rubber came into existence during World War II, when the war cut off supplies of rubber from Asia.

Since the 1950s growing concern about toxic waste produced by chemical industries has led to increased government regulation and the establishment of the Environmental Protection Agency (1972). The leakage of toxic chemicals at the Union Carbide plant in Bhopal, India (1984), was the worst industrial disaster in history and heightened public concern about lax environmental regulations for chemical companies in developing countries. Beginning in the 1980s, U.S. corporations faced expanding competition from foreign producers, including some Third World oil producers who have set up their own oil refining and petrochemical industries. In 1997 the U.S. chemical industry produced about $389 billion worth of products and employed 1,032,000 workers. It exported about $71 billion worth of chemicals.

Products

Polymers and plastics, especially polyethylene, polypropylene, polyvinyl chloride, polyethylene terephthalate, polystyrene and polycarbonate comprise about 80% of the industry's output worldwide.[citation needed] These materials are often converted to fluoropolymer tubing products and used by the industry to transport highly corrosive materials. Chemicals are used to make a wide variety of consumer goods, as well as thousands of inputs to agriculture, manufacturing, construction, and service industries. The chemical industry itself consumes 26 percent of its own output. Major industrial customers include rubber and plastic products, textiles, apparel, petroleum refining, pulp and paper, and primary metals. Chemicals are nearly a $3 trillion global enterprise, and the EU and U.S. chemical companies are the world's largest producers.

What does the chemical industry produce?

The products of the chemical industry can be divided into three categories:

- Basic chemicals
- Speciality chemicals
- Consumer chemicals

Basic Chemicals

Basic chemicals are divided into

- chemicals derived from oil, known as petrochemicals
- polymers
- basic inorganics

What are the challenges for the chemical industry today?

The chemical industry is undergoing huge changes worldwide. As we have seen above, one concerns the emergence of Middle Eastern countries and China, India and Brazil as manufacturers of chemicals on a mammoth scale, for their own consumption and also for export worldwide. Companies in these countries are also investing in plant in the US and Europe whilst US and European companies are investing in plant in these large emerging countries, making the industry as a whole totally international in the way it conducts business. The challenge for companies in the US and Europe is to cut their costs while ensuring that they conform to the best practice in protecting the environment. This concern about the environment is discussed in the separate units on individual chemicals.

A new revolution beckons. As oil and natural gas become ever scarcer and more expensive, chemists are searching for new feedstock's to supplement or even replace oil and natural gas. And they are rediscovering the virtues of coal (still in huge supply, even though it is a fossil fuel that cannot be replaced) and biomass.

Thus we are coming full circle. In the late 19th and the first part of the 20th centuries, the organic chemical industry was based largely on coal and biomass. Coal was heated strongly in the absence of air to form coal gas (a mixture of hydrogen, methane and carbon monoxide). A liquid (coal tar) was formed as a by-product which contained many useful organic chemicals, including benzene, and the solid residue was coke, an impure form of carbon. Coke was the source of what we now call synthesis gas. Steam was passed over it at high temperatures to yield carbon monoxide and hydrogen. Another source of organic chemicals was biomass. For example, the source of many C_2 chemicals was ethanol, produced by fermentation of biomass. C_3 and C_4chemicals such as propanone and butanol were also produced on a large scale by fermentation of biomass.

Importance of Chemical Industry

Chemical Industries are the prime factors to convert the raw materials into desired products that we use in our day-to-day life. This has brought a tremendous change in the way the things operate. It is very important for us to understand the importance of the chemical industry which has touched all our facets of life like Agriculture, Environment, Food, Hygiene, Décor, and Transportation. etc. It has also significantly used in re-cycling industries to curb the usage of virgin products. Re-cycling helps a lot in utilizing the waste materials, and gives one more life-cycle for the products.

- Chemicals play a major role in our food. The preservatives, taste enhancers and flavors helps the food to be palatable and increase the shelf life. Food Industry thrives for the reason, that the preservatives not only help them to maintain the quality of the food, but also helps them to import food to different parts of the world. Due to these advancements, we are able to enjoy fruits, canned food products, and ready-to-eat food products across the world.
- More than 80% of the chemical industry concentrates on producing polymers, and plastics. They are not only used in packing, but also in numerous other things, like wiring, furniture, clothing, home décor, prosthesis and electronics. PVC piping, water tanks, huge storage containers are made out of plastics.
- Fertilizers, and pesticides aids in the agriculture and development. The green revolution has happened only due to the advancement of chemical industry in India. The fertilizers and pesticides, not only increase the yield of the crop, but prevent from pest attacks. Apart from in-house usage of food products within our country, we are also exporting a lot of grains, fruits, flowers and ornamental stem to various parts of the world. The GDP of the country drastically increases with the industry.
- Pharma industries and lifesaving drugs are the fastest growing industry in India. Our country invites a lot of people for medical tour. Numerous laboratories are also set-up to study various drug for the endemic and epidemic diseases.

Before these laboratories in our country, we have been exporting a lot of chemicals, which was expensive and mostly unaffordable for a large sector of people.

- Toiletries like soaps, scents, and perfumes, deodorant are personnel products that we use every day, and we do not compromise on it. From the status of luxury products, it has come to a level of essentials. Other products like mosquito repelled, detergents, and cleaning agents have been tremendously increased.
- Advanced researches like bio-engineering, mutation, artificial human organ production and genetic-reengineering are made possible in India, only with the help of the proper chemical industries. India is taking a giant leap in research in these sectors.

Industrial Chemistry is part of applied chemistry that deals with the development, optimization and monitoring of fundamental chemical processes used in industry to produce chemicals and chemical products.

The main areas of research and teaching are on the catalyst and process development, mechanical and thermal unit operations and process of chemical reaction engineering. The Chemical Technology enables efficient production of basic, intermediate and end products. Between 1970 and 1980, the energy requirements for chemical reactions with the same production volume was reduced by about 40% through improvements of chemical processes.

The difference between classical and industrial chemistry

Before we define industrial chemistry, it may be helpful to know that the development of industrial chemistry started when a need to know how various chemicals are produced in much more than the laboratory scale, arose. Chemistry knowledge was applied to furnish the rapidly expanding chemical industries with ''recipes'' which we now call chemical processes. Industrial chemistry keeps up with the progress in science and technology. It incorporates other emerging disciplines such as biotechnology, microelectronics, and pharmacology and material science. The discipline is also concerned with economics and the need to protect the environment.

We define industrial chemistry as the branch of chemistry which applies physical and chemical procedures towards the transformation of natural raw materials and their derivatives to products that are of benefit to humanity. Classical chemistry (organic, inorganic and physical chemistry) is very essential for advancing the science of chemistry by discovering and reporting new products, routes and techniques. On the other hand industrial chemistry helps us to close the gap between classical chemistry as it is taught in colleges and universities, and chemistry as it is practiced commercially. The scope of industrial chemistry therefore includes:

- The exploitation of materials and energy in appropriate scale
- Application of science and technology to enable humanity experience the benefits of chemistry in areas such as food production, health and hygiene, shelter, protection, decoration, recreation and entertainment.

Classification of Industries

Industry is a general term that refers to all economic activities that deal with production of goods and services. Goods and services are key words when you talk of industry.

We then expect industry to include the following sectors:

- Manufacturing
- Agriculture
- Energy
- Transport
- Communication
- Education
- Tourism
- Building and construction
- Trade
- Finance
- etc

Classification of the Manufacturing Industry

The manufacturing industry is the area of focus in the study of this module. Manufacturing produces manufactured goods. This makes it distinct from other sectors like agriculture which also produce goods. In manufacturing, materials are transformed into other more valuable materials. We define manufacturing industry as follows: Manufacturing industry is a compartment of industry or economy which is concerned with the production or making of goods out of raw materials by means of a system of organized labor. Manufacturing industry can be classified into two major categories namely, heavy and light industry.

- Capital-intensive industries are classified as heavy while labor intensive industries are classified as light industries.
- Light industries are easier to relocate than heavy industries and require less capital investment to build.

Using the above classification criteria, examples of heavy industries include those that produce industrial machinery, vehicles and basic chemicals.

Other measures used to classify industries include the weight or volume of products handled and weight per cost of production. For example the weight of steel produced per dollar is more than the weight per dollar of a drug. In this case, steel industry is a heavy industry whereas drug manufacture is a light industry. Sometimes governments define heavy industry in terms of its impact on the environment. Many pollution control laws target heavy industries which in most cases pollute more than light industries. Therefore, pulp and paper industry is a heavy industry since its contribution to pollution is enormous. Both inorganic and organic chemical industry can be either heavy or light industry. For example the pharmaceutical industry which is basically organic is light industry. Petroleum refining is organic but heavy industry. Iron and steel industry is inorganic and heavy industry.

Manufacturing sub-sectors

Because the raw materials and the actual products manufactured are so varied, different skills and technologies are needed in manufacturing. Manufacturing is therefore divided into sub-sectors which typically deal with category of goods such as the following:

- Food, beverages and tobacco
- Textiles, wearing apparel, leather goods
- Paper products, printing and publishing
- Chemical, petroleum, rubber and plastic products
- Non-metallic mineral products other than petroleum products
- Basic metal products, machines and equipment.

The Chemical Industry

The chemical industry can also be classified according to the type of main raw materials used and/or type of principal products made. We therefore have industrial inorganic chemical industries and industrial organic chemical industries. Industrial inorganic chemical Industries extract inorganic chemical substances, make composites of the same and also synthesize inorganic chemicals. Heavy industrial organic chemical industries produce petroleum fuels, polymers, petrochemicals and other synthetic materials, mostly from petroleum. Light organic industries produce specialty chemicals which include pharmaceuticals, dyes, pigments and paints, pesticides, soaps and detergents, cosmetic products and miscellaneous products.

The Structure of the Global Chemical Industry

We normally put a value to something according to how much it has cost us. Some things are of high value while others are of low value. For low valued products, you need to produce them in large volumes to make significant profit. This means that the raw materials are cheap and easily accessible. There is also an existing, relatively simple, and easily accessible processing technology. To sell a large volume of product, there must be a large market. This brings stiff competition which also makes the price to remain low

Commodity Chemicals

The global chemical industry is founded on basic inorganic chemicals (BIC) and basic organic chemicals (BOC) and their intermediates. Because they are produced directly from natural resources or immediate derivatives of natural resources, they are produced in large quantities. In the top ten BIC, almost all the time, sulphuric acid, nitrogen, oxygen, ammonia, lime, sodium hydroxide, phosphoric acid and chlorine dominate. The reason sulphuric acid is always number one is because it is used in the manufacture of fertilizers, polymers, drugs, paints, detergents and paper. It is also used in petroleum refining, metallurgy and in many other processes. The top ranking of oxygen is to do with its use in the steel industry. Ethylene and propylene are usually among the top ten BOC. They are used in the production of many organic chemicals including polymers. BIC and BOC are referred to as commodity or industrial chemicals. Commodity chemicals are therefore defined as low-valued products produced in large quantities mostly in continuous processes. They are of technical or general purpose grade.

Specialty Chemicals

High-value adding involves the production of small quantities of chemical products for specific end uses. Such products are called specialty chemicals. These are high value-added products produced in low volumes and sold on the basis of a specific function. In this category are the so-called performance chemicals which are high value products produced in low volumes and used in extremely low quantities. They are judged by performance and efficiency. Enzymes and dyes are performance chemicals. Other examples of specialty chemicals include medicinal chemicals, agrochemicals, pigments, flavor and fragrances, personal care products, surfactants and adhesives. Specialty chemicals are mainly used in the form of formulations. Purity is of vital importance in their formulation. This calls for organic synthesis of highly valued pure chemicals known as fine chemicals.

Fine Chemicals

At times you will find that the raw materials for your product need to be very pure for the product to function as desired. Research chemicals are in this category as also are pharmaceutical ingredients. Such purified or refined chemicals are called fine chemicals. By definition they are high value-added pure organic chemical substances produced in relatively low volumes and sold on the basis of exact specifications of purity rather than functional characteristics

Raw material for the Chemical Industry

We have paid some attention to products from the chemical industry. But, since there would be no chemical industry without raw materials, the subject of raw materials is due for discussion at this stage. All chemicals are derived from raw materials available in nature. The price of chemicals depends on the availability of their raw materials. Major chemical industries have therefore developed around the most plentiful raw materials.

The natural environment is the source of raw materials for the chemical industry.

Raw materials from the atmosphere

The atmosphere is the field above ground level. It is the source of air from which six industrial gases namely N_2 , O_2 , N_e, Ar, Kr and Xe are manufactured. The mass of the earth's atmosphere is approximately 5x 1015 tons and therefore the supply of the gases is virtually unlimited.

Raw materials from the hydrosphere

Ocean water which amounts to about $1.5x\ 10^{21}$ litres contains about 3.5 percent by mass dissolved material. Seawater is a good source of sodium chloride, magnesium and bromine.

Raw materials from the lithosphere

The vast majority of elements are obtained from the earth's crust in the form of mineral ores, carbon and hydrocarbons. Coal, natural gas and crude petroleum besides being energy sources are also converted to thousands of chemicals.

Raw materials from the biosphere

Vegetation and animals contribute raw materials to the so-called agro-based industries. Oils, fats, waxes, resins, sugar, natural fibers and leather are examples of thousands of natural products.

Chemical Processes

Every industrial process is designed to produce a desired product from a variety of starting raw materials using energy through a succession of treatment steps integrated in a rational fashion. The treatments steps are either physical or chemical in nature

A Typical Industrial Process

Energy is an input to or output in chemical processes.

The layout of a chemical process indicates areas where:

- raw materials are pre-treated
- conversion takes place
- separation of products from by-products is carried out
- refining/purification of products takes place
- entry and exit points of services such as cooling water and steam

Units that make up a chemical process

A chemical process consists of a combination of chemical reactions such as synthesis, calcination, ion exchange, electrolysis, oxidation, hydration and operations based on physical phenomena such as evaporation, crystallization, distillation and extraction

A chemical process is therefore any single processing unit or a combination of processing units used for the conversion of raw materials through any combination of chemical and physical treatment changes into finished products.

Process flow diagram / flow sheet

Chemical plants are built from process flow drawings or flow sheets drawn by chemical engineers to communicate concepts and designs. Communication is impaired if the reader is not given clear and unmistakable picture of the design. Time is also wasted as reader questions or puzzles out the flow diagram. The reader may make serious mistakes based on erroneous interpretation of the flow diagram. Communication is improved if accepted symbols are used. The advantages of correct use of symbols include:

- the function being performed is emphasized by eliminating distractions caused by detail
- possibility of error that is likely to occur when detail is repeated many times is virtually done away with
- equipment symbols should neither dominate the drawing nor be too small for clear understanding

Flow sheet symbols are pictorial quick-to-draw, easy-to-understand symbols that transcend language barriers. Some have already been accepted as national standards while others are symbols commonly used in chemical process industries, which have been proven to be effective. Engineers are constantly devising their own symbols where standards do not exist. Therefore, symbols and presentation may vary from one designer or company to another.

Material Balances

From the law of conservation of mass, we know that mass can neither be created nor destroyed. However, in nuclear reactions, mass and energy can be converted into each other respectively. Because of this, we can write equations called "mass balances" or "material balances". Any process being studied must satisfy balances on the total amount of material, on each chemical component, and on individual atomic species.

CHAPTER 1

BIOSURFACTANTS: MULTIFUNCTIONAL BIOMOLECULES OF THE 21ST CENTURY

Danyelle Khadydja F. Santos, Raquel D. Rufino, Juliana M. Luna, Valdemir A. Santos and Leonie A. Sarubbo

Keywords: Biosurfactant, surface tension, critical micelle concentration, biodegradability, functional properties, physiology, kinetics, recovery, industrial applications

INTRODUCTION

Surfactants are amphipathic compounds with both hydrophilic and hydrophobic moieties that preferentially partition between liquid interfaces with different degrees of polarity and hydrogen bridges, such as oil/water or air/water interfaces. The apolar moiety is often a hydrocarbon chain, whereas the polar moiety may be ionic (cationic or anionic), non-ionic or amphoteric [1,2], as illustrated in Figure 1.

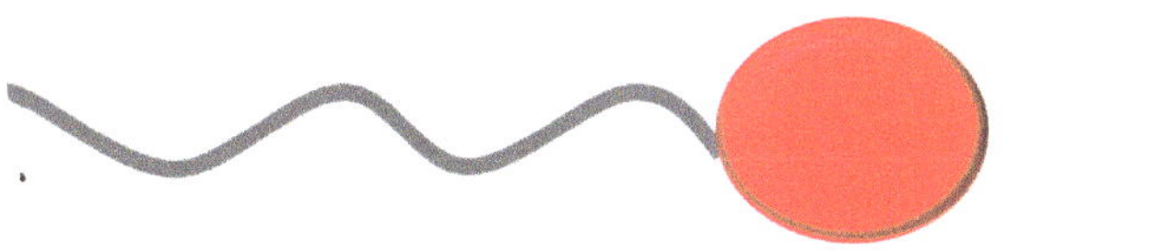

Figure 1: Surfactant molecule with apolar (hydrophobic) and polar (hydrophilic) moieties.

Surfactants increase the solubility of hydrophilic molecules, thereby reducing both surface and interfacial tensions at the oil/water interface [3]. The critical micelle concentration (CMC) is the concentration of surfactant at which organised molecular assemblies, known as micelles, are formed (Figure 2) and corresponds to the point at which the tensioactive agent achieves the lowest stable surface tension (Figure 3) [4].

Most currently produced surfactants are chemically derived from petroleum. However, such synthetic tensioactive agents are generally toxic and difficult to break down through

Citation: Danyelle Khadydja F. Santos, Raquel D. Rufino, Juliana M. Luna, Valdemir A. Santos , and Leonie A. Sarubbo, Biosurfactants: Multifunctional Biomolecules of the 21st Century, doi:10.3390/ijms17030401

the action of microorganisms. In recent years, such problems have motivated the scientific community to seek surfactants that are more environmentally friendly, such as those achieved through microbial production, known as biosurfactants [5]. Moreover, concerns regarding the environment on the part of consumers and new environmental control legislation have led to the development of natural surfactants as an alternative to existing products.

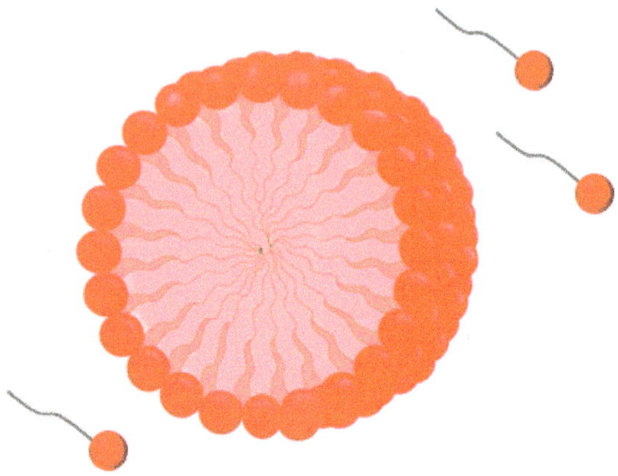

Figure 2: Schematic illustration of tensioactive agent and micelle formation.

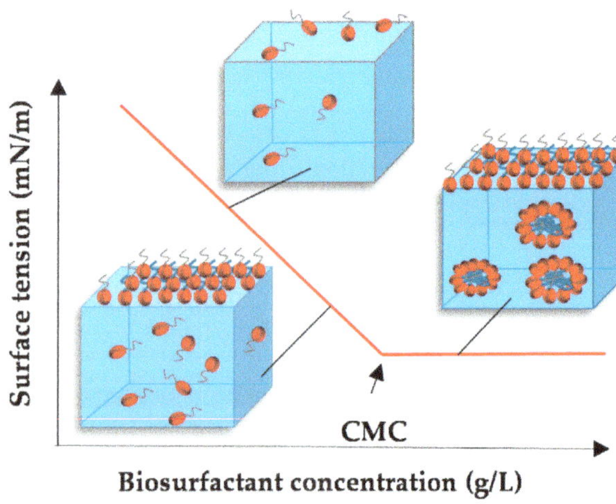

Figure 3: Illustration of regions in which micelle formation occurs (critical micelle concentration CMC).

Studies involving biosurfactants began in the 1960s and the use of these compounds has expanded in recent decades [2,6]. Biosurfactants have drawn the interest of different industries due to advantages such as structural diversity, low toxicity, greater biodegradability, ability to function in wide ranges of pH, temperature and salinity as well as greater selectivity, lower CMC and production involving renewable sources/industrial waste and industrial by-products [7,8,9]. The present review demonstrates the reasons for which biosurfactants are considered the multifunctional materials of the 21st century, with a description of concepts, properties, classification, modes of production, physiology and uses in the most diverse industries.

PRODUCING MICROORGANISMS

Microorganisms use a set of carbon sources and energy for growth. The combination of carbon sources with insoluble substrates facilitates the intracellular diffusion and production of different substances [10,11,12]. Microorganisms (yeasts, bacteria and some filamentous fungi) are capable of producing biosurfactants with different molecular structures and surface activities [4]. In recent decades, there has been an increase in scientific interest regarding the isolation of microorganisms that produce tensioactive molecules with good surfactant characteristics, such as a low CMC, low toxicity and high emulsifying activity [2].

The literature describes bacteria of the genera *Pseudomonas* and *Bacillus* as great biosurfactant producers [2]. However, most biosurfactants of a bacterial origin are inadequate for use in the food industry due to their possible pathogenic nature [13]. *Candida bombicola* and *Candida lipolytica* are among the most commonly studied yeasts for the production of biosurfactants. A key advantage of using yeasts, such as *Yarrowia lipolytica*, *Saccharomyces cerevisiae* and *Kluyveromyces lactis,* resides in their "generally regarded as safe" (GRAS) status. Organisms with GRAS status do not offer the risks of toxicity or pathogenicity, which allows their use in the food and pharmaceutical industries [4]. Table 1 displays a list of microorganisms that produce biosurfactants.

Table 1: Main classes of biosurfactants and respective producing microorganisms

Biosurfactant Class					
Glycolipids	**Polymeric Surfactants**	**Lipopeptides**	**Fatty Acids**	**Particulate Surfactant**	**Phospholipids**
Producer microorganisms					
Acinetobacter calcoaceticus *Alcanivorax borkumensis* *Arthrobacter paraffineus* *Arthrobacter* sp. *Candida antartica* *Candida apicola* *Candida batistae* *Candida bogoriensis* *Candida bombicola* *Candida ishiwadae* *Candida lipolytica* *Lactobacillus fermentum* *Nocardia* sp. *Pseudomonas aeruginosa* *Pseudomonas* sp. *Rhodococcus erythropolis* *Rhodotorula glutinus* *Rhodotorula graminus* *Serratia marcescens* *Tsukamurella* sp. *Ustilago maydis*	*Acinetobacter calcoaceticus* *Acinetobacter calcoaceticus* *Acinetobacter calcoaceticus* *Acinetobacter calcoaceticus* *Bacillus stearothermophilus* *Candida lipolytica* *Candida utilis* *Halomonas eurihalina* *Mycobacterium thermoautotrophium* *Sphingomonas paucimobilis*	*Acinetobacter* sp. *Bacillus licheniformis* *Bacillus pumilus* *Bacillus subtilis* *Candida lipolytica* *Gluconobacter cerinus* *Pseudomonas fluorescens* *Serratia marcescens* *Streptomyces sioyaensis* *Thiobacillus thiooxidans*	*Arthrobacter paraffineus* *Capnocytophaga* sp. *Corynebacterium insidibasseosum* *Corynebacterium lepus* *Nocardia erythropolis* *Penicillium spiculisporum* *Talaramyces trachyspermus*	*Acinetobacter calcoaceticus* *Cyanobacteria* *Pseudomonas marginalis*	*Acinetobacter* sp. *Aspergillus* *Corynebacterium lepus*

CLASSIFICATION

Most biosurfactants are either anionic or neutral, whereas those that contain amine groups are cationic. The hydrophobic moiety has long-chain fatty acids and the hydrophilic moiety can be a carbohydrate, cyclic peptide, amino acid, phosphate carboxyl acid or alcohol. The molar mass of biosurfactants generally ranges from 500 to 1500 Da [14]. Biosurfactants are generally categorised by their microbial origin and chemical composition, as follows [3,5,15].

Glycolipids

Rhamnolipids, sophorolipids and trehalolipids are the best known glycolipids [16]. Rhamnolipids were found as exoproducts of the pathogen *P. aeruginosa* and are a combination of α-l-rhamnopyranosyl-α-l-rhamnopyranosyl-β-hydroxydecanoyl-β-hydroxydecanoate (Rha-Rha-C10-C10) and α-l-rhamnopyranosyl-α-l-rhamnopyranosyl-β-hydroxydecanoate (Rha-Rha-C10) as well as their mono-rhamnolipid congeners (Rha-C10-C10 and Rha-C10) [17]. Sensitive analytical techniques have led to the discovery of rhamnolipid congeners and homologues (approximately 60) produced at different concentrations by species of *Pseudomonas* and bacteria belonging to other families, classes or even phyla [16]. For instance, various species of *Burkholderia* have been shown to produce rhamnolipids that have longer alkyl chains than those produced by *P. aeruginosa* [17,18,19]. Surface tensions values of 29 mN/m constitute a characteristic of such components, which can be produced using different substrates, such as alkanes, pyruvate, citrates, fructose, glycerol, olive oil and glucose [20]. Most studies involving rhamnolipids focus mainly on assessing the biodegradation efficiency of petroleum hydrocarbons [21,22]. Although researchers have found increased dissipation of target contaminant upon the addition of rhamnolipids, a decrease in biodegradation efficiency or no effect following rhamnolipid supplementation have also been reported [16,20]. The presence of surfactant molecules may induce changes in the microbial community, which, in turn, correspond to different degradation patterns. Interestingly, although rhamnolipids are considered biodegradable, few reports have demonstrated that these substances can be co-degraded or solely utilised as a carbon and energy source by various monocultures [18]. Rhamnolipids are described as potentially toxic to natural vegetation [23], but have also been found to reduce the toxicity of specific compounds by increasing hydrocarbon solubilisation, thereby facilitating biodegradation [24,25].

Sophorolipids are produced by yeasts that belong to the genus *Candida* [26,27]. These glycolipids have a dimeric carbohydrate sophorose linked to a long-chain hydroxyl fatty acid through a glycosidic bond. Sophorolipids and lactone form a sophorolipid that is preferable in many applications [28,29]. *C. bombicola* stands out among the different types of yeasts used in the production of these biosurfactants. Surface tension values of approximately 33 mN/m and a reduction in the surface tension of *n*-hexadecane and water from 40 to 5 mN/m has been recorded for these agents [30]. Mannosylerythritol lipids (MEL), which are yeast glycolipids, are one of the most promising biosurfactants known and are abundantly produced from vegetable oils by *Pseudozyma* (previously *Candida*) *antarctica* [31]. Trehalolipids are produced by species of *Mycobacterium*, *Nocardia* and *Corynebacterium*. Trehalolipids from *Arthrobacter* spp. and *Rhodococcus erythropolis* are able to lower surface and interfacial tensions in culture broth to 25–40 and 1–5 mN/m, respectively [5].

Fatty Acids, Phospholipids and Neutral Lipids

Different bacteria and yeasts produce large amounts of fatty acids and phospholipid surfactants during growth on n-alkanes. Phosphatidyl ethanolamine-rich vesicles are produced from *Acinetobacter* spp. and form optically clear microemulsions of alkanes in water. These biosurfactants are essential to medical applications. According to Gautam and Tyagi [28], phospholipid protein complex deficiency is the major cause of respiratory failure in the children born prematurely. The authors also suggest that the isolation and cloning of genes involved in the production of surfactants can be used in fermentative processes [28].

Polymeric Biosurfactants

Emulsan, lipomanan, alasan, liposan and other polysaccharide protein complexes are the best-studied polymeric biosurfactants. Emulsan is an emulsifier for hydrocarbons in water at concentrations as low as 0.001% to 0.01% [31,32]. Liposan is an extracellular water soluble emulsifier synthesised by *C. lipolytica* and is made up of 83% carbohydrates and 17% proteins. Chakrabarti [33] discuss the application of liposan as an emulsifier in the food and cosmetic industries.

Particulate Biosurfactants

Particulate biosurfactants partition extracellular membrane vesicles to form a microemulsion that exerts an influence on alkane uptake in microbial cells. The *Acinetobacter* spp. has vesicles with a diameter of 20 to 50 nm and a buoyant density of 1.158 cubic gcm composed of proteins, phospholipids and lipo-polysaccharides [5,33].

Figure 4 illustrates the chemical structure of the most studied microbial surfactants. Table 1 also displays biosurfactant classes and their producers.

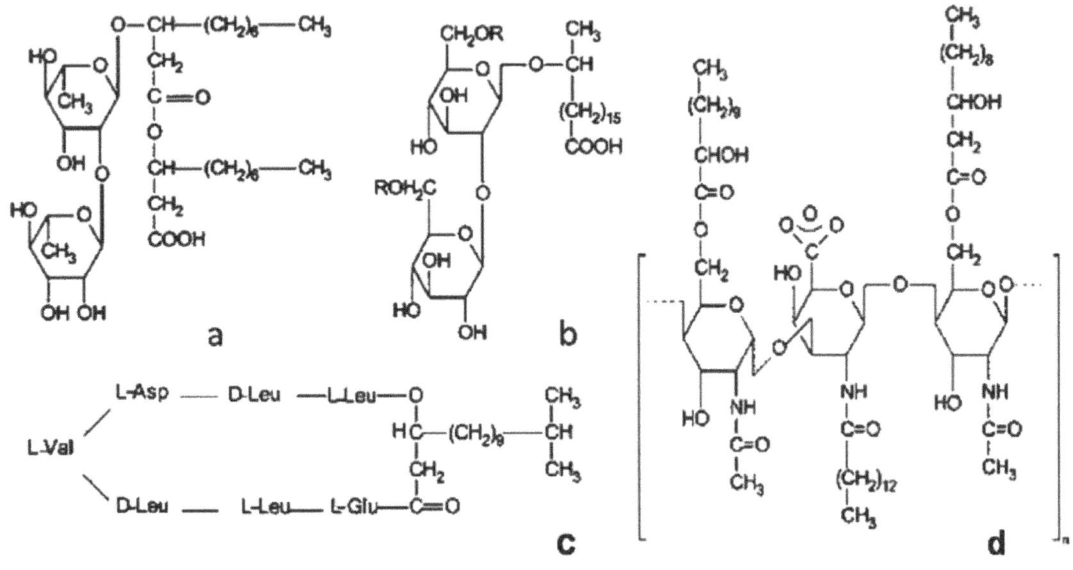

Figure 4. Chemical structure of most studied microbial surface-active compounds. (a) Rhamnolipid; (b) Sophorolipid; (c) Surfactin and (d) Emulsan.

PROPERTIES

It is necessary to submit a biosurfactant to conservation methods to evaluate its properties (surface tension and dispersion) over a period of 120 days to estimate the commercial validity of the product. Thus, heating methods are used separately or in combination with potassium sorbate, which is a conservative that inhibits the growth of mould that is widely used in the production and conservation of foods. Some characteristics are common to the majority of biosurfactants and have advantages over conventional surfactants, as described below [5].

Surface and Interfacial Activity

Efficiency and effectiveness are essential characteristics of a good surfactant. Efficiency is measured by the CMC, whereas effectiveness is related to surface and interfacial tensions [34]. The CMC of biosurfactants ranges from 1 to 2000 mg/L, whereas interfacial (oil/water) and surface tensions are respectively approximately 1 and 30 mN/m. Good surfactants are able to reduce water surface tension from 72 to 35 mN/m and the interfacial tension of *n*-hexadecane from 40 to 1 mN/m.

Tolerance to Temperature, pH and Ionic Strength

Many biosurfactants can be used at high temperatures and pH values ranging from 2 to 12. Biosurfactants also tolerate a salt concentration up to 10%, whereas 2% NaCl is enough to inactivate synthetic surfactants.

Biodegradability

Biosurfactants are easily degraded by microorganisms in water and soil, making these compounds adequate for bioremediation and waste treatment.

Low Toxicity

Low degree of toxicity allows the use of biosurfactants in foods, cosmetics and pharmaceuticals. Low toxicity is also of fundamental importance to environmental applications.

Biosurfactants can be produced from largely available raw materials as well as industrial waste.

Specificity

Biosurfactants are complex molecules with specific functional groups and therefore often have specific action. This is of particular interest in the detoxification of different pollutants and the de-emulsification of industrial emulsions as well as specific food, pharmaceutical and cosmetic applications.

Biocompatibility and Digestibility

These properties allow the use of biomolecules in different industries, especially the food, pharmaceutical and cosmetic industries.

Emulsion Forming/Breaking

Biosurfactants can be either emulsifiers or de-emulsifiers. An emulsion is a heterogeneous system consisting of an immiscible liquid dispersed in another liquid in the form of droplets, the diameter of which generally exceeds 0.1 mm. There are two basic types of emulsion: oil-in-water (o/w) and water-in-oil (w/o). Emulsions have minimal stability, but the addition of biosurfactants can lead to an emulsion that remains stable for months or even years [35]. Liposan, which is a water-soluble emulsifier synthesised by *C. Lipolytica*, has been used with edible oils to form stable emulsions. Liposan is commonly used in the cosmetic and food industries for producing stable oil/water emulsions [4,36].

FACTORS AFFECTING BIOSURFACTANT PRODUCTION

The production of biosurfactants can be either spontaneous or induced by the presence of lipophilic compounds, variations in pH, temperature, aeration and agitation speed or when cell growth is maintained under conditions of stress, such as a low concentration of nitrogen [37]. The various physicochemical factors are discussed below [38].

Carbon Source

The carbon source plays an important role in the growth and production of biosurfactants by microorganisms and varies from species to species. A very low yield was found when only either glucose or vegetable oil was used for the production of a biosurfactant by *T. bombicola*, but the yield increased to 70 g/L when both carbon sources were provided together [39]. At a concentration of 80 and 40 g/L of glucose and soybean oil, respectively, the maximum yield of sophorose lipids was obtained by *T. bombicola* [40]. Even higher yields of sophorolipids (120 g/L) were produced with *C. bombicola* in eight days when sugar and oil were used as carbon sources [41]. When canola oil and glucose were used as carbon sources at concentrations of 10% each, maximum yield of sophorolipids (8 g/L) was obtained from *C. lipolytica* [42]. Moreover, when industrial waste was used for the production of a biosurfactant by *C. lipolytica*, the yield of the protein-lipid-carbohydrate complex was 4.5 g/L, with a reduction in the surface tension of distilled water from 71 to 32 mN/m [43]. A high production of bioemulsifier was obtained with *C. lipolytica* when supplemented with 1.5% glucose (*w/v*) [44]. *C. antarctica* and *C. apicola*yielded 13.4 and 7.3 g/L of sophorolipids, respectively, when soapstock was used at a concentration of 5% (*v/v*) [45]. The resting cells of *Pseudozyma (C. antarctica)* were found to covert C_{12} to C_{18} n-alkanes into mannosylerythritol lipids (MEL); the yield was 140 g/L after four weeks and the biosurfactant was able to emulsify soybean oil [46]. A change in the fatty acid constitution of the final biosurfactant occurred when the fatty acid composition was changed in the fermentation medium containing *C. ingens* [47].

Nitrogen Sources

This is the second most important supplement for the production of biosurfactants by microorganisms. In fermentative processes, the C/N ratio affects the buildup of metabolites. High C/N ratios (*i.e.*, low nitrogen levels) limit bacterial growth, favouring cell metabolism towards the production of metabolites. In constrast, excessive nitrogen leads to the synthesis of cellular material and limits the buildup of products [48]. Different organic and inorganic nitrogen sources have been used in the production of biosurfactants. Santa Anna *et al.* [49] describe the importance of nitrogen for the production of a biosurfactant by *P. aeruginosa* cultivated in a mineral medium containing 3% glycerol. As $NaNO_3$ proved more effective than $(NH_4)_2SO_4$, nutritional limitations clearly guide the cell metabolism to the formation of the product. Mulligan and Gibbs [50] report that *P. aeruginosa* uses nitrates, ammonium and amino acids as nitrogen sources. Nitrates are first reduced to nitrite and then ammonium. Ammonium is assimilated either by glutamate dehydrogenase (EC 1.4.1.4) to form glutamate or glutamine synthetase (EC 6.3.1.2) to form glutamine. Glutamine and α-ketoglutarate are then converted to glutamine by l-glutamine 2-oxoglutarate aminotransferase (EC 1.4.1.13). However, lipid formation rather than sugar is the rate-determining factor in the biosynthesis of rhamnolipids and nitrogen limitation can lead to the accumulation of lipids. In comparison to ammonium, the assimilation of nitrate is slower and simulates nitrogen limitation, which is favourable to the production of rhamnolipids. High yields of sophorose lipids, which are biosurfactants produced by the fungi *T. bombicola* and *C. Bombicola*, have been achieved using yeast extract and urea as the nitrogen source [51]. Moreover, high yields of mannosylerythritol lipid by *Candida* sp. SY16, *C. lipolytica* and *C. glabrata* have been achieved with ammonium nitrate and yeast extract [42,43,46,52,53,54].

Growth Conditions

Growth conditions (temperature, pH, agitation speed and oxygen) also influence biosurfactant production [37]. Species of the genus *Candida* produce maximum biosurfactant yields in a wide pH range, such as pH 5.7 for *C. glabrata* UCP 1002, pH 7.8 for *Candida* sp. SY16, pH 5.0 for *C. Lipolytica* and pH 6.0 for *C. batistae* [52,54,55,56]. Moreover, *Pichia anamola* and*Aspergillus ustus* produce maximum biosurfactant yield at pH 5.5 and 7.0, respectively [57,58]. Different microbial processes are affected by even a small change in temperature. The most favourable temperature for the production of biosurfactants by different fungi is 30 °C, as observed for different species of *Candida*, viz. *Candida* sp. SY16, *C. bombicola, C. batistae* and *T. bombicola* [39,51,52,56]. In case of *C. lipolytica*, 27 °C has been found to be the best temperature. Incubation time also exerts a significant effect on biosurfactant production. Microorganisms produce biosurfactants in different time intervals. Maximum biosurfactant production by *Aspergillus ustus* was found after five days of incubation, whereas the incubation periods for *C. bombicola* were seven, eight and 11 days [59,60]. Maximum biosurfactant production by *C. bombicola* grown in animal fat was found after 68 h of incubation [49]. Moreover, an increase in agitation speed favoured the accumulation of a biosurfactant by*P. aeruginosa* UCP 0992 grown in glycerol [61]. Oliveira *et al.* [62] studied the effect of a change in agitation speed of cultures from 50 to 200 rpm on *P. alcaligenes* cultivated in palm oil. The authors found that the increase in rotation velocity favoured a reduction in the surface tension of the cell-free broth to 27.6 mN/m. In contrast, Cunha *et al.* [63] found that agitation had a negative effect regarding a reduction in surface tension using a biosurfactant from *Serratia* sp. SVGG16 grown in a hydrocarbon culture.

METABOLIC PATHWAYS OF BIOSURFACTANT PRODUCTION

Hydrophilic substrates are primarily used by microorganisms for cell metabolism and the synthesis of the polar moiety of a biosurfactant, whereas hydrophobic substrates are used exclusively for the production of the hydrocarbon portion of the biosurfactant [37,64]. Diverse metabolic pathways are involved in the synthesis of precursors for biosurfactant production and depend on the nature of the main carbon sources employed in the culture medium. For instance, when carbohydrates are the only carbon source for the production of a glycolipid, the carbon flow is regulated in such as way that both lipogenic pathways (lipid formation) and the formation of the hydrophilic moiety through the glycolytic pathway are suppressed by the microbial metabolism, as illustrated in Figure 5 [65].

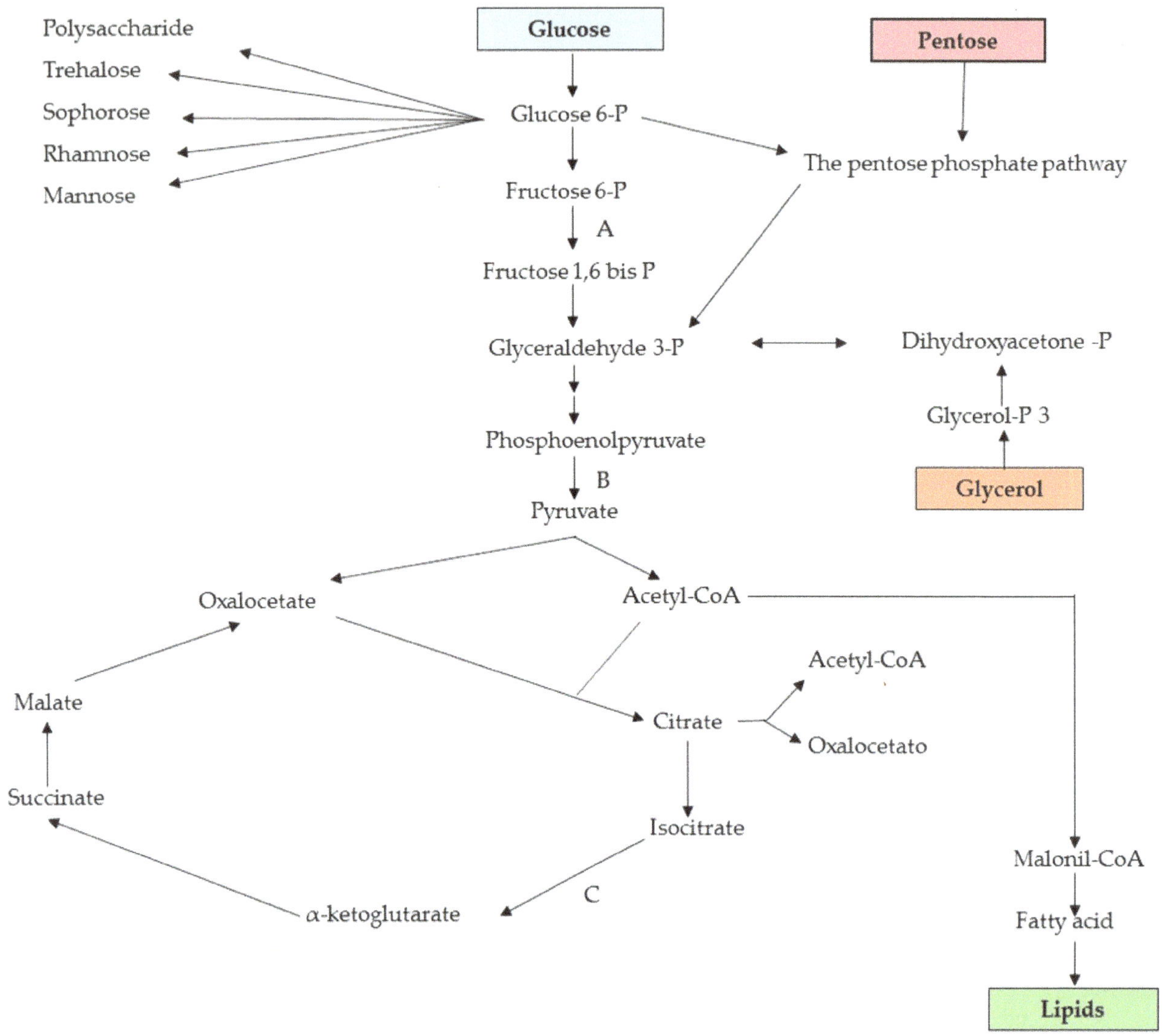

Figure 5: Intermediate metabolism related to synthesis of biosurfactant precursors with use of carbohydrates as substrate. Enzyme keys for control of carbon flow: (A) phosphofructokinase; (B) pyruvate kinase; (C) isocitrate dehydrogenase.

A hydrophilic substrate, such as glucose or glycerol, is degraded until forming intermediates of the glycolytic pathway, such as glucose 6-phosphate, which is one of the main precursors of carbohydrates found in the hydrophilic moiety of a biosurfactant. For the production of lipids, glucose is oxidised to pyruvate through glycolysis and pyruvate is then converted into acetyl-CoA, which produces malonyl-CoA when united with oxaloacetate, followed by conversion into a fatty acid, which is one of the precursors for the synthesis of lipids [66]. When a hydrocarbon is used as the carbon source, however, the microbial mechanism is mainly directed to the lipolytic pathway and gluconeogenesis (the formation of glucose through different hexose precursors), thereby allowing its use for the production of fatty acids or sugars. The gluconeogenesis pathway is activated for the production of sugars. This pathway consists of the oxidation of fatty acids through β-oxidation to acetyl-CoA (or propionyl-CoA in the case of odd chain fatty acids). Beginning with the formation of acetyl-CoA, the reactions involved in the synthesis of polysaccharide precursors, such as glucose 6-phosphate, are essentially the inverse of those involved in glycolysis. However, reactions catalysed by pyruvate kinase and phosphofructokinase-1 are irreversible. Thus, other enzymes exclusive to the process of gluconeogensis are required to avoid such reactions. Figure 6 illustrates the main reactions through to the formation of glucose 6-phosphate, which is the main precursor of polysaccharides and disaccharides formed for the production of the hydrophilic moiety of glycolipids [67].

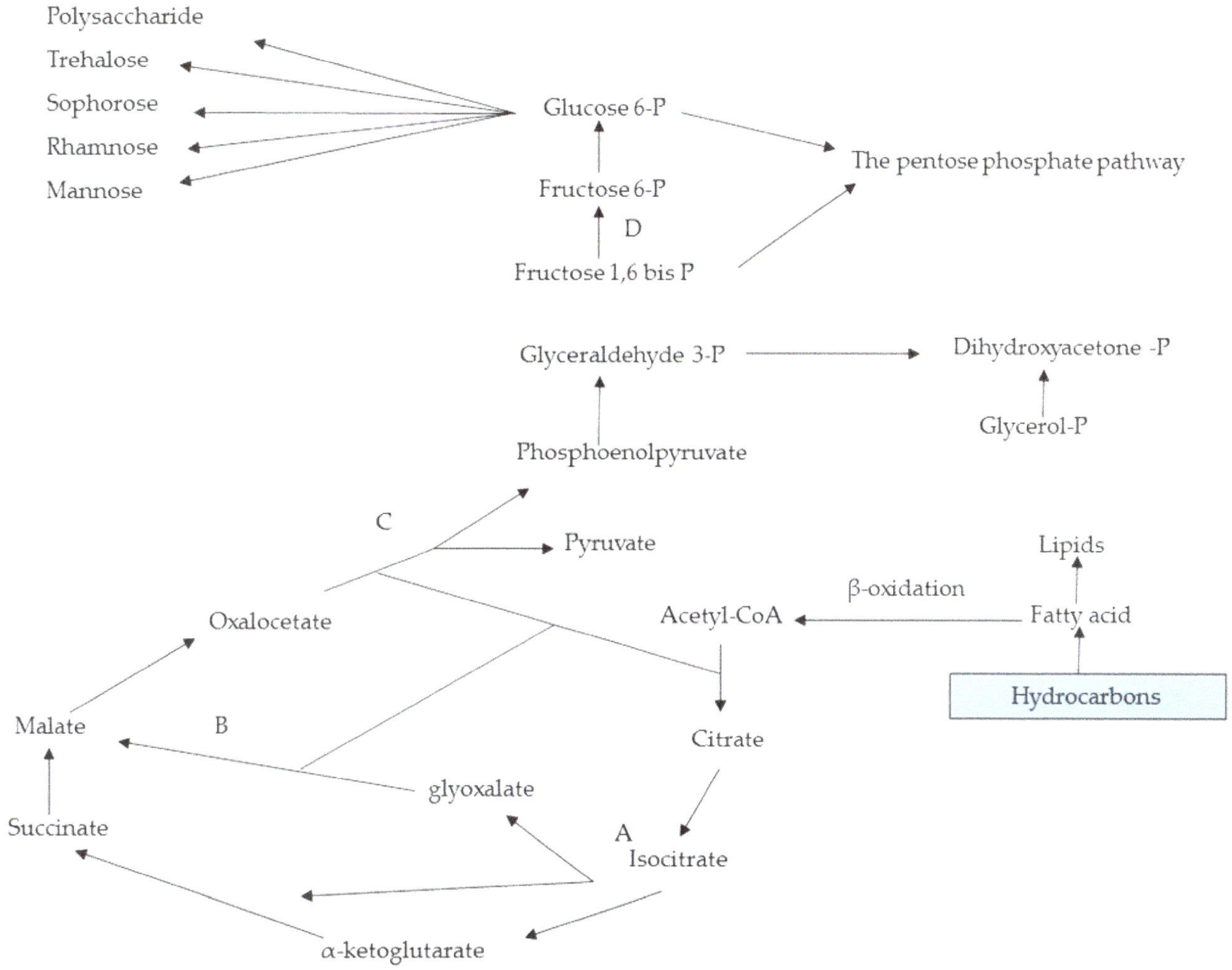

Figure 6: Intermediate metabolism related to synthesis of precursors of biosurfactant using hydrocarbons as substrate. Key enzymes: (A) isocitrate lyase; (B) malate synthase; (C) phosphoenolpyruvate; (D) fructose-1.

According to Sydatk and Wagner [68], the biosynthesis of a surfactant occurs through four different routes: (a) carbohydrate and lipid synthesis; (b) synthesis of the carbohydrate half while the synthesis of the lipid half depends on the length of the chain of the carbon substrate in the medium; (c) synthesis of the lipid half while the synthesis of the carbon half depends on the substrate employed; and (d) synthesis of the carbon and lipid halves, which are both dependent on the substrate. Therefore, the length of the n-alkane chain used as the carbon source alters the biosynthesis of a surfactant. Kitamoto *et al.* [46] studied the production of manosilerythritol lipid (MEL) by the yeast *C. antarctica* in the presence of different n-alkanes and found that this species does not grow or produce a biosurfactant in media containing C_{10} to C_{18}. However, production occurred when the species was grown in a medium containing C_{12} to C_{18} and octadekane as substrate led to the greatest yield. In contrast, production was minimal in media containing n-alkanes with more than 19 carbons.

PHYSIOLOGY

Biosurfactants are produced by microorganisms either through excretion or adhesion to cells, especially when cultivated on substrates that are insoluble in water. While the function in microbial cells is not yet fully understood, it has been speculated that biosurfactants are involved in the emulsification of insoluble substrates [37,69].

The main physiological role of biosurfactants is to allow microorganisms to grow on substrates that are insoluble in water through a reduction in surface tension between phases, making the substrate more available for uptake and metabolism. The uptake mechanisms of these substrates (such as, alkanes) are not yet fully clarified. The direct uptake of dissolved hydrocarbons in the aqueous phase, direct contact between cells and large hydrocarbon droplets, and the interaction with emulsified droplets (emulsion) have been described. Besides the emulsification of the carbon source, biosurfactants are also involved in the adhesion of microbial cells to hydrocarbons, as discussed in the following sections. Microorganism cell adsorption to insoluble substrates and the excretion of surfactant compounds allow growth on carbon sources [37].

FERMENTATION KINETICS

Biosurfactant production kinetics has considerable variation among different systems. For convenience, kinetic parameters are grouped as follows: (a) growth-associated production; (b) production under growth-limiting conditions; (c) production by resting or immobilised cells; and (d) production with precursor supplementation [37]. In growth-associated production, parallel relationships are found between growth, the use of the substrate and biosurfactant production. Production under growth-limiting conditions is characterised by an accentuated increase in biosurfactant concentration as a result of the limitation of one or more medium components. Production by resting or immobilised cells is a type of biosurfactant production in which there is no cell multiplication; the cells nonetheless continue to use the carbon source for biosurfactant synthesis. Investigators report that the addition of biosurfactant precursors to the medium leads to qualitative and quantitative changes in the final product.

RAW MATERIALS FOR BIOSURFACTANT PRODUCTION

Current society is characterised by an increase in expenditures, the need to reuse materials and environmental concerns. Consequently, greater emphasis has been given to recovery, recycling and reuse. Indeed, the need for environmental preservation has led to the reuse of different industrial wastes. This is particularly valid for the food production industry, the waste products, effluents and by-products of which can be reused [70]. Industrial waste had piqued the interest of researchers as a low-cost substrate for biosurfactant production [71]. The selection of waste products should ensure the proper balance of nutrients to allow microbial growth and consequent biosurfactant production. Industrial waste with a high content of carbohydrates or lipids is ideal for use as substrate [71]. According to Barros *et al.* [34], the use of agro-industrial waste is one of the steps towards the implantation of feasible biosurfactant production on an industrial scale, for which the optimisation of the different variables involved is required.

The literature describes a number of waste products employed in biosurfactant production, such as vegetable oils, oily effluents [42,72,73], starchy effluents [74,75], animal fat [51,76,77,78], vegetable fat [79], vegetable cooking oil waste [72,80,81,82], soapstock [76,83,84] molasses [85,86,87,88], dairy industry waste (whey) [89], corn steep liquor [43,71,90,91,92], cassava flour wastewater [93], oil distillery waste [43,90,94,95] and glycerol [61]. Some of the most commonly employed industrial waste products for biosurfactant production are detailed below.

Olive Oil Mill Effluent (OOME)

Olive Oil Mill Effluent (OOME) is a concentrated black liquor with a water-soluble portion of ripe olives and water that is used for the extraction of olive oil. OOME has polyphenols that represent a challenge in terms of the environment disposal. However, it also contains nitrogen compounds (12 to 24 g/L), sugars (20 to 80 g/L), residual oil (0.3 to 5 g/L) and organic acids (5 to 15 g/L). Mercade *et al.* [73] successfully employed OOME for the strain *Pseudomonas* sp. to produce rhamnolipids.

Animal Fat

Animal fat and lard can be obtained in large quantities from the meat processing industry and have been used as a medium for cooking foods. Recently, however, such fats have lost a large part of the market to vegetable oils due to the lower degree of harm to health caused by the latter [70]. Animal fat stimulates the production of sophorolipids by the yeast*C. bombicola* [51]. Using animal fat and corn steep liquor, Santos *et al.* [77,78] achieved maximum glycolipid production by the yeast *C. lipolytica* UCP 0988. The authors also report that the product has uses in bioremediation as well as oil mobilisation and recovery.

Frying Oils

Fry oil and edible fats are considered great carbon sources for biosurfactant production. Vegetable oils constitute a lipid carbon source and are mainly comprised of saturated or unsaturated fatty acids with chains of 16 to 18 carbon atoms [9]. Different oils are adequate substrates for biosurfactants. Babassu oil (5% *v*/*v*) with a carbon source (1% glucose *w*/*v*) is a good medium for biosurfactant production. Sarubbo *et al.* [96] found that two strains of *C. lipolytica* (1055 and 1120) produce biosurfactants toward the final of the exponential growth phase and onset of the stationary phase. Sunflower and olive oils have proven to be adequate energy and carbon sources for the production of biosurfactants. *P. aeruginosa* strains produce a biosurfactant on residue from corn, soybean and canola oil plants [97,98]. Canola oil residue and sodium nitrate has been reported adequate for microbial growth and the production of up to 8.50 g/L of rhamnoipids. The combination of glucose and canola oil has been used for the successful production of a biosurfactant by *C. lipolytica* [42].

Soapstocks

Oil cakes or soapstocks are produced from oilseed processing involving the refining of seed-based edible oils with the use of chemicals [76]. Soapstock has been used together with sunflower oil, olive oil or soybean oil as substates to produce rhamnolipids. Yields as high as 15.9 g/L have been reported using *P. aeruginosa* in a soapstock medium [83]. Soapstock and oil refinery wastes have been used with *C. antarctica* or *C. apicola* for biosurfactant production, achieving a greater yield than that in the medium without oil refinery residue [45]. Shabtai [84] also report the production of two extracellular heteropolysaccharides (emulsan and biodispersan) by *A. calcoaceticus* and *A. calcoaceticus*, respectively, using soapstock as a carbon source.

Molasses

Molasses is a by-product of sugarcane and beet processing. This inexpensive substrate has dry matter (75%), non-sugar organic matter (9%–12%), protein (2.5%), and potassium (1.5%–5.0%), as well as magnesium, phosphorus and calcium (≈1%). The inositol, biotin, thiamine and pantothenic acid contents (1%–3%) give molasses its thick consistency and brown colour. The high sugar content (48%–56%) makes molasses adequate for biosurfactant production by different microorganisms. Laboratories have used molasses for the production of different microbial metabolites. According to Makkar and Cameotra [88], *Bacillus subtilis* in a minimal medium supplemented with molasses as the carbon source produces a biosurfactant. Joshi*et al.* [99] used molasses as well as other carbon sources to produce biosurfactants from strains of *Bacillus*.

Whey

The dairy industry produces large quantities of whey, such as whey waste, cheese whey, curd whey and lactic whey, all of which can be used as substrates for the microbial production of metabolites [70,100,101,102]. A high amount of lactose (approximately 75%) is found in lactic whey. Other components, such as proteins, vitamins and organic acids, are good sources for microbial growth and biosurfactant production [75]. Moreover, whey disposal represents a major pollution problem, especially in countries that depend on a dairy economy [103]. Thus, the disposal of this by-product represents a waste of a widely available substrate and an environmental hazard.

Corn Steep Liquor

The agro-industry of corn-based products through wet processing results in both solid and liquid by-products, which, when disposed improperly, become a source of contamination and harm to the environment. Corn steep liquor is a by-product of the washing water and soaking of kernels as well as fractioning into starch and germen (oil) that contains 40% solid matter. This by-product consists of 21% to 45% proteins, 20% to 26% lactic acid, approximately 8% ash (containing Ca^{2+}, Mg^{2+}, K^{+}, *etc.*), approximately 3% carbohydrates and a low fat content (0.9% to 1.2%) [103,104,105]. Nut oil refinery residue and corn steep liquor are low-cost nutrients for the production of glycolipids by *C. sphaerica* (UCP 0995). The biosurfactant of this strain mobilises and removes up to 95% of motor oil on sand, making it useful for bioremediation [89,94,106]. Silva *et al.* [107] also report the production of a biosurfactant from *P. cepacia* grown in mineral medium supplemented with 2.0% corn steep liquor and 2.0% soybean waste frying oil.

Starchy Substrates

Abundant starch-based substrates also constitute renewable carbon sources. The potato processing industry produces significant amounts of starch-rich waste that are adequate for biosurfactant production. Besides the approximately 80% water content, potato waste has carbohydrates (17%), proteins (2%) and fats (0.1%) as well as inorganic minerals, trace elements and vitamins [103]. As an example, Fox and Bala investigated a commercially prepared potato starch in a mineral salt medium for the production of a biosurfactant by *B. subtilis* [74]. Cassava wastewater, which is another carbohydrate-rich waste product generated in large amounts, has been used for the production of surfactin by *B. Subtilis* 35. Other starchy wastes, such as rice water and cereal processing wastewater, have the potential to permit microbial growth and biosurfactant production [108].

RECOVERY OF BIOSURFACTANTS

The production of low-coast biosurfactants is unlikely due to the complicated recovery process. Process development is conduted in order to obtain biosurfactants that can be recovered easily and inexpensively. In many biotechnological processes, downstream processing accounts for 70%–80% of production costs. For economic reasons, most biosurfactant production processes need to involve spent whole-cell culture broths or other crude preparations [11,103,109]. Extraction with chloroform-methanol, dichloromethane-methanol, butanol, ethyl acetate, pentane, hexane, acetic acid, ether, *etc.* constitutes the most commonly used method in biosurfactant downstream processing. The most widely employed products are different ratios of chloroform and methanol, which facilitate the adjustment of the polarity of the extraction agent to the extractable target material. The disadvantages of using organic solvents for biosurfactant recovery include the large amount of solvent required and the increase in production costs due to the price of expensive solvents. Chloroform is a toxic chloro-organic compound that is harmful to human health and the environment. Thus, there is a need for inexpensive solvents with low toxicity for biosurfactant extraction processes that are suitable for industrial applications. Other product precipitation techniques have also been reported, such as precipitation with ammonium sulphate, centrifugation and adsorption. Biosurfactant recovery depends mainly on the ionic charge, water solubility and location (intracellular, extracellular or cell bond) [103,109]. Foam fractionation is a solvent-free method that separates biosurfactant molecules adsorbed to air bubbles in the culture medium. Biosurfactant production involves continuous foam formation due to the high surface activity. Foam in the broth interferes with mass and heat transfer processes, thereby affecting productivity. However, foam is beneficial to biosurfactant production, as it assists in the continuous removal of product, and therefore production and recovery processes can be accomplished in a single stage [110]. Continuous foam fractionation in the fermentation process helps prevent the accumulation of product that could otherwise inhibit biomass growth and product formation and also facilitates extended biosurfactant production in fed-batch or continuous mode operations. Moreover, biosurfactants do not readily undergo denaturation due to their small size and simple structure [111].

More research and development are required to optimise existing recovery processes to make such processes both commercially viable and more competitive [39,103]. Table 2 lists the most common biosurfactant recovery techniques and their advantages.

Table 2: Downstream processes for recovery of important biosurfactants and respective advantages

	Process	Biosurfactant Type	Biosurfactant Property Responsible for Separation	Advantages
Batch mode	Acid precipitation	Surfactin	Biosurfactants become insoluble at low pH values	Low cost, efficient in crude biosurfactant recovery
	Organic solvent extraction	Trehalolipids; Sophorolipids; Liposan	Biosurfactants are soluble in organic solvents due to the hydrophobic end	Efficient in crude biosurfactant recovery and partial purification, reusable nature
	Ammonium sulphate precipitation	Emulsan; Biodispersan; Lipopeptides	Salting-out of polymeric or protein-rich biosurfactants	Effective in isolation of certain type of polymeric biosurfactants
Continuous mode	Adsorption to wood-activated carbon	Rhamnolipids; Lipopeptides; Glycolipids; Mannosylerythritol Lipids (MEL)	Biosurfactants are adsorbed to activated carbon and can be desorbed using organic solvents	Highly pure biosurfactants, cheaper, reusability, recovery from continuous culture
	Adsorption to polystyrene resines	Rhamnolipids; Lipopeptides; Glycolipids; MEL	Biosurfactants are adsorbed to polystyrene resins and subsequently desorbed using organic solvents	Highly pure biosurfactants, cheaper, reusability, recovery from continuous culture
	Centrifugation	Glycolipids	Insoluble biosurfactants are precipitated due to centrifugal force	Reusable, effective in crude biosurfactant recovery
	Ion-exchange chromatography	Glycolipids	Charged biosurfactants are attached to ion-exchange resins and can be eluted with buffer	High purity, reusability, fast recovery
	Foam fractionation	Surfactin	Biosurfactant form and partition into foam	Useful in continuous recovery processes, high purity of product
	Ultrafiltration	Glycolipids	Biosurfactants form micelles above their critical micelle concentration (CMC), which are trapped by polymeric membranes	Fast, one-step recovery, high level of purity, reusability

INDUSTRIAL APPLICATIONS OF BIOSURFACTANTS

Biosurfactants have a wide range of biotechnological applications in petroleum, foods, beverages, cosmetics, detergents, textiles, paints, mining, cellulose, pharmaceutics and nanotechnology [112]. Currently, the main market is the petroleum industry. Biosurfactants can be used for oil residue recovery from storage tanks, other oil recovery processes, the cleanup of oil spills and the bioremediation of both soil and water [2,113]. Table 3 offers a summary of the uses of biosurfactants in different industries. The main biotechnological applications are detailed in the following sections.

Table 3: Applications of biosurfactants for industrial uses

Industry	Application	Role of Biosurfactants	References
Environment	Bioremediation; Oil spill cleanup operations; Soil remediation and flushing	Emulsification of oils, lowering of interfacial tension, dispersion of oils, solubilisation of oils, wetting, spreading, detergency, foaming, corrosion inhibition in fuel oils and equipment, soil flushing.	[2,8]
Petroleum	Enhanced oil recovery; De-emulsification	Emulsification of oils, lowering of interfacial tension, de-emulsification of oil emulsions, solubilisation of oils, viscosity reduction, dispersion of oils, wetting of solid surfaces, spreading, detergency, foaming, corrosion inhibition in fuel oils and equipment.	[8,120]
Mining	Heavy metal cleanup operations; Soil remediation; Flotation	Wetting and foaming, collectors and frothers, removal of metal ions from aqueous solutions, soil and sediments, heavy metals sequestrants, spreading, corrosion inhibition in oils.	[121]
Food	Emulsification and de-emulsification; Functional ingredient	Solubilisation of flavoured oils, control of consistency, emulsification, wetting agent, spreading, detergency, foaming, thickener.	[4]
Medicine	Microbiological; Pharmaceuticals and therapeutics	Anti-adhesive agents, antifungal agents, antibacterial agents, antiviral agents, vaccines, gene therapy, immunomodulatory molecules.	[20,122,123]
Agriculture	Biocontrol; Fertilisers	Wetting, dispersion, suspension of powdered pesticides and fertilisers, emulsification of pesticide solutions, facilitation of biocontrol mechanisms of microbes, plant pathogen elimination and increased bioavailability of nutrients for beneficial plant-associated microbes.	[124]
Cosmetics	Health and beauty products	Emulsification, foaming agents, solubilisation, wetting agents, cleansers, antimicrobial agents, mediators of enzyme action.	[5]
Cleaning	Washing detergents	Detergents and sanitisers for laundry, wetting, spreading, corrosion inhibition.	[3,5]
Textiles	Preparation of fibres; Dyeing and printing; Finishing of textiles	Wetting, penetration, solubilisation, emulsification, detergency and dispersion, wetting and emulsification in finishing formulations, softening.	[3,103]
Nanotechnology	Synthesis of nanoparticles	Emulsification, stabilisation.	[5,125]

Petroleum Recovery

Petroleum is an essential energy source and driving force of economic development. The US Department of Energy reports that fossil fuels constitute 83% of all primary energy sources in the country and petroleum accounts for 57% of such products. Indeed, 19.2 million cubic metres of petroleum were consumed per day in 2010 [114]. The USA produces 870,000 m^3 of crude oil from 530 thousand production wells, 35% of which produce 0.16 m^3/day and 79% produce < 1.59 m^3/day [115]. Through oil recovery processes, these oil wells produce only one third to half of the petroleum originally present at the sites. Oil residue in small pores within petroleum reservoirs accounts for 50% to 65% of oil and is trapped by high forces of capillarity as well as interfacial tension between the hydrocarbon and aqueous phases. Different reductions in interfacial tension are needed for the mobilisation of this hydrocarbon [116,117], which is only achieved with the use of surfactant concentrations significantly higher than that required for the formation of micelles [118,119]. In enhanced oil recovery, the use of heat, tensioactive agents, microbial processes and gas injection leads to the recovery of a significant portion of the retained oil. However, the high cost of chemical tensioactive agents hinders widespread use of surfactants in oil recovery processes. Thus, biosurfactants have been employed to reduce the interfacial tension between oil/water and oil/rock, which leads to a reduction in the capillary forces that impede oil from moving through rock pores (Figure 7). Biosurfactants also form an emulsion at the oil-water interface, which stabilises the desorbed oil in water and allows oil removal along with the injection water [1,8].

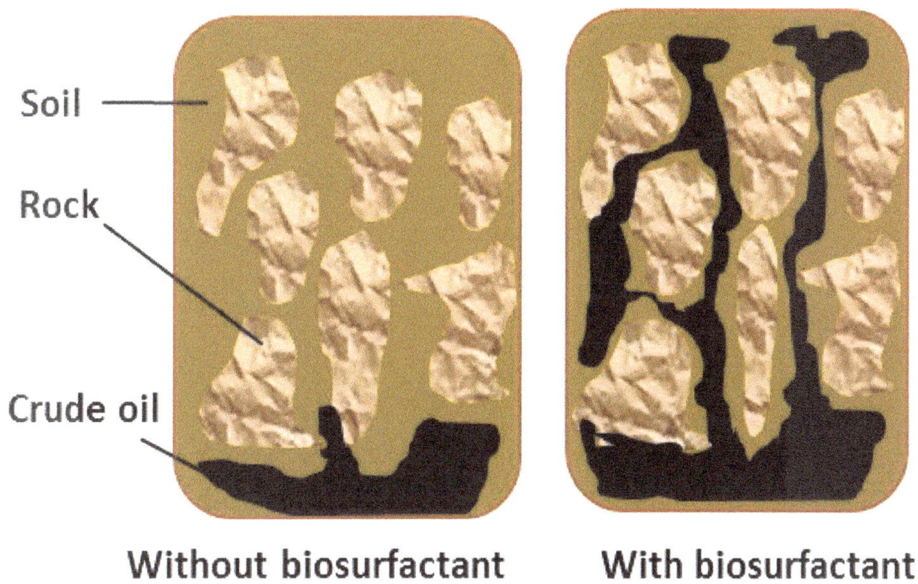

Figure 7: Enhanced oil recovery mechanism by biosurfactants.

Bioremediation

Oil spills occur during cargo transportation or in the form of industrial oil and by-product spills. Petroleum exerts a negative effect on cell membranes in living organisms, offering considerable risk of contamination to both marine and terrestrial ecosystems [2,114].

The US Environmental Protection Agency proposes different physical, chemical and biological technologies for the treatment of contaminated soil [126], one of the most studied of which is bioremediation. This process involves the natural degradation capacity of plants and microorganisms for either the partial conversion of contaminants into less toxic compounds or the complete conversion of such substances into carbon dioxide and water.

Larger degrading microorganism populations lead to a quicker, more efficient bioremediation processes. Therefore, this technique can be conducted through biostimulation, which consists of stimulating the growth of microorganisms present at the contaminated site. The process involves the introduction of specific electron receptors, oxygen and nutrients for the degradation of the contaminant as well as substances to correct the pH. Bioremediation can also be performed through bioaugmentation, in which indigenous (allochthonous) microorganisms are added to the contaminated environment to accelerate and complete the degradation of the pollutant [114].

Bioremediation played an important role in the cleanup of the 41 million litre oil spill caused by the oil tanker Exxon Valdez in the Gulf of Alaska in 1989, giving rise to the development of this technology and demonstrating that there are good reasons to believe in the effective application of this treatment method in future oil spills under the appropriate circumstances [114]. In the accident with the Exxon Valdez, the first measure taken was physical washing with high-pressure water. Chemical surfactants were then applied in polluted areas to accelerate the growth and activity of petroleum-degrading microorganisms. Two or three weeks later, the regions treated with surfactants were significantly cleaner than control areas. However, it was difficult to evaluate the exact effects of the treatment due to the heterogeneity of the contamination. Nonetheless, subsequent studies have demonstrated the importance of the use of surfactants to enhance the biodegradation of oil [114,127].

While bioremediation is an effective, environmentally friendly method, the time and costs involved make this process unviable for the treatment of large amounts of waste [128]. Thus, the use of biosurfactants emerges as a safe alternative for improving the solubility of hydrophobic compounds by allowing the desorption and solubilisation of hydrocarbons and facilitating the assimilation of these compounds by microbial cells [129].

The biodegradation of oil-derived hydrocarbons by biosurfactants occurs through two mechanisms. The first involves an increase in the bioavailability of the hydrophobic substrate to microorganisms, with a consequent reduction in surface tension of the medium around the bacterium as well as a reduction in interfacial tension between the cell wall and hydrocarbon molecules. The other mechanism involves the interaction between the biosurfactant and cell surface, leading to changes in the membrane, facilitating hydrocarbon adherence (increase in hydrophobicity) and reducing the lipopolysaccharide index of the cell wall without damaging the membrane. Thus, biosurfactants block the formation of hydrogen bridges and allow hydrophobic-hydrophilic interactions, which cause molecular rearrangements and reduce the surface tension of the liquid by increasing its surface area as well as promoting bioavailability and consequent biodegradability [130,131]. Figure 8 illustrates the action of biosurfactants in increasing the surface area of oil droplets as well as facilitating access to a greater number of bacteria and consequently producing a greater biomass.

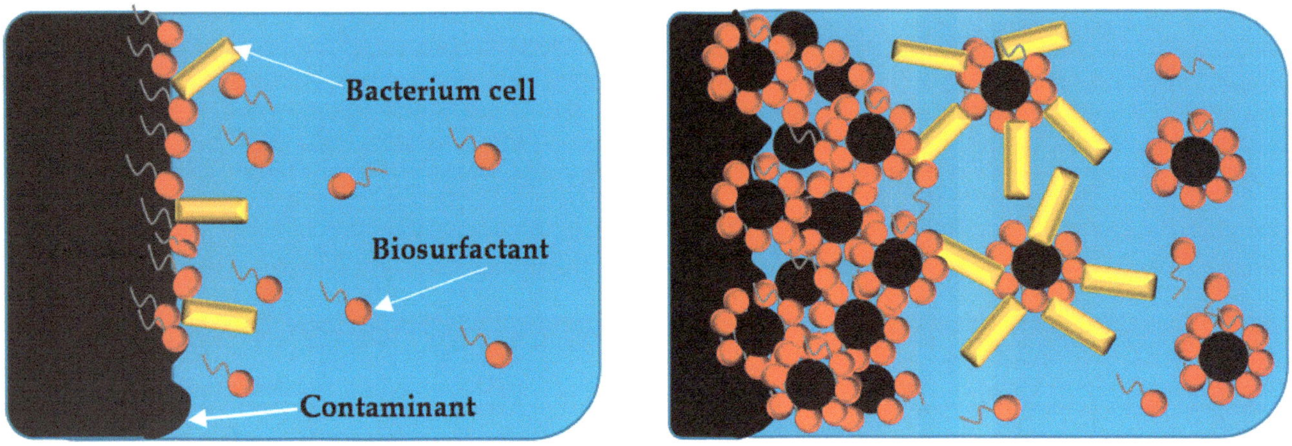

Figure 8: Illustration of biosurfactant action on petroleum.

Removal of Hydrophobic Organic Pollutants

The application of biosurfactants for the removal of contaminants from soil is less well known than the advanced application of these compounds in bioremediation processes, since removal efficiency is driven mainly by the physicochemical properties of the biosurfactant rather than the effects on metabolic activity or changes in the properties of the cell surface. However, the mechanisms that affect the mobilisation and solubilisation of hydrocarbons in soils are similar to those involved in the enhancement of bioavailability for bioremediation [131,132].

Biosurfactants enhance the removal of hydrocarbons through biodegradation, solubilisation, mobilisation or emulsification [8]. Solubilisation capacity depends on the ability of the surfactant to increase the solubility of hydrophobic components in the aqueous phase. A considerable increase in this capacity occurs above the CMC due to the partitioning of the hydrocarbon in the hydrophobic portion of the micelles. In this process, greater concentrations of surfactants are normally required, since the solubility of the hydrocarbon components in the solution depends wholly on the concentration of the surfactant [8]. Mobilisation occurs at concentrations below the CMC and is divided into displacement and dispersion. Displacement consists of the release of hydrocarbon droplets from the porous medium due to the reduction in interfacial tension. Using a theoretical explanation, hydrocarbon removal is possible when the interfacial tension between the aqueous and oil phases is sufficiently reduced to overcome the forces of capillarity that cause the formation of residual saturation. Dispersion is a process by which a hydrocarbon is dispersed in the aqueous phase as tiny emulsions. Emulsions are not generally thermodynamically stable, but may remain stable for significant periods of time due to kinetic restrictions. Dispersion is related to interfacial tension and surfactant concentration and differs from displacement, which is related only to interfacial tension between the aqueous and hydrophobic phases, with no formation of emulsion [132].

The efficiency of a surfactant in the removal of hydrophobic compounds also depends on the pH and ionic strength of the solution, which can alter the arrangement of the aggregated micelles and sorption of the surfactant to the soil, which, in turn, limits the transport of the hydrocarbon by the surfactant. Different biosurfactants have been tested for the removal of petroleum-derived products from contaminated soil and water. Rhaminolipids have been successfully used in biotechnological decontamination processes [2,7,17]. Other surfactants produced by species of *Pseudomonas* [110,133],*Bacillus* [5,134], and *Candida* [109,113,135,136,137], have also been successfully used in the remediation of soil.

Removal of Heavy Metals

Heavy metals and radionuclides are persistent soil contaminants. Increases in levels of heavy metals in soil have been reported in many industrialised countries. Metals and metalloids, such as chromium, cadmium, mercury and lead, can threaten ecosystems and human health through either the food chain or direct exposure to contaminated soil and water [1,12]. As different technologies can be used in combination for the treatment of organic pollutants and heavy metals, biosurfactants can be used in the removal of hydrophobic organic compounds and heavy metals [138,139,140]. Heavy metals mainly adsorb to the surface of soil in the form of ions or the precipitation of metal compounds. Unlike organic contaminants, heavy metals are removed from soil through surfactant-associated complexation [141] and ion exchange [142]. Therefore, surfactant-enhanced washing and surfactant-enhanced bio-extraction can be applied to the remediation of soils contaminated with heavy metals.

Surfactants in solutions facilitate the solubilisation, dispersion and desorption of contaminants and allow the reuse of the soil [143]. Decontamination tests have been performed with different synthetic surfactants [144,145], but the desire to replace such compounds with natural surfactants has led to research into the use of biosurfactants [119]. Studies have demonstrated the potential of surfactin, rhamnolipids and sophorolipids [146,147,148]. The ionic nature, biodegradability, low toxicity and excellent surface properties make biosurfactants adequate for the removal of heavy metals from sediment and soil. According to Mulligan [119], removal is possible with different concentrations of biosurfactants. Das *et al.* [149] found that the removal of cadmium using an aqueous solution also occurred at concentrations below the CMC, while a concentration fivefold greater than the CMC resulted in the virtually complete removal of 100 ppm of metal ions. Wen *et al.* [150] studied the degradation of a rhamnolipid in soils contaminated by cadmium and zinc and found that this compound could remain in the soil long enough to enhance the phytoextraction of the metals.

The removal of metals by ionic biosurfactants is thought to occur in the following order (Figure 9): (1) sorption of the biosurfactant to the soil surface and complexation with the metal; (2) detachment of the metal from the soil to the solution; and (3) association with micelles. Heavy metals are trapped within the micelles through electrostatic interactions and can be easily recovered through precipitation or membrane separation methods [151].

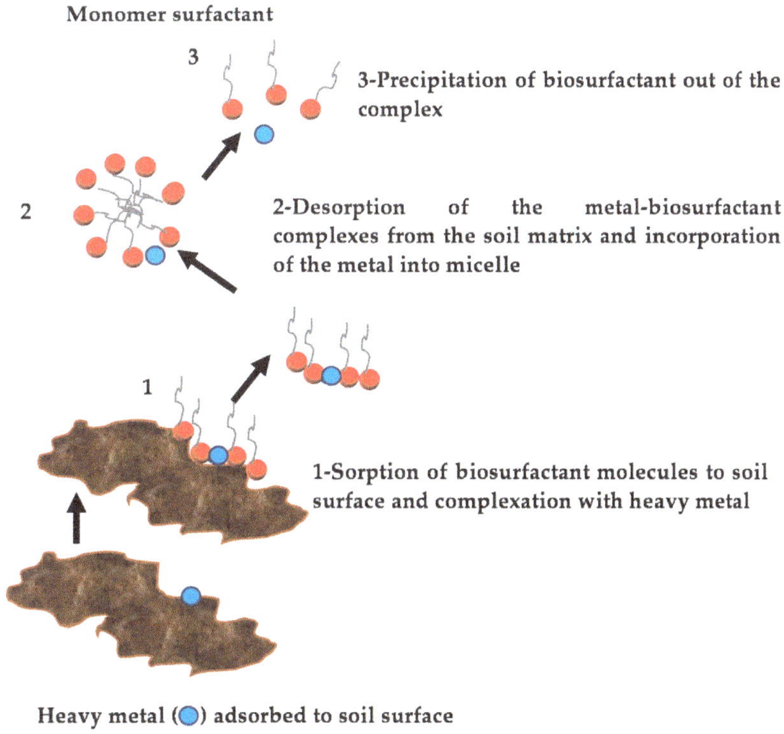

Figure 9: Mechanism of heavy metal removal by biosurfactants.

Anionic biosurfactants create non-ionic complexes with metals through ionic bonds. As such bonds are stronger than those between the metal and soil, the metal-biosurfactant complex is detached from the soil due to the reduction in interfacial tension. Cationic biosurfactants can replace similarly charged metal ions through ion exchange (competition for negatively charged surfaces). Surfactant micelles can also be used to remove metal ions from the soil surface [115].

Biosurfactants offer indisputable advantages, since surfactant-producing microorganisms do not need to survive in contaminated soil, although the continual addition of biosurfactant is required in the process [8]. Biosurfactants have also been applied in mining processes. Tensioactive compounds produced by *Pseudomonas* sp. and *Alcaligenes* sp. have been used for the floatation and separation of calcite and scheelite, with recovery rates of 95% for $CaWO_4$ and 30% for $CaCO_3$, whereas conventional chemical reagents are not capable of separating these two minerals [152]. Slizovskiy *et al.* [153] studied the enhanced remediation of soils contaminated with heavy metals using the cationic surfactant 1-dodecylpyridinium chloride (DPC), the non-ionic surfactant oleyl dimethyl benzyl ammonium chloride (trade name Ammonyx KP) and the ionic rhamnolipid biosurfactant (trade name JBR-425); the latter substance exhibited the best elution with regard to Zn (39%), Cu (56%), Pb (68%) and Cd (43%). Almeida *et al.* [154] studied the impact of surfactants on the removal of Cu by the salt marsh plant *Halimione portulacoides*. TX-100 and SDS (sodium dodecyl sulfate) were favourable to Cu harvesting and transportation in plant roots, but did not affect the transportation of Cu in the stem and leaves. The results of this study suggest that surfactants promote phytoremediation through a change in the membrane permeability of root cells. Therefore, surfactants can promote the desorption of metals and uptake by plants [1].

Rhamnolipids and soapberry-derived saponin have recently been found to assist in the removal of chromium and arsenic oxyanions from soils or the ore waste of mines [155,156]. Biosurfactants produced by yeast of the genus *Candida* have been successfully used in heavy metal flotation, demonstrating the ability to remove more than 90% of cations in columns and air-dissolved flotation processes [157,158]. *C. lipolytica* produces a biosurfactant that has also been used for the removal of heavy metals and petroleum byproducts using a soil barrier [90]. The biosurfactant significantly reduced soil permeability, thereby demonstrating its usefulness in reactive barriers, with the removal of approximately 96% of Zn and Cu as well as a reduction in Pb and Cd concentrations in groundwater.

Food Industry

Emulsification is important for the formation of consistency and texture in foods as well as phase dispersion and the solubilisation of aromas [4,159]. The general function of emulsifiers in food products is to stabilise the emulsion by controlling the agglomeration of fat globules and stabilising aerated systems [4,152]. An emulsion has at least one immiscible liquid (discontinuous internal phase) dispersed in another (continuous outer phase) in the form of droplets. The stability of this system is minimal, but can be enhanced by the addition of a surfactant, which reduces surface energy between the two phases by a reduction in interfacial tension, thereby preventing particle coalescence through the formation of steric and electrostatic barriers. Examples of processed foods that are emulsions include heavy cream, butter, mayonnaise, salad dressings, fillings, *etc.* [35]. Other uses for emulsifiers have been described, such as improving the texture and shelf life of products containing starch, the formation of complexes, altering the rheological properties of wheat flour and interactions with gluten as well as improving the consistency and texture of fat-based products through the control of polymorphism and the crystalline structure of fats [4].

Biosurfactants can also be used as emulsifiers in the processing of raw materials, the control of fat globule agglomeration, the stabilisation of aerated systems and an improvement in the consistency of fat-based products. The use of rhaminolipids to improve the emulsifying properties of frozen desserts, butter and croissants has also been reported [4,111]. For instance,*Candida utilis* produces a bioemulsifier used in processed salad dressings [160]. A manoprotein produced by *Saccharomyces cerevisiae* stabilises water/oil emulsions in cookies, mayonnaise and ice cream, *etc.* [13,161]. However, the food industry has not yet made widescale use of biosurfactants. Many of the properties of biosurfactants and their regulation as new ingredients for foods are pending approval.

Medicine

Biosurfactants have also been used in different biological (therapeutic) applications due to their fungicidal, bactericidal, insecticidal and anti-viral properties as well as use as anti-adhesive agents and enzyme inhibitors [111,116,117]. A number of rhaminolipids exhibit antibacterial activity. For instance, Abalos *et al.* [162] identified six rhaminolipids in cultures of *P. aeruginosa* AT10 grown on soybean oil refinery residue and evaluated the antimicrobial properties of the solution. These rhaminolipids exhibited excellent antifungal properties against different fungi at concentrations ranging from 16 to 32 μg/mL. *C. bombicola*-derived sophorolipids inhibited the growth of both Gram-negative and Gram-positive bacteria with a minimum inhibitory concentration of approximately 30 and 1 mg/mL in a contact time of 2 and 4 h, respectively, for *E. coli*(ATCC 8739) and *P. aeruginosa* (ATCC 9027) as well as 6 and 1 mg/mL in a contact time of 4 h for *S. aureus* (ATCC 6358) and*B. subtilis* (ATCC6633), respectively [163].

Despite the number of publications describing the antimicrobial activity of biosurfactants and patents related to their usage, real applications in the pharmaceutical, biomedical and health industries remains quite limited [164]. Some lipopeptides, such as daptomycin, have reached a commercial antibiotic status [165]. Daptomycin, is a branched cyclic lipopeptide isolated from *Streptomyces roseosporus* cultures and is produced by Cubist Pharmaceuticals under the name Cubicin® [166]. This drug was approved in 2003 for the treatment of skin infections caused by methicillin-resistant*Staphylococcus aureus* and other Gram-positive pathogens and approved in 2006 for the treatment of endocarditis and bacteraemia caused by *S. aureus*. Daptomycin had also been reported to display strong antibacterial activity against other important pathogens, such as penicillin-resistant *Streptococcus pneumoniae*, coagulase-negative *Staphylococci*, glycopeptide-intermediate-susceptible *S. aureus* and vancomycin-resistant *Enterococci* [166].

Biofilms are groups of bacteria and other organic matter that has colonised/accumulated on a given surface [117]. Bacterial adherence to the surface is the first step in the establishment of biofilm and is affected by various factors, such as the type of microorganism, hydrophobicity, electrical charges of the surface, environmental conditions and the ability of microorganisms to produce extracellular polymers that assist the cells in anchoring to the surface [167]. Anti-adherent activity, which is the ability to inhibit the adherence of pathogenic microorganisms to solid surfaces or infectious sites, has also been reported for biosurfactants, leading to a reduction in hospital infections with no need for drugs or synthetic chemical agents [112]. Meylheuc *et al.* [168] studied a biosurfactant obtained from *P. fluorescens* with inhibitory properties regarding the adherence of *Listeria monocytogenes* to stainless steel and polytetrafluoroethylene surfaces. *C. sphaerica*-derived lunasan inhibited the adherence of *P. aeruginosa, Streptococcus agalactiae* and *S. sanguis* between 80% and 92% at a concentration of 10 mg/mL [95], while the biosurfactant rufisan produced by *C. lipolytica* UCP 0988 demonstrated antimicrobial activities against *S. agalactiae, S. mutans, S. mutans* NS, *S. mutans* HG, *S. sanguis* 12, *S. oralis* J22 at a concentration above the CMC (0.3%). Moreover, the biosurfactant showed anti-adherent activity against most of the microorganisms tested [137].

Deficiency of lung surfactant, which is a protein-phopholipid complex, is responsible for respiratory failure in premature infants. Gene isolation for protein molecules in this surfactant and cloning in bacteria allow fermentative production for medical applications [108]. Sophorolipids from *C. bombicola* have been studied due to their spermicidal and cytotoxic activities as well as anti-HIV action that can reduce the proliferation of acquired immunodeficiency syndrome (AIDS). Sophorolipids have also been studied as anti-inflammatory agents for patients with immune diseases [108,169]. Iturin is a lipopeptide produced by *B. subtilis* that has demonstrated antifungal activity by affecting the morphology and structure of the cell membrane of yeasts. *In vitro* experiments have demonstrated that surfactin can effectively inactivate the virus that causes herpes as well as the retrovirus and other compact RNA and DNA viruses. The antiviral activity of surfactin has been determined for a broad spectrum of viruses. Moreover, surfactin has been found to exert an effect on insulin absorption in the lungs of laboratory rats [108].

Nanaotechnology

Biosurfactants have been used in nanotechnology and nanoparticle synthesis is emerging as part of green chemistry [118,170]. Nickel oxide (NiO) nano-rods can be produced by water-in-oil microemulsions [171]. In one experiment, two microemulsions were formed with the addition of a nickel chloride solution to a biosurfactant and heptane solution, with the addition of ammonium hydroxide to the same hydrocarbon mixture. The centrifuged microemulsions and ethanol was then used to wash the precipitates and remove the biosurfactant and heptane. The use of biosurfactants is a more ecofriendly approach [119]. Reddy *et al.* [172] found that silver nanoparticle synthesis could be stabilised for two months using surfactin, which is a biodegradable, renewable stabilising agent with low toxicity [3,119]. A biosurfactant produced by*P. Aeruginosa* grown in a low-cost medium has been employed to stabilise silver nanoparticles in the liquid phase [173]. The effect of a rhamnolipid on the electrokinetic and rheological behaviour of nano-zirconia particles has also been analysed, although this is not strictly an environmental application [174]. The biosurfactant increasingly adsorbs to zirconia with the increase in concentration and can serve as an ecofriendly product for the flocculation and dispersion of high solid contents of microparticles. New applications for biosurfactants are being developed in the field of nanotechnology. Future research should focus on the stabilisation of the nanoparticles by biosurfactants before addition during remediation processes [119].

FUTURE DIRECTIONS AND CONCLUDING REMARKS

The biosurfactant industry has demonstrated remarkable growth in recent decades, although the large-scale production of these biomolecules remains a challenge from the economic standpoint. This is mainly due to the enormous difference between the financial investment required and viable industrial production. Thus, the following are the main criteria to be considered for biosurfactant production to become truly viable: (a) type of raw materials; (b) continuous provision of the same composition of ingredients; (c) types of microorganisms; (d) the adequate design of industrial fermentors; (e) financial investments; (f) the target market; (g) purification processes; (h) biosurfactant properties; (i) production conditions, especially the time required for fermentation; (j) adequate production yields; and (k) the processing of recycled products (minimal or able to sell for more than the drop in value).

The target market is of fundamental importance to the implantation of an industrial biosurfactant production project. For cosmetic, medicinal and food products, production is only viable on a small-scale, as the column chromatography methods required to separate molecules are not economical on a large scale. Thus, the use of crude fermentation broths could be a viable solution, especially if the application is in an environmental context, as biosurfactants in such cases do not need to be pure and can be synthesized using a blend of inexpensive carbon sources, which would allow the creation of an economically and environmentally viable technology for bioremediation processes.

According to Hazra *et al.* [140], sophorolipids are offered as sophoron TM from Saraya (Osaka, Japan) and Soliance (Pomacle, France), whereas rhamnolipids are available from Ecover (Boulogne-sur-Mer, France), Jeneil Biosurfactant Inc. (Saukville, Wisconsin, USA) and Rhamnolipid Holdings Inc. (New york, USA). Sophorolipid production costs run from 2 to 5 €/kg. Rhamnolipid production costs US$ 20/kg at a volume of 20 m^3, but only US$ 5/kg when produced on a scale of 100 m^3, placing it closer to ethoxylate or alkyl polyglycoside (US$ 1 to 3/kg). The Exxon Company spent more than US$ 10 million in bioremediation studies between 1993 and 1997 after the spillage of petroleum (41 million litres) by the oil tanker Exxon Valdez in Alaska in 1989, leading to the generation of seven patents and making bioremediation second only to enhanced oil recovery within the initial years of use. Distribution in specific biosurfactant areas of the oil industry includes 17 patents for soil and water bioremediation as well as 20 for enhanced oil recovery [140,175].

Although improvements in biosurfactant technology have enabled a 10-to-20-fold increase in the production of these biomolecules, it is likely that further, significant advances (even if of a smaller magnitude) are needed to make this technology commercially viable.

ACKNOWLEDGMENTS

This study received funding from the Brazilian fostering agencies the State of Pernambuco Foundation for the Assistance to Science and Technology (FACEPE); the Research and Development Program of the Brazilian National Electrical Energy Agency (ANEEL); the National Council for Scientific and Technological Development (CNPq), and the Federal Agency for the Support and Evaluation of Graduate Education (CAPES).

AUTHOR CONTRIBUTIONS

Leonie A. Sarubbo proposed the theoretical frame; Danyelle Khadydja F. Santos, Raquel D. Rufino, Juliana M. Luna, Valdemir A. Santos and Leonie A. Sarubbo wrote the paper; Leonie A. Sarubbo performed manuscript editing and final improvement.

REFERENCES

1. Mao, X.; Jiang, R.; Xiao, W.; Yu, J. Use of surfactants for the remediation of contaminated soils: A review. *J. Hazard. Mater.* 2015, *285*, 419–435.
2. Silva, R.C.F.S.; Almeida, D.G.; Luna, J.M.; Rufino, R.D.; Santos, V.A.; Sarubbo, L.A. Applications of biosurfactants in the petroleum industry and the

remediation of oil spills. *Int. J. Mol. Sci.* 2014, *15*, 12523–12542.

3. Banat, I.M.; Franzetti, A.; Gandolfi, I.; Bestetti, G.; Martinotti, M.G.; Fracchia, L.; Smyth, T.J.; Marchant, R. Microbial biosurfactants production, applications. *Appl. Microbiol. Biotechnol.* 2010, *87*, 427–444.
4. Campos, J.M.; Stamford, T.L.M.; Sarubbo, L.A.; Luna, J.M.; Rufino, R.D.; Banat, I.M. Microbial biosurfactants as additives for food industries. *Biotechnol. Prog.* 2013, *29*, 1097–1108.
5. Vijayakumar, S.; Saravanan, V. Biosurfactants-types, sources and applications. *Res. J. Microbiol.* 2015, *10*, 181–192.
6. Cerqueira, V.S.; Hollenbach, E.B.; Maboni, F.; Vainstein, M.H.; Camargo, F.A.; do Carmo, M.; Peralba, R.; Bento, F.M. Biodegradation potential of oily sludge by pure and mixed bacterial cultures. *Bioresour. Technol.* 2011, *102*, 11003–11010.
7. Rosa, C.F.C.; Freire, D.M.G.; Ferraz, E.C. Biosurfactant microfoam: Application in the removal of pollutants from soil.*J. Environ. Chem. Eng.* 2015, *3*, 89–94.
8. Pacwa-Plociniczak, M.; Plaza, G.A.; Piotrowska-Seget, Z.; Cameotra, S.S. Environmental applications of biosurfactants: Recent advances. *Int. J. Mol. Sci.* 2011, *13*, 633–654.
9. Makkar, R.S.; Cameotra, S.S.; Banat, I.M. Advances in utilization of renewable substrates for biosurfactant production.*Appl. Microbiol. Biotechnol.* 2011, *1*, 1–19.
10. Deleu, M.; Paquot, M. From renewable vegetables resources to microorganisms: new trends in surfactants. *C. R. Chimie* 2004, *7*, 641–646.
11. Marchant, R.; Funston, S.; Uzoigwe, C.; Rahman, P.K.S.M.; Banat, I.M. Production of biosurfactants from nonpathogenic bacteria. In *Biosurfactants*; Taylor & Francis: New York, USA, 2014; pp. 73–81.
12. Chakraborty, J.; Das, S. Biosurfactant-based bioremeditaion of toxic metals. In *Microbial Biodegradation and Bioremediation*, 1st ed.; Das, S., Ed.; Elsevier: Rourkela Odisha, India, 2014; pp. 167–201.
13. Shepherd, R.; Rockey, J.; Shutherland, I.W.; Roller, S. Novel bioemulsifiers from microorganisms for use in foods. *J. Biotechnol.* 1995, *40*, 207–217.
14. Bognolo, G. Biossurfactants as emulsifying agents for hydrocarbons. *Coll. Surf. A Physicochem. Eng. Asp.* 1999, *152*, 41–52.
15. Pattanath, K.M.; Rahman, K.S.; Gakpe, E. Production, characterization and applications of biosurfactants—Review.*Biotechnology* 2008, *7*, 360–370.
16. Chrzanowski, L.; Ławniczak, L.; Czaczyk, K. Why do microorganisms produce rhamnolipids? *World J. Microbiol. Biotechnol.* 2012, *28*, 401–419.
17. Abdel-Mawgoud, A.M.; Lépine, F.; Deziel, E. Rhamnolipids: Diversity of structures, microbial origins and roles. *Appl. Microbiol. Biotechnol.* 2010, *86*, 1323–1336.
18. Hitsatsuka, K.; Nakahara, T.; Sano, N.; Yamada, K. Formation of a rhamnolipid by *Pseudomonas aeruginosa* and its function in hydrocarbon fermentation. *Agric. Biol. Chem.* 1971, *35*, 686–692.
19. Nitschke, M.; Costa, S.G.V.A.O.; Contiero, J. Rhamnolipids and PHAs: Recent reports on *Pseudomonas*-derived molecules of increasing industrial interest. *Process Biochem.* 2011, *46*, 621–630.
20. Ławniczak, L.; Marecik, R.; Chrzanowski, L. Contributions of biosurfactants to natural or induced bioremediation.*Appl. Microbiol. Biotechnol.* 2013, *97*, 2327–2339.
21. Chrzanowski, L.; Dziadas, M.; Ławniczak, L.; Cyplik, P.; Białas, W.; Szulc, A.; Lisiecki, P.; Jelen, H. Biodegradation of rhamnolipids in liquid cultures: Effect of biosurfactant dissipation on diesel fuel/B20 blend biodegradation efficiency and bacterial community composition. *Bioresour. Technol.* 2012, *111*, 328–335.
22. Szulc, A.; Ambrozewicz, D.; Sydow, M.; Ławniczak, L.; Piotrowska-Cyplik, A.; Marecik, R.; Chrzanowski, L. The influence of bioaugmentation and biosurfactant addition on bioremediation efficiency of diesel-oil contaminated soil: Feasibility during field studies. *J. Environ. Manag.* 2014, *132*, 121–128.
23. Marecik, M.; Wojtera-Kwiczor, J.; Ławniczak, L.; Cyplik, P.; Szulc, A.; Piotrowska-Cyplik, A.; Chrzanowski, L. Rhamnolipids increase the phytotoxicity of diesel oil towards four common plant species in a terrestrial environment.*Water Air Soil Pollut.* 2012, *223*, 4275–4282.
24. Chrzanowski, L.; Wick, L.Y.; Meulenkamp, R.; Kaestner, M.; Heipieper, H.J. Rhamnolipid biosurfactants decrease the toxicity of chlorinated phenols to *Pseudomonas putida* DOT-T1E. *Lett. Appl. Microbiol.* 2009, *48*, 756–762.
25. Chrzanowski, L.; Owsianiak, M.; Szulc, A.; Marecik, R.; Piotrowska-Cyplik, A.; Olejnik-Schmidt, A.K.; Staniewski, J.; Lisiecki, P.; Ciesielczyk, F.; Jesionowski, T.; *et al.* Interactions between rhamnolipid biosurfactants and toxic chlorinated phenols enhance biodegradation of a model hydrocarbon-rich effluent. *Int. Biodeterior. Biodegrad.* 2011, *65*, 605–611.
26. Cortés-Sánchez, A.J.; Sánchez, H.H.; Jaramillo-Flores, M.E. Biological activity of glycolipids produced by microorganisms: New trends and possible therapeutic alternatives. *Microbiol. Res.* 2013, *168*, 22–32.
27. Daverey, A.; Pakshirajan, K. Production, characterization, and properties of sophorolipids from the yeast *Candida bombicola* using a low-cost fermentative medium. *Appl. Biochem. Biotechnol.* 2009, *158*, 663–674.
28. Gautam, K.K.; Tyagi, V.K. Microbial Surfactants: A review. *J. Oleo Sci.* 2006, *55*, 155–166.
29. Hu, Y.; Ju, L.K. Purification of lactonic sophorolipids by crystallization. *J. Biotechnol.* 2001, *87*, 263–272.
30. Díaz De Rienzo, M.A.D.; Banat, I.M.; Dolman, B.; Winterburn, J.; Martin, P.J. Sophorolipid biosurfactants: Possible uses as antibacterial and antibiofilm agent. *New Biotechnol.* 2015.
31. Lang, S. Biological amphiphiles (microbial biosurfactants). *Curr. Opin. Coll. Interface Sci.* 2002, *7*, 12–20.

32. Hatha, A.A.M.; Edward, G.; Rahman, K.S.M.P. Microbial biosurfactants-review. *J. Mar. Atmos. Res.* 2007, *3*, 1–17.

33. Chakrabarti, S. Bacterial Biosurfactant: Characterization, Antimicrobial and Metal Remediation Properties. Ph.D. Thesis, National Institute of Technology, Surat, India, March 2012.

34. Barros, F.F.C.; Quadros, C.P.; Maróstica, M.R.; Pastore, G.M. Surfactina: Propriedades químicas, tecnológicas e funcionais para aplicações em alimentos. *Quím. Nova* 2007, *30*, 1–14.

35. Velikonja, J.; Kosaric, N. Biosurfactants in food applications. In *Biosurfactants: Production, Properties, Applications*; Kosaric, N., Sukan, F.V., Eds.; CRC Press: New York, NY, USA, 1993; pp. 419–448.

36. Cirigliano, M.C.; Carman, G.M. Purification and characterization of liposan, a bioemulsifier from *Candida lipolytica*.*Appl. Microbiol. Biotechnol.* 1985, *50*, 846–850.

37. Desai, J.D.; Banat, I.M. Microbial production of surfactants and their commercial potential. *Microbiol. Mol. Biol. Rev.*1997, *61*, 47–64.

38. Bhardwaj, G.; Cameotra, S.S.; Chopra, H.K. Biosurfactants from Fungi: A Review. *Pet. Environ. Biotechnol.* 2013, *4*, 1–6.

39. Cooper, D.G.; Paddock, D.A. Production of a biosurfactant from *Torulopsis bombicola*. *Appl. Environ. Microbiol.* 1984, *47*, 173–176.

40. Kim, S.Y.; Oh, D.K.; Lee, K.H.; Kim, J.H. Effect of soybean oil and glucose on sophorose lipid fermentation by*Torulopsis bombicola* in continuous culture. *Appl. Microbiol. Biotechnol.* 1997, *48*, 23–26.

41. Casas, J.A.; de Lara, S.G.; Garcia-Ochoa, F. Optimization of a synthetic medium for *Candida bombicola* growth using factorial design of experiments. *Enzyme Microb. Technol.* 1997, *21*, 221–229.

42. Sarubbo, L.A.; Farias, C.B.B.; Campos-Takaki, G.M. Co-utilization of canola oil and glucose on the production of a surfactant by *Candida lipolytica*. *Curr. Microbiol.* 2007, *54*, 68–73.

43. Rufino, R.D.; Sarubbo, L.A.; Campos-Takaki, G.M. Enhancement of stability of biosurfactant produced by *Candida lipolytica* using industrial residue as substrate. *World J. Microbiol. Biotechnol.* 2007, *23*, 729–734.

44. Sarubbo, L.A.; Porto, A.L.F.; Campos-Takaki, G.M. Bioemulsifier production in batch culture using glucose as carbon source by *Candida lipolytica*. *Appl. Biochem. Biotechnol.* 2001, *95*, 59–67.

45. Bednarski, W.; Adamczak, M.; Tomasik, J.; Plaszvzyk, M. Application of oil refinery waste in the biosynthesis of glycolipids by yeast. *Bioresour. Technol.* 2004, *95*, 15–18.

46. Kitamoto, D.; Ikegami, T.; Suzuki, G.T.; Sasaki, A.; Takeyama, Y.; Idemoto, Y.; Koura, N.; Yanagishita, H. Microbial conversion of n-alkanes into glycolipid biosurfactants, mannosylerythritol lipids by *Pseudozyma* (*Candida antartica*).*Biotechnol. Lett.* 2001, *23*, 1709–1714.

47. Amezcua-Veja, C.; Poggi-Varaldo, H.M.; Esparza-Garcia, F.; Rodriguez-Vazquez, R. Effect of culture conditions on fatty acids composition of a biosurfactant produced by *Candida ingens* and of surface tension of culture media.*Bioresour. Technol.* 2006, *98*, 237–240.

48. Robert, M.; Mercadé, M.E.; Bosch, M.P.; Parra, J.L.; Espiny, M.J.; Manresa, M.A.; Guinea, J. Effect of the carbon source on biosurfactant production by *Pseudomonas aeruginosa* 44T1. *Biotechnol. Lett.* 1989, *11*, 871–874.

49. Santa Anna, I.M.; Sebastian, G.V.; Pereira, N., Jr.; Alves, T.L.M.; Menezes, E.P.; Freire, D.M.G. Production of biosurfactant from a new and promissing strain of *Pseudomonas aeruginosa* PA1. *Appl. Biochem. Biotechnol.* 2002, *91*, 459–467.

50. Mulligan, C.N.; Gibbs, B.F. Correlation of nitrogen metabolism with biosurfactant production by *Pseudomonas aeruginosa*. *Appl. Environ. Microbiol.* 1989, *55*, 3016–3019.

51. Deshpande, M.; Daniels, L. Evaluation of sophorolipid biosurfactant production by *Candida bombicola* using animal fat.*Bioresour. Technol.* 1995, *54*, 143–150.

52. Kim, H.S.; Yoon, B.D.; Choung, D.H.; Oh, H.M.; Katsuragi, T. Characterization of a biosurfactant mannosylerythritol lipid produced from *Candida* sp. SY 16. *Appl. Microbiol. Biotechnol.* 1999, *52*, 713–721.

53. Rufino, R.D.; Sarubbo, L.A.; Benicio, B.N.; Campos-Takaki, G.M. Experimental design for the production of tensio-active agent by *Candida lipolytica*. *J. Ind. Microbiol. Biotechnol.* 2008, *35*, 907–914.

54. Sarubbo, L.A.; Luna, J.M.; Campos-Takaki, G.M. Production and stability studies of the bioemulsifier obtained from a new strain of *Candida glabrata* UCP 1002. *Electr. J. Biotechnol.* 2006, *9*, 400–406.

55. Cirigliano, M.C.; Carman, G.M. Isolation of a bioemulsifier from *Candida lipolytica*. *Appl. Environ. Microbiol.* 1984, *48*, 747–750.

56. Konishi, M.; Fukuoka, T.; Morita, T.; Imura, T.; Kitamoto, D. Production of new types of sophorolipids by *Candida batistae*. *J. Oleo Sci.* 2008, *57*, 359–369.

57. Kiran, G.S.; Hema, T.A.; Gandhimathi, R.; Selvin, J.; Thomas, T.A. Optimization and production of a biosurfactant from the sponge-associated marine fungus *Aspergillus ustus* MSF3. *Coll. Surf. B Biointerfaces* 2009, *73*, 250–256.

58. Thaniyavarn, J.; Chianguthai, T.; Sangvanich, P.; Roongsawang, N.; Washio, K.; Morikawa, M.; Thaniyavarn, S. Production of sophorolipid biosurfactant by *Pichia anomala*. *Biosci. Biotechnol. Biochem.* 2008, *72*, 2061–2068.

59. Cavalero, D.A.; Cooper, D.G. The effect of medium composition on the structure and physical state of sophorolipids produced by *Candida bombicola* ATCC 22214. *J. Biotechnol.* 2003, *103*, 31–41.

60. Felse, P.A.; Shah, V.; Chan, J.; Rao, K.J.; Gross, R.A. Sophorolipid biosynthesis by *Candida bombicola* from industrial fatty acid residues. *Enzyme Microb. Technol.* 2007, *40*, 316–323.

61. Silva, S.N.R.L.; Farias, C.B.B.; Rufino, R.D.; Luna, J.M.; Sarubbo, L.A. Glycerol as substrate for the production of biosurfactant by *Pseudomonas aeruginosa* UCP0992. *Coll. Surf. B Biointerfaces* 2010, *79*, 174–183.

62. Oliveira, F.J.S.; Vazquez, L.; de Campos, N.P.; de Franca, F.P. Production of rhamnolipids by a *Pseudomonas alcaligenes*strain. *Process Biochem.* 2009, *44*, 383–389.

63. Cunha, C.D.; Rosario, M.; Rosado, A.S.; Leite, G.F. *Serratia* sp. SVGG16: A promising biosurfactant producer isolated from tropical soil during growth with ethanol-blended gasoline. *Process Biochem.* 2004, *39*, 2277–2282.

64. Weber, L.; Doge, C.; Haufe, G.; Hommel, R.; Kleber, H.-P. Oxygenation of hexadecane in the biosynthesis of cyclic glycolipids in *Torulopsis apicola*. *Biocatal* 1992, *5*, 267–292.

65. Haritash, A.K.; Kaushik, C.P. Biodegradation aspects of Polycyclic Aromatic Hydrocarbons (PAHs): A review. *J. Hazard. Mater.* 2009, *169*, 1–15.

66. Hommel, R.K.; Huse, K. Regulation of sophorose lipid production by *Candida* (*Torulopsis*) *apicola*. *Biotechnol. Lett.* 1993,*15*, 853–858.

67. Tokumoto, Y.; Nomura, N.; Uchiyama, H.; Imura, T.; Morita, T.; Fukuoka, T.; Kitamoto, D. Structural characterization and surface-active properties of a succinoyl trehalose lipid produce by *Rhodococcus* sp. SD-74. *J. Oleo Sci.* 2009, *58*, 97–102.

68. Syldatk, C.; Wagner, F. Production of biosurfactants. In *Biosurfactants and Biotechnology*; Kosaric, N., Cairns, W.L., Gray, N.C.C., Eds.; Marcel Dekker: New York, NY, USA, 1987; pp. 89–120.

69. Tan, H.M. Biosurfactants and their roles in bioremediation. *Cheong Jit Kong* 2000, 1–12.

70. Banat, I.M.; Satpute, S.K.; Cameotra, S.S.; Patil, R.; Nyayanit, N. Cost effective technologies and renewable substrates for biosurfactants' production. *Front. Microbiol.* 2014, *5*.

71. Makkar, R.S.; Cameotra, S.S. An update on the use of uncoventional substrates for biosurfactant production and their new applications. *Appl. Micobiol. Biotechnol.* 2002, *58*, 428–434.

72. Batista, R.M.; Rufino, R.D.; Luna, J.M.; Souza, J.E.G.; Sarubbo, L.A. Effect of Medium components on the production of a biosurfactant from *Candida tropicalis* applied to the removal of hydrophobic contaminants in soil. *Water Environ. Res.*2010, *82*, 418–425.

73. Mercade, M.E.; Manresa, A.; Robert, M.; Espuny, M.J.; Deandres, C.; Guinea, J. Olive oil mill effluent (OOME). New substrat for biosurfactant production. *Bioresour. Technol.* 1993, *43*, 1–6.

74. Fox, S.I.; Bala, G.A. Production of surfactant from *Bacillus subtilis* ATTCC 21332 using potato substrates. *Bioresour. Technol.* 2000, *75*, 235–240.

75. Thompson, D.N.; Fox, S.L.; Bala, G.A. Biosurfactants from potato process effluents. *Appl. Biochem. Biotechnol.* 2000, *84*, 917–930.

76. Maneerat, S. Production of biosurfactants using substrates from renewable-resources. *Songklanakarin J. Sci. Technol.*2005, *27*, 675–683.

77. Santos, D.K.F.; Brandão, Y.B.; Rufino, R.D.; Luna, J.M.; Salgueiro, A.A.; Santos, V.A.; Sarubbo, L.A. Optimization of cultural conditions for biosurfactant production from *Candida lipolytica*. *Biocatal. Agric. Biotechnol.* 2014, *3*, 48–57.

78. Santos, D.K.F.; Rufino, R.D.; Luna, J.M.; Santos, V.A.; Salgueiro, A.A.; Sarubbo, L.A. Synthesis and evaluation of biosurfactant produced by *Candida lipolytica* using animal fat and corn steep liquor. *J. Pet. Sci. Eng.* 2013, *105*, 43–50.

79. Gusmão, C.A.B.; Rufino, R.D.; Sarubbo, L.A. Laboratory production and characterization of a new biosurfactant from*Candida glabrata* UCP 1002 cultivated in vegetable fat waste applied to the removal of hydrophobic contaminant.*World J. Microbiol. Biotechnol.* 2010, *26*, 1683–1692.

80. Alcântara, R.; Amores, J.; Canoira, L.; Fidalgo, E.; Franco, M.J.; Navarro, A. Catalytic production of biodiesel from soy-hean oil, used frying oil and tallow. *Biomass Bioenergy* 2000, *18*, 515–527.

81. Cvengros, Z.; Cvengrosova, Z. Used fruing oils and fats and their utilization in the production of methyl ester of higher fatty acids. *Biomass Bioenergy* 2004, *27*, 173–181.

82. Haba, E.; Espuny, M.J.; Busquets, M.; Manresa, A. Screening and production of rhamnolipids *Pseudomonas aeruginosa*47T2 NCIB 40044 from waste flying oils. *J. Appl. Microbiol.* 2000, *88*, 379–387.

83. Benincasa, M.; Contiero, J.; Manresa, M.A.; Moraes, I.O. Rhamnolipid production by *Pseudomonas aeruginosa* LBI growing on soapstock as the sole carbon source. *J. Food Eng.* 2002, *54*, 283–288.

84. Shabtai, Y. Production of exopolysaccharides by *Acinetobacter* strains in a controlled fed-batch fermentation process using soap stock oil (SSO) as carbon source. *Int. J. Biol. Macromol.* 1990, *12*, 145–152.

85. Ghurye, G.L.; Vipulananda, C.; Wilson, R.C. A practical approach to biosurfactant production using nonaseptic fermantation of mixed cultures. *Biotechnol. Bioeng.* 1994, *44*, 661–666.

86. Kalogiannis, S.; Iakovidou, G.; Liakopoulou-Kyriakides, M.; Kyriakidis, D.A.; Skaracis, G.N. Optimization of xanthan gum production by *Xanthomonas campestris* grown in molasses. *Process Biochem.* 2003, *39*, 249–256.

87. Lazaridou, A.; Roukas, T.; Biliaderis, C.G.; Vaikousi, H. Characterization of pullulan produced from beet molasses by*Aureobasidium pullulans* in a stirred tank reactor under varying agitation. *Enzyme Microbiol. Technol.* 2002, *31*, 122–132.

88. Makkar, R.S.; Cameotra, S.S. Utilization of molasses for biosurfactant production by two *Bacillus* strains at thermophilic conditions. *J. Am. Oil Chem. Soc.* 1997, *74*, 887–889.

89. Sudhakar-Babu, P.; Vaidya, A.N.; Bal, A.S.; Kapur, R.; Juwarka, A.; Khanna, P. Kinetics of biosurfactants production by *Pseudomonas aeruginosa* strain from industrial wastes. *Biotechnol. Lett.* 1996, *18*, 263–268.

90. Luna, J.M.; Rufino, R.D.; Albuquerque, C.D.C.; Sarubbo, L.A.; Campos-Takaki, G.M. Economic optimized medium for tenso-active agent production by *Candida sphaerica* UCP 0995 and application in the removal of hydrophobic contaminant from sand. *Int. J. Mol. Sci.* 2011, *12*, 2463–2476.

91. Rufino, R.D.; Luna, J.M.; Rodrigues, G.I.B.; Campos-Takaki, G.M.; Sarubbo, L.A.; Ferreira, S.R.M. Application of a yeast biosurfactant in the removal of heavy metals and hydrophobic contaminant in a soil used as slurry barrier. *Appl. Environ. Soil Sci.* 2011, *2011*.

92. Sobrinho, H.B.S.; Rufino, R.D.; Luna, J.M.; Salgueiro, A.A.; Campos-Takaki, G.M.; Leite, L.F.C.; Sarubbo, L.A. Utilization of two agroindustrial by-products for the production of a surfactant by *Candida sphaerica* UCP 0995. *Process Biochem.* 2008, *43*, 912–917.

93. Nitschke, M.; Ferraz, C.; Pastore, G.M. Selection of microorganisms for biosurfactant production using agroindustrial wastes. *Braz. J. Microbiol.* 2004, *435*, 81–85.

94. Luna, J.M.; Rufino, R.D.; Sarubbo, L.A.; Campos-Takaki, G.M. Characterisation, surface properties and biological activity of a biosurfactant produced from industrial waste by *Candida sphaerica* UCP 0995 for application in the petroleum industry. *Coll. Surf. B Biointerfaces* 2013, *102*, 202–209.

95. Luna, J.M.; Rufino, R.D.; Sarubbo, L.A.; Rodrigues, L.R.M.; Teixeira, J.A.C.; Campos-Takaki, G.M. Evaluation antimicrobial and antiadhesive properties of the biosurfactant Lunasan produced by *Candida sphaerica* UCP 0995.*Curr. Microbiol.* 2011, *62*, 1527–1534.

96. Sarubbo, L.A.; Marçal, M.C.; Neves, M.L.C.; Porto, A.L.F.; Campos-Takaki, G.M. The use of babassu oil as substrate to produce bioemulsifiers by *Candida lipolytica*. *Can. J. Microbiol.* 1999, *45*, 1–4.

97. Raza, Z.A.; Khan, M.S.; Khalid, Z.M.; Rehman, A. Production kinetics and tensioactive characteristics of biosurfactant from a *Pseudomonas aeruginosa* mutant grown on waste frying oils. *Biotechnol. Lett.* 2006, *28*, 1623–1631.

98. Raza, Z.A.; Rehman, A.; Khan, M.S.; Khalid, Z.M. Improved production of biosurfactant by a *Pseudomonas aeruginosa*mutant using vegetable oil refinery wastes. *Biodegradation* 2007, *18*, 115–121.

99. Joshi, S.; Bharucha, C.; Jha, S.; Yadav, S.; Nerurkar, A.; Desai, A.J. Biosurfactant production using molasses and whey under thermophilic conditions. *Bioresour. Technol.* 2007, *99*, 195–199.

100. Dubey, K.V.; Juwarkar, A.A. Distillery and curd whey wastes as viable alternative sources for biosurfactant production. *World J. Microbiol. Biotechnol.* 2001, *17*, 61–69.

101. Dubey, K.V.; Juwarkar, A.A.; Singh, S.K. Adsorption-desorption process using woodbased activated carbon for recovery of biosurfactant from fermented distillery wastewater. *Biotechnol. Prog.* 2005, *21*, 860–867.

102. Rodrigues, L.R.; Teixeira, J.A.; Oliveira, R. Low-cost fermentative medium for biosurfactant production by probiotic bacteria. *Biochem. Eng. J.* 2006, *32*, 135–142.

103. Helmy, Q.; Kardena, E.; Funamizu, N.; Wisjnuprapto, W. Strategies toward commercial scale of biosurfactant production as potential substitute for it's chemically counterparts. *Int. J. Biotechnol.* 2011, *12*, 66–86.

104. Akhtar, M.; Lentz, M.J.; Banchette, R.A.; Kirk, T.K. Corn steep liquor lowers the amount of inoculum for biopulping.*TAPPI J.* 1997, *80*, 161–164.

105. Cardinal, E.V.; Hedrick, L.R. Microbiological assay of corn steep liquor for amino acid content. *J. Biol. Chem.* 1948, *172*, 609–612.

106. Luna, J.M.; Rufino, R.D.; Jara, A.M.T.; Brasileiro, P.P.F.; Sarubbo, L.A. Environmental applications of the biosurfactant produced by *Candida sphaerica* cultivated in low-cost substrates. *Coll. Surf. A Physicochem. Eng. Asp.* 2015, *480*, 413–418.

107. Silva, N.M.P.R.; Rufino, R.D.; Luna, J.M.; Santos, V.A.; Sarubbo, L.A. Screening of *Pseudomonas* species for biosurfactant production using low-cost substrates. *Biocatal. Agric. Biotechnol.* 2014, *3*, 132–139.

108. Muthusamy, K.; Gopalakrishnan, S.; Ravi, T.K.; Sivachidambaram, P. Biosurfactants: Properties, commercial production and application. *Curr. Sci.* 2008, *94*, 736–747.

109. Marchant, R.; Banat, I.M. Biosurfactants: A sustainable replacement for chemical surfactants? *Biotechnol. Lett.* 2012, *34*, 1597–1605.

110. Winterburn, J.B.; Russell, A.B.; Martin, P.J. Integrated recirculating foam fractionation for the continuous recovery of biosurfactant from fermenters. *Biochem. Eng. J.* 2011, *54*, 132–139.

111. Rangarajan, V.; Sem, R. An inexpensive strategy for facilitated recovery of metals and fermentation products by foam fractionation process. *Coll. Surf. B Biointerfaces* 2013, *104*, 99–106.

112. Rodrigues, L.R.; Banat, I.M.; Teixeira, J.A.; Oliveira, R. Biosurfactants: Potential applications in medicine. *J. Antimicrob. Chemother.* 2006, *57*, 609–618.

113. Sobrinho, H.B.S.; Luna, J.M.; Rufino, R.D.; Porto, A.L.F.; Sarubbo, L.A. Assessment of toxicity of a biosurfactant from*Candida sphaerica* UCP 0995 cultivated with industrial residues in a bioreactor. *Electron. J. Biotechnol.* 2013, *16*.

114. US Energy Information Administration. Annual Energy Review, 2010. Available online: http://large.stanford.edu/courses/2012/ph241/druzgalski2/docs/aer.pdf (accessed on 05 September 2015).

115. Summers, D. Enhancing Oil Recovery: A Look at Stripper Wells. Available online: http//www.oilprice.com (accessed on 10 January 2011).

116. Austad, T.; Taugbøl, K. Chemical flooding of oil reservoirs 1. Low tension polymer flood using a polymer

gradient in the three-phase region. *Coll. Surf. A Phys. Eng. Asp.* 1995, *101*, 87–97.

117. West, C.C.; Harwell, J.H. Surfactant and subsurface remediation. *Environ. Sci. Technol.* 1992, *26*, 2324–2330.
118. Sabatini, D.A.; Harwell, J.H.; Knox, R.C. Surfactant selection criteria for enhanced subsurface remediation. In*Innovative Subsurface Remediation*; Brusseau, M.L., Sabatini, D.A., Gierke, J.S., Annable, M.D., Eds.; American Chemical Society: Washington, WA, USA, 1999; pp. 8–23.
119. Sabatini, D.A.; Knox, R.C.; Harwell, J.H.; Wu, B. Integrated design of surfactant enhanced DNAPL remediation: Efficient super solubilization and gradient systems. *J. Contam. Hydrol.* 2000, *45*, 99–121.
120. Souza, E.C.; Vessoni-Penna, T.C.; Souza Oliveira, R.P. Biosurfactant-enhanced hydrocarbon bioremediation: An overview. *Int. Biodeterior. Biodegrad.* 2014, *89*, 88–94.
121. Sarubbo, L.A.; Rocha, R.B., Jr.; Luna, J.M.; Rufino, R.D.; Santos, V.A.; Banat, I.M. Some aspects of heavy metals contamination remediation and role of biosurfactants. *Chem. Ecol.* 2015.
122. Mnif, I.; Ghribi, D. Microbial derived surface active compounds: Properties and screening concept. *World J. Microbiol. Biotechnol.* 2015, *31*, 1001–1020.
123. Banat, I.M.; De Rienzo, M.A.D.; Quinn, G.A. Microbial biofilms: Biosurfactants as antibiofilm agents. *Appl. Microbiol. Biotechnol.* 2014, *98*, 9915–9929.
124. Sachdev, D.P.; Cameotra, S.S. Biosurfactants in agriculture. *Appl. Microbiol. Biotechnol.* 2013, *97*, 1005–1016.
125. Mulligan, C.N. Recent advances in the environmental applications of biosurfactants. *Curr. Opin. Coll. Interface Sci.*2009, *14*, 372–378.
126. Environmental Protection Agency (EPA), U.S. *Treatment Technologies for Site Cleanup: Annual Status Report*, 10th ed.; Office of Solid Waste and Emergency Response: Cincinnati, Ohio, USA, 2001.
127. Satpute, S.K.; Banat, I.M.; Dhakephalkar, P.K.; Banpurkar, A.G.; Chopade, B.A. Biosurfactants, bioemulsifiers and exopolysaccharides from marine microorganisms. *Biotechnol. Adv.* 2010, *28*, 436–450.
128. Shavandi, M.; Mohebali, G.; Haddadi, A.; Shakarami, H.; Nuhi, A. Emulsification potential of a newly isolated biosurfactantproducing bacterium, *Rhodococcus* sp. strain TA6. *Coll. Surf. B Biointerfaces* 2011, *82*, 477–482.
129. Kuyukina, M.S.; Ivshina, I.B.; Makarov, S.O.; Litvinenko, L.V.; Cunningham, C.J.; Philip, J.C. Effect of biosurfactants on crude oil desorption and mobilization in a soil system. *Environ. Int.* 2005, *31*, 155–161.
130. Aparna, A.; Srinikethan, G.; Hedge, S. Effect of addition of biosurfactant produced by *Pseudomonas* ssp. on biodegradation of crude oil. In International Proceedings of Chemical, Biological & Environmental Engineering, Proceedings of the 2nd International Proceedings of Chemical, Singapore, Singapore, 26–28 February 2011; Volume 6, pp. 71–75.
131. Franzetti, A.; Caredda, P.; Ruggeri, C.; La Colla, P.; Tamburini, E.; Papacchini, M.; Bestetti, G. Potential applications of surface active compounds by *Gordonia* sp. strain BS29 in soil remediation technologies. *Chemosphere* 2009, *75*, 810–807.
132. Bai, G.; Brusseau, M.L.; Miller, R.M. Biosurfactant-enhanced removal of residual hydrocarbon from soil. *J. Contam. Hydrol.* 1997, *25*, 157–170.
133. Silva, R.C.F.S.; Rufino, R.D.; Luna, J.M.; Farias, C.B.B.; Filho, H.J.B.; Santos, V.A.; Sarubbo, L.A. Enhancement of biosurfactant production from *Pseudomonas cepacia* CCT6659 through optimisation of nutritional parameters using Response Surface Methodology. *Tenside Surf. Det.* 2013, *50*, 137–142.
134. Chaprão, M.J.; Ferreira, I.N.S.; Correa, P.F.; Rufino, R.D.; Luna, J.M.; Silva, E.J.; Sarubbo, L.A. Application of bacterial and yeast biosurfactants for enhanced removal and biodegradation of motor oil from contaminated sand. *Electron. J. Biotechnol.* 2015, *18*, 471–479.
135. Rufino, R.D.; Luna, J.M.; Campos Takaki, G.M.; Sarubbo, L.A. Characterization and properties of the biosurfactant produced by *Candida lipolytica* UCP 0988. *Electron. J. Biotechnol.* 2014, *17*, 34–38.
136. Rufino, R.D.; Luna, J.M.; Marinho, P.H.C.; Farias, C.B.B.; Ferreira, S.R.M.; Sarubbo, L.A. Removal of petroleum derivative adsorbed to soil by biosurfactant Rufisan produced by *Candida lipolytica*. *J. Pet. Sci. Eng.* 2013, *109*, 117–122.
137. Rufino, R.D.; Luna, J.M.; Sarubbo, L.A.; Rodrigues, L.R.M.; Teixeira, J.A.C.; Campos-Takaki, G.M. Antimicrobial and anti-adhesive potential of a biosurfactant Rufisan produced by *Candida lipolytica* UCP 0988. *Coll. Surf. B Biointerfaces*2011, *84*, 1–5.
138. Mulligan, C.N.; Wang, S. Remediation of a heavy metal contaminated soil by a rhamnolipid foam. In *Geoenvironmental Engineering: Integrated Management of Groundwater and Contaminated Land*; Thomas Telford: London, UK, 2004; pp. 544–551.
139. Singh, A.; Van-Hamme, J.D.; Ward, O.P. Surfactants in microbiology and biotechnology: Part 2. Application aspects.*Biotechnol. Adv.* 2007, *25*, 99–121.
140. Hazra, C.; Kundu, D.; Chaudhari, A. Biosurfactant-assisted bioaugmentation in bioremediation. In *Microorganisms in Environmental Management: Microbes and Environment*, 1st ed.; Satyanarayana, T., Johri, B.N., Prakash, A., Eds.; Springer: New York, NY, USA, 2012; pp. 631–664.
141. Ochoa-Loza, F.J.; Artiola, J.F.; Maier, R.M. Stability constants for the complexation of various metals with a rhamnolipid biosurfactant. *J. Environ. Qual.* 2001, *30*, 479–485.
142. Swarnkar, V.; Agrawal, N.; Tomar, R. Sorption of chromate and arsenate bysurfactant modified erionite (E-SMZ). *J. Dispers. Sci. Technol.* 2012, *33*, 919–927.

143. Hashim, M.A.; Mukhopadhyay, S.; Sahu, J.N.; Sengupta, B. Remediation technologies for heavy metal contaminated groundwater. *J. Environ. Manag.* 2011, *92*, 2355–2388.

144. Asçi, Y.; Nurbas, M.; Açikel, Y. A comparative study for the sorption of Cd(II) by soils with different clay contents and mineralogy and the recovery of Cd(II) using rhamnolipid biosurfactant. *J. Hazard. Mater.* 2008, *154*, 663–673.

145. Nash, J.; Traver, R.P.; Downey, D.C. Surfactant enhanced *in situ* soil washing. US.EPA HWERL: Edilsin, NJ, USA, 1987.

146. Herman, D.C.; Artiola, J.F.; Miller, R.M. Removal of cadmium, lead and zinc from soil by a rhamnolipid biosurfactant.*Environ. Sci. Technol.* 1995, *29*, 2280–2285.

147. Mulligan, C.N.; Yong, R.N.; Gibbs, B.F. Metal removal from contaminated soil and sediments by the biosurfactant surfactin. *Environ. Sci. Technol.* 1999, *33*, 3812–3820.

148. Ochoa-Loza, F.J.; Noordman, W.H.; Jannsen, D.B.; Brusseau, M.L.; Maier, R.M. Effect of clays, metal oxides, and organic matter on rhaminolipid biosurfactant sorpition by soil. *Chemosphere* 2007, *66*, 1634–1642.

149. Das, P.; Mukherjee, S.; Sen, R. Antiadhesive action of a marine microbial surfactant. *Coll. Surf. B Biointerfaces* 2009, *71*, 183–186.

150. Wen, J.; Stacey, S.P.; Mclaughlin, M.J.; Kirby, J.K. Biodegradation of rhamnolipid, EDTA and citric acid in cadmium and zinc contaminated soils. *Soil Biol. Biochem.* 2009, *41*, 2214–2221.

151. Kitamoto, D.; Isoda, H.; Nakahara, T. Functions and potential applications of glycolipid biosurfactants: From energy-saving materials to gene delivery carriers. *J. Biosci. Bioeng.* 2002, *94*, 187–201.

152. Nitschke, M.; Pastore, G.M. Biossurfactantes: Propriedades e aplicações. *Química Nova* 2002, *25*, 772–776.

153. Slizovskiy, I.B.; Kelsey, J.W.; Hatzinger, P.B. Surfactant-facilitated remediationof metal-contaminated soils: Efficacy and toxicological consequences to earth worms. *Environ. Toxicol. Chem.* 2011, *30*, 112–123.

154. Almeida, C.M.R.; Dias, A.C.; Mucha, A.P.; Bordalo, A.A.; Vasconcelos, M.T.S.D. Influence of surfactants on the Cu phytoremediation potential of a saltmarsh plant. *Chemosphere* 2009, *75*, 135–140.

155. Maity, J.P.; Huang, Y.M.; Fan, C.-W.; Chen, C.-C.; Li, C.-Y.; Hsu, C.-M.; Chang, Y.-F.; Wu, C.-I.; Chen, C.-Y.; Jean, J.-S. Evaluation of remediation process with soapberry derived saponin for removal of heavy metals from contaminated soils in Hai-Pu Taiwan. *J. Environ. Sci.* 2013, *25*, 1180–1185.

156. Ozturk, S.; Kaya, T.; Aslim, B.; Tan, S. Removal and reduction of chromium by *Pseudomonas* spp. and their correlation to rhamnolipid production. *J. Hazard. Mater.* 2012, *231*, 64–69.

157. Albuquerque, C.F.; Luna-Finkler, C.L.; Rufino, R.D.; Luna, J.M.; Menezes, C.T.B.; Santos, V.A.; Sarubbo, L.A. Evaluation of biosurfactants for removal of heavy metal ions from aqueous effluent using flotation techniques. *Int. Rev. Chem. Eng.* 2012, *4*, 156–161.

158. Menezes, C.T.B.; Barros, E.C.; Rufino, R.D.; Luna, J.M.; Sarubbo, L.A. Replacing synthetic with microbial surfactants as collectors in the treatment of aqueous effluent produced by acid mine drainage using the dissolved air flotation technique. *Appl. Biochem. Biotechnol.* 2011, *163*, 540–546.

159. Campos, J.M.; Stamford, T.L.M.; Sarubbo, L.A. Production of a Bioemulsifier with Potential Application in the Food Industry. *Appl. Biochem. Biotechnol.* 2014, *172*, 3234–3252.

160. Campos, J.M.; Stamford, T.L.M.; Rufino, R.D.; Luna, J.M.; Stamford, T.C.M.; Sarubbo, L.A. Formulation of mayonnaise with the addition of a bioemulsifier isolated from *Candida utilis*. *Toxicol. Rep.* 2015, *2*, 1164–1170.

161. Torabizadeh, H.; Shojaosadati, S.A.; Tehrani, H.A. Preparation and characterization of bioemulsifier from*Saccharomyces cerevisiae*. *Lebensm. Wiss. Technol.* 1996, *29*, 734–737.

162. Abalos, A.; Pinazo, A.; Infante, M.R.; Casals, M.; García, F.; Manresa, A. Physicochemical and antimicrobial properties of new rhamnolipids produced by *Pseudomonas aeruginosa* AT10 from soybean oil refinery wastes. *Langmuir* 2001, *17*, 1367–1371.

163. Joshi-Navare, K.; Prabhune, A. A biosurfactant sophorolipid acts in synergy with antibiotics to enhance their efficiency. *BioMed. Res. Int.* 2013, *2013*, 1–8.

164. Fracchia, L.; Banat, J.J.; Cavallo, M.; Ceresa, C.; Banat, I.M. Potential therapeutic applications of microbial surface-active compounds. *Bioengineering* 2015, *2*, 144–162.

165. Robbel, L.; Marahiel, M.A. Daptomycin, a bacterial lipopeptide synthesized by a nonribosomal machinery. *J. Biol. Chem.* 2010, *285*, 27501–27508.

166. Tally, F.P.; Zeckel, M.; Wasilewski, M.M. Daptomycin, A novel agent for Gram-positive infections. *Expert Opin. Investig. Drug* 1999, *8*, 1223–1238.

167. Zottola, E.A. Microbial attachment and biofilm formation: A new problem for the food industry? *Food Technol.* 1994,*48*, 107–114.

168. Meylheuc, T.; Van Oss, C.J.; Bellon-Fontaine, M.N. Adsorption of biosurfactants on solid surfaces and consequences regarding the bioahesion of *Listeria monocytogenes* LO28. *J. Appl. Microbiol.* 2001, *91*, 822–832.

169. Kitamoto, D.; Morita, T.; Fukuoka, T.; Konishi, M.; Imura, T. Self-assembling properties of glycolipid biosurfactants and their potential applications. *Curr. Opin. Coll. Interface Sci.* 2009, *14*, 315–328.

170. Kiran, G.S.; Sabu, A.; Selvin, J. Synthesis of silver nanoparticles by glycolipid biosurfactant produced from marine*Brevibacterium casei* MSA19. *J. Biotechnol.* 2010, *148*, 221–225.

171. Palanisamy, P. Biosurfactant mediated synthesis of NiO nanorods. *Mater. Lett.* 2008, *62*, 743–746.

172. Reddy, A.S.; Chen, C.-Y.; Baker, S.C.; Chen, C.-C.; Jean, J.-S.; Fan, C.-W.; Chen, H.-R.; Wang, J.-C. Synthesis of silver nanoparticles using surfactin: A biosurfactant stabilizing agent. *Mater Lett.* 2009, *63*, 1227–1230.

173. Farias, C.B.B.; Silva, A.F.; Rufino, R.D.; Luna, J.M.; Souza, J.E.; Sarubbo, L.A. Synthesis of silver nanoparticles using a biosurfactant produced in low-cost medium as stabilizing agent. *Electron. J. Biotechnol.* 2014, *17*, 122–125.

174. Biswas, M.; Raichur, A.M. Electrokinetic and rheological properties of nano zirconia in the presence of rhamnolipid biosurfactant. *J. Am. Ceram. Soc.* 2008, *91*, 3197–3201.

175. Santos, H.F.; Carmo, F.L.; Paes, J.E.S.; Rosado, A.S.; Peixoto, R.S. Bioremediation of mangroves impacted by petroleum. *Water Air Soil Pollut.* 2011, *216*, 329–350.

CHAPTER 2

ADVANCES IN MAGNETICALLY SEPARABLE PHOTOCATALYSTS: SMART, RECYCLABLE MATERIALS FOR WATER POLLUTION MITIGATION

Gcina Mamba and Ajay Mishra

Keywords: Magnetically separable, ferrites, recyclable photocatalysts, photocatalysis, nanocomposites

INTRODUCTION

Rapid development in terms of industrialisation and population growth over the years has put pressure on sustainable clean water supply. This is partly due to the ever increasing demand and the significant contribution towards compromised fresh water quality that comes with population growth and industrialisation. Similarly, climate change (drought conditions and elevated temperatures) has also impacted negatively on the amount of fresh water available. Consequently, the development of effective water treatment technologies remains an important area of research. This would ensure adequate utilisation of the available fresh water sources, which are constantly polluted through various domestic and industrial activities. Moreover, it would allow efficient treatment of industrial effluent water and enable it to be recycled which would lessen the burden on constant fresh water supply. This is particularly key to the smooth operations of the industries in water scarce localities where the amount of fresh water allocated per industry per day is highly restricted [1].

Over the years, a number of water treatment technologies such as membrane processes [1,2], biological methods [3,4], advanced oxidation processes (AOPs) [5,6], adsorption [7,8], electrochemical methods [9,10], etc. have been developed and explored for the destruction/removal of organic, inorganic, and microbial pollutants. These methods have also been explored in various combinations in order to improve their efficiency [11,12,13,14]. Despite the impressive level of success observed with these technologies in water treatment, they often suffer a number of setbacks, which relate to inefficiency, high operational costs, or both. For example, in membrane processes, fouling (organic, inorganic and biofouling) is a major problem that results in the alteration of the membrane flux and rejection properties [15,16,17]. Toxicity of the effluent matrix and the non-biodegradable nature of some pollutants often render biodegradation ineffective. Moreover, some of the pollutants may be transformed into more toxic

Citation: Gcina Mamba and Ajay Mishra, Advances in Magnetically Separable Photocatalysts: Smart, Recyclable Materials for Water Pollution Mitigation, doi:10.3390/catal6060079

compounds during biological treatment, resulting in secondary pollution [18,19,20]. The generation of highly concentrated sludge which poses further treatment and disposal challenges is a major setback associated with adsorption and coagulation–flocculation [21].

Advanced oxidation processes such as ozonation, hydrogen peroxide oxidation, and their combination with UV light have been extensively exploited in water treatment; however, problems such as the short half-life of ozone, activation by UV light, incomplete mineralisation of organic compounds, and the costs associated with these tools are usually cause for concern [12,22]. Semiconductor photocatalysis has rapidly emerged as the most attractive AOP in recent years for the destruction of organic, inorganic, and microbial pollutants due to a number of advantages it bears over other water treatment protocols. Such merits include its versatile nature in terms of the type of pollutants that can be removed (organic, inorganic, and microbial) and the medium in which the process is applicable (liquid and aqueous medium) [23,24,25,26,27]. Moreover, the semiconductor photocatalysts can also be exploited in other applications such as sensors [28], solar cells [29], pollutant adsorption [30], catalysts for organic synthesis [31], water splitting [32], etc. Apart from its versatile nature, semiconductor photocatalysis has the potential to completely mineralise organic pollutants into carbon dioxide, water, and inorganic ions, thereby eliminating the problem of sludge formation which otherwise cause secondary pollution. Semiconductor photocatalysis provides a stable reaction site/medium since the photocatalyst is not used up during the process and can be re-used several times, and the toxicity of some of the pollutants or their matrices does not affect their destruction by photocatalysis as would be the case with biodegradation [33,34]. In addition, photocatalysis has the advantage of being applicable at ambient conditions (room temperature and standard pressure).

Despite all the highlighted merits associated with semiconductor photocatalysis for water treatment, practical exploitation of this technology has remained a daunting challenge. The first challenge relates to the size of the band gap of the semiconductor since this determines the photoresponse of the photocatalyst. Some of the widely exploited semiconductors such as TiO_2, ZnO, SnO_2, etc. have wide band gaps of 3.0–3.2, 3.2, and 3.5 eV, respectively, which means these semiconductors can only be activated by UV light irradiation. However, artificial UV sources are costly, and UV light itself is toxic and requires protective gear when working under it. The sun does supply UV light but it only makes up about 4% of the solar spectrum, making practical exploitation unsustainable [35,36]. The second challenge in semiconductor photocatalysis is the fast recombination rate of the photogenerated charge carriers, which ultimately lowers the photocatalytic activity of the semiconductor [37,38]. Ideally, subsequent to light absorption and electron excitation in the semiconductor, the electrons, which are promoted to the conduction band, and the positive holes, which remain in the valence band of the semiconductor, must remain far apart so that they can react with oxygen and water or hydroxide ions to form the superoxide and hydroxyl radicals, respectively. These are the active species that together with the holes may be responsible for the degradation of the pollutants [39,40]. Thirdly, the slow degradation kinetics also hamper the full exploitation of photocatalysis in water treatment. The residence times are usually long and sometimes accompanied by low degradation efficiencies [41]. Incomplete mineralisation of organic pollutants is another problem usually observed in photocatalytic experiments and may lead to the formation of toxic by-products which are more toxic than the parent compound [42]. Lastly, the photocatalysts are usually applied in powdered form, which may undergo aggregation in the aqueous medium thereby lowering its activity. Moreover, after the treatment has been completed, the photocatalyst needs to be separated and recovered from the water, and this presents an expensive and time-consuming exercise, especially when it comes to nanometre-sized particles. This often leads to inefficient recovery of the photocatalyst, which not only affects its recyclability but also causes secondary pollution and escalates the treatment costs.

Over the past years, tremendous work has been done focusing on improving the visible light response of various photocatalysts, enhancing charge carrier separation and the overall photocatalytic activity. Efficient utilisation of visible light would enable the exploitation of sunlight as a source of energy, which consists of about 40% visible light. Generally, strategies such as doping the semiconductor with metal ions (alkali, alkaline earth, noble, rare earth, and transition metals) [43,44,45,46,47], non-metal ions (oxygen, sulphur, nitrogen, phosphorus, halogens, carbon, and boron) [48,49,50,51], codoping/multidoping (metal + metal, non-metal + metal, non-metal + non-metal) [52,53,54,55], coupling semiconductors with carbon nanomaterials (reduce graphene oxide (RGO), graphene oxide (GO), multiwalled carbon nanotubes (MWCNTs), single-walled carbon nanotubes (SWCNTs), carbon nanospheres (CNS), fullerenes, etc.) [56,57,58,59], sensitisation (dye, polymer and surface complex sensitisation) [60,61,62], and coupling two or more semiconductors [63,64,65] have been extensively investigated for improving the photocatalytic activity of various semiconductors. Encouraging results have been reported, especially for the nanocomposite photocatalysts, which showed improved visible light utilisation and efficient charge separation, which could be ascribed to the combined contribution of the individual components of the nanocomposite.

Engineering of materials with high visible light photocatalytic activity that are easily separable and recoverable from the treated water remains the ultimate goal of ongoing research. With the increased usage of nanophotocatalysts, efficient separation and recovery of these materials from water is becoming increasingly difficult. Immobilising the photocatalyst particles on solid supports such as glass, ceramics, polymers, zeolites, sand, etc. has been widely investigated and has successfully enabled easy separation of the photocatalyst [66,67,68,69,70,71]. However, in the process of attaining easy separation, the photocatalytic activity of the immobilised semiconductors decreased remarkably. This was due to a decrease in the surface area, poor interaction with the reaction medium, and limited exposure of the photocatalyst to light [72,73]. Since neither the photocatalytic activity nor the easy separation can be sacrificed, researchers turned their interests to magnetic nanoparticles as integral ingredients for developing photocatalysts, which were easily separated from water with an external magnetic field [74,75,76]. Magnetic nanoparticles such as haematite (α-Fe_2O_3), maghemite (γ-Fe_2O_3), magnetite (Fe_3O_4), ferrites (MFe_2O_4, M = Mg, Ni,

Zn, Cu, Co, Cd, etc.), etc. have been successfully coupled with various photocatalysts to induce magnetic behaviour in the composite material and allow easy magnetic separation [39,77,78,79]. Since the photocatalyst remains powdery upon incorporation of the magnetic nanoparticles, this does not compromise its surface area and the distribution of the photocatalyst in the aquatic medium. In addition, some of these magnetic nanoparticles have good visible light absorption and photocatalytic properties, which further enhance the overall pollutant degradation kinetics over the magnetic nanocomposite [39,40,64,80,81,82].

Considering the future prospects of magnetic photocatalysts, the design and preparation of magnetic visible-light-active nanocomposites has emerged as a hot topic in photocatalytic environmental pollution control. Therefore, it is necessary to bring into perspective the current developments in the tailoring and application of such materials in water treatment. A few reviews are available mainly on the synthesis, structure, and various applications of magnetic materials [83,84,85,86,87,88]. This review exclusively discusses the most recent developments in terms of the synthesis and exploitation of various magnetic nanocomposite photocatalysts in water pollution mitigation. Details on the synthesis routes and structure of the various magnetic nanoparticles have been well-reviewed and have been excluded in this review. Furthermore, this review only focusses on iron oxide magnetic nanoparticles resulting in four categories of magnetic nanocomposite photocatalysts: haematite, maghemite, magnetite, and ferrite-based magnetic photocatalysts. Under each category, we discuss the influence of the magnetic nanoparticle on optical properties, charge transfer mechanism, and the overall photocatalytic properties of the nanocomposites. We conclude the review by providing some summary remarks and highlight the future challenges towards tailoring and utilising magnetic photocatalysts in water treatment.

RECENT DEVELOPMENTS IN MAGNETIC PHOTOCATALYSTS

Haematite (α-Fe_2O_3)-Based Magnetic Photocatalysts

As a pristine material, haematite is a narrow band gap semiconductor (2–2.2 eV) with good visible light response, good chemical stability, low-cost, ferromagnetic behaviour, abundantly available, high resistance to corrosion, and environmentally friendly [89,90,91]. On its own, haematite shows poor photocatalytic activity due to the high recombination rate of the charge carriers and poor conductivity [90,91]. Despite its poor photocatalytic activity, the narrow band gap of haematite makes it an ideal sensitiser for improving the visible light response of various semiconductors. Zhao et al.developed a nanostructured α-Fe_2O_3-AgBr nonwoven cloth (Figure 1a,b) via a simple electrospinning-calcination method and studied its visible light photocatalytic behaviour towards rhodamine B (RhB) and para-chlorophenol (4-CP) decomposition. For both pollutants, the α-Fe_2O_3-AgBr nonwoven cloth showed superior activity over α-Fe_2O_3 and AgBr clothes individually, attaining 91.8% 60 min and 74.2% in 120 min for RhB and 4-CP decomposition, respectively [81]. Coupling AgBr with α-Fe_2O_3 resulted in a significant red-shift and resulted in the formation of heterojunctions, which facilitated charge separation, thereby improving the overall photocatalytic activity of the α-Fe_2O_3-AgBr nonwoven cloth. In addition, the cloth was easy to handle and recycle during photocatalysis by simple dipping and removal or by using an external magnet. It was found that the composite cloth was stable over four cycles with only a slight decrease in photocatalytic performance [81].

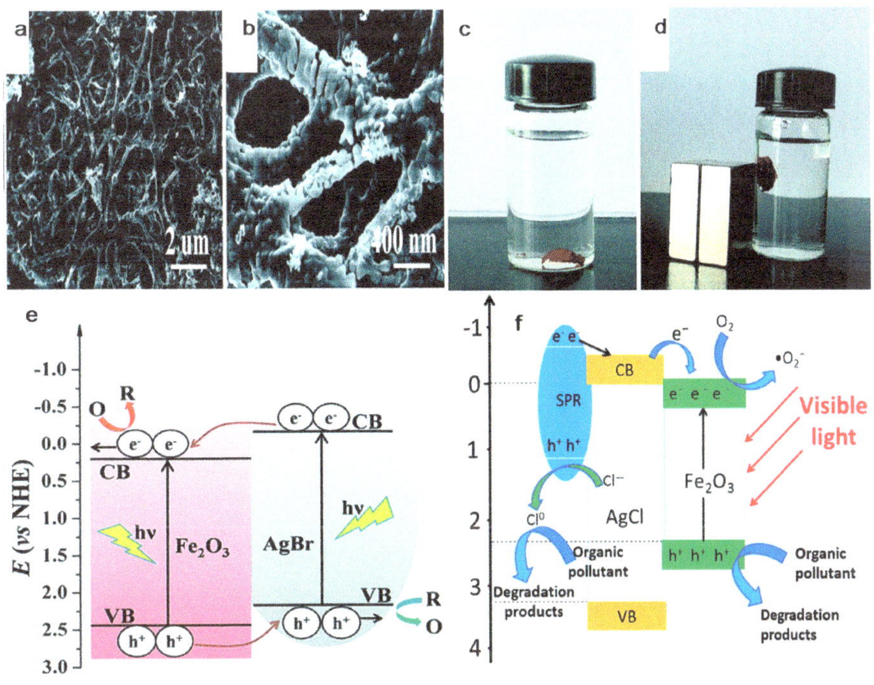

Figure 1: (a,b) SEM images of α-Fe_2O_3-AgBr nonwoven cloth; (c,d) illustration of magnetic separation of α-Fe_2O_3-AgBr cloth and charge transfer mechanism in (e) α-Fe_2O_3-AgBr cloth (Reproduced with permission from [81], Copyright 2015, RSC); and (f) Ag/AgCl/Fe_2O_3. Reproduced with permission from [39]. Copyright 2016, RSC.

Xue and coworkers fabricated magnetically separable α-Fe_2O_3/graphitic carbon nitride (g-C_3N_4) nanocomposite photocatalysts with varying α-Fe_2O_3 loadings (5%, 10%, 30%, 50%) and investigated their performance for degradation of RhB. The highest degradation efficiency (96.7% in 4 h) was recorded using the sample containing 10% α-Fe_2O_3, which suggests optimum composition [92]. The introduction of α-Fe_2O_3 nanoparticles into g-C_3N_4 matrix enhanced its visible light utilisation and charge separation efficiency. Moreover, the composite photocatalyst displayed good stability and could be easily separable and recycled using an external magnet [92]. Similarly, core-shell magnetically separable Ag/AgCl/Fe_2O_3showed remarkable photocatalytic activity towards bisphenol A (BPA) and E. coli under visible light exposure. At a 5% α-Fe_2O_3 loading, the composite photocatalyst reached the highest activity, which was 13 times that of Ag/AgCl for BPA degradation, and complete destruction of E. coli was observed after just 30 min of irradiation [39]. The nanocomposite displayed good stability and sufficient magnetic response to an external magnet.

In terms of charge transfer mechanism, the path followed by the electrons and holes depends on the semiconductor coupled with haematite and how their band structures match. In any case, under visible light illumination, haematite is excited and electrons accumulate in the conduction band while leaving positive holes in the valence band. In the case of α-Fe_2O_3-AgBr, both semiconductors are excited by visible light, and the electrons from AgBr transfer to the conduction band of α-Fe_2O_3, while the holes transfer in the opposite direction (Figure 1e). The holes will directly attack the organic pollutants to form the degradation products, while the electrons may react with adsorbed oxygen leading to the eventual formation of hydroxyl radicals, which are strong oxidising agents [81]. In the Ag/AgCl/Fe_2O_3 ternary nanostructure, the charge transfer mechanism (Figure 1f) is a bit complex. Only Ag and α-Fe_2O_3 are excited upon visible light absorption, and the plasmon-generated electrons in Ag will be injected into the conduction band of AgCl and further transfer to the conduction band of α-Fe_2O_3. Meanwhile, the holes in α-Fe_2O_3 can directly decompose the pollutants, while the holes left in Ag react with Cl^- ions to form Cl^0, which then attacks the pollutants [39].

Other haematite-based magnetically separable nanostructured photocatalysts such as graphene (GR)-α-Fe_2O_3-ZnO [93], α-Fe_2O_3/TiO_2 nanofibres [94], α-Fe_2O_3/RGO [95], and α-Fe_2O_3/ZnO [96] have been fabricated, have shown remarkable visible light utilisation, and have been easily separable using an external magnetic field. Coupling the various semiconductors with α-Fe_2O_3 resulted in synergy between the materials, leading to enhanced visible light response, improved charge separation, and an overall increase in photocatalytic activity. Moreover, the incorporation of haematite allowed for the easy separation of the photocatalysts after the water treatment. However, α-Fe_2O_3 generally shows weak ferromagnetic behaviour, and, most importantly, its magnetic properties largely depend on the synthesis method, which influences its particle size and shape [92]. Several α-Fe_2O_3-incorporating heterostructures such as α-Fe_2O_3/Ag_3VO_4 [91], g-C_3N_4/Ag/α-Fe_2O_3 [97], g-C_3N_4/α-Fe_2O_3 [98,99,100], α-Fe_2O_3/TiO_2 [101] Au/g-C_3N_4/α-Fe_2O_3 [102], and α-Fe_2O_3/CdS [90] have been fabricated, showing enhanced visible light photocatalytic properties, and no magnetic behaviour was reported. Therefore, when the main aim of incorporating the iron oxide nanoparticles is to induce magnetic response, haematite may not be the ideal choice.

Maghemite (γ-Fe_2O_3)-based Magnetic Photocatalysts

Maghemite is another important form of iron oxide but, unlike haematite, it shows good magnetic properties with saturation magnetisation values of up to 76 emu/g [103]. Consequently, the incorporation of maghemite into various photocatalyst matrices has been explored to fabricate magnetically separable composite photocatalysts for easy recovery and recycling [104,105]. Interestingly, despite the sensitivity to an external magnetic field, when the magnetic field is removed, the γ-Fe_2O_3-incorporating nanocomposites do not retain any significant residual magnetism that would otherwise cause problems by creating large clusters of magnetised particles during water treatment [106]. Maghemite shows good visible light response owing to its narrow band gap (2.2 eV) and possesses good adsorption properties. However, its photocatalytic activity is poor as a result of the high recombination rate of the photogenerated charge carriers and photodissolution [74,107]. Two approaches are usually adopted when coupling maghemite with other semiconductors. In the first approach, maghemite is directly coupled with other semiconductors to form heterojunctions, and it contributes towards visible light response of the nanocomposite photocatalyst as well as the overall charge transfer process [74,108,109]. According to the second approach, maghemite and the other semiconductor do not form heterojunctions (not in direct contact), but they are separated by an inert layer or shell usually silica (SiO_2). This is aimed at preventing photodissolution of the iron oxide and charge transfer between the two materials, which could otherwise result in the recombination on the maghemite surface. In this case, maghemite nanoparticles do not play any part in the photocatalysis process but function solely in the magnetic separation of the photocatalyst from the reaction medium [104,106,110].

A magnetically separable quaternary γ-Fe_2O_3/N and Fe-codoped TiO_2 nanocomposite was prepared using a sol-gel method (as illustrated in Figure 2a) and employed for the decomposition of reactive blue 4 (RB 4) under visible light illumination. Significantly higher photocatalytic activity was observed over the quaternary nanocomposite photocatalyst compared to N,Fe-TiO_2, N-TiO_2, and TiO_2, reaching 100% degradation in 180 min (Figure 2b) [111]. The remarkable improvement in activity of the nanocomposite could be explained in terms of the combined contribution of codoping $TiO_2$and coupling with γ-Fe_2O_3, which resulted in significant band gap narrowing, efficient visible light response, and efficient charge separation and transfer. Furthermore, the multicomponent photocatalyst demonstrated good stability and recyclability over four cycles without any significant loss in activity or magnetic properties, which allowed for easy magnetic separation [111]. Yu and coworkers probed the adsorption and UV photocatalytic oxidation of As(III) over TiO_2decorated on magnetic (γ-Fe_2O_3) mesoporous SBA-15 nanocomposite (γ-Fe_2O_3/SBA-15/TiO_2) derived from simple inner-pore hydrolysis combined with solvent evaporation route [112]. The nanocomposite photocatalyst displayed a dual function of adsorption and photocatalytic degradation. Firstly, As(III) ions were adsorbed onto the photocatalyst and then upon

UV irradiation, undergo photocatalytic oxidation to As(V), which is less toxic and also removable by the photocatalyst. In addition, the photocatalyst showed good stability and could be regenerated for five cycles using NaOH (0.01 M) and still maintained over 90% As(III) removal and good magnetic properties for easy separation [112]. The incorporation of the SBA-15 ensured good dispersibility of the nanoparticles and improved the surface area of the nanocomposite. All three components of the nanocomposite were equally important in the overall removal of As(III).

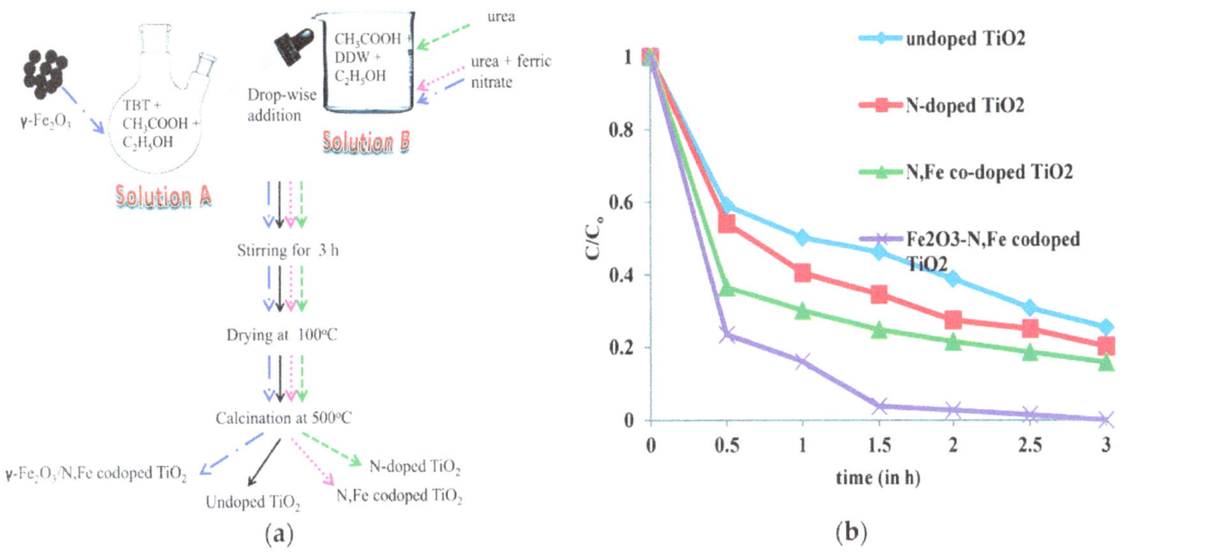

Figure 2: (a) Schematic illustration of the sol-gel preparation of γ-Fe_2O_3/N.Fe-TiO_2; and (b) photocatalytic degradation of RB 4 over γ-Fe_2O_3/N.Fe-TiO_2 and the controls. Reproduced with permission from [111]. Copyright 2015, RSC.

Improved optical and photocatalytic properties have also been reported for other magnetic photocatalysts such as ZnO/γ-Fe_2O_3 [113], g-C_3N_4/γ-Fe_2O_3 [109], and γ-Fe_2O_3/TiO_2 [74] towards the degradation of various pollutants including RhB, methylene blue (MB), and methyl orange (MO). Moreover, these nanostructures showed good stability and could be magnetically separated and recycled without a significant loss in activity. However, the maghemite nanoparticle loading needs to be carefully controlled in order to strike a good balance between sufficient magnetism to facilitate separation and the detrimental effect of recombination of the charge carriers on the maghemite nanoparticles. This means that, while enjoying the luxury of easy separation, the detrimental effect of recombination must not be ignored in order to positively benefit from the incorporation of the maghemite nanoparticles.

According to the proposed charge transfer route in γ-Fe_2O_3/N,Fe-TiO_2 (Figure 3a), both maghemite and codoped titania absorb visible light, and electrons are excited from the valence band to the conduction band in γ-Fe_2O_3, leaving holes in the valence band. Similarly, in the codoped titania, electrons are excited from the valence band of TiO_2 to the impurity states formed by Fe doping and also from N sub-band gap states to the conduction band of TiO_2. Subsequently, electrons transfer from the conduction band of maghemite to the Fe impurity states in TiO_2, while the holes migrate to the N states in TiO_2. The electrons and holes will be captured by molecular oxygen and water to form the superoxide and hydroxyl radicals, respectively [111]. These radical species are important oxidising agents for pollutant decomposition.

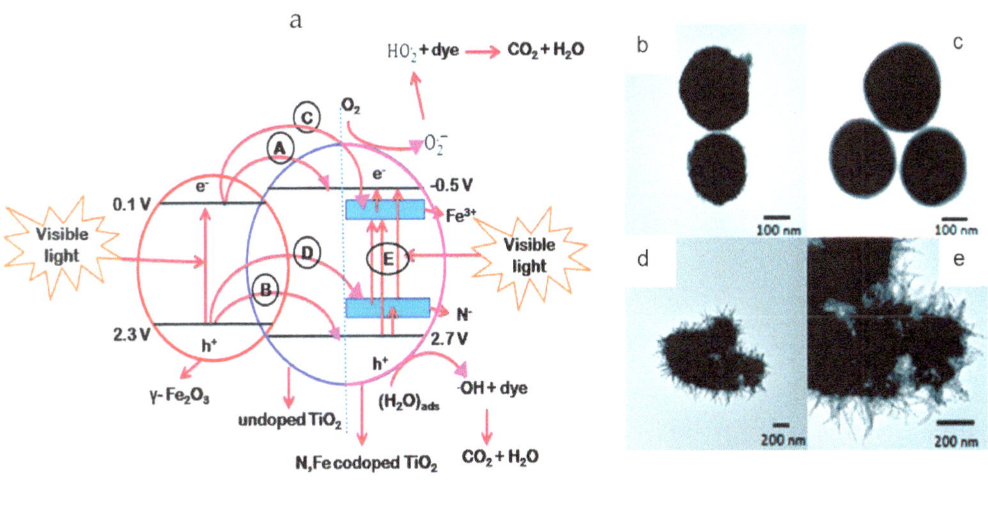

Figure 3: (a) Charge transfer mechanism in γ-Fe_2O_3/Fe,N-TiO_2 (Reproduced with permission from [111], Copyright 2015, RSC) and (b–e) SEM images of urchin-like γ-Fe_2O_3@SiO_2@TiO_2 composite microparticles. Reproduced with permission from [114]. Copyright 2015, RSC.

The strategy of preventing the formation of heterojunctions between maghemite and the semiconductor coupled with it has been widely explored using silica as the barrier/shell surrounding the magnetic core. For example, Szeto et al. used a layer-by-layer assembly route to fabricate urchin-like γ-Fe_2O_3@SiO_2@TiO_2 composite microparticles (Figure 3b–e) and observed enhanced phenol decomposition under UV irradiation [114]. The tailored composite photocatalyst was shown to possess a large surface area, a prolonged life of the charge carriers, remarkable permeability, and significantly higher photocatalytic activity compared to commercial P25. In addition, incorporation of the maghemite nanoparticles ensured easy and efficient separation of the photocatalyst using an external magnetic field [114]. Similar results have been reported for other magnetic heterostructures such as γ-Fe_2O_3@SiO_2@Ce-TiO_2 [110], γ-Fe_2O_3@SiO_2@AgBr:Ag [104], and γ-Fe_2O_3@SiO_2@TiO_2 [106], whereby a silica coating separates the maghemite nanoparticles from direct contact with the other components of the heterostructure. In these nanocomposites, the maghemite nanoparticles were only incorporated to facilitate the magnetic separation and play no part in the charge transfer process and overall photocatalytic decomposition of the pollutants. Although this approach is logical and has yielded some positive results, a critical comparison between the heterojunction type of photocatalyst and the core-shell structure of the same nanocomposite is rarely documented. This could map a way forward in terms of the most efficient approach towards engineering the maghemite-based magnetic photocatalyst and benefitting from the properties of this material.

Magnetite (Fe_3O_4)-based Magnetic Photocatalysts

Magnetite is by far one of the most widely exploited iron oxide magnetic nanoparticle in the fabrication of magnetic photocatalysts due to its low toxicity, good magnetic properties, biocompatibility, and remarkable adsorption properties [115,116]. The magnetisation saturation for magnetite can reach values of up to 92 emu/g, making it the ideal candidate for synthesis of magnetically separable photocatalysts [103]. However, unlike maghemite, magnetite shows no photocatalytic properties, and, like maghemite, the formation of heterojunctions with other semiconductors in nanocomposite photocatalysts is undesirable because magnetite may suppress the overall photocatalytic efficiency by acting as a recombination centre for the charge carriers [117,118]. Silica is commonly utilised to enclose the magnetite core, which is wrapped with the other semiconductor nanoparticles to form the magnetic composite photocatalysts. The silica barrier prevents electron transfer between magnetite and the other semiconductor, additionally preventing photodissolution of the iron oxide [119,120].

Despite the possible negative effect of magnetite in heterojunction nanocomposites, several reports on successful tailoring and improvement in photocatalytic activity over heterojunction nanocomposites have emerged, and careful control of the Fe_3O_4 loading in the nanocomposite is an important point of consideration. For example, Zhao et al. observed enhanced photocatalytic activity towards ampicillin (AMP) over Fe_3O_4/TiO_2/Ag with a sea urchin-like morphology, both under visible and UV light irradiation. The composite photocatalyst reached 98.7% and 91.5% AMP degradation after 360 min of UV and visible light illumination, respectively [116]. Furthermore, the nanocomposite possessed good magnetic behaviour with a magnetisation saturation of 26.5 emu·g^{-1} (Figure 4), which was high enough to facilitate magnetic separation (Figure 4insert). In addition to the degradation of AMP, the nanocomposite photocatalyst showed remarkable photocatalytic properties towards S. aureus (G^+), E. coli (G^-), and A. niger [116]. In another example, improved MO decomposition was realised over Fe_3O_4/N-TiO_2/Ag hollow nanospheres, reaching 99.5% in 80 min of visible light exposure when the Ag loading was 1.0% [121]. Ma and coworkers employed a low-temperature crystallisation method to wrap Fe_3O_4 spheres with TiO_2nanoparticles and subsequently immobilised Au nanoparticles on the core-shell Fe_3O_4/TiO_2 to yield a Fe_3O_4/TiO_2/Au plasmonic magnetic photocatalyst. RhB was used as a model pollutant to evaluate the photocatalytic properties of the nanocomposite, and it showed superior performance over Fe_3O_4 and Fe_3O_4/TiO_2, with good magnetic properties (magnetisation saturation (Ms), 44.6 emu·g^{-1}), and recyclability [122]. The heterojunctions formed between Fe_3O_4 and titania allow electrons to transfer between the two materials, which improves charge separation. However, the Fe_3O_4 loading needs to be kept in check as excess Fe_3O_4 will cause detrimental effects. Therefore, Fe_3O_4 nanoparticles have a dual function in the heterojunction nanostructures—electron trapping and enabling magnetic separation.

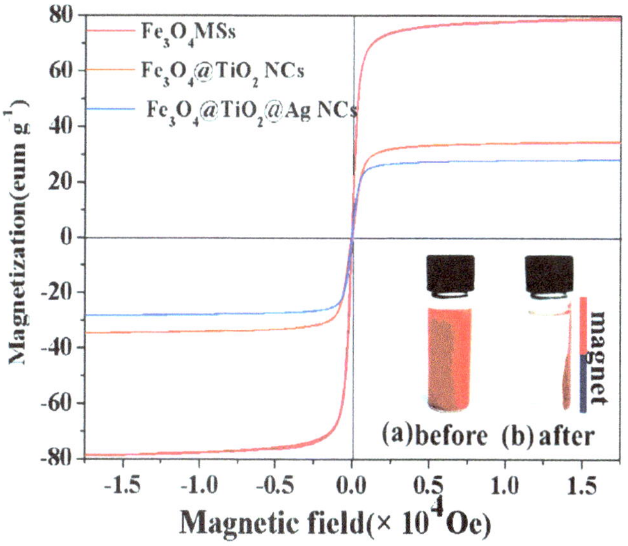

Figure 4: Magnetization curves and insert showing magnetic separation of $Fe_3O_4/TiO_2/Ag$. Reproduced with permission from [116]. Copyright 2016, RSC.

Li et al. fabricated Fe_3O_4 magnetic core coated with SiO_2 prior to the deposition of TiO_2 and surface modification with lysine, and the quaternary nanocomposite photocatalyst (Fe_3O_4@SiO_2@TiO_2-lysine) showed multifunctional properties as it was employed as an adsorbent, photocatalyst, and sensor for dissolved organic and inorganic phosphorus in seawater [123]. The photocatalyst displayed remarkable stability and could be magnetically separated and recycled for 10 cycles without any significant loss in its adsorption and photocatalytic properties [123]. Other core-shell Fe_3O_4@SiO_2 nanostructures decorated with TiO_2 have been fabricated and evaluated for the decomposition of various pollutants such as RhB, 2-chlorophenol (2-CP), phenol, MB, and MO. These materials showed improved photocatalytic properties, good stability, and sufficiently high magnetic response that enabled separation using an external magnetic field [124,125,126,127]. Table 1provides a summary of the synthesis and photocatalytic properties of various other magnetite-incorporating titania-based photocatalysts towards water treatment applications. In the magnetic core-shell nanostructures, the Fe_3O_4 nanoparticles encapsulated with the silica shell do not partake in the charge transfer process but only in the magnetic separation of the photocatalyst.

Table 1: Summary of the photocatalytic properties of Fe_3O_4-incorporating TiO_2-based photocatalysts

Sample	Preparation Method	Application	Catalyst Dose	Pollutant Concentration	Degradation Efficiency	Saturation Magnetisation	Reference
Fe_3O_4/GO/Ce-TiO_2	Low-temperature sol-gel/ ultrasonication	Degradation of tetracycline (TC)/Vis/300 W Xe lamp/400 nm UV filter	0.050 g	25 ppm	82.9% in 60 min	No data	[33]
Fe_3O_4/RGO/TiO_2	Reduction-deposition	Degradation of MB and tetrabromobisphenol A (TBBPA)/UV/230 W Hg lamp	0.025 g	10 ppm	99.5% in 60 min/TBBPA, 95.9% in 60 min/MB	1.14 emu/g	[128]
Fe_3O_4/P(MAA-DVB)/TiO_2	Magnetic field induced assembly/precipitation-polymerisation	Degradation of RhB/UV/250 W Hg lamp	0.040 g	4.43 ppm	71.5% in 120 min	35.2 emu/g	[129]
Fe_3O_4@SiO_2@N-TiO_2	Sol-gel	Degradation of phenol/Vis/15 W florescent lamp	0.800 g	100 ppm	46% in 480 min	~2 emu/g	[127]
Fe_3O_4/chitosan/TiO_2	Hydrothermal/crosslinking	Degradation of MB/UV/8 W UV lamp	1.00 g	1.28 ppm	93% in 40 min	4.2 emu/g	[130]
Fe_3O_4/TiO_2	Reverse microemulsion/sol-gel	Degradation of RhB/UV/50 W Xe lamp	0.004 g	22.12 ppm	100% in 100 min	21.68 emu/g	[131]
Ag_3PO_4/TiO_2/Fe_3O_4	*In situ* hydrolysis/deposition	Degradation of acid orange 7 (AO 7)/50 mW diode blue laser and *E. coli*/Vis/300 W Xe lamp/420 nm UV filter	0.100 g	5.25 ppm/AO7 and 10^7 CFU/mL/*E. coli*	~100% in 2.5 min/AO 7, 99.8% in 5 min/*E. coli*	No data	[38]
Fe_3O_4/TiO_2/Bi_2O_3	Sol-gel	Degradation of MO/Simulated solar light/350 W Xe lamp	0.200 g	No data	69% in 150 min	No data	[132]
Fe_3O_4/TiO_2/Au	Sol-gel/hydrothermal	Degradation of MB and *E. coli*/UV/4 × 9 W black lights	0.015 g/MB and 0.010 g/*E. coli*	8.00 ppm/MB and 10^8 CFU/mL/*E. coli*	78% in 4 h/MB, 89.3% in 60 min/ *E. coli*	No data	[133]
Ag@Fe_3O_4@SiO_2@TiO_2	Solvothermal	Degradation of MB and Cr(VI) reduction/Vis/500 W Xe lamp/ 425 UV filter	0.020 g	50 ppm/MB and 22.24 ppm/$K_2Cr_2O_7$	99.9% in 4 h/ Cr(VI), ~90 mg/g/MB	13.92 emu/g	[134]
Fe_3O_4@SiO_2@TiO_2@GO	Reverse microemulsion/sol-gel/amide conjugation	Degradation of RhB/UV/400 W column high pressure Hg lamp	0.050 g	8.85 ppm	92% in 120 min	16.90 emu/g	[120]
Fe_3O_4@C@TiO_2	Solvothermal/calcination	Degradation of RhB/UV/125 W high pressure Hg lamp	0.020 g	10 ppm	~100% in 80 min	6.04 emu/g	[135]
Fe_3O_4@SiO_2@TiO_2	Stobber process/sol-gel/chemical precipitation	Degradation of RhB/UV/50 W high pressure Hg lamp	0.010 g	8.85 ppm	100% in 10 min	30.60 emu/g	[136]
Fe_3O_4/TiO_2	Hydrothermal	Degradation of MB/UV/9 W UV lamp	0.002 g	1.00 ppm	100% in 10 min	No data	[75]
Fe_3O_4@TiO_2@GR	Sol-gel/assembly route	Degradation of 2,4-dichlorophenoxy-acetic acid (2,4-D)/simulated solar light/ 500 W Xe arc lamp	0.020 g	20 ppm	100% in 40 min	13.00 emu/g	[137]
WO_3/TiO_2/Fe_3O_4	Sol-gel	Degradation of direct blue 71 (DB 71)/Vis/200 W Xe lamp	0.030 g	50 ppm	98% in 35 min	18.20 emu/g	[138]

Chidambaram and coworkers prepared a novel plasmonic, magnetic Fe_3O_4/ZnO/Ag nanocomposite and employed it for the photocatalytic degradation of MB under visible light illumination. Complete MB removal was observed after 6 h of visible light exposure on Fe_3O_4/ZnO/Ag, while only 20% MB removal was realised over Fe_3O_4/ZnO during the same irradiation time. The improved activity could be explained in terms of improved charge separation and the surface plasmon resonance (SPR) effect of Ag nanoparticles, which is responsible for the visible light activity of the nanocomposite photocatalyst [139]. Fe_3O_4 nanoparticles played a dual role of facilitating charge separation and enabling magnetic separation.

In addition, the photocatalyst showed great stability over 10 cycles without much loss in activity or magnetic properties. Upon visible light irradiation, the SPR-generated electrons are injected into the conduction band of ZnO and further transfer to the Fe_3O_4 nanoparticles. Subsequently, the electrons reduce Fe^{3+} to Fe^{2+}, which in turn is oxidised back to Fe^{3+} upon reaction with adsorbed oxygen forming the superoxide radicals. Meanwhile, the SPR-generated holes in Ag could react with water to form the hydroxyl radicals and together with the superoxide radicals are responsible for pollutant degradation [139]. Yan et al. observed higher visible light photocatalytic performance over urchin-like Fe_3O_4@SiO_2@ZnO/CdS core-shell microspheres (Figure 5a,b) towards RhB compared to Fe_3O_4@SiO_2@ZnO microspheres. Coating Fe_3O_4 with SiO_2 prevented the formation of heterojunctions with ZnO and CdS, and the nanocomposite showed good magnetic properties with magnetisation saturation value of 18.97 emu/g [140]. On visible light exposure, only CdS nanoparticles were excited and played the role of a sensitiser for ZnO, with electrons transferred to the conduction band of ZnO while the holes remained in the valence band of CdS (Figure 5c), resulting in efficient charge separation and the formation of the oxidising species [140].

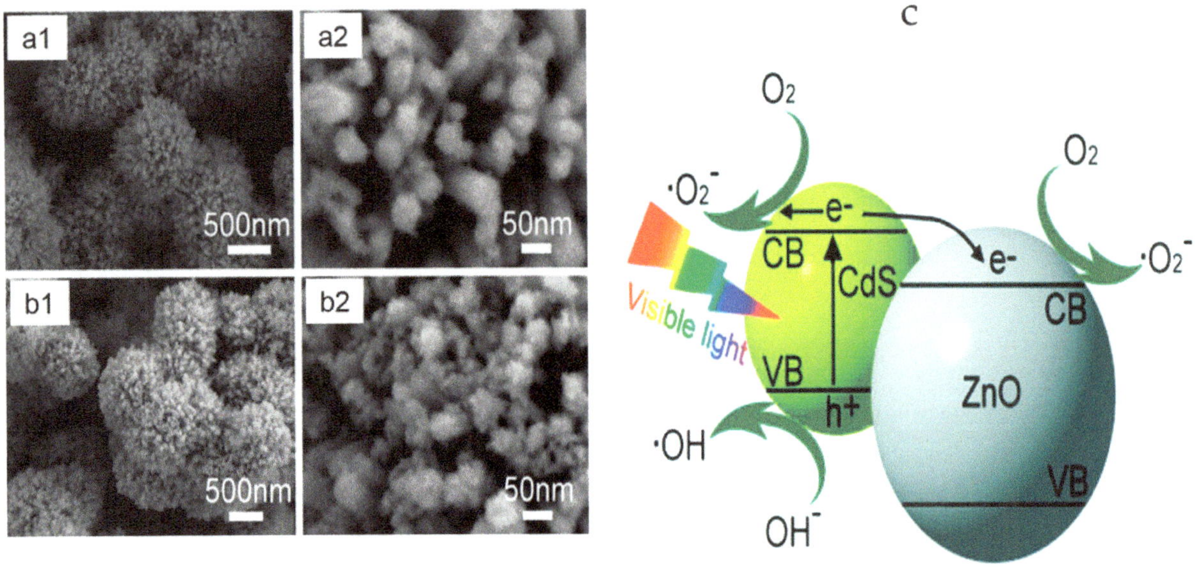

Figure 5: (a,b) SEM images of urchin-like Fe_3O_4@SiO_2@ZnO/CdS microspheres and (c) charge transfer route in Fe_3O_4@SiO_2@ZnO/CdS. Reproduced with permission from [140]. Copyright 2016, RSC.

Habibi-Yangjeh's group has recently prepared and investigated the visible light photocatalytic properties of various magnetic nanocomposites-based on Fe_3O_4 and ZnO such as ZnO/AgI/Fe_3O_4 [141], ZnO/AgBr/Fe_3O_4/Ag_3VO_4 [142], ZnO/Ag_3VO_4/Fe_3O_4 [143], Fe_3O_4/ZnO/AgCl [144], and Fe_3O_4/AgBr-ZnO [145] towards model pollutants such as RhB, phenol, MB, and MO. All of the nanocomposite photocatalysts showed good photocatalytic properties, which can be credited to efficient charge separation and improved visible light absorption as a result of the heterojunctions formed between ZnO and the co-catalysts. Moreover, it was shown that despite the formation of the composite photocatalysts, the presence of Fe_3O_4 nanoparticles ensured that the photocatalysts still retained sufficient magnetic response to allow for efficient separation using an external magnetic field. Obviously, the magnetisation saturation of Fe_3O_4 decreased sharply upon the formation of the nanocomposites due to their interaction and coverage by the non-magnetic components of the photocatalysts [141,142,143,144,145]. Under visible light irradiation, ZnO remains inactive and only the co-catalyst with a narrow band gap can be excited. For example, in ZnO/AgI/Fe_3O_4 and ZnO/AgBr/Fe_3O_4/Ag_3VO_4, only AgI, AgBr, and Ag_3VO_4, respectively, are excited, and electrons are promoted to the conduction band, while the holes remain in the valence band. Subsequently, the electrons transfer to the conduction band of ZnO where they react with adsorbed oxygen to form the radical species (superoxide and hydroxyl radicals). Accordingly, the holes are directly involved in the degradation of the pollutants since they are not positive enough to oxidise water for the hydroxyl radicals [141,142]. The electrons and holes occupied different sites in the nanocomposite photocatalysts, which ensured efficient charge separation and the formation of the reactive species for pollutant degradation.

Owing to its meta-free nature, polymeric structure, non-toxicity, good visible light response, good photocatalytic properties, and high thermal and chemical stability, tailoring graphitic carbon nitride (g-C_3N_4)-based magnetic photocatalysts is an important exploit towards potential practical application of these materials in water treatment. Using a combination of calcination and co-precipitation, Yang and coworkers prepared Fe_3O_4/g-C_3N_4 nanocomposite photocatalyst and probed its photocatalytic behaviour towards 2,4,6-trichlorophenol (2,4,6-TCP) under visible light illumination [77]. In a similar work, Jia et al. fabricated Fe_3O_4/g-C_3N_4 via a hydrothermal route and studied its visible light photocatalytic performance towards RhB decomposition [146]. In both studies, higher photocatalytic performance was observed over the nanocomposites compared to Fe_3O_4 and g-C_3N_4 individually, reaching 95.5% in 60 min [146] and 96.5% in 100 min [77] for RhB and 2,4,6-TCP, respectively. Coupling g-C_3N_4 with Fe_3O_4 improved charge separation since only g-C_3N_4 is excited by visible light, and the electrons in the conduction band quickly transfer to Fe_3O_4, where they are captured by oxygen to form the radical species. Meanwhile, the holes remain in the valence band of g-C_3N_4 and directly oxidise the pollutants to form the degradation products.

Therefore, the holes and electrons occupy different locations in the nanocomposite. Moreover, magnetic separation was made possible by the presence of the Fe_3O_4 nanoparticles [77,146].

Habibi-Yangjeh's group investigated the visible light photocatalytic performance of novel magnetic nanocomposites: g-C_3N_4/Fe_3O_4/BiOI [147], g-C_3N_4/Fe_3O_4/AgCl [148], and g-C_3N_4/AgBr/Fe_3O_4 [149] towards the decomposition of RhB, MO, and MB. Significantly higher photocatalytic activity was observed over the ternary heterostructures compared to single and binary nanocomposites, and this could be ascribed to the synergistic effect of the various components of the ternary nanocomposites, resulting in improved visible light response and charge separation efficiency. Moreover, the presence of Fe_3O_4 in the heterostructures endowed them with magnetic response, which allowed separation using an external magnetic field [147,148,149]. In terms of charge transfer, only g-C_3N_4 is excited by visible light in the case of g-C_3N_4/Fe_3O_4/AgCl, while both g-C_3N_4 and BiOI are excited in g-C_3N_4/Fe_3O_4/BiOI, and the electrons are promoted to their conduction bands, leaving positive holes in the valence bands. This is followed by the transfer of electrons from the conduction band of g-C_3N_4, which is more negative, to the conduction bands of AgCl and BiOI, which are less negative, whereby photoreduction reactions involving adsorbed oxygen take place and result in the formation of the radical species [147,148]. Accordingly, in both materials, the holes accumulate in the valence band of g-C_3N_4 where they directly oxidise the pollutants to form the degradation products.

Generally, the incorporation of Fe_3O_4 nanoparticles in the photocatalysts can serve as a means to prolong the life of the photogenerated charge carriers by acting as electron traps and induce magnetic response in the nanocomposites, which enables easy separation using an external magnetic field. In both magnetic core-shell and heterojunction nanostructures, the magnetic influence of the Fe_3O_4 nanoparticles was significant despite showing lower magnetisation saturation compared to neat Fe_3O_4 due to the presence of the non-magnetic components of the photocatalysts. In the heterojunction nanostructures, Fe_3O_4 nanoparticles played a dual role of charge separation and magnetic separation, but the detrimental effect of recombination becomes a problem if the optimum Fe_3O_4 loading is not carefully controlled. Meanwhile, the silica shell in the core-shell nanostructures serve as a barrier to prevent the formation of heterojunctions and charge transfer, limiting the role of the Fe_3O_4 nanoparticles only to magnetic separation of the photocatalysts. A host of other Fe_3O_4-based magnetic photocatalyst have been designed, prepared, and exploited in water treatment applications and are summarised in Table 2.

Table 2: Summary of the photocatalytic properties of various Fe_3O_4-based photocatalyst

Sample	Preparation Method	Application	Catalyst Dose	Pollutant Concentration	Degradation Efficiency	Magnetisation Saturation	Reference
Fe_3O_4@SiO_2@CdS	Stobber method/chelating assisted growth	Degradation of MB and TC/Vis	0.010 g	10 ppm/MB and 100 ppm/TC	80% in 21 min/TC, 94% in 140 min/MB	22.00 emu/g	[63]
Fe_3O_4(TAMH)/ZnO	Solvothermal	Degradation of phenol/Vis/ 575 W MSR metal halide lamp	0.325 g	20 ppm	71.3% in 150 min	No data	[150]
Fe_3O_4@mSiO_2@BiOBr	Solvothermal	Degradation of MB/Vis/500 W Xe lamp/420 nm UV filter	0.100 g	20 ppm	96% in 120 min	40.00 emu/g	[151]
Fe_3O_4/$BaTiO_3$	Solvothermal/sol-gel	Degradation of MO and orange II/UV/150 W UV lamp	0.002 g	10 ppm	71.2% in 20 h/MO, 43.7% in 20 h/ orange II	60.50 emu/g	[152]
Fe_3O_4/Cr_2O_3	Wet chemical/ultrasonication	Degradation of 4-CP/UV/12 W low-pressure Hg lamp	0.100 g	1.29 ppm	100% in 150 min	~20 emu/g	[153]
Cu_2O/chitosan/Fe_3O_4	Precipitation-reduction	Degradation of Reactive Brilliant red X-3B (X-3B)/Vis/500 W W-halogen lamp	0.100 g	50 ppm	99.7% in 50 min	15.1 emu/g	[154]
Fe_3O_4/AgBr	Precipitation route	Degradation of MO/Vis/300 W Xe arc lamp/420 nm UV filter	0.100 g	20 ppm	85% in 12 min	No data	[155]
RGO/Ag/AgCl/Fe_3O_4	Solvothermal/deposition-precipitation	Degradation of MB and RhB/Vis/500 W Xe arc/10 cm water filter	0.100 g	10 ppm	97.4% in 100 min/MB, 97.9% in 120 min/RhB	18.8 emu/g	[156]
Metalloporphyrin/Fe_3O_4	Covalent conjugation	Degradation of AO 7/Vis/125 W W-halogen lamp	0.444 g	17.5 ppm	69.0% in 5 h	61.45 emu/g	[157]
Au(Ag)/AgCl/Fe_3O_4@PDA@Au	Solvothermal/galvanic replacement	Degradation of MB/Vis	0.004 g	5 ppm	100% in 20 min	5.40 emu/g	[158]
Fe_3O_4@resorcinol-formaldehyde–Ag	*In situ* polymerisation/reduction	Inactivation of *S. aureus*, *E. coli* and degradation of 4-nitrophenol (4-NP), MB and RhB/Vis/No data on light source	0.010 g	16 ppm/ RhB/MB/4-NP and 10^7 CFU/mL/ bacteria	1.5 < OD in 6 h/ *E. coli*, ~1.0 OD in 6 h, *S. aureus*, ~100% in 3 min/ 4-NP, ~100% in 4 min/MB and ~100% in 3 min/RhB	~30 emu/g	[26]
Poly(p-phenylenediamine)–Fe_3O_4	Chemical oxidation polymerisation	Degradation of bromocresol green (BG), blue (BB), purple (BP), RhB, neutral red (NR), MB, Sudan III (SIII), MO and Congo red (CR)/UV (500 W Hg)/Vis (500 W Xe lamp/420 nm UV filter)	0.025 g	50 ppm	~95% in 1100 min/ BB/UV, ~80% in 1100 min/BB/Vis, ~100% in 1100 min/ BG/UV, ~90% in 1100 min/BG/Vis	No data	[159]
Polypyrrole (PPY)/ Fe_3O_4/ZnO	*In situ* polymerisation	Degradation of MB/UV/8 W germicidal lamp	0.200 g	10 ppm	85.2% in 4 h	No data	[160]
Fe_3O_4/GO/Ag_3PO_4	Co-precipitation/ ultrasonication	Degradation of MB/Vis/250 W halogen lamp/400 nm UV filter	0.025 g	20 ppm	100% in 10 min	12.56 emu/g	[161]
RGO/Fe_3O_4	*In situ* chemical synthesis	Degradation of MB/natural sunlight/bright sunny days/ 9.00 am to 2.00 pm	0.002 g	10 ppm	100% in 60 min	30.30 emu/g	[162]
ZrO_2/Fe_3O_4/chitosan	Co-precipitation/refluxing	Reduction of Cr(VI) and degradation of 4-CP/ natural sunlight	0.050 g/ Cr(III), 0.010 g/ 4-CP	70 ppm/$K_2Cr_2O_7$, 20 ppm/4-CP	88.6% in 180 min/4-CP, 90.2% in 180 min/Cr(VI)	42.00 emu/g	[115]
Cu_2O/Fe_3O_4	Solvothermal/precipitation	Degradation of MO/Vis/500 W Xe lamp/420 nm UV filter	0.100 g	30 ppm	90% in 90 min	41.70 emu/g	[163]
$FeWO_4$/Fe_3O_4	Hydrothermal	Degradation of MB/UV-Vis/ 500 W Xe lamp	0.020 g	20 ppm	97.1% in 60 min	9.00 emu/g	[164]
Fe_3O_4@carbon quantum dots (CQDs)	Hydrothermal	Degradation of MB/Vis75 W Xe lamp/420 UV filter	0.001 g	1.00 ppm	94.4% in 30 min	33.80 emu/g	[165]
BiOBr@SiO_2@Fe_3O_4	Hydrothermal/ Stobber method	2,2-bis(4-hydroxyphenyl)propane (BPA)/500 W Xe lamp/ 420 UV filter	0.100 g	20 ppm	87.0% in 50 min	No data	[119]

Ferrites (MFe_2O_4)-based Magnetic Photocatalysts

The spinel ferrites with the chemical formula MFe_2O_4, where M is a cation such as Mg^{2+}, Cu^{2+}, Zn^{2+}, Ni^{2+}, Co^{2+}, etc., are popular materials due to their narrow band gaps, which endow them with remarkable visible light response. Moreover, ferrites show good chemical, thermal, and ferromagnetic properties, numerous oxygen vacancies, and a high density of surface hydroxyl groups [166,167]. In terms of the crystal structure of ferrites, the oxygen atoms form a face-centred cubic (fcc) close packing, while the M^{2+} and Fe^{2+} cations occupy either octahedral or tetrahedral interstitial sites [168]. Generally, ferrites have attracted tremendous interests in various applications such as fabrication of memory devices, biotechnology, catalysts for organic reactions, water splitting, pollution remediation, lithium ion batteries, etc. [169,170,171,172]. Among the ferrites, zinc ferrite ($ZnFe_2O_4$) is one of the most attractive ferrite due to its photochemical and thermal stability, low toxicity, narrow band gap (1.9 eV), natural abundance, good magnetic properties, and easy preparation [64,173,174]. However, as a pristine photocatalyst, $ZnFe_2O_4$ show low photocatalytic activity (low quantum efficiency) due to the high recombination rate of the photogenerated charge carriers, unconvincing photoelectric conversion, and low valence band potential [64].

In an attempt to improve the photocatalytic performance of $ZnFe_2O_4$, laser ablation obtained $ZnO_x(OH)_x$ and FeO_xcolloids were used as precursors to uniformly deposit $ZnFe_2O_4$ nanoparticles on RGO sheets to form a magnetic RGO/$ZnFe_2O_4$ nanocomposite. The RGO/$ZnFe_2O_4$ nanocomposite exhibited improved charge separation and photocatalytic activity towards MB degradation under visible light irradiation [175]. In a similar work by Yang et al., $ZnFe_2O_4$ nanocrystals were confined within an interconnected graphene network, which served as dispersing agents, as transport channels for electrons, and as electron scavengers and reservoirs to minimise recombination. Enhanced visible light photocatalytic degradation of MB was observed over the nanocomposite photocatalyst [176]. In the RGO/$ZnFe_2O_4$ nanocomposite, electrons are excited by visible light from the valence band to the conduction band of $ZnFe_2O_4$, and these electrons quickly transfer to the RGO skeleton where they are trapped by adsorbed oxygen to form superoxide radicals (Figure 6a). The holes are confined to the valence band of $ZnFe_2O_4$ where they directly oxidise the pollutants since they are not positive enough to react with water to form the hydroxyl radicals [175,176]. More interestingly, the RGO/$ZnFe_2O_4$ nanocomposites showed good stability and recyclability and were easily separable using an external magnetic field owing to the good magnetic response (Figure 6b). The presence of RGO ensured that the electrons and holes are located at different parts of the nanocomposite, thereby ensuring efficient charge separation.

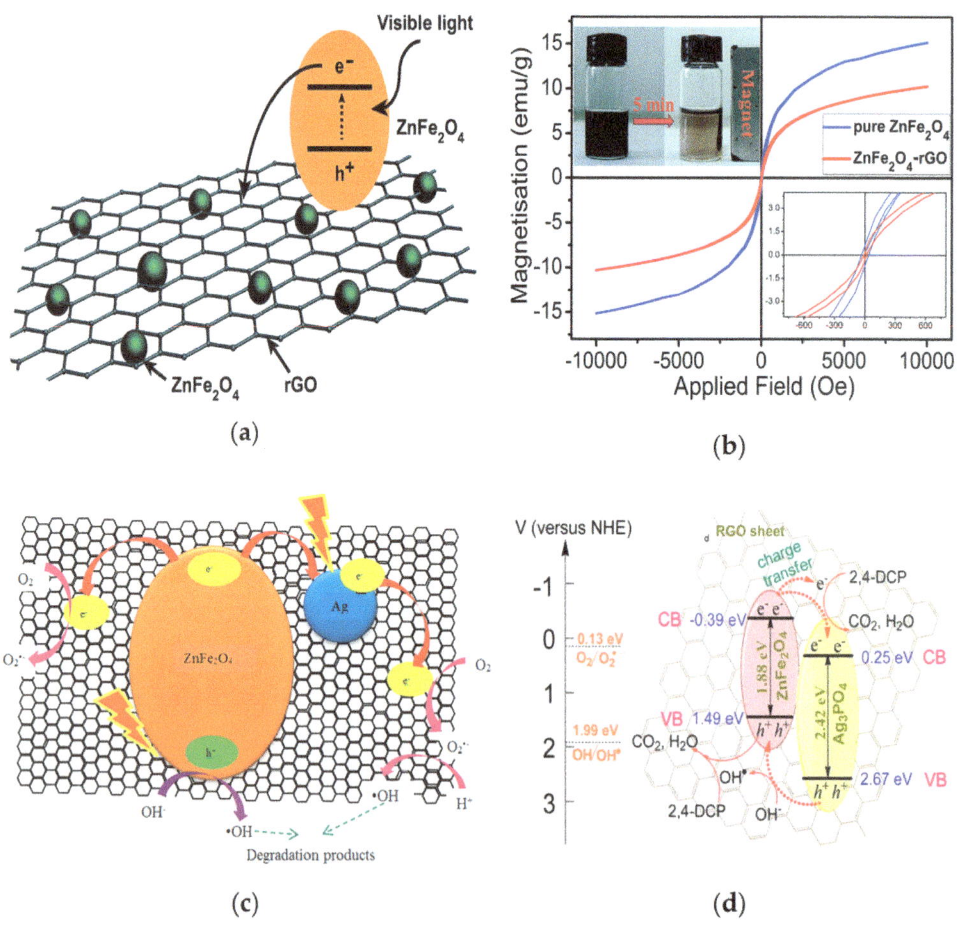

Figure 6: (a) Charge transfer route in RGO/$ZnFe_2O_4$; (b) magnetisation curves for pure $ZnFe_2O_4$ and RGO/$ZnFe_2O_4$, insert showing magnetic separation of RGO/$ZnFe_2O_4$(Reproduced with permission from [175], Copyright 2014, RSC); charge transfer mechanism in (c) $ZnFe_2O_4$–Ag/RGO (Reproduced with permission from [177], Copyright 2016, Wiley) and (d) RGO/$ZnFe_2O_4$/Ag_3PO_4. (Reproduced with permission from [178], Copyright 2016, ACS.

Enhanced visible light photocatalytic degradation of 17α-ethinylestradiol (EE 2) was observed over magnetic $ZnFe_2O_4$–Ag/RGO nanocomposite prepared via a hydrothermal method. In terms of degradation kinetics, the nanocomposite photocatalyst showed the highest activity, which was 14.6 and 5.6 times that of $ZnFe_2O_4$ and $ZnFe_2O_4$/RGO, respectively. This could be credited to the combined contribution of RGO, Ag, and $ZnFe_2O_4$ in the nanocomposite, resulting in an improved surface area, efficient charge separation and transportation, minimal aggregation, and improved visible light utilization [177]. Chen and coworkers used a combination of solvothermal and in situ precipitation routes to tailor a magnetic RGO/$ZnFe_2O_4$/Ag_3PO_4 nanocomposite and examined its photocatalytic behaviour towards 2,4-DCP under visible light exposure. The ternary nanocomposite exhibited significantly superior activity over pure Ag_3PO_4 [178]. In the $ZnFe_2O_4$–Ag/RGO heterostructure, electrons excited by visible light from the valence band of $ZnFe_2O_4$ to its conduction band transfer to the Ag nanoparticles and, further, to the RGO skeleton where they are captured by adsorbed oxygen to form the radical species (Figure 6c). The holes directly attack the pollutant to form the degradation products [177]. Similarly, in the RGO/$ZnFe_2O_4$/Ag_3PO_4 ternary heterostructure, a heterojunction type of charge transfer mechanism was the most plausible route (Figure 6d). Both $ZnFe_2O_4$ (band gap, 1.92 eV) and Ag_3PO_4 (band gap, 2.42 eV) are excited by visible light, and electrons are promoted to the conduction bands leaving positive holes in the valence bands. The conduction band potential (−0.39 eV) and valence band potential (+1.49 eV) of $ZnFe_2O_4$ are more negative than those of Ag_3PO_4; therefore, electrons transfer to the conduction band of Ag_3PO_4 and RGO, while the holes transfer in the opposite direction to the valence band of $ZnFe_2O_4$ [178]. This results in efficient charge separation and the formation of the reactive species.

Several other interesting $ZnFe_2O_4$-based magnetic nanocomposite photocatalysts such as $ZnFe_2O_4$/ZnO/Ag_3PO_4 [64], $ZnFe_2O_4$/ZnO [179], N-TiO_2/$ZnFe_2O_4$ [180], $ZnFe_2O_4$/Ag_3PO_4 [38], and Ag/ZnO/$ZnFe_2O_4$ [181] have recently been fabricated, and their photocatalytic properties were examined for the degradation of various pollutants in water. Notably, coupling the various semiconductors with $ZnFe_2O_4$ not only improved the overall visible response and photocatalytic properties of the photocatalysts but also introduced magnetic behaviour, which then allowed for easy separation using an external magnetic field.

Nickel ferrite ($NiFe_2O_4$) is another commonly exploited ferrite with a narrow band gap (1.7 eV), good magnetic properties, and high chemical and thermal stability. Structurally, $NiFe_2O_4$ has a typical inverse spinel structure and is ferromagnetic in nature. Its magnetism stems from the magnetic moment of antiparallel spins between Fe^{3+} cations occupying tetrahedral sites and Ni^{2+} cations sitting at octahedral sites [172,182,183]. Pure $NiFe_2O_4$ exhibits very low photocatalytic activity due to a high recombination rate of the charge carriers and the aggregation of the nanoparticles. However, $NiFe_2O_4$ is an attractive semiconductor to couple with other semiconductors or carbon nanomaterials and form nanocomposite photocatalysts with improved activity and, most importantly, could easily be separated using a magnet [167,184,185]. For example, magnetically separable AgBr/$NiFe_2O_4$ nanocomposite photocatalyst prepared via a combination of hydrothermal and an ultrasound-assisted precipitation method showed improved photocatalytic performance towards RhB under visible light illumination [186]. Ji and coworkers observed a significant improvement in the photocatalytic activity of magnetic g-C_3N_4/$NiFe_2O_4$ towards MB degradation, reaching 87% in 4 h of visible light irradiation [172]. In another example, magnetic Ag_3PO_4/$NiFe_2O_4$ nanocomposite tailored using an in situ precipitation method exhibited better visible light-driven photocatalytic activity towards MB compared to Ag_3PO_4 and $NiFe_3O_4$, individually [184]. $NiFe_2O_4$-based magnetic nanocomposites incorporating carbon nanomaterials such as Pd-$NiFe_2O_4$/RGO [27] and $NiFe_2O_4$/MWCNTs [187] have also shown remarkable enhancement in photocatalytic activity owing to the synergy between the carbon nanomaterials and $NiFe_2O_4$, which resulted in improved optical properties, charge separation, and transportation and specific surface area [27,187].

In all the $NiFe_2O_4$-incorporating photocatalysts, $NiFe_2O_4$ is excited upon the absorption of visible light. The charge transfer process depends on the other semiconductors involved in the nanocomposites. For example in AgBr/$NiFe_2O_4$ and Ag_3PO_4/$NiFe_2O_4$, in addition to $NiFe_2O_4$, AgBr and Ag_3PO_4 are also excited by visible light, and electrons are promoted to their conduction bands, leaving holes in the valence bands. The electrons transfer from the conduction band of $NiFe_2O_4$, since it is more negative, to the conduction bands of AgBr and Ag_3PO_4. Accordingly, the holes transfer from the valence bands of AgBr and Ag_3PO_4 to the valence band of $NiFe_2O_4$ where they are involved in the degradation of the pollutants [184,186]. Meanwhile, in g-C_3N_4/$NiFe_2O_4$ (both semiconductors are visible active), the electrons transfer from the conduction band of g-C_3N_4 to the conduction band of $NiFe_2O_4$, and the holes transfer in the opposite direction to the valence band of g-C_3N_4 (Figure 7a) [172]. Charge separation efficiency is greatly improved upon coupling $NiFe_2O_4$ with the other semiconductors since the holes and electrons are located at different sites in the nanocomposites. Moreover, the presence of $NiFe_2O_4$ nanoparticles in the nanocomposites endowed them with good magnetic response, which allowed for easy magnetic separation.

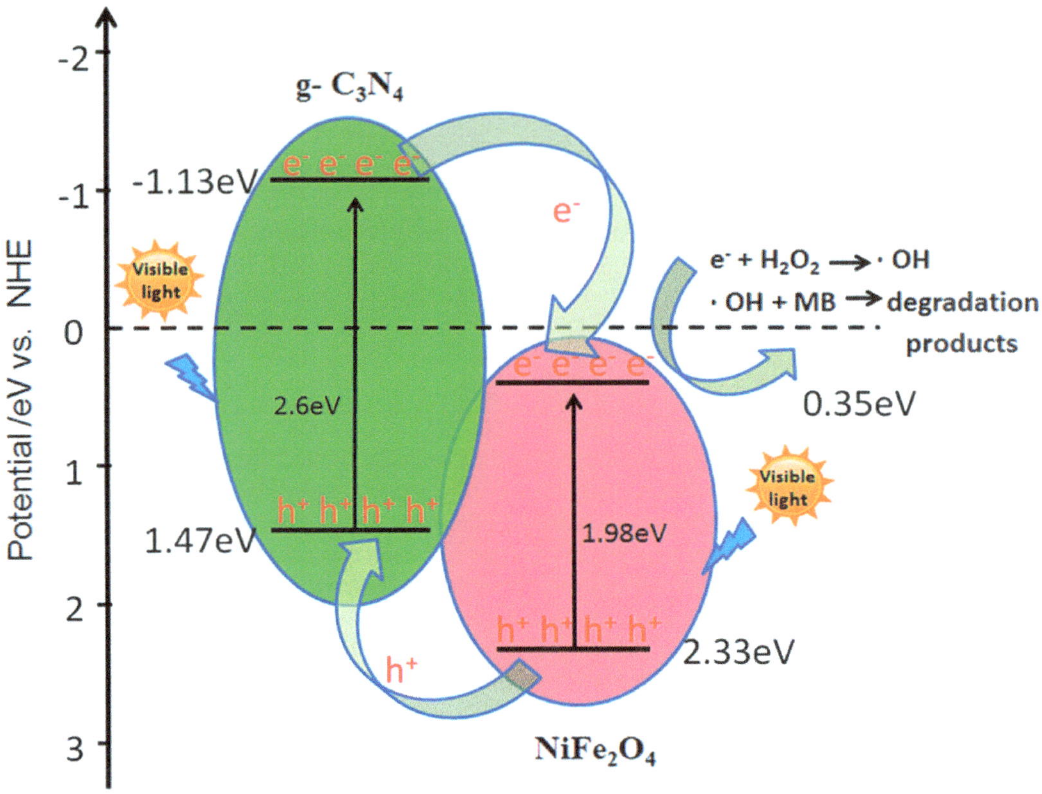

Figure 7: Charge transfer route in g-C_3N_4/$NiFe_2O_4$. Reproduced with permission from [172]. Copyright 2015, RSC.

Cobalt ferrite ($CoFe_2O_4$) is another important magnetic material with several attractive properties such as a crystallisation temperature similar to that of TiO_2, remarkable electromagnetic behaviour, high cubic magnetocrystalline anisotropy, good chemical stability, and high mechanical hardness [188,189,190,191]. Coupling $CoFe_2O_4$ with other semiconductors, carbon nanomaterials and conducting polymers are some of the strategies explored in order to benefit from its optical and magnetic properties. Pristine $CoFe_2O_4$ shows poor photocatalytic efficiency due to the high recombination rate of the charge carriers [61,192,193]. Kim et al. prepared hollow cobalt ferrite-polyaniline ($CoFe_2O_4$-PANI) nanofibres using the electrospinning method, calcination, and in situ chemical oxidative polymerisation, in succession. It was observed that the nanocomposite was magnetically separable and also showed significantly higher visible light photocatalytic degradation of MO, which was 80 times the activity of neat $CoFe_2O_4$ [61]. Hydrothermally prepared magnetic $CoFe_2O_4$/graphene (GR) nanocomposites with varying GR loadings exhibited improved visible light photocatalytic properties towards MB degradation compared to pristine $CoFe_2O_4$. At an optimum GR loading of 10%, the nanocomposite reached 100% MB degradation in 60 min, while the neat $CoFe_2O_4$ could only manage a low 34% degradation in twice the irradiation time [192]. The enhanced activity could be ascribed to the formation of heterojunctions between $CoFe_2O_4$ and GR, which allowed for the efficient separation of the charge carriers. Upon excitation by visible light, the electrons in the conduction band of $CoFe_2O_4$ transfer to GR, where they are trapped by molecular oxygen adsorbed on the surface [192]. A similar scenario was highlighted as responsible for the improved activity of $CoFe_2O_4$-PANI, whereby the electrons are transferred to the conduction band of $CoFe_2O_4$ from PANI while the holes transfer in the opposite direction to PANI [61]. The electrons and holes are kept apart, which ensures their availability to form the active species responsible for pollutant degradation.

Enhanced visible light photocatalytic decomposition of MB was reported over $CoFe_2O_4$ decorated CdS nanorods ($CoFe_2O_4$/CdS, prepared via a soft chemical method. The composite nanorods exhibited better activity than CdS and $CoFe_2O_4$, individually, and were magnetically separable [193]. Three-dimensional (3D) urchin-like TiO_2, prepared hydrothermally, was decorated with $CoFe_2O_4$ using a co-precipitation strategy to yield the nanocomposite photocatalyst $CoFe_2O_4$/TiO_2 with improved photocatalytic properties compared to the individual materials. Upon UV light illumination, the nanocomposite reached 98.9% MB degradation, while 79.9% MB decomposition was realised when the urchin-like TiO_2 was used as the photocatalyst [194]. Moreover, the nanocomposite photocatalyst showed good stability and recyclability, reaching 93.8% MB degradation in the fifth cycles while still maintaining good magnetic properties [194]. Impressive photocatalytic activity and versatility was observed for $CoFe_2O_4$/BiOX (X = Cl, Br and I) nanocomposites towards the UV and visible light photocatalytic decomposition of MO, RhB, MB, and their mixture (MO + MB + RhB) [195]. Similarly, Xu and coworkers examined the photocatalytic properties and versatile nature of the Ag/AgCl/$CoFe_2O_4$ nanocomposite by employing the photocatalyst for the decomposition of MO, bisphenol A (BPA), ciprofloxacin (CIP), and the deactivation of E. coli. The nanocomposite photocatalyst displayed outstanding photocatalytic performance towards both coloured (MO) and colourless (BPA and CIP) pollutants as well as microbial pollutants, reaching 93.4% in 150 min for MO and 100% in 90 min for CIP, and almost all of the E. coli were deactivated in just 30 min of visible light illumination [196]. In all the $CoFe_2O_4$-incorporating nanocomposites, the enhancement in activity could be ascribed to the formation of a heterojunction between $CoFe_2O_4$ and the other semiconductor, which allowed efficient charge separation and formation of the reactive species responsible for pollutant degradation. In addition, the presence of $CoFe_2O_4$ endowed the nanocomposite photocatalysts with sufficient magnetic sensitivity for easy separation using an external magnetic field [193,194,195,196]. Similar results have been reported for other $CoFe_2O_4$-based magnetic photocatalysts such as $CoFe_2O_4$/MCM-41/TiO_2 [197], Fe,N-TiO_2/$CoFe_2O_4$ [198], and BiOBr/$CoFe_2O_4$ [65].

A host of other ferrite-based magnetic nanocomposite photocatalysts (Table 3) have been designed and tailored for the removal of various pollutants in water, for which their photocatalytic properties were also examined. Enhanced photocatalytic performance has been observed upon the incorporation of the ferrites. Moreover, the presence of these ferrites induced magnetic response in the photocatalyst, which aids the separation process for easy recovery and recycling of the photocatalysts. Despite the decrease in magnetisation saturation of the nanocomposites compared to the neat ferrites, the photocatalyst retained enough magnetic response to enable separation using an external magnet.

Table 3: Summary of the photocatalytic properties of ferrite-based photocatalysts

Sample	Preparation Method	Application	Catalyst Dose	Pollutant Concentration	Degradation Efficiency	Magnetisation Saturation	Reference
P25/$CoFe_2O_4$/RGO	Hydrothermal	Degradation of MB, MO, neutral dark yellow (NDY)/Vis/500 W Xe lamp/420 nmUV filter	0.010 g	40 ppm	~90 mg/g in 60 min/MB, ~20 mg/g in 60 min and 10 mg/g in 60 min	~35 emu/g	[169]
$Mn_xMg_{1-x}Fe_2O_4$ ($0.0 \leq x \leq 0.5$)	Microwave-assisted combustion	Degradation of 4-CP/Vis/ 8 × 8 W medium pressure Hg lamps	0.030 g	200 ppm	~98% in 300 min	64.67 emu/g	[199]
Ag/TiO_2/$NiFe_2O_4$	Photodeposition/solvothermal	Degradation of MO/Vis/100 W LED	0.100 g	10 ppm	100% in 120 min	No data	[200]
TiO_2@SiO_2@Ni-Cu-$ZnFe_2O_4$	Chemical co-precipitation	Degradation of MB/simulated solar light/35 W Xe arc lamp	0.400 g	10 ppm	83.9% in 6 h	37.45 emu/g	[201]
$CoFe_2O_4$/TiO_2	Co-precipitation	Degradation of Reactive Red 120 (RR 120)/Vis/ 150 W W-halogen lamp	0.400 g	14.70 ppm	$4.98 \times 10^{-9}\ S^{-1}$	~0.2 emu/g	[202]
Ag/$NiFe_2O_4$	Combustion	Degradation of MB/Vis/300 W Xe lamp/450 nm UV filter	0.025 g	20 ppm	~80% in 120 min	~20 emu/g	[80]
$NiFe_2O_4$/TiO_2-SiO_2	Modified sol-gel/solvothermal	Degradation of cyanide/Vis/150 W blue fluorescent lamp/420 nm UV filter	No data	100 ppm	100% in 60 min	42.7 emu/g	[203]
$MgFe_2O_4$-ZnO	Solution method/chemical co-precipitation	Degradation of RhB/Vis/500 W Xe lamp/420 nm UV filter	0.050 g	4.40 ppm	100% in 120 min	21.37 emu/g	[204]
PANI-$CoFe_2O_4$-TiO_2	Hydrothermal/*in situ* chemical deposition	Degradation of MB/UV (500 W Xe lamp)/Vis/500 W Xe lamp/400 nm UV filter	0.100 g	50 ppm	0.0962/min/UV and 0.0110/min/Vis	11.4 emu/g	[205]
$Bi_{25}FeO_{40}$-RGO	Hydrothermal	Degradation of MB/Vis/500 W Xe lamp/400 nm UV filter	0.080 g	28.80 ppm	92.8% in 180 min	10.50 emu/g	[206]
BiOCl-$SrFe_{12}O_{19}$	Hydrothermal	Degradation of MB/UV/Vis/No data on light source	0.400 g	10 ppm	99% in 50 min/UV, 67.8% in 8 h/Vis	15.13 emu/g	[207]
$CuFe_2O_4$@C_3N_4	Self-assembly route	Degradation of orange II/Vis/500 W Xe lamp/ 420 nm UV filter	0.010 g	9.81 ppm	100% in 90 min	No data	[40]
$Co_{0.6}Zn_{0.4}Mn_xFe_{2-x}O_4$ ($x = 0.2$–1.0)	Sol-gel auto-combustion	Degradation of MO/Vis/400 W Hg lamp	0.100 g	30 ppm	94% in 60 min	39.51 emu/g	[208]
$Ni_{0.5}Zn_{0.5}Fe_2O_4$/$Zn_{0.95}Ni_{0.05}$	Egg albumen assisted sol-gel	Degradation of RhB/natural sunlight/sunny days/ 11 am to 3 pm.	0.010 g	20 ppm	~90% in 4 h	8.00 emu/g	[209]
$CdFe_2O_4$/GR	Hydrothermal	Degradation of MB/Vis/500 W Xe lamp/420 nm UV filter	0.100 g	10 ppm	89.2% in 4 h	14.26 emu/g	[82]
$Mn_xZn_{1-x}Fe_2O_4$-TiO_2 ($0.0 \leq x \leq 0.5$)	Auto-combustion	Degradation of 4-CP/UV/ 8 × 8 W low pressure Hg lamps	0.030 g	200 ppm	~98% in 270 min	45.17 emu/g	[210]
$Co_{0.6}Zn_{0.4}Cu_xFe_{2-x}O_4$	Sol-gel auto-combustion	Degradation of MO/Vis/400 W Hg lamp	0.100 g	30 ppm	86% in 60 min	38.02 emu/g	[211]
$Co_xZn_{1-x}Fe_2O_4$-GR	Chemical co-precipitation/ultrasonication	Degradation of MB/Vis/No data on light source	0.100 g	5 ppm	>95% in 60 min	No data	[212]
Sr-TiO_2/$Ni_{0.6}Zn_{0.4}Fe_2O_4$	Combustion/sol-gel	Degradation of BPA/UV (10 W low pressure Hg lamp)/Vis/500 W Xe arc lamp	0.150 g	10 ppm	100% in 4 h/UV, 90% in 4 h/Vis	19.04 emu/g	[213]
$CoFe_2O_4$-GR	Combustion	Degradation of MB/Vis/300 W Xe lamp/450 nm UV filter	0.025 g	20 ppm	100% in 120 min	5.3 emu/g	[166]
$MnFe_2O_4$/g-C_3N_4/TiO_2	Chemical impregnation	Degradation of MO/Simulated solar light/150 W Xe arc lamp	0.050g	10 ppm	99.3% in 180 min	0.065 emu/g	[214]
$Ni_{1-x}Co_xFe_2O_4$	Hydrothermal	Degradation of malachite green (MG)/natural sunlight (200–250 Wm^{-2})	0.025 g	0.365 ppm	~100% in 60 min	~50.7–64.2 emu/g.	[215]
$Mn_xZn_{1-x}Fe_2O_4$/β-Bi_2O_3	Dip-calcination	Degradation of RhB/Simulated solar light/300 W Xe lamp	0.200g	10 ppm	99.1% in 150 min	7.01 emu/g	[216]
$Zn_{1-x}Co_xFe_2O_4$	Hydrothermal	Degradation of MB/Natural sunlight	0.050 g	3.20 ppm	~88% in 270 min	~50 emu/g	[217]
$SrFeO_{3-x}$/g-C_3N_4	Sintering method	Degradation of chloramphenicol (CAP) and crystal violet (CV)/Vis/ 150 W Xe arc lamp	0.010 g	10 ppm	91.3% in 96 h/CAP, 99.9% in 12 h/CV	~0.17 emu/g	[41]
$CoFe_2O_4$/GR/CdS	Solvothermal	Degradation of MB/daylight/40 W daylight lamp	0.025 g	20 ppm	80% in 180 min	18.00 emu/g	[218]
ZnO/$CoFe_2O_4$	Co-precipitation	Degradation of direct blue 71 (BD 71)/150 W W-halogen lamp	0.160 g	38.64 ppm	~100% in 30 min	~0.040 emu/g	[219]

SUMMARY REMARKS AND FUTURE OUT-LOOK

The development of efficient visible light photocatalytic degradation of pollutants and easy recovery and recyclability of the photocatalyst is an exciting prospect in water treatment. The incorporation of magnetic nanoparticles such as haematite, maghemite, magnetite, and the ferrites into various photocatalyst matrices provides an attractive strategy to enhance the optical properties, charge separation efficiency, and the overall photocatalytic activity. Most interestingly, the presence of the magnetic nanoparticles endow the nanocomposite photocatalyst with sufficient magnetic response to enable separation of the photocatalyst using an external magnetic field. However, careful control of the magnetic nanoparticle, especially loading maghemite and magnetite, is important in ensuring a balance between good magnetic response and the detrimental effect of recombination that occurs on the surface of the nanoparticles. Alternatively, silica is used as a barrier between the magnetic core and the other semiconductor coupled with maghemite or magnetite. This prevents charge transfer between the coupled materials that could otherwise result in photodissolution and recombination. The incorporation of the ferrites in various photocatalysts provides more beneficial properties than just good magnetic response; they contribute towards visible light absorption, charge separation, and the photocatalytic performance of the nanocomposite photocatalyst. Therefore, the ferrites are the most attractive nanoparticles compared to haematite, maghemite, and magnetite, in terms of their photocatalytic performance and multifaceted contribution in the magnetic nanocomposites. However, a systematic, comparative study of the photocatalytic performance of the ferrites, haematite, magnetite, and maghemite is seldom reported, and this makes it difficult to map a way forward in terms of exploiting the best materials for photocatalytic applications.

Despite the promising results regarding visible-light-active magnetic photocatalysts, there are still some challenges that possibly hinder the practical exploitation of these materials. The most obvious problem relates to the reaction kinetics; most of the degradation experiments are slow (may take several hours) despite the improvements in visible light absorption of the photocatalyst, and this presents a massive challenge for practical applications. Secondly, work still needs to be done to develop synthesis routes that will ensure uniform shape and size of the magnetic particles as well as their uniform distribution within the nanocomposite matrix in order to ensure good magnetic response and efficient recovery. Strong contact between the magnetic nanoparticles and the other semiconductor particles is crucial in ensuring minimal leaching out and loss of the photocatalytic activity and magnetic response. In the magnetic core-shell structures, careful control of the thickness of the silica shell is important, as this will affect the magnetic response of the composite material. Extensive theoretical studies are needed in order to understand the interactions between the magnetic nanoparticles and the other semiconductor nanoparticles in the nanocomposites; this can help design and tailor materials with good heterojunctions that promote efficient charge separation. Moreover, theoretical studies can provide a good idea of the optimum loadings of the magnetic nanoparticles in the photocatalyst to ensure good magnetic response and minimal recombination. In most cases, the magnetisation saturation of the magnetic nanoparticles decrease significantly upon coupling with the non-magnetic components of the nanocomposite photocatalyst, which has a negative bearing on the separation efficiency and recovery of the photocatalyst. Work still needs to be done to ensure that nanocomposites retain sufficiently high magnetic response in order to enable efficient magnetic separation.

ACKNOWLEDGMENTS

Financial support from the College of Science, Engineering and Technology (CSET) of the University of South Africa is highly appreciated.

AUTHOR CONTRIBUTIONS

Gcina Mamba conceived the idea of writing a review on magnetically separable photocatalysts and wrote the introduction and Section 2.1, Section 2.2, and Section 2.3. Ajay Mishra is the project leader and was responsible for structuring the review, writing Section 2.4, and editing the entire review article.

REFERENCES

1. Fersi,C.;Gzara,L.;Dhahbi,M.Treatmentoftextileeffluents by membrane technologies. Desalination 2005, 185, 399–409.
2. Pendergast, M.M.; Hoek, E.M.V. A review of water treatment membrane nanotechnologies. Energy Environ. Sci. 2011,4, 1946–1971.
3. Wang, H.; Zheng, X.W.; Su, J.Q.; Tian, Y.; Xiong, X.J.; Zheng, T.L. Biological decolorization of the reactive dyes Reactive Black 5 by a novel isolated bacterial strain Enterobacter sp. EC3. J. Hazard. Mater. 2009, 171, 654–659.
4. Kumari, K.; Abraham, T.E. Biosorption of anionic textile dyes by nonviable biomass of fungi and yeast. Bioresour. Technol. 2007, 98, 1704–1710.
5. Soutsas, K.; Karayannis, V.; Poulios, I.; Riga, A.; Ntampegliotis, K.; Spiliotis, X.; Papapolymerou, G. Decolorization and degradation of reactive azo dyes via heterogeneous photocatalytic processes. Desalination 2010, 250, 345–350.
6. Wu, J.; Doan, H.; Upreti, S. Decolorization of aqueous textile reactive dye by ozone. Chem. Eng. J. 2008, 142, 156–160.
7. Harsha, N.; Krishna, K.V.S.; Renuka, N.K.; Shukla, S. Facile synthesis of g–Fe_2O_3 nanoparticles integrated $H_2Ti_3O_7$nanotubes structure as a magnetically recyclable dye-removal catalyst. RSC Adv. 2015, 5, 30354–30362.
8. Li, K.; Chai, F.; Zhao, Y.; Guo, X. Facile synthesis of magnetic Fe_3O_4/$CeCO_3OH$ composites with excellent adsorption capability for small cationic dyes. RSC Adv. 2015, 5, 94397–94404.
9. Song, S.; Fan, J.; He, Z.; Zhan, L.; Liu, Z.; Chen, J.; Xu, X. Electrochemical degradation of azo dye C.I. Reactive Red 195 by anodic oxidation on Ti/SnO_2-Sb/PbO_2 electrodes. Electrochim. Acta 2010, 55, 3606–3613.

10. Faouzi, A.M.; Nasr, B.; Abdellatif, G. Electrochemical degradation of anthraquinone dye Alizarin Red S by anodic oxidation on boron-doped diamond. Dyes Pigments 2007, 73, 86–89.
11. Rodrigues, C.S.D.; Madeira, L.M.; Boaventura, R.A.R. Synthetic textile dyeing wastewater treatment by integration of advanced oxidation and biological processes—Performance analysis with costs reduction. J. Environ. Chem. Eng. 2014,2, 1027–1039.
12. Basha, C.A.; Selvakumar, K.V.; Prabhu, H.J.; Sivashanmugam, P.; Lee, C.W. Degradation studies for textile reactive dye by combined electrochemical, microbial and photocatalytic methods. Sep. Purif. Technol. 2011, 79, 303–309.
13. Berberidou, C.; Avlonitis, S.; Poulios, I. Dyestuff effluent treatment by integrated sequential photocatalytic oxidation and membrane filtration. Desalination 2009, 249, 1099–1106.
14. Khouni, I.; Marrot, B.; Moulin, P.; Ben, R. Decolourization of the reconstituted textile effluent by different process treatments: Enzymatic catalysis, coagulation/flocculation and nanofiltration processes. Desalination 2011, 268, 27–37.
15. Trzcinski, A.P.; Stuckey, D.C. Inorganic fouling of an anaerobic membrane bioreactor treating leachate from the organic fraction of municipal solid waste (OFMSW) and a polishing aerobic membrane bioreactor. Bioresour. Technol.2016, 204, 17–25.
16. Schulz, M.; Soltani, A.; Zheng, X.; Ernst, M. Effect of inorganic colloidal water constituents on combined low-pressure membrane fouling with natural organic matter (NOM). J. Membr. Sci. 2016, 507, 154–164.
17. Hegab, H.M.; Elmekawy, A.; Barclay, T.G.; Michelmore, A.; Zou, L.; Saint, C.P.; Ginic-Markovic, M. Effective in-situ chemical surface modification of forward osmosis membranes with polydopamine-induced graphene oxide for biofouling mitigation. Desalination 2016, 385, 126–137.
18. Gottlieb, A.; Shaw, C.; Smith, A.; Wheatley, A.; Forsythe, S. The toxicity of textile reactive azo dyes after hydrolysis and decolourisation. J. Biotechnol. 2003, 101, 49–56.
19. Kariyajjanavar, P.; Jogttappa, N.; Nayaka, Y.A. Studies on degradation of reactive textile dyes solution by electrochemical method. J. Hazard. Mater. 2011, 190, 952–961.
20. Faouzi Elahmadi, M.; Bensalah, N.; Gadri, A. Treatment of aqueous wastes contaminated with Congo Red dye by electrochemical oxidation and ozonation processes. J. Hazard. Mater. 2009, 168, 1163–1169.
21. Mohammad, N.; Arami, M. Degradation and toxicity reduction of textile wastewater using immobilized titania nanophotocatalysis. J. Photochem. Photobiol. B 2009, 94, 20–24.
22. Aksu, Z.; Dönmez, G. Combined effects of molasses sucrose and reactive dye on the growth and dye bioaccumulation properties of Candida tropicalis. Process Biochem. 2005, 40, 2443–2454.
23. Yu, W.; Xu, D.; Peng, T. Enhanced photocatalytic activity of g-C_3N_4 for selective CO_2 reduction to CH_3OH via facile coupling of ZnO: A direct Z-scheme mechanism. J. Mater. Chem. A 2015, 3, 19936–19947.
24. Lu, D.; Zhang, G.; Wan, Z. Visible-light-driven g-C_3N_4/Ti^{3+}-TiO_2 photocatalyst co-exposed {001} and {101} facets and its enhanced photocatalytic activities for organic pollutant degradation and Cr(VI) reduction. Appl. Surf. Sci. 2015,358, 223–230.
25. Li, G.; Nie, X.; Chen, J.; Jiang, Q.; An, T.; Wong, P.K.; Zhang, H.; Zhao, H.; Yamashita, H. Enhanced visible-light-driven photocatalytic inactivation of Escherichia coli using g-C_3N_4/TiO_2 hybrid photocatalyst synthesized using a hydrothermal-calcination approach. Water Res. 2015, 86, 17–24.
26. Zhong, Y.; Ni, Y.; Li, S.; Wang, M. Chain-like Fe_3O_4@ resorcinol-formaldehyde resins-Ag composite microstructures: Facile construction and applications in antibacterial and catalytic fields. RSC Adv. 2016, 6, 15831–15837.
27. Li, Y.; Zhang, Z.; Pei, L.; Li, X.; Fan, T.; Ji, J.; Shen, J.; Ye, M. Multifunctional photocatalytic performances of recyclable Pd-$NiFe_2O_4$/reduced graphene oxide nanocomposites via different co-catalyst strategy. Appl. Catal. B Environ. 2016,190, 1–11.
28. Zeng, B.; Zhang, L.; Wan, X.; Song, H.; Lv, Y. Fabrication of α-Fe_2O_3/g-C_3N_4 composites for cataluminescence sensing of H_2S. Sens. Actuators Chem. 2015, 211, 370–376.
29. Wang, G.; Kuang, S.; Zhang, J.; Hou, S.; Nian, S. Graphitic carbon nitride/multiwalled carbon nanotubes composite as Pt-free counter electrode for high-efficiency dye-sensitized solar cells. Electrochim. Acta 2016, 187, 243–248.
30. Wang, M.; Yang, X.; Bi, W. Application of magnetic graphitic carbon nitride nanocomposites for the solid-phase extraction of phthalate esters in water samples. J. Sep. Sci. 2015, 38, 445–452.
31. Baig, R.B.N.; Verma, S.; Varma, R.S.; Nadagouda, M.N. Magnetic Fe@g-C_3N_4: A photoactive catalyst for the hydrogenation of alkenes and alkynes. ACS Sustain. Chem. Eng. 2016, 4, 1661–1664.
32. Preethi, V.; Kanmani, S. Photocatalytic hydrogen production using Fe_2O_3-based core shell nano particles with ZnS and CdS. Int. J. Hydrogen Energy 2014, 39, 1613–1622.
33. Cao, M.; Wang, P.; Ao, Y.; Wang, C.; Hou, J.; Qian, J. Visible light activated photocatalytic degradation of tetracycline by a magnetically separable composite photocatalyst: Graphene oxide/magnetite/cerium-doped titania. J. Colloid Interface Sci. 2016, 467, 129–139.
34. Bao, X.; Qiang, Z.; Chang, J.; Ben, W.; Qu, J. Synthesis of carbon-coated magnetic nanocomposite (Fe_3O_4@C) and its application for sulfonamide antibiotics removal from water. J. Environ. Sci. 2014, 26, 962–969.
35. Zhou, L.; Deng, J.; Zhao, Y.; Liu, W.; An, L.; Chen, F. Preparation and characterization of N–I co-doped nanocrystal anatase TiO_2 with enhanced photocatalytic

activity under visible-light irradiation. Mater. Chem. Phys. 2009, 117, 522–527.

36. Nolan, N.T.; Synnott, D.W.; Seery, M.K.; Hinder, S.J.; van Wassenhoven, A.; Pillai, S.C. Effect of N-doping on the photocatalytic activity of sol-gel TiO_2. J. Hazard. Mater. 2012, 211-212, 88–94.

37. Pany, S.; Parida, K.M. A facile in situ approach to fabricate N,S-TiO_2/g-C_3N_4 nanocomposite with excellent activity for visible light induced water splitting for hydrogen evolution. Phys. Chem. Chem. Phys. 2015, 17, 8070–8077.

38. Ge, M.; Chen, Y.; Liu, M.; Li, M. Synthesis of magnetically separable Ag_3PO_4/$ZnFe_2O_4$ composite photocatalysts for dye degradation under visible LED light irradiation. J. Environ. Chem. Eng. 2015, 3, 2809–2815.

39. Xu, Y.; Huang, S.; Xie, M.; Li, Y.; Jing, L.; Xu, H. Core-shell magnetic Ag/AgCl@Fe_2O_3 photocatalysts with enhanced photoactivity for eliminating bisphenol A and microbial contamination. New J. Chem. 2016, 40, 3413–3422.

40. Yao, Y.; Lu, F.; Zhu, Y.; Wei, F.; Liu, X.; Lian, C.; Wang, S. Magnetic core-shell $CuFe_2O_4$@C_3N_4 hybrids for visible light photocatalysis of Orange II. J. Hazard. Mater. 2015, 297, 224–233.

41. Lin, H.-P.; Chen, C.-C.; Lee, W.W.; Lai, Y.-Y.; Chen, J.; Chen, Y.; Fu, J.-Y. Synthesis of a $SrFeO_{3-x}$/g-C_3N_4heterojunction with improved visible-light photocatalytic activities in chloramphenicol and crystal violet degradation. RSC Adv. 2015, 6, 2323–2336.

42. Li, G.; Nie, X.; Gao, Y.; An, T. Can environmental pharmaceuticals be photocatalytically degraded and completely mineralized in water using g-C_3N_4/TiO_2 under visible light irradiation?—Implications of persistent toxic intermediates. Appl. Catal. B Environ. 2016, 180, 726–732.

43. Zhang, J.; Hu, S.; Wang, Y. A convenient method to prepare a novel alkali metal sodium doped carbon nitride photocatalyst with a tunable band structure. RSC Adv. 2014, 4, 62912–62919.

44. Jin, R.; Hu, S.; Gui, J.; Liu, D. A convenient method to prepare novel rare earth metal Ce-doped carbon nitride with enhanced photocatalytic activity under visible light. Bull. Korean Chem. Soc. 2015, 36, 17–23.

45. Akpan, U.G.; Hameed, B.H. Enhancement of the photocatalytic activity of TiO_2 by doping it with calcium ions. J. Colloid Interface Sci. 2011, 357, 168–178.

46. Inturi, S.N.R.; Boningari, T.; Suidan, M.; Smirniotis, P.G. Visible-light-induced photodegradation of gas phase acetonitrile using aerosol-made transition metal (V, Cr, Fe, Co, Mn, Mo, Ni, Cu, Y, Ce, and Zr) doped TiO_2. Appl. Catal. B Environ. 2014, 144, 333–342.

47. Li, Z.; Wang, J.; Zhu, K.; Ma, F.; Meng, A. Ag/g-C_3N_4 composite nanosheets: Synthesis and enhanced visible photocatalytic activities. Mater. Lett. 2015, 145, 167–170.

48. Hu, S.; Ma, L.; You, J.; Li, F.; Fan, Z.; Wang, F.; Liu, D.; Gui, J. A simple and efficient method to prepare a phosphorus modified g-C_3N_4 visible light photocatalyst. RSC Adv. 2014, 4, 21657–21663.

49. Sun, H.; Wang, S.; Ang, H.M.; Tadé, M.O.; Li, Q. Halogen element modified titanium dioxide for visible light photocatalysis. Chem. Eng. J. 2010, 162, 437–447.

50. Dong, G.; Ai, Z.; Zhang, L. Efficient anoxic pollutant removal with oxygen functionalized graphitic carbon nitride under visible light. RSC Adv. 2014, 4, 5553–5560.

51. Sharotri, N.; Sud, D. Ultrasound-assisted synthesis and characterization of visible light responsive nitrogen-doped TiO_2 nanomaterials for removal of 2-Chlorophenol. Desalin. Water Treat. 2016, 57, 8776–8788.

52. Reszczyńska, J.; Grzyb, T.; Sobczak, J.W.; Lisowski, W.; Gazda, M.; Ohtani, B.; Zaleska, A. Visible light activity of rare earth metal doped (Er^{3+}, Yb^{3+} or Er^{3+}/Yb^{3+}) titania photocatalysts. Appl. Catal. B: Environ. 2015, 163, 40–49.

53. Kaur, N.; Kaur, S.; Singh, V. Preparation, characterization and photocatalytic degradation kinetics of Reactive Red dye 198 using N, Fe codoped TiO_2 nanoparticles under visible light. Desalination Water Treat. 2016, 57, 9237–9246.

54. Xu, X.; Zhou, X.; Zhang, L.; Xu, L.; Ma, L.; Luo, J.; Li, M.; Zeng, L. Photoredox degradation of different water pollutants (MO, RhB, MB, and Cr(VI)) using Fe-N-S-tri-doped TiO_2 nanophotocatalyst prepared by novel chemical method. Mater. Res. Bull. 2015, 70, 106–113.

55. Antonopoulou, M.; Vlastos, D.; Konstantinou, I. Photocatalytic degradation of pentachlorophenol by N-F-TiO_2: Identification of intermediates, mechanism involved, genotoxicity and ecotoxicity evaluation. Photochem. Photobiol. Sci.2015, 14, 520–527.

56. Mamba, G.; Mbianda, X.Y.; Mishra, A.K. Photocatalytic degradation of the diazo dye naphthol blue black in water using MWCNT/Gd,N,S-TiO_2 nanocomposites under simulated solar light. J. Environ. Sci. 2015, 33, 219–228.

57. Mamba, G.; Mamo, M.A.; Mbianda, X.Y.; Mishra, A.K. Nd,N,S-TiO_2 decorated on reduced graphene oxide for a visible light active photocatalyst for dye degradation: Comparison to its MWCNT/Nd,N,S-TiO_2 analogue. Ind. Eng. Chem. Res. 2014, 53, 14329–14338.

58. Mamba, G.; Mbianda, X.Y.; Mishra, A.K. Enhanced visible light photocatalytic degradation of eriochrome black T and eosin blue shade in water using tridoped titania decorated on SWCNTs and MWCNTs: Effect of the type of carbon nanotube incorporated. Mater. Chem. Phys. 2015, 149–150, 734–742.

59. Apostolopoulou, V.; Vakros, J.; Kordulis, C.; Lycourghiotis, A. Preparation and characterization of [60] fullerene nanoparticles supported on titania used as a photocatalyst. Colloids Surf. A Physicochem. Eng. Asp. 2009, 349, 189–194.

60. Shang, J.; Zhao, F.; Zhu, T.; Li, J. Photocatalytic degradation of rhodamine B by dye-sensitized TiO_2 under visible-light irradiation. Sci. China Chem. 2011, 54, 167–172.

61. Nam, K.; Jung, H.; Lee, W. Hollow cobalt ferrite-polyaniline nanofibers as magnetically separable visible-light photocatalyst for photodegradation of methyl orange. J. Photochem. Photobiol. A Chem. 2016, 321, 257–265.

62. Kim, G.; Choi, W. Charge-transfer surface complex of EDTA-TiO_2 and its effect on photocatalysis under visible light.Appl. Catal. B Environ. 2010, 100, 77–83.

63. Shi, W.; Lu, D.; Wang, L.; Teng, F.; Zhang, J. Core-shell structured Fe_3O_4@SiO_2@CdS nanoparticles with enhanced visible-light photocatalytic photocatalytic activities. RSC Adv. 2015, 5, 106038–106043.

64. Li, J.; Liu, Z.; Zhu, Z. Enhanced photocatalytic activity in $ZnFe_2O_4$-ZnO-Ag_3PO_4 hollow nanospheres through the cascadal electron transfer with magnetical separation. J. Alloys Compd. 2015, 636, 229–233.

65. Jiang, R.; Zhu, H.; Li, J.; Fu, F.; Yao, J.; Jiang, S.; Zeng, G.-M. Fabrication of novel magnetically separable BiOBr/$CoFe_2O_4$ microspheres and its application in the efficient removal of dye from aqueous phase by an environment-friendly and economical approach. Appl. Surf. Sci. 2016, 364, 604–612.

66. Mi, J.; Hwan, Y.; Keun, B.; Ho, S.; Soo, B.; Bin, J. Novel immobilization of titanium dioxide (TiO_2) on the fluidizing carrier and its application to the degradation of azo-dye. J. Hazard. Mater. 2006, 134, 230–236.

67. Fernhndez, A.; Lassaletta, G.; Jimknez, V.M.; Justo, A. Preparation and characterization of TiO_2 photocatalysts supported on various rigid supports (glass, quartz and stainless steel). Comparative studies of photocatalytic activity in water purification. Appl. Catal. B Environ. 1995, 7, 49–63.

68. Carlson, P.J.; Pretzer, L.A.; Boyd, J.E. Solvent deposition of titanium dioxide on acrylic for photocatalytic application.Ind. Eng. Chem. Res. 2007, 46, 7970–7976.

69. Arabatzis, I.M.; Antonaraki, S.; Stergiopoulos, T.; Hiskia, A. Preparation, characterization and photocatalytic activity of nanocrystalline thin film TiO_2 catalysts towards 3,5-dichlorophenol degradation. J. Photochem. Photobiol. A Chem.2002, 149, 237–245.

70. Anpo, M.; Zhang, S.G.; Mishima, H.; Matsuoka, M.; Yamashita, H. Design of photocatalysts encapsulated within the zeolite framework and cavities for the decomposition of NO into N_2 and O_2 at normal temperature. Catal. Today 1997,39, 159–168.

71. Rao, N.N.; Chaturvedi, V. Photoactivity of TiO_2-coated pebbles. Ind. Eng. Chem. Res. 2007, 46, 4406–4414.

72. Gad-allah, T.A.; Kato, S.; Satokawa, S.; Kojima, T. Treatment of synthetic dyes wastewater utilizing a magnetically separable photocatalyst (TiO_2/SiO_2/Fe_3O_4): Parametric and kinetic studies. Desalination 2009, 244, 1–11.

73. Wang, W.; Yao, J. Synthesis of magnetically separable Sn doped magnetite/silica core-shell structure and photocatalytic property. Mater. Res. Bull. 2010, 45, 710–716.

74. Belessi, V.; Lambropoulou, D.; Konstantinou, I.; Zboril, R.; Tucek, J.; Jancik, D. Structure and photocatalytic performance of magnetically separable titania photocatalysts for the degradation of propachlor. Appl. Catal. B Environ.2009, 87, 181–189.

75. Linley, S.; Leshuk, T.; Gu, F.X. Synthesis of magnetic rattle-type nanostructures for use in water treatment. ACS Appl. Mater. Interfaces 2013, 5, 2540–2548.

76. Kumar, S.; Kumar, B.; Baruah, A.; Shanker, V. Synthesis of magnetically separable and recyclable g-C_3N_4–Fe_3O_4hybrid nanocomposites with enhanced photocatalytic performance under visible-light irradiation. J. Phys. Chem. C2013, 117, 26135–26143.

77. Yang, J.; Chen, H.; Gao, J.; Yan, T.; Zhou, F.; Cui, S.; Bi, W. Synthesis of Fe_3O_4/g-C_3N_4 nanocomposites and their application in the photodegradation of 2,4,6-trichlorophenol under visible light. Mater. Lett. 2016, 164, 183–189.

78. Shih, Y.; Su, C.; Chen, C.; Dong, C. Synthesis of magnetically recoverable ferrite (MFe_2O_4, M = Co, Ni and Fe)-supported TiO_2 photocatalysts for decolorization of methylene blue. Catal. Commun. 2015, 72, 127–132.

79. Pang, Y.L.; Lim, S.; Ong, H.C.; Chong, W.T. Synthesis, characteristics and sonocatalytic activities of calcined α-$Fe_2O_3$and TiO_2 nanotubes/α-Fe_2O_3 magnetic catalysts in the degradation of Orange G. Ultrason. Sonochem. 2016, 29, 317–327.

80. Zhang, D.; Pu, X.; Du, K.; Moon, Y.; Jeong, J.; Cai, P.; Kim, S., II; Seo, H.J. Combustion synthesis of magnetic Ag/$NiFe_2O_4$ composites with enhanced visible-light photocatalytic properties. Sep. Purif. Technol. 2014, 137, 82–85.

81. Zhao, H.; Zhang, L.; Gu, X.; Li, S.; Li, B.; Wang, H. Fe_2O_3-AgBr nonwoven cloth with hierarchical nanostructures as efficient and easily recyclable macroscale photocatalysts. RSC Adv. 2015, 5, 10951–10959.

82. Zhang, D.; Wang, Q.; Wang, L.; Zhang, L. Magnetically separable $CdFe_2O_4$/graphene catalyst and its enhanced photocatalytic properties. J. Mater. Chem. A 2015, 3, 3576–3585.

83. Liu, J.; Qiao, S.Z.; Hu, Q.H.; Qing, G.; Lu, M. Magnetic nanocomposites with mesoporous structures: Synthesis and applications. Small 2011, 7, 425–443.

84. Polshettiwar, V.; Luque, R.; Fihri, A.; Zhu, H.; Bouhrara, M.; Basset, J. Magnetically recoverable nanocatalysts. Chem. Rev. 2011, 111, 3036–3075.

85. Zhang, D.; Zhou, C.; Sun, Z.; Wu, L.-Z.; Tung, C.-H.; Zhang, T. Magnetically recyclable nanocatalysts (MRNCs): A versatile integration of high catalytic activity and facile recovery. Nanoscale 2012, 4, 6244–6255.

86. Wang, D.; Astruc, D. Fast-growing field of magnetically recyclable nanocatalysts. Chem. Rev. 2014, 114, 6949–6985.

87. Linley, S.; Leshuk, T.; Gu, F.X. Magnetically separable water treatment technologies and their role in future advanced water treatment: A patent review. CLEAN Soil Air Water 2013, 41, 1152–1156.

88. Yao, H.; Fan, M.; Wang, Y. Magnetic titanium dioxide based nanomaterials: Synthesis, characteristics, and photocatalytic application in pollutant degradation. J. Mater. Chem. A 2015, 3, 17511–17524.

89. Liu, D.; Li, Z.; Wang, W.; Wang, G.; Liu, D. Hematite doped magnetic TiO_2 nanocomposites with improved photocatalytic activity. J. Alloy Compd. 2016, 654, 491–497.

90. Shi, Y.; Li, H.; Wang, L.; Shen, W.; Chen, H. Novel α-Fe_2O_3/CdS cornlike nanorods with enhanced photocatalytic performance. ACS Appl. Mater. Interfaces 2012, 4, 4800–4806.

91. Xu, Y.; Jing, L.; Chen, X.; Ji, H.; Xu, H.; Li, H. Novel visible-light-driven Fe_2O_3/Ag_3VO_4 composite with enhanced photocatalytic activity toward organic pollutants degradation. RSC Adv. 2016, 6, 3600–3607.

92. Xu, Y.; Huang, S.; Xie, M.; Li, Y.; Xu, H.; Huang, L.; Zhang, Q.; Li, H. Magnetically separable Fe_2O_3/g-C_3N_4 catalyst with enhanced photocatalytic activity. RSC Adv. 2015, 5, 95727–95735.

93. Kumar, S.V.; Huang, N.M.; Yusoff, N.; Lim, H.N. High performance magnetically separable graphene/zinc oxide nanocomposite. Mater. Lett. 2013, 93, 411–414.

94. Wang, H.; Fei, X.; Wang, L.; Li, Y.; Xu, S. Magnetically separable iron oxide nanostructures-TiO_2 nanofibers hierarchical heterostructures: Controlled fabrication and photocatalytic activity. New J. Chem. 2011, 35, 1795–1802.

95. Muthukrishnaraj, A.; Vadivel, S.; Kamalakannan, V.P.; Balasubramanian, N. α-Fe_2O_3/reduced graphene oxide nanorod as efficient photocatalyst for methylene blue degradation. Mater. Res. Innov. 2016, 19, 258–264.

96. Wu, W.; Zhang, S.; Xiao, X.; Zhou, J.; Ren, F.; Sun, L.; Jiang, C. Controllable synthesis, magnetic properties, and enhanced photocatalytic activity of spindle-like mesoporous α-Fe_2O_3/ZnO core-shell heterostructures. ACS Appl. Mater. Interfaces 2012, 4, 3602–3609.

97. Xue, J.; Ma, S.; Zhou, Y.; Zhang, Z.; Liu, X. Fabrication of porous g-C_3N_4/Ag/Fe_2O_3 composites with enhanced visible light photocatalysis performance. RSC Adv. 2015, 5, 58738–58745.

98. Liu, X.; Jin, A.; Jia, Y.; Jiang, J.; Hu, N.; Chen, X. Facile synthesis and enhanced visible-light photocatalytic activity of graphitic carbon nitride decorated with ultrafine Fe_2O_3 nanoparticles. RSC Adv. 2015, 5, 92033–92041.

99. Theerthagiri, J.; Senthil, R.A.; Priya, A.; Madhavan, J.; Michael, R.J.V.; Ashokkumar, M. Photocatalytic and photoelectrochemical studies of visible-light-active α-Fe_2O_3-g-C_3N_4 nanocomposites. RSC Adv. 2014, 4, 38222–38229.

100. Liu, Y.; Yu, Y.X.; Zhang, W. Photoelectrochemical study on charge transfer properties of nanostructured Fe_2O_3modified by g-C_3N_4. Int. J. Hydrogen Energy 2014, 39, 9105–9113.

101. Peng, L.; Xie, T.; Lu, Y.; Fan, H.; Wang, D. Synthesis, photoelectric properties and photocatalytic activity of the Fe_2O_3/TiO_2 heterogeneous photocatalysts. Phys. Chem. Chem. Phys. 2010, 12, 8033–8041.

102. Pawar, R.C.; Pyo, Y.; Ahn, S.H.; Lee, C.S. Photoelectrochemical properties and photodegradation of organic pollutants using hematite hybrids modified by gold nanoparticles and graphitic carbon nitride. Appl. Catal. B Environ. 2015, 176–177, 654–666.

103. Yamauraa, M.; Camilo, R.L.; Sampaio, L.C.; Macedo, M.A.; Nakamurad, M.; Toma, H.E. Preparation and characterization of (3-aminopropyl) triethoxysilane-coated magnetite nanoparticles. J. Magn. Magn. Mater. 2004, 279, 210–217.

104. Tian, B.; Wang, T.; Dong, R.; Bao, S.; Yang, F.; Zhang, J. Core-shell structured–Fe_2O_3@SiO_2@AgBr: Ag composite with high magnetic separation efficiency and excellent visible light activity for acid orange 7 degradation. Appl. Catal. B Environ. 2014, 147, 22–28.

105. Idris, A.; Hassan, N.; Rashid, R.; Ngomsik, A. Kinetic and regeneration studies of photocatalytic magnetic separable beads for chromium (VI) reduction under sunlight. J. Hazard. Mater. 2011, 186, 629–635.

106. Yu, X.; Liu, S.; Yu, J. Superparamagnetic γ-Fe_2O_3@SiO_2@TiO_2 composite microspheres with superior photocatalytic properties. Appl. Catal. B Environ. 2011, 104, 12–20.

107. Lin, Y.; Chang, C. Magnetic mesoporous iron oxide/carbon aerogel photocatalysts with adsorption ability for organic dye removal. RSC Adv. 2014, 4, 28628–28631.

108. Zhang, Y.; Zhang, Y.; Tan, J. Novel magnetically separable AgCl/iron oxide composites with enhanced photocatalytic activity driven by visible light. J. Alloy. Compd. 2013, 574, 383–390.

109. Ye, S.; Qiu, L.-G.; Yuan, Y.-P.; Zhu, Y.-J.; Xia, J.; Zhu, J.-F. Facile fabrication of magnetically separable graphitic carbon nitride photocatalysts with enhanced photocatalytic activity under visible light. J. Mater. Chem. A 2013, 1, 2929–3188.

110. He, M.; Li, D.; Jiang, D.; Chen, M. Magnetically separable γ-Fe_2O_3@SiO_2@Ce-doped TiO_2 core-shell nanocomposites: Fabrication and visible-light-driven photocatalytic activity. J. Solid State Chem. 2012, 192, 139–143.

111. Kaur, N.; Shahi, S.K.; Singh, V. Synthesis, characterization and photocatalytic activity of magnetically separable γ-Fe_2O_3/N,Fe codoped TiO_2 heterojunction for degradation of Reactive Blue 4 dye. RSC Adv. 2015, 5, 61623–61630.

112. Yu, L.; Yang, X.; Wang, D. TiO_2 incorporated in magnetic mesoporous SBA-15 by a facile inner-pore hydrolysis process toward enhanced adsorption-photocatalysis performances for As(III). J. Colloid Interface Sci. 2015, 448, 525–532.

113. Liu, Y.; Yu, L.; Hu, Y.; Guo, C.; Wen, X.; Lou, D. A magnetically separable photocatalyst based on nest-like γ-Fe_2O_3/ZnO double-shelled hollow structures with enhanced photocatalytic activity. Nanoscale 2012, 4, 183–187.

114. Szeto, W.; Li, J.; Huang, H.; Xuan, J.; Leung, D.Y.C. Novel urchin-like Fe_2O_3@SiO_2@TiO_2 microparticles with magnetically separable and photocatalytic properties. RSC Adv. 2015, 5, 55363–55371.

115. Kumar, A.; Guo, C.; Sharma, G.; Pathania, D.; Naushad, M.; Kalia, S.; Dhiman, P. Magnetically recoverable ZrO_2/Fe_3O_4/chitosan nanomaterials for enhanced sunlight driven photoreduction of carcinogenic Cr(VI) and dechlorination and mineralization of 4-chlorophenol from simulated waste water. RSC Adv. 2016, 6, 13251–13263.

116. Zhao, Y.; Tao, C.; Xiao, G.; Wei, G.; Li, L.; Liu, C.; Su, H. Controlled synthesis and photocatalysis of sea urchin-like Fe_3O_4@TiO_2@Ag nanocomposites. Nanoscale 2016, 8, 5313–5326.

117. Li, G.; Wong, K.H.; Zhang, X.; Hu, C.; Yu, J.C.; Chan, R.C.Y.; Wong, P.K. Degradation of Acid Orange 7 using magnetic AgBr under visible light: The roles of oxidizing species. Chemosphere 2009, 76, 1185–1191.

118. Álvarez, P.M.; Jaramillo, J.; López-pi, F.; Plucinski, P.K. Preparation and characterization of magnetic TiO_2nanoparticles and their utilization for the degradation of emerging pollutants in water. Appl. Catal. B Environ. 2010,100, 338–345.

119. Zhang, L.; Wang, W.; Sun, S.; Sun, Y.; Gao, E.; Zhang, Z. Elimination of BPA endocrine disruptor by magnetic BiOBr@SiO_2@Fe_3O_4 photocatalyst. Appl. Catal. B Environ. 2014, 148-149, 164–169.

120. Chen, F.; Yan, F.; Chen, Q.; Wang, Y.; Han, L.; Chen, Z.; Fang, S. Fabrication of Fe_3O_4@SiO_2@TiO_2 nanoparticles supported by graphene oxide sheets for the repeated adsorption and photocatalytic degradation of rhodamine B under UV irradiation. Dalton Trans. 2014, 43, 13537–13544.

121. Wang, H.; Wang, W. Synthesis of Fe_3O_4/N-TiO_2/Ag hollow nanospheres and their application as recyclable photocatalysts. J. Cluster Sci. 2016, 27, 403–415.

122. Ma, J.; Guo, S.; Guo, X.; Ge, H. A mild synthetic route to Fe_3O_4@TiO_2-Au composites: Preparation, characterization and photocatalytic activity. Appl. Surf. Sci. 2015, 353, 1117–1125.

123. Li, S.; Liang, W.; Zheng, F.; Zhou, H.; Lin, X. Lysine surface modified Fe_3O_4@SiO_2@TiO_2 microspheres-based preconcentration and photocatalysis for in situ selective determination of nanomolar dissolved organic and inorganic phosphorus in seawater. Sens Actuators B Chem. 2016, 224, 48–54.

124. Hongfei, L.I.U.; Shengfu, J.I.; Yuanyuan, Z.; Ming, L.I. Porous TiO_2-coated magnetic core-shell nanocomposites: Preparation and enhanced photocatalytic activity. Chin. J. Chem. Eng. 2013, 21, 569–576.

125. Ma, J.; Guo, S.; Guo, X.; Ge, H.-G. Liquid-phase deposition of TiO_2 nanoparticles on core-shell Fe_3O_4@SiO_2 spheres: Preparation, characterization, and photocatalytic activity. J. Nanopart. Res. 2015, 17, 1–11.

126. Rashid, J.; Barakat, M.A.; Ruzmanova, Y.; Chianese, A. Fe_3O_4/SiO_2/TiO_2 nanoparticles for photocatalytic degradation of 2-chlorophenol in simulated wastewater. Environ. Sci. Pollut. Res. 2015, 22, 3149–3157.

127. Hamzezadeh-Nakhjavani, S.; Tavakoli, O.; Akhlaghi, S.P. Efficient photocatalytic degradation of organic pollutants by magnetically recoverable nitrogen-doped TiO_2 nanocomposite photocatalysts under visible light irradiation. Environ. Sci. Pollut. Res. 2015, 22, 18859–18873.

128. Cao, M.; Wang, P.; Ao, Y.; Wang, C.; Hou, J.; Qian, J. Photocatalytic degradation of tetrabromobisphenol A by a magnetically separable graphene-TiO_2 composite photocatalyst: Mechanism and intermediates analysis. Chem. Eng. J.2015, 264, 113–124.

129. Li, C.; Tan, J.; Fan, X.; Zhang, B.; Zhang, H.; Zhang, Q. Magnetically separable one dimensional Fe_3O_4/P(MAA-DVB)/TiO_2 nanochains: Preparation, characterization and photocatalytic activity. Ceram. Int. 2015, 41, 3860–3868.

130. Xiang, Y.; Wang, H.; He, Y.; Song, G. Efficient degradation of methylene blue by magnetically separable Fe_3O_4/chitosan/TiO_2 nanocomposites. Desalin. Water Treat. 2015, 55, 1018–1025.

131. Su, W.; Zhang, T.; Li, L.; Xing, J.; He, M.; Zhong, Y.; Li, Z. Synthesis of small yolk-shell Fe_3O_4@TiO_2 nanoparticles with controllable thickness as recyclable photocatalysts. RSC Adv. 2014, 4, 8901–8906.

132. Dong, X.; Shao, Y.; Zhang, X.; Ma, H.; Zhang, X.; Shi, F.; Ma, C.; Xue, M. Synthesis and properties of magnetically separable Fe_3O_4/TiO_2/Bi_2O_3 photocatalysts. Res. Chem. Intermed. 2014, 40, 2953–2961.

133. Li, C.; Younesi, R.; Cai, Y.; Zhu, Y.; Ma, M.; Zhu, J. Photocatalytic and antibacterial properties of Au-decorated Fe_3O_4@mTiO_2 core-shell microspheres. Appl. Catal. B Environ. 2014, 156-157, 314–322.

134. Su, J.; Zhang, Y.; Xu, S.; Wang, S.; Ding, H.; Pan, S. Highly efficient and recyclable triple-shelled Ag@Fe_3O_4@SiO_2@TiO_2 photocatalysts for degradation of organic pollutants and reduction of hexavalent chromium ions. Nanoscale 2014, 6, 5181–5192.

135. Chen, L.; Li, L.; Wang, T.; Zhang, L.; Xing, S. A novel strategy to fabricate multifunctional Fe_3O_4@C@TiO_2 yolk-shell structures as magnetically recyclable photocatalysts. Nanoscale 2014, 6, 6603–6608.

136. Chi, Y.; Yuan, Q.; Li, Y.; Zhao, L.; Li, N.; Li, X.; Yan, W. Magnetically separable Fe_3O_4@SiO_2@TiO_2-Ag microspheres with well-designed nanostructure and enhanced photocatalytic activity. J. Hazard. Mater. 2013, 262, 404–411.

137. Tang, Y.; Zhang, G.; Liu, C.; Luo, S.; Xu, X.; Chen, L.; Wang, B. Magnetic TiO_2-graphene composite as a high-performance and recyclable platform for efficient photocatalytic removal of herbicides from water. J. Hazard. Mater.2013, 252-253, 115–122.

138. Shojaei, F.A.; Shams-Nateri, A.; Ghomashpasand, M. Comparative study of photocatalytic activities of magnetically separable WO_3/TiO_2/Fe_3O_4 nanocomposites and TiO_2, WO_3/TiO_2 and TiO_2/Fe_3O_4 under visible light irradiation.Superlattice Microstruct. 2015, 88, 211–224.

139. Chidambaram, S.; Pari, B.; Kasi, N.; Muthusamy, S. ZnO/Ag heterostructures embedded in Fe_3O_4 nanoparticles for magnetically recoverable photocatalysis. J. Alloy Compd. 2016, 665, 404–410.

140. Yang, J.; Wang, J.; Li, X.; Wang, D.; Song, H. Synthesis of urchin-like Fe_3O_4@SiO_2@ZnO/CdS core-shell microspheres for the repeated photocatalytic degradation of rhodamine B under visible light. Catal. Sci. Technol. 2016.

141. Shekofteh-Gohari, M.; Habibi-Yangjeh, A. Ultrasonic-assisted preparation of novel ternary ZnO/AgI/Fe_3O_4nanocomposites as magnetically separable visible-light-driven photocatalysts with excellent activity. J. Colloid Interface Sci. 2016, 461, 144–153.

142. Shekofteh-Gohari, M.; Habibi-Yangjeh, A. Novel magnetically separable ZnO/AgBr/Fe_3O_4/Ag_3VO_4 nanocomposites with tandem n–n heterojunctions as highly efficient visible-light-driven photocatalysts. RSC Adv. 2015, 6, 2402–2413.

143. Shekofteh-Gohari, M.; Habibi-Yangjeh, A. Ternary ZnO/Ag_3VO_4/Fe_3O_4 nanocomposites: Novel magnetically separable photocatalyst for efficiently degradation of dye pollutants under visible-light irradiation. Solid State Sci.2015, 48, 177–185.

144. Shekofteh-Gohari, M.; Habibi-Yangjeh, A. Novel magnetically separable Fe_3O_4@ZnO/AgCl nanocomposites with highly enhanced photocatalytic activities under visible-light irradiation. Sep. Purif. Technol. 2015, 147, 194–202.

145. Shekofteh-Gohari, M.; Habibi-Yangjeh, A. Facile preparation of Fe_3O_4@AgBr-ZnO nanocomposites as novel magnetically separable visible-light-driven photocatalysts. Ceram. Int. 2015, 41, 1467–1476.

146. Jia, X.; Dai, R.; Sun, Y.; Song, H.; Wu, X. One-step hydrothermal synthesis of Fe_3O_4/g-C_3N_4 nanocomposites with improved photocatalytic activities. J. Mater. Sci. 2016, 27, 3791–3798.

147. Mousavi, M.; Habibi-Yangjeh, A. Magnetically separable ternary g-C_3N_4/Fe_3O_4/BiOI nanocomposites: Novel visible-light-driven photocatalysts based on graphitic carbon nitride. J. Colloid Interface Sci. 2016, 465, 83–92.

148. Akhundi, A.; Habibi-Yangjeh, A. Novel magnetic g-C_3N_4/Fe_3O_4/AgCl nanocomposites: Facile and large-scale preparation and highly efficient photocatalytic activities under visible-light irradiation. Mater. Sci. Semicond. Process.2015, 39, 162–171.

149. Akhundi, A.; Habibi-Yangjeh, A. Novel magnetically separable g-C_3N_4/AgBr/Fe_3O_4 nanocomposites as visible-light-driven photocatalysts with highly enhanced activities. Ceram. Int. 2015, 41, 5634–5643.

150. Feng, X.; Lou, X. The effect of surfactants-bound magnetite (Fe_3O_4) on the photocatalytic properties of the heterogeneous magnetic zinc oxides nanoparticles. Sep. Purif. Technol. 2015, 147, 266–275.

151. Li, W.; Tian, Y.; Li, P.; Zhang, B.; Zhang, H.; Geng, W.; Zhang, Q. Synthesis of rattle-type magnetic mesoporous and investigation of its photoactivity in the degradation of methylene blue. RSC Adv. 2015, 5, 48050–48059.

152. Li, X.; Wang, G.; Cheng, Y. Preparation and characteristics of Fe_3O_4-$BaTiO_3$ heterostructural nanocomposite as photocatalyst. Res. Chem. Intermed. 2015, 41, 3031–3039.

153. Singh, K.K.; Senapati, K.K.; Borgohain, C.; Sarma, K.C. Newly developed Fe_3O_4-Cr_2O_3 magnetic nanocomposite for photocatalytic decomposition of 4-chlorophenol in water. J. Environ. Sci. 2015.

154. Cao, C.; Xiao, L.; Chen, C.; Cao, Q. Magnetically separable Cu_2O/chitosan–Fe_3O_4 nanocomposites: Preparation, characterization and visible-light photocatalytic performance. Appl. Surf. Sci. 2015, 333, 110–118.

155. Cao, Y.; Li, C.; Li, J.; Li, Q.; Yang, J. Magnetically separable Fe_3O_4/AgBr hybrid materials: Highly efficient photocatalytic activity and good stability. Nanoscale Res. Lett. 2015, 10, 952.

156. Zhong, S.; Jiang, W.; Han, M.; Liu, G.; Zhang, N.; Lu, Y. Graphene supported silver@silver chloride and ferroferric oxide hybrid, a magnetically separable photocatalyst with high performance under visible light irradiation. Appl. Surf. Sci. 2015, 347, 242–249.

157. Jing, J.; Zhang, Y.; Feng, J.; Li, W.; Yu, W.W. Facile preparation and high performance of magnetically separable metalloporphyrin. Chem. Eng. J. 2015, 263, 385–391.

158. Wang, B.; Zhang, M.; Li, W.; Wang, L.; Zheng, J.; Gan, W. Fabrication of Au(Ag)/AgCl/Fe_3O_4@PDA@Au nanocomposites with enhanced visible-light-driven photocatalytic activity. Dalton Trans. 2015, 44, 17020–17025.

159. Yang, S.; Ye, C.; Song, X.; He, L.; Liao, F. Theoretical calculation based synthesis of a poly(p-phenylenediamine)–Fe_3O_4composite: A magnetically recyclable photocatalyst with high selectivity for acid dyes. RSC Adv. 2014, 4, 54810–54818.

160. An, L.; Wang, G.; Shi, X.; Su, M.; Gao, F.; Cheng, Y. Recyclable Fe_3O_4/ZnO/PPy composite photocatalyst: Fabrication and photocatalytic activity. Russ. J. Phys. Chem. A 2014, 88, 2419–2423.

161. Wang, C.; Cao, M.; Wang, P.; Ao, Y.; Hou, J.; Qian, J. Preparation of a magnetic graphene oxide-Ag_3PO_4 composite photocatalyst with enhanced photocatalytic activity under visible light irradiation. J. Taiwan Inst. Chem. Eng. 2014, 45, 1080–1086.

162. Peik-See, T.; Pandikumar, A.; Ngee, H.; Ming, N. Magnetically separable reduced graphene oxide/iron oxide nanocomposite materials for environmental remediation. Catal. Sci. Technol. 2014, 4, 4396–4405.

163. Li, Z.; Wen, Y.; Shang, J.; Wu, M.; Wang, L.; Guo, Y. Magnetically recoverable Cu_2O/Fe_3O_4 composite photocatalysts: Fabrication and photocatalytic activity. Chin. Chem. Lett. 2014, 25, 287–291.

164. Cao, X.; Chen, Y.; Jiao, S.; Fang, Z.; Xu, M.; Liu, X.; Li, L.; Pang, G.; Feng, S. Magnetic photocatalysts with a p–n junction: Fe_3O_4 nanoparticle and $FeWO_4$ nanowire heterostructures. Nanoscale 2014, 6, 12366–12370.

165. Wang, H.; Wei, Z.; Matsui, H.; Zhou, S. Fe_3O_4/carbon quantum dots hybrid nanoflower for highly active and recyclable visible-light driven. J. Mater. Chem. A 2014, 2, 15740–15745.

166. Zhang, D.; Pu, X.; Gao, Y.; Su, C.; Li, H.; Li, H. One-step combustion synthesis of $CoFe_2O_4$-graphene hybrid materials for photodegradation of methylene blue. Mater. Lett. 2013, 113, 179–181.

167. Zhu, H.; Jiang, R.; Huang, S.; Yao, J.; Fu, F.; Li, J. Novel magnetic $NiFe_2O_4$/multi-walled carbon nanotubes hybrids: Facile synthesis, characterization, and application to the treatment of dyeing wastewater. Ceram. Int. 2015, 41, 11625–11631.

168. Bai, S.; Shen, X. One-pot solvothermal preparation of magnetic reduced graphene oxide-ferrite hybrids for organic dye removal. Carbon 2012, 50, 2337–2346.

169. Sun, J.; Fu, Y.; Xiong, P.; Sun, X.; Xu, B.; Wang, X. A magnetically separable P25/$CoFe_2O_4$/graphene catalyst with enhanced adsorption capacity and visible-light-driven photocatalytic activity. RSC Adv. 2013, 3, 22490–22497.

170. Xiong, P.; Zhu, J.; Wang, X. Cadmium sulfide-ferrite nanocomposite as a magnetically recyclable photocatalyst with enhanced visible-light-driven photocatalytic activity and photostability. Ind. Eng. Chem. Res. 2013, 52, 17126–17133.

171. Fu, Y.; Chen, Q.; He, M.; Wan, Y.; Sun, X.; Xia, H.; Wang, X. Copper ferrite-graphene hybrid: A multifunctional heteroarchitecture for photocatalysis and energy storage. Ind. Eng. Chem. Res. 2012, 51, 11700–11709.

172. Ji, H.; Jing, X.; Xu, Y.; Yan, J.; Li, H.; Li, Y.; Huang, L.; Zhang, Q.; Xu, H.; Li, H. Magnetic g-C_3N_4/$NiFe_2O_4$ hybrids with enhanced photocatalytic activity. RSC Adv. 2015, 5, 57960–57967.

173. Zhang, W.; Wang, M.; Zhao, W.; Wang, B. Magnetic composite photocatalyst $ZnFe_2O_4$/$BiVO_4$: Synthesis, characterization, and visible-light photocatalytic activity. Dalton Trans. 2013, 42, 15464–15474.

174. Yao, Y.; Cai, Y.; Lu, F.; Qin, J.; Wei, F.; Xu, C.; Wang, S. Magnetic $ZnFe_2O_4$-C_3N_4 hybrid for photocatalytic degradation of aqueous organic pollutants by visible light. Ind. Eng. Chem. Res. 2014, 53, 17294–17302.

175. Wu, S.; Wang, P.; Cai, Y.; Liang, D.; Ye, Y.; Tian, Z.; Liu, J.; Linag, C. Reduced graphene oxide anchored magnetic $ZnFe_2O_4$ nanoparticles with enhanced visible-light photocatalytic activity. RSC Adv. 2015, 5, 9069–9074.

176. Yang, D.; Feng, J.; Jiang, L.; Wu, X.; Sheng, L.; Jiang, Y.; Wei, T.; Fan, Z. Photocatalyst interface engineering: Spatially confined growth of $ZnFe_2O_4$ within graphene networks as excellent visible-light-driven photocatalysts. Adv. Funct. Mater. 2015, 25, 7080–7087.

177. Khadgi, N.; Li, Y.; Upreti, A.R.; Zhang, C.; Zhang, W.; Wang, Y.; Wang, D. Enhanced photocatalytic degradation of 17 α-ethinylestradiol exhibited by multifunctional $ZnFe_2O_4$-Ag/rGO nanocomposite under visible light. Photochem. Photobiol. 2016, 92, 238–246.

178. Chen, X.; Dai, Y.; Guo, J.; Liu, T.; Wang, X. Novel magnetically separable reduced graphene oxide (RGO)/$ZnFe_2O_4$/Ag_3PO_4 nanocomposites for enhanced photocatalytic performance toward 2,4-dichlorophenol under visible light. Ind. Eng. Chem. Res. 2016, 55, 568–578.

179. Kulkarni, S.D.; Kumbar, S.; Menon, S.G.; Choudhari, K.S.; Santhosh, C. Magnetically separable core-shell $ZnFe_2O_4$@ZnO nanoparticles for visible light photodegradation of methyl orange. Mater. Res. Bull. 2016, 77, 70–77.

180. Yao, Y.; Qin, J.; Chen, H.; Wei, F.; Liu, X.; Wang, J.; Wang, S. One-pot approach for synthesis of N-doped TiO_2/$ZnFe_2O_4$ hybrid as an efficient photocatalyst for degradation of aqueous organic pollutants. J. Hazard. Mater.2015, 291, 28–37.

181. Wu, S.; Shen, X.; Zhu, G.; Zhou, H.; Ji, Z.; Chen, K. Synthesis of ternary Ag/ZnO/$ZnFe_2O_4$ porous and hollow nanostructures with enhanced photocatalytic activity. Appl. Catal. B Environ. 2016, 184, 328–336.

182. Ren, A.; Liu, C.; Hong, Y.; Shi, W.; Lin, S.; Li, P. Enhanced visible-light-driven photocatalytic activity for antibiotic degradation using magnetic $NiFe_2O_4$/Bi_2O_3 heterostructures. Chem. Eng. J. 2014, 258, 301–308.

183. Fu, Y.; Chen, H.; Sun, X.; Wang, X. Graphene-supported nickel ferrite: A magnetically separable photocatalyst with high activity under visible light. Am. Inst. Chem. Eng. J. 2012, 58, 3298–3305.

184. Patil, S.S.; Tamboli, M.S.; Deonikar, V.G.; Umarji, G.G.; Ambekar, J.D.; Kulkarni, M.V.; Kolekar, S.S.; Kale, B.B.; Patil, D.R. Magnetically separable Ag_3PO_4/$NiFe_2O_4$ composites with enhanced photocatalytic activity. Dalton Trans. 2015,44, 20426–20434.

185. Kim, S.H.; Kim, D.; Kwak, S.B.; Han, B.G.; Um, M.; Kang, M. Synthesis of magnetically separable core@shell structured $NiFe_2O_4$@TiO_2 nanomaterial and its use for photocatalytic hydrogen production by methanol/water splitting. Chem. Eng. J. 2014, 243, 272–279.

186. Ge, M.; Hu, Z. Novel magnetic AgBr/$NiFe_2O_4$ composite with enhanced visible light photocatalytic performance. Ceram. Int. 2016, 42, 6510–6514.

187. Xiong, P.; Fu, Y.; Wang, L.; Wang, X. Multi-walled carbon nanotubes supported nickel ferrite: A magnetically recyclable photocatalyst with high photocatalytic activity on degradation of phenols. Chem. Eng. J. 2012, 195-196, 149–157.

188. Li, C.; Wang, J.; Wang, B.; Ru, J.; Lin, Z. A novel magnetically separable TiO_2/$CoFe_2O_4$ nanofiber with high photocatalytic activity under UV-vis light. Mater. Res. Bull. 2012, 47, 333–337.

189. Wilson, A.; Mishra, S.R.; Gupta, R.; Ghosh, K. Preparation and photocatalytic properties of hybrid core-shell reusable $CoFe_2O_4$-ZnO nanospheres. J. Magn. Magn. Mater. 2012, 324, 2597–2601.

190. Pongwan, P.; Inceesungvorn, B.; Phanichphant, S.; Wetchakun, N. Synthesis and characterization of a magnetically separable $CoFe_2O_4$/TiO_2 nanocomposite

for the photomineralization of formic acid. Ferroelectrics 2013, 453, 133–140.

191. Wetchakun, N.; Chaiwichain, S.; Wetchakun, K. Synthesis and characterization of novel magnetically separable $CoFe_2O_4/CeO_2$ nanocomposite photocatalysts. Mater. Lett. 2013, 113, 76–79.

192. Gan, L.; Shang, S.; Wah, C.; Yuen, M.; Jiang, S.; Hu, E. Hydrothermal synthesis of magnetic $CoFe_2O_4$/graphene nanocomposites with improved photocatalytic activity. Appl. Surf. Sci. 2015, 351, 140–147.

193. Singh, S.; Khare, N. Magnetically separable, $CoFe_2O_4$ decorated CdS nanorods for enhanced visible light driven photocatalytic activity. Mater. Lett. 2015, 161, 64–67.

194. Haw, C.; Chiu, W.; Rahman, S.A.; Khiew, P.; Radiman, S.; Shukor, R.A.; Hamid, M.A.A.; Ghazali, N. The design of new magnetic-photocatalyst nanocomposites ($CoFe_2O_4$-TiO_2) as smart nanomaterials for recyclable-photocatalysis applications. New J. Chem. 2016, 40, 1124–1136.

195. Choi, Y.I.; Kim, Y.; Cho, D.W.; Kang, J.; Leung, K.T.; Sohn, Y. Recyclable magnetic $CoFe_2O_4$/BiOX (X = Cl, Br and I) microflowers for photocatalytic treatment of water contaminated with methyl orange, rhodamine B, methylene blue, and a mixed dye. RSC Adv. 2015, 5, 79624–79634.

196. Xu, Y.; Zhou, T.; Huang, S.; Xie, M.; Li, H.; Xu, H. Preparation of magnetic Ag/AgCl/$CoFe_2O_4$ composites with high photocatalytic and antibacterial ability. RSC Adv. 2015, 5, 41475–41483.

197. Shi, B.; Wan, J.; Liu, C.; Yu, X. Synthesis of $CoFe_2O_4$/MCM-41/TiO_2 composite microspheres and its performance in degradation of phenol. Mater. Sci. Semicond. Process. 2015, 37, 241–249.

198. Gaikwad, P.N.; Hankare, P.P.; Wandre, T.M.; Garadkar, K.M.; Sasikala, R. Photocatalytic performance of magnetically separable Fe,N co-doped TiO_2-cobalt ferrite nanocomposite. Mater. Sci. Eng. B 2016, 205, 40–45.

199. Manikandan, A.; Durka, M.; Seevakan, K.; Antony, S.A. A novel one-pot combustion synthesis and opto-magnetic properties of magnetically separable spinel $Mn_xMg_{1-x}Fe_2O_4$ ($0.0 \leq x \leq 0.5$) nanophotocatalysts. J. Supercond. Novel Magn. 2015, 28, 1405–1416.

200. Tio, A.; Nife, O.; Nicola, D.; Vanags, M. Ag sensitized TiO_2 and $NiFe_2O_4$ three-component nanoheterostructures: Synthesis, electronic structure and strongly enhanced visible light. RSC Adv. 2016, 6, 18834–18842.

201. Chen, C.-C.; Fu, Y.-P.; Hu, S.-H. Characterizations of TiO_2/SiO_2/Ni-Cu-Zn ferrite composite for magnetic photocatalysts. J. Am. Chem. Soc. 2015, 98, 2803–2811.

202. Sathishkumar, P.; Viswanathan, R. $CoFe_2O_4$/TiO_2 nanocatalysts for the photocatalytic degradation of Reactive Red 120 in aqueous solutions in the presence and absence of electron acceptors. Chem. Eng. J. 2013, 220, 302–310.

203. Kadi, M.W.; Mohamed, R.M. Environmental remediation of aqueous cyanide by photocatalytic oxidation using a $NiFe_2O_4$/TiO_2-SiO_2 core-shell nanocomposite. Desalin. Water Treat. 2016, 56, 1940–1948.

204. Su, R.N.; Lv, P.; Li, M.; Zhang, X.; Li, M.; Niu, J. Fabrication of $MgFe_2O_4$-ZnO heterojunction photocatalysts for application of organic pollutants. Mater. Lett. 2014, 122, 201–204.

205. Leng, C.; Wei, J.; Liu, Z. Facile synthesis of PANI-modified $CoFe_2O_4$-TiO_2 hierarchical flower-like nanoarchitectures with high photocatalytic activity. J. Nanopart. Res. 2013, 15, 1643.

206. Sun, A.; Chen, H.; Song, C.; Jiang, F.; Wang, X.; Fu, Y. Magnetic $Bi_{25}FeO_{40}$-graphene catalyst and its high visible-light photocatalytic performance. RSC Adv. 2013, 3, 4332–4340.

207. Xie, T.; Xu, L.; Liu, C.; Yang, J.; Wang, M. Magnetic composite BiOCl-$SrFe_{12}O_{19}$:A novel p–n type heterojunction with enhanced photocatalytic activity. Dalton Trans. 2014, 43, 2211–2220.

208. Bhukal, S.; Bansal, S.; Singhal, S. Magnetic Mn substituted cobalt zinc ferrite systems: Structural, electrical and magnetic properties and their role in photo-catalytic degradation of methyl orange azo dye. Physica B 2014, 445, 48–55.

209. Qasim, M.; Asghar, K.; Raj, B.; Prathapani, S.; Khan, W.; Naqvi, A.H.; Das, D. Magnetically recyclable $Ni_{0.5}Zn_{0.5}Fe_2O_4$/$Zn_{0.95}Ni_{0.05}O$ nano-photocatalyst: Structural, optical, magnetic and photocatalytic properties.Spectrochim. Acta A 2015, 137, 1348–1356.

210. Manikandan, A.; Antony, S.A. Magnetically separable $Mn_xZn_{1-x}Fe_2O_4$; ($0.0 \leq x \leq 0.5$) nanostructures: Structural, morphological, opto-magnetic and photocatalytic properties. Synth. React. Inorg. Met-Org. Nano-Met. Chem. 2016.

211. Shakir, I.; Sarfraz, M.; Ali, Z.; Aboud, M.F.A. Magnetically separable and recyclable graphene-$MgFe_2O_4$nanocomposites for enhanced photocatalytic applications. J. Alloy. Compd. 2016, 660, 450–455.

212. Bhukal, S.; Singhal, S. Magnetically separable copper substituted cobalt-zinc nano-ferrite photocatalyst with enhanced photocatalytic activity. Mater. Sci. Semicond. Process. 2014, 26, 467–476.

213. Liu, F.; Xie, Y.; Yu, C.; Liu, X.; Dai, Y. Novel hybrid Sr-doped TiO_2/magnetic $Ni_{0.6}Zn_{0.4}Fe_2O_4$ for enhanced separation and photodegradation of organics under visible light. RSC Adv. 2015, 5, 24056–24063.

214. Vignesh, K.; Suganthi, A.; Min, B.-K.; Kang, M. Photocatalytic activity of magnetically recoverable $MnFe_2O_4$/g-C_3N_4/TiO_2 nanocomposite under simulated solar light irradiation. J. Mol. Catal. A Chem. 2014, 395, 373–383.

215. He, H.-Y. Photocatalytic degradations of dyes on magnetically separable $Ni_{1-x}Co_xFe_2O_4$ nanoparticles synthesized by a hydrothermal process. Part. Sci. Technol. 2016, 34, 143–151.

216. Zhang, Z.; Xu, L.; Liu, C. Preparation and characterization of composite magnetic photocatalyst $Mn_xZn_{1-x}Fe_2O_4$/β-Bi_2O_3. RSC Adv. 2015, 5, 79997–80004.

217. He, H.; Yan, Y.; Huang, J.; Lu, J. Rapid photodegradation of methyl blue on magnetic $Zn_{1-x}Co_xFe_2O_4$ nanoparticles synthesized by hydrothermal process. Sep. Purif. Technol. 2014, 136, 36–41.

218. Shi, Y.; Zhou, K.; Wang, B.; Jiang, S.; Qian, X.; Gui, Z.; Yuen, R.K.K.; Hu, Y. Ternary graphene-$CoFe_2O_4$/CdS nanohybrids: Preparation and application as recyclable photocatalysts. J. Mater. Chem. A 2014, 2, 535–544.

219. Sathishkumar, P.; Pugazhenthiran, N. ZnO supported $CoFe_2O_4$ nanophotocatalysts for the mineralization of Direct Blue 71 in aqueous environments. J. Hazard. Mater. 2013, 252–253, 171–179.

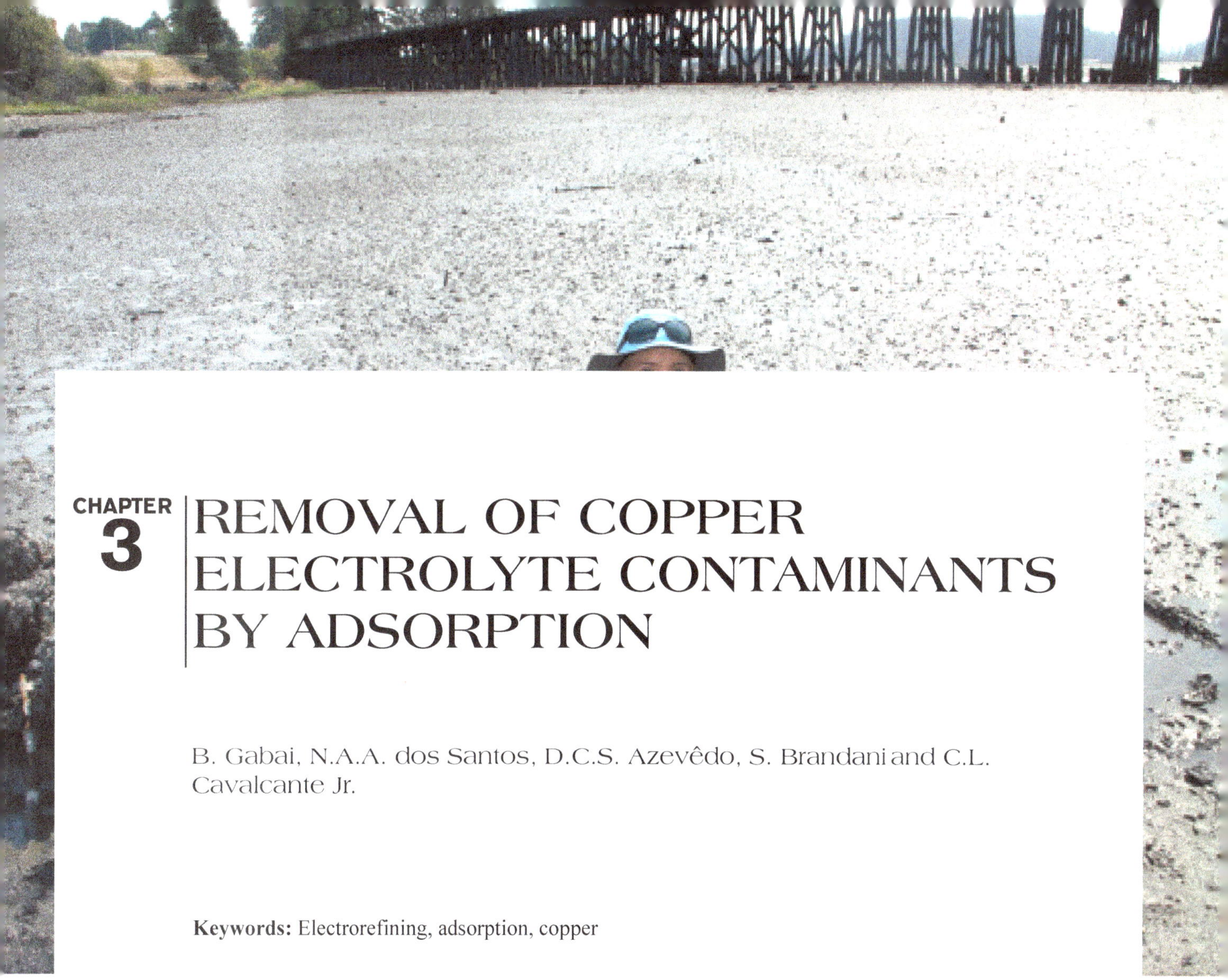

CHAPTER 3

REMOVAL OF COPPER ELECTROLYTE CONTAMINANTS BY ADSORPTION

B. Gabai, N.A.A. dos Santos, D.C.S. Azevêdo, S. Brandani and C.L. Cavalcante Jr.

Keywords: Electrorefining, adsorption, copper

INTRODUCTION

During the electrorefining of copper, many impurities are introduced to the electrolyte through the dissolution of an impure copper anode. Contaminants such as antimony, bismuth and arsenic may be found as dissolved metals in the anode slime which forms on the electrode surface, accumulated in the electrolyte or incorporated into the refined copper on the cathode by deposition. The removal of these impurities by conventional techniques involves the use of electrolysis processes in liberating tanks or bleeding the electrolyte in multi-stage electrowinning. In general, the mixture of deposited copper-arsenic-antimony-bismuth is recycled back into the smelter, to recover most of the copper from the slime. However, the use of these conventional methods often results in an insufficient reduction of contaminant concentrations. Besides, it presents disadvantages such as the need to reduce and control the copper concentration level in the electrolyte to facilitate the removal of impurities that may damage the quality of the final product (refined copper).

Recent studies have demonstrated the possibility of using adsorption on separation of heavy metals in several industrial processes, including copper refining, to remove electrolyte contaminants, with better results than those obtained with conventional techniques. The adsorbents mainly cited in the literature for metals removing processes are chelating resins and activated carbons.

This paper studies the possibility of removing antimony and bismuth from copper refinery electrolyte by selective adsorption. As adsorbents, activated carbon, synthetic zeolites, and chelating resins were tested to select the more adequate adsorbent to this separation, as well as the best operating temperature for the process.

EXPERIMENTAL

Resistance to Electrolyte Acidity

Before initiating any of the experiments, resistance to the acidity of the electrolyte (pH @ 0) was checked for some of the adsorbents. Although zeolites are known to resist only to pHs as low as 2 (Ruthven, 1988), some samples of zeolites were tested as well. Pellets of Y, 4A, USY (dealuminized) zeolites, silicalite, some samples ofactivated carbon and chelating resins were chosen for this test. A synthetic electrolyte, composed of $CuSO_4.5H_2O$ (Cu^{+2}@42 g/l) and H_2SO_4 (185 g/l), was prepared. Then, small samples of the adsorbents (@ 0.5 g each) were

Citation: B. Gabai, N.A.A. dos Santos, D.C.S. Azevêdo, S. Brandani and C.L. Cavalcante Jr, Removal of Copper Electrolyte Contaminants By Adsorption, http://dx.doi.org/10.1590/S0104-66321997000300002

soaked in 100 ml of electrolyte and maintained for 24 hours at 60°C or 27°C (room temperature). Table 1shows the results obtained in which only the activated carbons, the USY zeolite and the chelating resins samples remained intact.

Tested Adsorbents

Table 2 summarizes some physical properties of the adsorbents that were used in the finite bath experiments. Zeolite Ultra-Stable Y was provided by Degussa, activated carbon by Carbomafra S/A and chelating resins by Unitika and Rohm & Haas.

Finite Bath

The finite bath immersion method (Azevêdo, 1993), which is widely used to screen adsorbents and temperatures in liquid phase sorption processes, was chosen for the adsorption studies. This method, with a properly designed adsorption cell, illustrated in Figure 1, allows experiments by adsorbent immersion in a finite volume of liquid (containing the sorbate) and satisfies the following conditions:

- Operational ease in set up, control over the experimental starting conditions and sample collection;
- Thermostatization of the cell to assure isothermality of the system;
- Lowest variation in system volume during the experiment;
- Minimal mechanical damage to the particles to maintain their volume constant over the entire experiment.

The experimental system is sketched in Figure 2 and consists basically of a glass cell immersed in a thermostatic bath. A properly design shaker is used to assure good mixing of the solid/liquid system. Samples are taken with syringes during the experiments without disturbing the overall conditions of the system.

Table 1: Results of the Test of Adsorbent Resistance to Electrolyte Acidity

Adsorbent	Behaviour
Activated Carbon	*Remained undamaged.*
UR-3300S Resin	*Remained undamaged.*
Duolite C-467 Resin	*Remained undamaged.*
Zeolite Y	*Structure softened, degenerating easily when shaken by a glass stick.*
Zeolite 4A	*Solubilized as soon as there was contact*
Zeolite USY	*Resisted, apparently undamaged. However, it's possible that the structure may have been weakened.*
Silicalite	*Solubilized.*

Table 2: Some Physical Properties of Adsorbents Used

Adsorbent	Manufacturer	Typical Shape	Main Dimension (mm)
Zeolite Ultra-Stable Y	*Degussa*	*Cylindric pellets*	r_p *@ 1; L @ 5*
Activated Carbon	*Carbomafra S/A*	*Random*	r_p *@ 0.6*
Duolite C-467 Resin	*Rohm & Haas*	*Spherical*	r_p *@ 0.28*
UR-3300S	*Unitika*	*Spherical*	r_p *@ 0.3*

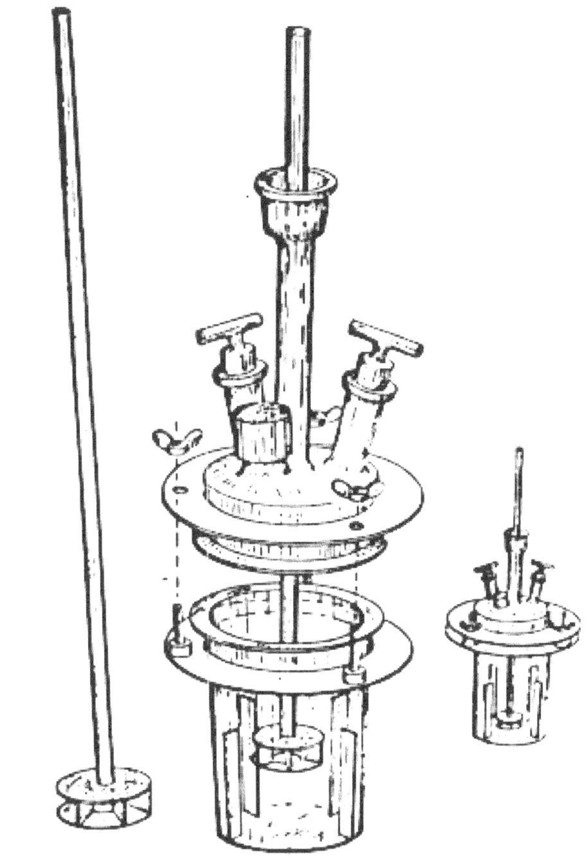

Figure 1: Detail of the Glass Cell (ports on top are for extracting samples with the syringe, and introducing the components and the shaker).

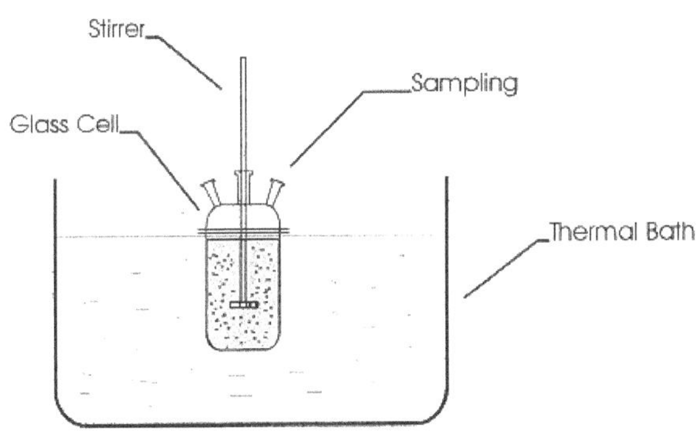

Figure 2: Finite Bath System.

The experimental procedure consists in bringing the adsorbent into contact with a limited volume of electrolyte in the glass cell, under controlled temperature, and continuously monitoring the variations in metals concentrations. The analyses of contaminant concentrations were done by atomic absorption spectrophotometry (AAS) (Interlab/Varian Mod. AA 1475), with hollow cathode lamps of Sb and Bi and air-acetylene flame. In regard to the antimony analysis, copper and nickel appear as interferences in the air-acetylene flame, decreasing the absorption signal, specially in reducing flame. This effect, however, may be eliminated without problems with an oxidizing flame. There is no risk of interference with the other electrolyte components in the analysis for bismuth.

The synthetic electrolyte prepared for the experiments was composed of a solution of $CuSO_4 . 5H_2O$ and H_2SO_4(in concentrations given above) with the addition of dissolved Sb^{+3} (@ 296g/l) and Bi^{+3} (@ 142,5g/l) salts.

Before each experimental series, the adsorbent was previously activated. Activated carbon and zeolites were dried in an oven at 150°C to remove moisture occasionally present into the pores. The chelating resins, however, were activated by soaking in deionized water for at least 24 hours.

The experiment began with the addition of the electrolyte to the cell. Then, the adsorbent was dropped into the electrolyte and immediately agitation was initiated to maximize the contact between the liquid and the adsorbent. At predetermined intervals, samples were taken from the electrolyte (@ 3.5 ml) with a syringe, to observe the decrease in concentrations (adsorption kinetics) of antimony and bismuth in the electrolyte during the process. A typical experiment lasted about 3 to 4 hours.

For the equilibrium experiments, the cell with adsorbent and electrolyte was left still for 24 hours (or more) in a thermostatic bath at a constant temperature. Then, a sample of the electrolyte was taken for concentration analysis. For the experiments with activated carbon or zeolite, the electrolyte with the adsorbent was maintained in balloons after the tests. A few samples were extracted several days later, for equilibrium concentration analysis.

After each run with resins, the used resin was separated from the electrolyte and rinsed with deionized water. Then, elution was done with Hcl 6M, followed by a new rinse with deionized water. The resin was maintained in this water to be reused.

Diffusion Model

Ruthven (1984) shows the solution for the uptake curve of a sorbate in a finite bath system:

$$\frac{m_t}{m_\infty} = 1 - 6\sum_{n=1}^{\infty} \frac{\exp(-D_e P_n t / r_p^2)}{\frac{9\Lambda}{1-\Lambda} + (1-\Lambda) P_n^2} \quad (1)$$

where P_n is given by the non-zero roots of:

$$\tan P_n = \frac{3P_n}{3 + (1/\Lambda - 1)P_n^2} \quad (2)$$

and

$$\Lambda \equiv (C_0 - C_\infty)/C_0 \quad (3)$$

This model was employed for evaluation of the effective diffusivities of antimony and bismuth in the chelating resins.

RESULTS AND DISCUSSION

Activated Carbon

Experimental runs with activated carbon, were performed at temperature of 60°C and 27°C (room temperature). The results showed a very smooth or almost negligible decrease in the contaminants in the electrolyte while under agitation during 4 hours. However, after left still for a few days, there was a noticeable, although only partial, reduction in the level of contaminants. As expected, increased values of solid/liquid ratio decreased the level of contaminants. But even with higher values of amount adsorbed (g metal /g adsorbent), the adsorption kinetics of the activated carbon was too slow, usually taking several days to achieve equilibrium. As previously observed by Reed and Nonavinakere (1992), there was no adsorbate competition between the two adsorbing metals. However, some selectivity favoring Sb over Bi was observed. Using fixed bed adsorption, Toyabe et al. (1987) also observed high Sb sorption preference, with somewhat higher selectivity values. Table 3 shows a summary of experimental results for activated carbon adsorption equilibrium.

USY Zeolite

No reference was found in the literature to the removal of copper electrolyte contaminants using zeolites. However, the removal of heavy metals from industrial wastes using zeolites has been reported (Zamzow and Murphy., 1992; Rupp et al., 1995). Based on these studies, room temperature was initially chosen for the experiments. The results showed that even after several days of the experiment, there was no significant variation in the Sb and Bi concentrations in the electrolyte and the amount of adsorbed metals in the zeolite was still very low (Table 4).

We may infer as well that, due to their chemical structure, zeolites are strongly affected by the very low pH of the electrolyte, and their physical resistance to high acidity hinders any possibility of usage with copper electrolyte at process conditions.

Chelating Resins

Chelating resins have been reported as viable adsorbents for heavy metal removal (Oda et al.,1986; Shibata et al., 1987; Abe and Takasawa, 1987; Sasaki et al., 1991; Dreisinger et al., 1994; Dreisinger and Scholey, 1995; Yang et al., 1995), even for copper electrolyte conditions. In this study, Duolite C-467 and Unitika UR-3300S were tested at temperatures of 45°C and 55°C. Both resins showed high levels of contaminant removal, as may be seen in Table 5.

The sorption kinetics was essentially the same for both resins, as may be observed in Figures 3 and 4. Increasing temperature accelerated the overall adsorption process, as expected (Figures 5–8), which indicates a diffusion-controlled process. Applying the prposed diffusion model (equation 1), we could estimate values for the diffusivities by matching the theoretical curves against the observed experimental data. Some curves are shown in Figures 9–11

The estimated values for the effective diffusion coefficients are shown in Table 6. These again confirm that the two resins have very similar behaviors and may be applied in practice for the removal of Sb and Bi. The diffusivity values for Sb are slightly higher than those for Bi at the same temperature.

Table 3: Experimental Results for the Adsorption of Antimonyand Bismuth with Activated Carbon (Equilibrium)

Adsorbent		Activated Carbon	
T (°C)		*27*	*60*
Mass (g) for 100 ml of electrolyte		*0.3 - 20*	*10 - 20*
Initial Concentration	*Sb*	*261 - 304*	*225*
(mg/l)	*Bi*	*128 - 141*	*124*
Final Concentration	*Sb*	*149 - 207*	*130-170*
(mg/l)	*Bi*	*70 - 109*	*92 - 125*
Capacity	*Sb*	*60*	*0.4 - 0.6 (*)*
(1000 ′ g/g Ads.)	*Bi*	*22*	*0.1 - 0.3*
Time to reach	*Sb*	*(days)*	*48 hours*
90% equilib. (min)	*Bi*	*(days)*	*48 hours*

Table 4: Experimental Results for the Adsorption of Antimony and Bismuth with Zeolites (Equilibrium)

Adsorbent		Zeolite
T (°C)		*27*
Mass (g) for 100 ml of electrolyte		*3.5*
Initial Concentration	*Sb*	*285*
(mg/l)	*Bi*	*131*
Final Concentration	*Sb*	*203*
(mg/l)	*Bi*	*105*
Capacity	*Sb*	*2.2*
(1000′ g/g Ads.)	*Bi*	*0.8*
Time to reach	*Sb*	*(days)*
90% equilib. (min)	*Bi*	*(days)*

Table 5: Experimental Results for the Adsorption of Antimony and Bismuth with Chelating Resins (Equilibrium)

Adsorbents		Resin UR-3300S		Resin Duolite C-467	
T (°C)		*45*	*55*	*45*	*55*
Mass (g) for 100 ml of electrolyte		*4.3 - 10*		*4.3 - 10*	
Initial Concentration	*Sb*	*227-232*	*227-237*	*216*	*216-233*
(mg/l)	*Bi*	*127-139*	*130*	*118*	*118-138*
Final Concentration	*Sb*	*24-32.6*	*21.6-36*	*18.7-26*	*17.1-30*
(mg/l)	*Bi*	*7.8-10.2*	*6-9*	*6.5-12*	*5.6-16.5*
Capacity	*Sb*	*3-5*	*1-5*	*2-5*	*1-5*
(1000g/g Ads.)	*Bi*	*2-3*	*1-3*	*1-3*	*1-3*
Time to reach	*Sb*	*30*	*20*	-	*15-30*
90% equilib. (min)	*Bi*	*25*	*15*	-	*10-20*

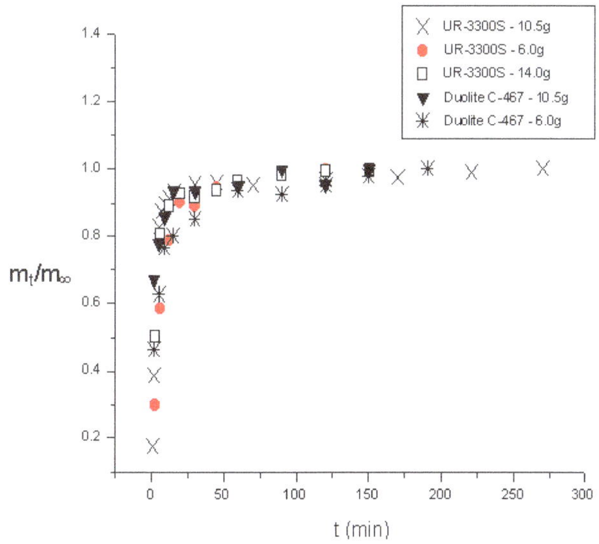

Figure 3: Adsorption Kinetics of Sb on resins UR-3300S and Duolite C-467 at 55°C.

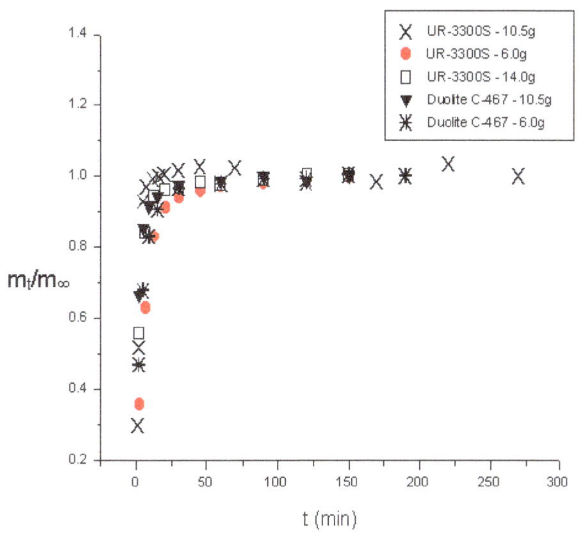

Figure 4: Adsorption Kinetics of Bi on resins UR-3300S and Duolite C-467 at 55°C.

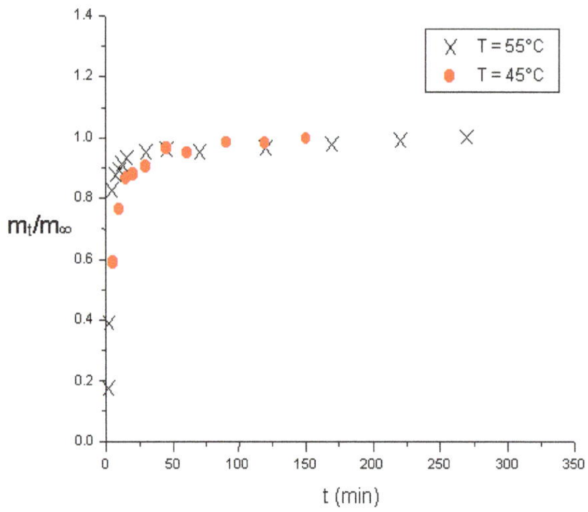

Figure 5: Adsorption Kinetics of Sb on resin UR-3300S (m = 10.5g) at 45°C and 55°C

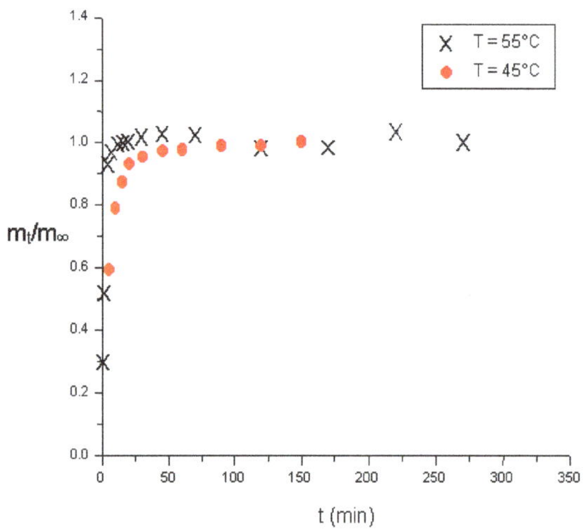

Figure 6: Adsorption Kinetics of Bi on resin UR-3300S (m = 10.5g) at 45°C and 55°C.

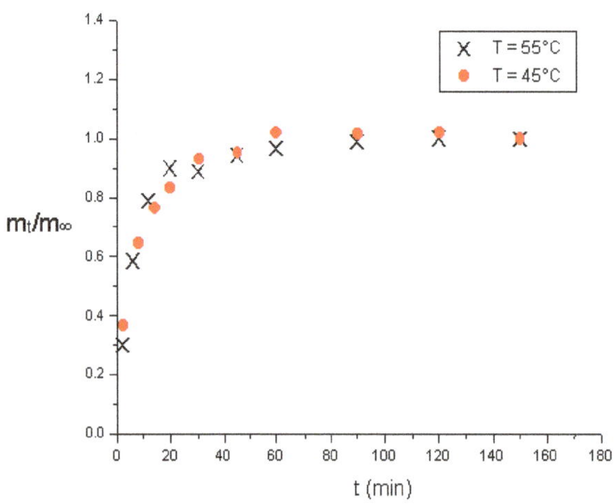

Figure 7: Adsorption Kinetics of Sb on resin UR-3300S (m = 6.0g) at 45°C and 55°C.

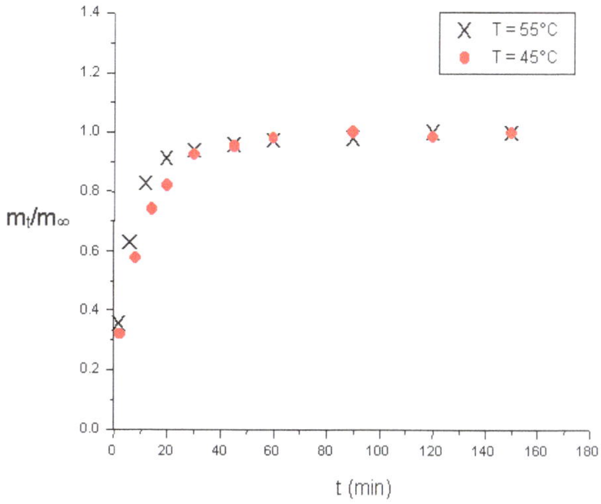

Figure 8: Adsorption Kinetics of Bi on resin UR-3300S (m = 6.0g) at 45°C and 55°C.

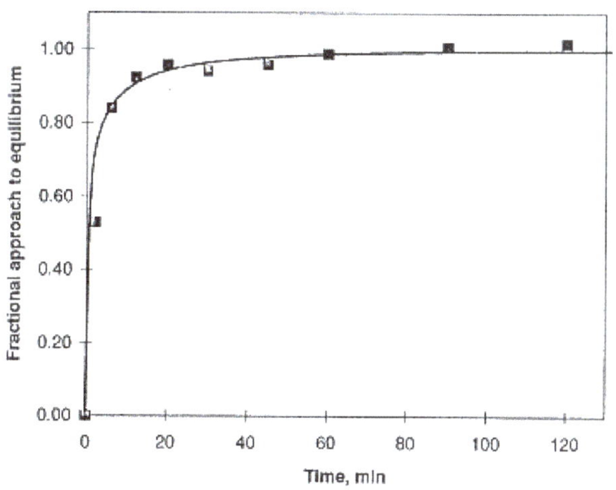

Figure 9: Theoretical diffusion model (full line) versus experimental results (squares) for sorption kinetics of Sb on UR-3300S at 55°C.

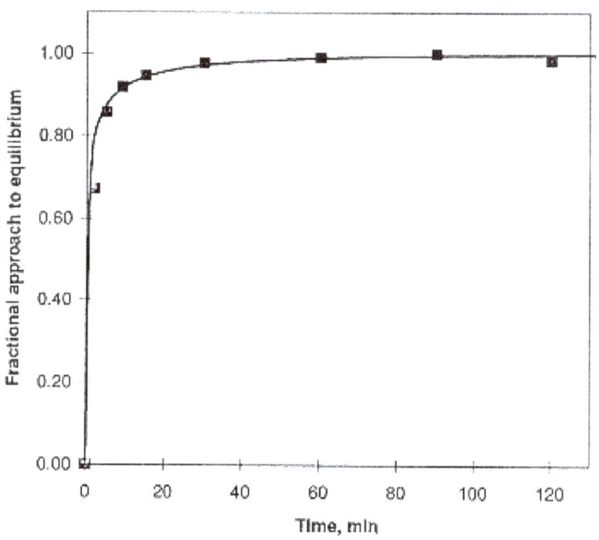

Figure 10: Theoretical diffusion model (full line) versus experimental results (squares) for sorption kinetics of Bi on UR-3300S at 55°C.

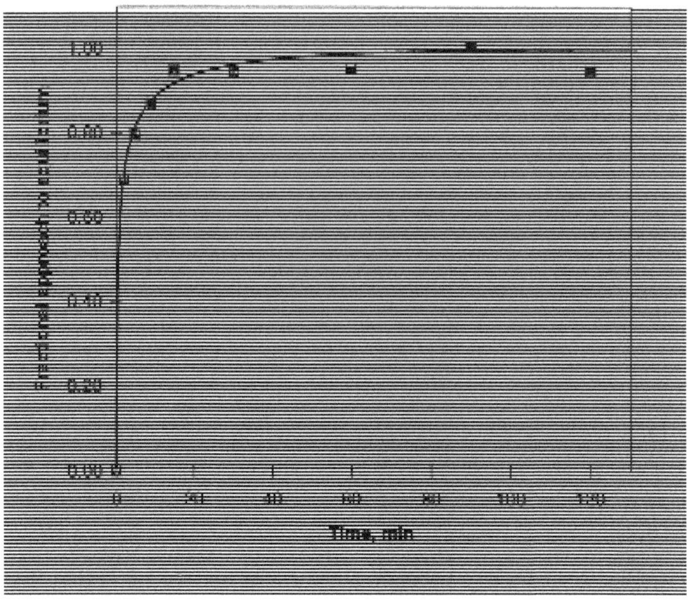

Figure 11: Theoretical diffusion model (full line) versus experimental results (squares) for sorption kinetics of Sb on Duolite C-467 at 55°C.

Table 6: Estimated Effective Diffusivity Coefficients for Chelating Resins

D_e (cm^2/min)			
Resin		UR-3300S	Duolite C-467
45°C	*Sb*	$6.48\ '\ 10^{-7}$	–
	Bi	$3.60\ '\ 10^{-7}$	–
55°C	*Sb*	$1.69\ '\ 10^{-6}$	$1.57\ '\ 10^{-6}$
	Bi	$7.50\ '\ 10^{-7}$	$1.04\ '\ 10^{-6}$

CONCLUSIONS

Several adsorbents were tested for Sb and Bi removal from a copper electrolyte using the finite bath experimental method.

Activated carbons are selective for antimony and bismuth with reasonable adsorption capacities (g metal/g adsorbent). However, the adsorption kinetics for these samples of activated carbon was too slow (15 or more days to reach the equilibrium). So, it may be concluded that these adsorbents are not recommended for industrial applications of this process.

Several zeolites did not physically resist the process conditions, as was expected. For USY zeolite, the results indicated very little selectivity towards the contaminants, with very slow kinetics. So zeolites are not adequate adsorbents for this application either.

Chelating resins definitely presented the best results, as compared to zeolites and activated carbon. They reduced the contamination level considerably and in a short time (about 3 hours under shaking and 24 hours or fewer at rest).

The sorption rate in both resins increased as temperature increased. Both resins, UR-3300S Unitika and Duolite C-467 Rohm & Haas, achieved similar results under all the conditions that were studied. The effective difusion coefficients were very similar for both resins and on the order of 10^{-6} to 10^{-7} cm^2/min in the temperature range of 45 to 55°C.

ACKNOWLEDGMENTS

The support of Caraíba Metais S/A, Departamento de Solos/ Universidade Federal do Ceará and ASTEF (Associação Técnico Científica Engenheiro Paulo de Frontin) is gratefully acknowledged.

REFERENCES

1. Abe, S. and Takasawa, Y., Prevention of Floating Slimes Precipitation in Copper Electrorefining. The Electrorefining and Winning of Copper, Hoffmann et al. (Eds.), TMS-AIME, Warrendale, PA, USA, pp. 87-98 (1987).
2. Azevêdo, D.C.S., Estudo Cinético e Termodinâmico de Adsorção para o Sistema Etanol-Água Sobre Zeolita 3A. Msc. Thesis, Universidade Federal de São Carlos-São Paulo, Brazil (1993).
3. Dreisinger, D.B., Leong, B.J.Y. and Grewal, I., Recent Advances in the Solvent Extraction and Ion Exchange of As, Sb, and Bi from Copper Refinery Electrolytes. 24thannual hydrometallurgical Meeting, Harris et al. (Eds.), Canada, pp. 71-89 (1994).
4. Dreisinger, D.B. and Scholey, B.J.Y., Ion Exchange Removal of Antimony and Bismuth from Copper Refinery Electrolytes. Electrorefining and Hydrometallurgy of Copper, Proceedings of Copper '95, Volume III, Cooper

et al. (Eds.), The Metallurgical Society of CIM, pp. 305-314 (1995).

5. McLaughlin, H. S., Regenerate Activated Carbon Using Organic Solvents. Chemical Engineering Progress, July, pp. 45-53 (1995).

6. Oda, Y., Hashiuchi, M., Kato, T. and Yonekawa, H., Tamano Chelating Resin Adsorption Process. Metallurgical Review of MMIJ, 3 (2), Nov., Symposium Proceedings, pp. 155-163 (1986).

7. Reed, B.E. and Nonavinakere, S.K., Metal Adsorption by Activated Carbon: Effect of Complexing Ligands, Competing Adsorbates, Ionic Strength, and Background Electrolyte. Separation Science and Technology, Hassan et al. (Eds.), USA, 27 (14), pp. 1985-2000 (1992).

8. Rupp, M.T.C, Russo, C., Sousa-Aguiar, E.F. and Imenes, E.S., Avaliação dos Parâmetros Dinâmicos na Remoção de Metais Pesados Através da Utilização da Zeólita Mineral Clinoptilolita. Paper, XXIII ENEMP, Maringá, PR, Brazil, (14) (1995).

9. Ruthven, D. M., Principles of Adsorption and Adsorption Processes, Wiley, New York, NY, USA (1984).

10. Ruthven, D.M., Zeolites as Selective Adsorbents. Chemical Engineering Progress, Feb, pp. 42-50 (1988).

11. Sasaki, Y., Kawai, S., Takasawa, Y. and Furuga, S., Development of Antimony Removal Process for Copper Electrolyte. Hydrometallurgy and Electrometallurgy of Copper. Proceedings of Copper '91, Volume III, Cooper et al. (Eds.), Elmsford, N.Y., USA, pp. 245-254 (1991).

12. Shibata, T., Hashiuchi, M. and Kato, T., Tamano Refinery's New Processes for Removing Impurities from Electrolyte. The Electrorefining and Winning of Copper, Hoffmann et al. (Eds.), TMS-AIME, Warrendale, PA, USA, pp. 99-116 (1987).

13. Tarasenko, Yu. A., Bagreev, A.A., Reznik, G.V. and Strelko, V.V., Reductive Sorption Methods for Extraction of Noble Metals from Solution. Hydrometallurgy '94, Jul., Cambridge, England, pp. 517-526 (1994).

14. Toyabe, K., Segawa, C. and Sato, H., Impurity Control of Electrolyte at Sumitomo Niihama Copper Refinery. The Electrorefining and Winning of Copper, Hoffmann et al. (Eds.), TMS-AIME, Warrendale, PA, USA, pp. 117-128 (1987).

15. Yang, J.H.K., Burban, J.H. and Cussler, E.L., Copper Selective Adsorption with a Microemulsion-Based Resin. AIChE Journal, 41(5), May, pp. 1165-1170 (1995).

16. Zamzow, M.J. and Murphy, J.E., Removal of Metal Cations from Water Using Zeolites. Separation Science and Tecnology, Hassan et al. (Eds.), USA, 27, pp. 1969-1984 (1992).

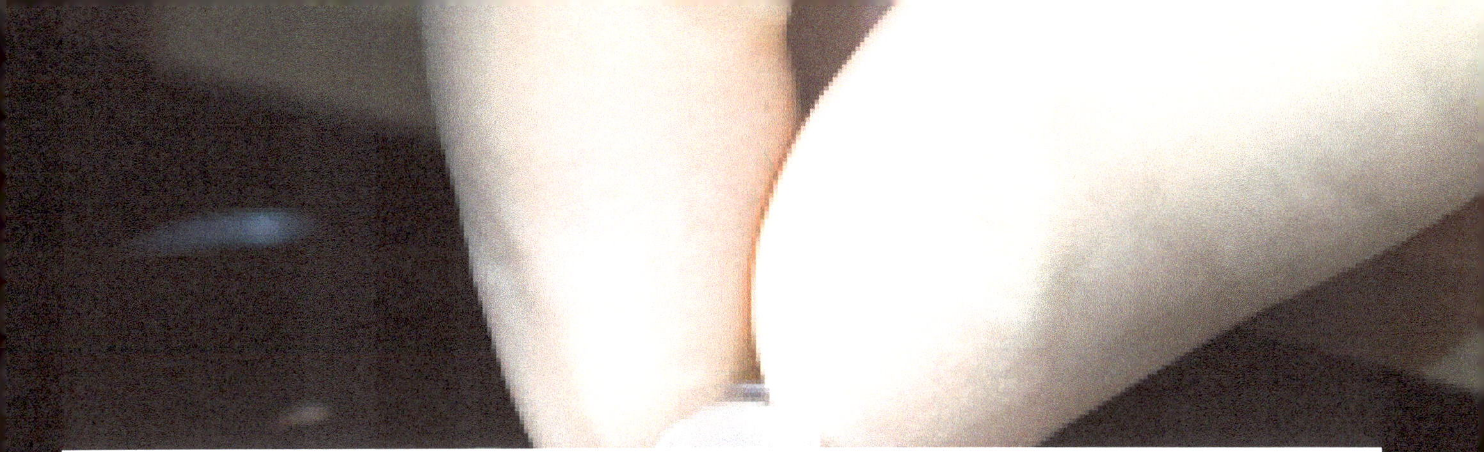

CHAPTER 4

FIRST FIFTY YEARS OF CHEMORESISTIVE GAS SENSORS

Giovanni Neri

Keywords: Chemosensors, gas sensor history, sensing materials, gas sensors perspectives

INTRODUCTION: HISTORICAL OVERVIEW

Chemoresistive gas sensors were introduced for the first time fifty years ago. At the beginning of the 1960s, Seyama, using ZnO thin film as a sensing layer, was able to demonstrate that gas sensing is possible with simple electrical devices [1]. He used a simple chemoresistive device based on ZnO thin films operating at the temperature of 485 °C. The response of the detection system to propane was about 100 times higher compared to the thermal conductivity detector used at that time.

In the early 1950s, Brattain and Bardeen, two scientists working at Bell Laboratories, gave the first demonstration that some semiconductor materials such as Ge modify their resistance, depending on the atmosphere they are in contact with [2]. Later, Heiland additionally described [3] that metal oxides such as ZnO modify their semiconducting properties with a change in the partial pressure of oxygen or other gases in the surrounding atmosphere. However, these discoveries were not investigated further.

After Seyama's work, Shaver [4] in 1967 described effects achievable with oxide semiconductors modified by the addition of noble metals (e.g., Pt, Pd, Ir, Rh). Since that time, the sensitivity and selectivity of semiconductor sensing devices have been enhanced significantly, and the search of new formulations for sensing materials has been intensified.

At the beginning of the 1970s, Taguchi fabricated and patented the first chemoresistive gas sensor device for practical applications using tin dioxide (SnO_2) as the sensitive material [5].

Indeed, after investigating many metal oxides, he found that SnO_2 has many advantageous properties such as higher sensitivity, low operating temperature, and a thermal stable structure. Its first generation of thick film sensors was based on a mixture of tin chloride ($SnCl_4$) and stearic acid that was painted on the substrate and fired at 700 °C in air. Firing burned off the organic component, leaving behind a porous SnO_2 layer. In order to increase the sensitivity, selectivity and stability, Taguchi used also Pd as a metal catalyst. The main application of these devices commercialized by Figaro Inc. was as alarms to prevent accidents and fire in domestic residences by monitoring the presence of hazardous levels of explosive gases. Intense efforts in this direction resulted in widespread application of semiconductor gas sensors.

Citation: Giovanni Neri, First Fifty Years of Chemoresistive Gas Sensors, doi:10.3390/chemosensors3010001

In the late 1980s, the field of semiconductor gas sensors underwent a significant expansion and became one of the most active research areas within the sensor community. The demand for high performance gas sensors with high sensitivity and selectivity, faster response, together with low power consumption and high device reliability, generated intensive efforts in order to develop new sensing materials. The rapid growth of materials chemistry, the broader field of materials science, has led to a dramatic increase in the number of new sensing materials available, subsequently enlarging the basis for a wider choice and favoring the development of high performance solid state gas sensors.

Seyama presumably would not have been able to anticipate that, half a century after the publication of his seminal paper, research on chemoresistive devices would assume an important role in daily life. Indeed, over the last five decades, due to their simplicity, low cost, small size and ability to be integrated into electronic devices, chemical sensors have seen an increase in their application to a variety of fields, including industrial emission control, household security, vehicle emission control and environmental monitoring, agricultural, biomedical, *etc.*, [6,7,8,9].

Today, the development of semiconducting sensing materials is adjunctively reliant on opportunities provided by new nanoscale technologies. Nanoscience, enabling controllable manipulation of matter at the molecular level, has become a fundamental generator for innovations in materials processing. Furthermore, emerging nanotechnologies promise dramatic changes in sensor designs and capabilities. A substantially smaller size, lower weight, more modest power requirements, greater sensitivity, and better specificity are just a few of the improvements we will see in sensor design.

It is not the intention here to do a comprehensive research survey within this short review article. There are already many excellent review articles on this topic in which readers can obtain a complete picture of chemoresistive gas sensors research from various perspectives [8,9,10,11,12,13,14,15,16,17,18,19]. In this brief review, a chronological approach is taken in order to not only examine the past, but also to identify key concepts, new materials and technologies, as well as predict innovative ideas for the future.

CHEMORESISTIVE GAS SENSORS

The Device

Detection of target gases by chemoresistive gas sensors has received impulse great deal of attention because of its many advantages over other sensing technologies. Table 1 illustrates the main characteristics of electrochemical, optical and chemoresistive gas sensors.

Table 1: omparison of three gas-sensing technologies

Characteristic	Electrochemical	Optical	Chemoresistive
Cost	*low*	*high*	*low*
Lifetime	*short*	*long*	*long*
Sensitivity	*high*	*high*	*high*
Selectivity	*good*	*excellent*	*poor*
Response time	*fast*	*fast*	*fast*
Size	*medium*	*large*	*small*

Electrochemical sensors are becoming unpopular as they have a short lifetime, rendering them unacceptable for some applications. Optical sensors show excellent characteristics of sensitivity, selectivity, adequate lifetime, and fast response; however, they have a high cost and large size. Although the chemoresistive sensors are largely unselective, the low cost and fabrication simplicity of chemoresistive sensors are the main factors contributing to their widespread use.

A chemoresistive sensor is based on a sensitive material, in bulk or deposited on a suitable support, upon which the molecular recognition process takes place. The analyte recognition process occurs either at the surface of the sensing element or in the bulk of the material, leading to a concentration-dependent change property that can be transformed into an electrical signal by the appropriate transducer. This simple transduction mechanism allowed the fabrication of devices with different configurations.

Classical Taguchi sensors have a sensitive material in the form of sintered porous ceramic body (see Figure 1). Other examples of chemoresistive sensors bearing different forms and dimensions are shown in Figure 2.

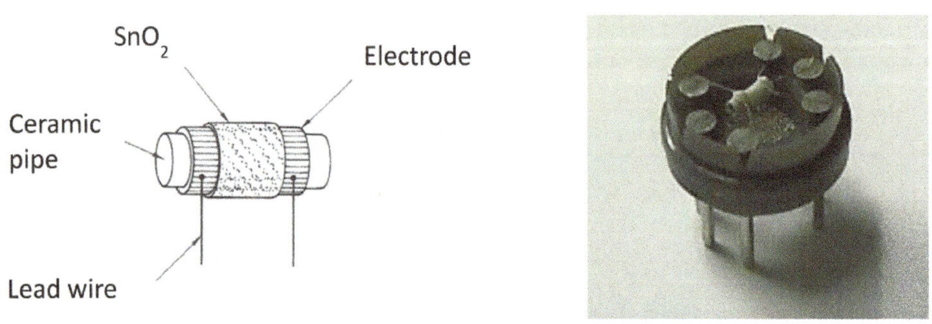

Figure 1: Schematization and picture of Taguchi chemoresistive gas sensor.

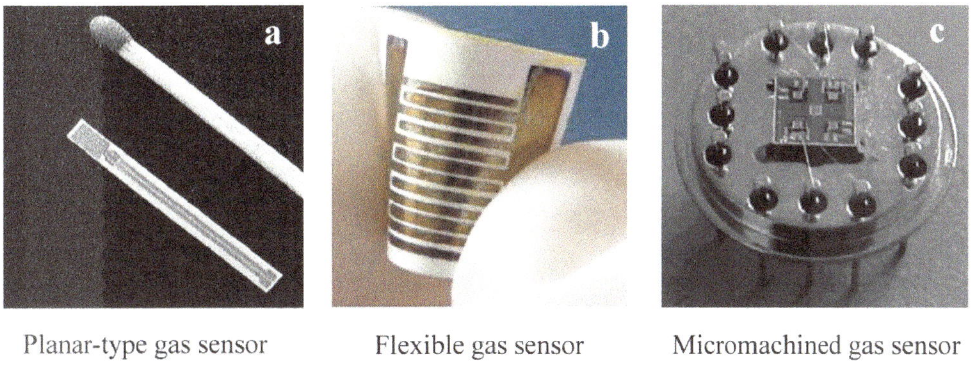

Figure 2: Chemoresistive gas sensors with different configurations.

Planar-type gas sensors are constituted of a sensing thick/thin layer deposited by chemical or physical methods onto a ceramic substrate with interdigited electrodes (Figure 2a). The sensing layer can be deposited also on a plastic support, thereby allowing the fabrication of flexible chemoresistive gas sensors (Figure 2b).

Although these sensors are small in size, further miniaturization of sensors is demanded by a variety of applications. In particular, compatibility with integrated circuit (IC) technology is now largely pursued. Fully integrated devices have been fabricated using top-down processes, but the top-down approach typically has the advantage of higher throughput and amenability for large-scale integration. By these techniques, micromachined chemoresistive sensors can be fabricated on a chip substrate, enabling easy interfacing with conventional silicon microelectronics (Figure 2c).

The continual drive towards even finer microstructures will lead to the fabrication of sensing nanodevices. Already now there are existing mature industrial technologies (e.g., technologies based on thin film semiconductors), but long-term stability could be a serious problem for these sensing nanodevices. For example, sensing nanomaterials may be quite unstable under thermal conditions due to their high surface energy leading to severe grain growth, which may result in the degradation of the device performances. Under such operations, the mechanical deterioration of electrodes can also lead to performance degradation due to the possible formation of micro-cracks on the electrode structure.

Therefore, unless there are significant cost and/or performance benefits from using nanosensors, it will be difficult to displace the existing bulk sensor technologies in the near future.

Sensing Materials

From the first tin dioxide sensor developed in the 1970s for domestic gas alarms, there has been an increase in the demand for high performance solid-state gas sensors. Results of a search study on semiconductor metal oxides used as sensing materials for chemoresistive gas sensors, including both the n-type and p-type oxides, are summarized in the graph shown in Figure 3.

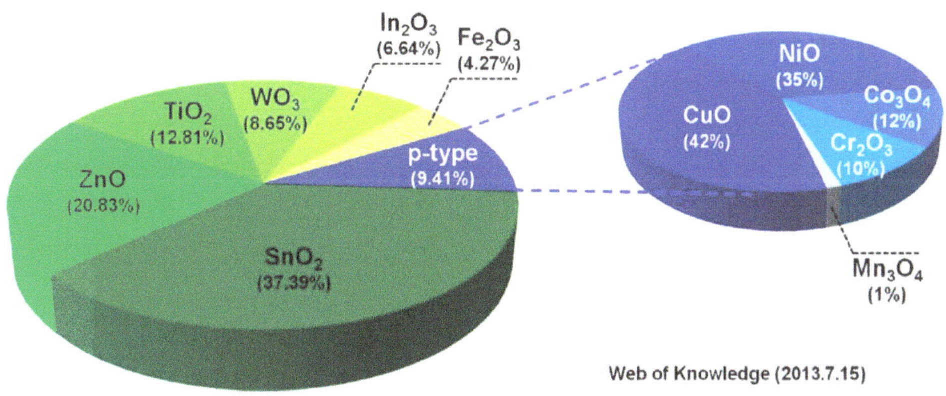

Figure 3: Studies on n- and p-type oxide semiconductor gas sensors (internet search of Web of Knowledge). Reprinted with permission from [15].

Metal oxides stand out as one of the most common, diverse and, most likely, largest class of materials due to their extensive structural, physical and chemical properties and functionalities. The most common metal oxides utilized as sensing layer in chemoresistive devices are binary oxides such as SnO_2, ZnO, TiO_2, *etc.*; however, ternary and more complex oxides are also applied in practical gas sensors [14].

SnO_2 is the most extensively studied metal oxide and is widely applied in practical commercial devices. Tin dioxide is a wide-band gap (3.6 eV) semiconductor with interesting electrical properties [20]. Due to its high sensitivities for different gas species, tin oxide-based sensors allow the detection of low concentration levels of gaseous species, though it unfortunately suffers from the lack of selectivity. Nevertheless, strategies devoted to enhance the SnO_2-based devices performance (synthesis and treatment conditions, and nano- micro- and nano-structure, addition of dopants) have been extensively studied [21].

Zinc oxide is a II–VI semiconductor showing a wide-band gap (3.37 eV), with the dominant defects identified as O vacancies [22]. ZnO has attracted much attention in the gas-sensing field because of its high mechanical and chemical stability, suitability to doping, non-toxicity and low cost. The sensing property of ZnO is strongly influenced by the microstructural features, such as the grain size, geometry and connectivity between the grains.

Titania is particularly attractive for gas sensors because of its lower cross-sensitivity to humidity than other metal oxides [23]. Among the other applications, TiO_2 has been largely investigated as a sensing layer in resistive oxygen gas sensors operating at medium-high temperatures for automotive air/fuel ratio control. At medium temperatures (400–600 °C), oxygen detection is mainly due to reactions that takes place on the surface, whereas at high temperatures (700–1000 °C), oxygen detection is mainly due to diffusion of oxygen ions in the bulk of the material [24].

A lot of factors affect the gas sensor performance of metal oxide chemosensors which, in turn, is determined by the reception and transduction function along with the fabrication procedure [16]. The synthesis procedure, crystal size and shape, the addition of foreign elements with the role of sensitizers or conductivity modifiers are some of the factors influencing the sensor response [17,18,19,20,21,22,23,24,25].

Doping the metal oxide layer with suitable promoters (metal particles, foreign metal oxide, ions) is a common way of enhancing the sensing characteristics of chemoresistive gas sensors. A comprehensive review dealing with all aspects of metal doping can be found in [26]. The modification of the sensing properties of the sensing material by the introduced additives depends on the nature of the latter. For example, Pt is known to promote the gas-sensing reaction by the spill-over of sample gas (chemical sensitization), whereas Pd is known to promote the gas-sensing reaction by electronic interaction between Pd and sensing materials (electronic sensitization) [27].

The grain-size reduction at nanometric level is another of the main factors enhancing the detection properties of metal oxides. It is in fact well recognized that by reducing the particle size of the sensing material in the nanometer range the sensitivity of chemoresistive gas sensors is greatly improved both for the large specific surface offered and for the influence in reducing the surface charge density [21,28,29,30,31,32,33,34,35]. Furthermore, in this size range, a large fraction of the atoms (up to 50%) are present at the surface or the interface region; therefore, the chemical and electronic of nanoparticles are different from those of the bulk, consequently contributing to an increase in the sensing properties.

Over the last fifty years, novel sensing materials others than metal oxides have been proposed. In 1983, the gas-sensing properties of conducting polymers were first reported. The sensor involved the use of doped polypyrrole (PPy), functioning as an ammonia sensor [36]. In fact, as the conductivity of the pure conducting polymer is rather low, in order to achieve the high conductivity suitable for sensing applications, a doping process was necessary (Figure 4). A wide variety of polymers of this type is now available, including substituted polypyrroles, polythiophenes, polyindoles, and polyanilines.

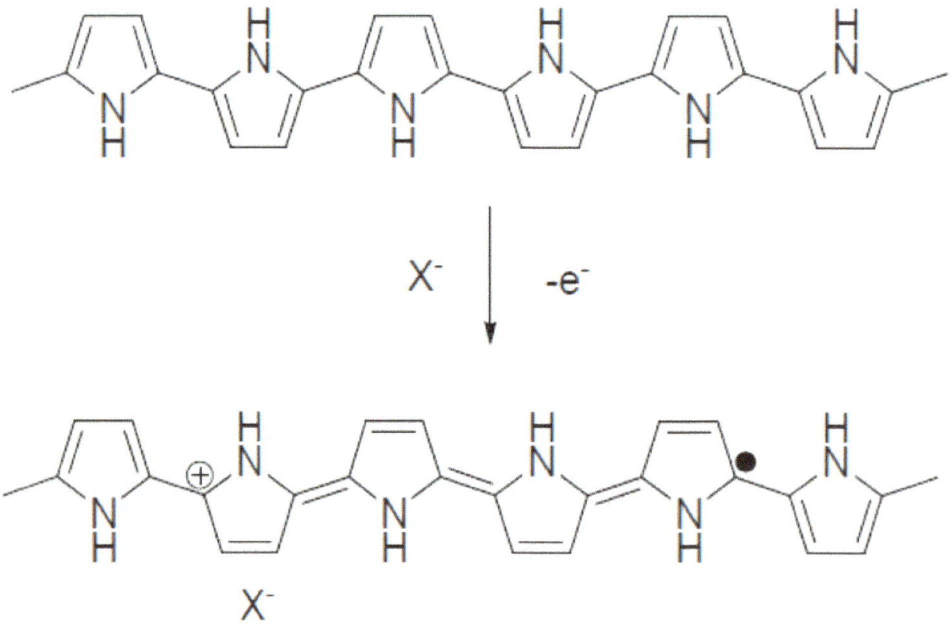

Figure 4: Molecular structure of polypyrrole (PPy) and related doping process.

Organic materials are much more easily modified than inorganic materials with respect to such characteristics as sensitivity, working temperature, and selectivity. Long-term instability is a main drawback of the sensors based on conducting polymers as they are thermally unstable, so it is often impossible to use them at temperatures at which gas-solid interactions proceed rapidly and reversibly.

To fully use all the potential advantages of organic and inorganic materials, hybrid composites have also been introduced as sensing elements for resistive sensors [37]. These hybrid organic/metal oxide composites are not merely the sum of the individual components, but rather new materials with new functionalities and properties. From the viewpoint of structure, organic support can induce the nucleation, growth and formation of fine metal oxide nano-/microstructures with uniform dispersion and controlled morphology, thereby avoiding the agglomeration of metal oxides. If the organic component is a good electrical conductor (e.g., conductive carbon, carbon nanotubes or graphene), the resulting composites can form a perfect integrated structure with a developed electron conductive network and shortened current transport paths, improving the poor electrical properties and charge transfer of pure metal oxides. Significant synergistic effects, such as room-temperature sensing capability when exposed to low-concentration gases such as NO_2, H_2, and CO, often occur in hybrid composites because of size effects and interfacial interactions in contrast to the high-temperature operation required for metal oxides alone.

The possibility of synthesized materials having dimensions in the order of nanometers has provided enormous advantages for gas sensing because their extremely high surface-to-volume ratio is ideal for gas molecule adsorption. One important example is that of carbon nanotubes (CNTs), which are currently receiving a great deal of interest in gas sensing. Dai *et al.* first demonstrated the variety that they can offer for practical applications in highly sensitive gas sensors [38]. The high surface-to-volume ratio, high conductivity and mechanical stability make CNTs very attractive for gas-sensing applications. Carbon nanotubes can now be synthesized by different methods in large quantities and with high purity, such as the chemical vapor deposition (CVD) technique, and can be doped with B- and/or N-groups, thus rendering them very sensitive to a wide range of gas vapors [39]. In a similar way, functionalizing the nanotube surface with the polar COOH group attached, the sensors will give stronger responses towards the volatile organic compounds (VOCs) as their absorption efficiency with these volatile organic molecules will be increased due to the dipole-dipole interactions (mainly hydrogen bonding) between the COOH and the polar organic molecules [40].

CNTs are also often used as carriers of supported metal oxide particles [41]. CNTs provide high surface area, then help the dispersion of the sensing materials on the nanotube walls (Figure 5a). The better performance of these hybrid sensors are also attributed to the effective electron transfer between the metal oxide particles and the highly conductive carbon nanotube network [42,43,44]. CNTs could also be coated with metal oxides of controlled thickness [45]. By combining the non-aqueous sol-gel route with the atomic layer deposition, metal oxides have been grown from the respective metal alkoxide precursors at low temperatures (Figure 5b). Moreover, the surface reaction leading to the M−O−M bond formation is self-limited and allows then the deposition of films with well-controlled thickness on the internal and external surface of CNTs.

The successful synthesis of nanoparticles, nanowires, nanotubes and other shapes has generated a lot of work to use these nanostructured materials in gas sensing [46,47,48,49,50]. In particular, low-dimensional structures of metal oxides were found to possess characteristics and enhanced gas-sensing properties. In 2002, the Sberveglieri and Yang groups initiated the investigation of gas-sensing properties of one-dimensional nanostructures [51,52]. The shape and morphology of the key one-dimensional nanostructures in chemoresistive sensors are shown in Figure 6.

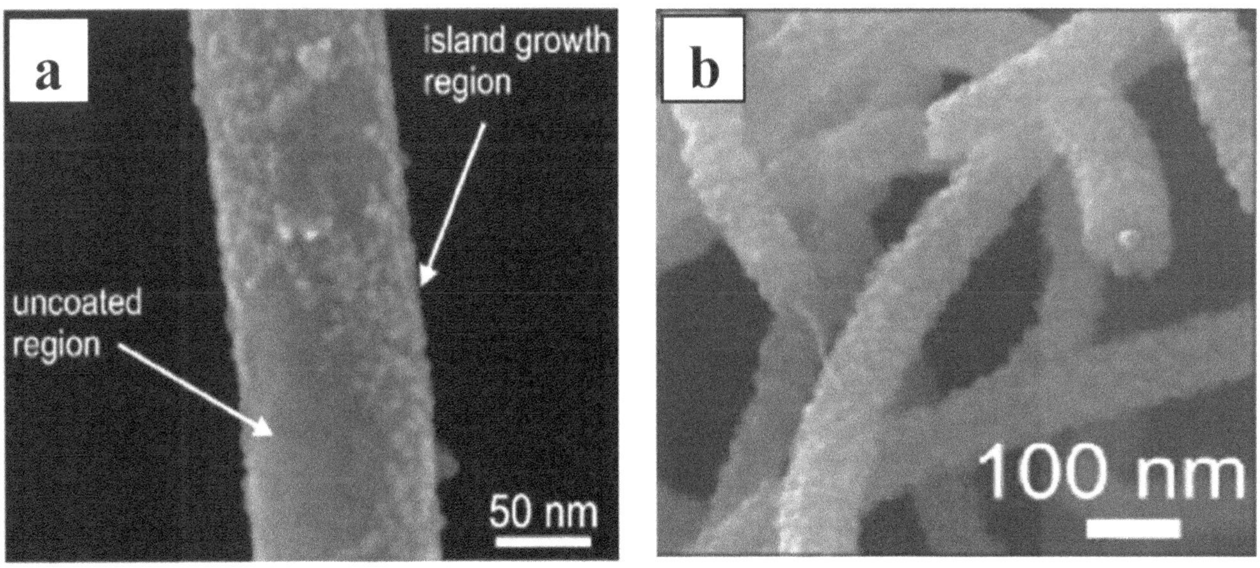

Figure 5: SEM images showing metal oxide nanoparticles, visible as white spots, dispersed on CNT: (**a**) irregularly coated CNT; (**b**) uniformly coated CNT. Reprinted with permission from [45].

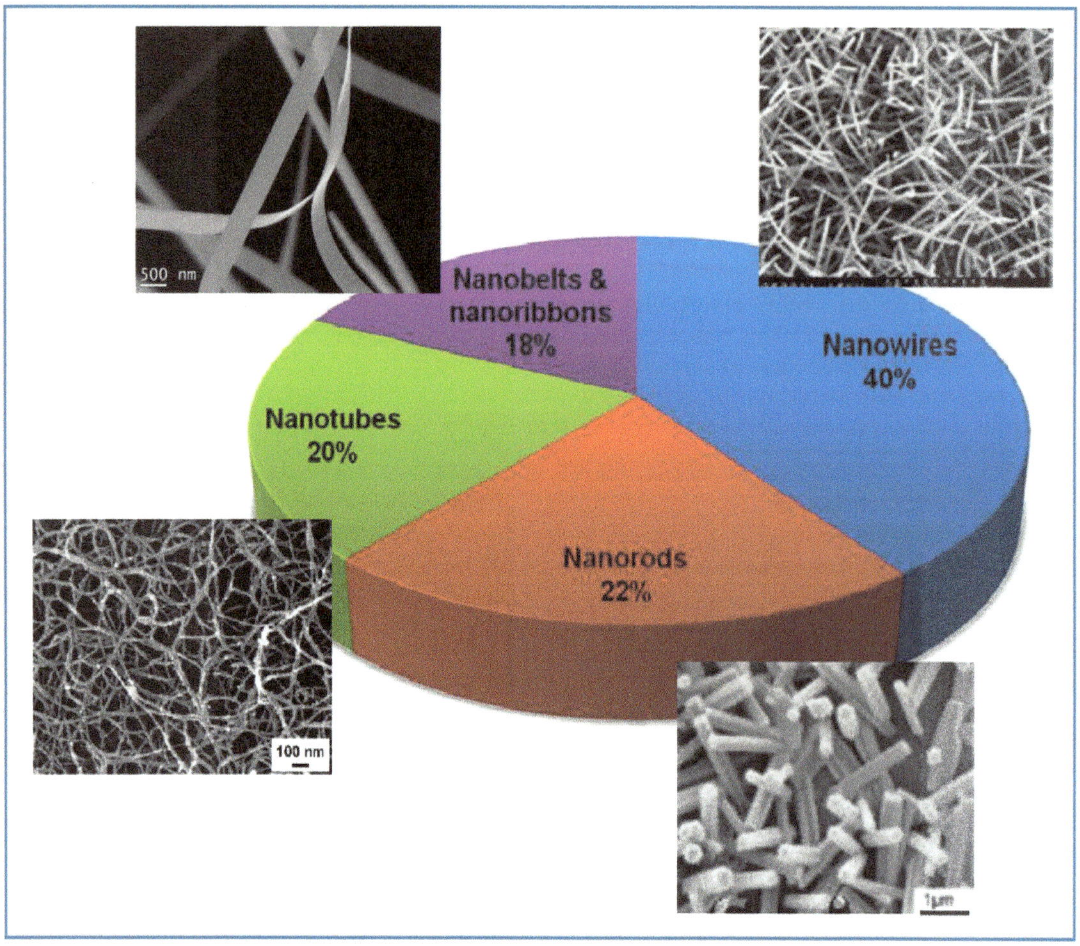

Figure 6: Morphology of the key one-dimensional nanostructures in gas chemiresitors (internet search of Web of Knowledge). Adapted from [53].

Hierarchical and hollow oxide nanostructures are also very promising gas sensor materials due to well-aligned nanoporous structures with less agglomerated configurations. The literature data clearly show that these peculiar nanostructures can increase both the gas response and response speed simultaneously and substantially [54]. This can be explained by the rapid and effective gas diffusion toward the entire sensing surfaces via the highly porous structures.

In 2007, a two-dimensional monolayer of carbon atoms, named graphene, has been identified to be a promising sensing material because it has unique and excellent electrical and mechanical properties [55]. Graphene appears to be more suitable for integration into devices than CNTs because its planar nanostructure makes graphene advantageous for use in standard microfabrication techniques. Furthermore, the sheet structure of graphene (Figure 7a) makes it suitable for depositing and stabilizing small metal and/or metal oxide nanoparticles (Figure 7b), which is of remarkable interest for use in gas sensors.

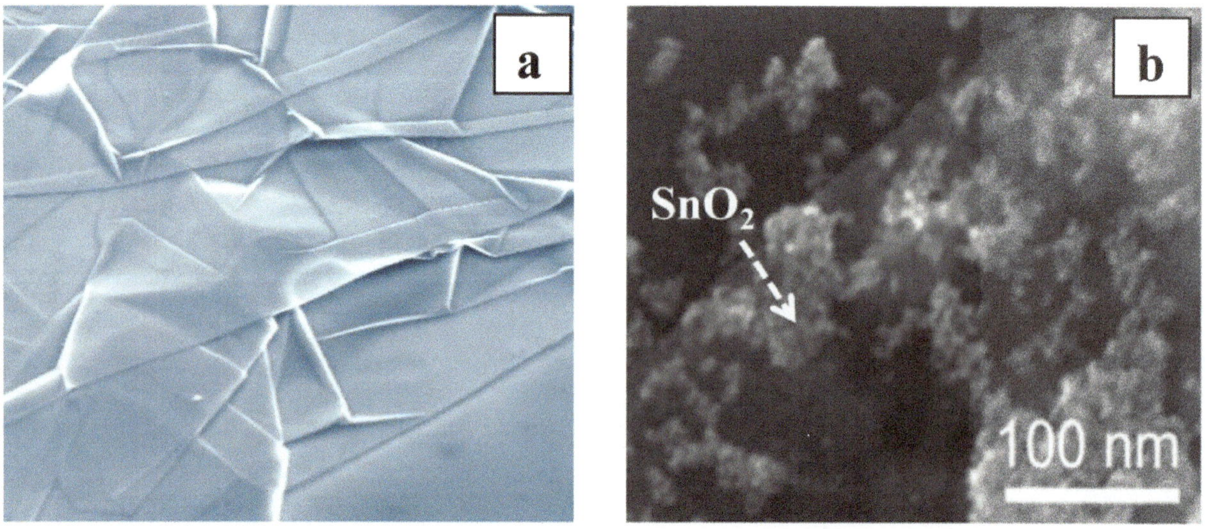

Figure 7. (a) Graphene sheets; (b) picture of metal oxide nanoparticles deposited on graphene sheets.

Working Mechanism

Along with the improvement of the characteristics and properties of the devices, a great deal of effort has been also made during these years to improve knowledge of the sensing mechanism and understanding the related processes. The description of mechanisms of chemoresistive sensors based on metal oxides was first depicted through the application of electron theory of chemisorption and catalysis on semiconductors originally formulated by Wolkenstein [56].

Later, Morrison, Yamazoe and Gopel especially contributed to the description of conditions of transport of electric charges through the metal oxide semiconducting layer in the presence of oxygen and reactive gases [57,58,59]. Following these theories, the sensing mechanism of MOS sensors relies on reactions which occur between adsorbed oxygen species and the probed gas on the surface of the sensing layer. Details on this subject can be found in many books and reviews [10,12,60,61]. In first approximation, oxygen adsorbed on the surface of n-type metal oxide semiconductors plays a key role, trapping free electrons because of its high electron affinity, and forming a potential barrier at the grain boundaries. This potential barrier restricts the flow of electrons, causing the electric resistance to increase. When the sensor is exposed to an atmosphere containing reducing gases, e.g., CO, the gas molecules adsorbs on the surface and reacts with active oxygen species, e.g., O^-, which liberates free electrons in the bulk. This lowers the potential barrier allowing electrons to flow more easily, thereby reducing the electrical resistance. With oxidizing gases such as NO_2 and ozone, the adsorption process increases instead the surface resistance [62]. The converse is true for p-type oxides, where electron exchange due to the gas interaction leads either to a reduction (reducing gas) or an increase (oxidizing gas) in electron holes in the valence band [63]. However, to give a complete description of the gas-sensing mechanism for these sensors, it is necessary to take into account all elementary steps (adsorption, reaction, desorption, *etc.*) governing surface-gas target interactions leading to charge transfer [64].

Yamazoe first demonstrated that reduction in crystal size would significantly increase the sensor performance [7]. In particular, the sensitivity of metal oxides is drastically promoted when the crystallite size, D, is in the nanometer range (Figure 8). The sensing properties are partially associated with the depth of the surface space charge region which is affected by the gas adsorption and depend on the particle size. Specifically, the sensor's response increases significantly if the crystallite size is about twice the adsorption depth, 2L (L is the depth of the space-charge layer), of oxygen adsorbates. This means that the sensor performance improves not only when D decreases, but also if L increases, since a major proportion of material takes place in the reaction of oxygen adsorbates with the target gas even if the grain size is not excessively small. For a typical sensing material, SnO_2, with different grain sizes between 5 and 80 nm, simulations showed that the conductivity increases linearly with decreasing trapped charge densities and the sensitivity to the gas-induced variations in the trapped charge density increases too, in agreement with experimental findings [65].

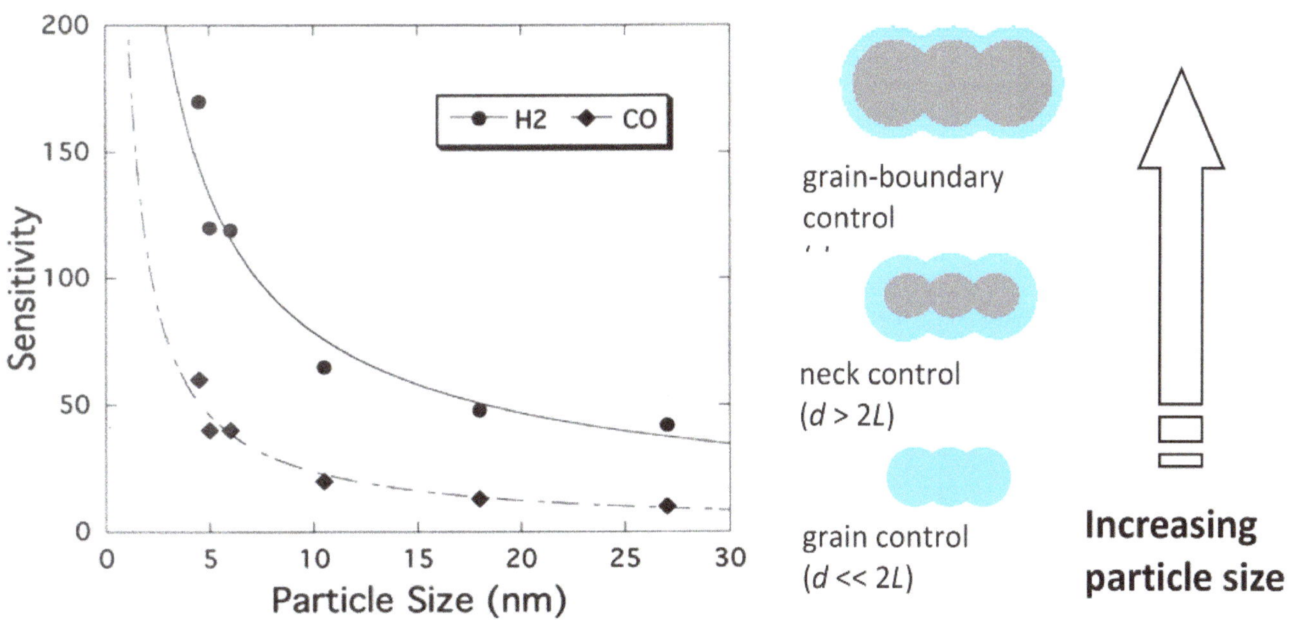

Figure 8: Sensitivity to H_2 and CO of SnO_2 *vs.* particle size and sensing mechanisms associated. Adapted with permission from [7].

When dealing with wire with lateral dimensions to the order of hundreds of nanometers, gas adsorption creates a surface depletion layer, consequently reducing the conducting channel thickness [50,66]. The mobility dependence on surface coverage can be neglected because electron diffusion length (about 1 nm) is much shorter than the diameter (tens of nanometers). Electrical transport changes when the thickness of the wire is small like in nano-wires with low lateral dimensions low compared to the Debye length, the space charge region extends through the entire wire cross-section and all electrons are trapped in surface states.

A model of the sensing mechanism of hybrid heterostructures such as metal oxide/CNTs has been attempted. The junctions existing between the nanotubes of the networking at their crossing points (Figure 9), as well as the different depletion regions existing as a function of the presence of the target gas, have been taken into account. Their role in the sensing mechanism has been established by several authors [67,68].

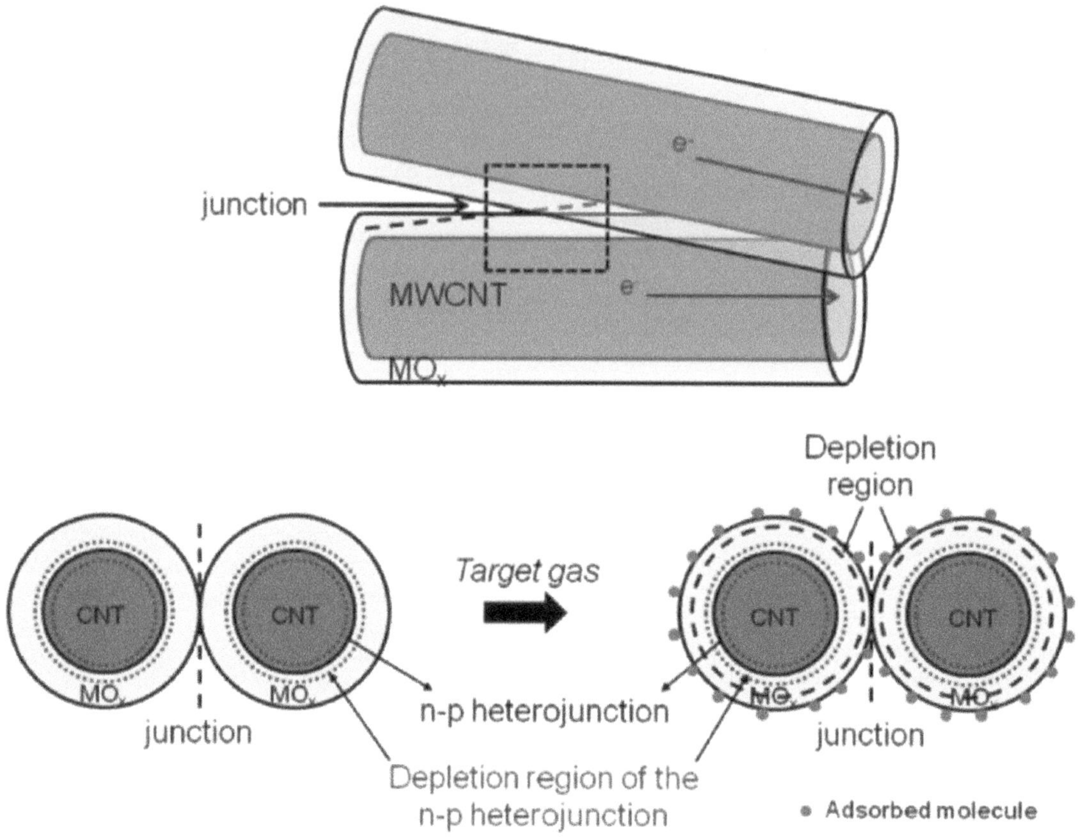

Figure 9: Schematic illustration of the junctions involved in the sensing mechanism of metal oxide/CNTs. Adapted with permission from [68].

APPLICATIONS OF CHEMORESISTIVE GAS SENSORS

While the previous sections focused on the scientific breakthroughs achieved by constantly pushing the fundaments of sensing materials and devices, focus in the following is addressed towards the applicative aspects of chemoresistive gas sensors.

Rapid, comprehensive, and reliable information regarding the chemical state of a gaseous system is currently indispensable in many high-technology fields. Solid-state gas sensors are thus generating tremendous interest because of their widespread applications in industry, environmental monitoring, space exploration, biomedicine, and pharmaceutics. Generally, basic criteria for these practical gas-sensing devices are: (i) high sensitivity and selectivity; (ii) fast response time and recovery time; (iii) low power consumption; (iv) low operating temperature and temperature independence; (v) high stability.

For most of the practical applications, a relevant problem is not only to estimate the target gas concentration, but also to identify it in the real mixture (outdoor and indoor ambient, exhaust gases, breath). Unfortunately, chemoresistive gas sensors are very sensitive but not selective. Such a task can therefore not be performed by a single sensor. E-nose, an artificial olfactory system consisting of an array of sensors, then needs to be used. Each sensor shows an individual response to all (or to a certain subset) of the target gases which are identified by pattern recognition of the sensor signals, e.g., that developed by Cyrano Sciences is based on the concept of using multiple semi-selective sensors combined with electronic computation, first proposed by Gardner and Bartlett [69].

In the following, some of the most important applicative fields of semiconductor gas sensors are listed.

Environmental Applications

In environmental applications, field-portable monitors based on chemoresistive sensor devices are particularly suited to comply with the newly reinforced environmental regulations, and give a valuable alternative to conventional analytical techniques; they are accurate but more expensive and time consuming [70]. Metal oxide semiconductor sensors are relatively inexpensive compared to other sensing technologies; they are also robust, lightweight, long lasting and benefit from high material sensitivity and quick response times. These properties could lead to them becoming ever more important tools in environmental monitoring with respect to the other type of sensors. They can be used to measure and monitor trace amounts of environmentally important gases such as carbon monoxide and nitrogen dioxide [71]. Today, it is also highly desirable to develop simple and inexpensive sensors to measure atmospheric carbon dioxide in order to monitor indoor air quality [72].

Automotive Applications

The ability of chemosensors based on refractory metal oxides to withstand high temperatures, the robust nature, efficient packaging procedures and small size, are strong technical merits for automotive applications. Indeed, usually high temperatures are reached in the exhausts, and sensor devices should operate in the range 600–900 °C in a harsh ambient requiring advanced packaging. Other requirement for these sensors are reaction measurement speed of the order of fraction of seconds and 5–10 year service life. As a result of the introduction of emergent technologies in the automotive industry, new vehicle diagnostic sensors are at present necessary to control motor functioning, monitor a range of emission gases (NO, NO_2, CO, CO_2, HC, O_2, *etc.*), and detect high pollution levels in the vehicle cabin [73].

Chemoresistive sensors can be used to minimize the emission of CO, hydrocarbons, and nitrogen oxides coming from combustion engines. It is important for the engine to operate with the proper air-to-fuel ratio (A/F) so that combustion parameters are optimized. The first titania gas sensors were developed in the late 1970s and early 1980s. They were primarily used to detect the stoichiometric A/F [74]. In these instances, the sensor resistance increased by orders of magnitude around the stoichiometric A/F, making it a very useful device for these applications.

Chemoresistive sensors will have high importance also in future cars. In fact, CO sensors have been identified as a critical need for proton exchange membrane (PEM) fuel cell systems, which offer a viable approach to improve efficiency of power generation from fossil fuels, while reducing emissions of pollutants and greenhouse gases [75]. The primary uses of such sensors include measurement of the CO content of the reformate gas entering the PEM fuel cell (*i.e.*, for fuel cell protection) and measurement of the CO content of the reformate gas at various catalytic stages of the fuel processor (for fuel processor feedback and control). A further requirement is that these sensors must not exhibit cross-sensitivities to the gaseous components of the reformed gases (e.g., hydrogen, carbon dioxide, and/or humidity) or to other potential contaminants.

Biomedical Applications

Biomedical uses of chemoresistive gas sensors (e.g., in breath test) require instead high sensitivity to detect very low concentrations of target gases (in the ppt–ppm range), coming from biochemical processes occurring in the human body and used as markers of for several pathologies [76]. Furthermore, the target gas is to be analyzed and quantified in a complex mixture, *i.e.*, in the presence of many other interferent gases and a high humidity content [77]. Such sensors can complement or serve as an alternative to more sophisticated spectrometric systems for breath analysis and speciation, especially for clinical diagnostics and monitoring [78].

Miscellaneous

During recent years, single or array of metal oxide semiconductor (MOS) sensors have found applications in many other different fields than those listed above, including odor landfill, chemical agent warfare detection, food quality control, *etc.*[79]. These technologies enable rapid detection and identification of substances based on their chemical profile. They find applications in the monitoring of some medical conditions as well as industrial applications generally related to quality control or contamination detection. A new multi-chamber electronic nose (MCE-nose) which overcomes, to a great extent, one of the major disadvantages of the use of MOS technology for e-noses—the long recovery period needed after each gas exposure that severely restricts its use in applications where the gas concentrations may change rapidly—has been also proposed [80]. It comprises several identical sets of MOS sensors accommodated in separate chambers which alternate between sensing and recovery states, providing, as a whole, a device capable of sensing more quickly changes in chemical concentrations.

NEW OPPORTUNITIES

Although resistive gas sensors are still commercialized intensively as bulk devices, there is a strong demand for novel, smaller, integrated, high performance and reliable devices for advanced applications. Technological advances will assist gas sensor device scale of economies by exploring more efficient fabrication methods that may include material and component assembly into the final device.

At the material level, technological advances will arise from multidisciplinary contributions, which will hopefully open up new areas of nanoscience research. In the past few years, large efforts have been expended in the development of new synthetic approaches for sensing materials at the nanoscale. Advances in recent activities concerning the synthesis, characterization and properties of novel materials with potential in sensing applications can be found in many reviews [81,82]. Among the obvious challenges are improving reproducibly and control over nanoparticle structure, surface chemistry and dispersion.

Future work will continue to focus also on generating improved component assembly and device fabrication strategies. For small or medium-scale batch production, a hybrid design whereby component parts are created separately for subsequent assembly into a complete system is an attractive option, as it removes many of the restrictions imposed by the need for process compatibility.

Integrated gas sensor with a micro-hotplate (MHP) demonstrates better sensitivity, faster response, and lower power consumption than traditional thick film devices. For the detection of a target gas, a gas-sensitive layer is applied on a suspended micro-hotplate that is mounted on top of a silicon chip. Any gas-induced change of the gas-sensitive layer's surface potential is detected and digitally processed by the integrated electronics.

Sensor platforms robust enough for integration into standard, industrial weaving processes have been proposed [83]. Such sensors were fabricated on a thin, flexible plastic stripe (e.g., made of Kapton®, DuPont™, Wilmington, De, USA) and woven into a cotton fabric by using a standard weaving machine, to create a smart fabric able to detect ambient gas. The fabrication process is actually very simple and compatible with large-scale roll-to-roll fabrication. Inkjet-printing can be used as an effective, low-cost alternative, suitable for the fabrication of long sensor stripes. Bending tests performed on the devices showed that they can be successfully woven within a cotton fabric using a standard weaving machine without being damaged.

From a processing point of view, employing modern microelectronics technology to manufacture both the sensing element and the signal conditioning circuitry on a single silicon chip, is now possible to make low-power, low-cost smart gas sensors in high volume and this should result in a new generation of miniaturized gas sensors [84]. Full-integration of micro-electronic and micro-mechanical components on a single wafer has been achieved commercially using silicon processing technology. Figure 10 shows a commercial micro electro mechanical system (MEMS) gas sensor for indoor air quality monitoring [85].

Figure 10: TGS 8100 MEMS gas sensors by FIGARO Eng. Inc., Osaka, Japan.

The sensing element is comprised of a sensing chip and an integrated heater fabricated on a silicon substrate using MEMS technology, and a metal oxide semiconductor layer on the sensing chip. Due to miniaturization of the sensing chip, the sensor requires a heater power consumption of only 15 mW and is suitable for low-power equipment and battery-operated instruments. We expect many areas will benefit from these miniaturized sensors, including transportation (land, sea, air, and space), buildings and facilities (homes, offices, factories), humans (especially for health and medical monitoring) and robotics of all types.

Recent technological advances are quickly changing the landscape of gas sensors. As these devices are becoming smaller in size, require less support infrastructure than currently used equipment, and are capable of operating autonomously, new opportunity for gas sensors could come from their integration into smartphones [86]. Chemical sensors, integrated within smartphones and other wireless and wearable devices, provide a vast array of functionality, from informing people about their environment, air quality and safety, to diagnosing health conditions through breath analysis. For example, background apps on a smartphone (see Figure 11) can monitor CO or CO_2 levels and provide Safe/Warning/Alert indication to the user about possible dangerous changes in their environment [87].

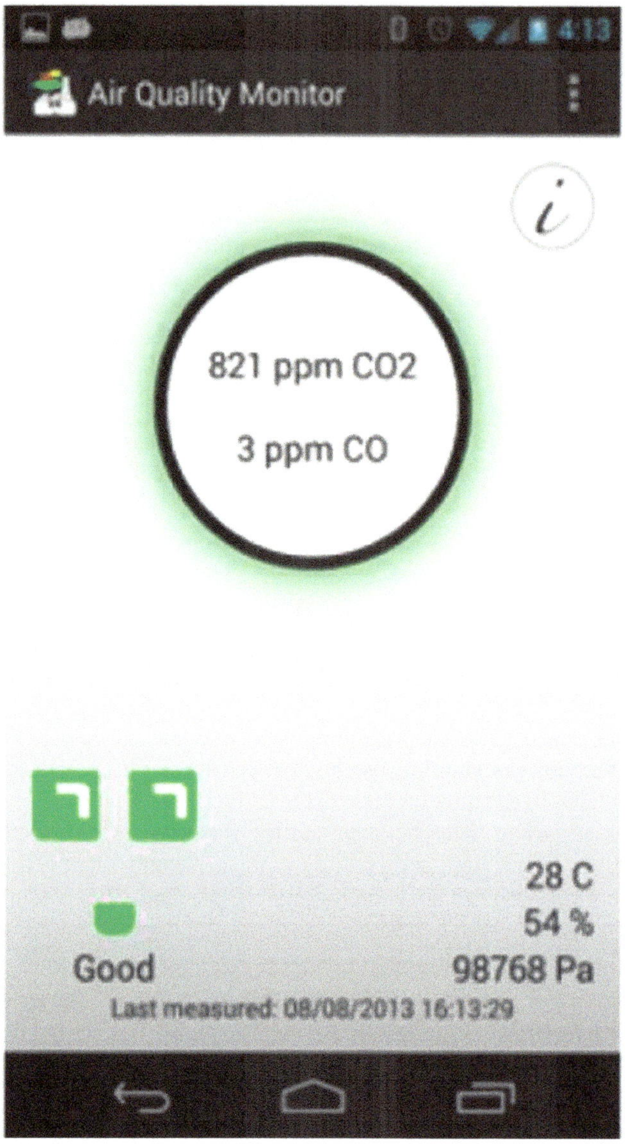

Figure 11. App on smartphone for providing CO or CO_2 levels from gas sensors.

Furthermore, monitoring devices can become simpler and cheaper as the computation is pushed to the cloud. This enables users to buy off-the-shelf devices and access customized monitoring applications via cloud-based services.

CONCLUSIONS

In this paper the historical development of semiconductor gas sensors accomplished during the past five decades has been briefly reviewed. Since early 1970s, after the pioneering work of Seyama and Taguchi, a number of research teams in both academia and industry have been engaged to investigate gas sensing mechanism and propose gas sensor devices featuring a wide offering of functionalities. With the advent of the latest technological breakthroughs, recent progress in gas sensor devices has been stimulated by inter-disciplinary perspectives at the intersection of materials science, microelectronics, nanotechnology and signal processing/pattern recognition in an attempt to overcome limitations of early systems. The effect of microstructure, size feature of the material, the advantages of novel nano-structured materials, have been widely pursued to improve the sensors' performance as well as to reduce their power consumption. With time, sensors migrated to process control market, then industrial controls, automotive and medical markets. Further, as embedded system technologies continues to growth at very fast rates offering great functionality at lower costs, efforts in gas sensing area seek for system solutions based on miniaturized smart sensors. Taking this into account, new generation of semiconductor gas sensors are now appearing in cloth fabrics, smart phones, media tablets, *etc.*, and in the near future will permeate the lives of consumers in ways unimaginable 50 years ago.

REFERENCES

1. Seiyama, T.; Kato, A. A new detector for gaseous components using semiconductor thin film. *Anal. Chem.* 1962, *34*, 1502–1503.
2. Brattain, W.H.; Bardeen, J. Surface properties of germanium. *Bell Syst. Tech. J.* 1952, *32*, 1–41.
3. Heiland, G. Zum einfluss von wasserstoff auf die elektrische leitfähigkeit von ZnO-kristallen. *Z. Phys.* 1954, *138*, 459–464.
4. Shaver, P.J. Activated tungsten oxide gas detectors. *Appl. Phys. Lett.* 1967, *11*, 255–257.
5. Taguchi, N. Gas Detecting Devices. U.S. Patent 3,631,436, 28 December 1971.
6. Seiyama, T. Chemical sensors—Current status and future outlook. In *Chemical Sensor Technology*; Seiyama, T., Ed.; Elsevier: Amsterdam, The Netherlands, 1988; Volume 1.
7. Yamazoe, N. New approaches for improving semiconductor gas sensors. *Sens. Actuators B Chem.* 1991, *5*, 7–19.
8. Shimizu, Y.; Egashira, M. Basic aspects and challenges of semiconductor gas sensors. *MRS Bull.* 1999, *24*, 18–24.
9. Yamazoe, N. Toward innovations of gas sensor technology. *Sens. Actuators B Chem.* 2005, *108*, 2–14.
10. Williams, D.E. *Solid State Gas Sensors*; Moseley, P.T., Tofield, B.C., Eds.; Adam Hilger: Bristol, UK, 1987.
11. Gerblinger, J.; Hardtl, K.H.; Meixner, H.; Aigner, R. High-Temperature Microsensors. In *Sensors Set: A Comprehensive Survey*; Gopel, W., Ed.; VCH-Verlag: Weinheim, Germany, 1995; p. 181.
12. Sberveglieri, G. *Gas Sensors*; Kluwer Academic Publishing: Berlin, Germany, 1992.
13. Capone, S.; Siciliano, P. *Encyclopedia of Nanoscience and Nanotechnology*; Nalwa, H.S., Ed.; American Scientific Publishers: Valencia, CA, USA, 2004; Volume 3, p. 769.
14. Gurlo, A.; Bârsan, N.; Weimar, U. Gas sensors based on semiconductiong metal oxides. In *Metal Oxides: Chemistry and Applications*; Fierro, J.L.G., Ed.; CRC Press: Boca Raton, FL, USA, 2006; p. 683.
15. Kim, H.-J.; Lee, J.-H. Highly sensitive and selective gas sensors using p-type oxide semiconductors: Overview. *Sens. Actuators B Chem.* 2014, *192*, 607–627.
16. Yamazoe, N.; Shimanoe, K. Receptor function and response of semiconductor gas sensor. *J. Sens.* 2009, *2009*, 875704.
17. Aleixandre, M.; Gerboles, M. Review of small commercial sensors for indicative monitoring of ambient gas. *Chem. Eng. Trans.* 2012, *30*, 169–174.
18. Huang, X.-J.; Choi, Y.-K. Chemical sensors based on nanostructured materials. *Sens. Actuators B Chem.* 2007, *122*, 659–671.
19. Jimenez-Cadena, G.; Riu, J.; Xavier Rius, F. Gas sensors based on nanostructured materials. *Analyst* 2007, *132*, 1083–1099.
20. Ogawa, H.; Nishikawa, M.; Abe, A. Hall measurement studies and an electrical conduction model of tin oxide ultrafine particle films. *J. Appl. Phys.* 1982, *53*, 4448–4455.
21. Korotcenkov, G. Gas response control through structural and chemical modification of metal oxide films: State of the art and approaches. *Sens. Actuators B Chem.* 2005, *107*, 209–232.
22. Mahmood, M.R.; Soga, T.; Mamat, M.H.; Khusaimi, Z.; Nor, A.M. A Review on zinc oxide nanostructures: Doping and gas sensing. *Adv. Mater. Res.* 2013, *667*, 329–332.
23. Tricoli, A.; Righettoni, M.; Pratsinis, S.E. Minimal cross-sensitivity to humidity during ethanol detection by SnO_2-TiO_2solid solutions. *Nanotechnology* 2009, *20*, 315502.
24. Ramamoorthy, R.; Dutta, P.K.; Akbar, S.A. Oxygen sensors: Materials, methods, designs and applications. *J. Mater. Sci.*2003, *38*, 4271–4282.
25. Malagù, C.; Guidi, V.; Carotta, M.C.; Martinelli, G. Unpinning of Fermi level in nanocrystalline semiconductors. *Appl. Phys. Lett.* 2004, *84*, 4158–4160.
26. Neri, G. *Encyclopedia of Sensors*; Grimes, C.A., Dickey, E.C., Pishko, M.V., Eds.; American Scientific Publishers: Valencia, CA, USA, 2006; Volume 6, p. 1.
27. Cabot, A.; Arbiol, J.; Morante, J.R.; Weimar, U.; Barsan, N.; Gopel, W. Analysis of the noble metal catalytic additives introduced by impregnation of as obtained SnO_2 sol-gel nanocrystals for gas sensors. *Sens. Actuators B Chem.* 2000, *70*, 87–100.
28. Baraton, M.-I.; Merhari, L. Influence of the particle size on the surface reactivity and gas sensing properties of SnO_2nanopowders. *Mater. Trans.* 2001, *42*, 1616–1622.
29. Xu, C.; Tamaki, J.; Miura, N.; Yamazoe, N. Grain size effects on gas sensitivity of porous SnO_2-based elements. *Sens. Actuators B Chem.* 1991, *3*, 147–155.
30. Dolbec, R.; el Khakani, M.A.; Serventi, A.M.; Saint-Jacques, R.G. Influence of the nanostructural characteristics on the gas sensing properties of pulsed

laser deposited tin oxide thin films. *Sens. Actuators B Chem.* 2003, *93*, 566–571.

31. Gurlo, A.; Ivanovskaya, M.; Barsan, N.; Schweizer-Berberich, M.; Weimar, U.; Gopel, W.; Dieguez, A. Grain size control in nanocrystalline In_2O_3 semiconductor gas sensors. *Sens. Actuators B Chem.* 1997, *44*, 327–333.
32. Ansari, S.G.; Boroojerdian, P.; Sainkar, S.R.; Karekar, R.N.; Aiyer, R.C.; Kulkarni, S.K. Grain size effects on H_2 gas sensitivity of thick film resistor using SnO_2 nanoparticles. *Thin Solid Films* 1997, *295*, 271–276.
33. Dieguez, A.; Romano-Rodriguez, A.; Morante, J.R.; Weimar, U.; Schweizer-Berberich, M.; Gopel, W. Morphological analysis of nanocrystalline SnO_2 for gas sensor applications. *Sens. Actuators B Chem.* 1996, *31*, 1–8.
34. Korotcenkov, G. The role of morphology and crystallographic structure of metal oxides in response of conductometric-type gas sensors. *Mater. Sci. Eng. R* 2008, *61*, 1–39.
35. Rothschild, A.; Komem, Y. Metal oxide gas sensors with nanosized grains. *J. Electroceram.* 2004, *13*, 697–701.
36. Nylander, C.; Armgarth, M.; Lundstrom, L. An ammonia detector based on a conducting polymer. *Anal. Chem. Symp. Ser.* 1983, *17*, 203–207.
37. Jiang, T.; Wang, Z.; Li, Z.; Wang, W.; Xu, X.; Liu, X.; Wang, J.; Wang, C. Synergic effect within n-type inorganic–p-type organic nano-hybrids in gas sensors. *J. Mater. Chem. C* 2013, *1*, 3017–3025.
38. Kong, J.; Franklin, N.R.; Zhou, C.; Chapline, M.G.; Peng, S.; Cho, K.; Dai, H. Nanotube molecular wires as chemical sensors. *Science* 2000, *287*, 622–625.
39. Peng, S.; Cho, K. Ab initio study of doped carbon nanotube sensors. *Nano Lett.* 2003, *3*, 513–517.
40. Sin, M.L.Y.; Chow, G.C.T.; Wong, G.M.K.; Li, W.J.; Leong, P.H.W.; Wong, K.W. Ultralow-power alcohol vapor sensors using chemically functionalized multiwalled carbon nanotubes. *IEEE Trans. Nanotechnol.* 2007, *6*, 571–577.
41. Sun, Y.P.; Fu, K.F.; Lin, Y.; Huang, W.J. Functionalized carbon nanotubes: Properties and applications. *Acc. Chem. Res.*2002, *35*, 1096–1104.
42. Lu, G.; Ocola, L.E.; Chen, J. Room-temperature gas sensing based on electron transfer between discrete tin oxide nanocrystals and multiwalled carbon nanotubes. *Adv. Mater.* 2009, *21*, 1–5.
43. Sun, Y.-F.; Liu, S.-B.; Meng, F.-L.; Liu, J.-Y.; Jin, Z.; Kong, L.-T.; Liu, J.-H. Metal oxide nanostructures and their gas sensing properties: A review. *Sensors* 2012, *12*, 2610–2631.
44. Willinger, M.G.; Neri, G.; Rauwel, E.; Bonavita, A.; Micali, G.; Pinna, N. Vanadium oxide sensing layer grown on carbon nanotubes by a new atomic layer deposition process. *Nano Lett.* 2008, *8*, 4201–4204.
45. Marichy, C.; Pinna, N. Carbon-nanostructures coated/decorated by atomic layer deposition: Growth and applications.*Coord. Chem. Rev.* 2013, *257*, 3232–3253.
46. Lu, J.G.; Chang, P.; Fan, Z. Quasi-one-dimensional metal oxide materials—Synthesis, properties and applications. *Mater. Sci. Eng. R* 2006, *52*, 49–91.
47. Pan, Z.W.; Dai, Z.R.; Wang, Z.L. Nanobelts of semiconducting oxides. *Science* 2001, *291*, 1947–1949.
48. Li, C.; Zhang, D.; Liu, X.; Han, S.; Tang, T.; Han, J.; Zhou, C. In_2O_3 nanowires as chemical sensors. *Appl. Phys. Lett.*2003, *82*, 1613–1615.
49. Cao, G. *Nanostructures and Nanomaterials*; IC Press: London, UK, 2004.
50. Comini, E. Metal oxide nano-crystals for gas sensing. *Anal. Chim. Acta* 2006, *568*, 28–40.
51. Comini, E.; Faglia, G.; Sberveglieri, G. Stable and highly sensitive gas sensors based on semiconducting oxide nanobelts. *Appl. Phys. Lett.* 2002, *81*, 1869–1871.
52. Law, M.; Kind, H.; Messer, B.; Kim, F.; Yang, P.D. Photochemical sensing of NO_2 with SnO_2 nanoribbon nanosensors at room temperature. *Angew. Chem. Int. Ed.* 2002, *41*, 2405–2408.
53. Choi, K.J.; Jang, H.W. One-dimensional oxide nanostructures as gas-sensing materials: Review and issues. *Sensors*2010, *10*, 4083–4099.
54. Lee, J.-H. Gas sensors using hierarchical and hollow oxide nanostructures: Overview. *Sens. Actuators B Chem.* 2009,*140*, 319–336.
55. Schedin, F.; Geim, A.K.; Morozov, S.V.; Hill, E.W.; Blake, P.; Katsnelson, M.I.; Novoselov, K.S. Detection of individual gas molecules adsorbed on graphene. *Nat. Mater.* 2007, *6*, 652–655.
56. Wolkenstein, F.F. Effect of ionizing radiation on the adsorptive and catalytic properties of semiconductors. *Discuss. Faraday Soc.* 1961, *31*, 209–218.
57. Morrison, R.S. Semiconductor gas sensors. *Sens. Actuators* 1982, *2*, 329–341.
58. Yamazoe, N.; Kurokawa, Y.; Seiyama, T. Effects of additives on semiconductor gas sensors. *Sens. Actuators B Chem.*1983, *4*, 283–289.
59. Gopel, W.; Schierbaum, K. SnO_2 sensors: Current status and future prospects. *Sens. Actuators B Chem.* 1995, *26/27*, 1–12.
60. Azad, A.M.; Akbar, S.A.; Mhaisalkar, S.G.; Birkefeld, L.D.; Goto, K.S. Solid-State Gas Sensors: A Review. *J. Electrochem. Soc.* 1992, *139*, 3690–3704.
61. Barsan, N.; Schweizer-Berberich, M.; Göpel, W. Fundamental and practical aspects in the design of nanoscaled SnO_2gas sensors: A status report. *Fresenius. J. Anal. Chem.* 1999, *365*, 287–304.
62. Leblanc, E.; Perier-Camby, L.; Thomas, G.; Gibert, R.; Primet, M.; Gelin, P. NO_x adsorption onto dehydroxylated or hydroxylated tin dioxide surface. Application to SnO_2-based sensors. *Sens. Actuators B Chem.* 2000, *62*, 67–72.
63. Li, Y.; Wlodarski, W.; Galatsis, K.; Moslih, S.H.; Cole, J.; Russo, S.; Rockelmann, N. Gas sensing properties of p-type semiconducting Cr-doped TiO_2 thin films. *Sens. Actuators B Chem.* 2002, *83*, 160–163.

64. Yamazoe, N.; Shimanoe, K. Theory of power laws for semiconductor gas sensors. *Sens. Actuators B Chem.* 2008, *128*, 566–573.

65. Rothschild, A.; Komem, Y. The effect of grain size on the sensitivity of nanocrystalline metal-oxide gas sensors. *J. Appl. Phys.* 2004, *95*, 6374–6380.

66. Chen, X.; Wong, C.K.Y.; Yuan, C.A.; Zhang, G. Nanowire-Based Gas Sensors. *Sens. Actuators B Chem.* 2013, *177*, 178–195.

67. Wei, B.-Y.; Hsu, M.-C.; Su, P.-G.; Lin, H.-M.; Wu, R.-J.; Lai, H.-J. A novel SnO_2 gas sensor doped with carbon nanotubes operating at room temperature. *Sens. Actuators B Chem.* 2004, *101*, 81–89.

68. Marichy, C.; Russo, P.A.; Latino, M.; Tessonnier, J.-P.; Willinger, M.-G.; Donato, N.; Neri, G.; Pinna, N. Tin dioxide-carbon heterostructures applied to gas sensing: Structure-dependent properties and general sensing mechanism. *J. Phys. Chem. C* 2013, *117*, 19729–19739.

69. Wilson, A.D.; Baietto, M. Applications and advances in electronic-nose technologies. *Sensors* 2009, *9*, 5099–5148.

70. Lee, D.-D.; Lee, A.-S. Environmental gas sensors. *IEEE Sens. J.* 2001, *1*, 214–224.

71. Fine, G.F.; Cavanagh, L.M.; Afonja, A.; Binions, R. Metal oxide semi-conductor gas sensors in environmental monitoring. *Sensors* 2010, *10*, 5469–5502.

72. Marsal, A.; Dezanneau, G.; Cornet, A.; Morante, J.R. A new CO_2 gas sensing material. *Sens. Actuators B Chem.* 2003, *95*, 266–270.

73. Pijolat, C.; Pupier, C.; Sauvan, M.; Tournier, G.; Lalauze, R. Gas detection for automotive pollution control. *Sens. Actuators B Chem.* 1999, *59*, 195–202.

74. Cederquist, A.; Gibbons, E.; Meitzler, A. Characterization of Zirconia and Titania Engine Exhaust Gas Sensors for Air/Fuel Feedback Control Systems. *SAE Tech. Paper* 1976.

75. Holt, C.T.; Azad, A.-M.; Swartz, S.L.; Rao, R.R.; Dutta, P.K. Carbon monoxide sensor for PEM fuel cell systems. *Sens. Actuators B Chem.* 2002, *87*, 414–420.

76. D'Amico, A.; di Natale, C.; Paolesse, R.; Macagnano, A.; Martinelli, E.; Pennazza, G.; Santonico, M.; Bernabei, M.; Roscioni, C.; Galluccio, G.; *et al.* Olfactory systems for medical applications. *Sens. Actuators B Chem.* 2008, *130*, 458–465.

77. *Disease Markers in Exhaled Breath*; Marczin, N., Kharitonov, S.A., Yacoub, M.H., Barnes, P.J., Eds.; Marcel Dekker: New York, NY, USA, 2002.

78. Righettoni, M.; Amann, A.; Pratsinis, S.E. Breath analysis by nanostructured metal oxides as chemoresistive gas sensors. *Mater. Today* 2014.

79. Casalinuovo, I.; Pierro, D. Application of electronic noses for disease diagnosis and food spoilage detection. *Sensors*2006, *6*, 1428–1439.

80. Gonzalez-Jimenez, J.; Monroy, J.G.; Blanco, J.L. The multi-chamber electronic nose—An improved olfaction sensor for mobile robotics. *Sensors* 2011, *11*, 6145–6164.

81. Neri, G. Better sensors through chemistry: Some selected examples. *Lect. Notes Electr. Eng.* 2011, *91*, 19–30.

82. Carbone, L.; Cozzoli, P.D. Colloidal heterostructured nanocrystals: Synthesis and growth mechanisms. *Nano Today*2010, *5*, 449–493.

83. Ataman, C.; Kinkeldei, T.; Mattana, G.; Vásquez Quintero, A.; Molina-Lopez, F.; Courbat, J.; Cherenack, K.; Briand, D.; Tröster, G.; de Rooij, N.F. A robust platform for textile integrated gas sensors. *Sens. Actuators B Chem.* 2013, *177*, 1053–1061.

84. Gardner, J.W.; Guha, P.K.; Udrea, F.; Covington, J.A. CMOS interfacing for integrated gas sensors: A review. *IEEE Sens. J.* 2010, *10*, 1833–1848.

85. MEMS Technical Data. Available online: http://www.memsgassensor.com (accessed on 28 February 2014).

86. Oletic, D.; Bilas, V. Empowering smartphone users with sensor node for air quality measurement. *J. Phys. Conf. Ser.*2013, *450*, 012028.

87. Sensorcon. Available online: http://sensorcon.com (accessed on 10 December 2014).

CHAPTER 5 SUSTAINABLE HIGH QUALITY RECYCLING OF AGGREGATES FROM WASTE-TO-ENERGY

Philip Van den Heede, Niels Ringoot, Arno Beirnaert, Andres Van Brecht, Erwin Van den Brande, Geert De Schutter and Nele De Belie

Keywords: Municipal solid waste incineration, bottom ash, concrete, aggregate replacement, prefabricated Lego brick

INTRODUCTION

Yearly, 3,331,000 tons of municipal solid waste are generated in Flanders. Around 880,000 tons are incinerated to gain energy and reduce the waste volume with 90% (Waste-to-Energy) [1,2]. The incineration results in bottom ash (BA), boiler ash and air pollution control residues. BA is mostly disposed of as landfill. The high waste volume, cost of landfill and lack of disposal sites force the industry to recycle the BA as secondary raw material [3]. This would also be more in agreement with European Directive 2008/98/EC [4], as it saves valuable natural resources and gives waste a financial value [5]. To obtain an end-of-waste status *cf.* ED 2008/98/EC [4], several criteria need to be fulfilled. The material should serve a purpose. The technical requirements and legislation of this purpose should be met. There needs to be a market for it and there should be no negative environmental effects [4]. In this context, companies involved in municipal solid waste incineration (MSWI) intend to recover the BA as aggregates for concrete. The successful production of so-called concrete Lego bricks is seen as the minimal aim. These are prefabricated unreinforced blocks which are used to build modular partition walls at industrial and rural sites. For several years now, the waste management company Indaver nv in Belgium has been trying to tune the properties of BA at its incineration plant to meet this goal. In the past, this already resulted in the publication of papers on the proper characterization and treatment of the Belgian BA (e.g., Vandecasteele *et al.* [6], Van Gerven *et al.*[7]).

According to literature, BA from Belgium [8,9], the Netherlands [10], France [11], Germany [5], Italy [3,12], Spain [2] and South-Korea [13] show quite some similarities in composition. They usually consist of bottle glass (around 15%, composed of Na_2O, CaO, $CaCO_3$ and SiO_2 [2]), as well as metals (ferrous and non-ferrous, such as Al, Cu and Zn), ceramics (brick, plaster and mortar) and organic residues (bone fragments, charcoal, plant fibers and polymers). The latter elements have only a small contribution to the total mass (each with an amount of about 2%) [5,10]. Evidently, the variety in constituting materials implies a heterogeneous chemical composition of the BA: 30%–70% silica, 10%–15% aluminum oxide, 20% sodium oxide and 10%–15% calcium oxide. A smaller yet still important fraction of (heavy) metals is also present [5,10]. BA is lighter than gravel (2210 kg/m^3 *vs.* 2470 kg/m^3). The water absorption is 4–10 times the value of gravel (4%–10%). Both characteristics are due to a high porosity [3,11]. BA is highly angular and has a weaker abrasion strength

Citation: Philip Van den Heede , Niels Ringoot, Arno Beirnaert, Andres Van Brecht , Erwin Van den Brande , Geert De Schutter, and Nele De Belie, Sustainable High Quality Recycling of Aggregates from Waste-to-Energy, doi:10.3390/ma9010009

than limestone [14]. BA concrete is usually characterized by a lower compressive strength, a lower Young's modulus, a lower workability and a higher porosity than concrete with limestone as aggregate. Nevertheless, even without modifications and with full replacement of the aggregates by BA, a concrete strength class of C20/25 can be assured [15].

In the fresh state, extensive aggregate replacement by BA may have a negative effect on the concrete workability. This is mainly due to the higher water absorption of the BA. However, when using rather high water-to-cement (W/C) ratios, workability may not be a real issue. According to Pera *et al.*, a W/C ratio of 0.60–0.65 is ideal to achieve both an acceptable workability and a sufficient mechanical performance [11]. Moreover, the use of a superplasticizer (SP) can easily improve the workability if needed [16].

In hardened state, most damage phenomena are induced by the expansive reaction of metallic Al in an alkaline environment, which increases the porosity and reduces the strength. This effect exists for all hydroxide-forming elements present in the BA [5,11]. Alkali-silica reaction (ASR) is considered not detrimental because of the higher porosity of the BA concrete. A literature survey on other expected damage phenomena (e.g., freeze-thaw, acid attack, *etc.*) provides little or no information on the performance of BA concrete.

Literature on the leaching behavior of concrete containing BA from MSWI is rather scarce. Since the material is immobilized within the cementitious matrix of the concrete, leaching values are normally lower than when the individual grains are in direct contact with the leaching solution. Regarding the leaching of (heavy) metals, Sorlini and Ginés found values that meet the national criteria (Italy for Sorlini and Spain for Ginés) [17,18].

In the past, tests were mainly performed in view of a partial replacement of traditional aggregates by BA [10,11,16,17] and they included variations in cement type [17], W/C ratio [8,12], workability and compaction [8,19]. The concrete quality can also be improved by optimizing the BA properties using opto-mechanical glass separation [11], reactive washing with NaOH (=Lye treatment) [11], vitrification [12] and sintering [17].

RESEARCH RELEVANCE AND METHODOLOGY

In this research, the focus was on obtaining a qualitative concrete with replacement of traditional crushed limestone aggregates by BA for the production of prefabricated concrete products. The main novel character of the research lies in the fact that full instead of partial aggregate replacement is aimed for. The main benefit of such a practice is rather evident. If we are able to incorporate more BA from MSWI in concrete, less of the material will get landfilled, its valorization potential will increase significantly and the concrete industry will have to rely less on natural resources to ensure its production capacity. In other words, there will be considerable advantages from both an economic and environmental point of view. However, these benefits only exist if the production of this BA concrete is practically feasible in the casting stage and if the hardened concrete is sufficiently strong and durable for its intended field of application. Therefore, the following research approach was adopted. After careful characterization of the physical and chemical properties of the BA, several test mixtures with varying W/C ratios were made to find an optimal balance between workability and strength of the concrete. The effects of pre-treating the BA in various ways were carefully evaluated. A dedicated study of the expected long-term behavior for one potentially suitable mixture was done and a final optimized concrete mix design was proposed. Prefabricated Lego bricks were made using both mix designs and were compared with a Lego brick made of traditional concrete.

MATERIALS AND METHODS

Concrete Constituents

The inert fraction of concrete traditionally consists of sand and aggregates. For the aggregates, it is often useful to use both a fine and a coarse fraction to achieve a better packing. In this research, the aim was a full replacement of the traditional fine and coarse aggregates by BA processed at the MSWI plant of Indaver nv. In terms of phase composition, this consisted of glass remains, a stony fraction (ceramics, porcelain, bricks, *etc.*), amorphous slags and some ferrous (e.g., steel) and non-ferrous (e.g., Al, Cu, Pb, Zn) metals that cannot be removed in the treatment installation. The glass remains (bottle glass, grains of glass with devitrification products) normally represent the largest fraction. The proportioning with the other fractions is somewhat seasonally bound.

Different types of fine and coarse BA aggregates were considered, *i.e.*, sieved and aged BA 0/6 and 2/6, crushed BA 0/6, sieved and washed BA 0/6 and 6/20, and crushed BA 6/20. The crushed and sieved and washed fractions originated from the same 6/50 fraction which was obtained after an initial processing stage at Indaver nv. The other fractions were the result of a natural aging process of a 2/6 batch. More details on how the 6/50 fraction and the aged 2/6 batch were obtained can be found in Vandecasteele *et al.* [6]. The crushing, sieving, washing and aging techniques that were applied afterwards to obtain the 0/6, 2/6 and 6/20 fractions were similar to the ones mentioned in that previous research. Crushing was done by means of a commercial mobile cone crusher that allows for setting the required maximum aggregate size. A mobile installation was also used for sieving the material to the desired sieve aperture. The material to be sieved was usually wet due to its outdoor unsheltered storage. No additional water was added during sieving. Washing operations were done in most cases with process water that resulted from drying the filter cake at the Indaver incineration plant. Logically, this process water contains fines that adhere somewhat to the BA. The aging process consisted of piling the BA on 5–10 m high heaps in open air on a paved floor at the incineration plant with the collection of percolating rainwater, exposing them to wind and rain.

The BA-based aggregates should replace the traditional crushed limestone 2/6 and 6/20 which are commonly used for concrete manufacturing in Belgium. The remaining portion of the inert fraction, *i.e.*, the sand, was a natural river sand 0/4.

Two different cement types were used for concrete manufacturing, either a traditional Portland cement CEM I 52.5 N MF or a low alkali (LA) high sulfate resistant CEM I 52.5 N HES LA HSR. The total cement content for each concrete composition amounted to 350 kg/m^3.

The mixing water consisted of tap water at a temperature of 20 °C. The amount of water used for the concrete varied with the applied W/C ratio: 0.65, 0.60 or 0.55. When lowering the water content substantially, the use of a SP was often imperative. Initially, polycarboxylic ether-based MasterGlenium 51 con. 35% (dry matter mass percentage: 35%, density at 20 °C: 1100 kg/m^3) was used for this purpose. Later on, a switch was made to polycarboxylic ether-based MasterGlenium ACE 30 con. 30% SPL (dry matter mass percentage: 35%, density at 20 °C: 1070 kg/m^3) to reduce the delay in setting time.

Physical and Chemical Characterization of the Aggregates

Water Absorption over 24 h and Mass Density

The water absorption over 24 h (WA_{24}) of all studied aggregates was measured in accordance with NBN EN 1097-6 using a calibrated pycnometer. The apparent mass density ρ_a (kg/m^3) and the relative mass density after oven drying ρ_{rd} (kg/m^3) were determined in accordance with the same standard.

Water Absorption as a Function of Time

To evaluate the water absorption as a function of time, a hydrostatic weighing technique similar to the one proposed by Garcia-Gonzalez *et al.* and Tegguer was applied [20,21]. Per aggregate type, a sample of at least 1000 g was inserted in a cylindrical strainer. This strainer was hung from a balance and submerged in a basin filled with demineralized water. The weight increase was monitored over a time period of 10 days.

Particle Size Distribution

The particle size distributions of the aggregates and sand were determined *cf.* NBN EN 933-1. The sieving column consisted of standardized sieves with following apertures: 63, 45, 40, 31.5, 22.4, 20, 16, 14, 12.5, 10, 8, 6.3, 4, 2, 1, 0.5, 0.25, 0.125 and 0.063 mm. The cumulative percentage of material passing through each sieve was calculated to plot the particle size distributions.

Aggregate Crushing Value

The strength of an aggregate can be characterized by its crushing value. In this research, this parameter—also known as the static compressive strength—was determined in accordance with Belgian standard NBN B11-205, for both the BA and the limestone aggregates. A cylindrical metal mold with an inner diameter of 150 mm was filled with a dried aggregate sample of a certain particle size range. After adding the sample, the mold was fitted with a no-friction plunger which was then loaded to 400 kN in 4 min. Before unloading again, this maximum load onto the sample was maintained for 2 min. Finally, the crushed fraction of the aggregate <2 mm was separated from the sample by means of a sieving operation. The mass ratio of the sample without the crushed fraction over the initial sample before loading counts as the static compressive strength of the aggregate. Normally, this test is to be conducted on coarse aggregate fractions only (minimum aggregate diameter: 6 mm). Nevertheless, the static compressive strengths of the fine BA and limestone aggregates were determined in the same manner. Per type of sample, the test was repeated three times.

Chemical Composition

The chemical properties of the BA were determined using Inductively Coupled Plasma (ICP), Ion Chromatography (IC) and Flow Injection Mercury System (FIMS) techniques available at the laboratories of Indaver nv.

Concrete Mix Design and Production

Once the total cement content (=350 kg/m^3) and the W/C ratio (0.65, 0.60 or 0.55) were set for the concrete compositions under investigation, the required amounts of sand, fine and coarse aggregates were determined. This was done by approximating the optimal particle size distribution curve of Fuller using the least-of-squares method. This approach normally results in a minimal spacing between the different particles of the inert fraction.

In the initial research stage only the fine aggregates were replaced by BA. In the next stage, the same was done for the coarse aggregates. As such, the individual effects of each BA could be studied unambiguously. In the final stage, both the fine and coarse limestone aggregates were replaced.

Concrete was produced in accordance with NBN EN 12390-2. First, the fine and coarse aggregates, the sand and the cement, were mixed for 1 min. Then, the water was added whereupon mixing continued for 2 min. After this step, the workability of the concrete was evaluated by means of the slump and flow test (NBN EN 12350-2, NBN EN 12350-5) and the required amount of SP was determined. This amount was added while mixing for another 2 min. The SP dosing step was repeated until the proper slump and flow class were obtained. Finally, the test samples were cast and optimally cured in a climate chamber at 20 ± 2 °C and 95% ± 5% relative humidity (RH).

Strength Performance

The compressive strength was determined in accordance with NBN EN 12390-2. This was done after seven days, 28 days and 56 days of optimal curing. Per testing age three cubic specimens (n = 3) with a 150 mm side were subjected to the test. The characteristic value of the compressive strength f_{ck} at 28 days was obtained in correspondence with EN 1990 *cf.* Equation (1).

$$f_{ck} = \overline{x} - k_n \times s\,(N/mm) \tag{1}$$

where x is the mean value of the compressive strength of n samples (N/mm^2); k_n is a factor equal to 3.37 for n = 3 and a variation coefficient V_X unknown from prior knowledge; and s is the standard deviation on the individual values of the compressive strength of n samples. With the characteristic compressive strength known, the proper compressive strength class as specified in NBN EN 206-1 can be assigned.

Susceptibility to Expansion

The susceptibility to expansion was evaluated by monitoring the height of the cubic samples intended for the compressive strength tests as a function of time (after 1, 7, 28 and 56 days) and relative to the their original height just after casting (=150 mm). Note that this is only a rudimentary assessment approach to get a first idea on the expansive behavior. The change in length of only one cube side reflects the total expansion only to a certain extent since the initial early age expansion in other directions is hindered by the cubic mold. Moreover, the duration of dihydrogen gas generation after reaction with the metallic Al and the setting time of the concrete also govern the amount of expansion as well. Further investigation on these influencing parameters is certainly still necessary in the future.

Porosity

The porosity was determined using a vacuum saturation technique followed by hydrostatic weighing *cf.* NBN B 05-201. Since the amount of water absorbed during the vacuum saturation is mainly limited to the pores in contact with the outer sample surface, the resulting porosity only reflects the open/permeable porosity. Per concrete mixture, the test was performed on nine cylindrical samples (diameter: 100 mm, height: 50 mm) taken from cubes (side: 150 mm). After 28 days of optimal curing, the samples were weighed and dried at a temperature of 40 ± 5 °C until constant mass (Δm after 24 h < 0.1%). Next, the samples were placed in a normalized vacuum tank. A vacuum with a residual pressure of 2.7 kPa was applied for 2.5 h. While keeping this vacuum condition, water was introduced at a rate of 50 mm/h until complete immersion of the samples. Subsequently, the air pressure was restored and the samples were kept under water for 24 h. Afterwards, they were weighed hydrostatically as well as above water. After the weighing, they were dried at 105 ± 5 °C until constant mass (Δm after 24 h < 0.1%) whereupon the whole procedure was repeated. Then, the open/permeable porosity φ was calculated for the two pre-drying temperatures (Equation (2)).

$$\varphi = \frac{m_s - m_d}{m_s - m_l} \times 100\,(\%) \tag{2}$$

where m_d is the oven dry mass after pre-drying at 40 °C or 105 °C (g); m_s is the water saturated mass after vacuum saturation (g); and m_l is the mass under water after vacuum saturation (g).

Long-Term Behavior

Susceptibility to Leaching

After 28 days of optimal curing, six cubes (side: 100 mm) were conserved in a closed tank filled with demineralized water. Conform CMA/2/II/A, the volume of the reagents should be two to five times the sample volume. Conform NEN 7345, this is four times the sample volume. In this research, the latter volume ratio was chosen. A water sample was taken after 30 days of immersion and examined using ICP, IC and FIMS.

Resistance to ASR

The susceptibility to ASR was evaluated using the modified Oberholster test *cf.* the Walloon technical guideline STM D424 ST6. This method was chosen because, in contrast with the original Oberholster test or South-African NBRI (National Building Research Institute) accelerated test method, it allows for a test directly on the concrete mixtures under investigation instead of on mortar bars containing the aggregates to be studied after an initial processing (crushing and sieving). Moreover, the modified Oberholster test holds the advantage of being a rather fast assessment technique (exposure period: 20 days) in comparison with some of the other methods designed for concrete (exposure periods of up to 12 months *cf.* RILEM TC 106-AAR [22]). Per mixture, six cylinders (diameter: 50 mm, height: 150 mm) taken from cubes (side: 150 mm) were tested. After 28 days of optimal curing the samples were immersed for 24 h in water at 80 °C. Afterwards, the samples were stored in a 40 g NaOH per L water solution at 80 °C. The expansion was continuously monitored on dial gauges in contact with the upper surface of the cylinders. Normally, an expansion exceeding 0.1% implies a potential ASR sensitivity.

Resistance to Acid Attack

Six cylinders (diameter: 113 mm, height: 150 mm) were drilled and cut from six cubes (side: 150 mm). After 28 days, the cylinders were exposed to 15 cycles involving intermittent exposure to acid. One cycle started by keeping the samples 24 h in a dry environment (minimal 60% ± 5% RH) and continued with 24 h of immersion in an acid solution (30 g acetic acid and 30 g lactic acid per liter), all taking place at a temperature of 20 ± 2 °C and 60% ± 5% RH. After each drying period, the mass of the samples was determined. After the last cycle, the mass was determined once again. The damage was evaluated based on the observed mass decrease and visual appearance.

Resistance to Freeze-Thaw Attack

Four cubes (side: 150 mm), optimally cured for 23 days, were subjected to a freeze-thaw test in compliance with NBN B15-231, which in turn refers to NBN B 05-203. The samples were submerged in demineralized water at a temperature of 20 ± 2 °C until constant mass (Δm 24 h < 0.05%) *cf.* NBN B 15-215. After saturation was observed, the samples were placed in a laboratory freezing chamber and were subjected to 14 freeze-thaw cycles which consisted of: (i) cooling down till 0 °C in 1–2 h; (ii) progressive freezing from 0 °C to −15 °C in 5 h (at a mean rate of 3 ± 0.5 °C/h and always in the range of 2–4 °C/h); (iii) maintaining a constant freezing temperature of −15 ± 2 °C for 10–11 h; (iv) defrosting in water; (v) keeping the samples in water until the end of the cycle at 24 h.

Two non-destructive test methods were used to evaluate the freeze-thaw resistance: a visual check-up and ultrasonic pulse velocity measurements using a commercial device. The visual damage assessment was done in accordance with NBN B 27-009/A1.

Microscopic Characterization on Thin Sections

To explain the porosity, a microscopic investigation on thin sections was necessary. These thin sections were prepared as follows: 45 × 30 × 15 mm^3 prisms were cut from the concrete and 45 × 30 mm^2 faces were glued onto glass slides with a thickness of 2.9 mm. Then, the samples were cut and polished until the height of the concrete specimens and the glass equaled 10.1 mm. In a next step, the samples were impregnated under vacuum with a fluorescent epoxy. After impregnation, the excess epoxy was ground away and an object glass was glued onto the polished surface. In a final step, the glass slides were cut off and the concrete samples on the object glasses were polished until thin sections with a 25 μm thickness were obtained. A cover glass was glued onto them for protection. All thin sections were examined with a Leica DM LP polarizing microscope. Images were taken with a Leica DFC295 camera in fluorescent light mode.

Optimization of the Bottom Ash

Pre-Wetting

To limit the high water absorption during concrete mixing and the resulting workability issues, the BA could be pre-wetted. In total, three pre-wetting techniques were examined. All of them were based on their expected water absorption as a function of time (Section 3.2.2). The first technique focused on pre-wetting the BA to a level corresponding with the deflection point in the absorption curve. This level is met as soon as the initial high water uptake rate ends. The time needed for this counts as the required immersion time span of the aggregates prior to the concrete mixing. After immersion, a self-made strainer bucket was used to drain the water. Note that this method was only applied for the fine BA. The second and third methods also relate to the water absorption behavior as a function of time. For these two methods, the expected contact time with water was the decisive criterion and not the change in water absorption rate. As the concrete mixing process and specimen casting takes about 6 min after adding the water, it makes sense to pre-wet the BA with the amount of water that is expected to be absorbed within the 6 min timeframe. This can be done in two ways. Either the aggregates can be pre-wetted in the mixer (this second method is commonly used when manufacturing concrete with lightweight aggregates), or the additional amount of water can be added afterwards as a surplus to the mixing water (the third method). Both the fine and coarse BA were treated like this.

Reactive Washing

The applied reactive washing procedure was based upon literature findings [23] and existing know-how at Indaver nv. To passivate Al inclusions, the BA was washed with 1 M NaOH. The BA was completely immersed in this solution for two weeks. The metallic Al reacts with the NaOH solution and forms non-active Al and H_2. The mixture was stirred from time to time to ensure an optimal interaction between the BA and the solution and to guarantee a proper release of H_2.

To evaluate the effectiveness of this reactive washing, the amount of reactive Al before and after the treatment was determined. Therefore, the BA samples were subjected to a test described in CUR recommendation 116 [24]. Per sample, the BA was combined with the NaOH solution in a closed system flask. The H_2 being released was captured in a second flask initially full of water. Since a portion of the water is displaced by the H_2, the amount of metallic Al can be calculated from Equation (3) after measuring the initial and final water weight.

$$v_{H_2} = \frac{3}{2}\frac{m_{Al}}{M_{Al}} v_0 \text{ (L)} \tag{3}$$

where v_{H2} is the volume of entrapped H_2 (L); M_{Al} is the molar mass of Al (27 g/mol); v_0 is the molar volume (22.4 L/mol); and m_{Al} is the mass of reactive Al (g). Before applying the BA after the reactive washing in concrete, they were first rinsed with water until the pH had decreased sufficiently. For the fine and coarse BA fractions, the pH decreased from 12.8 to 9.0–10.0 and from 12.8 to 8.0, respectively.

Manufacturing of the Lego Bricks

Three 1600 × 800 × 400 mm^3 Lego bricks were cast: one made of traditional concrete, one made of the initial BA concrete with full replacement of both the fine and coarse aggregate fractions and one made of the optimized BA concrete, also with full aggregate replacement. The mold for the Lego brick was provided by CB Recycling BV. The concrete was manufactured in a concrete mixer with a total volume capacity of only 200 L. As a consequence, the Lego bricks had to be cast in three subsequent stages. Each layer of concrete was carefully compacted using a vibration needle.

RESULTS AND DISCUSSION

Physical and Chemical Characterization of the Bottom Ash

Particle Size Distributions

The particle size distributions of crushed limestone 2/6 and BA 0/6 are within the same range (Figure 1). Only the fraction <1 mm is more present in BA 0/6, whereas the limestone has a steeper grading curve, indicating a pure and narrow range of grain diameters. The crushed fraction has more fines than the sieved and washed fraction. This is probably due to the dust generation during crushing.

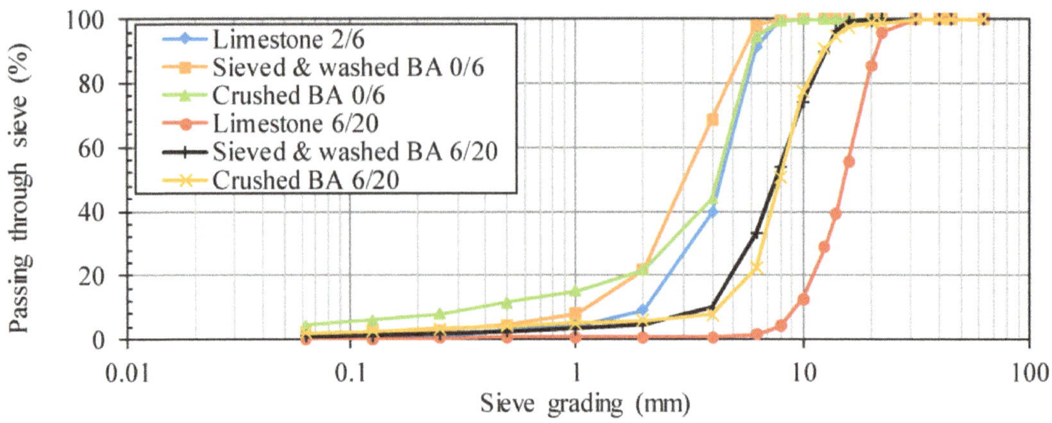

Figure 1: Influence of aggregate type (limestone *vs.* bottom ash (BA)), fraction (2/6 or 0/6 *vs.*6/20) and BA pre-treatment (sieving and washing *vs.* crushing) on the particle size distribution.

When looking at the particle size distributions of the coarse fractions, a clear shift to the left can be observed between limestone 6/20 and BA 6/20. In accordance with the proper definitions for minimum and maximum aggregate size *cf.* NBN EN 12620, this BA fraction rather meets the requirements of an aggregate 2/14 instead of those of an aggregate 6/20. Comparison between the crushed and sieved and washed BA 6/20 reveals only very limited differences in particle size distribution.

Note that there are several factors with an influence on the difference in particle size distributions, e.g., the grading and strength of the original untreated material that is subjected to the processing as well as the effectiveness of the equipment used for crushing, sieving and washing.

Aggregate Crushing Value

When making the comparison between the recorded aggregate crushing values or static compressive strengths for the sieved and washed and crushed BA and the limestone aggregates, it is immediately clear that the BA clearly shows a lower mechanical performance (Table 1).

Table 1: Influence of aggregate type (limestone *vs.* BA), fraction (2/6 *vs.* 6/10 and 6/20) and BA pre-treatment (sieving and washing *vs.* crushing) on the static compressive strength

Limestone 2/6	**Limestone 6/20**	
Fraction 2/6	Fraction 6/10	Fraction 10/20
63.8% ± 0.9%	75.1% ± 0.7%	82.7% ± 0.3%
Sieved & washed BA 0/6	**Sieved and washed BA 6/20**	
Fraction 2/6	Fraction 6/10	Fraction 10/20
42.2% ± 0.2%	55.5% ± 1.0%	63.3% ± 0.5%
Crushed BA 0/6	**Crushed BA 6/20**	
Fraction 2/6	Fraction 6/10	Fraction 10/20
50.3% ± 0.6%	58.8% ± 0.3%	65.8% ± 0.8%

This statement holds true for both the fine and coarse aggregates. The difference ranges between 13.6% and 21.6%, depending on the aggregate type and fraction. It is worth mentioning that the crushed BA fractions are all characterized by a significantly higher static compressive strength than the sieved and washed fractions. It is likely that the crushing operations performed in the processing installations of Indaver nv reduced the weaker particles in size, to values lower than the particle sizes considered during the static compressive strength tests. Thus, only the stronger aggregates remain. In the case of the sieved and washed BA there was no elimination of the weaker elements as such.

Mass Density and Water Absorption over 24 h

From the apparent mass densities recorded (Table 2), it is clear that BA is a lighter aggregate than the limestone. This statement holds true for both the fine and the coarse fractions. The relation between the surface dry mass density ρ_{rd} and the apparent mass density ρ_a gives an indication of the material's porosity. As such, the third column of the table clearly shows that the BA is more porous than the reference material. In case of the fine fraction, the crushed BA has a significantly lower porosity than the sieved and washed BA. On the other hand, the porosities of the crushed and sieved and washed coarse fractions are not significantly different from each other.

Table 2: Influence of aggregate type (limestone *vs.* BA), fraction (2/6 or 0/6 *vs.* 6/20) and BA pre-treatment (sieving and washing *vs.* crushing) on the apparent mass density (ρ_a), surface dry mass density (ρ_{rd}) and water absorption after 24 h (WA_{24})

Aggregate Type	ρ_a (kg/m^3)	ρ_{rd} (kg/m^3)	$(\rho_a-\rho_{rd})/\rho_a \times 100$ (%)	WA_{24} (%)
Limestone 2/6	2750 ± 31	2644 ± 37	3.85 ± 0.28	1.5 ± 0.1
Sieved and washed BA 0/6	2246 ± 13	1793 ± 13	20.17 ± 0.44	11.2 ± 0.3
Crushed BA 0/6	2536 ± 6	2144 ± 11	15.46 ± 0.59	7.2 ± 0.3
Limestone 6/20	2762 ± 18	2710 ± 16	1.87 ± 0.06	0.7 ± 0.0
Sieved and washed BA 6/20	2649 ± 7	2285 ± 11	13.72 ± 0.20	6.0 ± 0.1
Crushed BA 6/20	2697 ± 22	2329 ± 23	13.65 ± 0.25	5.8 ± 0.1

More or less similar conclusions can be drawn from the water absorption after 24 h (Table 2: WA_{24}). This parameter is to a high extent correlated to the ratio of apparent and surface dry mass density. It is immediately clear that the BA has a higher inherent water absorption than the corresponding reference limestone aggregates. The value is about eight times the value of crushed limestone. Fine BA obtained by crushing has a substantially lower water absorption than the sieved and washed fine BA. Its value is comparable to the ones of the coarse BA. As indicated by the static compressive strength values (Section 4.1.2), crushing removes more of the weaker elements in the fine BA. These are probably the more porous elements responsible for the high water absorption. Thus, the crushing operation as pre-treatment seems beneficial for the fine fraction in order to control that property. In general, the fine fraction still absorbs more water than the coarse one. This is a recurring phenomenon for all types of aggregate materials due to their higher surface area. Nevertheless, in case of a highly absorptive BA aggregate, this particular behavior of the fine fraction may result in workability issues when used in concrete (see Section 4.2.1).

Water Absorption as a Function of Time

Figure 2 shows the water absorption as a function of time. The reference limestone aggregate materials barely absorb water within the considered timeframe of around 240 h. The fine and coarse BA fraction are much more absorptive and show a clear deflection point after around 2.5 h of immersion in water. At this point, the fine fraction has already reached 70% of its total absorption capacity, while the coarse BA reaches 50% of its capacity. Note that the fine BA already absorbs over 60% of its total absorption capacity within the first 15 min which is a very relevant timeframe within the concrete manufacturing process.

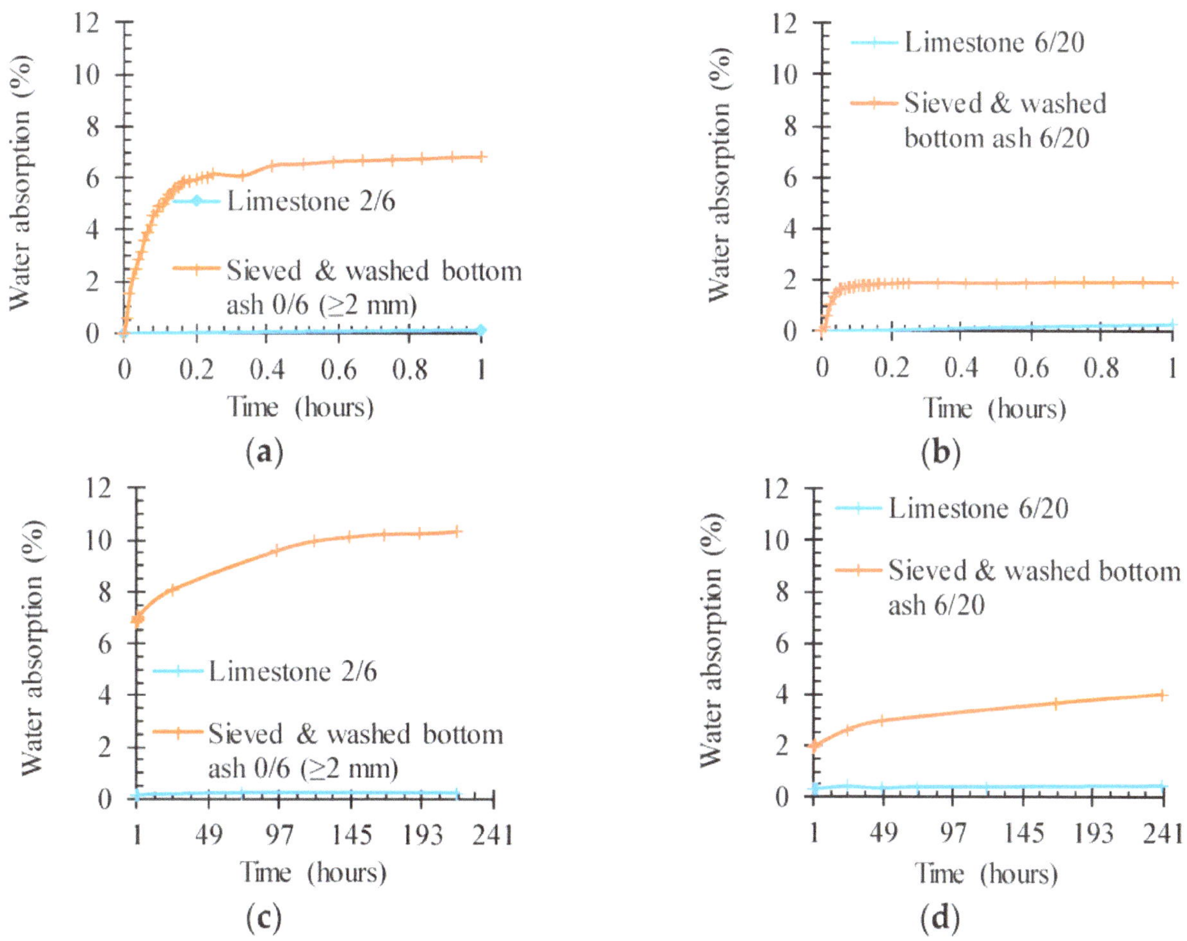

Figure 2. The higher water absorption as a function of time for sieved and washed BA 0/6 (≥2 mm) and 6/20 in comparison with limestone 2/6 and 6/20. (a) fine, between 0 and 1 h; (b) coarse, between 0 and 1 h; (c) fine, between 1 and 241 h; (d) coarse, between 1 and 241 h.

Chemical Composition

The presence of (heavy) metals is of relevance in view of the concrete performance (e.g., the susceptibility to expansion, seeSection 4.2.3) and leaching-related environmental problems (see Section 4.4.1). The chemical composition was determined for the sieved and washed BA 0/6, the crushed BA 0/6 and the sieved and washed BA 6/20 (Table 3). Unfortunately, there was no crushed BA 6/20 available at the time of the chemical analysis at the laboratories of Indaver nv. The (heavy) metal content is higher for sieved and washed BA 0/6 than for crushed BA 0/6 and sieved and washed BA 6/20. Also, the amount of Al is lower in case of the latter two BA aggregates, which indicates a lower susceptibility to Al-induced expansion. On the other hand, Fe, Cu and Zn are less present in the sieved and washed 0/6 fraction. This either indicates that these elements were less present in the MSW that was incinerated to obtain this batch of BA or that the separation of these ferrous and non-ferrous elements was more successful.

Table 3: Influence of the fraction (0/6 *vs.* 6/20) and pre-treatment (sieving and washing *vs.* crushing) on the chemical composition of the BA, especially in view of the (heavy) metal content

Element	Sieved and Washed BA 0/6	Crushed BA 0/6	Sieved and Washed BA 6/20
Al (μg/L)	31,590	5757	3920
Ba (μg/L)	61	148	58
Cu (μg/L)	61	213	117
Zn (μg/L)	13	1332	167
Fe (mg/L)	<0.4	3.0	0.7
Ca (mg/L)	156	137	146
K (mg/L)	45	51	58
Na (mg/L)	85	51	58

Concrete Mixtures

Mixture Proportions and Fresh State Properties

Lowering the W/C ratio normally results in the application of a higher SP dosage to achieve a sufficient workability (slump class S3–S4). For the reference concrete (Table 4: mixtures REF1–REF4), SP only needed to be added in case of a W/C ratio of 0.55. In case of a replacement of the fine limestone 2/6 by fine BA (Table 4: mixtures BA1–BA12 and BA24), the concrete workability was more problematic. Much higher SP dosages (up to 8 mL/kg cement) were necessary and still it remained difficult to achieve the same workability as for the reference concrete, even if the W/C ratio amounted to 0.60 or 0.65. The use of crushed BA instead of sieved and washed BA turned out less problematic somehow. The insufficient workability can be attributed to the fact that the BA has a high porosity and will thus absorb a lot of water in the beginning. Nevertheless, when also considering mixtures BA13–BA16 with replacement of only the coarse fraction by BA 6/20 (Table 4), which is also quite porous (see Table 2), workability was much less of an issue. Still, when considering full replacement of the limestone by fine and coarse BA (Table 4: mixtures BA17–BA23 and BA25–BA28), the detrimental effect of the fine fraction kept on dominating the workability and the required SP dosage. For the W/C ratio of 0.55, this dosage amounted to no less than 36 mL/kg cement, which is far above the dosage allowed for that type of SP. Moreover, the high dosages would increase the price of these mixtures for rather low-value applications (*i.e.*, prefabricated Lego bricks) quite dramatically.

Strength Performance

For the reference concrete (REF1–REF4), adequate strength results (>33 N/mm²) are already obtained after only seven days of curing (Table 5), including for the highest W/C ratio considered (=0.65). For the lower W/C ratios of 0.60 and 0.55, these initial strength results amount to no less than 38 and 44 N/mm², respectively. This strength performance is still bound to increase because of the still-ongoing hydration of the cement. The most significant strength increase could be observed between 7 and 28 days of curing. As a consequence, the 28-day compressive strength class of this traditional concrete type is at least C30/37, which makes it very suitable for prefabricated Lego brick production, which requires a strength class of only C20/25.

In case only the fine aggregate fraction has been replaced with BA (Table 5: mixtures BA1–BA12 and BA24), the concrete performed worse in terms of compressive strength. Still, the 28-day strength class was at least C20/25 in most cases. Mixture BA10 was the only exception. This can mainly be attributed to the 20 min immersion of the fine BA in water prior to the mixing process. The pre-treatment with water was very beneficial for the concrete workability (Table 4: slump class S4, without SP addition). However, the strength decrease caused by it (Table 5: C16/20) proved rather unacceptable given the intended purpose of the concrete. Lowering the W/C ratio from 0.65 over 0.60 to 0.55 obviously turned out beneficial for the strength performance. For these W/C ratios, a 20 min immersion in water as a pre-treatment still resulted in concrete with strength class C20/25 (Table 5: mixtures BA11–BA12).

The replacement of only the coarse limestone by BA 6/20 (Table 5: mixtures BA13–BA15) clearly had a less detrimental effect on the strength. Just as for the reference concrete mixtures, the compressive strength is already higher than 30 N/mm² after only seven days of optimal curing. Their 28-day strength class was at least C25/30 and can reach C30/37 for W/C ratios of 0.60 and 0.55. As such, a more or less similar strength performance to the reference concrete can be assumed. Only the strength gain between 28 days and 56 days of optimal curing is more pronounced for the BA concrete.

Table 4: Influence of the mixture proportions (ratio sand, fine and coarse aggregates, water-to-cement (W/C) ratio and superplasticizer (SP) dosage), replacement of limestone by BA and BA pre-treatment on the concrete workability

Mixture	Natural Aggregates			BA		Water	Cement CEM I 52.5 N	SP	W/C Ratio	Workability	
	Sand 0/4	Limestone 2/6	Limestone 6/20	BA 0/6/BA 2/6	BA 6/20					Slump	Flow
	(kg/m³)	(kg/m³)	(kg/m³)	(kg/m³)	(kg/m³)	(kg/m³)	(kg/m³)	(mL/kg Cement)	-	-	-
REF1	439.00	596.00	695.00	-	-	227.50	350.00	-	0.65	S4	-
REF4	439.00	596.00	695.00	-	-	227.50	350.00	-	0.65	S4	F5
REF3	451.00	613.00	714.00	-	-	210.00	350.00	-	0.60	S4	F5
REF2	464.00	627.00	733.00	-	-	192.50	350.00	1 [a]	0.55	S3	F4
BA2	333.00	-	674.00	664.00 [b]	-	227.50	350.00	3 [c]	0.65	S3	-
BA3	282.00	-	700.00	690.00 [b, d]	-	227.50	350.00	3 [c]	0.65	S3	-
BA4	471.00	–	686.00	519.00 [e]	-	227.50	350.00	2 [c]	0.65	S4	-
BA6	436.00	-	717.00	525.00 [e]	-	227.50	350.00	1 [c]	0.65	S2	-
BA7	436.00	-	717.00	525.00 [e]	-	227.50	350.00	1 [a]	0.65	S2	-
BA10	436.00	-	717.00	525.00 [b]	-	227.50 [f]	350.00	-	0.65	S4	-
BA1	342.00	-	350.00	683.00 [b]	-	210.00	350.00	2 [c]	0.60	S1	-
BA5	342.00	-	665.00	683.00 [b]	-	210.00	350.00	8 [c]	0.60	S4	-
BA8	448.00	-	737.00	539.00 [e]	-	210.00	350.00	3 [a]	0.60	S2	-
BA11	448.00	-	737.00	539.00 [b]	-	210.00 [f]	350.00	-	0.60	S3	-
BA24	397.00	-	703.00	557.00 [g]	-	210.00	350.00	-	0.60	S4	F4
BA9	460.00	-	757.00	554.00 [e]	-	192.50	350.00	6 [a]	0.55	S3	-
BA12	460.00	-	757.00	554.00 [b]	-	192.50 [f]	350.00	-	0.55	S2	-
BA13	613.00	444.00	-	-	633.00 [h]	227.50	350.00	-	0.65	S3	F4
BA14	630.00	456.00	-	-	650.00 [h]	210.00	350.00	-	0.60	S3	F4
BA16	617.00	540.00	-	-	594.00 [i]	210.00	350.00	-	0.60	S4	F3
BA15	647.00	469.00	-	-	668.00 [h]	192.50	350.00	6 [a]	0.55	S3	F4
BA17	517.00	-	-	347.00 [j]	797.00 [h]	227.50	350.00	6 [a]	0.65	S3	F4
BA20	517.00	-	-	347.00 [j]	797.00 [h]	227.50	350.00	5 [a]	0.65	S3	F4
BA21	517.00	-	-	347.00 [j]	797.00 [h]	227.50 [k]	350.00	-	0.65	S4	F5
BA22	517.00	-	-	347.00 [j]	797.00 [h]	227.50 [l]	350.00	-	0.65	S2	F4
BA23	517.00	-	-	347.00 [j]	797.00 [h]	227.50	350.00 [m]	3 [a]	0.65	S2	F4
BA26	488.00	-	–	324.00 [j]	793.00 [h*]	227.50 [n]	350.00	-	0.65	S4	F4
BA27	488.00	-	-	324.00 [j]	793.00 [h*]	227.50 [o]	350.00	6 [a]	0.65	S2	F1
BA28	543.00	-	-	343.00 [j, p]	727.00 [h, p]	227.50	350.00	2 [a]	0.65	S2	F4
BA18	531.00	-	-	357.00 [j]	819.00 [h]	210.00	350.00	10 [a]	0.60	S2	F3
BA25	547.00	-	-	400.00 [g]	717.00 [i]	210.00	350.00	5 [a]	0.60	S3	F4
BA19	545.00	-	-	366.00 [j]	841.00 [h]	192.50	350.00	36 [a]	0.55	S3	F2
Lego brick 1	488.00	-	-	366.00 [j]	808.00 [h*]	227.50	350.00	10 [a]	0.65	S2	F1
Lego brick 2	439.00	596.00	695.00	-	-	227.50	350.00	-	0.65	S3	F5
Lego brick 3	667.00	-	-	382.00 [g, p]	566.00 [h*]	227.50 [n]	350.00	-	0.65	S4	F5

Table 5: Influence of W/C ratio, replacement of limestone by BA and pre-treatment of the BA on the compressive strength after 7, 28 and 56 days and the strength class

Mixture	W/C (-)	Mix Details	$f_{c\ 7\ days}$ (N/mm²)	$f_{c\ 28\ days}$ (N/mm²)	$f_{c\ 56\ days}$ (N/mm²)	Strength Class (-)
REF1	0.65	-	33.49 ± 1.38	41.00 ± 0.43	44.85 ± 0.41	C30/37
REF4	0.65	-	33.61 ± 0.42	40.45 ± 0.61	41.72 ± 0.92	C30/37
REF3	0.60	-	38.54 ± 0.81	48.60 ± 1.14	49.37 ± 1.18	C30/37
REF2	0.55	a	44.29 ± 0.55	52.51 ± 1.43	55.15 ± 0.75	C35/45
BA2	0.65	b, c	23.43 ± 0.87	29.18 ± 0.71	31.26 ± 1.01	C20/25
BA3	0.65	b, c, d	29.69 ± 1.16	32.85 ± 0.41	33.26 ± 0.54	C25/30
BA4	0.65	c, e	32.75 ± 1.26	37.23 ± 0.29	38.32 ± 1.07	C25/30
BA6	0.65	c, e	28.62 ± 1.53	34.49 ± 0.46	38.30 ± 1.32	C25/30
BA7	0.65	a, e	28.69 ± 1.21	34.73 ± 0.42	40.46 ± 0.91	C25/30
BA10	0.65	b, f	18.88 ± 0.45	22.98 ± 0.68	26.66 ± 0.53	C16/20
BA1	0.60	b, c	25.52 ± 0.63	31.35 ± 0.81	34.95 ± 0.69	C20/25
BA5	0.60	b, c	30.39 ± 1.54	36.51 ± 1.54	38.10 ± 0.50	C25/30
BA8	0.60	a, e	33.40 ± 0.46	37.81 ± 0.53	42.16 ± 1.42	C25/30
BA11	0.60	b, f	20.82 ± 0.80	25.73 ± 0.14	29.24 ± 0.34	C20/25
BA24	0.60	g	30.34 ± 0.32	36.93 ± 1.38	39.10 ± 2.05	C25/30
BA9	0.55	a, e	36.62 ± 0.97	42.15 ± 1.28	43.20 ± 1.26	C30/37
BA12	0.55	b, f	22.35 ± 0.67	27.71 ± 0.70	31.83 ± 0.67	C20/25
BA13	0.65	h	34.40 ± 0.56	40.90 ± 2.13	46.32 ± 1.24	C25/30
BA14	0.60	h	37.40 ± 0.71	42.82 ± 1.35	50.30 ± 0.84	C30/37
BA16	0.60	i	35.23 ± 1.01	39.85 ± 0.86	47.42 ± 0.80	C25/30
BA15	0.55	a, h	38.34 ± 0.93	40.97 ± 0.53	47.73 ± 2.08	C30/37
BA17	0.65	a, h, j	32.93 ± 1.74	38.72 ± 0.61	40.18 ± 1.30	C25/30
BA20	0.65	a, h, j	26.26 ± 0.71	31.51 ± 1.10	33.15 ± 0.55	C20/25
BA21	0.65	h, j, k	22.03 ± 0.70	29.86 ± 0.92	30.55 ± 0.51	C20/25
BA22	0.65	h, j, l	25.05 ± 0.13	33.11 ± 1.06	34.80 ± 1.71	C20/25
BA23	0.65	a, h, j, m	26.67 ± 0.30	34.81 ± 0.21	34.92 ± 0.74	C25/30
BA26	0.65	h*, j, n	26.23 ± 0.74	32.47 ± 1.17	34.59 ± 0.10	C20/25
BA27	0.65	a, h*, j, o	21.74 ± 0.37	26.03 ± 0.94	28.11 ± 0.23	C20/25
BA28	0.65	a, h, j, p	30.23 ± 1.14	37.02 ± 0.46	38.62 ± 0.40	C25/30
BA18	0.60	a, h, j	28.38 ± 0.66	34.53 ± 0.76	33.67 ± 1.91	C25/30
BA25	0.60	a, g, i	32.15 ± 0.18	37.38 ± 0.95	38.98 ± 0.88	C25/30
BA19	0.55	a, h, j	29.72 ± 2.90	35.80 ± 1.11	36.18 ± 4.37	C25/30

In case of a full replacement of both the fine and coarse aggregate fraction by BA (Table 5: mixtures BA17–BA23 and BA25–BA28), the earlier observed rather negative strength effects inherent to the fine fraction keep on dominating the overall strength performance of the concrete. Nevertheless, it remains possible to achieve a strength class for the concrete of at least C20/25, even when additional water is added to the BA before mixing to achieve a sufficiently workable concrete without adding SP (mixtures BA21–BA22). Lowering the W/C ratio from 0.65 to 0.60 and 0.55 makes it easier to obtain a strength class of C25/30.

Note that the lower strength performance of the BA concrete can mainly be attributed to the lower aggregate crushing value or static compressive strength of the BA aggregates (Section 4.1.2) and the higher porosity of the material. A more in-depth evaluation of the latter property follows in Section 4.2.3.

Susceptibility to Expansion

All concrete specimens with BA showed signs of expansion (Figure 3a). This behavior is attributed to the presence of metallic Al in the BA. When placed in an alkaline environment (pH > 7), the metallic Al dissolves and emits dihydrogen gas [11]. Obviously, the cement matrix (pH 13) provides these required alkaline conditions. The dihydrogen-induced expansion leads to the formation of longitudinal voids, which were, in this study, mainly concentrated near the troweled surface of the test samples and perpendicular to the casting direction (Figure 3b).

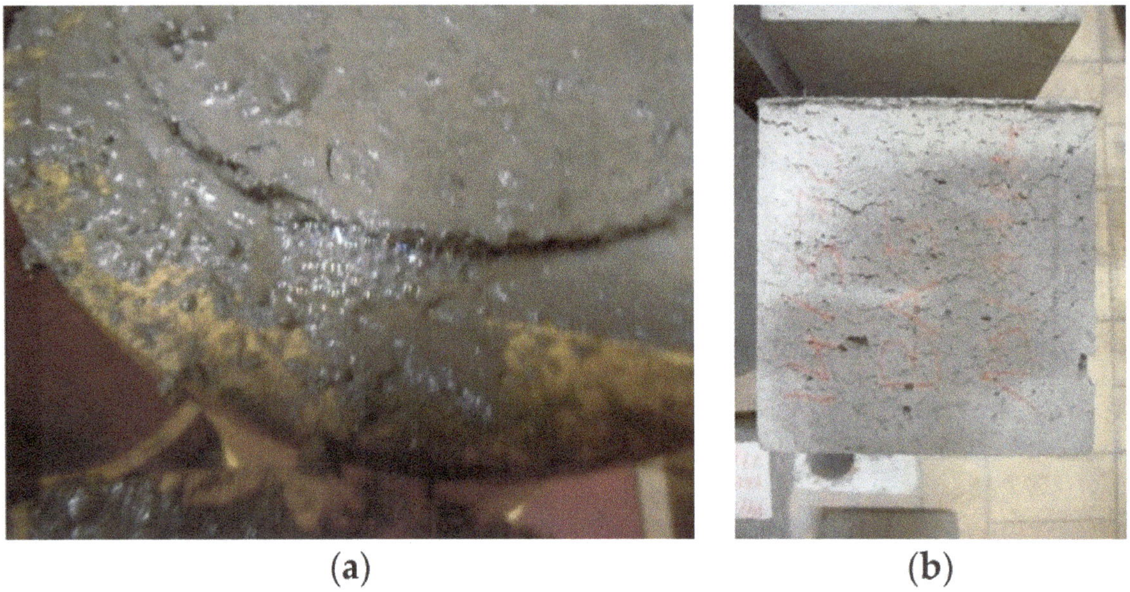

Figure 3: Reactive Al induced expansion near the troweled surface of a concrete cylinder (a); longitudinal expansion voids observed on a concrete test cube (b) when containing BA.

There is a clear difference in expansive behavior between the reference concrete and concrete in which the fine and/or the coarse aggregate fractions were replaced with BA. For all three W/C ratios (Figure 4a: 0.55, Figure 4b: 0.60, Figure 4c: 0.65), the expansive behavior of the BA concrete is evident. Whereas the reference concrete never shows a significant change in length for the distance between the bottom and the troweled surface of the test cubes, the concrete with replacement of the fine aggregate by BA is characterized by an expansion of a little less than 4%. Replacement of only the coarse aggregate fraction by BA 6/20 gives less expansion (around 1%), be it only for W/C ratios of 0.65 and 0.60. For a W/C ratio of 0.55, the expansion also equaled around 4%. The recorded expansion in case of a full replacement of the limestone by the appropriate proportioning of fine and coarse BA represents an accumulative effect of the two alternative aggregate fractions. This accumulative effect is the most pronounced for the lowest W/C ratio considered, *i.e.*, 0.55.

The BA concrete with the highest W/C ratio probably shows less expansion because it requires much less SP. The presence of SP usually tends to postpone the time of setting. This means that there is more time for the dihydrogen-induced longitudinal voids to form and, thus, also more time for the resulting expansion to develop. Nevertheless, further testing is still needed to find further confirmation for this.

Selection of the Most Suitable Test Mixture

The selection of the most suitable BA concrete mixture for further testing was mainly governed by the achievable balance between workability (without extensive use of expensive SP) and strength performance. Compressive strength testing revealed that even with a W/C ratio of 0.65, the minimum required strength class for prefabricated Lego bricks (C20/25) could still be met (Section 4.2.2). Knowing this, it seems no problem to stick to this high W/C ratio as it also helps to compensate for the workability issues inherent to the presence of mainly the fine BA (Section 4.2.1). Sieved and washed BA were used for mixture BA20. This particular pre-treatment does not result in a significantly higher strength or resistance to expansion as opposed to the crushed BA. Nevertheless, sieving and washing normally leads to a somewhat lower workability than crushing. Still, the former pre-treatment was chosen since a crushing of the aggregates is more energy intensive and it does not improve the concrete workability that much. Instead, a suitable pre-wetting procedure for the BA was investigated in order to lower the required SP dosage for an adequate workability (see Section 4.5.1). Mixture BA20, still without pre-wetting of the BA aggregates, was manufactured for a further in-depth evaluation of the concrete mainly regarding its porosity (Section 4.3) and expected behavior in the long term (Section 4.4). Within Section 4.5.2, an additional reactive washing step was considered for the BA to overcome the expansion problem.

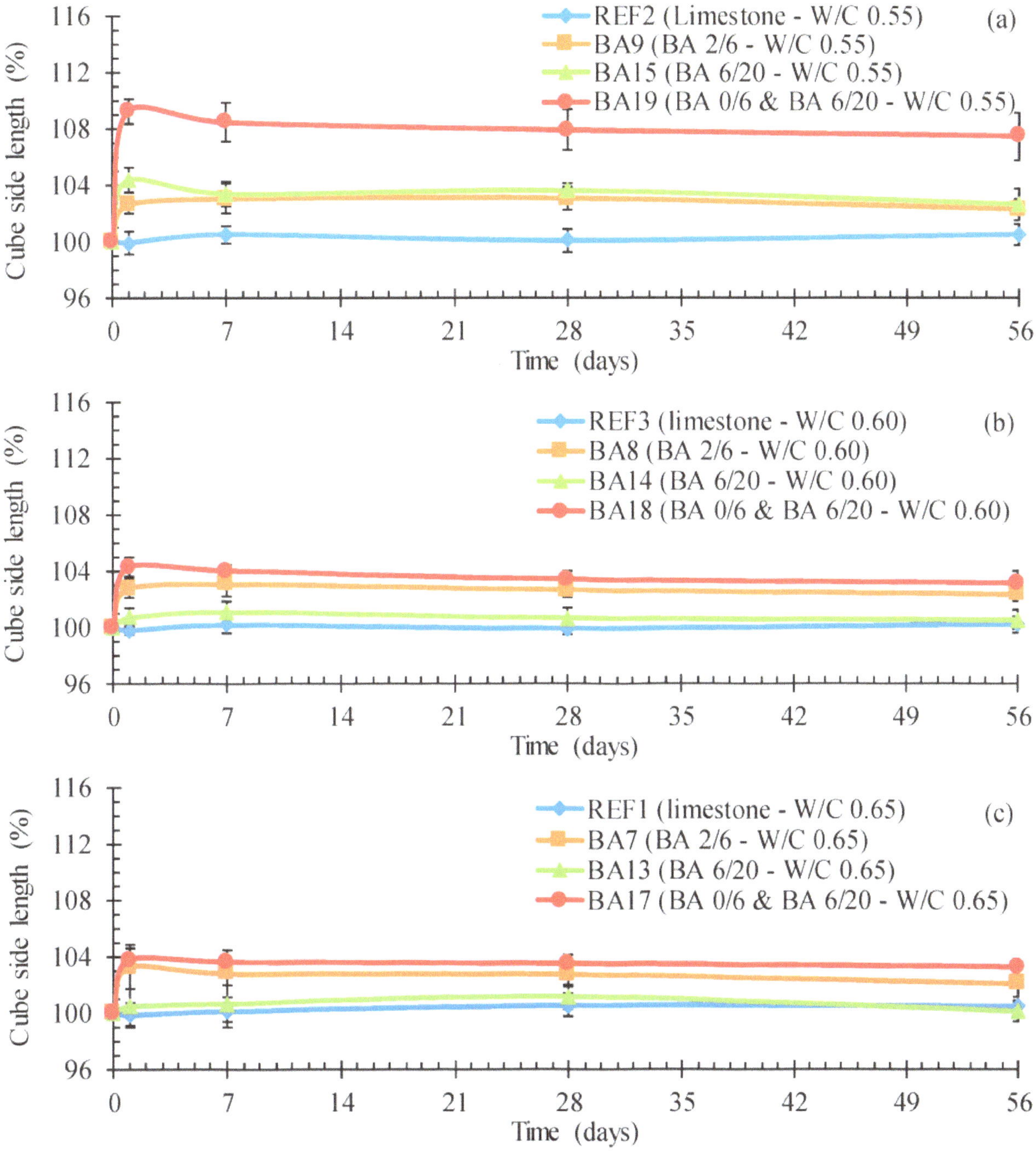

Figure 4: The higher susceptibility of BA concrete toward expansion in comparison with limestone concrete, an expansion which decreases with increasing W/C ratio 0.55 (a); 0.60 (b); 0.65 (c).

Porosity

From the vacuum saturation experiments, it can be concluded that the BA concrete is clearly more porous than the reference concrete (Table 6). The pronounced porosity difference between the BA and the reference concrete exists both after pre-drying at 40 °C and 105 °C.

Table 6. The higher open/permeable porosity of concrete BA20 *vs.* REF4 both after pre-drying at 40 °C (capillary porosity) and 105 °C (total porosity) and their similar calculated gel porosity

Mixture	Capillary Porosity φ (%) after Pre-Drying at (40 ± 5) °C	Total Porosity φ (%) after Pre-Drying at (105 ± 5) °C	Gel Porosity φ (%)
REF4	11.1 ± 0.4	16.1 ± 0.6	5.1 ± 0.4
BA20	17.6 ± 0.4	22.5 ± 0.5	4.9 ± 0.1

In literature, the porosity obtained after pre-drying at 40 °C is said to correspond with the capillary porosity (pore diameter: 1–10 μm), while the difference in porosity obtained after pre-drying at 105 °C and 40 °C should represent the gel porosity (pore diameter: 10 nm) [25]. The capillary pores are the result of the fact that water and cement in unbound conditions take a larger volume than the hydrated cement. The gel pores relate to the gel water which is physically adsorbed by the hydration products and the monomolecular water film in between the flat crystallization products of interlayer water. When considering the theory of the capillary and gel porosity, one could state that the gel porosity is equal for both mixtures, while the capillary porosity is significantly different. However, since for both mixtures the same amount of water and cement was used, there is no reason to conclude that the capillary porosity should be different. Now, since the BA aggregates absorb a lot of water during the mixing process, it can be stated that the BA concrete has a lower effective water/cement ratio and thus also a lower capillary porosity. The increase in porosity after pre-drying at 40 ± 5 °C should be assigned to the existence of other voids, *i.e.*, the longitudinal expansion voids at the surface and the pores in the aggregates, if capillary pores reach these aggregates. The expansion voids will not directly attract water as their size (around 100 μm in width) does not really accommodate capillary suction. Still, these voids enlarge the possible water penetration surface from which more capillary pores can be reached. If a porous aggregate is in contact with them, an extra amount of water would be stored in the concrete as well.

Note that the capillary pores and the longitudinal expansion voids are influencing the strength and permeability of the concrete. The higher porosity after pre-drying at 40 ± 5 °C is a measure for the extra voids at the surface, but not of the inner structure. However, one could state that if these voids could be minimized, permeability will be diminished and strength will increase.

For a better insight on the inner porosity, a microscopic analysis was performed as well (Figure 5).

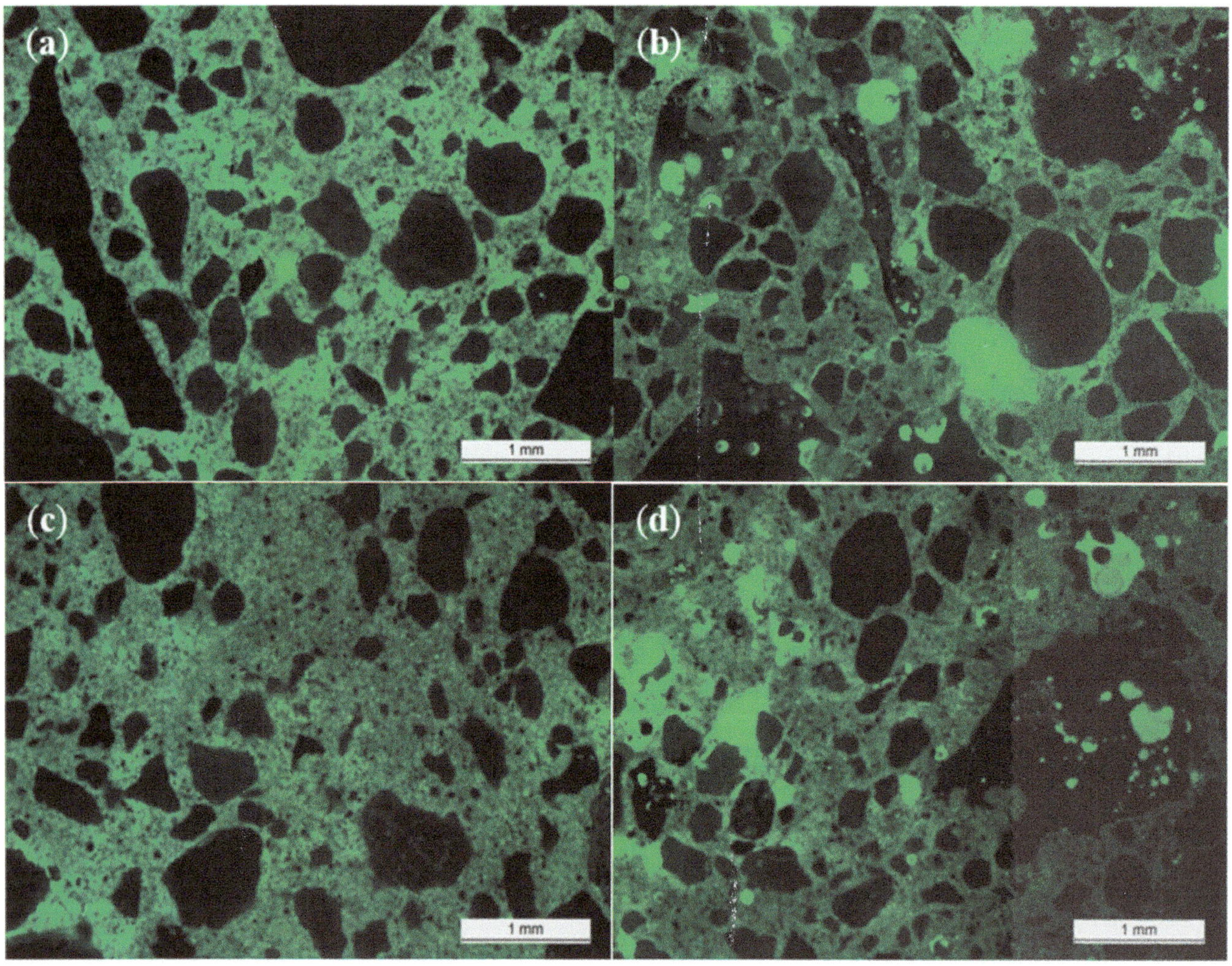

Figure 5: The darker green color of the thin sections of concrete BA20 *vs.* limestone concrete REF4 in fluorescent light mode, indicating a lower effective W/C ratio of the former. (a) REF4, thin section 1; (b) BA20, thin section 1; (c) REF4, thin section 2; (d) BA20, thin section 2.

When analyzing the thin sections in fluorescent light mode, it can be seen that the cement matrix of the BA concrete has a darker green color than the one of the reference concrete. This normally indicates that the effective W/C ratio of the BA concrete is lower. The concrete's effective water content was also calculated. In accordance with NBN EN 206-1, this is the reduced amount of mixing water available given the water absorption (WA_{24}) of the concrete's inert fraction. The calculation revealed that when aiming at a theoretical W/C ratio of 0.65, the effective W/C ratio due to the high water absorption of the BA would only be 0.56. It also means that the higher open/permeable porosity for the BA mixture after pre-drying at 40 °C cannot be assigned to the presence of more capillary pores. Hence, it must indeed be caused by the presence of the longitudinal expansion voids and the highly absorptive aggregates.

Long-Term Behavior

Susceptibility to Leaching

Leaching values stayed below the limiting values imposed by the Flemish VLAREMA standards for the normalized elements, except for bromine (Table 7). Copper was much more present in the leaching solution of the BA concrete as opposed to the one of the reference concrete (28 µg/L *vs.* <6 µg/L). Still, the measured concentration was acceptable (<80 µg/L). There is a two-fold explanation for the adequate leaching behavior. First of all, the BA itself already meets the VLAREMA criteria. Leaching components usually come from the bottle glass and metal fragments present in the BA, but in this study these fractions apparently had no problematic effects. Secondly, once in concrete they are immobilized in the cement matrix which reduces the leaching even more. Leaching data published by Vandecasteele *et al.* on the original non-processed BA 6/50 further confirm both effects [6]. The highest leaching value was recorded for Al (5302 µg/L), for which there are no VLAREMA criteria. It was about the 10-fold of the Al leaching value of the reference concrete.

Table 7: Leaching values for concrete BA20 and REF4 relative to demineralized water, indicating a fulfillment of most VLAREMA criteria, yet with a much higher Al leaching value for BA20

Normalized	BA20	REF4	Demineralized Water	Vlarema Criterion	Non-Normalized	BA20	REF4	Demineralized Water
As (µg/L)	<12	<12	<12	80	Al (µg/L)	5302	620	<100
Ba (µg/L)	<4	<4	<4	200	B (µg/L)	480	182	49
Cd (µg/L)	<2	<2	<2	3	Mn (µg/L)	<6	<6	<6
Co (µg/L)	<2	<2	<2	50	Tl (µg/L)	<15	<15	<15
Cr (µg/L)	51	53	<4	260	Fe (mg/L)	<0.4	<0.4	<0.4
Cu (µg/L)	28	<6	<6	80	Ca (mg/L)	15	6.4	<1
Mo (µg/L)	49	31	<10	3000	K (mg/L)	169	136	<1
Ni (µg/L)	<2	<2	<2	75	Mg (mg/L)	<1	<1	<1
Pb (µg/L)	<20	<20	<20	130	Na (mg/L)	160	96	<1
Sb (µg/L)	<12	<12	<12	100	P (mg/L)	<0.02	0.06	<0.02
Se (µg/L)	<10	<10	<10	200	S (mg/L)	8.8	10.7	<0.02
Sn (µg/L)	<10	<10	<10	100	NO_2 (mg/L)	<1	<1	<1
V (µg/L)	165	169	<10	250	NO_3 (mg/L)	<1	<1	<1
Zn (µg/L)	<8	<8	<8	280	PO_4 (mg/L)	<5	<5	<5
F (mg/L)	<5	<5	<5	5.5	I (mg/L)	<5	<5	<5
Cl (mg/L)	<100	<100	<100	100				
Br (mg/L)	<4	<4	<4	2				
SO_4 (mg/L)	24	30	<10	220				
Hg (µg/L)	<1	<1	<1	2				

Resistance to ASR

After the modified Oberholster test, substantial expansion and cracking were observed for the BA concrete. The length increase was around twice the value of the reference concrete (Figure 6: 0.36% ± 0.16% *vs.* 0.18% ± 0.05%). Especially for the BA mixture, the recorded length increase exceeded 0.1% by far. This normally indicates a serious sensitivity to ASR. Surprisingly, this criterion was also not met for the reference limestone concrete. Note that the length change of the BA concrete is also characterized by a much larger standard deviation on the individual values. It demonstrates that the ASR reactivity of the BA depends on its specific composition and its often high heterogeneity.

The observed expansion during the modified Oberholster test for the BA concrete is rather in contrast with literature which normally reports an acceptable ASR resistance. Perhaps the accelerated testing conditions and the fact that the exposure solution is NaOH, which also reacts with the metallic Al present in the BA, may explain this unexpected behavior. To see whether the experiment indeed resulted in expansion and cracking due to ASR, additional microscopic thin section analysis on the test samples from the Oberholster test is needed. It would also be useful to perform an additional Oberholster test on concrete with BA that was subjected to the reactive washing operation. That investigation is, for the moment, still ongoing.

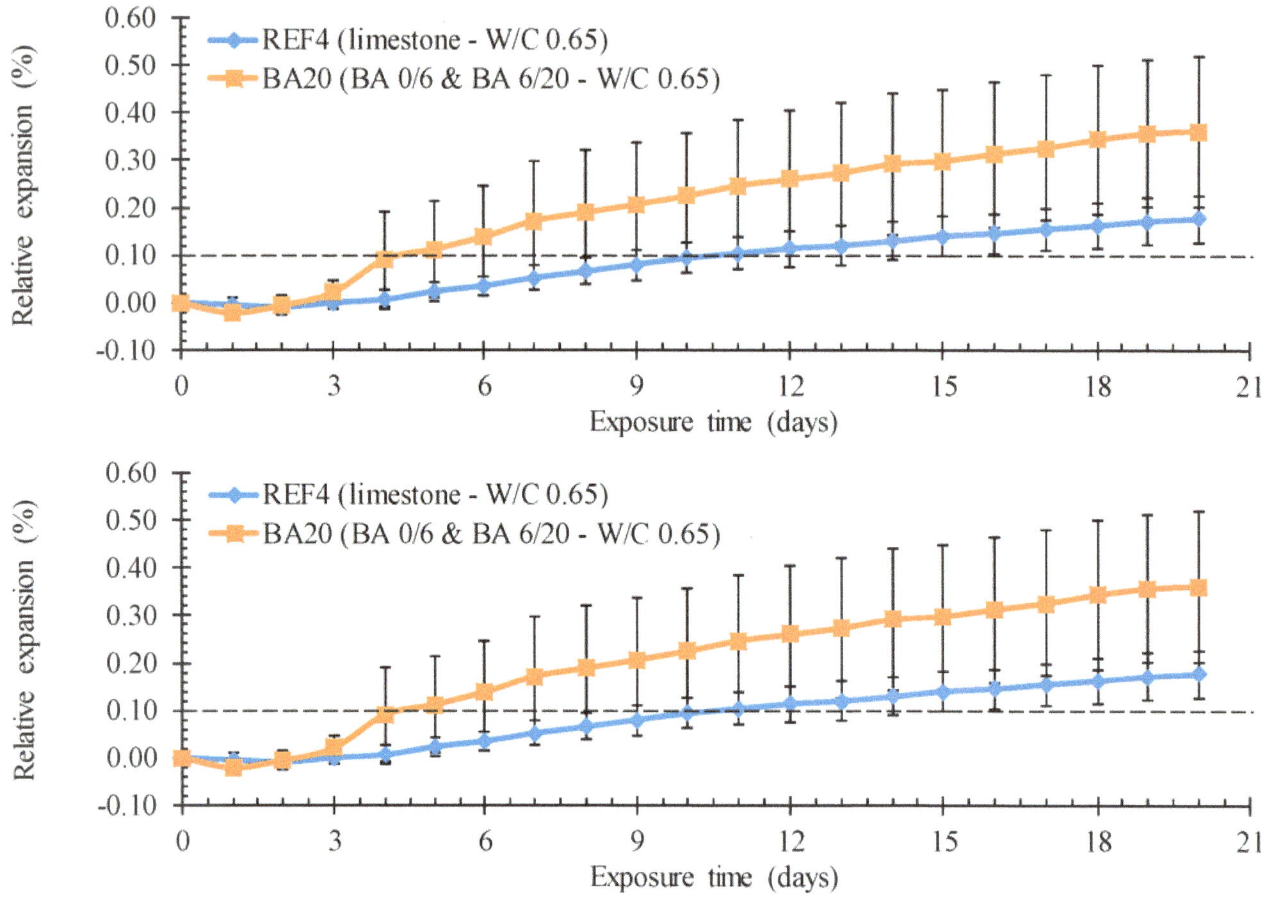

Figure 6: Higher mean relative expansion of concrete BA20 *vs.* limestone concrete REF4 after the modified Oberholster test.

Resistance to Acid Attack

From the recorded relative mass changes, concrete BA20 seems to be more resistant to cyclic lactic and acetic acid exposure than the reference limestone concrete REF4 (Figure 7). However, a visual inspection of the samples shows that one should state this conclusion rather carefully. Apparently, the reference concrete eroded rather homogenously and the sample surface remained smooth. The BA concrete showed a quite different behavior. There, only the cement matrix eroded in between the BA aggregates. This resulted in a rough sample surface and some loss of individual, mainly spherical and smooth-shaped BA grains (Figure 8). As the mass loss of individual aggregates can vary considerably per test sample, it is no surprise that the relative mass change of the BA concrete is characterized by a larger standard deviation on the individual values than the reference concrete (Figure 7).

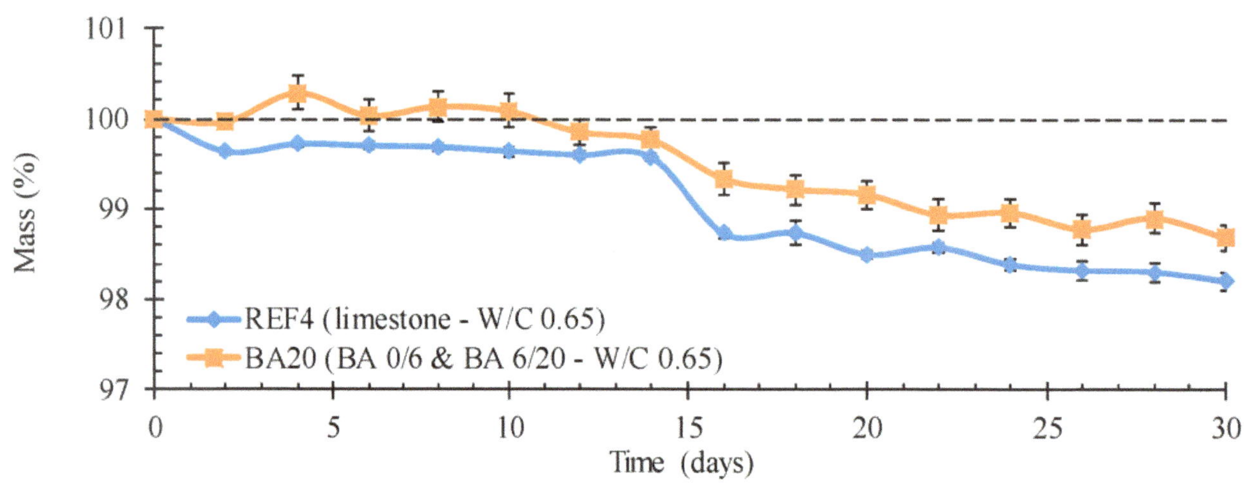

Figure 7: Lower mass loss of concrete BA20 *vs.* limestone concrete REF4 during cyclic exposure to a mixture of lactic and acetic acid.

Figure 8: Higher superficial deterioration of concrete BA20 (**a**) *vs.* limestone concrete REF4 (**b**) after cyclic exposure to lactic and acetic acid.

The difference in sample deterioration between the two concrete types can be explained as follows. The lactic and acetic acid react with both the calcium hydroxide present in the cement matrix and with the limestone aggregates which are, in essence, $CaCO_3$. In both cases, the reaction products are CO_2 and H_2O as well as either calcium acetate or lactate, depending on the acid. Thus, in the case of the reference concrete, both the cement matrix and the aggregate fraction deteriorated. It explains the still rather smooth appearance of the concrete surface after the acid exposure. As the calcium content of the BA-based aggregates is much less (around 10%–15%), it evidently contains much less of the vulnerable $CaCO_3$ in comparison with the limestone aggregates of the reference concrete. In other words, just the cement matrix of the BA concrete is susceptible to the acid attack. As this cement matrix is also more porous due to the presence of more longitudinal expansion voids (Section 4.3), it is also more accessible for the acidic exposure solution. As a consequence, it will deteriorate more than the cement matrix of the reference concrete. This, in contrast with the BA-based aggregates remaining more or less intact, resulted in a rather inhomogeneous deterioration of the BA concrete. The corresponding very distinctive roughening of the concrete surface is probably as undesirable as the higher mass loss of the reference concrete.

Resistance to Freeze-Thaw Attack

No problems were observed regarding the freeze-thaw resistance of the BA concrete. This conclusion follows from both the visual inspection and the ultrasonic testing. Visually, no damage phenomena such as cracks were observed. Secondly, the ultrasonic pulse velocity before freeze-thaw exposure (3951 ± 93 m/s) is more or less equal to the one after 14 freeze-thaw cycles (3994 ± 88 m/s). The ultrasonic pulse velocities recorded for the reference mixture did not change either (before: 4369 ± 82 m/s, after: 4320 ± 160 m/s). The difference in ultrasonic pulse velocity between the BA mixture and the reference mixture is due to the higher porosity and lower strength of the former concrete type. Note that the influence of de-icing salts on the freeze-thaw resistance was not investigated.

Optimization of the Bottom Ash Characteristics

Optimization regarding Workability and Water Absorption

Initially, it seemed that the large amount of fines was causing the high water absorption of the BA. Therefore, its water absorption with exclusion of the fraction <2 mm was determined and compared to the value of the non-sieved BA. Surprisingly, it was found that the water absorption was not really affected by this sieving operation (Table 8). Also, washing the BA with tap water instead of with industrial process water (containing more fine particles) did not result in a less absorptive behavior. Although the sieving operation influenced the concrete workability in a positive manner (Table 4), this is not attributable to the omission of the fine fraction.

Table 8: The limited influence of excluding the fraction <2 mm and washing with tap water instead of process water on the water absorption of the fine BA

BA Treatment	BA 0/6 Sieved and Aged	BA 2/6 Sieved and Aged	BA 0/6 Sieved and Washed (Process Water)	BA 0/6 Sieved and Washed (Tap Water)
WA_{24} (%)	10.3 ± 1.6	11.8 ± 0.1	11.2 ± 0.3	11.0 ± 0.2

The high water absorption was believed to be caused by the high porosity of the BA rather than by the amount of fines. Attempts were made to separate the porous from the non-porous elements. However, this turned out rather difficult. The porous elements could also not be detected visually. For instance, quite some metal parts within the BA looked porous because they were covered by with porous-looking layers due to the sintering process during incineration. Floating tests based on the difference in density between the non-porous and porous elements were also unsuccessful. However, it could be demonstrated that crushing the aggregates in a laboratory ball mill had a positive effect on the water absorption. As can be seen in Table 9, this crushing operation mainly influenced the smaller fractions of the bottom ash (BA 0/6 (≥2 mm) and BA 6/20 (<4 mm)). In view of concrete manufacturing, it is thus advantageous to use the crushed material because the weaker porous elements are crushed. The slightly higher water absorption induced by having a somewhat greater fine fraction this way appears not to be comparable to the diminished absorption that can be obtained by removing the porous parts.

Table 9: The pronounced influence of a crushing operation on the reduction in water absorption of BA 0/6 (≥2 mm) and BA 6/20 (<4 mm) *vs.* BA 6/20 (≥4 mm)

BA Fraction	BA 0/6 (≥2 mm)		BA 6/20 (≥4 mm)		BA 6/20 (<4 mm)	
Crushing	Before	After	Before	After	Before	After
WA_{24} (%)	9.5 ± 0.9	6.2 ± 0.9	3.4 ± 0.2	3.0 ± 0.0	12.4 ± 1.1	7.1 ± 0.4

Also pre-wetting of the BA had a distinct positive influence. However, a lot depends on the method applied. The first technique described in Section 3.9.1 did not give satisfying results. Putting the BA in contact with water to achieve a 60% saturation resulted in a too-wet, non-surface-dry aggregate. The mixtures made with this material (Table 4: mixtures BA10–BA12) showed a good workability without SP addition. Still, mixture BA10 with the highest W/C ratio (=0.65) was on the verge of segregating. The overdose of water also affected the compressive strength in a negative way (Table 5). Its strength class was only C16/20, which is lower than the required C20/25.

The second and third pre-wetting methods were more promising in terms of gaining a sufficient workability without losing too much strength performance (Table 4, Table 5: mixtures BA21–BA22). It is recommended to pre-wet the BA in advance (third method) rather than to add the same surplus water content simultaneously with the mixing water (second method). The pre-wetting establishes a water-cement film around the grains in advance which prevents further absorption by the BA aggregates. As such, the aggregates do not get saturated, a higher effective water/cement ratio is reached and the workability is less of an issue. Thus, the sequence of adding the different components during concrete manufacturing is important. After testing several amounts of water, 1 m% of the coarse BA 6/20 and 2.5 m% of the fine BA 0/6 appeared to be the optimal pre-wetting content of water. This more or less corresponds with the amount of water absorbed in the first 2 min after the addition of the mixing water. Thus, it is not necessary to consider the whole 6 min mixing timeframe.

Optimization regarding the Susceptibility to Expansion

To diminish the expansion, LA cement could be used. From the length change measurements performed on mixture BA23, it could be concluded that the expansion was reduced from 3%–4% to 2.5%, but it was not completely eliminated (Figure 9: mixture BA23).

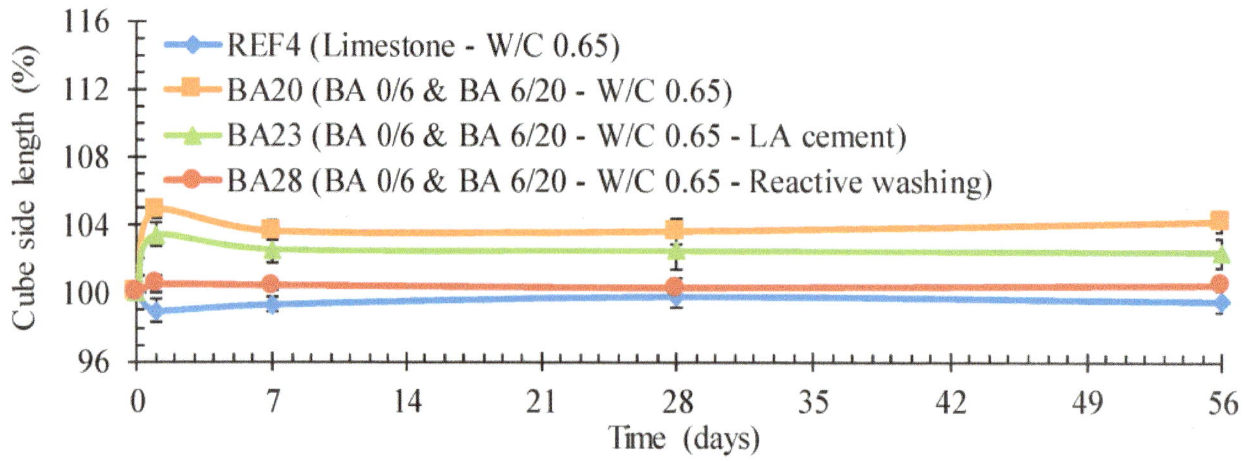

Figure 9: he achievable lower relative expansion of BA concrete with LA cement as binder or after reactive washing of the BA aggregates.

On the other hand, a reactive washing of the aggregates with 1 M NaOH was very effective and eliminated the problem completely (Figure 9: Mixture BA28). As shown in Table 10, reactive washing resulted in a significant decrease in Al reactivity after 14 days, especially in case of the fine BA. After only 4 h, the change in Al reactivity for both BA fractions was still non-significant.

Table 10: The pronounced effectiveness of a reactive washing operation on the Al reactivity of the fine and coarse BA, especially after 14 days

Time	Al Reactivity (%)	
	BA 0/6	BA 6/20
Before reactive washing	2.22	0.16
After 4 h	1.93	0.21
After 14 days	0.73	0.02

Almost no expansion was recorded for concrete BA28 containing BA 0/6 and 6/20 washed with NaOH (Figure 9). There was a concern regarding the strength of this concrete because the surface showed signs of peeling. However, this appeared to be not an issue (Table 5: C25/30). Although this reactive washing can solve the expansion problem, one should remain aware of the fact that this chemical processing will increase the production cost of prefabricated Lego bricks significantly. Moreover, there will be some additional environmental issues. Further research is needed to see whether recycled NaOH obtained from other industrial processes can be used for this purpose.

Evaluation of the Lego Bricks

Applied Concrete Mixtures

Lego brick 1 was cast using a concrete with the same composition as mixture BA20. Lego brick 2 must be seen as the reference. Its composition was the same as mixtures REF 1 and REF 4. Finally, an optimized BA concrete composition was used for the production of Lego brick 3. It contained crushed BA 0/6 (with somewhat better workability properties) instead of the sieved and washed BA 0/6. For BA 6/20, there was no difference in water absorption between the sieved and washed and the crushed aggregates, so the former was used again. Both the BA 0/6 and 6/20 were sieved afterwards to exclude the fractions smaller than 1 mm and 4 mm, respectively. This enabled a better approximation of the particle size distribution of the reference limestone 2/6 and 6/20 aggregates. Since the BA 0/6 was the most reactive due to its higher metallic Al content, this fine aggregate was subjected to the reactive washing with 1 M NaOH for two to three weeks. Prior to concrete mixing, the aggregates were pre-wetted with 2.9 l water per 180 L of concrete (1 m% of BA 6/20 and 2.5 m% of BA 0/6). A LA cement CEM I 52.5 N HES LA HSR was used as binder. The predefined W/C ratio amounted to 0.65. The compositions of all Lego bricks were included in Table 4.

Observations Made in Fresh and Hardened State

The unadjusted BA concrete mixture (*cf.* BA20) that was used for Lego brick 1 was characterized by a poor workability. Even the addition of 10 mL SP per kg cement could not overcome the problem. This behavior can be explained by the fact that another delivery of coarse BA was used for this concrete which was characterized by a higher water absorption. Nevertheless, with some extra compaction efforts, it was possible to cast the Lego brick. All protrusions were filled in a proper manner and no difficulties were encountered during demolding. However, in hardened state, surface laitance, clear traces of the vibration needle, longitudinal expansion voids and compaction voids could be seen. The expansion itself was comparable to what was observed earlier for mixture BA20 on small concrete cubes. Despite the larger dimensions of the Lego brick, the expansion was not excessive. This can be explained by the fact that the ratio of the troweled surface over the total concrete surface available (troweled and cast surfaces) was lower for the Lego brick than for the cubes with a 150 mm side. The mechanical performance of the concrete used for Lego brick 1 was inadequate since its strength class was only C16/20. This was mainly due to the difficult compaction of the poorly workable concrete which resulted in more voids.

None of these issues were encountered when casting reference Lego brick 2. The only problem was that the concrete was too workable. Without SP addition the concrete was almost segregating. Soon after casting, a distinct water layer appeared on the surface and bleeding occurred. Apparently, applying a W/C ratio of 0.65 for a large concrete volume was too much. This is in contrast with the smaller batches of this concrete that were produced earlier on (REF1 and REF4).

Once Lego brick 3 was produced, the influence of the different optimization techniques could be evaluated. The use of crushed BA 2/6 and pre-wetting the aggregates solved the workability problem. No SP was needed anymore. The reactive washing of the fine BA 2/6 again appeared to be very effective in eliminating expansion and the longitudinal voids. The use of LA cement probably also helped to achieve this. The sieving operations performed on the BA to eliminate the excess in fine material and to approximate the particle size distributions of the traditional limestone aggregates resulted in a more evenly distributed use of sand and BA in accordance with the Fuller approach. This contributed to the better strength (C20/25) and workability (Table 4: slump: S4, flow: F5) performance as well. Figure 10 shows pictures of all three Lego bricks.

Figure 10: The three Lego bricks (3: the optimized BA-based Lego brick; 2: the reference limestone-based Lego brick; and 1: the BA-based unadjusted Lego brick).

CONCLUSIONS

It is possible to produce a proper concrete in which both the fine and coarse aggregates have been replaced with BA resulting from MSWI. The concrete seems suitable for casting prefabricated Lego bricks. The minimum required strength class (=C20/25) can be assured even with W/C ratios of no less than 0.65 and without using an expensive SP.

However, a dedicated processing of the BA prior to concrete manufacturing is imperative. Excluding the excess of fines to approximate the particle size distributions of the traditional limestone aggregates is advised. This results in a better packing and a higher strength of the concrete. A crushing operation before the sieving eliminates porous elements in the BA which absorb a lot of water. As such, a better workability of the concrete can be obtained. A controlled pre-wetting of the BA just before concrete mixing in relation to the time-dependent water absorption curve also helps to improve the concrete workability. Given these techniques, it will probably be possible to manufacture a sufficiently workable BA concrete with a design W/C ratio of only 0.60 in the future.

The significant presence of metallic Al in the BA makes the concrete susceptible to the formation of longitudinal voids and expansion. The use of a LA cement can somewhat reduce this problem. A reactive washing of the BA with 1 M NaOH eliminates the problem completely. Still, the latter procedure is not ideal from both an economic and an environmental point of view. Further research on more appropriate reactive washing techniques is still needed.

No problems regarding leaching and freeze-thaw were observed for the BA concrete. On the other hand, the concrete shows a pronounced surface roughening after acetic and lactic acid exposure and does not pass the modified Oberholster test for concrete. The expansion (>0.1%) after the latter test indicates ASR sensitivity. However, further microscopic analysis is needed to see whether ASR is indeed causing the problem. Moreover, the tested BA concrete was not fully optimized yet (e.g., with exclusion of the reactive washing stage). Additional tests on the finally optimized mixture used for Lego brick 3 are, for the moment, still ongoing.

ACKNOWLEDGMENTS

The authors would like to thank Ghent University for the financial support and the technicians of the Magnel Laboratory for Concrete Research for their assistance during the experiments. Acknowledgments also go to Indaver nv for providing and treating the BA and performing the ICP, IC and FIMS analyses as well as the Al reactivity tests at their laboratories. A word of thanks also goes to CB Recycling BV for providing the mold for the manufacturing of the Lego bricks.

AUTHOR CONTRIBUTIONS

Niels Ringoot and Arno Beirnaert worked on this research topic for their master dissertation. Together with their supervisors, Philip Van den Heede, Andres Van Brecht and Erwin Van den Brande, they conceived and designed the research plan and all experiments. They performed all tests and analyzed the data which they then compiled in their master dissertation. Philip Van den Heede wrote the paper which was revised and approved by all co-authors. Nele De Belie and Geert De Schutter were the promoters of the master dissertation and hence provided input at all stages of the research.

REFERENCES

1. European Commission. *Being Wise with Waste: The EU's Approach to Waste Management*; Publications Office of the European Union: Luxembourg, 2010.
2. Chimenos, J.M.; Segarra, M.; Fernandez, M.A.; Espiell, F. Characterization of the bottom ash in municipal solid waste incinerator. *J. Hazard. Mater.* 1998, *64*, 211–222.
3. Siddique, R. Use of municipal waste ash in concrete. *Resour. Conserv. Recycl.* 2010, *55*, 83–91.
4. European Union. Directive 2008/98/EC of the European parliament and of the council of 19 November 2008 on waste and repealing certain directives. *Off. J. Eur. Union* 2008, *312*, 3–20.
5. Müller, U.; Rübner, K. The microstructure of concrete made with municipal waste incinerator bottom ash as an aggregate component. *Cem. Concr. Res.* 2006, *36*, 1434–1443.
6. Vandecasteele, C.; Wauters, G.; Arickx, S.; Jaspers, M.; van Gerven, T. Integrated municipal solid waste treatment using a grate furnace incinerator: The Indaver case. *Waste Manag.* 2007, *27*, 1366–1375.

7. Van Gerven, T.; van Keer, E.; Arickx, S.; Jaspers, M.; Wauters, G.; Vandecasteele, C. Carbonation of MSWI bottom ash to decrease heavy metal leaching, in view of recycling. *Waste Manag.* 2005, *25*, 291–300.
8. Courard, L.; Degeimbre, R.; Darimont, A.; Laval, A.L.; Dupont, L.; Bertrand, L. Use of municipal solid waste incineration residues in the fabrication of concrete paving. *Mater. Struct.* 2002, *35*, 365–372. (In French).
9. Van Rooy, S. Feasibility Study of MSWI Bottom Ash Aggregate as Secondary Resource for Ground and Road Works. Master's Thesis, Hogeschool Antwerpen, Antwerpen, Belgium, 2007. (In Dutch).
10. Keulen, A.; Florea, M.; Brouwers, J. *Upgrading MSWI Bottom Ash as Building Material for Concrete Mixes*; Eindhoven University of Technology: Eindhoven, The Netherlands, 2012.
11. Pera, J.; Coutaz, L.; Ambroise, J.; Chababbet, M. Use of incinerator bottom ash in concrete. *Cem. Concr. Res.* 1997, *27*, 1–5.
12. Abbá, A.; Collivignarelli, M.C.; Sorlini, S.; Bruggi, M. On the reliability of reusing bottom ash from municipal solid waste incineration as aggregate in concrete. *Composites B* 2014, *58*, 502–509.
13. Song, G.J.; Kim, K.H.; Seo, Y.C.; Kim, S.C. Characteristics of ashes from different locations at the MSW incinerator equipped with various air pollution control devices. *Waste Manag.* 2003, *24*, 99–106.
14. Forteza, R.; Far, M.; Seguí, C.; Cerdá, V. Characterization of bottom ash in municipal solid waste incinerators for its use in road base. *Waste Manag.* 2004, *24*, 899–909.
15. Rübner, K.; Haamkens, F.; Linde, O. *Use of Municipal Solid Waste Incinerator Bottom Ash as Aggregate in Concrete*; European Geosciences Union General Assembly: Vienna, Austria, 2007.
16. Van der Wegen, G.; Hofstra, U.; Speerstra, J. Upgraded MSWI bottom ash as aggregate in concrete. *Waste Biomass Valoris.*2013, *4*, 737–743.
17. Sorlini, S.; Abba, A.; Collivignarelli, C. Recovery of MSWI and soil washing residues as concrete aggregates. *Waste Manag.*2010, *31*, 289–297.
18. Ginés, O.; Chimenos, J.M.; Vizcarro, A.; Formosa, J.; Rosell, J.R. Combined use of MSWI bottom ash and fly ash as aggregate in concrete formulation: Environmental and mechanical considerations. *J. Hazard. Mater.* 2009, *169*, 643–650.
19. Berg, E.R.; Neal, J.A. Municipal solid waste bottom ash as Portland cement concrete ingredient. *J. Mater. Civ. Eng.* 1998, *10*, 168–173.
20. Garcia-Gonzalez, J.; Rodriguez-Robles, D.; Juan-Valdes, A.; Moran-del Pozo, J.M.; Guerra-Romero, M.I. Pre-saturation technique of the recycled aggregates: Solution to the water absorption drawback in the recycled concrete manufacture.*Materials* 2014, *7*, 6224–6236.
21. Tegguer, D.A. Determining the water absorption of recycled aggregates utilizing hydrostatic weighing approach. *Constr. Build. Mater.* 2012, *27*, 112–116.
22. RILEM TC 106-AAR: Alkali-aggregate reaction. *Mater. Struct.* 2000, *33*, 283–293.
23. Fabien, A.; de Boom, A.; Lenormand, T.; Lambrette, N.; Degrez, M.; Staquet, S. Reuse of municipal solid waste incinerator fly ash into cement-based materials. In Proceedings of the 1st Concrete Innovation Conference (CIC2014), Oslo, Norway, 11–13 June 2014; Justnes, H., Ed.; Sintef Building and Infrastructure: Trondheim, Norway, 2014.
24. CUR Bouw & Infra. *CUR Recommendation 116, Processed Municipal Incinerator Bottom Ash as Aggregate for Concrete*; CUR Bouw & Infra: Gouda, The Netherlands, 2012. (In Dutch)
25. Audenaert, K. Transport Mechanisms of Self-Compacting Concrete Related to Carbonation and Chloride Penetration. Ph.D. Thesis, Ghent University, Ghent, Belgium, 2006. (In Dutch).

CHAPTER 6

CATALYSIS OF A NANOMETRE SOLID SUPER ACID OF SO42-/TIO2 ON THE THERMAL DECOMPOSITION OF AMMONIUM NITRATE

Xiaolan Song, Yi Wang, Dan Song, Chongwei An and Jingyu Wang

Keywords: Nanoparticles, Solid Super Acid, Ammonium Nitrate, Thermal Decomposition, Mechanism

INTRODUCTION

Compared with ammonium perchlorate (AP) propellants, ammonium nitrate (AN) propellants have advantages such as nontoxic products, a low signature and insensitivity. These advantages are mainly attributed to the chloride-free composition and low water content in their combustion products. In addition, AN is a type of energetic material with super low insensitivity. For example, pure AN cannot be ignited by flame or heated in the air, and it has a remarkable resistance to stimulation such as friction, impacts, shock waves, etc. Nevertheless, AN has not served as the main oxidizer in propellant formulation and is used only as an ingredient in gas-generating agent [1,2]. This is because AN has three major defects [3]: (1) serious hygroscopicity, (2) phase transition at storage temperature and (3) poor ignition and combustion performance. In particular, the third defect is fatal, i.e., AN-based propellants burn so slowly that their combustion cannot provide sufficient thrust to boost the missile (or rocket) to fly at high speed. Therefore, introducing a catalyst to improve the combustion performance of AN-based propellants is crucial.

At present, with the solving of the first two defects, AN is frequently used in studies of high energy and low vulnerability (HELV) in the field of military science and technology [4-6]. Therefore, over the past five years, there has been an increase in the number of studies about improving the performance of AN-based propellants. In particular, many studies concerning better combustion performance were conducted but the results were not satisfactory. In studies about AP, researchers often find that a good catalyst can always decrease the decomposition temperature by more than 100°C. However, please note that this type of result was almost impossible in studies concerning AN because of its special decomposition mechanism. According to all reported information, if a catalyst can decrease the decomposition temperature of AN by more than 10°C, we consider it to exhibit good catalysis ability. Researchers have improved the thermal decomposition and combustion performance of AN as much as possible but the results remain notably limited. For example, Hasue studied the combustion of phase-stabilized ammonium nitrate (PSAN) mixed with BTA-NH_3 and found that the mixture could be ignited under 1 MPa (improved ignition performance), but the burning rate remained notably low (<1 mm s^{-1}) [7]. Sinditskii studied the effects of NaCl, $BaCl_2$ and NaF as catalysts on the thermal decomposition and combustion properties of AN. The results indicated that the AN/GAP propellant that was

Citation: Xiaolan Song, Yi Wang, Dan Song, Chongwei An and Jingyu Wang. Catalysis of a Nanometre Solid Super Acid of SO42-/TiO2 on the Thermal Decomposition of Ammonium Nitrate. Nanomater Nanotechnol, 2016, 6:23. doi: 10.5772/62633

doped with 7% NaCl continuously combusted under 0.5 MPa, but the burning rate remained less than 1 mm s^{-1} [8]. Miyata studied the combustion of the AN/AGAT mixture (50/50) and found that, although the mixture could be sustained under 0.5 MPa, the burning rate was not significantly improved (only approximately 1 mm s^{-1}) [9]. Naya investigated the catalysis of MnO_2 in the thermal decomposition of AN [10]. The results indicated that the catalyst did not function at a high heating rate (20°C min^{-1}). When the heating rate was decreased to 2°C min^{-1}, its catalysis activity was demonstrated, i.e., the DSC peak temperature of 4%MnO_2-doped AN decreased by 16°C compared with that of pure AN. Nevertheless, the propellant with added MnO_2 as the catalyst could not be ignited at low pressure (≤1 MPa). Vargeese studied the catalytic effect of CuO, TiO_2 and LiF on the thermal decomposition of AN. The results showed that the addition of TiO_2 did not change the activation energy of the AN decomposition, CuO had limited catalytic activity and LiF had a negative effect [11].

The number of studies on AN-based propellants has increased since 2011. In particular, researchers from Russia, Japan and India have devoted the most attention to this subject [12-15]. Moreover, it is worth noting that the catalytic effects of nano materials on the thermal decomposition and combustion of AN have not been reported. In this paper, after studying the decomposition mechanism of AN in detail, we propose that solid super acids may exhibit a good catalysis ability because their super high acidity may strongly promote decomposition. Super acids are solid acids whose acidities are much higher than that of 100% H_2SO_4. Generally, their *pKa* values are less than -11.93 [16,17]. In industry, many organic synthesis reactions must use acid as a catalyst. Compared with liquid acids, solid acids have advantages such as high catalysis, high selectivity, lack of pollution and ease of separation from the reaction system. Thus, their application would be much suited to the organic industry. The studies of Guo [18] and Wang [19] considered that, due to the strong inductive effects of the S=O group, the adsorbed proton contributes as a Brønsted acid centre. The surface acid sites are associated with metal ions, whose acidic strength can be strongly enhanced by the induction effect of the S=O groups. With these results in mind, we decided to prepare this type of new catalyst and attempt to determine whether it would help to promote the decomposition of ammonium nitrate.

EXPERIMENTAL

Materials and Instruments

In this study, analytically pure ammonium nitrate (AN) was produced by Guangzhou Hengyu Chemical Co. Ltd. Analytically pure titanium tetrachloride ($TiCl_4$) was produced by Shanghai Huzhen Industrial Co. Ltd. Strong ammonia (25~28%) was supplied by Jining Huide Chemical Co. Ltd. Concentrated sulphuric acid (98%) was provided by Chongqing Zhaohui Chemical Factory. Analytically pure ethanol was provided by Concord Tianjin Branch Co., Ltd. Deionized water was provided by Taiyuan Meijiayuan Beverage Co. Ltd. A muffle furnace model LHT04/16 (30°C~3000°C) was made by Germany RETSCH Company. A vacuum pump model SHZ-CD (≤0.1 MPa) was made by Henan the Great Wall Instrument Factory.

Preparation of Nano So_4^{2-}/Tio_2

Strong ammonia (25%) was diluted to ammonia solution with a concentration of 6 mol/L, and ethanol was subsequently introduced into the solution as a dispersant. Then, 23.75 g of $TiCl_4$ was slowly added into 84 mL of the ammonia solution and the pH value was adjusted to 9~10, using ammonia to ensure that Ti^{4+} completely converted to $Ti(OH)_4$. After 24 h of aging, the precipitation was washed with H_2O three times, with ethanol two times and once with acetone. After washing, the precipitate was dried at 50°C. The dried powder was carefully grinded in an agate mortar and tiny particles of amorphous TiO_2 were obtained. With ultrasonic stirring, 2 g of amorphous TiO_2 was put into the sulphuric acid solution and dipped in 30 min. After dipping, the amorphous TiO_2 was dried at 80°C. The dried powders were grinded and calcined at 300°C, 400°C, 500°C or 800°C, and SO_4^{2-}/TiO_2solid super acids were subsequently obtained.

To investigate the catalysis of SO_4^{2-}/TiO_2, 0.97 g of AN was blended with 0.03 g of SO_4^{2-}/TiO_2 by careful manual grinding and the obtained sample was tagged as [AN+3%SO_4^{2-}/TiO_2]. For comparison, an additional 0.97 g of AN was blended with 0.03 g of nano TiO_2 by careful manual grinding and the obtained sample was tagged as [AN+3%TiO_2].

Sample Characterization

Morphology was observed with a field-emission scanning electron microscope (SEM, JEOL JSM-7500) and a Philips-Tecnai-12 transmission electron microscope (TEM). The phases of the samples were investigated using an X-ray diffractometer (XRD, Bruker Advance D8) with Cu K_α radiation at 40 kV and 30 mA. XPS analysis was performed using X-ray photoelectron spectroscopy (XPS) and a PHI5000 Versa-Probe (ULVAC-PHI). TG-DSC analysis and DSC-IR analysis were performed using a thermal analyser system (TG/DSC, Mettler Toledo), which was coupled with a Fourier transform infrared spectrometer.

RESULTS AND DISCUSSION

Morphology and Structure

The calcined powders were analysed using XRD and the results are shown in Fig. 1. The results indicate that the amorphous TiO_2 transformed to anatase-phase TiO_2 after it was calcined at 300, 400 or 500°C. With the increase in temperature, the intensity of the diffraction peak increased, which indicates that the crystallization degree of the sample gradually increased. In addition, the diffraction peak presented a width phenomenon, which implies that the crystal grain may have had a nano-scale size. According to the Debye-Scherrer equation ($D=K\lambda/(\beta\cos\theta)$), the average grain size of SO_4^{2-}/TiO_2(baked at 500°C) was calculated to be 25.2 nm. Via calcination at 800°C, the anatase-phase TiO_2transformed into rutile-phase TiO_2, and its average grain size was more than 100 nm. In Fig. 1b, the XRD patterns of TiO_2 and SO_4^{2-}/TiO_2 that were calcined at 500°C were compared. The result shows that TiO_2 had a stronger peak intensity than SO_4^{2-}/TiO_2, which implies that surface acidification can delay the crystallization of TiO_2. The average grain size of TiO_2 was 29.7 nm, which is larger than that of SO_4^{2-}/TiO_2.

Fig. 2 shows the XPS spectrum of SO_4^{2-}/TiO_2 calcined at 500°C. Three elements (i.e., S, Ti and O) were detected in the spectrum. The O and Ti peaks are located at the binding energies of 529.9 eV and 458.7 eV (463.8 eV), respectively. These peaks should reflect the elements in TiO_2 because their intensity was notably strong. The peak at the binding energy of 168.5 eV is related to the 2p electron transition of the S element. The S element should originate from the SO_4^{2-} anion on the surface of SO_4^{2-}/TiO_2, which implies that SO_4^{2-} was fixed on the surface of TiO_2 after acidification.

Fig. 3 shows the morphology and particle size of the non-acidification samples and acidification samples that were calcined at 500°C. Fig. 3a indicates that pure TiO_2 (with a particle size of 25~40 nm) showed a certain agglomeration. Fig. 3b shows that SO_4^{2-}/TiO_2 (with a particle size of 20~30 nm) also massively agglomerated. Therefore, we had sufficient reason to reconsider the drying process in the fabrication of SO_4^{2-}/TiO_2 nanoparticles. Freeze-drying or supercritical drying may be more feasible. The powder was added into ethanol, and strong ultrasonic was used to disperse the agglomerate particles. Then, the obtained suspension was subjected to a TEM analysis. The TEM images are shown in Figs. 3c and d, and the particle sizes were measured and recorded. The specific size of each particle in Figs. 3c and d was carefully measured, and the statistical data are displayed in Figs. 3e and f. Fig. 3 shows that the average particle sizes ($\bar{X}$) of TiO_2 and SO_4^{2-}/TiO_2 were 33 nm and 29 nm, respectively, i.e., SO_4^{2-}/TiO_2 were somewhat smaller than TiO_2 in terms of particle size, which is consistent with the XRD analysis results, i.e., surface acidification slightly delays the growth of particles (or grains).

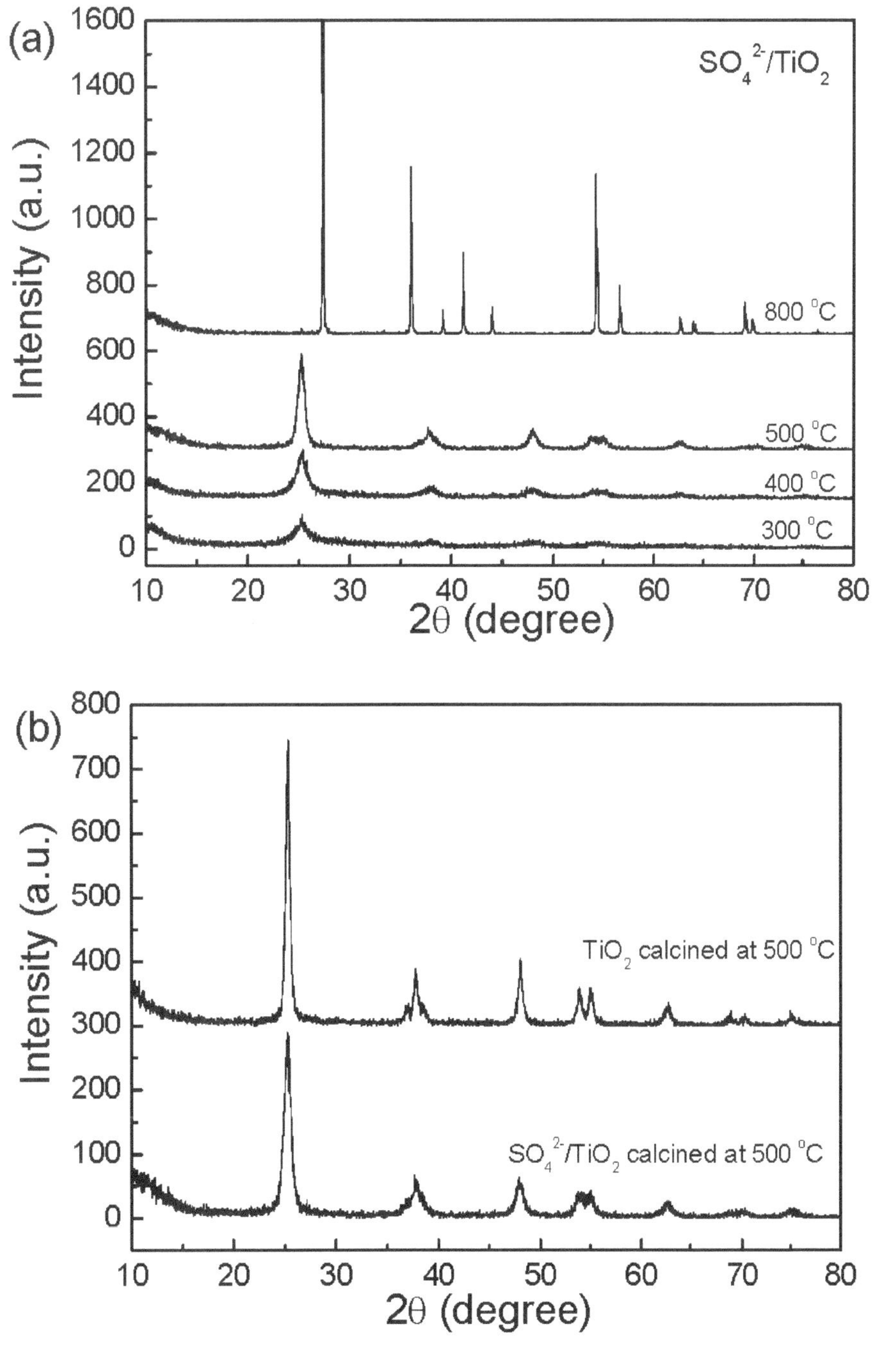

Figure 1: XRD patterns of samples calcined at different temperatures.

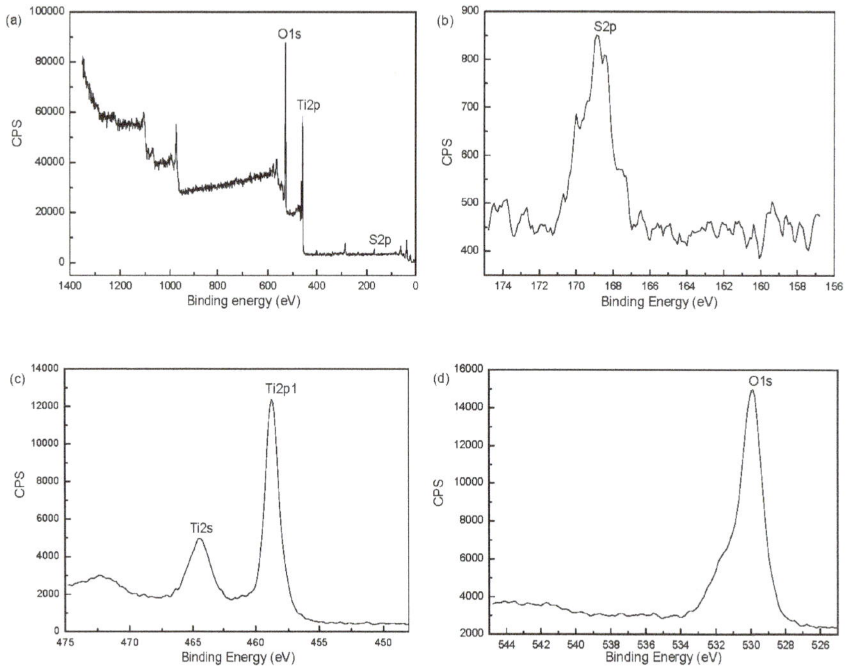

Figure 2: XPS spectrum of nano SO_4^{2-}/TiO_2 super acid.

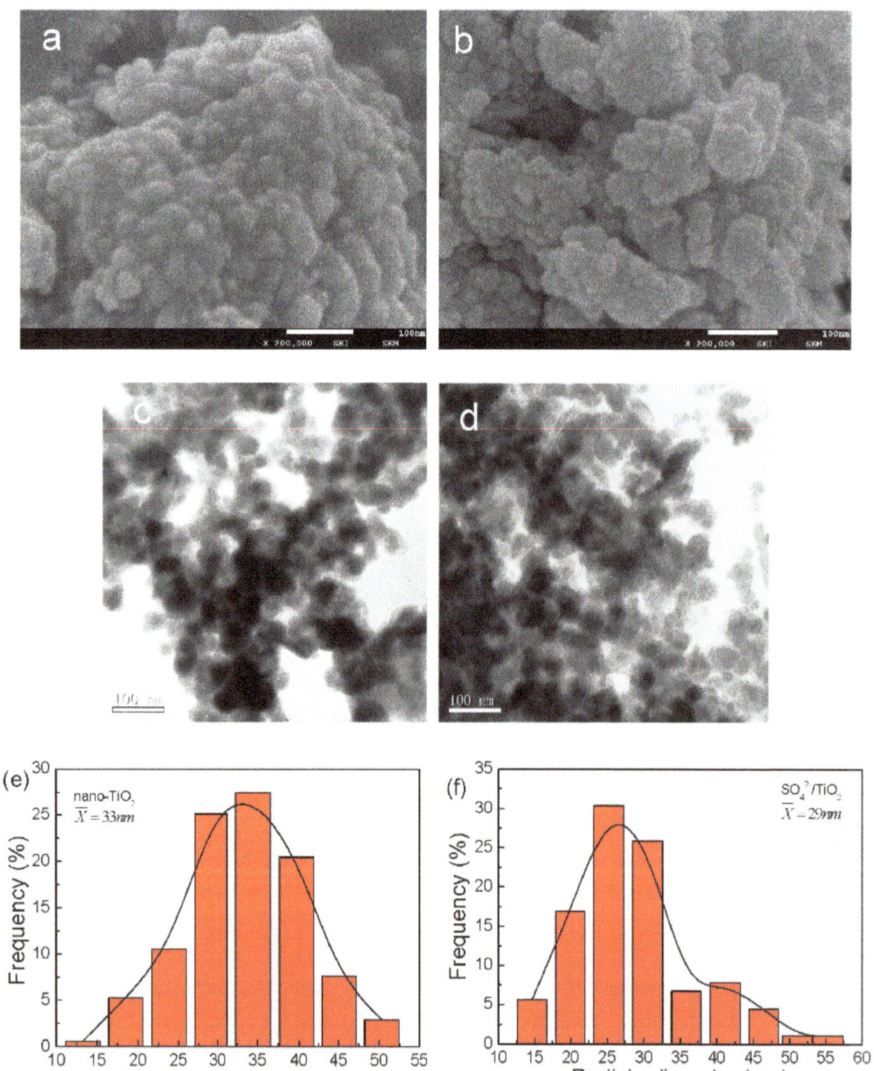

Figure 3: SEM and TEM images of the samples: (a, c) nano TiO_2; (b, d) nano SO_4^{2-}/TiO_2; (e, f) size distribution of nano TiO_2 and nano SO_4^{2-}/TiO_2, which was calculated from Fig. 3 (c and d).

Thermal Analysis

Thermal analyses were performed using the TG-DSC technique and the results are shown in Fig. 4. Pure AN has no exothermic peak and four endothermic peaks. The first phase transition occurs at 52.9°C, which corresponds to the transformation from phase III to phase II (ΔH=13.9J g^{-1}). The second phase transition (at 126.1°C) is the transformation from phase II to phase I (ΔH=40.1J g^{-1}). The third endothermic peak (at 167.6°C, ΔH=55.4J g^{-1}) reflects the melting of AN. The strongest endothermic peak depicts the decomposition process of AN. The decomposition begins at 239.2°C and has its peak point at 277.4°C (ΔH=888.9J g^{-1}). In theory, liquid $NH_4^+NO_3^-$ dissociates via a proton transformation (from NH_4^+ to NO_3^-) to generate NH_3(g) and HNO_3(g) (Eq. 1), which is the first decomposition step of almost all onium salts. This dissociation course does not spontaneously proceed (ΔG=+90.6 kJ mol^{-1}) but it can occur after 2.18 kJ g^{-1} of heat is absorbed [20]. After the dissociation, the redox reactions (exothermic) between the oxidizer and the reducer begin. However, the amount and rate of heat release from the redox reactions are notably low. Thus, the presented decomposition course in the DSC traces is endothermic.

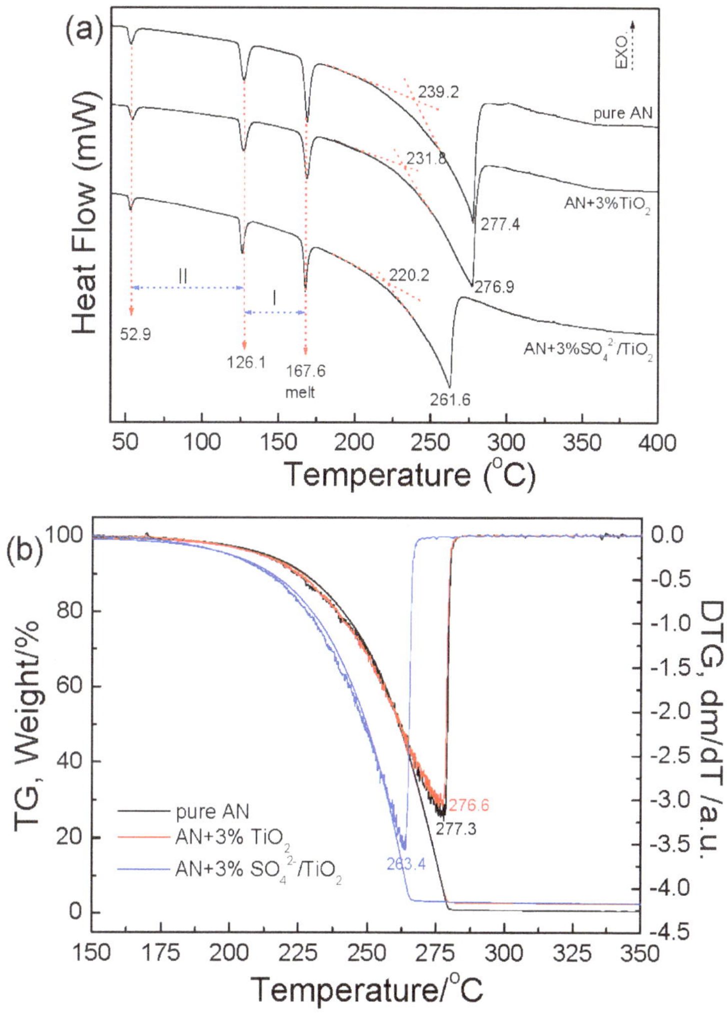

Figure 4: DSC, TG and DTG traces of the samples.

The results in Fig. 4a also show that the main decomposition process of AN did not change with the doping of either SO_4^{2-}/TiO_2 or TiO_2. The presented processes in the DSC traces remained endothermic, which implies that the addition of a nano catalyst cannot increase the reaction heat in thermodynamics. However, the onset and peak temperatures (*To* and *Tp*) decreased with the addition of nano catalysts. The catalytic effect of TiO_2 is notably limited because, compared with pure AN, *To*decreased by only 7.4°C, and *Tp* seldom changed. In comparison, SO_4^{2-}/TiO_2 had a much stronger catalytic effect than TiO_2, where *To* and *Tp* decreased by 19°C and 15.8°C, respectively. This result may imply that the SO_4^{2-} ions on the surface of the catalyst particles are favourable for the thermal decomposition of AN. Fig. 4b shows the TG-DTG curves of pure AN, [AN+3%TiO_2] and [AN+3%SO_4^{2-}/TiO_2]. In addition to the onset temperature, there was no obvious distinction among these TG curves. After the derivation calculus, the DTG curves were obtained and show the decomposition rates of the samples. These DTG peaks are similar to the DSC peaks.

Catalytic Mechanism

The decomposition products of [AN+3%SO_4^{2-}/TiO_2] were detected using the DSC-IR technique and the results are shown in Fig. 5. Fig. 5a shows the stack plots of the gas products and Fig. 5b shows the infrared spectra of the gas products at the times of 528.9 s, 706.6 s, 770.5 s, 856.9 s and 920.8 s. It is clear that the decomposition products are mostly N_2O and a small amount of H_2O. This result indicates that the redox reactions between the dissociation products of 3% SO_4^{2-}/TiO_2-doped ammonium nitrate completely proceeded because no $NH_3(g)$ or $HNO_3(g)$ was detected. In theory, the thermal decomposition of AN has two mechanisms: ionization reaction (Eqs. 1-4) and radical reaction (Eqs. 1and 5-7). Note that the final product of both channels is N_2O, which is consistent with our experimental results. At a low temperature, the decomposition reaction should proceed according to the ionization mechanism because the rupture of the HO-NO_2 bond (i.e., Eq. 5) only occurs at elevated temperatures (approximately 1300°C).

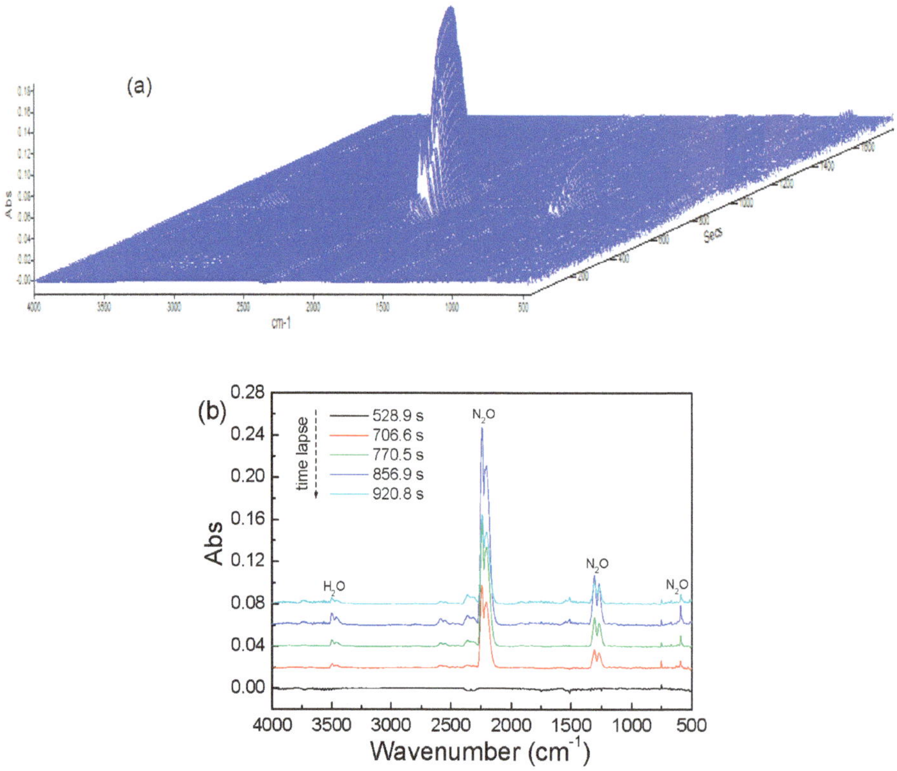

Figure 5: Decomposition products of the AN+3%SO_4^{2-}/TiO_2 mixture: (a) stack plots of gas products; (b) IR spectra of the gas products intercepted at different times.

Generally, the reactions (Eqs. 2 and 5) are controlled steps for ionization reaction and radical reaction, respectively, because their rate constant is notably small. Hence, the issue of promoting AN decomposition may be addressed by finding a method to accelerate the limiting steps. Herein, we attempted to promote decomposition using simple chemical processes, based on the principle that a stronger acid can replace a weak acid. First, a type of acid, "HA", was introduced ("H" means proton; "A" means anion), whose acidity was quantified by pK_α. When we added HA into the decomposition system of AN, the reaction $NH_3 + H^+ \rightleftharpoons NH_4^+$ would occur if pK_α(HA) was less than $pK_\alpha(NH_3)$. This process helped to fix $HN_3(g)$ as NH_4^+ to avoid the cessation of the decomposition by NH_3 poisoning. Likewise, if $pK_\alpha(HA) < pK_\alpha(HNO_3)$, the reaction $NO_3^- + H_3O^+ \rightleftharpoons HNO_3 + H_2O$ would begin. This process considerably increases the concentration of HNO_3 in the decomposition system and accelerates the controlled steps (Eqs. 2 and 5). In particular, when $pK_\alpha(HA) << pK_\alpha(HNO_3)$, the decomposition of HNO_3 is changed (see Eq. 8). Note that the decomposition of HNO_3 is important for the thermolysis of ammonium nitrate [12].

$$NH_4^+NO_3^- \rightarrow NH_3 + HNO_3 + 2.18 kJ/g \tag{1}$$

$$2HNO_3 \rightleftharpoons [NO_2^+NO_3^-] + H_2O \text{ (slow, rate limiting step)} \tag{2}$$

$$[NO_2^+NO_3^-] \rightleftharpoons NO_2^+ + NO_3^- \tag{3}$$

$$NO_2^+ + NH_3 \rightleftharpoons [NH_3 \cdot NO_2^+] \rightarrow N_2O + H_3O^+ \tag{4}$$

$$HNO_3 \rightarrow \cdot OH + \cdot NO_2 \text{ (slow, rate limiting step)} \tag{5}$$

$$NH_3 + \cdot OH \rightarrow \cdot NH_2 + H_2O \tag{6}$$

$$\cdot NH_2 + \cdot NO_2 \rightarrow [NH_2NO_2] \rightarrow N_2O + H_2O \tag{7}$$

$$HNO_3 + H^+ \rightleftharpoons [H_2ONO_2]^+ \rightarrow NO_2^+ + H_2O \text{ (faster than Eq. 2)} \tag{8}$$

In fact, many Lewis acid sites and Brønsted acid sites coexist on the surface of SO_4^{2-}/TiO_2 (Fig. 6). The studies of Guo [18] and Wang [19] considered that, due to the strong inductive effects of the S=O group, the proton contributes as a Brønsted acid centre. The surface acid sites are associated with metal ions, whose acidic strength can be strongly enhanced by the induction effect of S=O groups. Hence, we believe that, instead of Lewis acid sites, Brønsted acid sites are important in this process. The experimental result confirms that TiO_2 did not present catalysis in the AN decomposition. Please note that there are many Lewis acid sites on the surface of TiO_2 (Fig. 6). Some papers have reported that Lewis acid sites (on the surface of anatase titanium dioxide) could catalyse the decomposition of ammonia ($2NH_3 \rightarrow N_2+3H_2$) [21-23]. According to the results, although the decomposition of NH_3(catalysed by Lewis acid sites) is a feasible process in thermodynamics, the reaction(s) processed notably slowly in kinetics. Meanwhile, please note that this dissociation of NH_3 is endothermic (exactly like the pyrolysis of ammonia in industry). Moreover, for ammonia pyrolysis in industry, a high reaction temperature and a noble metal catalyst are indispensable.

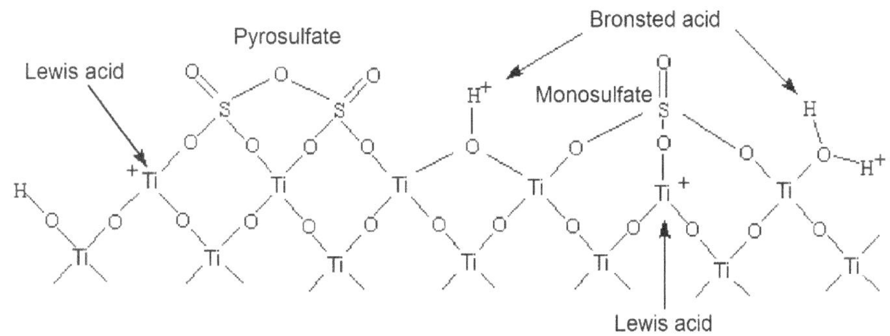

Figure 6: Lewis acid sites and Bronsted acid sites on the surface of nano SO_4^{2-}/TiO_2.

Molecular H_2O, which is generated from AN decomposition and absorbed on the surface of SO_4^{2-}/TiO_2, can enable the transformation (Lewis acid transforms to Brønsted acid). However, liquid H_2O can also cause a loss of SO_4^{2-} from the surface because of the lixiviation of sulphate groups. Thus, in the condensed phase, H_2O can weaken the catalysis of SO_4^{2-}/TiO_2. However, this factor is not important because the temperature of the AN decomposition is more than 210°C. In fact, Sun and MacNeil studied the catalysis of some inorganic acids in the AN decomposition [24, 25]. They found that hydrochloric acid had a distinctly higher catalysis ability than sulphuric acid [24]. Their reported catalysis mechanism of HCl and H_2SO_4 is not repeated here. Briefly, they used the function of the anion of the acid, whereas we used the function of the proton of the acid. In addition, a rocket designer would not introduce liquid acids into the formulation of any type of propellant because they would seriously hinder the preparation process. Hence, using solid acid may be a better choice. Moreover, in terms of acidity, solid super acids have a much lower pK_α (<-12) than sulphuric acid (pK_α=-3.0) and hydrochloric acid (pK_α=-8.0).

CONCLUSION

To promote the thermal decomposition of ammonium nitrate, we prepared a nanometre solid super acid (SO_4^{2-}/TiO_2) as the catalyst. The best baking temperature of 500°C was determined using XRD analysis. SEM and TEM analyses show that the average size of SO_4^{2-}/TiO_2 particles is 29 nm. There is an S element on the catalyst surface, as confirmed via XPS analysis.

The thermolysis properties of pure AN and AN doped with 3%SO_4^{2-}/TiO_2 were probed using TG/DSC technology. Meanwhile, for comparison, the thermolysis of AN doped with 3%TiO_2 was also analysed. The results indicate that SO_4^{2-}/TiO_2 helps to decrease the decomposition temperature by 15.8°C, but TiO_2 did not show catalysis activity in the decomposition. Using DSC-IR technology, we conclude that the main decomposition products of [AN+3%SO_4^{2-}/TiO_2] are massive amounts of N_2O and few H_2O. Please note that there was no NH_3 in the products. In the mechanism analysis, the function of SO_4^{2-}/TiO_2 was proposed based on the principle that a stronger acid can replace a weak acid. Due to the significantly low pK_a value, the protons on the surface of SO_4^{2-}/TiO_2 cause the reactions $NH_3+H^+ \rightleftharpoons NH_4^+$ and $NO_3^-+H_3O^+ \rightleftharpoons HNO_3+H_2O$ to occur. As a result, the concentrations of NH_4^+ ions and HNO_3 increase on the surface of the catalyst, which promotes AN decomposition.

ACKNOWLEDGEMENTS

This research was supported by the National Natural Science Foundation of China (Grant No.: 51206081).

REFERENCES

1. Blomquist H R (2000) Ammonium Nitarte Gas Generating Composition. US patent US006149746A.
2. Blomquist H R (2001) Cool Burning Ammonium Nitrate Based Gas Generating Composition. US patent US006231702B1.
3. Oommen C, Jain S R (1999) Ammonium Nitrate: A Promising Rocket Propellant Oxidizer. J. Hazard. Mater. A67: 253-281.
4. Xiong X T, Liu Z L (2013) Improvement of the Hygroscopicity of AN by Modified Paraffin. Chin. J. Explos. Propell. 36: 50-52.
5. Zhang X D, Li J M, Yang R J (2011) Study on Ammonium Nitrate Reactively Coated by Amines and Their Application in Azido Polyether Propellants. J. Solid Rocket Technol. 34: 86-91.
6. Erickson J D (2004) Phase-Stabilized Ammonium Nitrate. US Patent US20040149363A1.
7. Kazuo Hasue (2014) A Burning Rate Equation as a Function of Pressure and Temperature for a BTA NH3/PSAN Mixture. J. Energetic Mater. 32: 199-206.
8. Sinditskii V P, Egorshev V Yu, Levshenkov A I (2005) Ammonium Nitrate: Combustion Mechanism and the Role of Additives. Propell. Explos. Pyrot. 30: 269-280.
9. Miyata Y, Hasue K (2011) Burning Characteristics of Aminoguanidinium 5,5′-Azobis-1H-Tetrazolate/Ammonium Nitrate Mixture—Effects of Particle Size

and Composition Ratio on Burning Rate. J. Energetic Mater. 29: 344-359.

10. Naya T, Kohga M (2013) Burning Characteristics of Ammonium Nitrate-based Composite Propellants Supplemented with MnO2. Propell. Explos. Pyrot. 38: 87-94.
11. Vargeese A A, Mija S J, Muralidharan K (2014) Effect of Copper Oxide, Titanium Dioxide, and Lithium Fluoride on the Thermal Behavior and Decomposition Kinetics of Ammonium Nitrate. J. Energetic Mater. 32: 146-161.
12. Sinditskii V P, Egorshev V Yu, Serushkin V V (2012) Combustion of Energetic Materials Controlled by Condensed-Phase Reactions. Combust. Explo. Shock Waves 48: 81-99.
13. Naya T, Kohga M (2015) Thermal Decomposition Behaviors and Burning Characteristics of AN/Nitramine-Based Composite Propellant. J. Energetic Mater. 33: 73-90.
14. Kohga M, Okamoto K (2011) Thermal Decomposition Behaviors and Burning Characteristics of Ammonium Nitrate/Polytetrahydrofuran/Glycerin Composite Propellant. Combust. Flame 158: 573-582.
15. Chaturvedi S, Dave P N (2013) Review on Thermal Decomposition of Ammonium Nitrate. J. Energetic Mater. 31: 1-26.
16. Li Q Y (2005) Steam Method of Determining Acid Strength of Solid Superacids. Petrochemical Technol. 4: 75-77.
17. Zhan R R (2009) Research of Nanosolid Superacid Catalyst SO42-/ZrO2-CeO2-SiO2 Modified by Rare Earth. Chinese Rear Earths. 3: 84-88.
18. Guo H F (2008) Influences of Introducing Al on the Solid Super Acid SO42−/SnO2. Mater. Chem. Phys. 112: 1065-1068.
19. Wang Y (2009) Lewis and Brønsted Acids in Super-acid Catalyst SO42-/ZrO2-SiO2. J. Mater. Sci. 44: 6736-6740.
20. Gunawan R, Zhang D K (2009) Thermal Stability and Kinetic of Decomposition of Ammonium Nitrate in the Presence of Pyrite. J. Hazard. Mater. 165: 751-758.
21. Amores J M, Escribano V S, Ramis G (1997) An FT-IR Study of Ammonia Adsorption and Oxidation over Anatase-Supported Metal Oxides. Appl. Catal. B 13: 45-58.
22. Chuang C C, Shiu J S, Lin J L (2000) Interaction of Hydrazine and Ammonia with TiO2. Phys. Chem. Chem. Phys. 2: 2629-2633.
23. Chang J G, Ju S P, Chang C S (2009) Adsorption Configuration and Dissociative Reaction of NH3 on Anatase (101) Surface with and Without Hydroxyl Groups. J. Phys. Chem. C 113: 6663–6672.
24. Sun J H, Sun Z H, Wang Q S (2005) Catalytic Effects of Inorganic Acids on the Decomposition of Ammonium Nitrate. J. Hazard. Mater. 127: 204-210.
25. Joseph H. MacNeil (1997) Catalytic Decomposition of Ammonium Nitrate in Superheated Aqueous Solutions. J. Am. Chem. Soc. 119: 9738-9744.

CHAPTER 7 STRUCTURE-BASED ENGINEERING INCREASED THE CATALYTIC TURNOVER RATE OF A NOVEL PHENAZINE PRENYLTRANSFERASE

Georg Zocher, Orwah Saleh, Joel B. Heim, Dominik A. Herbst, Lutz Heide , Thilo Stehle

INTRODUCTION

The transfer of prenyl entities is a common enzymatic reaction found in many primary and secondary metabolic processes, for example during terpenoid biosynthesis, attachment of prenyl anchors to proteins, the biosynthesis of lipoquinones, or the production of prenylated secondary metabolites. [1], [2], [3] All of these reactions are catalyzed by prenyltransferases (PTs). A new superfamily of PTs, the soluble aromatic PT family, was identified with the discovery of two PTs involved in the biosynthetic pathway of secondary metabolites, [4], [5],[6] the enzymes CloQ and NphB from *Streptomyces*. The members of the soluble aromatic PT family are found in certain taxa of bacteria and fungi [7], [8] and catalyzes prenylation reactions of diverse aromatic substrates in secondary metabolism. The PT EpzP from*Streptomyces cinnamonensis* DSM 1042 reported here is a member of a new type of bacterial prenyltransferases, as it catalysis the prenylation of a phenazine substrate. Previously described examples of PTs are the fungal enzymes FgaPT2, [9] FtmPT1, [10] CdpNPT [11] and AnaPT [12] and the bacterial proteins NphB, [5] Fnq26, [13] NovQ, [14] Fur7, [15] and CymD. [16] Crystal structure determinations of NphB, [5] CloQ, [17] FgaPT2 [18] and FtmPT1[19] revealed that all four enzymes share a unique PT-barrel fold. The soluble aromatic PTs have been termed ABBA PTs, [8] due to the presence of five repetitive αββα-motifs that assemble into a ten-stranded antiparallel β-barrel. Furthermore, structures of NphB, [5] FgaPT2[18] and FtmPT1 [19] in complex with substrates and substrate analogs established the reaction mechanism, which is based on a Friedel-Crafts alkylation. In the first step, the prenyl diphosphate is bound to the enzyme and an electrophilic carbocation is generated, which subsequently attacks the nucleophilic aromatic substrate. In a second step, the resulting Wheland-complex is deprotonated and the product is released from the enzyme. Such a sequential order of substrate binding was determined for NphB [20] and likely also occurs in the homologous PTs.

By sequence comparison, the superfamily of soluble aromatic PTs can be divided into two families. [7], [17] The CloQ/NphB family catalyzes the prenylation of phenols, naphthalenes and phenazines, while the DMATS/CymD family catalyzes the prenylation of indole derivatives. Members of both families are found in eubacteria and fungi. The family of CloQ/NphB enzymes comprises both magnesium-independent PTs (e.g. CloQ) and magnesium-dependent PTs (e.g. NphB). Detailed insight into the catalytic mechanism of indole prenylation by FgaPT2 [18] and FtmPT1 [19] is available, where regioselectivity of the reaction is achieved by a

---

Citation: Zocher G, Saleh O, Heim JB, Herbst DA, Heide L, Stehle T (2012) Structure-Based Engineering Increased the Catalytic Turnover Rate of a Novel Phenazine Prenyltransferase. PLoS ONE 7(10): e48427. doi:10.1371/journal.pone.0048427

modifiable indole binding site inside the β-barrel. By contrast, knowledge about mechanistic aspects of the prenylation of phenols, naphthalenes and phenazines by CloQ/NphB PTs remains limited, in part because no structural data of such a PT in complex with both of its genuine substrates are available. Although a structure of the magnesium-dependent enzyme NphB in complex with geranyl diphosphate (GPP) and the non-natural substrate 1,6-dihydroxy naphthalene revealed insight into naphthalene prenylation, [5] a more sophisticated understanding of the prenylation reaction, and especially of the determinants of substrate selectivity, regioselectivity and turnover rates, is desired in the group of CloQ/NphB enzymes, in order to fully exploit their large potential as biocatalysts.

Recently, the first phenazine PT, PpzP, [21] was identified in *Streptomyces annulatus* 9663 and classified as a member of the CloQ/NphB group. [7] Gene inactivation as well as biochemical studies showed that PpzP catalyzes the regioselective prenylation at the C9 position of 5,10-dihydrophenazine-1-carboxylate (dhPCA) with the use of dimethylallyl diphosphate (DMAPP), yielding the product 5,10-dihydroendophenazine A (Figure 1). [21]Natural phenazine products are secondary metabolites isolated mainly from *Streptomyces* and *Pseudomonas,* and their pharmacological properties include broad-spectrum antibiotic, antitumor, antimalaria and antiparasitic activities. [22], [23] Several phenazine compounds derived from *Streptomyces* strains feature an *N*- or a *C*-prenylated side chain. Prenylation of phenazines contributes to biological diversity [24] and results in a wide range of biological activity. [23] For example, the *N*-prenylated phenazines benthocyanin A–C act as radical scavengers and inhibit lipid peroxidation in rat microsomes, [25], [26] whereas the *N*-prenylated phenazinomycin has antitumor activity against murine tumors and cytotoxic activity against human tumor cells. [27] Antimicrobial activities have been reported by agar plate diffusion assays for the *C*-isoprenylated compounds endophenazine A and C, and it has been shown that these compounds are active against seven out of eight tested Gram-positive bacteria and some fungi. [28].

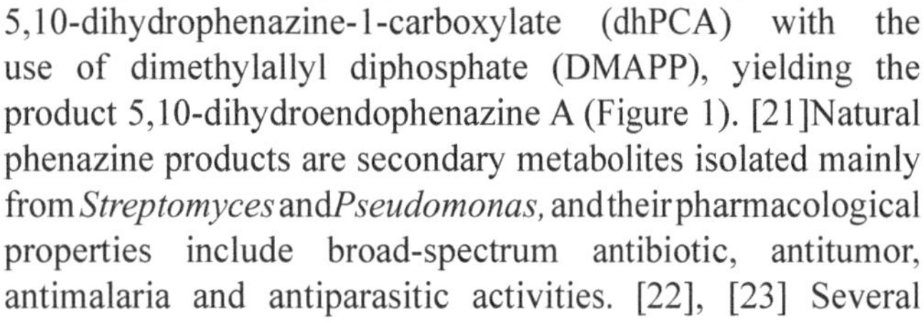

Figure 1: EpzP and PpzP catalyze the regioselective C-prenylation of 5,10-dihydrophenazine-1-carboxylate (dhPCA) to yield 5,10-dihydroendophenazine A.

The phenazine PT EpzP [29] was discovered in *Streptomyces cinnamonensis* DSM 1042. The phenazine biosynthetic gene cluster in this strain is divided into two different chromosomal loci, and previous studies revealed that isoprenoid building blocks utilized by PTs are derived mostly via the mevalonate pathway, but are additionally synthesized via the methyl-erythritol phosphate MEP pathway. [30] Biochemical characterization showed that EpzP [29] catalyzes a prenylation reaction that is identical to that described for PpzP, albeit with substantially lower turnover rates. [21], [29] We have now determined the crystal structure of EpzP from *Streptomyces cinnamonensis* DSM 1042, which is the first structure of a phenazine PT. We report structures of the enzyme in its native state as well as in complex with *S*-thiolodiphosphate (SPP). *In silico* docking of the aromatic substrate dhPCA provides insight into a likely catalytic mechanism, which is also supported by extensive site-directed mutagenesis experiments. Furthermore, based on our structural data and a comparison with PzpP we were able to engineer EpzP to yield a 14-fold higher turnover-rate compared to the wild-type enzyme with the introduction of a single point mutation.

METHODS

Protein Expression and Purification

Expression of wildtype EpzP from *Streptomyces cinnamonensis* DSM 1042 and mutant proteins was performed using an EpzP-pET28a construct. For protein production, *E. coli*Rosetta2 cells harboring the EpzP plasmid were grown at 37°C and 120 rpm in Luria-Bertani medium containing kanamycin (50 μg/mL) and chloramphenicol (34 μg/mL). 30 min before cell density reached OD600 of 0.6 the temperature was reduced to 20°C. Isopropyl-β-D-thiogalactopyranoside was added to a final concentration of 0.5 mM to start EpzP production. Cells were harvested after 16 h by centrifugation (5440*g, 10 min, 4°C). Cells were resuspended in His-A-buffer (20 mM Tris-HCl pH 8.5, 500 mM NaCl) and placed on ice. Cell lysis was performed by sonication. Subsequent centrifugation (25579*g, 30 min, 4°C) yielded supernatant that was passed through a 0.45 μm filter. The protein solution was loaded onto a 5 mL HisTrap FF column (GE Healthcare) followed by extensive washing with His-A-buffer (10 column volumes). A step gradient with His-B-buffer (20 mM Tris-HCl pH 8.5, 500 mM NaCl, 500 mM imidazole) was applied to remove impurities, and pure EpzP was eluted with 30% of His-B-buffer. Fractions containing pure EpzP were pooled and concentrated. For protein used in activity assays, the buffer was exchanged to SEC buffer (HEPES-NaOH pH 7.5, 150 mM NaCl). Protein folding of EpzP and point mutants of EpzP was verified by circular dichroism spectroscopy. For structure determination the His-Tag of EpzP was removed by thrombin cleavage. This was achieved by exchanging the buffer (20 mM Tris pH 8, 200 mM NaCl, 2.5 mM $CaCl_2$), adding thrombin (0.5 U per milligram of EpzP) and incubating for 12 h at 4°C. Uncleaved EpzP, as well as the cleaved His_6-tag were removed by passing the solution through a 5 mL HisTrap FF column (GE Healthcare) coupled with a benzamidine column. The flowthrough was pooled and concentrated for size-exclusion chromatography using SEC buffer and a Superdex S200/16/60 column. The elution volume of EpzP corresponds to a monomeric state of the enzyme. The pure protein was either used for crystallization trials or methylated prior to crystallization. Initial crystallization trials with the wildtype enzyme failed. Therefore, we performed methylation of the protein sample as 19 lysine residues were found in the sequence of EpzP (302 amino acids). Methylation alters the size and mobility of the lysine side chains and can in some cases promote crystallization.

For methylation of lysine residues the protein sample was diluted to 0.4 mg/mL and placed on ice. Freshly prepared dimethylamine-borane (1 M, 20 μL/mg of EpzP) and formaldehyde solution (1 M, 40 μL/mg of EpzP) were added, and the mixture was extensively stirred. The addition of dimethylamine-borane and formaldehyde was repeated after 3 h and incubated over night at 4°C. Afterwards, aggregates were removed by centrifugation (25579*g, 20 min, 4°C), and the supernatant was concentrated and applied to size exclusion chromatography (SEC), as described above. Successful methylation was determined by mass spectroscopy, which showed that 18 out of 19 lysine residues had been methylated in EpzPm. Proper folding of all EpzP proteins was analyzed with circular dichroism spectroscopy (J-720 CD spectrometer, JASCO; wavelength 205–250 nm), using samples with a concentration of 0.3 mg/mL. CD spectra of the EpzP variants did not show any substantial deviation compared to the CD spectrum of the wildtype enzyme and therefore indicate correct folding of all EpzP variants. All protein samples were flash-frozen in liquid nitrogen until crystallization experiments were performed.

Crystallization and Structure Determination

All crystallization experiments of EpzP were performed by sitting drop vapor diffusion method. Therefore, 600 nL of the protein solution (30 mg/mL) was mixed 1:1 with the crystallization buffer and placed on a 96-well plate above the reservoir solution (100 µL). Initially, crystals of methylated EpzP (EpzPm-nat) that diffract to 1.33 Å resolution were obtained in crystallization buffer I (200 mM magnesium chloride, 30% (w/v) PEG4000, Tris hydrochloride pH 8.5) at 4°C. A second crystal form of methylated EpzP grew in crystallization buffer II (200 mM $(NH_4)_2SO_4$, 30% (w/v) PEG2000MME, 100 mM sodium acetate 4.6) at 20°C. To incorporate dimethylallyl S-thiolodiphosphate (DMSPP), these crystals were soaked with 5 mM DMSPP for 30 min at 20°C to give EpzPm-SPP. Cryoprotection was achieved by transferring crystals into the crystallization solution containing 12% (v/v) PEG400. EpzPwt was crystallized at 20°C (EpzPwt). The protein solution (0.6 µL, 25 mg/mL) was mixed 1:1 with crystallization buffer III (2.4 M $(NH_4)_2SO_4$, 100 mM HEPES/NaOH pH 7.5, 200 mM sodium acetate, 0.6 µL). Crystals grew to a final size of 200×100×100 µm³ within three months. They were then harvested, mounted in loops, and flash-frozen in liquid nitrogen until data collection.

Data sets of EpzPm-SPP, EpzPm-nat and EpzPwt were recorded at beam line X06DA at the Swiss Light Source in Villigen, Switzerland. Data reduction was performed with XDS and XSCALE. [31] Crystals of EpzPwt belong to orthorhombic space group $P2_12_12_1$, and contain a single monomer in the asymmetric unit. Two copies of the protein were found in the asymmetric units of EpzPm-SPP and EpzPm-nat. Protein phases resulted from molecular replacement using a model of CloQ [17] that had been modified with CHAINSAW [32]. To avoid model bias, the models were either rebuilt using ARP/wARP [33], or simulated annealing was performed in PHENIX. [34] The final models were obtained after several cycles of manual building with COOT, [35] followed by refinement with REFMAC5. [36] Water molecules were placed using COOT:Find_waters. *S*-thiolodiphosphate in EpzPm-SPP, sulfate and PEG molecules in EpzPwt and EpzPm-nat were inserted with COOT and verified after refinement. The final refinement step involved TLS parameterization [37] using a single TLS group per protomer for EpzPwt and multi-group TLS refinement for EpzPm-SPP and EpzPm-nat. The TLS parameters were obtained from the TLSMD server. [38] The geometries of the final models were analyzed with RAMPAGE [39] and SF-CHECK. [32] Figures were generated using POVscript+ [40] and POVRAY (http://povray.org) or pymol. [41].

Substrate Modeling

For DMSPP modeling, the diphosphate groups of EpzPm-SPP and NphB (PDB code:1ZB6) were superposed and kept as fixed anchors. The dimethylallyl moiety was placed manually to fit the binding mode of GPP in NphB. The coordinates of the substrate dhPCA were obtained from the PRODRG server [42] and used for *in silico* modeling into the structure of EpzPwt by the use of Vina. [43] For docking calculations, all solvent molecules inside the cavity were removed, with the exception of two structurally conserved water molecules (W1 and W2). The side chain atoms were kept in fixed positions, with the exception of R267, which was known to be flexible, and the search box for the ligand docking experiment was extended to cover the entire active site. Conformations that exhibit the lowest binding energy were investigated for meaningful chemical and biochemical properties of the catalyzed prenylation reaction.

Mutagenesis

Site-directed mutagenesis was performed to generate site-specific mutations on EpzP. To confirm the active site and the docking calculation, mutant primer pairs were designed to generate the appropriate EpzP mutants: R62N_fwd (5′-GAC CTC GAC TGC AAC TTC ACG ATG CTC CCG AAG GGC CTC GAC CCG-3′) and R62N_rev (5′-GAG CAT CGT GAA GTT GCA GTC GAG GTC GTC CGC GCT GCT GCC GGT-3′); R267Q_fwd (5′-C ACC GAG GGC CAG AAC TAC GTC TAC GGC ATC GCG TCG ACG CCG-3′) and R267Q_rev (5′-GTA GAC GTA GTTCTG GCC CTC GGT GTC GTA CGG CGC GCT CTT GAT G-3′); V270F_fwd (5′-GC CGC AAC TAC TTT TAC GGC ATC GCG TCG ACG CCG AAG GGC GAA TAC-3′) and V270F_rev (5′-GC GAT GCC GTA AAA GTA GTT GCG GCC CTC GGT GTC GTA CGG CGC-3′); S236A_fwd (5′-GG TTC ACG TAC GCG GTC ATG ACG CCG GAG CCG CTG GGC CTC-3′) and S236A_rev (5′-GG CGT CAT GAC CGC GTA CGT GAA CCG GTT GAT CTT CGG CGA G-3′); T64Y_fwd (5′-C TGC CGC TTC TAT ATG CTC CCG AAG GGC CTC GAC CCG TAC GCC CGC-3′) and T64Y_rev (5′-CTT CGG GAG CAT ATA GAA GCG GCA GTC GAG GTC GTC CGC GCT GCT GCC-3′); Y287F_fwd (5′-G ATC GCG TCG TTC TAC CAG TGG CAG AAG CGC GTG GAG AAG CTG-3′) and Y287F_rev (5′-G CCA CTG GTA GAA CGA CGC GAT CTT GTG GTA TTC GCC CTT CGG C-3′); A285L_fwd (5′-C CAC AAG ATC CTG TCG TAC TAC CAG TGG CAG AAG CGC GTG GAG AAG-3′) and A285L_rev (5′-G GTA GTA CGA CAG GAT CTT GTG GTA TTC GCC CTT CGG CGT CGA CG-3′); A285Q_fwd (5′-C CAC AAG ATC CAG TCG TAC TAC CAG TGG CAG AAG CGC GTG GAG AAG-3′) and A285Q_rev (5′-G GTA GTA CGA CTG GAT CTT GTG GTA TTC GCC CTT CGG CGT CGA CG-3′);V218G_fwd (5′-G CGC TCG TTC GGCATC TAC GTC ACC CTG AGC TGG GAC TCG C-3′) and V218G_rev (5′-GT GAC GTA GATGCC GAA CGA GCG CTC GCA GAA CTT GAG CAG CTC-3′); G272V_fwd (5′-C TAC GTC TAC GTG ATC GCG

TCG ACG CCG AAG GGC GAA TAC CAC-3′) and G272V_rev (5′-CGT CGA CGC GAT CAC GTA GAC GTA GTT GCG GCC CTC GGT GTC GTA C-3′); dM65_fwd (5′-GC CGC TTC ACG CTC CCG AAG GGC CTC GAC CCG TAC GCC CGC-3′) and dM65_rev (5′-CC CTT CGG GAG CGT GAA GCG GCA GTC GAG GTC GTC CGC GCT G-3′). The PCR mixture (50 μL) contained dNTP (1 μL, 25 mM), polymerase buffer (5 μL, 200 mM Tris-HCl (pH 8.8), 100 mM $(NH_4)_2SO_4$, 100 mM KCl, 1 mg/mL BSA, 1% (v/v) Triton X-100, 20 mM $MgSO_4$), Pfu DNA Polymerase (1 μL, Fermentas, 2.5 U/μL), EpzP-pET28a-Plasmid (25 ng), primer premix (1 μL, forward and reverse primer, 100 pM) and DMSO (1 μL). The PCR was performed using a typical thermal cycling (1 cycle: 95°C for 5:00 min, 18 cycles: 95°C for 1:00 min; 65.7°C for 1:00 min; 72°C for 14:00 min; 1 cycle: 72°C for 20:00 min). The template DNA was digested by adding Dpn I (Fermentas,1 μL, 10 U/μL) and incubated for 2 h at 37°C. Precipitation of DNA was carried out by adding 1.1 volumes of isopropanol and 0.14 volumes of sodium acetate (3M, pH=5.2). The mixtures were stored for 2 h at −80°C and then centrifuged (10 min, 13000 g, 4°C). The supernatant was removed, the DNA pellet was dried and diluted in 10 μL water. 5 μL of the DNA were applied for transformation in XL10-Gold® Ultracompetent Cells according to the manual (Stratagene). Successful mutagenesis was verified by sequencing. The mutant proteins were expressed and purified as described above. Correct folding of all EpzP mutants was determined by circular dichroism spectroscopy (J-720 CD spectrometer, JASCO).

Enzyme Assays

In the prenyltransferase assays, product formation was quantified by HPLC analysis using UV detection at 365 nm. The substrate dihydro-PCA was prepared from commercially available PCA by reduction with dithionite. The enzymatic product dihydroendophenazine A is rapidly oxidized and was therefore converted to the more stable endophenazine A by use of sodium persulfate before extraction from the aqueous reaction mixture by ethyl acetate. The enyzme assay (100 μL) contained 100 mM Na-TAPS (pH 7.5) (Sigma), 0.4 mM freshly prepared dhPCA, 1 mM DMAPP, and 5 μg/mL protein. For the preparation of dhPCA, 90 μL of 50 mM freshly dissolved sodium dithionite (Merck) were mixed with 10 μL of 100 mM PCA dissolved in Tris-HCl 1 M, pH 8.0. 4 μL of this mixture was added to the incubation mixture. After incubation for 5 min at 30°C, 10 μL of 100 mM sodium persulfate (Sigma) were added to oxidize dihydro-PCA and dihydroendophenazine A into PCA and endophenazine A, respectively. The mixture was immediately extracted with 100 μL of ethyl acetate/formic acid (975:25). After vortexing and centrifugation, 75 μL of the organic phase were evaporated. The residue was dissolved in 100 μL of methanol, and 35 μL thereof were analyzed by HPLC-UV and HPLC-MS. The HPLC-UV analysis was carried out using an Eclipse XDB-C18 column (4.6×150 mm, 5 μm; Agilent) at a flow rate of 1 mL/min with a linear gradient from 40 to 100% of solvent B in 20 min (solvent A: water/formic acid (999:1); solvent B, methanol) and detection at 252 and 365 nm. Additionally, a UV-spectrum from 200 to 400 nm was logged by a photodiode array detector. For quantitative analysis, authentic reference sample of PCA as external standards was used. The HPLC-MS analysis was carried out using a Nucleosil 100-C18 column (3 μm, 100×2 mm) coupled to an ESI mass spectrometer (LC/MSD Ultra Trap System XCT 6330; Agilent Technology). Analysis was carried out at a flow rate of 0.4 mL/min with a linear gradient from 10 to 100% of solvent B in 15 min (solvent A: water/formic acid (999:1); solvent B: acetonitrile/formic acid (999.4:0.6)). Detection was carried out at 230, 260, 280, 360, and 435 nm. Electrospray ionization (positive and negative ionization) in Ultra Scan mode with capillary voltage of 3.5 kV and heated capillary temperature of 350°C was used for LC-MS analysis. Enzymatic activities were calculated in nmol product s^{-1} mg^{-1} enzyme. Activities of EpzP variants were expressed relative to the activity of wildtype EpzP, which was set as 1.

RESULTS

Structures of EpzP

Crystallization of unmodified, wild-type EpzP proved initially difficult. However, methylation of the EpzP lysine residues (see Methods) allowed us to grow needle-like crystals (200 mM magnesium chloride, 30% (w/v) PEG4000, Tris hydrochloride pH 8.5) of methylated EpzP (EpzPm). These crystals diffracted to very high resolution of 1.33 Å, belong to the monoclinic space group $P2_1$, and were used to assemble a data set for the unliganded protein (EpzPm-nat). Unfortunately, the EpzPm-nat crystals were not suitable for soaking with substrates because the active site is blocked by a symmetry-related protomer from one site, whereas the aromatic binding site is blocked by a PEG molecule (Figure 2b). Therefore, a second crystal form for EpzPm that grew in an acidic crystallization condition (200 mM $(NH_4)_2SO_4$, 30% (w/v) PEG2000MME, 100 mM sodium acetate 4.6) was used to prepare a complex by soaking with 5 mM of the substrate analog dimethylallyl *S*-thiolodiphosphate (DMSPP). These crystals belonged to the orthorhombic space group $P2_12_12_1$, diffracted to 1.67 Å resolution and were used to determine the structure of EpzP in complex with DMSPP (EpzPm-SPP). Inspection of the electron density map for EpzPm-SPP revealed the presence of a *S*-thiolodiphosphate (SPP) moiety in the diphosphate binding site, implying that hydrolysis of DMSPP had taken place at the acidic pH of 4.5 used during crystallization. Subsequent crystallization screening of non-methylated EpzP eventually yielded a third crystal form (2.4 M $(NH_4)_2SO_4$, 100 mM HEPES/NaOH pH 7.5, 200 mM sodium acetate) that also belonged to space group $P2_12_12_1$and diffracted to 1.93 Å resolution. Structure analysis of this crystal from (EpzPwt) and comparison with the two earlier structures confirmed that lysine methylation had not led to structural changes in the protein. Moreover, the EpzPwt structure showed that, in contrast to EpzPm-SPP and EpzPm-nat, the *C*-terminal portion of the enzyme was visible in the electron density and found to seal one entrance of the β-barrel (Figure 2a). All structures of EpzP were refined to the best possible quality and possess low free R-factors, plausible stereochemistry, and exhibit small root-mean-square deviations (r.m.s.d.) from ideal values for bond lengths and bond angles (Table 1). However, as crystals of EpzPwt were not reproducible and other crystal forms could not be obtained, we were unable to prepare a ternary complex. In order to gain insight into the mode of substrate binding, we therefore decided to perform *in silico* based substrate modeling.

Table 1: Data collection and refinement statistics[a]

	EpzPwt	EpzPm-nat	EpzPm-SPP
Data collection statistics			
Resolution (Å)	[illegible]	[illegible]	[illegible]
Spacegroup	$P2_12_12_1$	$P2_1$	[illegible]
Unit cell (Å)	[illegible]	[illegible]	[illegible]
No. of reflections			
Measured	[illegible]	[illegible]	[illegible]
Unique	57232 (4072)	[illegible]	65469 (4725)
R_{merge} (%)	[illegible]	[illegible]	[illegible]
Completeness (%)	98.6 (94.4)	93.5 (83.7)	99.3 (99.6)
Multiplicity	[illegible]	[illegible]	[illegible]
$I/\sigma(I)$	10.7 (2.0)	18.7 (3.2)	10.7 (2.9)
Wilson B Factor (Å²)	24.1	16.5	20.1
Mosaicity (°)	0.291	0.205	0.212
Refinement statistics			
R_{work}/R_{free} (%)	17.3/19.8	17.9/20.1	18.6/22.4
No. of atoms			
Chain A/B	3115/-	2295/2325	[illegible]
Ions/Water	[illegible]	[illegible]	[illegible]
Average B-Factors (Å²)			
Chain A/B	[illegible]	13.2/13.4	[illegible]
Ions/Water	43.6/31.1	20.1/18.7	27.7/32.8
Rmsd bond length (Å)	[illegible]	[illegible]	[illegible]
Rmsd bond angle (°)	1.529	[illegible]	1.487
Ramachandran angles			
Favored (%)	97.7	[illegible]	98.0
Allowed (%)	[illegible]	[illegible]	[illegible]
Outliers (%)	0.3	0.0	0.4

[a]Values in parentheses are for the highest resolution shell. [illegible]

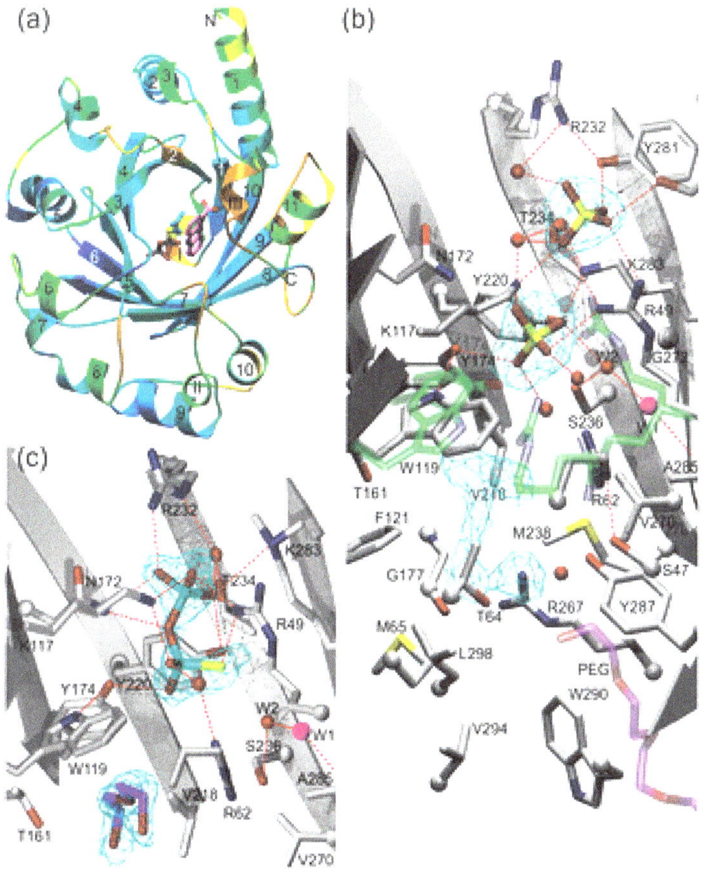

Figure 2: Structures of EpzP.

a) Overall structure of EpzPwt reveals an ABBA PT-fold. Secondary structural elements are colored according to their r.m.s. deviations of C_α-atom positions compared to the most homologous structure of NphB (PDB-Code: 1ZB6) from blue (zero r.m.s.d.) to orange (r.m.s.d. above 3 Å). The substrates dhPCA and DMAPP are modeled and shown in stick representation. b) View into the active site of EpzPwt. Side chains which showed a different orientation in EpzPm-nat are depicted in transparent green color. R267 points towards the cavity in EpzPwt and away from it in the EpzPm-nat and EpzPm-SPP structures. Water molecule W1, which is proposed to deprotonate the Wheland complex, is shown in magenta. Two sulfate ions occupy the diphosphate binding site of EpzPwt. The (F_{obs}-F_{calc})-omit map (cyan) is shown at σ-level of 3.2 for both sulfate ions and inside the barrel. The remaining electron density in the proposed substrate binding site of EpzPwt could not be explained; it does not fit with any molecule used in downstream purification and crystallization experiments and may represent a molecule inserted into the enzyme during protein production. c) The active site of EpzPm-SPP viewed along the same axis. The (F_{obs}-F_{calc})-omit map (cyan) is shown at σ-level of 3.0 and clearly reveals the presence of the *S*-thiolodiphosphate moiety (cyan) and a PEG molecule (purple). Hydrogen bonds and hydrophobic interactions are represented with red and green lines, respectively.

Overall Structure and Comparison to ABBA Prenyl Transferases

EpzP is a soluble, monomeric ABBA PT consisting of 302 amino acids with a molecular mass of 33.2 kDa. The overall structure of EpzP belongs to the PT-fold family [7] and is characterized by ten antiparallel β-strands that assemble into a circular β-barrel. The β-barrel houses a solvent-filled reaction chamber in which the substrates bind and the regioselective Friedel-Crafts prenylation of dhPCA takes place. The β-barrel is surrounded by eleven α-helices that constitute the major fraction of solvent accessible surface of the enzyme (Figure 2a). Additionally, three 3_{10}-helices are present in the structure. One of these, the *C*-terminal 3_{10}-helix in EpzPwt, seals the reaction chamber of the aromatic binding site of dhPCA.

As expected, structural comparisons using DALI [44] and PDBeFold [45] revealed highest similarities to the four known soluble ABBA PT structures. The bacterial PTs NphB from*Streptomyces sp. strain* CL190 [5] and CloQ from *Streptomyces roseochromogenes* [17] are very similar in size (307 and 324 amino acids, respectively). With C_α-rmsd values of 1.8 Å (288 of 301 aligned amino acids) and 2.2 Å (282 of 301 aligned amino acids), respectively, NphB and CloQ represent the closest structural homologs of EpzP (Figure 2a). The largest deviations among the enzymes occur at the *C*-termini of the enzymes (Figure 3a–c). The dimeric, fungal prenyltransferases FgaPT2 [18] and FtmPT1, [19] both from *Aspergillus fumigatus*, are significantly larger (459 residues for FgaPT2, 464 residues for FtmPT1) but their core regions align well with the ten-stranded antiparallel β-barrel of EpzP, as shown by reasonable C_α-rmsd values of 3.2 Å for each FgaPT2 and FtmPT1 (231 amino acids were aligned in each case). The additional residues in FgaPT2 and FtmPT1 are found in α-helical and loop regions that connect the core domain. Except for the *C*-termini, these residues do not contribute to the formation of the reaction chamber. Although the β-barrel structures of the NphB/CloQ and DMATS/CymD family of enzymes superimpose well, the architectures of the active centers differ substantially between both families. Since EpzP is a member of the NphB/CloQ family, [7] we will compare the EpzP structure with these two enzymes.

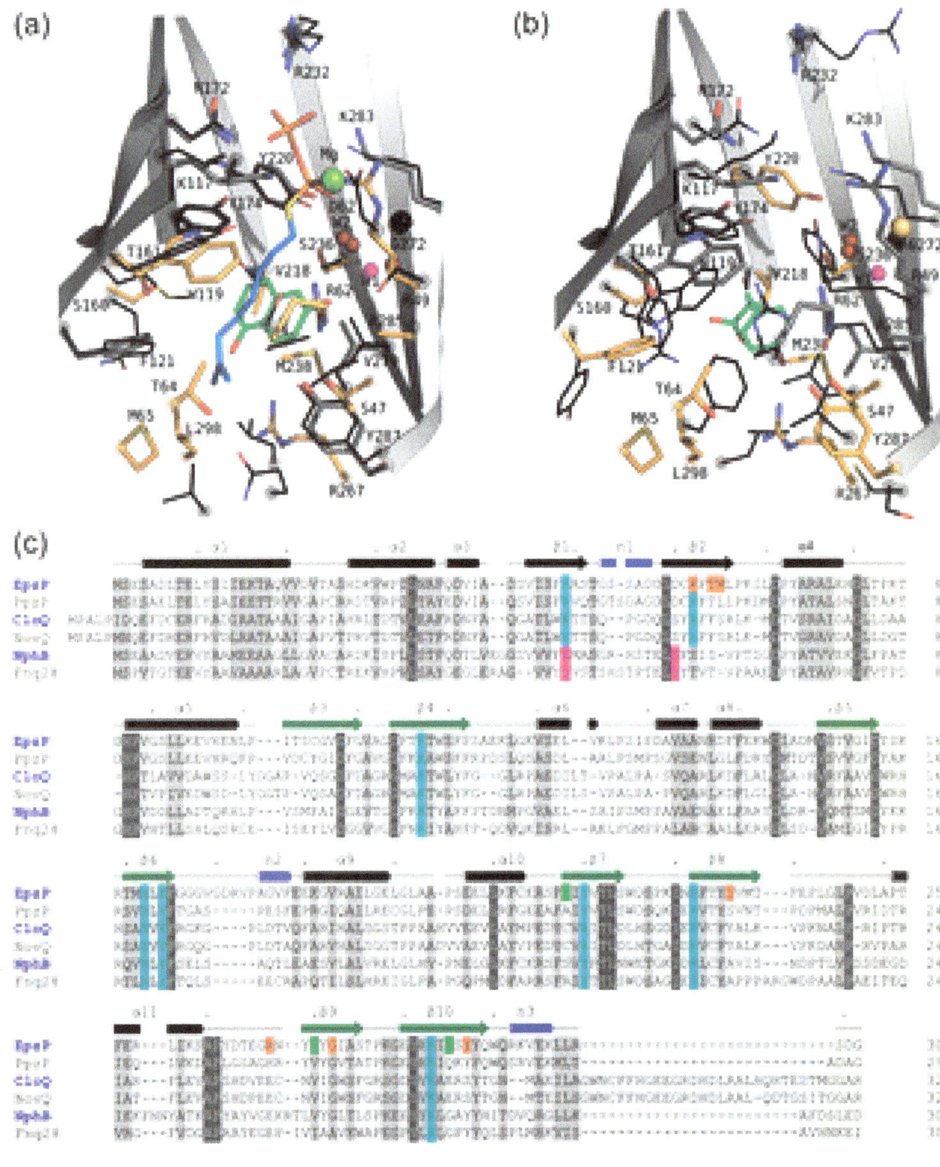

Figure 3: Comparison of EpzP structure and sequence with other PTs of the NphB/CloQ family. a)

Superposition of NphB (black lines) and EpzP. b) Superposition of CloQ (black lines) with EpzP. The superpositions reveal variability in structure and sequence in the aromatic binding site at the C-terminal part of the enzymes. Side chain residues of EpzP that differ from NphB or CloQ are shown orange, whereas conserved residues are shown grey. The catalytic water molecules W1 and W2 are shown as spheres in magenta and red, respectively. The substrates of NphB and CloQ are shown in stick representation (green). c) Structure-based sequence comparison of EpzP, CloQ and NphB and sequence comparison of PpzP, NphB and Fnq28. NphB and Fnq28 are magnesium-dependent enzymes, whereas the others are not. Amino acids of the diphosphate binding site (cyan), mutations that confirm the *in silico* docking model (orange), and mutations that modify the enzymatic turnover (green) are emphasized.

In comparison to EpzP, the aromatic binding site of NphB (Figure 3a and c) is wider and reveals a substantial difference in the geranyl binding cavity. Compared to EpzP, M65 of EpzP is deleted in NphB, which results in a reorientation of strand β2 to facilitate binding of the geranyl entity. A structure comparison of the active center of EpzP with CloQ revealed a substantially smaller cavity for the aromatic substrate of CloQ (Figure 3 and c). The size of cavity in CloQ is restricted mainly by F68, F161, Y233 and W295. All residues are replaced with less voluminous side chains (T64, T161, S236, L298) in EpzP, in agreement with the elongated shape of dhPCA. In summary, the binding site for the diphosphate entity is moderately conserved in structure and sequence, whereas the aromatic binding sites of EpzP, NphB and CloQ differ significantly. A similar observation has been made for the FgaPT2/CymD family, but variations in the aromatic groove are more subtle in that family as substrates share the indole entity as a core feature for substrate binding. [19].

The Active Center of EpzP

The active site within the β-barrel can be subdivided into the aromatic substrate site and the DMAPP binding site. When the C-terminal 3_{10}-helix is defined as the bottom of the β-barrel, the diphosphate binding pocket is found in the upper part of the β-barrel, whereas the aromatic binding site of dhPCA is found in its lower part. The diphosphate binding site is formed by four basic residues (R49, K117, R232 and K283) that compensate for the negative charge of the diphosphate entity (Figure 2b and c). Furthermore, two tyrosines (Y174 and Y220) and an asparagine (N172) contribute to the diphosphate binding as hydrogen bond donors. Additionally, two water-mediated interactions link R62 and T234 to the diphosphate moiety. All basic residues are highly conserved within the family of magnesium-independent PTs. The positively-charged diphosphate binding groove of magnesium-dependent PTs differ at two residues (R49 and R62 for EpzP) (Figure 3a–c). While R62 is not involved in diphosphate binding, R49 is replaced with an aspartate (D51) in NphB. The loss of the positive charge of the R49 side chain in the magnesium-independent PT EpzP is compensated in NphB by a magnesium ion that is coordinated via the aspartate side chain.

The aromatic binding pocket of dhPCA is located opposite to the diphosphate binding site, within a large cavity inside EpzPwt. The pocket is lined with predominantly hydrophobic residues (V47, M65, W119, F121, G177, V218, M238, V270, A285, L298, W290). Only a few polar side chains (S47, T64, T161, Y287, S236) are present, and these could conceivably contribute to substrate binding via a small number of hydrogen bonds (Figure 2b and c). This situation agrees well with the mainly hydrophobic character of dhPCA, a substrate that offers limited options for hydrogen bond formation. Additionally, two positively charged side chains (R62 and R267) are present in this cavity. While R62 contributes to interactions with SPP, R267 does not. Furthermore, the R267 side chain is located in a mostly hydrophobic environment. It is thus tempting to speculate that R267, and perhaps also R62, might interact with the carboxyl function of dhPCA. Interestingly, we found substantial electron density in the cavity of EpzPwt, in an area near the predicted binding site of dhPCA (Figure 2c). The density is indicative of the presence of an ordered compound. Although we were not able to unambiguously identify the nature of this compound, we conclude that it is inserted into the enzyme during protein production, as the shape of the density does not fit with any molecules used in downstream purification and crystallization experiments.

Conformational Changes in EpzP

In contrast to EpzPwt, which contains a fully ordered *C*-terminus that closes the reaction chamber, the structures of EpzPm-SPP and EpzPm-nat have disordered *C*-termini that cannot be traced beyond residue K292. The lack of an ordered C-terminus renders the aromatic binding site solvent-accessible, and both methylated structures furthermore contain PEG molecules that occupy the cavity (Figure 2b). Interestingly, the opening of the aromatic site is linked to a change in the conformation of R267. In the closed structure of EpzPwt, R267 points towards the active center, whereas in the opened structure of EpzPm-SPP R267 is orientated to the solvent. It is unlikely that this conformational change is related to the methylation of the protein, as K292 (the only lysine residue at the *C*-terminus) is not involved in intramolecular interactions, and no other lysine residue is involved in contact formation with the terminal 3_{10}-helix. Furthermore, the flexibility of the C-terminus is confirmed by B-factor analysis, which reveals that the average B-factor for the main chain atoms of residues 292 to 302 is increased by a factor of two. Therefore, we suggest that R267 might trigger a conformational change of EpzP by "scanning" the solvent for the negatively charged substrate and pulling it into the active center resulting in a closed up active site. An important role of R267 for catalysis is furthermore established by *in silico* docking, as well as site-directed mutagenesis.

Model of Catalysis

Although we were able to incorporate SPP into the binding site of EpzP (EpzPm-SPP), we could not establish a ternary complex of EpzP with its substrate analogue DMSPP and the aromatic substrate dhPCA in our crystallization experiments. Therefore, we modeled the prenyl moiety into the active site and performed *in silico* docking experiments with dhPCA to gain insight into catalysis.

Based on the structures of EpzPm-SPP (Figure 2c) and NphB in complex with GPP (seeMethods), we modeled DMSPP into the EpzP binding site, utilizing the SPP group as fixed anchors. The dimethyallyl entity faces towards T64 and F121, in such a way that the π-system can interact with the π-orbitals of W119 (Figure 4a and b). Furthermore, F121, T64, R62 and Y174 shield the dimethylallyl moiety. This is comparable with the situation found in the NphB-GPP complex. [5] Here, the geranyl moiety can interact with the aromatic system of Y121 to shield a nascent carbocation during catalysis. Interestingly, W122 in CloQ [17] occupies a different side chain conformation, which would preclude π-π-interaction of the substrate to the aromatic ring system. This difference is most likely a consequence of the absence of sulfate, phosphate or diphosphate groups in the diphosphate binding sites of all structures of CloQ. These ions presumably mimic substrate binding. Comparison of EpzPm-nat, which lacks diphosphate or sulfate in the binding pocket, with EpzPwt and EpzPm-SPP revealed different side chain conformations of W119. In the unbound state, the side chain of W119 is rotated by 180°. Furthermore, residues R62 and R49, which both participate in DMAPP binding, adopt alternative conformations when the anion binding site is not fully occupied (Figure 3b). A substrate-induced fit in EpzP is furthermore supported by the reorientation of Y174 towards the phosphate group during phosphate binding, as this would trigger a side chain flip of W119 as a result of the formation of a hydrogen bond of Y174 to W119 (Figure 2b). The induced fit therefore yields a protected environment that stabilizes the formation of the carbocation during the first step of the catalytic cycle.

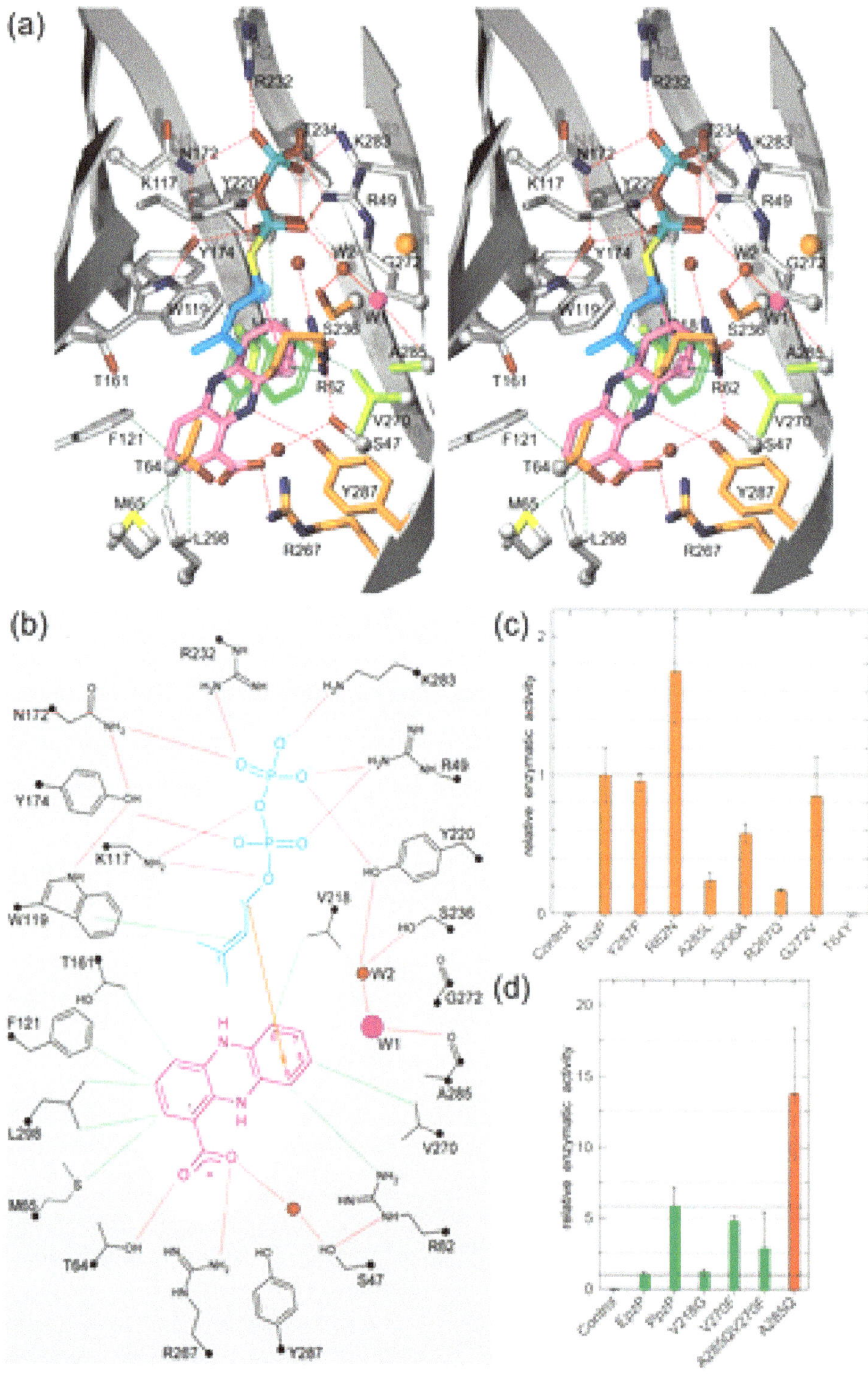

Figure 4: Model of catalysis.

a) Model of phenazine prenylation based on crystal structures, *in silico* docking and site-directed mutagenesis. The reactions centers of DMAPP (blue) and dhPCA (purple) are emphasized with spheres. Water molecule W1 (magenta) deprotonates the intermediate Wheland complex. EpzP mutants that confirm the model of catalysis or have been engineered to modify turnover-rates are colored orange and green, respectively. 1,6-dihydroxy naphthalene (green), the substrate of NphB, occupies essentially the same binding pocket as dhPCA in EpzP. **b)** Schematic view of substrate binding in EpzP. **c)** and **d)** Relative enzymatic activities of EpzP variants compared to the wild-type EpzP.

The aromatic substrate dhPCA was docked *in silico* into the active site of EpzPwt (seeMethods) by the use of Vina. [43] Two structurally conserved water molecules were kept in the dhPCA binding pocket for the docking calculations, whereas all other solvent molecules were removed from the pocket. A planar configuration of dhPCA was used for docking. Structure determination of phenazine complexes in the oxidized and reduced state revealed that the energy barrier between a bent and planar conformation of the tricyclic ring system is low and that dihydrophenazine predominantly adopts a planar conformation. [46] A model of the ternary complex is shown in Figure 4a and b. We selected this model for two reasons. First, it exhibited the lowest binding energy in our docking calculations and revealed meaningful interactions between dhPCA and the protein environment. Secondly, the model is in good agreement with our biochemical data, [29] as it explains the regioselectivity of the catalyzed prenylation reaction. The docking solution is also supported by superpositions of NphB [5](pdb-code: 1ZB6) and CloQ [17] (pdb-code: 2XLQ) with EpzP, which reveal that dhPCA occupies the same binding pocket as found for the aromatic substrate 1,6-dihydroxy naphthalene in NphB (Fig. 4a) and 2-keto-3-(4-hydroxyphenyl)propanoic acid for CloQ. Furthermore, the distance between the reacting carbon atoms is 4.8 Å and is slightly shorter compared to the distance of 6.2 Å found in the ternary complex of NphB with the non-natural substrate 1,6-dihydroxy naphthalene. [5].

The dhPCA moiety fits tightly into a pocket formed by residues R62, Y174, V218, Y220, S236, R267, V270, Y287 and A285. As a result, the C1-position of DMAPP is oriented towards C9 of dhPCA. An electrophilic attack to C6 of dhPCA seems impossible as steric collisions would preclude the formation of the nascent Wheland complex. It has been shown for CloQ that E281 can act as base to abstract the proton of the sigma complex in order to regain the aromaticity of the product. This has been demonstrated by a point mutation E281Q, which exhibited only 1% of the wild-type activity. [17] Residue A285 of EpzP, which is equivalent to E281 in CloQ, is not able to act as a base. However, two structurally conserved water molecules are located near A285, and both of these are held in place by hydrogen bonds. Water molecule W1 (Figures 4a and b), which we propose acts as base, is fixed via hydrogen bonds to the main chain carbonyl of V270 and the second water molecule, W2, which has tight interactions with the hydroxyl groups of S236 and Y220. Interestingly, both water molecules are also present in the NphB structure, and a similar proton relay for deprotonation has been described in that case. [5] Furthermore, this proton transfer was confirmed in NphB on the basis of quantum mechanics and molecular mechanics simulation. [47].

Site-directed Mutagenesis of Active-site Residues

In order to provide experimental support for the docking solution, we performed site-directed mutagenesis of residues that are likely to play a role in catalysis. We purified these EpzP variants and determined their catalytic activities as described in a previous study. [29] The substrate dihydro-PCA was prepared from commercially available PCA by reduction with sodium dithionite. The enzymatic product dihydro-endophenazine A is rapidly oxidized and was therefore converted to the more stable endophenazine A by use of sodium persulfate before extraction from the aqueous reaction mixture and HPLC analysis. A summary of the results is depicted in Figures 4c and d. In these figures, reaction velocities observed under catalysis of the different EpzP variants are expressed relative to the reaction velocity under catalysis of wildtype EpzP (0.93 nmol product • s^{-1} • mg^{-1} enzyme). All mutants were purified as soluble proteins and are properly folded as analyzed by CD spectroscopy.

The DMAPP binding site of indole prenyltransferases is highly conserved, and substrate specificity and regioselectivity for the prenylation reaction is obtained via a modifiable reaction chamber at the bottom of the β-barrel of fungal indole prenyltransferases. [19] A similar situation is observed for aromatic PTs in the CloQ/NphB family, which includes EpzP. Structure-based sequence alignments of different members in this family revealed no significant sequence diversities within the binding sites for the prenyl diphosphate, with the exception of the above-described differences between magnesium-dependent and magnesium-independent enzymes. Therefore, a single mutant T64Y was prepared to verify the correct modeling of the dimethylallyl entity of DMAPP. Mutation of T64 to tyrosine abrogates enzymatic activity (Figure 4c), as the increased size of the tyrosine side chain would result in a steric collision with the dimethylallyl entity of the substrate DMAPP. The enzyme cannot avoid this collision, as the side chain of tyrosine is not able to adopt different conformations due to steric restrains in its surrounding. Therefore, no detectable turnover is observed. This result directly supports the placement of DMAPP and also the proposed catalytic mechanism.

Two positively charged side chains, R62 and R267, are found in the dhPCA binding site of EpzPwt. In principle, both could participate in substrate binding by compensating the negative charge of the carboxyl group of dhPCA. Our docking solution for dhPCA suggests that R267 is involved in binding of the carboxylate group, while R62 mostly contributes to surface complementary, as well as to the water-mediated diphosphate stabilization as described above. These observations are in full agreement with site-directed mutagenesis experiments, as we observe a drastic reduction of catalysis when R267Q is introduced in EpzP (Figure 4c). The reaction velocity drops to 17% of that achieved with the wild-type protein (i.e. to 0.16 nmol s^{-1} mg^{-1}), which clearly indicates that the substrate is stabilized by interactions with R267 that cannot be replicated with the shorter glutamine side chain of the mutated enzyme. In contrast, the R62N mutation leads to a significant increase of the velocity (1.63 nmol s^{-1} mg^{-1}). The positively charged arginine residue faces towards the hydrophobic phenazine ring system. The exchange to asparagine would avoid such an unfavorable situation and additionally retain water-mediated DMAPP binding. The docking suggested substrate stabilization via a weak hydrogen bond (3.3 Å) from dhPCA to the hydroxyl group of Y287. Enzymatic activity of Y287F remained at wild-type level (0.88 nmol s^{-1} mg^{-1}), showing that the hydroxyl group of Y287 is not critical for dhPCA binding. R267 thus appears to be sufficient in fixing the carboxyl group of the phenazine substrate. To investigate the suggested proton relay for deprotonation of the Wheland complex, we prepared a S236A mutant, which should reduce the catalytic activity as the conserved water molecule W2 loses one of its hydrogen binding partners. Indeed, we found that the reaction velocity of this mutant is decreased to approximately 50% of the wildtype enzyme (0.54 nmol s^{-1} mg^{-1}). Furthermore, we produced a G272V mutant to determine whether the proton transfer over two water molecules is necessary for deprotonation of the σ-complex. This mutation would be expected to displace the water molecule W2, which lies adjacent to W1. The enzymatic activity remained at wildtype level (0.79 nmol s^{-1} mg^{-1}) demonstrating that a proton transfer to the diphosphate moiety is not crucial for deprotonation of the Wheland complex. We investigated the binding mode of the catalytic water W1 to design point mutations which would exclusively displace this water molecule. However, this is not possible as W1 is fixed exclusively by the contact to W2 and a main chain interaction with the carbonyl of V270 (Figure 4c). Next, we prepared EpzP variant A285L. Although, a repulsion of the water molecule W1 and the leucine residue of mutation A285L is possible, which might result in a conformation unfavored for deprotonation, we expected a decrease of catalytic activity of A285L mainly due to steric collision to dhPCA. Indeed, the replacement of alanine in mutant A285L leads to a decrease of the relative reaction velocity to 24% (0.22 nmol s^{-1} mg^{-1}) and therefore supports proper docking of the substrates.

Structure-based Engineering of EpzP

On the basis of our docking solution and the proposed reaction mechanism, we investigated whether modifications of EpzP can be designed that increase the turnover rate of the enzyme to yield higher catalytic turnover rate for chemoenzymatic synthesis. Examination of the docking solution revealed that dhPCA stabilization might be achieved by introduction of V270F. The aromatic side chain of V270F can adopt a geometry that arranges the phenyl ring of the V270F mutant and the substrate dhPCA in a π-stacking conformation. Indeed, the V270F mutant showed a five-fold increased enzymatic activity (4.43 nmol s^{-1} mg^{-1}) compared to the wildtype enzyme (Figure 4d). Remarkably, the introduction of this single point mutation led to approximately the same catalytic reaction rate as found for PpzP (5.42 nmol s^{-1} mg^{-1}). EpzP and PpzP catalyze an identical regioselective prenylation of dhPCA. Although the affinity to DMAPP and dhPCA were reported to be similar for both enzymes, [21]· [29] EpzP exhibits substantially lower turnover rates. To determine the structural differences that could underlie these differences in catalytic reaction rate, we prepared a model of PpzP (57% sequence identity) by mapping the sequence of PpzP onto the structure of EpzPwt using MODELLER.[48] We investigated the cavity of the resulting PCA binding site of the PpzP model and found that replacement of four key residues would suffice to essentially replace the dhPCA cavity of EpzP into that of PpzP. EpzP residues T161, V218, R267 and A285 are replaced with V162, A214, H263 and Q281, respectively, in PpzP. A remarkable difference is the replacement of A285 in EpzP with Q281 in PpzP. Therefore, we decided to prepare point mutations A285Q and A285QV270F to investigate the catalytic activity of these substitutions. Instead of V218A, we prepared EpzP variant V218G as substitution of valine to alanine as found in PpzP is proposed to have minor effects on catalysis due to moderate change in side chain size. On the basis of the *in silico* model, no obvious reason for an enhancement of the reaction velocity could be obtained for V218G variant and indeed, the activity remained nearly unchanged when compared with the wildtype enzyme (1.06 nmol s^{-1} mg^{-1}; Figure 4d). The enzymatic activity was not further increased in A285QV270F (3.65 nmol s^{-1} mg^{-1}) compared to V270F, which indicates that substitution of two voluminous side chain in close vicinity may lead to steric restriction for dhPCA binding. However, we achieved an approximately 14-fold activity for the A285Q mutant (12.89 nmol s^{-1} mg^{-1}) compared to wildtype EpzP (0.93 nmol s^{-1} mg^{-1}), and a two-fold enhancement compared to PpzP (5.42 nmol s^{-1} mg^{-1}). The structure suggests that this boost in catalysis is achieved by providing additional interactions with water molecule W1, which we identified as the likely catalytic base. These additional interactions would be expected to contribute to the rigidity of the dhPCA cavity. The observed drastic enhancement of the reaction velocity of EpzP-A285Q is remarkable, as nature tends to optimize enzymatic reactions and this mutant enzyme is an even more efficient catalyst than the already optimized enzyme PpzP.

CONCLUSION AND SUMMARY

Aromatic PTs are valuable enzymatic catalysts as they offer the possibility of a regioselective C-C-, C-N- or C-O-functionalisation [49] of aromatic substrates. A relaxed substrate specificity has been observed for several enzymes of this family, for example NphB, [20] Fnq26 [13] or 7-DMATS. [50] The possibility to design their regio- or chemoselectivity by protein engineering[19] renders them attractive candidates for chemoenzymatic synthesis of bioactive substrates, as demonstrated for the production of prenylated indole derivatives. [51] Here we report the structure of EpzP, a magnesium-independent ABBA prenyltransferase and member of the CloQ/NphB family. This is the first structure for a prenyltransferase that is capable of catalyzing the prenylation of phenazine derivatives, and detailed information about its reaction mechanism broadens the spectrum of aromatic substrates for regioselective prenylations by phenazine PTs as well as all ABBA PTs.

In combination with computational docking, the crystallographic data result in a model for the reaction mechanism that is in good agreement with the observed prenylation reaction and is also supported by the results from site-directed mutagenesis experiments. We have been able to identify amino acids that are crucial for catalysis, as demonstrated by EpzP variants T64Y or R267Q. Furthermore, we provide insight into the reaction mechanism of the phenazine prenylation and identify a catalytic water molecule that could act as a base for the deprotonation of the Wheland complex. EpzP variants S236A, G272V and A285L confirmed that this water molecule (W1) nearby the reaction center is important for catalysis and that the proposed proton relay (W2) is not necessary for the deprotonation of the Wheland complex. Such a deprotonation had already been suggested for the catalytic mechanism of NphB, [5]but lacked experimental verification.

We were able to interpret our model for the catalysis and to use it as a template for the design of EpzP modifications. As the reaction velocity for EpzP is reduced 6-fold compared to that of the homologous enzyme PpzP, we probed the differences between the enzymes by analyzing amino acid substitutions in their active sites. Based on this analysis, we established EpzP variants that significantly increased the reaction rate. This impressive result also serves as further confirmation for our docking calculation, especially as mutation V218G, which was designed on the basis of sequence similarity to PpzP but was predicted to have small effects based on the model, indeed showed no substantial change in turnover rates. In contrast EpzP variant V270F increased catalytic turnover rate to PpzP levels, and the mutation A285Q boosted enzymatic performance, in line with our structural model.

In contrast to other PTs, which revealed relaxed substrate specificity, both phenazine PTs are specific for their substrate. [21], [29] We are confident that our structure/function analysis of EpzP can serve as a template for the interpretation of structural and mechanistic properties of other phenazine PTs, including their regioselectivity, and that it can furthermore provide a pathway for the design of optimized phenazine PTs as well as other ABBA-type PTs. Thus, the data presented here elucidate the basis of enzymatic phenazine prenylation and therefore form a platform from which the synthesis of valuable phenazine products by chemoenzymatic methods can be launched.

ACCESSION CODES

The coordinates and structure factor amplitudes were deposited in the PDB data base under accession codes 4EE6, 4EE7 and 4EE8 for EpzPm-nat, EpzPm-SPP and EpzPwt, respectively.

ACKNOWLEDGMENTS

We thank Andreas Maurer for performing the MS experiment to verify the methylation of EpzP. We gratefully acknowledge the Swiss light source beamline X06DA for offering beam time.

AUTHOR CONTRIBUTIONS

Conceived and designed the experiments: GZ TS LH. Performed the experiments: GZ OS JBH DAH. Analyzed the data: GZ OS LH TS. Contributed reagents/materials/analysis tools: TS LH. Wrote the paper: GZ LH TS.

REERENCES

1. Sacchettini JC, Poulter CD (1997) Creating isoprenoid diversity. Science (New York, NY) 277: 1788–1789. doi: 10. 1126/science. 277. 5333. 1788
2. Liang P-H, Ko T-P, Wang AHJ (2002) Structure, mechanism and function of prenyltransferases. European Journal of Biochemistry/FEBS 269: 3339–3354. doi: 10. 1046/j. 1432-1033. 2002. 03014. x
3. Saleh O, Haagen Y, Seeger K, Heide L (2009) Prenyl transfer to aromatic substrates in the biosynthesis of aminocoumarins, meroterpenoids and phenazines: the ABBA prenyltransferase family. Phytochemistry 70: 1728–1738. doi: 10. 1016/j. phytochem. 2009. 05. 009
4. Pojer F, Wemakor E, Kammerer B, Chen H, Walsh CT, et al. (2003) CloQ, a prenyltransferase involved in clorobiocin biosynthesis. Proceedings of the National Academy of Sciences of the United States of America 100: 2316–2321. doi: 10. 1073/pnas. 0337708100
5. Kuzuyama T, Noel JP, Richard SB (2005) Structural basis for the promiscuous biosynthetic prenylation of aromatic natural products. Nature 435: 983–987. doi: 10. 1038/nature03668
6. Heide L (2009) Prenyl transfer to aromatic substrates: genetics and enzymology. Curr Opin Chem Biol 13: 171–179. doi: 10. 1016/j. cbpa. 2009. 02. 020
7. Bonitz T, Alva V, Saleh O, Lupas AN, Heide L (2011) Evolutionary Relationships of Microbial Aromatic Prenyltransferases. PLoS ONE 6: e27336–e27336. doi: 10. 1371/journal. pone. 0027336
8. Tello M, Kuzuyama T, Heide L, Noel JP, Richard SB (2008) The ABBA family of aromatic prenyltransferases: broadening natural product diversity. Cellular and Molecular Life Sciences: CMLS 65: 1459–1463. doi: 10. 1007/s00018-008-7579-3
9. Steffan N, Unsöld IA, Li S-M (2007) Chemoenzymatic synthesis of prenylated indole derivatives by using a 4-dimethylallyltryptophan synthase from Aspergillus fumigatus. Chembiochem: A European Journal of Chemical Biology 8: 1298–1307. doi: 10. 1002/cbic. 200700107
10. Grundmann A, Li S-M (2005) Overproduction, purification and characterization of FtmPT1, a brevianamide F prenyltransferase from Aspergillus fumigatus. Microbiology (Reading, England) 151: 2199–2207. doi: 10. 1099/mic. 0. 27962-0
11. Yin W-B, Ruan H-L, Westrich L, Grundmann A, Li S-M (2007) CdpNPT, an N-prenyltransferase from Aspergillus fumigatus: overproduction, purification and biochemical characterisation. Chembiochem: A European Journal of Chemical Biology 8: 1154–1161. doi: 10. 1002/cbic. 200700079
12. Yin W-B, Grundmann A, Cheng J, Li S-M (2009) Acetylaszonalenin biosynthesis in Neosartorya fischeri. Identification of the biosynthetic gene cluster by genomic mining and functional proof of the genes by biochemical investigation. The Journal of Biological Chemistry 284: 100–109. doi: 10. 1074/jbc. m807606200
13. Haagen Y, Unsöld I, Westrich L, Gust B, Richard SB, et al. (2007) A soluble, magnesium-independent prenyltransferase catalyzes reverse and regular C-prenylations and O-prenylations of aromatic substrates. FEBS Letters 581: 2889–2893. doi: 10. 1016/j. febslet. 2007. 05. 031
14. Ozaki T, Mishima S, Nishiyama M, Kuzuyama T (2009) NovQ is a prenyltransferase capable of catalyzing the addition of a dimethylallyl group to

both phenylpropanoids and flavonoids. The Journal of Antibiotics 62: 385–392. doi: 10. 1038/ja. 2009. 48

15. Kumano T, Tomita T, Nishiyama M, Kuzuyama T (2010) Functional characterization of the promiscuous prenyltransferase responsible for furaquinocin biosynthesis: identification of a physiological polyketide substrate and its prenylated reaction products. J Biol Chem 285: 39663–39671. doi: 10. 1074/jbc. m110. 153957
16. Schultz AW, Lewis CA, Luzung MR, Baran PS, Moore BS (2010) Functional characterization of the cyclomarin/cyclomarazine prenyltransferase CymD directs the biosynthesis of unnatural cyclic peptides. J Nat Prod 73: 373–377. doi: 10. 1021/np9006876
17. Metzger U, Keller S, Stevenson CEM, Heide L, Lawson DM (2010) Structure and mechanism of the magnesium-independent aromatic prenyltransferase CloQ from the clorobiocin biosynthetic pathway. Journal of Molecular Biology 404: 611–626. doi: 10. 1016/j. jmb. 2010. 09. 067
18. Metzger U, Schall C, Zocher G, Unsöld I, Stec E, et al. (2009) The structure of dimethylallyl tryptophan synthase reveals a common architecture of aromatic prenyltransferases in fungi and bacteria. Proceedings of the National Academy of Sciences of the United States of America 106: 14309–14314. doi: 10. 1073/pnas. 0904897106
19. Jost M, Zocher G, Tarcz S, Matuschek M, Xie X, et al. (2010) Structure-function analysis of an enzymatic prenyl transfer reaction identifies a reaction chamber with modifiable specificity. Journal of the American Chemical Society 132: 17849–17858. doi: 10. 1021/ja106817c
20. Kumano T, Richard SB, Noel JP, Nishiyama M, Kuzuyama T (2008) Chemoenzymatic syntheses of prenylated aromatic small molecules using Streptomyces prenyltransferases with relaxed substrate specificities. Bioorganic & Medicinal Chemistry 16: 8117–8126. doi: 10. 1016/j. bmc. 2008. 07. 052
21. Saleh O, Gust B, Boll B, Fiedler H-P, Heide L (2009) Aromatic prenylation in phenazine biosynthesis: dihydrophenazine-1-carboxylate dimethylallyltransferase from Streptomyces anulatus. The Journal of Biological Chemistry 284: 14439–14447. doi: 10. 1074/jbc. m901312200
22. de Andrade-Neto VF, Goulart MOF, da Silva Filho JF, da Silva MJ, Pinto MdCFR, et al. (2004) Antimalarial activity of phenazines from lapachol, beta-lapachone and its derivatives against Plasmodium falciparum in vitro and Plasmodium berghei in vivo. Bioorganic & Medicinal Chemistry Letters 14: 1145–1149. doi: 10. 1016/j. bmcl. 2003. 12. 069
23. Laursen JB, Nielsen J (2004) Phenazine Natural Products: Biosynthesis, Synthetic Analogues, and Biological Activity. ChemInform 35: no-no-no-no.
24. Mavrodi DV, Peever TL, Mavrodi OV, Parejko JA, Raaijmakers JM, et al. (2010) Diversity and evolution of the phenazine biosynthesis pathway. Applied and Environmental Microbiology 76: 866–879. doi: 10. 1128/aem. 02009-09
25. Shinya K, Furihata K, Teshima Y, Hayakawa Y, Seto H (1993) Benthocyanins B and C, new free radical scavengers from Streptomyces prunicolor. J Org Chem 58: 4170–4172. doi: 10. 1021/jo00067a069
26. Shin-ya K, Furihata K, Hayakawa Y, Seto H, Kato Y, et al. (1991) The structure of benthocyanin A. A new free radical scavenger of microbial origin. Tetrahedron Letters 32: 943–946. doi: 10. 1016/s0040-4039(00)92126-9
27. Omura S, Eda S, Funayama S, Komiyama K, Takahashi Y, et al. (1989) Studies on a novel antitumor antibiotic, phenazinomycin: taxonomy, fermentation, isolation, and physicochemical and biological characteristics. The Journal of Antibiotics 42: 1037–1042. doi: 10. 7164/antibiotics. 42. 1037
28. Gebhardt K, Schimana J, Krastel P, Dettner K, Rheinheimer J, et al. (2002) Endophenazines A-D, new phenazine antibiotics from the arthropod associated endosymbiont Streptomyces anulatus. I. Taxonomy, fermentation, isolation and biological activities. The Journal of Antibiotics 55: 794–800. doi: 10. 7164/antibiotics. 55. 794
29. Seeger K, Flinspach K, Haug-Schifferdecker E, Kulik A, Gust B, et al. (2011) The biosynthetic genes for prenylated phenazines are located at two different chromosomal loci of Streptomyces cinnamonensis DSM 1042. Microbial Biotechnology 4: 252–262. doi: 10. 1111/j. 1751-7915. 2010. 00234. x
30. Bringmann G, Haagen Y, Gulder TAM, Gulder T, Heide L (2007) Biosynthesis of the isoprenoid moieties of furanonaphthoquinone I and endophenazine A in Streptomyces cinnamonensis DSM 1042. The Journal of Organic Chemistry 72: 4198–4204. doi: 10. 1021/jo0703404
31. Kabsch W (1988) Automatic indexing of rotation diffraction patterns. Journal of Applied Crystallography 21: 67–72. doi: 10. 1107/s0021889887009737
32. Winn MD, Ballard CC, Cowtan KD, Dodson EJ, Emsley P, et al. (2011) Overview of the CCP4 suite and current developments. Acta Crystallographica Section D, Biological Crystallography 67: 235–242. doi: 10. 1107/s0907444910045749
33. Langer G, Cohen SX, Lamzin VS, Perrakis A (2008) Automated macromolecular model building for X-ray crystallography using ARP/wARP version 7. Nature Protocols 3: 1171–1179. doi: 10. 1038/nprot. 2008. 91
34. Adams PD, Afonine PV, Bunkóczi G, Chen VB, Echols N, et al. (2011) The Phenix software for automated determination of macromolecular structures. Methods (San Diego, Calif) 55: 94–106. doi: 10. 1016/j. ymeth. 2011. 07. 005
35. Emsley P, Lohkamp B, Scott WG, Cowtan K (2010) Features and development of Coot. Acta Crystallographica Section D, Biological Crystallography 66: 486–501. doi: 10. 1107/s0907444910007493
36. Murshudov GN, Skubák P, Lebedev AA, Pannu NS, Steiner RA, et al. (2011) REFMAC5 for the refinement of macromolecular crystal structures. Acta

Crystallographica Section D, Biological Crystallography 67: 355–367. doi: 10. 1107/s0907444911001314

37. Winn MD, Murshudov GN, Papiz MZ (2003) Macromolecular TLS refinement in REFMAC at moderate resolutions. Methods in Enzymology 374: 300–321. doi: 10. 1016/s0076-6879(03)74014-2
38. Painter J, Merritt EA (2006) Optimal description of a protein structure in terms of multiple groups undergoing TLS motion. Acta Crystallographica Section D, Biological Crystallography 62: 439–450. doi: 10. 1107/s0907444906005270
39. Lovell SC, Davis IW, Arendall WB, 3rd, de Bakker PIW, Word JM, et al (2003) Structure validation by Calpha geometry: phi,psi and Cbeta deviation. Proteins 50: 437–450.
40. Fenn TD, Ringe D, Petsko GA (2003) POVScript+ : a program for model and data visualization using persistence of vision ray-tracing. Journal of Applied Crystallography 36: 944–947. doi: 10. 1107/s0021889803006721
41. Schrodinger LLC (2010) The PyMOL Molecular Graphics System, Version 1. 3r1.
42. Schüttelkopf AW, van Aalten DMF (2004) PRODRG: a tool for high-throughput crystallography of protein-ligand complexes. Acta Crystallographica Section D, Biological Crystallography 60: 1355–1363. doi: 10. 1107/s0907444904011679
43. Trott O, Olson AJ (2010) AutoDock Vina: improving the speed and accuracy of docking with a new scoring function, efficient optimization, and multithreading. Journal of Computational Chemistry 31: 455–461. doi: 10. 1002/jcc. 21334
44. Holm L, Kääriäinen S, Wilton C, Plewczynski D (2006) Using Dali for structural comparison of proteins. Current Protocols in Bioinformatics/Editoral Board, Andreas D Baxevanis [et Al] Chapter 5: Unit 5. 5-Unit 5. 5.
45. Krissinel E, Henrick K (2004) Secondary-structure matching (SSM), a new tool for fast protein structure alignment in three dimensions. Acta Crystallographica Section D Biological Crystallography 60: 2256–2268. doi: 10. 1107/s0907444904026460
46. Thalladi VR, Smolka T, Gehrke A, Boese R, Sustmann R (2000) Role of weak hydrogen bonds in the crystal structures of phenazine, 5,10-dihydrophenazine and their 1:1 and 3:1 molecular complexes. New J Chem 24: 143–147. doi: 10. 1039/a908744c
47. Yang Y, Miao Y, Wang B, Cui G, Merz KM (2012) Catalytic Mechanism of Aromatic Prenylation by NphB. Biochemistry.
48. Eswar N, Webb B, Marti-Renom MA, Madhusudhan MS, Eramian D, et al. . (2007) Comparative protein structure modeling using MODELLER. Current Protocols in Protein Science/Editorial Board, John E Coligan [et Al] Chapter 2: Unit 2. 9–Unit 2. 9.
49. Zou HX, Xie X, Zheng XD, Li SM (2011) The tyrosine O-prenyltransferase SirD catalyzes O-, N-, and C-prenylations. Appl Microbiol Biotechnol 89: 1443–1451. doi: 10. 1007/s00253-010-2956-x
50. Yu X, Li SM (2011) Prenylation of Flavonoids by Using a Dimethylallyltryptophan Synthase, 7-DMATS, from Aspergillus fumigatus. ChemBioChem 12: 2280–2283. doi: 10. 1002/cbic. 201100413
51. Li SM (2010) Prenylated indole derivatives from fungi: structure diversity, biological activities, biosynthesis and chemoenzymatic synthesis. Nat Prod Rep 27: 57–78. doi: 10. 1039/b909987p

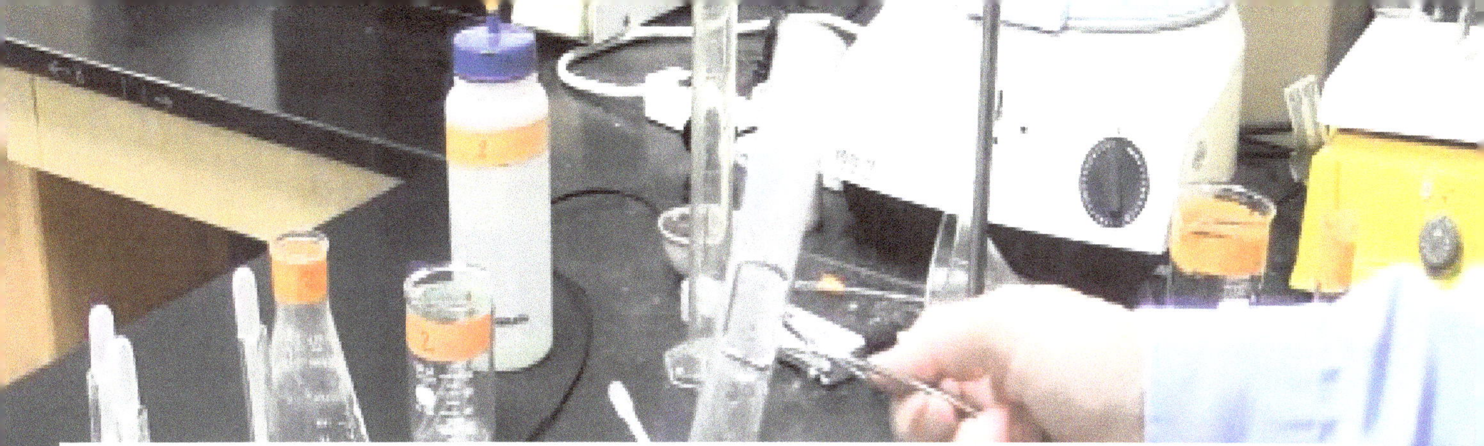

CHAPTER 8

GRAPHENE-BASED NANOMATERIALS AS HETEROGENEOUS ACID CATALYSTS

Bhaskar Garg, Tanuja Bisht and Yong-Chien Ling

Keywords: heterogeneous acid catalysis; nanomaterials; graphene; graphene oxide; sulfonation; sulfation; green chemistry

INTRODUCTION

The prime source of energy on Earth for the present and foreseeable future is found in chemical bonds. Catalysis affords the means of changing the rates at which chemical bonds are formed and broken making it possible to elicit a desirable product over an undesired one, the phenomenon of chemical specificity. Among catalysis of substantial interest, acid catalysis is quite prevalent and probably the most routine operation in both industrial processes and academic research worldwide. The petroleum (biodiesel production as alternative fuels), chemical (production of most chemicals and materials), plastic (production of polymers and plasticizers), and pharmaceutical industries (production of new heterocycles as "wonder drugs"), essential to a healthy economy, rely heavily on acid catalysis. Not surprisingly, therefore, it lies at the heart of our quality of life.

In industry, with foremost focus on economic benefits, a large class of liquid-phase reactions are routinely executed with sulfuric acid (H_2SO_4), hydrofluoric acid (HF), and methanesulfonic acid (CH_3SO_3H) as bulk commodity catalysts. Unquestionably, the unique spatial separation (similar to that of enzymes) as well as the self-similarity of structures between the active sites in these homogeneous molecular acids not only allow consistent energetic interactions between each active site and reaction substrate (even at ambient conditions) but also make them responsive to most spectroscopic characterization techniques. Despite this, these catalysts require challenging processes for their separation from the homogeneous reaction mixture, resulting in abundant non-recyclable acid waste.

The shift towards an emphasis on catalyst recoverability and reusability has resulted in extensive research efforts devoted to the development of solid or heterogeneous acid catalysts, which are easier to separate, recover and reuse in desired reactions. In this milieu, endeavors have been made using, to name but a few, cation exchange resins (Amberlyst™-15, Nafion-H, *etc.*), solid superacids (SO_4^{2-}/ZrO_2), (supported)metal oxides/sulfides/triflates (Al_2O_3, CdS, $Yb(OTf)_3$), zeolites (H-ZSM-5), molecular sieves, silica, natural clays (bentonite, montmorillonite), and heteopolyacids to complement liquid-acid catalysts in a broad spectrum of acid-catalyzed reactions [1,2,3,4,5,6,7,8]. Although the mentioned solid catalysts certainly overcome some shortcomings, most of them are inappropriate owing to their deactivation (leaching or undesired deposits on active sites) [9,10], internal mass transfer (such as in zeolites) [11] or high mass transfer resistance (as in clay, molecular sieves *etc.*), unfavorable side reactions, high cost, and most importantly, toxicity issues associated with the presence of certain metals.

Citation: Garg, Tanuja Bisht and Yong-Chien Ling, Graphene-Based Nanomaterials as Heterogeneous Acid Catalysts, doi:10.3390/molecules190914582

The pursuit of heterogeneous acid catalysts capable of meeting most of the abovementioned challenges has resulted in the consideration of carbon nanomaterials (CNMs), which can function as metal-free catalysts. Indeed, with the added advantage of classical nanofabrication techniques, the CNMs, which include mesoporous/amorphous carbon [12,13] and rich polytypes of carbon such as carbon nanotubes (CNTs), have produced considerable advancements in heterogeneous acid catalysis [14,15]. Notably, until the 20th century, the carbon family was only dominated by the well-known materials like graphite, diamond, fullerenes, and CNTs. However, this "aura" was lessened with the first isolation of free standing 2D graphene in 2004 [16] (in order to avoid unnecessary complexity to potential readers, we have adopted herein the common names currently in use in graphene-based catalysis research, yet we would also like to recommend the readers to consult the guest editorial published in the journal *Carbon* [17] for an exhaustive and productive account of the nomenclature of sp^2carbon nanoforms).

Graphene could be considered as a potential building block of 3D graphite, 0D fullerenes (in a broader context or to some extent, it is reasonable to declare graphene as "mother" of all carbon nanostructures, as increasingly highlighted by many research groups; certainly, however, it is a misconception that graphene can produce fullerenes–notably, fullerenes have alternate pentagonal and hexagonal arrangements, whereas in graphene only hexagons should be present), and 1D CNTs (Figure 1) [18,19] and possess many exceptional properties that have surpassed those observed with its predecessor, CNTs. Not surprisingly, therefore, graphene-based nanomaterials (G-NMs) are (and will continue to be) increasingly used in optoelectronic [18], energy conversion and storage [18,20], photocatalysis/photovoltaic [21], switching and information storage [22], bacteria killing [23], biosensors [24,25] and other technological applications [26,27,28] in the imminent carbon age. On another front, the concept of carbocatalysis within the graphene framework is a relatively new but rapidly emerging research area at the interface of organic, material, green and sustainable chemistry [29].

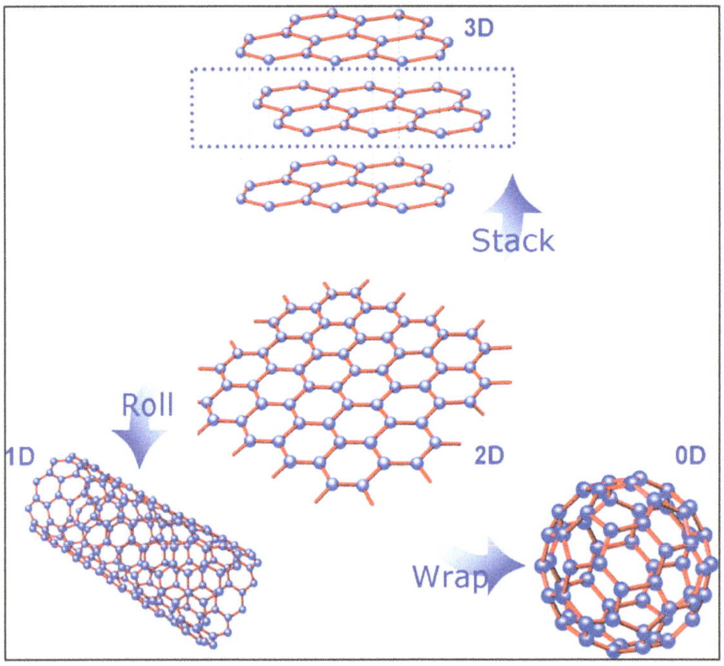

Figure 1: Graphene: the basic building block for other carbon allotropes, graphite (3D), fullerene (0D), and CNT (1D). Reprinted with permission from [18]. Copyright (2012) American Chemical Society.

It is worth commenting here that at certain instances, "carbocatalysis" can act as a bridge toward homogeneous catalysis (where both the reactants and catalysts are in a single phase) due to the existence of indefinitely persistent colloidal dispersions of carbocatalysts. In such cases, the process should be referred as "pseudo-homogeneous" because during the reaction, the catalyst is ostensibly in the same phase as reactants and products but can be recovered by means of centrifugation or membrane filtration after the reaction. Nevertheless, in order to avoid any confusion and also taking into account the recoverable features of G-NMs (lacking molecular entities), herein, we use the term "heterogeneous" throughout for graphene-based acid catalysis (G-AC).

After the seminal discovery by Bielawski and coworkers [30] of the use of graphene oxide (GO) as catalyst in alcohol oxidation, unquestionably, an appreciable range of G-NMs are currently available as catalysts for a variety of useful transformations. However, since the focus of this review article strongly lies on G-AC, interested readers are recommended to consult other review articles [29,31,32,33,34] for the corresponding unique and productive details.

The current state-of-the-art in G-AC can be largely categorized into graphite/graphene oxide (GO) and their sulfated (-O-SO_3H) or sulfonated (-SO_3H) derivatives. Whereas the former are obtained by chemical oxidation of natural graphite and/or subsequent exfoliation, the latter rely on the fundamental idea of surface functionalization of GO with sulfonating agents. Both classes exhibit unique acid activity and hold great promise in replacing traditional homogeneous and heterogeneous acid catalysts in an environmentally benign manner (Table 1). Thus far, to the best of our knowledge, G-AC has not been reviewed by any research group and we would like to be the first to shed light on this fast-growing field.

Table 1: A comparison between the characteristic features of sulfuric acid, Amberlyst™-15 and G-NMs in acid catalysis

Features	Sulfuric Acid	Amberlyst™-15	G-NMs
Physical state	*Liquid*	*Solid*	*Solid*
Acid activity	*High*	*High*	*High*
Surface Area	-	*Relatively low*	*High*
Operating conditions	*Typically harsh*	*Relatively Mild*	*Variable* [a]
Diffusion/Leaching	*None*	*Possible*	*Negligible*
Recoverability	*Difficult*	*Easy* [b]	*Easy* [b]
Reusability/cycles	*Difficult*	*Easy/Less cycles* [c]	*Easy/More cycles* [c]
Water tolerance/stability	*High*	*Relatively low*	*High*
Cost-effectiveness	*Cost-effective*	*Relatively expensive*	*Variable* [a]
Mass transfer	-	*Relatively low*	*High*
Reaction selectivity	*High*	*Variable* [a]	*Variable* [a]
Eco-friendly use	*No*	*Yes*	*Yes*
Industrial use	*Frequent*	*Selected* [d]	*Awaiting*

As such, this review will begin with a brief discussion on the synthesis and structure of GO and its precursor, graphite oxide. Then, we will focus on unconventional structural models that have emerged only recently and are ascribed to be responsible for the inherent acidic properties of graphite oxide/GO. Afterwards, we will focus on the description of various synthetic routes to obtain sulfonated and sulfated analogues of graphene. In this regard, systematization according to the type of reducing and sulfonating agents is established. An equal emphasis is placed while discussing the properties and characterization data of G-NMs based on elemental analysis (EA) and Raman spectroscopy. Lastly, the catalytic applications of G-NMs are categorized based on: (i) graphite oxide; (ii) GO; and (iii) sulfated or sulfonated derivatives and discussed in conjunction with other acid catalysts. It should be noted that in some applications, both GO and its sulfonated counterparts have been examined together and thus the organization of this review should be regarded as a flexible guide to the classification of these applications.

GRAPHITE OXIDE AND GRAPHENE OXIDE (GO)

Synthesis and Structure

Historically, graphite oxide was discovered as a way to measure the molecular weight of graphite. Indeed, this was the intention of Brodie when in 1859 [35] he heated a mixture of graphite and fuming nitric acid in the presence of potassium chlorate. The repeated oxidation (four replicates) of the as-obtained product afforded the very first sample of graphite oxide as a light yellow solid which was termed "graphic acid", based on its dispersion behavior at different pH values. In a subsequent follow-up work in 1898 [36,37], Staudenmaier improved the oxidation protocol by adding potassium chlorate in small portions and acidifying the mixture with concentrated H_2SO_4, which completely avoided the four repetitions of the oxidation steps and was thus ascertained more practical. Over the span of 60 years, in 1958 [38], Hummers introduced potassium permanganate as an oxidant in a mixture of sodium nitrate and concentrated H_2SO_4 resulting in a more heavily oxygenated form of graphite oxide. This protocol was proved safer as it not only allowed *in situ* generation of nitric acid but also avoided the use of highly corrosive fuming nitric acid. In a bid to replace *in situ* production of nitric acid with less corrosive phosphoric acid, a new oxidation protocol (Tour's method) is reported recently [39], yet Hummers method has been well appreciated among scientific community and increasingly adopted to date by many researchers worldwide.

Aside from the operative oxidation procedures, the chemical structure of graphite oxide has been the subject of extensive research over the years. Consequently, several structural models of graphite oxide have been surfaced from time to time. In this context, Hofmann-Holst [40], Ruess [41], Clauss [42], Scholz-Boehm [43], Nakajima [44], Nakajima-Matsuo [45], Lerf-Klinowski (LK) [46,47,48], Dékány [49], and Ajayan [50] models are especially notable. However, due to the variable oxidation procedures leading to sample-to sample variability, nonstoichiometric atomic composition and lack of a "hallmarked" characterization technique, the precise chemical structure of graphite oxide remains elusive and still a subject of further discussion. Overall, the LK model [48] of graphite oxide is perhaps the most popular and widely accepted among graphene research community.

As such, it is assumed that the basal plane of heavily oxygenated graphite oxide is highly populated with hydroxyl (-OH) and epoxide (C-O-C) functions, while the edge-plane consists of carbonyl (C=O) and carboxylic acid (-COOH) groups. The coverage of the oxygen-rich functional groups in graphite oxide substantiates hole defects (nanovoids and vacancies) throughout the surface and deviates it from the state of pristine graphite. Indeed, the presence of such features on the graphite oxide surface has recently been observed [51] using high-resolution transmission electron microscopy (HRTEM).

Graphite oxide is chiefly composed of carbon, oxygen, and hydrogen atoms retaining a C/O ratio in between 1.8 to 2.5, while the most common C/O ratio for different samples is ~2 [40,42,49,52]. In this context, it should be noted that the stoichiometric ratio between the constituent elements of graphite oxide is not fixed and varies depending upon the oxidation level. Consequently, regardless of the synthesis protocol followed [36,38] the composition and structure of graphite oxide do not change significantly even after the addition of excess oxidizing agent. GO, an exfoliated form of graphite oxide, is another important intermediate between graphite and graphene and usually attained via mechanical (vigorous stirring or ultrasonication), chemical (ionic liquids) or thermal treatment of graphite oxide. At this juncture, it is critical to understand the technical point of differentiation between graphite oxide and GO. Whereas graphite oxide represents the lamellar structure (consisting of layers of carbon from the original graphene lattice), GO largely exists as mono-, bi- or at most a few layers (3 to 4) of graphene sheets. Nevertheless, despite being structurally different, both are chemically similar, having a complex cocktail of oxygen functionalities (Figure 2) rendering them hydrophilic in nature. As a consequence, water molecules can readily intercalate between the layers resulting in an increase of the interlayer distance and change of hybridization state (in oxidized carbons) from sp^2 (planar) to sp^3 (tetrahedral).

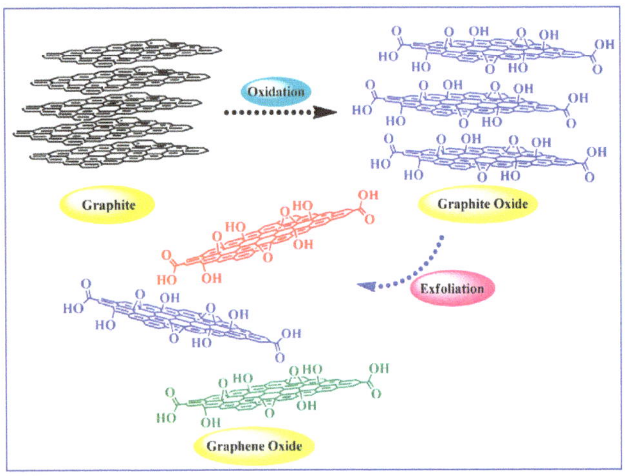

Figure 2: Synthesis of graphite oxide and graphene oxide (GO) from natural graphite.

Acidic Properties of Graphite Oxide and GO

Aqueous solutions of graphite oxide have significant acidity, with a pH in between 3 to 4. It has long been demonstrated that 100 g of graphite oxide contains 500–800 mmol of active acid sites [40,42] *i.e.*, one acid site is present for every 6–8 carbon atoms that can participate in cation exchange reactions. In order to explain such high acidity of graphite oxide, Clauss *et al.* [42] suggested the presence of enolic groups and proposed the empirical formula $C_8O_2(OH)_2$ for the composition of graphite oxide. However, the authors were unable to provide any experimental data in support of this suggestion.

In yet a more recent scenario, it is commonly assumed that graphite oxide obtained by the (modified)Hummer's method [38,48] contains lots of recoverable Brønsted acid sites (though benchmark experimental evidences are still in question) including -OH and -COOH groups on the basal plane and flake edges, respectively. Accordingly, the acidic character of graphite oxide is likely due to the synergistic effect of these functionalities. However, controlled experiments by Dhakshinamoorthy *et al.* [53] with GO, H_2SO_4, *p*-toluenesulfonic acid (*p*-TSA), and glacial acetic acid have clearly indicated that the active acid sites of GO are not indeed -COOH groups but probably HSO_4 groups as evidenced by elemental analysis (1.16 wt % of sulfur atoms). In fact, the thermal treatment of GO at 200 °C removed most of the HSO_4 groups and resulted in catalytically inactive, partially reconstituted, GO having a negligible sulfur content. A similar observation (at relatively high temperature; 400 °C) has also been reported recently while using GO as an acid catalyst in the conversion of carbohydrates into 5-ethoxymethylfurfural [54].

In an elegant work, Dimiev *et al.* [55] have recently proposed a new explanation for the graphite oxide acidity that may allow one to add critical details to existing structural models of graphite oxide with higher confidence. It is well known, as increasingly highlighted in most published work, that graphite oxide has a light yellow appearance during its first quenching with water. However, with cumulative water washings, this color gradually changes from light yellow to dark brown or even black, which indicates that significant chemical transformations occur in the course of the purification of graphite oxide. Taking this into account, the group [55] decided to avoid exposing the graphite oxide material to water. Thus, by introducing non aqueous solvents such as methanol, isopropanol, ethyl acetate and glacial/trifluoro- acetic acid during the quenching and purification steps of graphite oxide, they synthesized different samples of pristine graphite oxide (pGO) having a significant amount of vacancy defects terminated by ketone groups. The products were extensively characterized by Fourier transform infrared spectroscopy (FTIR), UV-visible and X-ray photoelectron spectroscopy (XPS), thermogravimetric analysis (TGA), ^{13}C solid state nuclear magnetic resonance (SSNMR), and scanning electron microscopy (SEM).

It was postulated that the oxidized areas (sp^3) of as-synthesized pGO are dominated by epoxide groups while covalent sulfates (primarily existing as cyclic sulfates **1**) and -OH groups are present in small amounts. When pGO was exposed to water, hydrolysis occurred in two steps. In the first step, cyclic sulfates are cleaved to monosulfate 2, while 1,2-diol 3 and H_2SO_4 are formed in the second step of hydrolysis (Scheme 1). The as-produced 2 and H_2SO_4 were ascribed to be responsible, in part, for the highly acidic properties of graphite oxide. Furthermore, as evidenced by experimental data, it was proposed that acidic properties of graphite oxide are significantly affected by stepwise conversion of tertiary alcohols into edge-pane ketones, which remain in equilibrium with their hydrated form.

The group further examined the effect of water on GO acidity and proposed an unconventional view of GO chemistry [56], a "dynamic structural model" (DSM). As per this theory, "GO does not exist as a static structure with a given set of functional groups, but rather generates them and constantly changes its structure through interaction with water". As such, the GO acidity was brought about by CC bond cleavage, formation of vinylogous carboxylic acid 5 (rendering the -OH groups acidic) (Scheme 2), and the generation of protons. In this milieu, it should be noted that the formation of an electrical double layer at the GO interface in aqueous solutions is the main driving force for the observed GO acidity. Equally important is to understand that on prolonged exposure to water, GO flakes gradually degrade and attain a humic acid-like structure.

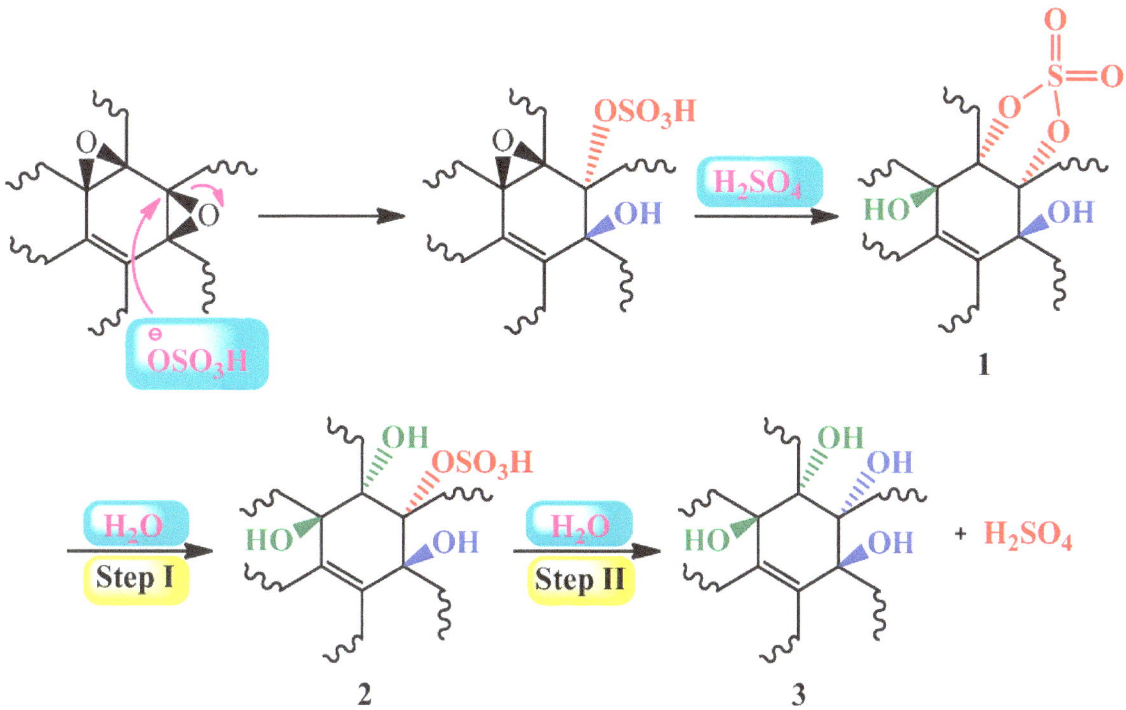

Scheme 1: Hydrolysis of cyclic sulfate 1 into hydrosulfate 2 (Step I) and sulfuric acid (Step II). Adapted with permission from [55]. Copyright (2012) American Chemical Society.

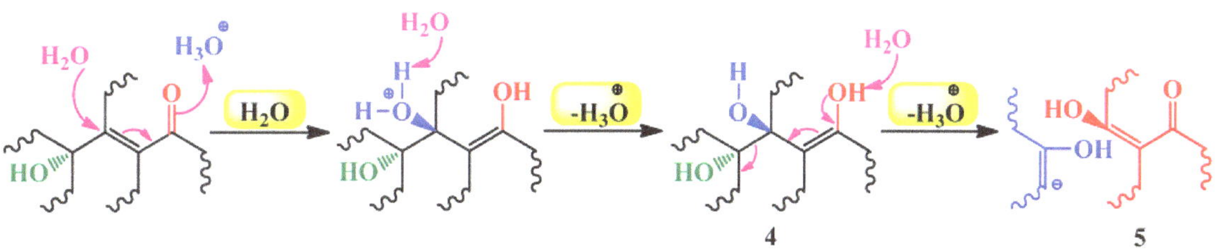

Scheme 2: A schematic representing the formation of vinylogous carboxylic acid 5 via C-C bond cleavage of enol 4, which in turn was obtained via nucleophilic attack of water molecule on a GO fragment having one hydroxyl and ketone group. Adapted with permission from [56]. Copyright (2013) American Chemical Society.

In a more recent report, Eigler *et al.* [57] undertook a systematic study in order to determine the sulfur species in GO and its base-treated counterparts, the so called GO-NaOH, GO-$NaHSO_4$, and GO-Na_2SO_3. In contrast to the previous work on graphite oxide hydrolysis [55], the group postulated that only the first step of hydrolysis takes place, while mono/organosulfates **2** are fairly stable and remain bound on GO (see Scheme 1), a key aspect for graphite oxide/GO acidity. As evidenced by thermogravimetry-mass spectroscopy (TGA-MS), it was concluded that organosulfates are the integral part of any GO and decompose at about 250 °C. On the other hand, the weight loss step in between 700–800 °C could be attributed to the pyrolysis of inorganic sulfur as sulfates or sulfite in GO-$NaHSO_4$, and GO-Na_2SO_3, respectively (Figure 3a). Moreover, a TGA-MS analysis using sodium dodecylbenzenesulfonate (decomposition temperature of ~500 °C) as a reference completely excluded the presence of sulfonic acid groups on the GO surface. Accordingly, the structural model of GO was extended by adding organosulfate in addition to epoxy, and hydroxyl groups, which are predominantly located above and below the carbon skeleton (Figure 3b). It was estimated that, in addition to hydroxyl and epoxy groups, one organosulfate group per 20 carbon atoms is present in GO and must be located on both sides of carbon lattice as well as the edges of GO flakes. The observation brought about by this study that the basic washing procedures could apparently be effective in GO purification, especially in removing sulfur species, is quite appealing.

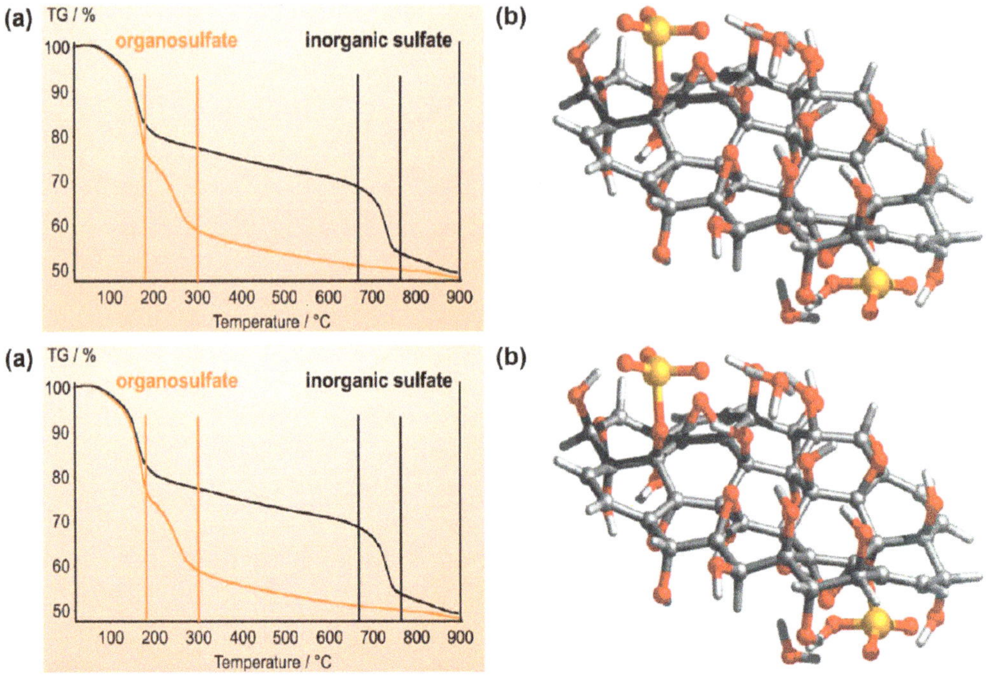

Figure 3: (a) The TGA data of GO-$NaHSO_4$ revealing two main decomposition steps for sulfur-containing species. (b) The proposed structural model of GO terminated with epoxy, hydroxyl, and organosulfate as dominant functional groups. Reproduced with permission from [57]. Copyright (2013) Wiley-VCH Verlag GmbH & Co.

In light of the above discussion, it may be concluded that -COOH groups have nothing to do with the acidic nature of graphite oxide/GO while the presence of sulfur as organosulfates, an integral part of oxidation protocols, is largely responsible for the observed acidity. As a final note, we anticipate that the exhaustive characterization of GO is a prerequisite prior to surface functionalization, especially, in sulfonation reactions as the presence of organosulfates in GO may significantly influence the characterization data of the resulting NMs and lead to misinterpretations.

G-NMS *VS.* SULFONATION AND SULFATION

The integration of sulfonic acid functional groups (-SO_3H) into CNMs is well known and emerged as a powerful strategy to produce promising solid acids for a range of acid-catalyzed reactions [12,13,14,15,58,59]. However, the sulfonation chemistry of G-NMs has developed only recently and continues to expand for diverse applications. At this juncture, it is critical to differentiate between sulfonation and sulfation reactions (Scheme 3) as the two are often confused but the structural differences are quite significant, making the use of separate terms more appropriate.

In sulfonation, sulfur trioxide (SO_3), the active species, reacts for example with an alkyl benzene and forms a C-S bond. Notably, the resultant alkyl benzenesulfonic acid is a stable molecule. Sulfation, on the other hand, involves the formation of a C-O-S bond and the resultant alcohol sulfuric acid is somewhat less stable. Unless neutralized, it decomposes to form sulfuric acid and the starting alcohol. This stability difference in the products of SO_3 reaction also has an intense impact on the choice of process used to produce sulfonates or sulfates.

In general, graphene-based sulfonation can be achieved in two steps: (a) reduction of GO to reduced graphene oxide (rGO) and (b) sulfonation of rGO. However, the direct sulfonation of GO and two-steps reduction protocols have also been undertaken at certain instances. The concrete methods, reducing and sulfonating agents are classified, summarized, and discussed below.

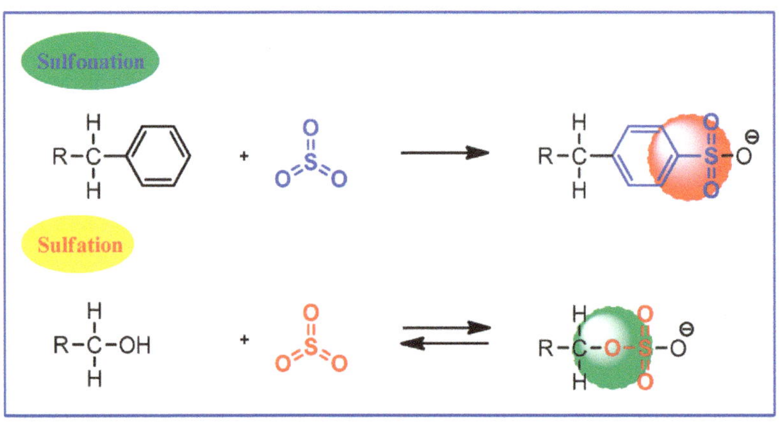

Scheme 3: Sulfonation *vs.* sulfation.

Reduction of GO

To date, one of the most important reactions of GO is its reduction leading to rGO, similar to that of pristine graphene. Although the reduction of GO can be achieved through thermal/H_2 [60,61,62], electrochemical [63], or chemical means [19,64,65,66,67,68], the latter is widely used in the graphene research and will only be discussed here. The emphasis is given to C/O ratio as well as Raman spectroscopy.

The aqueous colloidal dispersions of GO can be effectively reduced by a range of reducing agents having an established background in synthetic chemistry (see the recent review [69] for a comprehensive table of reducing agents for GO). The conversion of GO into rGO is usually achieved by an increased aggregation or hydrophobicity (due to the removal of oxygen containing groups) manifested in a color change from brown to black. Based on elemental analysis (EA), a decrease in elemental oxygen contents thereby resulting an increase of C/O ratio certainly signify the efficacy of the employed reducing agents.

Raman spectroscopy is a handy tool to characterize the structure and quality of G-NMs including graphite, graphite oxide, GO, rGO, and other carbonaceous materials [70,71]. In this context, the two most prominent features in (r)GO are the D and G vibrational bands (1100–1800 cm^{-1}). Whereas the former gives an idea about ordered or disordered structure of carbon, the latter is an indicator of the stacked carbon domain. Also of interest is the 2D band, a dominant sharp peak always observed for single layer graphene. It has been demonstrated that Raman scattering can be used as a fingerprint for single, bi, and a few layer graphene [72,73]. The ratio of intensities of two bands (I_D/I_G) can be used as a means of determining the structural changes within graphene samples. For instance, a high I_D/I_G ratio clearly indicates a certain amount of disorders/edges or exfoliation [71].

In general, graphite displays a prominent G peak at 1580 cm^{-1}, attributable to the first-order scattering of the E_{2g} mode. On the contrary, GO shows two peaks at 1363 and 1594 cm^{-1} corresponding to the D and G bands, respectively. The G band is associated with the vibrations of sp^2 carbon atoms in the 2D graphitic lattice whereas the D band is related to the vibrations of sp^3 carbon atoms of defects or disorder as mentioned above [71,73]. After the reduction, the I_D/I_G ratio changes considerably, as will be discussed below taking hydrazine (N_2H_4), sodium borohydride ($NaBH_4$), and lithium aluminum hydride ($LiAlH_4$) as representative reducing agents.

Hydrazine (N_2H_4)

N_2H_4 is known to scavenge oxygen while itself being broken down into water and nitrogen, making it an attractive option where most reducing agents exhibit moderate to strong reactivity toward water. $N_2H_4{\cdot}H_2O$ was firstly used as reducing agent by Stankovich *et al.* [74] in an effort to obtain individual sheets of graphene from GO. Under the optimized conditions, the number of polar functionalities on GO's surface was decreased significantly and resulted in aggregated layers of graphene sheets, rGO. Based on the extensive characterization data, it was proposed that the reduction of GO proceeds via a nucleophilic attack of N_2H_4 on an epoxide function. The EA results were in accord to expectations and a high C/O ratio (10.3/1) was measured for rGO over GO (2.7/1). Likewise, the Raman spectra indicated an increase in I_D/I_G ratio of as-obtained rGO which was suggested to be the result of a decrease in the size of sp^2 carbon domains (by the formation of sp^3carbons) upon reduction. Despite its good efficiency, the severe toxicity coupled with the introduction of heteroatomic impurities somewhat limits the practical use of N_2H_4.

Sodium Borohydride ($NaBH_4$)

$NaBH_4$ is one of the most used hydrides in research laboratories, as it can be readily solubilized in aqueous and alcoholic media. The first example of $NaBH_4$ usage in GO reduction was reported by Muszynski *et al.* [75] to achieve the physisorption of Au nanoparticles on amine grafted graphene. Later on, Gao *et al.* [50] performed the total reduction of GO using $NaBH_4$ as the first of a three-step protocol. The C/O ratio of as-obtained rGO was almost twice (4.78) that of GO (2.44). In addition, the Raman spectra of rGO exhibited an increased I_D/I_G ratio (1.91) compared to GO (0.95), which is consistent with what is observed with N_2H_4, as mentioned above. Unlike N_2H_4, $NaBH_4$ generates additional alcohols as principal impurities [76], a possible cause of a low C/O ratio relative to that of N_2H_4.

Lithium Aluminum Hydride ($LiAlH_4$)

$LiAlH_4$ (LAH) is another strong reducing agent of the hydride family that remained unused in GO reduction for a long time, presumably, due to its strong reactivity towards water (production of H_2 gas) and capability to reduce carbonyl, epoxy, ester, and carboxylic functions (unselective nature) to hydroxyl groups. The first LAH-triggered reduction of GO, in conjunction with N_2H_4 and $NaBH_4$, was reported by Ambrosi *et al.* [77]. The strong reducing strength of LAH was evidenced by a highest C/O ratio (12) while the reduction of GO (C/O ratio 3.4) with N_2H_4 and $NaBH_4$ provided relatively low C/O ratios of 11.5 and 9.5, respectively. The FTIR spectrum of as-obtained rGO indicated the presence of $-CH_2$ groups due to the alkene hydrogenation of unsaturated carbonyls. The GO obtained by graphite oxidation provided a considerably high I_D/I_G ratio (1.10) and was postulated to be due to the extensive disorder in the graphitc structure. On the contrary, upon reduction, the I_D/I_G ratio decreased to 0.61, 0.97, and 0.63 for $NaBH_4$, N_2H_4 and LAH, respectively. A decrease in I_D/I_Gratio clearly indicates that the disorder associated with the amorphous GO diminishes. There are a handful of other characterization techniques (XRD, XPS, EDS elemental mapping) for differentiating GO and rGO. Besides this, the AFM can be used to determine the single layer morphology of the desired sample. We recommend that readers consult the original literature for additional details, as mentioned previously.

Sulfonation of rGO and GO

As discussed above, the chemical reduction of oxygen functionalities on GO sheets produces rGO with the reintroduction of large aromatic domains. Consequently, rGO has increasingly used for the non-covalent adsorption of polymers [78,79] and aromatic species [80,81] via π–π stacking and van der Waals interactions. On the other hand, the covalent functionalization of rGO is quite uncommon and not performed significantly. In this milieu, the surface modifications of (r)GO with $-SO_3H$ groups have provided a handle to narrow down this huge gap. This section will deal the concrete methodologies used for sulfonation reactions of (r)GO along with a brief on characterization data. For the sake of clarity, sulfonation reactions are classified according to the type of sulfonating agents (Table 2).

Table 2: List of sulfonating agents for (r)GO to G-AC

Sulfonating Agents	Reducing Agents [a]	Reaction Conditions [b]	Acid Density [c]	Applications [d]	Ref.
4-Benzenediazonium sulfonate	$N_2H_4{\cdot}H_2O$	$EtOH/H_2O$, H_3PO_2, 3–5 °C, 1.5 h	1.55 mmol H^+ g^{-1}	Hydrolysis of ethyl acetate	[82]
Fuming sulfuric acid	$NaBH_4$	Autoclave, 180 °C, 24 h	1.2 mmol H^+ g^{-1}	(1) Esterification reactions (2) Peckmann reaction (3) Hydration reaction	[83]
Chlorosulfonic acid	*n*-butyl lithium	$ClSO_3H$, 0 °C, RT [e], overnight, NaOH, HCl, H_2O	-	Ester exchange reactions	[84]
4-Benzenediazonium sulfonate	$NaBH_4$	0 °C, 2 h, Hydrazine, 100 °C, 24 h	-	Synthesis of xanthenes and benzoxanthenes	[85]
Chlorosulfonic acid	-	$CHCl_3$, 70 °C, 4 h	1.2 mmol H^+ g^{-1}	Chemical conversion of biomass derived carbohydrates	[86]
4-Benzenediazonium sulfonate	(i) $NaBH_4$ (ii) N_2H_4	RT [e], overnight	0.5–1.7 mmol H^+ g^{-1}	Dehydration of xylose to furfural	[87]
Ammonium sulfate	-	235 °C, 30 min, argon atmosphere	EW [f] = 725 ± 5 g/mol	As support for metal nanocatalysts	[88]
Sulfuric acid	Microwave irradiation, 190 °C	160 °C, 5 h nitrogen atmosphere	1.9 mmol H^+ g^{-1}	Production of biofuels	[89]

Diazonium Salt of Sulfanilic Acid; 4-Benzenediazonium Sulfonate

Diazonium salts represent a class of organic compounds having a general formula of $RN_2^+X^-$ where R may be an alkyl or aryl residue and X is an (in)organic anion. The diazonium salts having an aryl group are usually stable in aqueous solutions at temperatures below 5 °C (the N_2 group does not tend to be lost as nitrogen gas) and serve as important intermediates in the organic synthesis of azo dyes [91,92,93,94]. The nitrosation of primary aromatic amines with HNO_2, which itself generated *in situ* from $NaNO_2$ and a mineral acid, is the most common method for the preparation of diazonium salts. However, being non-nucleophilic, the diazonium tetrafluoroborate salts can be isolated and are stable at room temperature. The process of forming diazonium compounds has been given a variety of names, including diazotation, diazoniation, and diazotization.

The usage of diazonium chemistry in the graphene domain was first reported by Lomeda *et al.* [95] to achieve the high dispersibility of graphene platelets in a range of polar aprotic solvents. However, the first attempt towards graphene sulfonation was performed by Si and Samulski [96] following a three step protocol as shown in Scheme 4. It was postulated that prereduction (step 1) is crucial both to enable sulfonation reaction (step 2) and to achieve complete reduction (step 3). The EA indicated that after prereduction of GO, a S/C ratio of 1:35 could be achieved via diazotization. However, only a 1:148 ratio is possible without step 1. The as-obtained sulfonated graphene exhibited an electrical conductivity of 1250 S m^{-1}implying that an extended conjugated sp^2 network is restored in the sulfonated graphene.

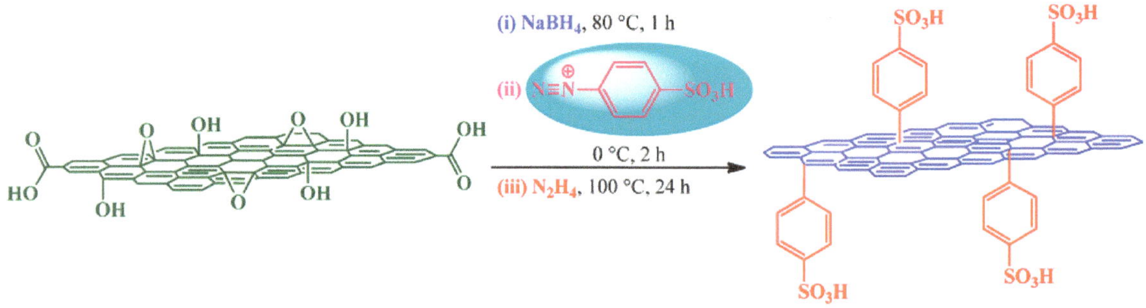

Scheme 4: Sulfonation of GO with 4-benzenediazonium sulfonate.

On another note, Ji *et al.* [82] introduced a two-step protocol avoiding the prereduction step. Specifically, GO was first reduced with $N_2H_4{\cdot}H_2O$ and dialyzed for 2 days. The diazotization of sulfanilic acid followed by sulfonation of purified rGO with 4-benzenediazonium sulfonate in the presence of hypophosphorous acid (H_3PO_2) afforded sulfonated graphene (S-graphene) having crumpling features. The solid state ^{13}C-NMR spectrum of S-graphene exhibited a small peak at 136 ppm, which was assigned to the carbons in the covalently attached phenyl-SO_3H groups. The Raman spectra of S-graphene indicated a slightly higher I_D/I_G ratio than that of rGO, which was likely a result of introduction of abundant $-SO_3H$ groups to the sp^2 carbon network. It was further evidenced by EDS mapping where sulfur was found to be homogeneously distributed on the whole surface. The neutralization titrations of S-graphene indicated an ion exchange capacity (IEC) of 1.55 mmol H^+ g^{-1}, which was significantly higher than commercially available Nafion NR50 (0.80 mmol H^+ g^{-1}) solid acid catalyst.

Following Samulski's approach, Zhao *et al.* [97] obtained highly dispersed sheets of sulfonated graphene (SG) and applied them as adsorbents for persistent organic aromatic pollutants such as naphthalene and 1-naphthol from aqueous solutions.

A similar, three-step preparation method was also carried out by Lam *et al.* [87] during their investigations on xylose dehydration. The BET surface area of resulting NMs was measured to be 318 (GO), 680 (SGO), and 634 (SG) $m^2 \cdot g^{-1}$; quite higher than that of Hu's work, even though the S/C ratio was nearly comparable (1:135 *vs.* 1:140) for SG. Of further interest is the clear discrepancy in the Raman data. The I_D/I_G ratio was increased from GO (0.89) to SG (1.00) in the previous work [97], whereas a significant decrease was observed in I_D/I_G ratio from GO (1.16) to SG (0.78) in the latter [87]. However, the product of step 2 (SGO; obtained after prereduction of GO followed by sulfonation) showed a slight increase in I_D/I_G ratio from GO (1.16) to SGO (1.27). Although no explanation has been given for this difference, we assume that the synthesis and purification steps of graphite oxide or GO, as mentioned previously, can significantly affect its further functionalization and experimental properties. Indeed, this assumption may be linked to the EA analysis in the same work. Surprisingly, all three GO, SGO, and SG NMs indicated the presence of 0.6% of sulfur, even though their IEC values were significantly different; 2.0, 1.7, and 0.5 mmol H^+ g^{-1} for GO, SGO, and SG, respectively. This data suggests that the presence of oxygen-functionalities, in particular, -OH groups are responsible for the acidity of G-NMs, which indeed is a misconception as mentioned above.

Zhang *et al.* [98] prepared SG sheets as described above [96] and used them as a sorbent in micro-solid-phase extraction (μ-SPE) coupled with gas chromatography-mass spectrometry for the determination of polycyclic aromatic hydrocarbons in water. Based on Samulski's pioneering work [96] on GO sulfonation, a number of other groups have also prepared SGs and used them for different applications [85,99,100].

Fuming Sulfuric Acid

Fuming sulfuric acid or oleum commonly refers to a solution of sulfur trioxide (SO_3) dissolved in sulfuric acid. It can be described by the general formula $H_2SO_4 \cdot xSO_3$ where x is the free SO_3 molar content and is often known as 30% oleum. It is produced in the contact process in which S is oxidized to SO_3 and subsequently dissolved in concentrated H_2SO_4. Owing to its high enthalpy of hydration, it serves as a central intermediate in the production of H_2SO_4. In typical organic chemistry, oleum is used in the nitration of nitrobenzene and the manufacture of explosives. Due to its relatively low cost (~US$ 0.153 per pound of reactive SO_3), oleum sulfonation is one of the industrially preferred strategies, especially, for versatility in feedstock selection to build materials that are truly application specific. In general, oleum sulfonation is an equilibrium and hydrothermal process, which can be operated both as a batch and continuous process.

In graphene research, the oleum-assisted sulfonation was first achieved by Liu *et al.* [83] by following the route shown inScheme 5. The resulting material was termed sulfated graphene ($G-SO_3H$). However, we would like to rename it sulfonated graphene, which is more appropriate as there is C-S bonding rather than O-S. The TEM and AFM images indicated the sheet like structure of $G-SO_3H$ with an average thickness of 0.8–3.0 nm. Compared with GO (0.89), $G-SO_3H$ gave a relatively highI_D/I_G ratio (1.02), which can be attributed to a decrease in the average size of the sp^2 domain as described above. The high resolution S2p XPS spectrum of $G-SO_3H$ displayed a peak at 168.3 eV (associated with S-O bonds); rather lower than those of SO_3H-functionalized ordered mesoporous carbon ($OMC-SO_3H$; 168.8 eV) and Amberlyst-15 (168.9 eV), which was likely a result of electron transfer from graphene to the $-SO_3H$ groups [16]. In this milieu, it is important to stress that S2p binding energy is a direct measure and quite sensitive to the acidic strength [101].

The thermal stability of $G-SO_3H$ was evaluated by TGA in association with $OMC-SO_3H$. Specifically, $G-SO_3H$ revealed two main decomposition steps, centered at 268 and 568 °C, which were assigned to the degradation of $-SO_3H$ groups and carbon framework, respectively. In contrast, weight loss temperatures in $OMC-SO_3H$ were centered at 237 °C and 390 °C for $-SO_3H$ groups and carbon framework, respectively. The better stability of carbon framework and $-SO_3H$ groups in $G-SO_3H$ over $OMC-SO_3H$ was attributed to its non-amorphous characteristic leading to stronger interaction with $-SO_3H$ groups.

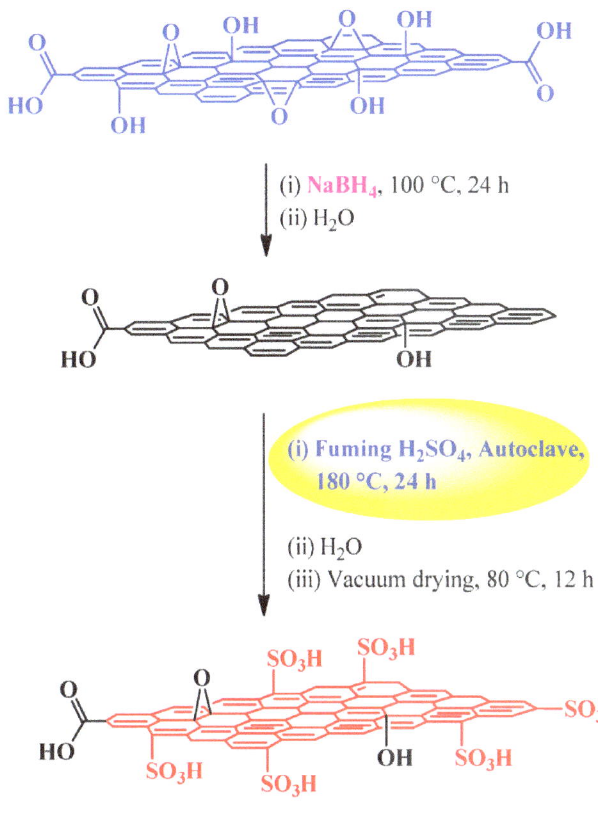

Scheme 5: Oleum-assisted sulfonation of GO into $G-SO_3H$.

It would be worthwhile to remind readers again here that TGA results are in sharp contrast to Hirsch's experimental findings [57], where it is shown that -SO_3H groups dissociate at ~500 °C. Furthermore, the observed weight loss at 268 °C in G-SO_3H is quite close enough to that of organosulfates, which may be attributed to the presence of sulfur species in the parent GO. The acid content of G-SO_3H was evaluated by acid-base titration and was found closer (1.2 mmol H^+ g^{-1}) to OMC-SO_3H (1.3 mmol H^+ g^{-1}) but far below that of Amberlyst-15 (4.7 mmol H^+ g^{-1}).

While diazotization-based GO sulfonation involves tedious multi-step synthesis, oleum-assisted hydrothermal sulfonation does not, making it an attractive strategy for sulfonation of GO. Despite this, the use of autoclaves with the added disadvantage of relatively high temperature conditions (≥180 °C) make this process quite expensive as well as energy inefficient. With these realizations, we decided to reinvestigate this reaction under relatively mild conditions. Indeed, combined with a knowledge of acid functional ionic liquids [102,103,104] and their exceptional product-switching ability in tetrapyrrolic organic systems [102], we were able to synthesize sulfonated graphene with simple processing and handling, one of the important tenets of "green chemistry". In particular, the oleum-assisted hydrothermal sulfonation of rGO in a standard fume cupboard at 120 °C for 3 days afforded sulfonated graphene exhibiting an IEC of 1.55 mmol H^+ g^{-1}. The as-obtained sulfonated graphene was successfully used in the synthesis of ester plasticizers, and other industrially important chemicals. The details of these findings will be reported elsewhere.

In yet another work, Liu *et al.* [90] used oleum to develop single-layer sheets of sulfated GO (GO-OSO_3H) (though oleum-assisted sulfation is not commonly practiced) as a hole-extraction layer in high performance polymer solar cells. It was postulated that on treatment with oleum, epoxy and hydroxyl groups on the basal plane of GO are either removed or transformed into sulfate groups as shown in Scheme 6. This is quite reasonable because strong acids [50,105] such as oleum exhibit a high dehydration tendency and can lead to the partial reduction of GO. The oleum-induced basal plane reduction of GO was evidenced by comparing the XRD patterns for GO and GO-OSO_3H. In particular, a peak shift from 2θ = 11.50° (d= 7.65 Å) for GO to 2θ = 26.40° (d = 3.91 Å) was recognized for GO-OSO_3H, a characteristic feature of partially restacked graphene domain. These results were further verified by XPS, UV-visible absorption, and Raman spectroscopy, the latter of which showed a slightly higher I_D/I_G ratio for GO-OSO_3H (0.88) than that for GO (0.78).

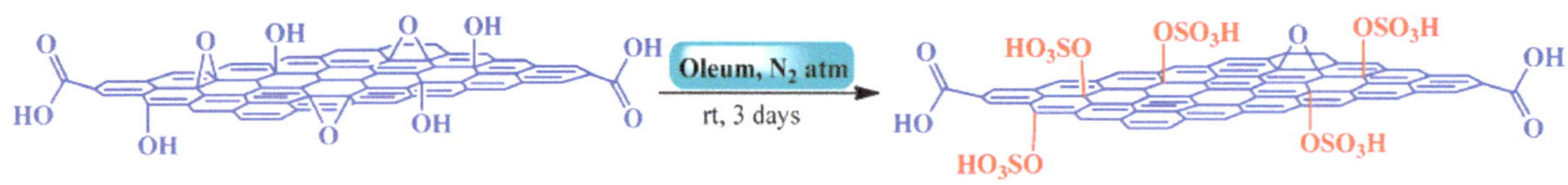

Scheme 6: Oleum-assisted sulfation of GO into GO-OSO_3H.

The acid density of GO-OSO_3H was evaluated to be 5.86 mmol H^+ g^{-1}, which was relatively higher than that of GO (3.15 mmol H^+ g^{-1}). Such a high acidity (even higher than that of Amberlyst™-15) of GO-OSO_3H was attributed to both the presence of -OSO_3H and -COOH groups and may be linked to high S content (4.49 wt %) as determined by EA. Furthermore, the acidity of GO was interpreted in terms of -COOH and -OH groups, however, no EA data was presented by the authors [90].

Chlorosulfonic Acid

Chlorosulfonic acid ($ClSO_3H$) or sulfuric chlorohydrin is another strong acid widely used to produce primarily alcohol sulfates, alcohol ether sulfates, and dye intermediates. $ClSO_3H$ is an expensive source of SO_3 (~US$ 0.255 per pound of reactive SO_3), highly toxic, corrosive, rapid, and stoichiometric reactant that can be used as a dehydrating, oxidizing, and chlorinating agent. During its reaction, it liberates hydrochloric acid as a by-product, which is normally neutralized with water or dilute basic solution. In graphene research, $ClSO_3H$ has been used as a solvent for high concentration dispersion and expansion/exfoliation of graphitic domains under ambient conditions [106,107,108].

Taking into account the excellent nucleophilicity of organolithium compounds toward epoxy and carbonyl functions, Wang *et al.* [84] introduced a novel synthesis methodology for sulfated graphene under relatively mild conditions. As such, the sulfation steps of epoxide, hydroxyl and carbonyl functions were suggested to follow the mechanism shown in Scheme 7. These include pretreatment of GO, organolithiation of rGO with n-butyl lithium and sulfation of as-obtained chemically converted graphene (CCG).

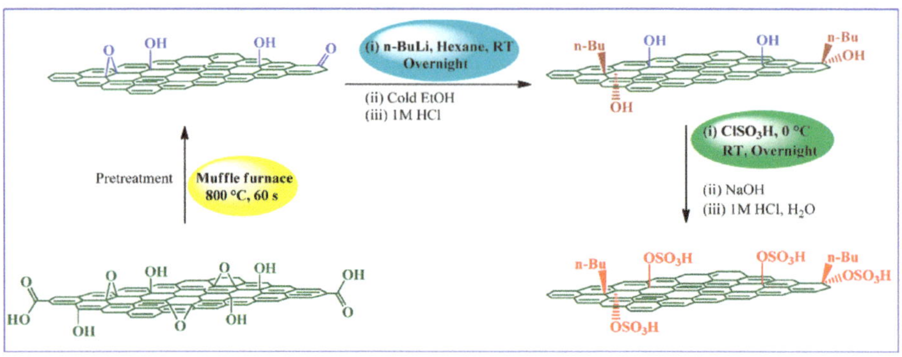

Scheme 7: $ClSO_3H$-induced sulfation of GO.

The formation of O-S bonds in the as-prepared sulfated graphene was evidenced by FTIR spectroscopy, where a peak at 3691 cm^{-1} corresponding to the O-H stretching vibrations in CCG disappeared and reappeared (at 3450 cm^{-1}) along with an additional peak at 1006 cm^{-1} (S=O stretching vibrations). The Raman spectra of rGO, CCG, and sulfated graphene indicated an I_D/I_G ratio of 0.92, 0.84, and 0.94, respectively. The thermal stability of all samples was assayed by TG analysis and determined to be: rGO > sulfated graphene > CCG. The quantitative XPS analysis of sulfated graphene showed 2.36% S by mass, however, IEC was not taken into account by the authors [84].

In another work, Upare *et al.* [86] employed $ClSO_3H$ to functionalize commercial samples (Sigma-Aldrich, Germany) of GO and activated carbon with $-SO_3H$ groups. It has been claimed that as-prepared GO-SO_3H catalyst can be used for the selective decomposition of hexose sugars into levulinic acid and other value-added derivatives. However, this report is somewhat difficult to harmonize in view of the reported experimental data. For instance, XRD pattern of GO are reported to be at $2\theta = 26°$; GO exhibits a sharp diffraction peak at $2\theta = 10°–13°$. We think that perhaps rGO is misinterpreted as GO in this report, as only the former gives diffraction patterns in the range from $2\theta = 21.5°$ to $27.5°$. Our assumption may be further linked to EA results demonstrating a very high C/O ratio of 14.12 for GO, which is quite unlikely. In fact, this value is very close to that observed (13.4) with $NaBH_4$-induced reduction of graphite oxide [76].

Combined with a knowledge of sulfonation or sulfation chemistry, we can make an educated guess that the reactivity of $ClSO_3H$ towards -OH functions of GO should even be higher than that of with oleum. Nevertheless, chlorosulfonation reactions cannot be ruled out completely.

Sulfuric Acid

Sulfuric acid (H_2SO_4), historically known as oil of vitriol, is a well-known strong mineral acid having significant oxidizing and dehydrating properties. In the chemical industry it has a central importance and is increasingly used for variety of applications such as in oil refining, fertilizer manufacturing, mineral and waste water processing, acidic drain cleaning, as well as in acid-catalyzed reactions as mentioned previously. In typical organic chemistry, it is used for sulfonation of benzene through an electrophilic aromatic substitution under reflux conditions. However, the sulfonation with H_2SO_4 can be considered as a special case of oleum sulfonaion. In present scenario, the H_2SO_4 sulfonation is principally used for production of hydrotropes by azeotropic reaction with toluene, xylene, or benzene.

Recently, Antunes *et al.* [89] have reported the use of H_2SO_4 to introduce sulfur-containing acid groups onto the graphene surface. Unlike previous studies, GO (in conjunction with CNTs and carbon black (CB)) was reduced by high temperature (190 °C) treatment with benzyl alcohol under microwave irradiation. The as-obtained rGO (coined as RGO in the original report) was subjected to H_2SO_4 (97 wt %) treatment at 160 °C for 5 h and this process was repeated twice more. The BET surface area of the as-obtained material (S-RGO) was determined to be 12 $m^2 \cdot g^{-1}$; far below than those of S-CNTs (44 $m^2 \cdot g^{-1}$) and S-CB (228 $m^2 \cdot g^{-1}$). As described above, the Raman results were consistent and an increase in I_D/I_G ratio was observed in all CNMs after H_2SO_4 treatment.

The acid strength of S-RGO in combination with rGO, Amberlyst-15, and S-CNTs was assayed by ^{31}P magic-angle spinning (MAS) NMR spectroscopy using triethylphosphine oxide (TEPO) as a basic probe. As shown in Figure 4, the appearance of signals at 55–60 ppm in both rGO and S-RGO were attributed to the presence of non-sulfur acid groups while the latter gave additional signals at higher chemical shifts (71–80 ppm), possibly due to the presence of sulfur species. As expected, the sulfonic acid-based cation exchange resin, Amberlyst™-15 (referred to as Amberlyst-15 in ref. [89]), displayed a single sharp resonance ~90 ppm demonstrating its stronger acidity than that of carbocatalysts.

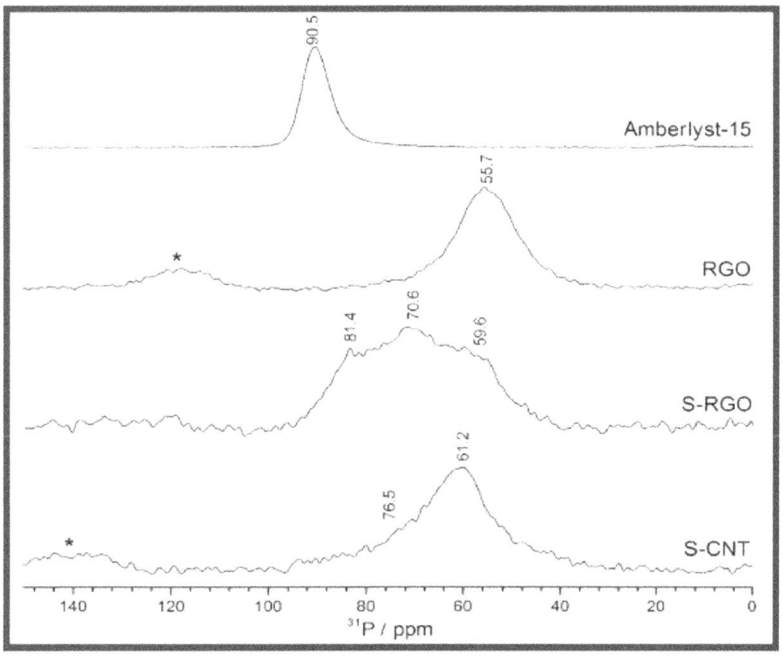

Figure 4: The ^{31}P MAS NMR spectra of TEPO adsorbed onto carbocatalysts and Amberlyst-15. Reproduced with permission from [89]. Copyright (2014) Wiley-VCH Verlag GmbH & Co.

The total acid content of S-RGO as measured by acid-base titrations was determined to be 2.2 H^+ mmol·g^{-1} and was attributed to the presence of -COOH and -OH groups. On the other hand, EA results indicated relatively low sulfur content (1.08 mmol·g^{-1}) which was credited to the presence of sulfonic acids. However, the TGA outcome is in contrast to this hypothesis. In particular, a second weight loss in TGA is observed in between 230–320 °C; far below the decomposition temperature (~500 °C) of sulfonic acid groups [57]. As such, one can assume that rather than sulfonic these are hydrosulfate groups that may be generated during oxidative treatment of rGO with H_2SO_4. In this regard, a comparison between S-RGO and GO, especially, ^{31}P MAS NMR experimental data is not only desirable but also would be praiseworthy.

Aside from the above-described sulfonating agents, He *et al.* [88] have reported the use of ammonium sulfate ($(NH_4)_2SO_4$) in the simultaneous reduction and sulfonation of GO. It was demonstrated that when a mixture of GO and $(NH_4)_2SO_4$ (in mass ratio of 5:1) was dried by lyophilization and subsequently heated at 235 °C in an inert atmosphere, the sulfonic acid groups could be grafted onto GO, whereas GO could be simultaneously reduced. The as-obtained material (S-rGO) was characterized by SEM, EDX, IR and Raman spectroscopy and used as support for metal nanocatalysts. The acid-density of S-rGO in terms of equivalent weight was evaluated to be 727 ± 5 g·mol^{-1}.

It is worthy of mention here that despite the efforts of many chemists since the 18th century [109,110,111,112], the exact mechanism of thermal decomposition of $(NH_4)_2SO_4$ remains elusive. Nevertheless, the use of $(NH_4)_2SO_4$ as a sulfonating agent, as far as we are aware, is not common in practice, presumably, due to its most feasible two-steps decomposition pathway [113] as shown in Equations (1) and (2):

$$2(NH_4)_2SO_4 \rightarrow (NH_4)_2S_2O_7 + 2NH_3 + H_2O \quad (1)$$

$$3(NH_4)_2S_2O_7 \rightarrow 2NH_3 + 6SO_2 + 2N_2 + 9H_2O \quad (2)$$

$$(NH_4)_2SO_4 \rightarrow 2NH_3 + SO_3 + H_2O \quad (3)$$

The ammonium pyrosulphate ($(NH_4)_2S_2O_7$), the primary condensation phase product (Equation (1)) is generally obtained at temperatures in between 100–250 °C while the reaction shown in Equation (2) occurs at relatively high temperature (>250 °C). Thus, the only possibility of generating SO_3 is the likely reaction of SO_2 with oxygen at relatively high temperature. Though a similar $(NH_4)_2SO_4$-induced sulfonation protocol was also reported previously [114], we believe that characterization data is insufficient, leading to confusion in understanding the nature of this reaction. Furthermore, the acidity is commonly presumed and interpreted from the perspective of sulfur content in $-SO_3H$ groups. Now, if we assume that the decomposition pathway does exist in accordance with Equation (3), is this process gentle enough to leave other molecules, such as ammonia, untouched?

Indeed, the thermal decomposition of $(NH_4)_2SO_4$ within graphene domain appears to us quite interesting and it is quite reasonable to expect that some interesting materials can be formed simply by switching the experimental conditions.

APPLICATIONS OF G-NMS IN HETEROGENEOUS ACID CATALYSIS

Acid-catalyzed reactions are classified according to the type of acid catalysts. The catalytic performances of G-NMs in conjunction with other solid/liquid catalysts have been discussed in terms of catalyst loadings, products yield, recoverability, and recyclability at certain instances.

Graphite Oxide as Acid Catalyst

Michael-Type Friedel-Crafts Addition

Intrigued by the strong acidity of graphite oxide, Kumar and Rao [115] used it as catalyst in the Friedel-Crafts addition of indoles to α,β-unsaturated ketones as well as nitrostyrenes (Scheme 8). The superiority of graphite oxide over other acid catalysts including HCl, H_2SO_4, acidic Al_2O_3, nano MgO/CuO/Fe_2O_3, (sulfo)β-cyclodextrins, activated charcoal, and *p*-TSA was demonstrated by addition of indole to methyl vinyl ketone in THF-H_2O mixtures at ambient conditions. Whereas graphite oxide (20 wt %) afforded excellent yield (92%), the other catalysts only gave poorer to moderate yields (5%–67%) of addition product. For the same reaction, graphite oxide could be recycled up to five times without apparent loss in catalytic activity. Indoles with electron-donating and electron-withdrawing substituents underwent reactions smoothly in the addition to methyl vinyl ketone with no polymerized or dimerized side products. However, no reaction was observed with carboxylic acid/ester substituents, even after 24 h. Furthermore, the reaction of indole with electron deficient nitrostyrene took quite a longer time to afford products in moderate yields.

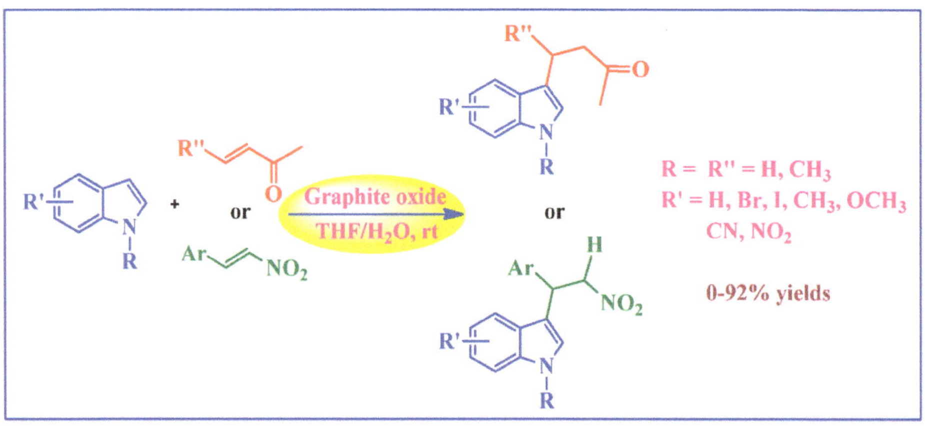

Scheme 8: Graphite oxide catalyzed Michael-type Friedel-Crafts addition of indoles.

Polymerization

Inspired by their initial discovery of GO's acidity [30], Dreyer *et al.* [116] used graphite oxide (2.5–20 wt % loadings) to catalyze the ring opening polymerization of various cyclic lactones and lactams to their corresponding carbon-reinforced, high molecular weight polyesters, or polyamide composites in good to excellent yields (39%–100%). In contrast to most other methods, which often rely on clumsy two-step protocols–preparing the polymer followed by mixing of the carbon filler, the use of a graphite oxide combines the polymerization and dispersion of the filler into a single expeditious step.

Synthesis of Dipyrromethanes

Chauhan *et al.* [117] examined the reactions of pyrrole and ketones using graphite oxide as acid catalyst under a variety of conditions. Specifically, moderate to high yields of 5,5-dialkyldipyrromethanes were achieved along with linear oligomers and cyclic tetramers under ambient conditions.

Graphene Oxide (GO) as Acid Catalyst

Hydration of Alkynes

The first use of GO [30] as an acid catalyst was demonstrated in the hydration of various alkynes to their corresponding ketones (Scheme 9a). In particular, in the presence of GO, low to excellent conversions of alkynes (27%–98%) were achieved at 100 °C as monitored by ^{1}H-NMR spectroscopy. However, the need to use significantly high GO loadings (200 wt %) in these reactions somewhat limits the practical utility of this protocol.

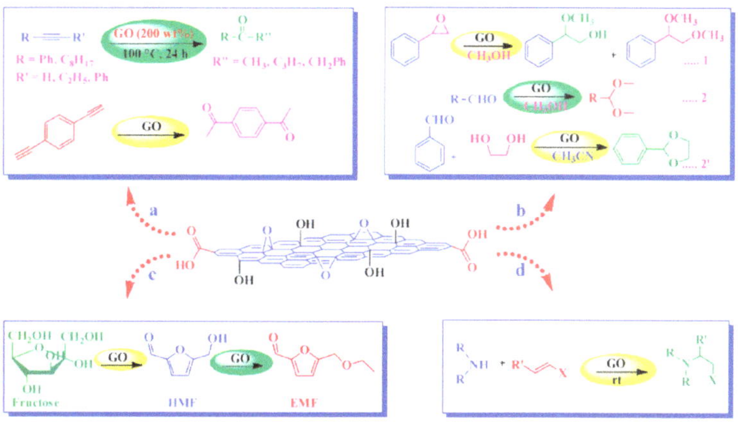

Scheme 9: GO catalyzed (a) Hydration of alkynes. (b) Ring opening of epoxides and acetalization of aldehydes. (c) Conversion of carbohydrates into 5-ethoxymethylfurfural (EMF). (d) Aza-Michael addition.

Aza-Michael Addition

The Aza-Michael addition of amines to activated alkenes in the presence of GO catalyst furnishing β-amino compounds was reported by Verma *et al.* [118] (Scheme 9d). The reactions were found to proceed under relatively mild conditions and to afford excellent product yields (70%–97%) within short reaction times. The GO was recyclable up to nine runs with the consistent catalytic activity.

Ring Opening of Epoxides and Acetalization of Aldehydes

Dhakshinamoorthy *et al.* [53] employed GO as an acid catalyst in the ring opening of epoxides with methanol and other primary alcohols as nucleophile and solvents. Under ambient conditions, the product selectivity could be switched by changing the reaction time. However, most of the reactions exhibited high regioselectivity rendering the product expected from S_N1 mechanism. Interestingly, the conversion and selectivity in the presence of GO were found comparable to those of H_2SO_4 and *p*-TSA and much higher than Norit A, activated carbon, as demonstrated by styrene oxide with methanol as a model reaction (Scheme 9b-1). The high activity of GO was attributed to the presence of hydrogen sulfate groups on GO surface. The catalyst could be easily recovered after the reaction and successfully reused for up to three cycles. In their subsequent work [119], the group further examined the activity of GO for the acetalization of aldehydes in methanol (Scheme 9b-2). The GO-induced reaction of a variety of aromatic and aliphatic aldehydes with methanol afforded the corresponding acetalized products in moderate to excellent yields as determined by GC. In the case of 4-methylbenzaldehyde, the low yield (10%) was attributed to the poisoning of GO due to the formation of toluic acid, while 4-pyridylcarboxaldehyde and acetophenone were completely unreactive. The protocol was further extended successfully to the acetalization of benzaldehyde with ethylene glycol (Scheme 9b-2').

Conversion of Carbohydrates into 5-Ethoxymethylfurfural (EMF)

Wang *et al.* [54] discovered that GO can be used as a facile acid catalyst for the one-step conversion of fructose-based carbohydrates into high-heating value liquid biofuels (Scheme 9c). In this regard, one-pot conversions of sucrose, inulin, and fructose into EMF using DMSO-ethanol mixtures (1:2.3, v/v) were achieved with yields of 34%, 66%, and 71%, respectively. Interestingly, GO-catalyzed direct etherification of 5-hydroxymethyl furfural (HMF), a dehydration product of carbohydrates, in ethanol afforded high EMF yield of 92% as determined by LC/MS. On the contrary, traditional liquid and solid acid catalysts such as Amberlyst-15, *p*-TSA, and H_2SO_4 gave only moderate yields of EMF (54%–61%) even though with 100% conversions as determined by

NMR. Though a high yield of EMF (82%) was detected with $H_3PW_{12}O_{40}$, it was dissolved in the reaction mixture and couldn't be separated. The high activity of GO was attributed to the presence of hydrogen sulfate groups as mentioned above and found consistent even after several runs.

Sulfonated/Sulfated Graphene and Graphene Oxide (GO) as Acid Catalysts

Sulfonated Graphene and GO as Acid Catalysts Prepared by Diazonium Salt

Hydrolysis of Ethyl Acetate

Ji *et al.* [82] used sulfonated graphene in the water participating reaction, hydrolysis of ethyl acetate. To evaluate its activity and stability, H_2SO_4 and Nafion NR50 catalysts were employed as controls. Whereas Nafion NR50 showed poorer activity, the activity of sulfonated graphene was comparable to that of H_2SO_4 and remain unchanged even after five consecutive runs.

Synthesis of Xanthenes and Benzoxanthenes

Shaabani *et al.* [85] used sulfonated graphene in conjunction with GO for the one-pot condensation of aldehydes, 1,3-diketones, and naphthols into the corresponding xanthenes and benzoxanthenes (Scheme 10).

Unlike $InCl_3$, P_2O_5, HY zeolite, and $Sr(OTF)_2$, the G-NMs exhibited high catalytic activity in water and could be reused several times; the best results were obtained with sulfonated graphene.

Dehydration of Xylose

Lam *et al.* [87] applied sulfonated GO (SGO) and graphene in conjunction with unfunctionalized GO and graphene for the dehydration of xylose to furfural (Scheme 11). Out of the four tested catalysts (2 wt% loadings, 35 min), SGO afforded highest yield of furfural (62%) with 75% selectivity as averaged over three runs. On the other hand, yields determined in the absence of catalyst as well as in the presence of graphene, GO, and sulfonated graphene were in the following order: sulfonated graphene (55%) > GO (53%) > graphene (51%) > no catalyst (44%). Interestingly, at only 0.5 wt % loading *vs.*xylose, SGO could be reused over multiple runs maintaining an average yield of 61%, despite the presence of accumulated byproducts on its surface. It has been suggested that strongly acidic aryl sulfonic acid groups in SGO are the key active sites for the high temperature production of furfural in water.

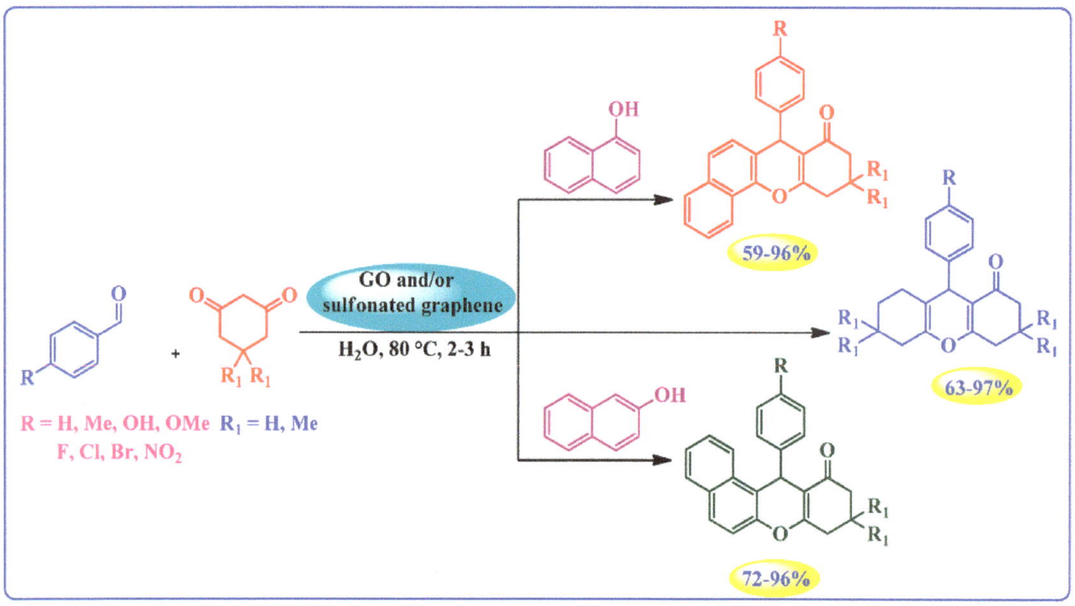

Scheme 10: GO and/or sulfonated graphene catalyzed synthesis of xanthenes and benzoxanthenes in water.

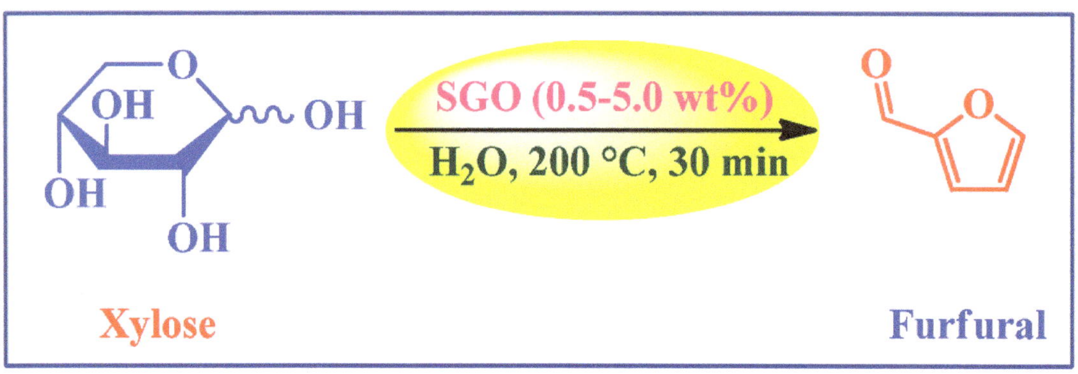

Scheme 11: Sulfonated GO catalyzed dehydration of xylose into furfural.

Sulfonated Graphene as Acid Catalyst Prepared by Fuming Sulfuric Acid or Oleum

Esterification, Peckmann, and Hydration Reactions

Liu *et al.* [83] employed sulfonated graphene (G-SO_3H) as solid catalyst in a number of acid-catalyzed reactions including esterification, hydration, and Peckmann reactions (Scheme 12). The esterification reactions of acetic acid with cyclohexanol and *n*-butanol in the presence of G-SO_3H (after pretreatment with 0.1 M H_2SO_4) afforded the corresponding esters with the highest conversions of ~79% and 89%, respectively (Scheme 12a,a'). Furthermore, G-SO_3H exhibited almost similar activity when reactions were performed under static conditions. On the contrary, the conversions were significantly low (<53% and <76% for cyclohexanol and *n*-butanol, respectively) in the presence of other solid catalysts such as OMC-SO_3H, rGO and SBA-15-SO_3H. Nevertheless, Amberlyst-15 gave second highest conversions for the same. Notably, in contrast to available reports, GO showed poorer catalytic activity with as low as <22% conversions. Surprisingly, no acid sites were detected in GO as measured by acid-base titrations and S analysis.

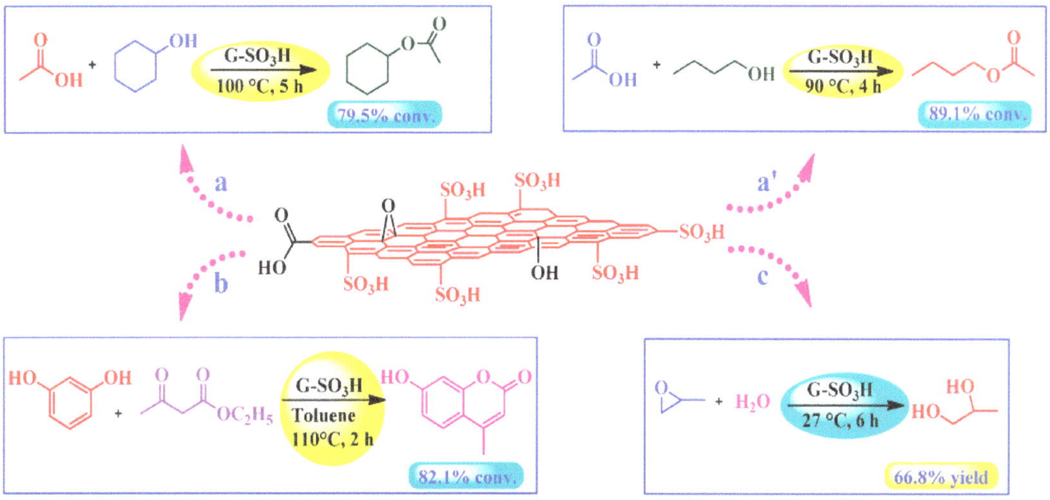

Scheme 12: G-SO_3H catalyzed (a) Esterification of acetic acid with cyclohexanol. (a') Esterification of acetic acid with *n*-butanol. (b) Peckmann reaction of resorcinol with ethyl acetoacetate. (c) Hydration of propylene oxide.

In yet another example, G-SO_3H catalyzed Peckmann reaction of resorcinol with ethyl acetoacetate in toluene afforded corresponding coumarin with ~82% conversion (Scheme 12b). On the other hand, conversions were decreased significantly in the presence of other catalysts and determined in the following order: Amberlyst-15 (~75%) > OMC-SO_3H > SBA-15-SO_3H >> GO > rGO (~11%). A similar trend was recognized in the hydration of propylene oxide (Scheme 12c), however, GO and rGO were found completely inactive.

The recyclability of G-SO_3H was examined in association with OMC-SO_3H. After being recycled five times, G-SO_3H gave relatively higher conversions than OMC-SO_3H. The activity loss in G-SO_3H was 1.5%–2.8% whereas OMC-SO_3H had a significant reduction in activities (~7%–10%) after five cycles. The high stability and activity of G-SO_3H were attributed to the presence of stable sulfonic acid groups on its surface and unlimited mass transfer owing to its unique sheet structure.

Sulfated Graphene as Acid Catalyst Prepared by $ClSO_3H$

Ester-Exchange Reactions

Wang *et al.* [84] evaluated the catalytic efficiency of sulfated graphene for the ester-exchange reaction between ethyl acetate and a variety of alcohols. Specifically, sulfated graphene catalyzed reaction of alcohols gave corresponding esters in ≥96% isolated yield; somewhat superior to that of Dowex 50 W × 2 (50–100 mesh; 95% yield). It was postulated that nucleophilicity of alcohols play a significant role in ester-exchange reactions. The catalyst could be recycled up to three times.

Chemical Conversion of Carbohydrates to Levulinic Acid (LA)

Upare *et al.* [86] used -SO_3H functional GO catalyst in conjunction with activated carbon for the selective decomposition of glucose and fructose into levulinic acid (LA) (Scheme 13). The graphene-based catalyst gave high yield of LA (78%) and showed good reusability with reliable performance. It was postulated that chemical transformation patterns for hexose sugar decompositions are affected by the type of catalyst, the density of acid sites as well as temperature. In addition, the presence of other functional groups including -COOH and -OH groups on catalyst surface help in enhancing the adsorption of sugar for the reaction.

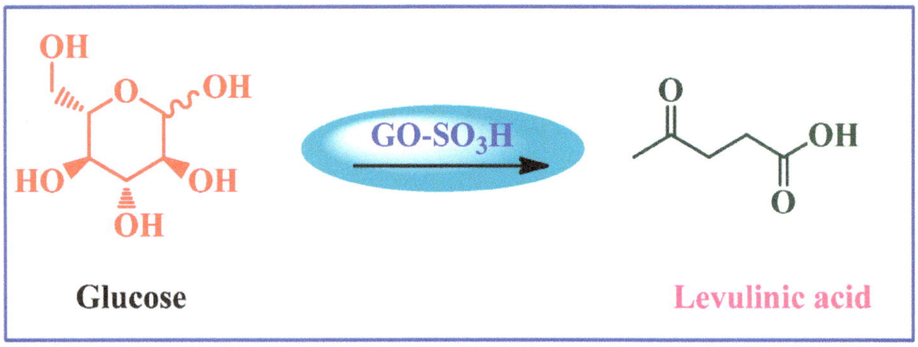

Scheme 13: Chemical conversion of glucose to levulinic acid (LA).

Sulfated rGO as Acid Catalyst Prepared by H_2SO_4

Conversion of HMF into Biofuels

Antunes *et al.* [89] used H_2SO_4-treated rGO (S-RGO) as potential catalyst in the conversion of 5-hydroxymethylfurfural (HMF) to biofuels (Scheme 14) such as 5-ethoxymethylfurfural (EMF), 5-(ethoxymethyl)furfural diethyl acetal (EMF da), and ethyl levulinate (EL). The catalytic performance of S-RGO was found superior to that of H_2SO_4 treated carbon black, CNTs as well as Amberlyst-15 and was attributed to the different acid sites and 2D structure of the catalyst. The catalyst could be reused up to three cycles without significant loss in catalytic activity.

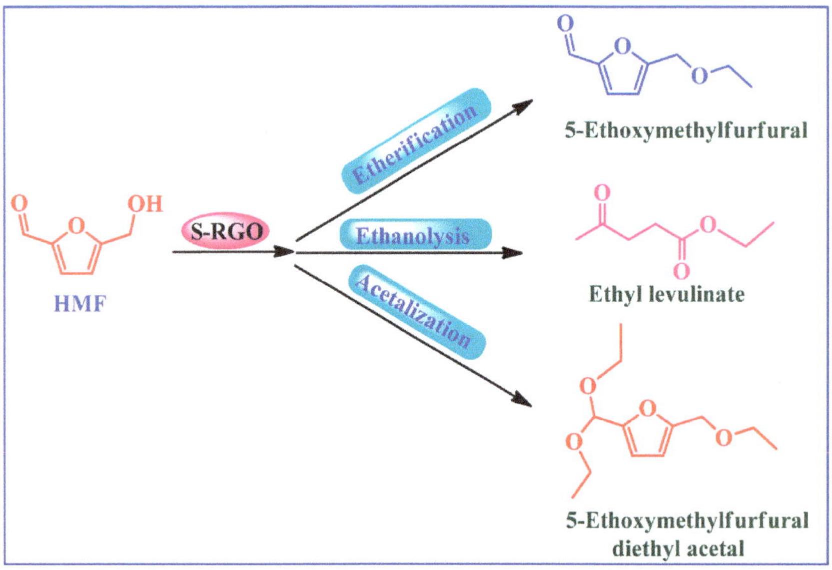

Scheme 14: Conversion of 5-hydroxymethylfurfural (HMF) into value-added biofuels.

CONCLUSIONS: FINAL REMARKS

As the building block of CNMs of all dimensions, graphene possesses many unique properties and physicochemistry. In this comprehensive review, we have shed light on graphite oxide, GO, and their sulfonated or sulfated derivatives acting as main performers in heterogeneous acid catalysis. The observation that graphite oxide and GO, prepared by a modified Hummer's method, convey inherent acidic character has motivated scientists to get new insights into the structural properties of graphite oxidation products. In this regard, new theories and approaches are constantly being explored and somewhat mystifying the commonly accepted LK model. Another important property of GO is its easy functionalization rendering it useful in preparing a library of catalytically active sulfonated or sulfated NMs. Indeed, the heterogeneous acid catalysis using these NMs has the potential to undergo a renaissance and to address many industrial processes not previously accessible to it.

Despite this, a basic understanding of working actions of sulfonating agents on (r)GO is still in its infancy, which have resulted in conceptual vagueness in many aspects. Furthermore, due to the lack of comparative yet solid experimental data of starting (GO) and final products (sulfonated or sulfated G-NMs), a more precise acid activity has not been ascertained. These shortcoming are of utmost importance since two key concerns may plague future scientists interested in using sulfonation methods, mainly whether a sulfonating agent of interest would: (i) specifically target aromatic carbon (sulfonation) or -OH functions (sulfation) in a certain mode of reaction and (ii) undergo a random mode of reaction. In this context, a fundamental understanding of sulfonation chemistry combined with routine industrial experience can only act as a guide for advancements in this area.

As a final note, while increasingly tight legislation on the control of chemical effluents currently at play, heterogeneous acid catalysis with G-NMs holds great promise to achieve the objectives of industrial catalysis following the principles of sustainable and green chemistry. We postulate that by the mutual cooperation between chemists, physicists and material scientists, graphene-based heterogeneous acid catalysis may lead to new disciplines in the 21st century. It is just a matter of patience and constant efforts to discover new catalysts or novel methodologies; after all, they are also patiently waiting to be discovered.

ACKNOWLEDGMENTS

This work was supported by National Tsing Hua University (102N1807E1) and the Ministry of Science and Technology (NSC101-2113-M-007-006-MY3) of Taiwan.

AUTHOR CONTRIBUTIONS

B. Garg and Y.-C. Ling were responsible for the subject of the review. T. Bisht performed bibliographic researches, participated in the discussions and also played an important role in preparing graphics or art work. The manuscript was designed, organized and written by B. Garg while the final editing was done by Y.-C. Ling. All authors read and approved the final manuscript. B. Garg was also responsible for the revisions and the proofs.

REFERENCES

1. Mizuno, N.; Misono, M. Heterogeneous catalysis. *Chem. Rev.* 1998, *98*, 199–217.
2. Pal, R.; Sarkar, T.; Khasnobis, S. Amberlyst-15 in organic synthesis. *ARKIVOC* 2012, 570–609.
3. Okuhara, T. Water-tolerant solid acid catalysts. *Chem. Rev.* 2002, *102*, 3641–3666.
4. Harmer, M.A.; Sun, Q. Solid acid catalysis using ion-exchange resins. *Appl. Catal. A: Gen.* 2001, *221*, 45–62.
5. Harmer, M.A.; Farneth, W.E.; Sun, Q. High surface area nafion resin/silica nanocomposites: A new class of solid acid catalyst. *J. Am. Chem. Soc.* 1996, *118*, 7708–7715.
6. Shylesh, S.; Thiel, W.R. Bifunctional acid–base cooperativity in heterogeneous catalytic reactions: Advances in silica supported organic functional groups. *ChemCatChem* 2011, *3*, 278–287.
7. Harmer, M.A. Industrial processes using solid acid catalysts. In *Handbook of Green Chemistry and Technology*; Clark, J.H., Macquarris, D.J., Eds.; Blackwell Publishers: London, UK, 2002; pp. 86–117.
8. Chauhan, S.M.S.; Garg, B.; Bisht, T. Syntheses of calix[4] pyrroles by amberlyst-15 catalyzed cyclocondensations of pyrrole with selected ketones. *Molecules* 2007, *12*, 2458–2466.
9. Suwannakarn, K.; Lotero, E.; Goodwin, J.G., Jr.; Lu, C. Stability of sulfated zirconia and the nature of the catalytically active species in the transesterification of triglycerides. *J. Catal.* 2008, *255*, 279–286.
10. Onda, A.; Ochi, T.; Yanagisawa, K. Selective hydrolysis of cellulose into glucose over solid acid catalysts. *Green Chem.*2008, *10*, 1033–1037.
11. Suwannakarn, K.; Lotero, E.; Goodwin, J.G., Jr. A comparative study of gas phase esterification on solid acid catalysts.*Catal. Lett.* 2007, *114*, 3–4.
12. Wang, X.; Liu, R.; Waje, M.M.; Chen, Z.; Yan, Y.; Bozhilov, K.N.; Feng, P. Sulfonated ordered mesoporous carbon as a stable and highly active protonic acid catalysts. *Chem. Mater.* 2007, *19*, 2395–2397.
13. Nakajima, K.; Hara, M. Amorphous carbon with SO_3H groups as a solid Brønsted acid catalyst. *ACS Catal.* 2012, *2*, 1296–1304.
14. Peng, F.; Zhang, L.; Wang, H.; Lv, P.; Yu, H. Sulfonated carbon nanotubes as a strong protonic acid catalyst. *Carbon*2005, *43*, 2405–2408.
15. Yu, H.; Jin, Y.; Li, Z.; Peng, F.; Wang, H. Synthesis and characterization of sulfonated single-walled carbon nanotubes and their performance as solid acid catalyst. *J. Solid State Chem.* 2008, *181*, 432–438.
16. Novoselov, K.S.; Geim, A.K.; Morozov, S.V.; Jiang, D.; Zhang, Y.; Dubonos, S.V.; Grigorieva, I.V.; Firsov, A.A. Electric field effect in atomically thin carbon films. *Science* 2004, *306*, 666–669.
17. Guest Editorial. Nomenclature of sp^2 carbon nanoforms. *Carbon* 2012, *50*, 741–747.
18. Wan, X.; Huang, Y.; Chen, Y. Focusing on energy and optoelectronic applications: A journey for graphene and graphene oxide at large scale. *Acc. Chem. Res.* 2012, *45*, 598–607.
19. Compton, O.C.; Nguyen, S.T. Graphene oxide, highly reduced graphene oxide, and graphene: Versatile building blocks for carbon-based materials. *Small* 2010, *6*, 711–723.
20. Dai, L.; Chang, D.W.; Baek, J.-B.; Lu, W. Carbon nanomaterials for advanced energy conversion and storage. *Small*2012, *8*, 1130–1166.
21. Huang, X.; Qi, X.; Boey, F.; Zhang, H. Graphene-based composites. *Chem. Soc. Rev.* 2012, *41*, 666–686.
22. Chen, Y.; Zhang, B.; Liu, G.; Zhuang, X.; Kang, E.-T. Graphene and its derivatives: Switching ON and OFF. *Chem. Soc. Rev.* 2012, *41*, 4688–4707.
23. Wu, M.C.; Deokar, A.R.; Liao, J.H.; Shih, P.Y.; Ling, Y.-C. Graphene-based photothermal agent for rapid and effective killing of bacteria. *ACS Nano* 2013, *7*, 1281–1290.
24. Guo, C.X.; Ng, S.R.; Khoo, S.Y.; Zheng, X.; Chen, P.; Li, C.M. RGD-peptide functionalized graphene biomimetic live-cell sensor for real-time detection of nitric oxide molecules. *ACS Nano* 2012, *6*, 6944–6951.
25. Li, Y.; Wen, T.; Xue, C.; Han, Q.; Wang, Y.; Hong, J.; Zhou, X.; Jiang, H. RGO LBL modified biomimetic electrochemical sensor for detection of Sildenafil in herbal sexual health products. *Biosens. Bioelectron.* 2013, *42*, 287–292.
26. Ganesh, G.; Chang, C.C.; Ling, Y.-C. Facile synthesis of smart magnetic graphene for safe drinking water: Heavy metal removal and disinfection control. *ACS Sustain. Chem. Eng.* 2013, *1*, 462–471.
27. Liu, J.Y.; Chang, H.Y.; Truong, Q.D.; Ling, Y.-C. Synthesis of nitrogen-doped graphene by pyrolysis of

ionic-liquid-functionalized graphene. *J. Mater. Chem. C* 2013, *1*, 1713–1716.

28. Ganesh, G.; Ling, Y.-C. Magnetic and fluorescent graphene for dual modal imaging and single light induced photothermal and photodynamic therapy of cancer cells. *Biomaterials* 2014, *35*, 4499–4507.
29. Garg, B.; Ling, Y.-C. Versatilities of graphene-based catalysts in organic transformations. *Green Mater.* 2013, *1*, 47–61.
30. Dreyer, D.R.; Jia, H.-P.; Bielawski, C.W. Graphene oxide: A convenient carbocatalyst for facilitating oxidation and hydration reactions. *Angew. Chem. Int. Ed.* 2010, *49*, 6813–6816.
31. Su, C.; Loh, K.P. Carbocatalysts: Graphene oxide and its derivatives. *Acc. Chem. Res.* 2013, *46*, 2275–2285.
32. Machado, B.F.; Serp, P. Graphene-based materials for catalysis. *Catal. Sci. Technol.* 2012, *2*, 54–75.
33. Dreyer, D.R.; Todd, A.D.; Bielawski, C.W. Harnessing the chemistry of graphene oxide. *Chem. Soc. Rev.* 2014, *43*, 5288–5301.
34. Navalon, S.; Dhakshinamoorthy, A.; Alvaro, M.; Garcia, H. Carbocatalysis by graphene-based materials. *Chem. Rev.*2014, *114*, 6179–6212.
35. Brodie, B.C. On the atomic weight of graphite. *Philos. Trans. R. Soc. Lond.* 1859, *149*, 249–259.
36. Staudenmaier, L. Verfahren zür darstellung der graphitsäure. *Ber. Dtsch. Chem. Ges.* 1898, *31*, 1481–1487.
37. Staudenmaier, L. Verfahren zür darstellung der graphitsäure. *Ber. Dtsch. Chem. Ges.* 1899, *32*, 1394–1399.
38. Hummers, W.S., Jr.; Offeman, R.E. Preparation of graphitic oxide. *J. Am. Chem. Soc.* 1958, *80*, 1339–1339.
39. Marcano, D.C.; Kosynkin, D.V.; Berlin, J.M.; Sinitskii, A.; Sun, Z.; Slesarev, A.; Alemany, L.B.; Lu, W.; Tour, J.M. Improved Synthesis of Graphene Oxide. *ACS Nano* 2010, *4*, 4806–4814.
40. Hofmann, U.; Holst, R. Über die säurenatur und die methylierung von graphitoxyd. *Ber. Dtsch. Chem. Ges. A/B* 1939,*72*, 754–771.
41. Ruess, G. Uber das graphitoxyhydroxyd (graphitoxyd). *Monatshefte Chem.* 1947, *76*, 381–417.
42. Clauss, A.; Plass, R.; Boehm, H.-P.; Hofmann, U. Untersuchungen zür Struktur des Graphitoxyds. *Z. Anorg. Allg. Chem.* 1957, *291*, 205–220.
43. Scholz, W.; Boehm, H.P. Untersuchungen am Graphitoxid. VI. Betrachtungen zür Struktur des Graphitoxids. *Z. Anorg. Allg. Chem.* 1969, *369*, 327–340.
44. Nakajima, T.; Mabuchi, A.; Hagiwara, R. A new structure model of graphite oxide. *Carbon* 1988, *26*, 357–361.
45. Nakajima, T.; Matsuo, Y. Formation process and structure of graphite oxide. *Carbon* 1994, *32*, 469–475.
46. He, H.; Riedl, T.; Lerf, A.; Klinowski, J. Solid-state NMR studies of the structure of graphite oxide. *J. Phys. Chem.* 1996,*100*, 19954–19958.
47. Lerf, A.; He, H.; Riedl, T.; Forster, M.; Klinowski, J. ^{13}C and ^{1}H MAS NMR studies of graphite oxide and its chemically modified derivatives. *Solid State Ion.* 1997, *101–103 Part 2*, 857–862.
48. Lerf, A.; He, H.; Forster, M.; Klinowski, J. Structure of graphite oxide revisited. *J. Phys. Chem. B* 1998, *102*, 4477–4482.
49. Szabó, T.; Berkesi, O.; Forgó, P.; Josepovits, K.; Sanakis, Y.; Petridis, D.; Dékány, I. Evolution of surface functional groups in a series of progressively oxidized graphite oxides. *Chem. Mater.* 2006, *18*, 2740–2749.
50. Gao, W.; Alemany, L.B.; Ci, L.J.; Ajayan, P.M. New insights into the structure and reduction of graphite oxide. *Nat. Chem.* 2009, *1*, 403–408.
51. Erickson, K.; Erni, R.; Lee, Z.; Alem, N.; Gannett, W.; Zettl, A. Determination of the local chemical structure of graphene oxide and reduced graphene oxide. *Adv. Mater.* 2010, *22*, 4467–4472.
52. Szabó, T.; Tombácz, E.; Illés, E.; Dékány, I. Enhanced acidity and pH-dependent surface charge characterization of successively oxidized graphite oxides. *Carbon* 2006, *44*, 537–538.
53. Dhakshinamoorthy, A.; Alvaro, M.; Concepción, P.; Fornés, V.; Garcia, H. Graphene oxide as an acid catalyst for the room temperature ring opening of epoxides. *Chem. Commun.* 2012, *48*, 5443–5445.
54. Wang, H.; Deng, T.; Wang, Y.; Cui, X.; Qi, Y.; Mu, X.; Hou, X.; Zhu, Y. Graphene oxide as a facile acid catalyst for the one-pot conversion of carbohydrates into 5-ethoxymethylfurfural. *Green Chem.* 2013, *15*, 2379–2383.
55. Dimiev, A.; Kosynkin, D.V.; Alemany, L.B.; Chaguine, P.; Tour, J.M. Pristine graphite oxide. *J. Am. Chem. Soc.* 2012,*134*, 2815–2822.
56. Dimiev, A.; Alemany, L.B.; Tour, J.M. Graphene oxide. Origin of acidity, its instability in water, and a new dynamic structural model. *ACS Nano* 2013, *7*, 576–588.
57. Eigler, S.; Dotzer, C.; Hof, F.; Bauer, W.; Hirsch, A. Sulfur species in graphene oxide. *Chem. Eur. J.* 2013, *19*, 9490–9496.
58. Toda, M.; Takagaki, A.; Okamura, M.; Kondo, J.N.; Hayashi, S.; Domen, K.; Hara, M. Green chemistry: Biodiesel made with sugar catalyst. *Nature* 2005, *438*, 178–178.
59. Liu, X.-Y.; Hunag, M.; Ma, H.-L.; Zhang, Z.-Q.; Gao, J.-M.; Zhu, Y.-L.; Han, X.-J.; Guo, X.-Y. Preparation of a carbon-based solid acid catalyst by sulfonating activated carbon in a chemical reduction process. *Molecules* 2010, *15*, 7188–7196.
60. McAllister, M.J.; Li, J.-L.; Adamson, D.H.; Schniepp, H.C.; Abdala, A.A.; Liu, J.; Herrera-Alonso, M.; Milius, D.L.; Car, R.; Prud'homme, R.K.; *et al.* Single sheet functionalized graphene by oxidation and thermal expansion of graphite.*Chem. Mater.* 2007, *19*, 4396–4404.
61. Schniepp, H.C.; Li, J.-L.; McAllister, M.J.; Sai, H.; Herrera-Alonso, M.; Adamson, D.H.; Prud'homme, R.K.; Car, R.; Saville, D.A.; Aksay, I.A. Functionalized single graphene sheets derived from splitting graphite oxide. *J. Phys. Chem. B*2006, *110*, 8535–8539.

62. Kaniyoor, A.; Baby, T.T.; Ramaprabhu, S. Graphene synthesis via hydrogen induced low temperature exfoliation of graphite oxide. *J. Mater. Chem.* 2010, *20*, 8467–8469.
63. Zhou, M.; Wang, Y.; Zhai, Y.; Zhai, J.; Ren, W.; Wang, F.; Dong, S. Controlled synthesis of large-area and patterned electrochemically reduced graphene oxide films. *Chem. Eur. J.* 2009, *15*, 6116–6120.
64. Dreyer, D.R.; Park, S.; Bielawski, C.W.; Ruoff, R.S. The chemistry of graphene oxide. *Chem. Soc. Rev.* 2010, *39*, 228–240.
65. Amarnath, C.A.; Hong, C.E.; Kim, N.H.; Ku, B.-C.; Kuila, T.; Lee, J.H. A; Hong, C.E.; Kim, N.H; Ku, B.-C.; Kuila, T.; Lee, J.H. Efficient synthesis of graphene sheets using pyrrole as a reducing agent. *H. Efficient synthesis of graphene sheets using pyrrole as a reducing agent. Carbon* 2011, *49*, 3497–3502.
66. Liu, J.; Tang, J.; Gooding, J.J. Strategies for chemical modification of graphene and applications of chemically modified graphene. *J. Mater. Chem.* 2012, *22*, 12435–12452.
67. Pei, S.; Cheng, H.-M. The reduction of graphene oxide. *Carbon* 2012, *50*, 3210–3228.
68. Liu, Y.; Zhang, Y.; Ma, G.; Wang, Z.; Liu, K.; Liu, H. Ethylene glycol reduced graphene oxide/polypyrrole composite for supercapacitor. *Electrochim. Acta* 2013, *88*, 519–525.
69. Chua, C.K.; Pumera, M. Chemical reduction of graphene oxide: A synthetic chemistry viewpoint. *Chem. Soc. Rev.* 2014,*43*, 291–312.
70. Kudin, K.N.; Ozbas, B.; Schniepp, H.C.; Prud'homme, R.K.; Aksay, I.A.; Car, R. Raman spectra of graphite oxide and functionalized graphene sheets. *Nano Lett.* 2008, *8*, 36–41.
71. Das, A.; Chakraborty, B.; Sood, A.K. Raman spectroscopy of graphene on different substrates and influence of defects.*Bull. Mater. Sci.* 2008, *31*, 579–584.
72. Gupta, A.; Chen, G.; Joshi, P.; Tadigadapa, S.; Eklund, P.C. Raman scattering from high-frequency phonons in supported n-graphene layer films. *Nano Lett.* 2006, *6*, 2667–2673.
73. Graf, D.; Molitor, F.; Ensslin, K.; Stampfer, C.; Jungen, A.; Hierold, C.; Wirtz, L. Spatially resolved Raman spectroscopy of single- and few-layer graphene. *Nano Lett.* 2007, *7*, 238–242.
74. Stankovich, S.; Dikin, D.A.; Piner, R.D.; Kohlhaas, K.A.; Kleinhammes, A.; Jia, Y.; Wu, Y.; Nguyen, S.T.; Ruoff, R.S. Synthesis of graphene-based nanosheets via chemical reduction of exfoliated graphite oxide. *Carbon* 2007, *45*, 1558–1565.
75. Muszynski, R.; Seger, B.; Kamat, P.V. Decorating graphene sheets with gold nanoparticles. *J. Phys. Chem. C* 2008, *112*, 5263–5266.
76. Shin, H.-J.; Kim, K.K.; Benayad, A.; Yoon, S.-M.; Park, H.K.; Jung, I.-S.; Jin, M.H.; Jeong, H.-K.; Kim, J.M.; Choi, J.-Y.; *et al.* Efficient reduction of graphite oxide by sodium borohydride and its effect on electrical conductance. *Adv. Funct. Mater.* 2009, *19*, 1987–1992.
77. Ambrosi, A.; Chua, C.K.; Bonanni, A.; Pumera, M. Lithium aluminium hydride as reducing agent for chemically reduced graphene oxides. *Chem. Mater.* 2012, *24*, 2292–2298.
78. Stankovich, S.; Piner, R.D.; Chen, X.; Wu, N.; Nguyen, S.T.; Ruoff, R.S. Stable aqueous dispersions of graphitic nanoplatelets via the reduction of exfoliated graphite oxide in the presence of poly(sodium 4-styrenesulfonate). *J. Mater. Chem.* 2006, *16*, 155–158.
79. Bai, H.; Xu, Y.; Zhao, L.; Li, C.; Shi, G. Non-covalent functionalization of graphene sheets by sulfonated polyaniline.*Chem. Commun.* 2009, 1667–1669.
80. Hao, R.; Qian, W.; Zhang, L.; Hou, Y. Aqueous dispersions of TCNQ-anion-stabilized graphene sheets. *Chem. Commun.* 2008, 6576–6578.
81. Xu, Y.; Bai, H.; Lu, G.; Li, C.; Shi, G. Flexible graphene films via the filtration of water-soluble noncovalent functionalized graphene sheets. *J. Am. Chem. Soc.* 2008, *130*, 5856–5857.
82. Ji, J.; Zhang, G.; Chen, H.; Wang, S.; Zhang, G.; Zhang, F.; Fan, X. Sulfonated graphene as water-tolerant solid acid catalyst. *Chem. Sci.* 2011, *2*, 484–487.
83. Liu, F.; Sun, J.; Zhu, L.; Meng, X.; Qi, C.; Xiao, F.-S. Sulfated graphene as an efficient solid catalyst for acid-catalyzed liquid reactions. *J. Mater. Chem.* 2012, *22*, 5495–5502.
84. Wang, L.; Wang, D.; Zhang, S.; Tian, H. Synthesis and characterization of sulfonated graphene as a highly active solid acid catalyst for the ester-exchange reaction. *Catal. Sci. Technol.* 2013, *3*, 1194–1197.
85. Shaabani, A.; Mahyari, M.; Hajishaabanha, F. The synthesis of xanthenes and benzoxanthenes on graphene oxide and sulfated graphene nanosheets in water. *Res. Chem. Intermed.* 2013.
86. Upare, P.P.; Toon, J.-W.; Kim, M.Y.; Kang, H.-Y.; Hwang, D.W.; Hwang, Y.K.; Kung, H.H.; Chang, J.-S. Chemical conversion of biomass-derived hexose sugars to levulinic acid over sulfonic acid-functionalized graphene oxide catalysts. *Green Chem.* 2013, *15*, 2935–2943.
87. Lam, E.; Chong, J.H.; Majid, E.; Liu, Y.; Hrapovic, S.; Leung,A.C.W.; Luong, J.H.T. Carbocatalytic dehydration of xylose to furfural in water. *Carbon* 2012, *50*, 1033–1043.
88. He, D.; Kou, Z.; Xiong, Y.; Cheng, K.; Chen, X.; Pan, M.; Mu, S. Simultaneous sulfonation and reduction of graphene oxide as highly efficient supports for metal nanocatalysts. *Carbon* 2014, *66*, 312–319.
89. Antunes, M.M.; Russo, P.A.; Wiper, P.V.; Veiga, J.M.; Pillinger, M.; Mafra, L.; Evtuguin, D.V.; Pinna, N.; Valente, A.A. Sulfonated graphene oxide as effective catalyst for conversion of 5-(hydroxymethyl)-2-furfural into biofuels.*ChemSusChem* 2014, *7*, 804–812.
90. Liu, J.; Xue, Y.; Dai, L. Sulfated graphene oxide as a hole-extraction layer in high performance polymer solar cells.*Phys. Chem. Lett.* 2012, *3*, 1928–1933.
91. Garg, B.; Bisht, T.; Chauhan, S.M.S. Synthesis and anion binding properties of novel 3,12- and 3,7-bis(4'-

nitrophenyl)-azo-calix[4]pyrrole receptors. *New J. Chem.* 2010, *34*, 1251–1254.

92. Chauhan, S.M.S.; Garg, B.; Bisht, T. Synthesis and anion binding of 2-arylazo-*meso*-octamethylcalix[4]pyrroles. *Supramol. Chem.* 2009, *21*, 394–400.

93. Chauhan, S.M.S.; Bisht, T.; Garg, B. 1-arylazo-5,5-dimethyl Dipyrromethanes: Versatile chromogenic probes for anions. *Sens. Actuators B: Chem.* 2009, *141*, 116–123.

94. Garg, B.; Bisht, T.; Chauhan, S.M.S. 2,2'-Diaminoazobenzene, a potential scaffold for the synthesis of bis-ureas and thioureas: Solution phase anion sensing and binding studies. *Sens. Actuators B: Chem.* 2012, *168*, 318–328.

95. Lomeda, J.R.; Doyle, C.D.; Kosynkin, D.V.; Hwang, W.-F.; Tour, J.M. Diazonium functionalization of surfactant-wrapped chemically converted graphene sheets. *J. Am. Chem. Soc.* 2008, *130*, 16201–16206.

96. Si, Y.; Samulski, E.T. Synthesis of water soluble graphene. *Nano Lett.* 2008, *8*, 1679–1682.

97. Zhao, G.; Jiang, L.; He, Y.; Li, J.; Dong, H.; Wang, X.; Hu, W. Sulfonated graphene for persistent aromatic pollutant management. *Adv. Mater.* 2011, *23*, 3959–3963.

98. Zhang, H.; Low, W.P.; Lee, H.K. Evaluation of sulfonated graphene sheets as sorbent for micro-solid-phase extraction combined with gas chromatography-mass spectrometry. *J. Chromatogr. A* 2012, *1233*, 16–21.

99. Coşkun, E.; Zaragoza-Contreras, E.A.; Salavagione, H.J. Synthesis of sulfonated graphene/polyaniline composites with improved electroactivity. *Carbon* 2012, *50*, 2235–2243.

100. Ankamwar, B.; Surti, F. Water soluble graphene synthesis. *Chem. Sci. Trans.* 2012, *1*, 500–507.

101. Geim, A.K. Graphene: Status and prospects. *Science* 2009, *324*, 1530–1534.

102. Garg, B.; Ling, Y.-C. Highly efficient synthesis of N-confused *meso*-tetraspirocyclohexyl calix[4] pyrrole using Brønsted acidic ionic liquids as catalysts. *Tetrahedron Lett.* 2012, *53*, 5674–5677.

103. Garg, B.; Lei, S.-L.; Liu, S.-C.; Bisht, T.; Liu, J.-Y.; Ling, Y.-C. Rapid identification of trimethyl and triethyl amines using sulfonic acidic ionic liquids: A TOF-SIMS study of fragmentation reactions. *Anal. Chim. Acta* 2012, *757*, 48–55.

104. Garg, B.; Ling, Y.-C. One-pot green synthesis of azides from alcohols using Brønsted acidic ionic liquid [HMIM][BF_4] as solvent and catalyst. *J. Chin. Chem. Soc.* 2014, *61*, 737–742.

105. Pei, S.; Zhao, J.; Du, J.; Ren, W.; Cheng, H. Direct reduction of graphene oxide films into highly conductive and flexible graphene films by hydrohalic acids. *Carbon* 2010, *48*, 4466–4474.

106. Behabtu, N.; Lomeda, J.R.; Green, M.J.; Higginbotham, A.L.; Sinitskii, A.; Kosynkin, D.V.; Tsentalovich, D.; Parra-Vasquez, A.N.G.; Schimidt, J.; Kesselman, E.; *et al.* Spontaneous high-concentration dispersions and liquid crystals of graphene. *Nat. Nanotechnol.* 2010, *5*, 406–411.

107. Lu, W.; Liu, S.; Qin, X.; Wang, L.; Tian, J.; Luo, Y.; Asiri, A.M.; Al-You, A.O.; Sun, X. High-yield, large-scale production of few-layer graphene flakes within seconds: Using chlorosulfonic acid and H_2O_2 as exfoliating agents. *J. Mater. Chem.*2012, *22*, 8775–8777.

108. Mutlay, I.; Tudoran, L.B. Chlorosulfonic acid-based room temperature chemical expansion route for the bulk production of graphite nanoplatelets. *Fuller. Nanotub. Carbon Nanostruct.* 2013, *21*, 149–157.

109. Dixon, P. Formation of sulphamic acid during the decomposition of ammonium sulphate. *Nature* 1944, *154*, 706–706.

110. Kiyoura, R.; Urano, K. Mechanism, kinetics, and equilibrium of thermal decomposition of ammonium sulfate. *Ind. Eng. Chem. Process Des. Dev.* 1970, *9*, 489–494.

111. Li, J.-H.; Zhang, G.-E. Investigation of the kinetics and mechanism of decomposition of ammonium hydrogen sulfate.*Acta Phys. Chim. Sin.* 1992, *8*, 123–127.

112. Li, J.; Zhang, G.; Wang, J. The identification of the mechanism function and the kinetic investigation of the energy-storing reaction of ammonium hydrogen sulfate. *Thermochem. Acta* 1992, *207*, 219–225.

113. Halstead, W.D. Thermal decomposition of ammonium sulphate. *J. Appl. Chem.* 1970, *20*, 129–132.

114. Xu, Z.; Qi, Z.; Kaufman, A. Superior catalysts for proton exchange membrane fuel cells sulfonation of carbon-supported catalysts using sulfate salts. Electrochem. *Solid-State Lett.* 2005, *8*, A313–A315.

115. Kumar, A.V.; Rao, K.R. Recyclable graphite oxide catalyzed Friedel-Crafts addition of indoles to α,β-unsaturated ketones. *Tetrahedron Lett.* 2011, *52*, 5188–5191.

116. Dreyer, D.R.; Jarvis, K.A.; Ferreirab, P.J.; Bielawski, C.W. Graphite oxide as a carbocatalyst for the preparation of fullerene-reinforced polyester and polyamide nanocomposites. *Polym. Chem.* 2012, *3*, 757–766.

117. Chauhan, S.M.S.; Mishra, S. Use of graphite oxide and graphene oxide as catalysts in the synthesis of dipyrromethane and calix[4]pyrrole. *Molecules* 2011, *16*, 7256–7266.

118. Verma, S.; Mungse, H.P.; Kumar, N.; Choudhary, S.; Jain, S.L.; Sain, B.; Khatri, O.P. Graphene oxide: An efficient and reusable carbocatalyst for aza-Michael addition of amines to activated alkenes. *Chem. Commun.* 2011, *47*, 12673–12675.

119. Dhakshinamoorthy, A.; Alvaro, M.; Puche, M.; Fornes, V.; Garcia, H. Graphene oxide as catalyst for the acetalization of aldehydes at room temperature. *ChemCatChem* 2012, *4*, 2026–2030.

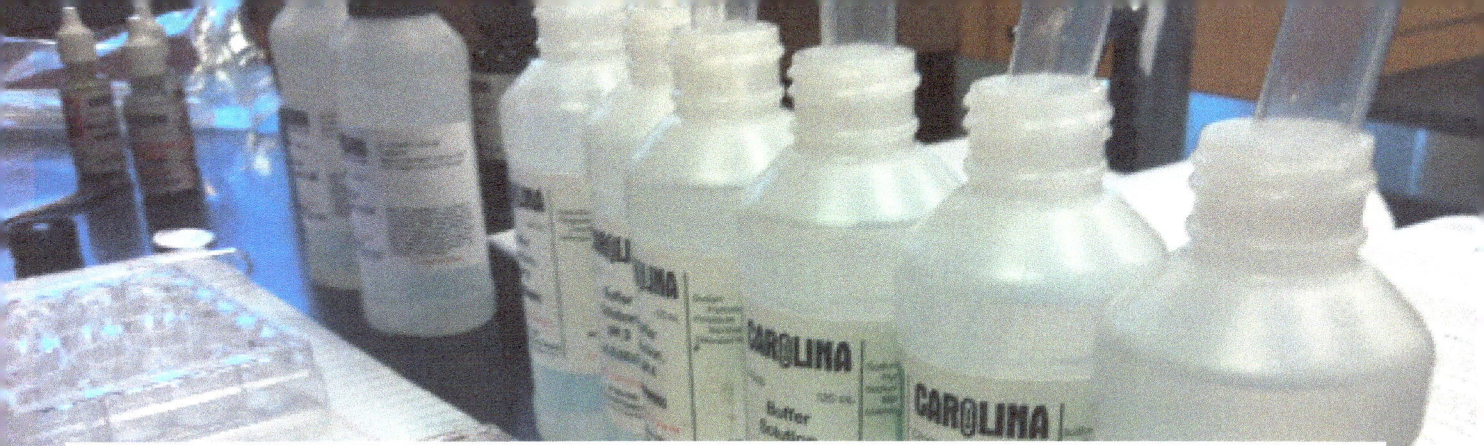

CHAPTER 9

PERIODIC MESOPOROUS ORGANOSILICA FUNCTIONALIZED WITH SULFONIC ACID GROUPS AS ACID CATALYST FOR GLYCEROL ACETYLATION

Els De Canck, Inmaculada Dosuna-Rodríguez, Eric M. Gaigneaux and Pascal Van Der Voort

Keywords: Periodic Mesoporous Organosilica, sulfonic acid groups, catalysis, glycerol acetylation

INTRODUCTION

The discovery of Periodic Mesoporous Organosilicas (PMOs) [1,2,3] with organic bridging groups incorporated in their silica framework has been the start of a fascinating research area which provides materials with huge potential [4,5,6]. Different organic bridges have been employed for very diverse applications, such as heterogeneous catalysts [7,8], bio-sensors [9,10], chromatographic packing materials [11,12], low-k materials [13,14], adsorbents of pollutants [15] and controlled drug delivery systems [16,17,18,19]. PMOs are highly porous materials with large specific surface areas, pore volumes and narrow pore size distributions. Furthermore, they exhibit a high thermal and mechanical stability [20,21,22], especially in comparison with other porous silica materials [23]. This type of material is synthesized with structure directing agents such as the non-ionic triblock copolymer P123. Around this template, a silica source is condensed in basic or acid aqueous environment. Usually, an organo bis-silane $(R'O)_3$–Si–R–Si–$(OR')_3$ is used where R represents the organic bridging group and R′ usually a methyl or ethyl group. Already many reports have appeared on different bridging groups (R) like phenylene, ethylene, ethenylene and ethylbenzene but also more complex and flexible organic functionalities have been described. Furthermore the bridging group can be modified to fine-tune the material for a specific application such as solid acid catalysis [4].

Concerning this topic, some very promising results have already been published regarding the incorporation of an acid functionality such as a sulfonic acid group and its catalytic activity. Several diverse methods have been applied to prepare sulfonic acid containing PMO materials. These strategies include the direct sulfonation of the phenylene bridge [24,25,26], as first attempted by Inagaki et al. [27], and the cocondensation of an organo bis-silane with (3-mercaptopropyl) trimethoxysilane (MPTMS) [28,29,30,31] followed by an oxidation of the thiol functionality. The latter can also be achieved by the in-situ oxidation of the thiol functionality by the addition of H_2O_2 during the cocondensation process of tetraethoxyorthosilicate (TEOS) and MPTMS [32,33]. Other silanes have been used in cocondensation processes with an organo bis-silane such as 2-(4-chlorosulfonylphenyl)-trimethoxysilane [34] and perfluorinated alkylsulfonic acid silanes [35,36,37].

Citation: Els De Canck, Inmaculada Dosuna-Rodríguez, Eric M. Gaigneaux and Pascal Van Der Voort, Periodic Mesoporous Organosilica Functionalized with Sulfonic Acid Groups as Acid Catalyst for Glycerol Acetylation, doi:10.3390/ma6083556

In the specific case of ethenylene bridged PMOs, $-SO_3H$ moieties can be acquired by the direct sulfonation of the C=C bond [38]. However, the sulfonic acid group can detach from the material, depending on the environment used during catalysis. Another explored route is the use of a Diels Alder reaction where the ethene bond acts as dienophile. Kondo et al.[39,40] described the cycloaddition of the ethene bond with benzocyclobutene and subsequently the resulting phenylene moiety is sulfonated to obtain a heterogeneous catalyst. These authors tested this material for several catalytic reactions (esterification of acetic acid with ethanol, the Beckmann and pinacole-pinacolone rearrangement) and the catalyst showed excellent conversion results. This Diels Alder process has been further used to expand the functionalization possibilities [41].

Another example of modifying the surface of an ethenylene bridged PMO material has been reported by the research group of Kaliaguine [42]. First, the surface silanols were end-capped with hexamethyldisilazane after which a Friedel-Crafts alkylation with benzene was made, and subsequently the benzene moiety was sulfonated with concentrated sulfuric acid. The $-SO_3H$ containing material exhibited a high catalytic activity in the self-condensation of heptanal.

In this study, the pure trans-ethenylene bridged PMO material [43] is chosen as support material. Thiol functionalities were incorporated according to a procedure previously described by our research group [44] and subsequently oxidized in order to obtain $-SO_3H$. This material was thoroughly characterized and the solid acid was tested in the acetylation of glycerol. Furthermore, the reusability of this catalyst was investigated.

RESULTS AND DISCUSSION

Starting from the trans-ethenylene bridged PMO (EP), a sulfonic acid modified PMO material is prepared, characterized and its catalytic activity is studied in the acetylation of glycerol. A general overview of the preparation method, including the starting material, is presented in Figure 1. Firstly, an ethenylene bridged Periodic Mesoporous Organosilica is synthesized starting from E-1,2-bis(triethoxysilyl)ethene (Figure 1, pathway A) [43,45].

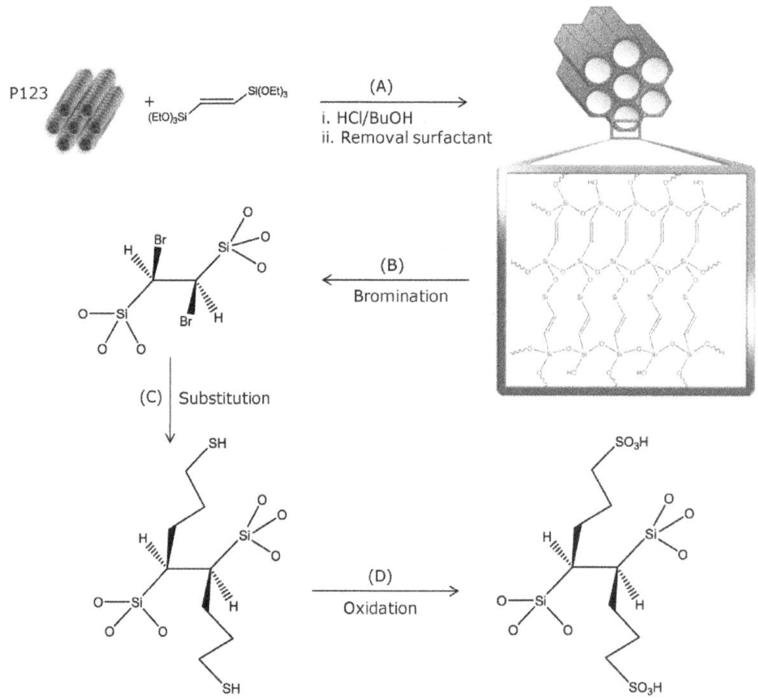

Figure 1: Summary of the synthetic pathways followed in this study. (A) Preparation of trans-ethenylene bridged Periodic Mesoporous Organosilica [EP]; (B) Bromination of EP [BEP]; (C) Substitution of the bromine with Grignard reagent of 3-chloro-1-propanethiol [EP–$(CH_2)_3$–SH] and (D) Oxidation with sulfuric acid [EP–$(CH_2)_3$–SO_3H].

Afterwards, the ethene bridge can be further functionalized to incorporate sulfonic acid groups in the material. When pathway B is followed, the material EP is firstly brominated and subsequently the bromine can be substituted via an SN2 reaction by the Grignard reagent of 3-chloro-1-propanethiol (Figure 1, pathway C). A material with a propylthiol group is obtained (EP–$(CH_2)_3$–SH) [44,46]. Oxidizing the thiol moiety is performed with H_2SO_4 or other oxidizing agents such as HNO_3 [28,29,30,31] or H_2O_2 [25,47,48]. In this study, a treatment of EP–$(CH_2)_3$–SH with sulfuric acid and a thorough washing step was selected by which it converted the –SH into a $-SO_3H$ group (Figure 1, pathway D). This route resulted in the material EP–$(CH_2)_3$–SO_3H.

Characterization of the Solids

Nitrogen sorption measurements were performed to examine the porosity of the different materials obtained by the reaction pathway shown in Figure 1 (EP, BEP, EP–$(CH_2)_3$–SH and EP–$(CH_2)_3$–SO_3H). The nitrogen adsorption and desorption isotherms are shown in Figure 2. The type IV isotherms with the condensation step at relative pressures between 0.55 and 0.75 and the H1 hysteresis of the solids clearly indicate that the materials are mesoporous and possess cylindrical pores with a narrow pore size distribution.

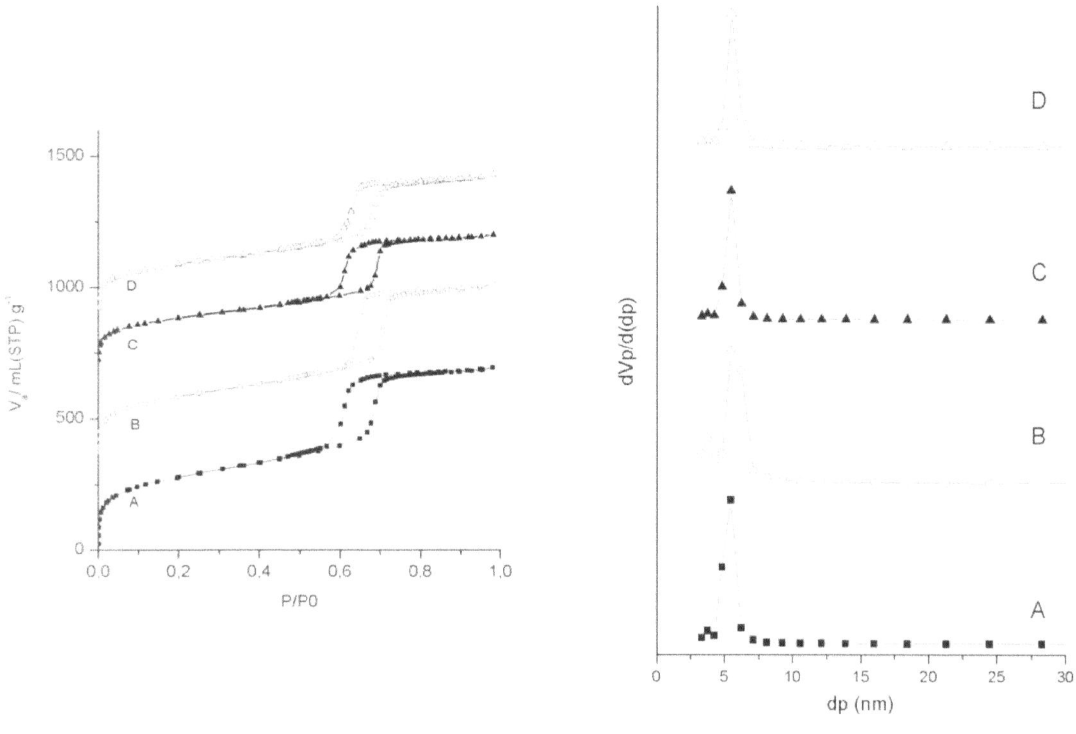

(a) (b)

Figure 2: The nitrogen adsorption and desorption isotherms (**a**) and pore size distributions (**b**) of (A) EP; (B) BEP; (C) EP–$(CH_2)_3$–SH and (D) EP–$(CH_2)_3$–SO_3H. The isotherms of BEP, EP–$(CH_2)_3$–SH and EP–$(CH_2)_3$–SO_3H are vertically offset for clarity by 350, 700 and 900 mL (STP) g^{-1} (Standard Temperature and Pressure), respectively.

A summary of the properties of these materials is shown in Table 1. The materials exhibit high specific surface areas (S_{BET}) ranging from 850 to 523 m^2 g^{-1} and large total pore volumes around 0.84 mL g^{-1}. The S_{BET} decreases when the material is functionalized due to the decoration of the pore walls with the bromine and later on with the propylthiol functionality but also due to the overall weight gain of the functionalized materials. The pore diameter of all the materials lies in the range of 6 to 5 nm. Only a minor shift to smaller pore diameters and a slight broadening of the pore size distribution is observed (Figure 2 and Table 1). The structural characteristics of the commercially available resin Amberlyst-15 are also presented inTable 1 for comparison. This ethenylbenzenesulfonic acid polymer is a strong acid ion exchange resin with unordered macropores. The material is also prone to swelling.

Table 1: Overview of the structural characteristics of the materials compared in this study

Sample	Path	S_{BET} [a] (m^2 g^{-1})	V_p [b] (mL g^{-1})	d_p [c] (nm)
EP	A	850	1.03	5.8
BEP	B	663	0.84	5.6
EP–$(CH_2)_3$–SH	C	523	0.59	5.3
EP–$(CH_2)_3$–SO_3H	D	688	0.72	5.4
Amberlyst-15	–	50	–	300

The XRD patterns of the materials in Figure 3 reveal three well-resolved signals originating from the low angle (100) and second-order (110) and (200) reflections. This evidently indicates that the materials possess a 2D-hexagonal ordered structure and thus retain their P6mm space group ordering throughout the syntheses. Only a slight broadening can be observed at the patterns of sample BEP, EP–$(CH_2)_3$–SH and EP–$(CH_2)_3$–SO_3H.

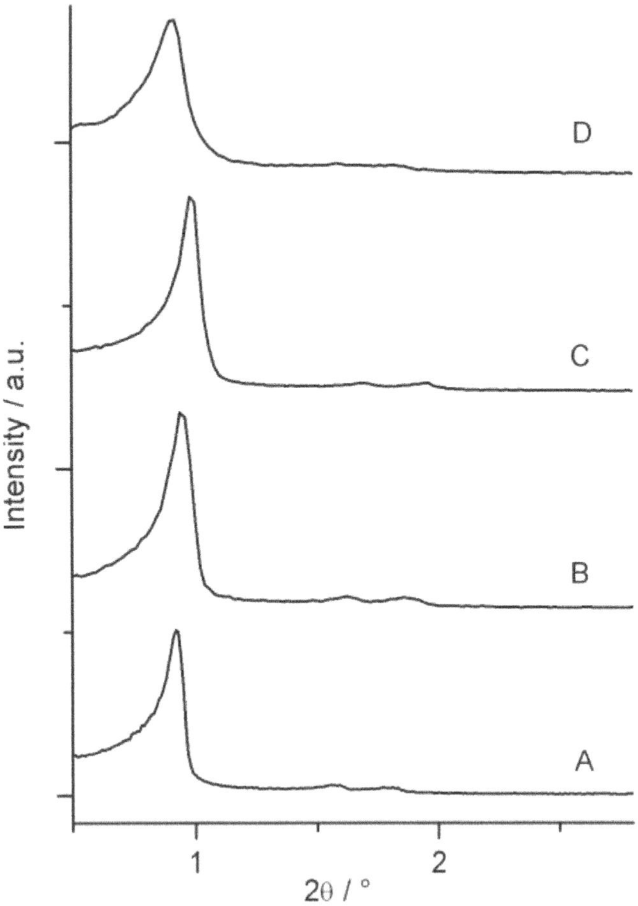

Figure 3: The powder X-ray diffraction patterns of (A) EP; (B) BEP; (C) EP–$(CH_2)_3$–SH; and (D) EP–$(CH_2)_3$–SO_3H.

It is quite remarkable that all the materials discussed in this study, show outstanding structural stability. The materials retain porosity and ordering after three consecutive reactions as can be seen from the nitrogen sorption and XRD data. These results also confirm the reported stability of Periodic Mesoporous Organosilicas [20,23].

Table 2 presents an overview of the chemical characterization of the solids after the different synthetic procedures. The bromination of the ethene bridge is a very straightforward reaction and approximately 25%–30% of the double bonds are brominated as the remaining fraction of double bonds are buried inside the walls and are unavailable for further reaction [45]. The subsequent substitution with the Grignard reagent results in thiol functionalities. The amount of thiol functionalities is determined using a silver titration [44,46]. After oxidation of the thiol groups using sulfuric acid and a thorough washing step, a total amount of 0.60 mmol H^+ per gram of material has been observed. This also includes the intrinsic acidity of the PMO material originating from the surface silanols (~0.15 mmol g^{-1}), as we described earlier [49]. It is clear that the conversion of the thiol containing PMO into the sulfonic acid containing-material has occurred via the oxidation process. This is also confirmed by Raman spectroscopy by the appearance of two signals in the region between 1160 and 1190 cm^{-1} (See figure S1 in supplementary information). Also, the thiol titration after oxidation showed a zero concentration of remaining thiol groups. Amberlyst-15 exhibits a high acidity of 4.7 mmol H^+ g^{-1}.

Table 2: Overview of the chemical characteristics of the materials compared in this study

Sample	Functionality	mmol g^{-1}
BEP	–Br [a]	2.39
EP–$(CH_2)_3$–SH	–SH [b]	0.40
EP–$(CH_2)_3$–SO_3H	–SO_3H [c]	0.60 [d]

Catalytic Experiments and Recyclability

The catalytic ability of the sulfonic acid functionalized PMO material has been explored for an esterification reaction, i.e., the glycerol acetylation reaction (Figure 4). The activity of EP–$(CH_2)_3$–SO_3H is compared with a commercially available catalyst Amberlyst-15 and moreover the catalysts' reusability is explored.

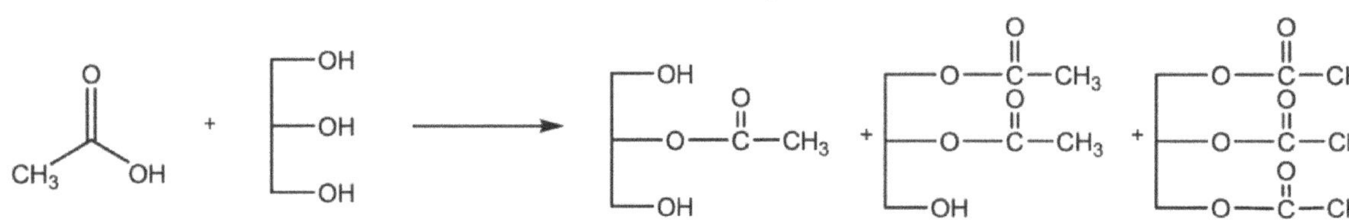

Figure 4: The esterification reaction: the acetylation of glycerol with the formation of glycerol monoacetate, glycerol diacetate and glycerol triacetate.

In this study the esterification of glycerol is probed due to its economic importance. Glycerol is an important by-product of first generation biodiesel and is produced in a relative large quantity [50]. This overproduction of glycerol can be used in order to develop second generation biodiesel which uses glycerol as a raw product. As carboxylic acid, acetic acid is probed as shown in the general reaction (Figure 4). Three products may in principle be obtained from this reaction: glycerol monoacetate (MAG), glycerol diacetate (DAG) and glycerol triacetate (TAG). However, experimentally, only MAG (~94%) and DAG (~6%) are formed using the specific catalytic conditions described in the experimental part [51].

The catalytic activity of EP–$(CH_2)_3$–SO_3H for the esterification of acetic acid with glycerol is presented in Figure 5 where the total acetylation yield is shown as a function of time. The total acetylation yield is defined according to the equation below:

$$Yield_t(\%) = \frac{[P]_t}{[HAC]_0}\frac{\upsilon_{HAC}}{\upsilon_p}\times 100 \tag{1}$$

where $[P]_t$ and $[HAc]_0$ represent the product and acetic acid concentration at a certain reaction time and at t = 0, respectively. Furthermore, υ_{HAc} and υ_P represent the stoichiometric coefficients of the acetic acid and the ester formed, i.e., 1 for mono-substituted, 2 for di-substituted and 3 for fully substituted products, respectively. Also, as acetic acid contains acid protons which can induce a self-catalyzed process, the reaction in absence of any solid catalyst was monitored. Corresponding data are shown in Figure 5. It is clear that the sulfonated PMO possesses a significant catalytic activity with a yield of almost 80% for this reaction after ~300 min; whereas the blank test (without any solid catalyst involved) yielded only ~50% of esters after ~300 min. The conversion of Amberlyst-15 is shown in the same figure. Comparing the two materials clearly shows that EP–$(CH_2)_3$–SO_3H exhibits a similar catalytic activity as Amberlyst-15, which is a well-performing catalyst in this type of reaction.

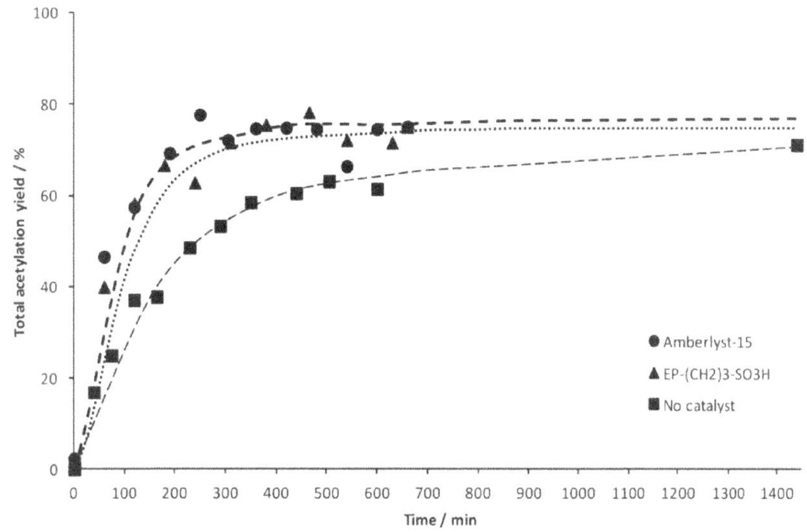

Figure 5: The total acetylation yield for the catalytic reaction with EP–$(CH_2)_3$–SO_3H and Amberlyst-15. Also the blank reaction is represented for clarity. A catalyst loading of 0.25 g per 40 mL of glycerol was used. The lines are intended as visual aids only.

Furthermore, the recyclability of the sulfonic acid containing PMO material is studied for three consecutive runs. First, the initial rate of the catalytic reaction is studied for each run by focusing on the first hour of the acetylation (Figure 6). These experiments are all performed in the same catalytic set-up as the standard catalytic experiment. The solid is filtered after 1 hour and re-used without any further treatment in the subsequent run with a fresh reaction medium (run 2); this being performed again for two additional consecutive runs (runs 3 and 4). As one can see from the figure, the material still possesses catalytic activity for the acetylation after four runs. However, recycling of the material in the consecutive runs results in a slight decrease of the initial reaction rate.

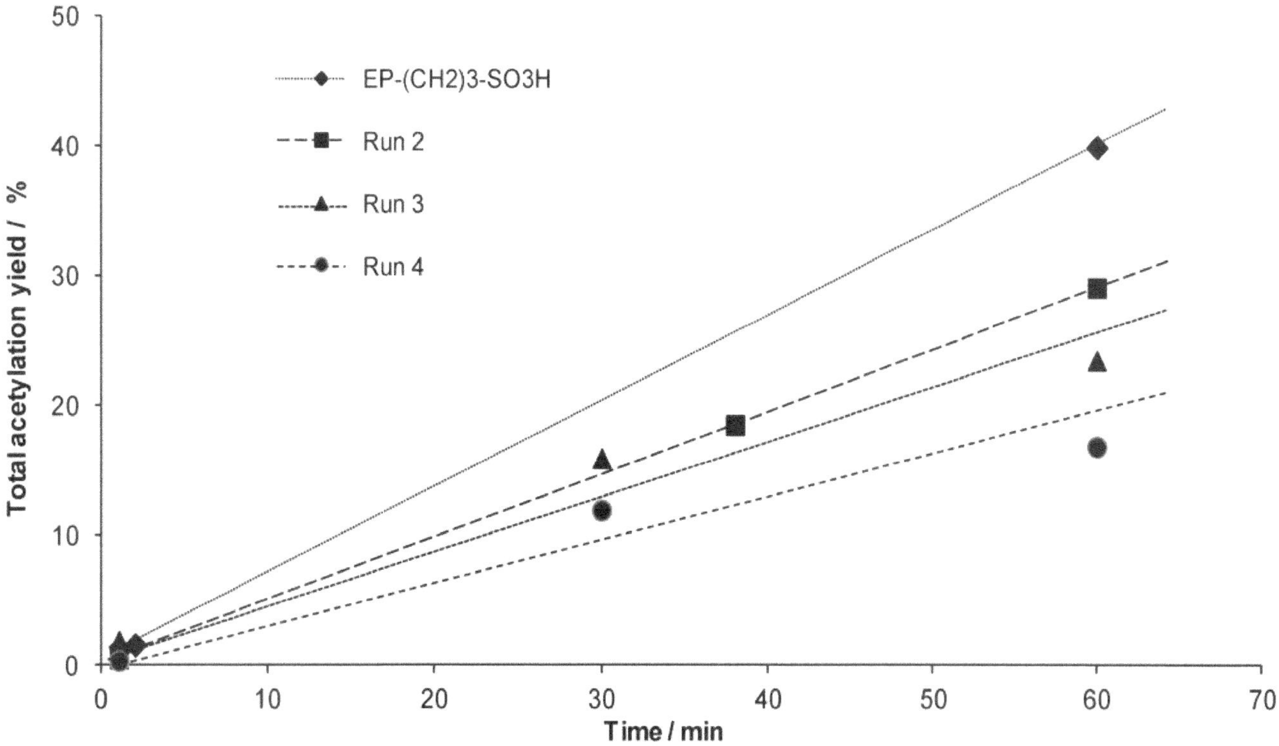

Figure 6: Recyclability experiments for EP–$(CH_2)_3$–SO_3H with several runs during the first hour of the reaction. The lines are intended as visual aids only.

After the last catalytic run, i.e., run 4, the total acetylation yield is monitored for approximately 10 h in order to compare it with the acetylation yield of the pristine sulfonated PMO material, EP–$(CH_2)_3$–SO_3H (Figure 7). Although a decrease in the initial reaction rate was observed as already mentioned, at the third consecutive run, the material still reaches the equilibrium after approximately 5 h and finally results in the same acetylation level as the fresh pristine material.

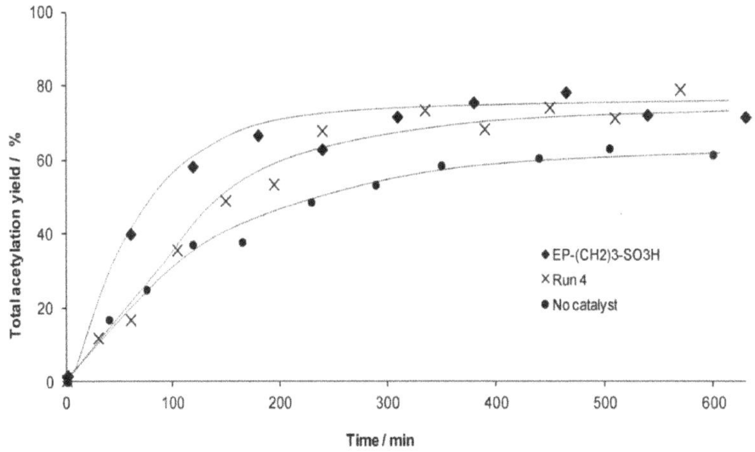

Figure 7: Recyclability experiment for EP–$(CH_2)_3$–SO_3H: a comparison between the catalytic activity of the pristine material and the third catalytic run. The blank reaction is presented for clarity. The lines are intended as visual aids only.

Moreover, additional tests are performed to evaluate the actual heterogeneous character of the observed catalytic activity. Therefore, the solids are removed from the liquid after 1 h along the first and second runs, and the corresponding recovered solutions are kept under the catalytic conditions to follow the occurrence of a further evolution of the total acetylation yield without solid catalyst in the system anymore. As can be seen from Figure 8, it is clear that the acetylation occurs much slower, i.e., in the range of the blank, when the catalysts are removed of the reaction media, than when the catalyst is maintained in the reactor for the whole test duration.

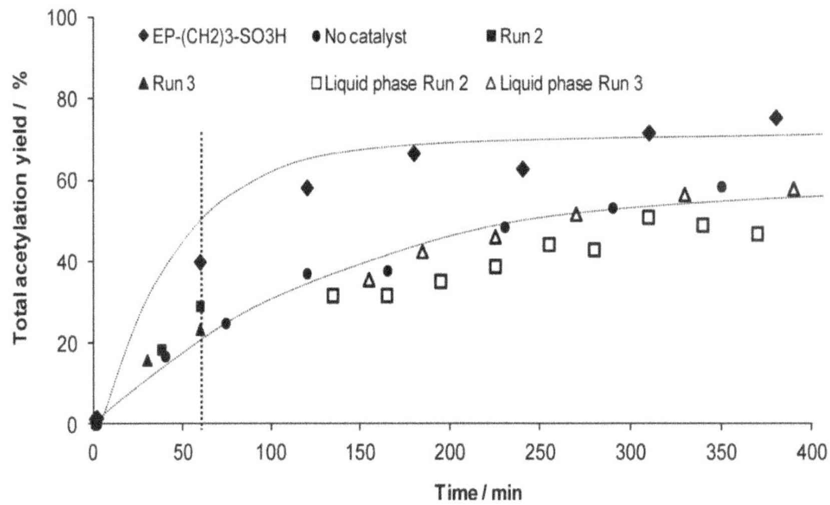

Figure 8: The total acetylation yield for the EP–$(CH_2)_3$–SO_3H, the second and third run. After 60 min (represented by the vertical line) the liquid is separated from the catalyst and the catalytic activity of the liquid phase of run 2 and 3 is further followed in function of time (open square and triangle). The blank reaction is presented for clarity. The dotted lines are intended as visual aids only.

EXPERIMENTAL SECTION

Materials

The following reagents were used for the synthesis and characterization: Grubbs' first generation catalyst $(PCy_3)Cl_2Ru{=}CH{-}Ph$, pluronic P123 $(PEO_{20}PPO_{70}PEO_{20})$, 1-butanol, bromine, 3-chloro-1-propanethiol (98%), magnesium turnings (98%), sodium chloride, potassium thiocyanate, ammonium iron sulfate dodecahydrate, nitric acid (65%), iron (III) chloride (98%; anhydrous), toluene (anhydrous), acetonitrile, sulfuric acid (H_2SO_4), silver nitrate were obtained from Sigma-Aldrich (Bornem, Belgium). Acetone (>99.5%) was acquired from VWR (Belgium). Tetrahydrofurane (rotidry) and hydrochloric acid (37%; p.a.) were purchased at Carl Roth (Karlsruhe, Germany). Vinyltriethoxysilane (97%) was acquired from ABCR (Karlsruhe, Germany). The following reagents were used for the catalytic reaction: acetic acid (puriss. p.a., ACS reagens, ≥99.8%, GC/T), glycerol (puriss. p.a., ACS reagens, anhydrous, dist., ≥99.5%) were purchased from Fluka (Bornem, Belgium). 1-Propanol, 1-heptanol (98%) and o-xylene (98%) were obtained from Sigma Aldrich. All chemicals were used as received.

Synthesis of Pure E-1,2-Bis(Triethoxysilyl) Ethene (E-BTSE)

The diastereoisomerically pure E-isomer of the ethenylene precursor was synthesized according to a procedure described by our research group [43,45]. An amount of 0.0535 g of the Grubbs' first generation catalyst and 42.95 mL of vinyltriethoxysilane (VTES) were mixed together under inert atmosphere. The mixture was stirred for 1 h at room temperature and subsequently refluxed. After 3 h, a distillation was performed to remove remaining VTES. Afterwards, the colorless E-1,2-bis(triethoxysilyl)ethene (E-BTSE) was distilled off under vacuum (~0.1 Pa).

Synthesis of Ethenylene Bridged Periodic Mesoporous Organosilica (Figure 1; Pathway A)

In a typical synthesis, reported by our group [45], a total amount of 1.00 g of P123 was placed in a flask and 47.80 mL of distilled H_2O, 3.42 mL concentrated HCl and 2.45 mL 1-butanol were added. The solution was stirred for 1.5 h at room temperature and subsequently 1.86 mL of E-BTSE was added to the mixture. Next, the mixture was vigorously stirred at 35 °C. After 4 h, the temperature was raised to 90 °C and aged for 16 h. Afterwards, the mixture was left to cool down, and the white solid was filtered and washed with acetone and distilled water. Finally, P123 was removed via Soxhlet extraction with acetone for 5 h. The white solid, denoted as EP, was dried at 120 °C under vacuum (~0.1 Pa).

Synthesis of BEP and EP-Propylthiol (Figure 1; Pathways B and C)

The synthesis of this material was extensively described by our research group [38,46]. A certain amount of EP was first brominated with bromine gas under vacuum for 3 h. The resulting material was then dried at 90 °C for 16 h. Magnesium (0.74 g), iron(III) chloride (0.54 g) and dry tetrahydrofuran (THF) (30 mL) were mixed under inert atmosphere and stirred for 30 min at 50 °C. Next, 3-chloro-1-propanethiol (0.22 mL) was added and left to stir for 4 h at room temperature. Subsequently, the solution was added to 0.70 g of dry brominated EP. The resulting mixture was stirred for an additional 24 h at 40 °C. Then, the solid was filtered and washed several times with THF, 2 mol L^{-1} HCl, H_2O and acetone. Finally, the material was dried at 90 °C for 16 h under vacuum (~0.1 Pa) and denoted as EP–$(CH_2)_3$–SH.

Synthesis of EP-Propylsulfonic Acid (Figure 1; Pathway D)

A volume of 25 mL H_2SO_4 (2.5 mol L^{-1}) was added to 0.50 g of EP–$(CH_2)_3$–SH and stirred at room temperature. After 1 h, the solids were filtered and washed thoroughly with water and acetone. Finally, the material was dried at 90 °C for 16 h under vacuum (~0.1 Pa) and referred to as EP–$(CH_2)_3$–SO_3H.

Determination of the Amount of Reachable Thiols

A titration is performed as described in the literature [44]. A solution of silver nitrate with a known concentration is added to 0.050 g of material and left to stir until the equilibrium is reached. The excess of silver is titrated with potassium thiocyanate and $FeNH_4(SO_4)_2 \cdot 12H_2O$ in 0.3 mol L^{-1} HNO_3 as indicator.

Determination of the Total Acidity

The material containing sulfonic acid groups was stirred for 24 h in 20 g of 2 mol L^{-1} NaCl. Next, an acid-base titration was achieved with NaOH and phenolphtaleine as indicator.

Esterification of Acetic Acid with Glycerol

The catalytic reaction was carried out at 105 °C and at atmospheric pressure in a round bottom flask reactor equipped with a magnetic stirrer. The sulfonic acid containing material was dried prior to use at 105 °C under vacuum (~1.33 Pa) for 18 h. The following concentrations of the reagents were used: 100 g acetic acid/L of glycerol and the catalyst concentration was 6.25 g/L of glycerol. First, the catalyst and glycerol are added together to the flask and are brought to reaction temperature. When the temperature is reached, the acetic acid is injected. Before analysis the samples are extracted with 1-heptanol and using o-xylene as an internal standard. The total yield of the esters was determined with gas chromatography. After one minute a sample of the reaction mixture is taken and analyzed with gas chromatography. This sample represents the starting point of the catalytic experiment. The recyclability experiments were carried out under identical conditions as the catalytic tests. Therefore, the catalyst was separated from the liquid phase after 1 h and reused in a consecutive run without further treatment. The liquid phase was kept under reaction conditions and further followed as a function of time. Gas chromatography was performed with a Varian CP-3800 using a Cp-Sil 8CD column (Varian, Palo Alto, CA, USA) with a FID detector.

Characterization

Nitrogen sorption measurements were carried out by a Belsorp-mini II gas analyzer at 77 K. The specific surface area (S_{BET}) was determined by the BET equation (p/p_0 = 0.05–0.15). The pore size distribution was determined from the desorption branch of the isotherm using the Barrett-Joyner-Halenda (BJH) theory. The samples were pretreated at 90 °C while degassing (~0.1 Pa). X-ray diffraction (XRD) measurements were conducted with an ARL X'tra X-ray diffractometer of Thermo Scientific (Waltham, MA, USA) equipped with a Cu $K_{\alpha 1}$ tube and a Peltier cooled lithium drifted silicon solid stage detector. Raman spectra are recorded using a Raman type FRA106/S spectrometer of Bruker (Karlsruhe, Germany), equipped with a Nd-YAG laser (λ = 1064 nm).

CONCLUSIONS

Periodic Mesoporous Organosilica functionalized with sulfonic acid groups has been successfully synthesized and characterized. An ethenylene bridged PMO material was chosen as starting material and this was further post-modified in several steps. A bromination and subsequent substitution reaction was used followed by an oxidation turning the material into a solid acid catalyst. This material has been investigated for the acetylation of glycerol. The material showed an equal activity as Amberlyst-15; the latter being usually considered as the reference most-efficient catalyst material for such kind of reactions. Recyclability experiments showed that the sulfonated Periodic Mesoporous Organosilica is reusable and showed the same total acetylation yield after three runs. Homogeneous tests showed that the acetylation of glycerol indeed occurs heterogeneously, suggesting the absence of leaching, e.g., of sulfonate species, in the medium, and thus the high stability of our material under working conditions.

ACKNOWLEDGMENTS

EDC and PVDV thank Ghent University for financial support and Tom Planckaert from the department of Inorganic and Physical Chemistry (Ghent University) for nitrogen sorption and X-ray diffraction measurements. IDR and EMG acknowledge the support of the Région Wallonne (DG06, convention LIGNOFUEL No. 716721).

REFERENCES

1. Inagaki, S.; Guan, S.; Fukushima, Y.; Ohsuna, T.; Terasaki, O. Novel mesoporous materials with a uniform distribution of organic groups and inorganic oxide in their frameworks. J. Am. Chem. Soc. 1999, 121, 9611–9614.
2. Asefa, T.; MacLachan, M.J.; Coombs, N.; Ozin, G.A. Periodic mesoporous organosilicas with organic groups inside the channel walls. Nature 1999, 402, 867–871.
3. Melde, B.J.; Holland, B.T.; Blanford, C.F.; Stein, A. Mesoporous sieves with unified hybrid inorganic/organic frameworks. Chem. Mater. 1999, 11, 3302–3308.
4. Van der Voort, P.; Esquivel, D.; de Canck, E.; Goethals, F.; van Driessche, I.; Romero-Salguero, F.J. Periodic Mesoporous Organosilicas: from simple to complex bridges; A comprehensive overview of functions, morphologies and applications. Chem. Soc. Rev. 2013, 42, 3913–3955.
5. Mizoshita, N.; Tani, T.; Inagaki, S. Syntheses, properties and applications of periodic mesoporous organosilicas prepared from bridged organosilane precursors. Chem. Soc. Rev. 2011, 40, 789–800.
6. Hoffmann, F.; Froba, M. Vitalising porous inorganic silica networks with organic functions-PMOs and related hybrid materials. Chem. Soc. Rev. 2011, 40, 608–620.
7. Sasidharan, M.; Fujita, S.; Ohashi, M.; Goto, Y.; Nakashima, K.; Inagaki, S. Novel synthesis of bifunctional catalysts with different microenvironments. Chem. Commun. 2011, 47, 10422–10424.
8. Ohashi, M.; Kapoor, M.P.; Inagaki, S. Chemical modification of crystal-like mesoporous phenylene-silica with amino group. Chem. Commun. 2008, 7, 841–843.
9. Johnson-White, B.; Zeinali, M.; Shaffer, K.M.; Patterson, C.H.; Charles, P.T.; Markowitz, M.A. Detection of organics using porphyrin embedded nanoporous organosilicas. Biosens. Bioelectron. 2007, 22, 1154–1162.
10. Johnson-White, B.; Zeinali, M.; Malanoski, A.P.; Dinderman, M.A. Sunlight-catalyzed conversion of cyclic organics with novel mesoporous organosilicas. Catal. Commun. 2007, 8, 1052–1056.
11. Huang, L.L.; Lu, J.; Di, B.; Feng, F.; Su, M.X.; Yan, F. Self-assembled highly ordered ethane-bridged periodic mesoporous organosilica and its application in HPLC. J. Sep. Sci. 2011, 34, 2523–2527.
12. Zhang, Y.P.; Jin, Y.; Yu, H.; Dai, P.C.; Ke, Y.X.; Liang, X.M. Pore expansion of highly monodisperse phenylene-bridged organosilica spheres for chromatographic application. Talanta 2010, 81, 824–830.
13. Goethals, F.; Baklanov, M.R.; Ciofi, I.; Detavernier, C.; van der Voort, P.; van Driessche, I. A new procedure to seal the pores of mesoporous low-k films with precondensed organosilica oligomers. Chem. Commun. 2012, 48, 2797–2799.
14. Goethals, F.; Ciofi, I.; Madia, O.; Vanstreels, K.; Baklanov, M.R.; Detavernier, C.; van der Voort, P.; van Driessche, I. Ultra-low-k cyclic carbon-bridged PMO films with a high chemical resistance. J. Mater. Chem. 2012, 22, 8281–8286.
15. Walcarius, A.; Mercier, L. Mesoporous organosilica adsorbents: Nanoengineered materials for removal of organic and inorganic pollutants. J. Mater. Chem. 2010, 20, 4478–4511.
16. Parambadath, S.; Rana, V.K.; Moorthy, S.; Chu, S.W.; Park, S.K.; Lee, D.; Sung, G.; Ha, C.S. Periodic mesoporous organosilicas with co-existence of diurea and sulfanilamide as an effective drug delivery carrier. J. Solid State Chem.2011, 184, 1208–1215.
17. Parambadath, S.; Rana, V.K.; Zhao, D.Y.; Ha, C.S. N,N'-diureylenepiperazine-bridged periodic mesoporous organosilica for controlled drug delivery. Microporous Mesoporous Mater. 2011, 141, 94–101.
18. Vathyam, R.; Wondimu, E.; Das, S.; Zhang, C.; Hayes, S.; Tao, Z.M.; Asefa, T. Improving the adsorption and release capacity of organic-functionalized mesoporous materials to drug molecules with temperature and synthetic methods.J. Phys. Chem. C 2011, 115, 13135–13150.
19. Moorthy, M.S.; Park, S.S.; Fuping, D.; Hong, S.H.; Selvaraj, M.; Ha, C.S. Step-up synthesis of amidoxime-functionalised periodic mesoporous organosilicas with an amphoteric ligand in the framework for drug delivery. J. Mater. Chem. 2012, 22, 9100–9108.
20. Goethals, F.; Vercaemst, C.; Cloet, V.; Hoste, S.; van der Voort, P.; van Driessche, I. Comparative study of ethylene- and ethenylene-bridged periodic mesoporous organosilicas. Microporous Mesoporous Mater. 2010, 131, 68–74.
21. Kruk, M.; Jaroniec, M.; Guan, S.Y.; Inagaki, S. Adsorption and thermogravimetric characterization of mesoporous materials with uniform organic-inorganic frameworks. J. Phys. Chem. B 2001, 105, 681–689.
22. Esquivel, D.; Jimenez-Sanchidrian, C.; Romero-Salguero, F.J. Comparison of the thermal and hydrothermal stabilities of ethylene, ethylidene, phenylene and biphenylene bridged periodic mesoporous organosilicas. Mater. Lett. 2011, 65, 1460–1462.
23. Burleigh, M.C.; Markowitz, M.A.; Jayasundera, S.; Spector, M.S.; Thomas, C.W.; Gaber, B.P. Mechanical

and hydrothermal stabilities of aged periodic mesoporous organosilicas. J. Phys. Chem. B 2003, 107, 12628–12634.

24. Esquivel, D.; Jimenez-Sanchidrian, C.; Romero-Salguero, F.J. Thermal behaviour, sulfonation and catalytic activity of phenylene-bridged periodic mesoporous organosilicas. J. Mater. Chem. 2011, 21, 724–733.
25. Rac, B.; Hegyes, P.; Forgo, P.; Molnar, A. Sulfonic acid-functionalized phenylene-bridged periodic mesoporous organosilicas as catalyst materials. Appl. Catal. A Gen. 2006, 299, 193–201.
26. Kapoor, M.P.; Kasama, Y.; Yanagi, M.; Yokoyama, T.; Inagaki, S.; Shimada, T.; Nanbu, H.; Juneja, L.R. Cubic phenylene bridged mesoporous hybrids from allylorganosilane precursors and their applications in Friedel-Crafts acylation reaction. Microporous Mesoporous Mater. 2007, 101, 231–239.
27. Inagaki, S.; Guan, S.; Ohsuna, T.; Terasaki, O. An ordered mesoporous organosilica hybrid material with a crystal-like wall structure. Nature 2002, 416, 304–307.
28. Yang, Q.H.; Kapoor, M.P.; Shirokura, N.; Ohashi, M.; Inagaki, S.; Kondo, J.N.; Domen, K. Ethane-bridged hybrid mesoporous functionalized organosilicas with terminal sulfonic groups and their catalytic applications. J. Mater. Chem. 2005, 15, 666–673.
29. Yang, Q.H.; Kapoor, M.P.; Inagaki, S.; Shirokura, N.; Kondo, J.N.; Domen, K. Catalytic application of sulfonic acid functionalized mesoporous benzene-silica with crystal-like pore wall structure in esterification. J. Mol. Catal. A Chem.2005, 230, 85–89.
30. Yang, Q.H.; Liu, J.; Yang, J.; Kapoor, M.P.; Inagaki, S.; Li, C. Synthesis, characterization, and catalytic activity of sulfonic acid-functionalized periodic mesoporous organosilicas. J. Catal. 2004, 228, 265–272.
31. Yang, Q.H.; Kapoor, M.P.; Inagaki, S. Sulfuric acid-functionalized mesoporous benzene-silica with a molecular-scale periodicity in the walls. J. Am. Chem. Soc. 2002, 124, 9694–9695.
32. Melero, J.A.; Vicente, G.; Morales, G.; Paniagua, M.; Moreno, J.M.; Roldan, R.; Ezquerro, A.; Perez, C. Acid-catalyzed etherification of bio-glycerol and isobutylene over sulfonic mesostructured silicas. Appl. Catal. A Gen. 2008, 346, 44–51.
33. Melero, J.A.; Bautista, L.F.; Morales, G.; Iglesias, J.; Sanchez-Vazquez, R. Biodiesel production from crude palm oil using sulfonic acid-modified mesostructured catalysts. Chem. Eng. J. 2010, 161, 323–331.
34. Rat, M.; Zahedi-Niaki, M.H.; Kaliaguine, S.; Do, T.O. Sulfonic acid functionalized periodic mesoporous organosilicas as acetalization catalysts. Microporous Mesoporous Mater. 2008, 112, 26–31.
35. Dube, D.; Rat, M.; Shen, W.; Nohair, B.; Beland, F.; Kaliaguine, S. Perfluorinated alkylsulfonic acid functionalized periodic mesostructured organosilica: A new acidic catalyst. Appl. Catal. A Gen. 2009, 358, 232–239.
36. Dube, D.; Rat, M.; Shen, W.; Beland, F.; Kaliaguine, S. Perfluoroalkylsulfonic acid-functionalized periodic mesostructured organosilica: a strongly acidic heterogeneous catalyst. J. Mater. Sci. 2009, 44, 6683–6692.
37. Shen, W.; Dube, D.; Kaliaguine, S. Alkylation of isobutane/1-butene over periodic mesoporous organosilica functionalized with perfluoroalkylsulfonic acid group. Catal. Commun. 2008, 10, 291–294.
38. De Canck, E.; Vercaemst, C.; Verpoort, F.; van der Voort, P. A New Sulphonic Acid Functionalized Periodic Mesoporous Organosilica as A Suitable Catalyst. In Scientific Bases for the Preparation of Heterogeneous Catalysts: Proceedings of the 10th International Symposium; Gaigneaux, E.M., Devillers, M., Hermans, S., Jacobs, P.A., Martens, J.A., Ruiz, P., Eds.; Elsevier Science BV: Amsterdam, The Netherlands, 2010; Volume 175, pp. 365–368.
39. Nakajima, K.; Tomita, I.; Hara, M.; Hayashi, S.; Domen, K.; Kondo, J.N. A stable and highly active hybrid mesoporous solid acid catalyst. Adv. Mater. 2005, 17, 1839–1842.
40. Nakajima, K.; Tomita, I.; Hara, M.; Hayashi, S.; Domen, K.; Kondo, J.N. Development of highly active SO_3H-modified hybrid mesoporous catalyst. Catal. Today 2006, 116, 151–156.
41. Esquivel, D.; de Canck, E.; Jimenez-Sanchidrian, C.; van der Voort, P.; Romero-Salguero, F.J. Formation and functionalization of surface Diels-Alder adducts on ethenylene-bridged periodic mesoporous organosilica. J. Mater. Chem. 2011, 21, 10990–10998.
42. Dube, D.; Rat, M.; Beland, F.; Kaliaguine, S. Sulfonic acid functionalized periodic mesostructured organosilica as heterogeneous catalyst. Microporous Mesoporous Mater. 2008, 111, 596–603.
43. Vercaemst, C.; Ide, M.; Allaert, B.; Ledoux, N.; Verpoort, F.; van der Voort, P. Ultra-fast hydrothermal synthesis of diastereoselective pure ethenylene-bridged periodic mesoporous organosilicas. Chem. Commun. 2007, 22, 2261–2263.
44. De Canck, E.; Lapeire, L.; de Clercq, J.; Verpoort, F.; van der Voort, P. New ultrastable mesoporous adsorbent for the removal of mercury ions. Langmuir 2010, 26, 10076–10083.
45. Vercaemst, C.; Ide, M.; Wiper, P.V.; Jones, J.T.A.; Khimyak, Y.Z.; Verpoort, F.; van ver Voort, P. Ethenylene-Bridged Periodic Mesoporous Organosilicas: From E to Z. Chem. Mater. 2009, 21, 5792–5800.
46. Gao, Y.; de Canck, E.; Leermakers, M.; Baeyens, W.; van der Voort, P. Synthesized mercaptopropyl nanoporous resins in DGT probes for determining dissolved mercury concentrations. Talanta 2011, 87, 262–267.
47. Hamoudi, S.; Royer, S.; Kaliaguine, S. Propyl- and arene-sulfonic acid functionalized periodic mesoporous organosilicas. Microporous Mesoporous Mater. 2004, 71, 17–25.
48. Hamoudi, S.; Kaliaguine, S. Sulfonic acid-functionalized periodic mesoporous organosilica. Microporous Mesoporous Mater. 2003, 59, 195–204.

49. Ide, M.; El-Roz, M.; de Canck, E.; Vicente, A.; Planckaert, T.; Bogaerts, T.; van Driessche, I.; Lynen, F.; van Speybroeck, V.; Thybault-Starzyk, F.; van der Voort, P. Quantification of silanol sites for the most common mesoporous ordered silicas and organosilicas: total versus accessible silanols. Phys. Chem. Chem. Phys. 2013, 15, 642–650.

50. Nigam, P.S.; Singh, A. Production of liquid biofuels from renewable resources. Prog. Energy Combust. Sci. 2011, 37, 52–68.

51. Dosuna-Rodriguez, I.; Gaigneaux, E.M. Glycerol acetylation catalysed by ion exchange resins. Catal. Today 2012, 195, 14–21.

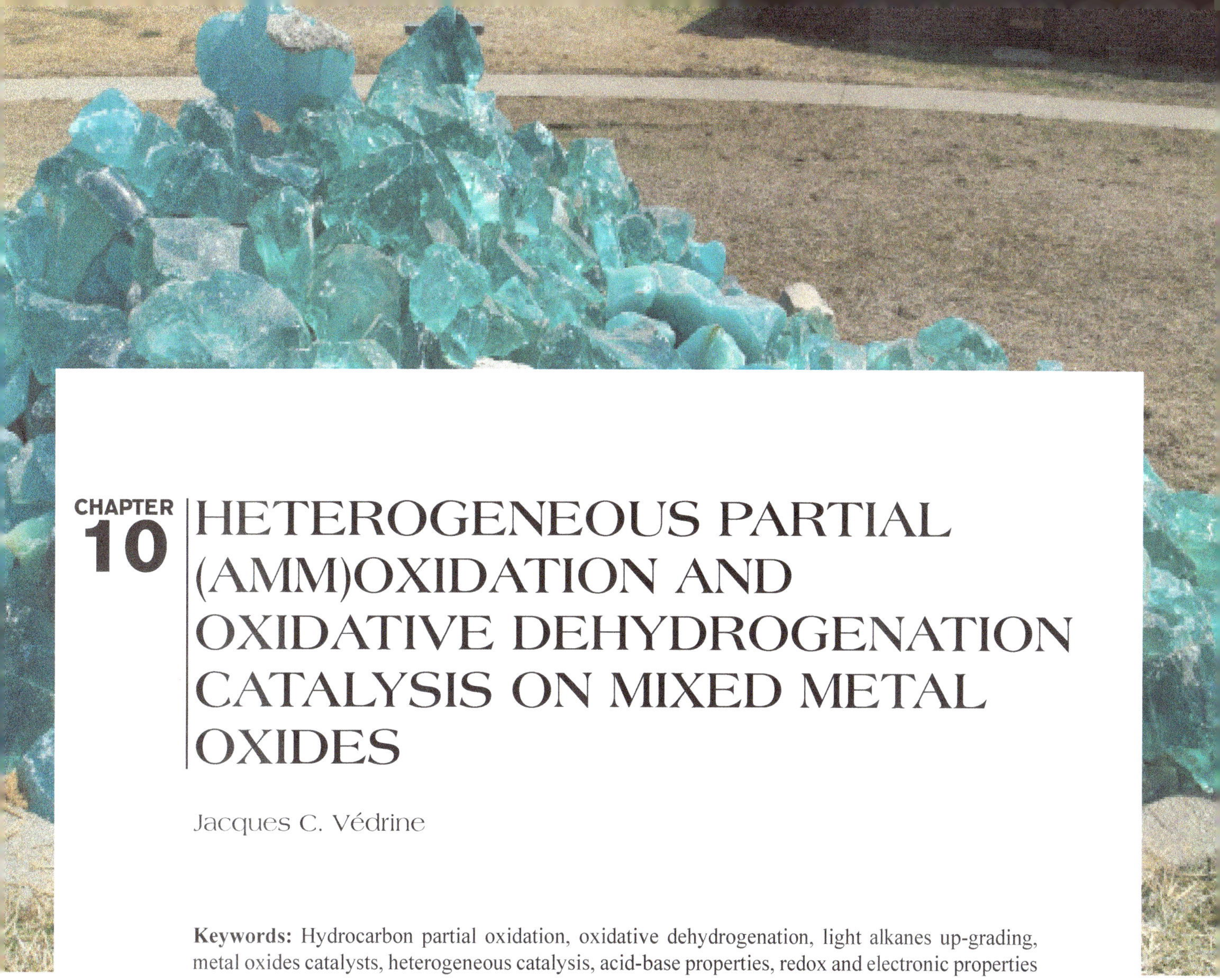

CHAPTER 10

HETEROGENEOUS PARTIAL (AMM)OXIDATION AND OXIDATIVE DEHYDROGENATION CATALYSIS ON MIXED METAL OXIDES

Jacques C. Védrine

Keywords: Hydrocarbon partial oxidation, oxidative dehydrogenation, light alkanes up-grading, metal oxides catalysts, heterogeneous catalysis, acid-base properties, redox and electronic properties

INTRODUCTION AND GENERAL FEATURES OF PARTIAL OXIDATION REACTIONS

In the chemical industry, hydrocarbons' partial oxidation reactions correspond to important processes, playing a key role in numerous industrial, environmental and energy applications [1]. The majority of monomers and ~25% of all catalytic reactions are obtained by heterogeneous oxidation of hydrocarbons over mainly metal oxide catalysts, and even ~50% of bulk chemicals if one includes the synthesis of NO (NH_3 oxidation over a Pt-based catalyst) and of SO_3 (oxidation of SO_2over a V-based catalyst). For instance, C_2 to C_8 hydrocarbons lead to monomers such as ethylene, vinyl chloride, ethylene oxide, propene, acrolein, acrylic acid, acrylonitrile, butenes, methacrylic acid and methacrylates, maleic and phthalic anhydrides, *etc.* In the past, the majority of studies have dealt with olefin selective (amm)oxidation, in particular propene to acrolein over bismuth molybdate based catalysts and acrolein to (acrylonitrile)acrylic acid over Bi and V molybdates based catalysts by SOHIO group as early as in the 1960s [2], but also on light alkanes such as C_2 to C_5 activation by oxidative dehydrogenation (ODH) to olefins ($C_2^=$ to $C_5^=$) or by oxidation to carboxylic acids on different oxides such as Bi molybdates, antimonates or vanadates, metal ions exchanged heteropolyacids of Keggin-type, *etc.* In the past 30 years, direct alkane selective oxidation has been the focus of many efforts in industry and academia [3], because of its higher abundance and lower price compared to the corresponding olefins. For instance, direct oxidation of ethane to acetic acid and/or ethylene on MoNbTeV-O or Nb doped NiO, direct (amm)oxidation of propane to (acrylonitrile)acrylic acid on MoNbTe(Sb)V–O, oxidation of butane to maleic anhydride on VPO catalysts, oxidative dehydrogenation of isobutyric acid to methacrylic acid on Fe hydroxyl-oxy-phosphates, *etc.*, have led to the development of new industrial processes, as it will be discussed in the text below. So far, the most used solid catalysts for oxidation reactions are metals, metal oxides or metal complexes immobilized in zeolites, silica, alumina or polymeric resins [4] and, more recently, in metal organic frameworks (MOFs) for selective and chiral oxidation catalysis [5].

However, it is worth noting that alkanes' ODH transformation is still a challenging task due to the low intrinsic chemical reactivity of the alkanes, which demands a high energy input to activate them. Note also that direct alkanes' non oxidative dehydrogenation (DH) to olefins on noble metal supported catalysts is still employed industrially as being more efficient than alkane ODH although showing some major disadvantages, namely high endothermicity and

Citation: Jacques C. Védrine, Heterogeneous Partial (amm)Oxidation and Oxidative Dehydrogenation Catalysis on Mixed Metal Oxides, doi:10.3390/catal6020022

high tendency to coking and, consequently, leading to short catalyst lifetime [6].

A main aspect of a partial oxidation reaction comprises the first C–H bond activation for oxidative dehydrogenation (ODH) or oxyfunctionalisation, considered as the rate determining step in the overall process. Usually, monomeric and dimeric V=O bonds are considered as active sites for light alkane activation in ODH reactions and larger V–O moieties for acrylic acid or acrylonitrile formation, as it will be discussed below (Section 3).

Complex metal oxides and multicomponent oxides catalysts represent the most important family of solid catalysts for partial heterogeneous oxidation catalysis as active phases or as supports. They include mixed metal oxides and hetero polyoxometallates [7]. The three main features of these oxides, which are essential for their application in catalysis, are (i) the coordination environment of surface atoms, (ii) the redox and, subsequently, acid-base and electron transfer properties and (iii) the oxidation states of the surface cations.

Seven key factors, designated as pillars by Grasselli [8], have been proposed to be satisfied for partial oxidation reactions to occur, namely: (i) nature of lattice oxygen anions: nucleophilic (selective) rather than electrophilic (total oxidation); (ii) redox properties of the metal oxide (removal of lattice oxygen and its rapid reinsertion); (iii) host structure (permits redox mechanism to take place without structure collapse); (iv) phase cooperation in multicomponent catalyst or supported catalyst (epitaxial growth and synergetic effects); (v) multifunctionality (e.g., α-H abstraction and O–/NH– insertion); (vi) active site isolation (to avoid a too high lattice O surface mobility and thus overoxidation); (vii) M-O bond strength (to be not too weak (total oxidation) nor too strong (inactivity) (Sabatier principle).

Point (i) of active oxygen species was discussed a long time ago in two review papers by Tench and Che [9], who described different types of oxygen species. For the redox mechanism one may write: $O_2 + e^- \rightarrow O_2^-$; $O_2^- + e^- \rightarrow O_2^{2-} \rightarrow 2O^-$; $2O^- + 2e^- \rightarrow 2O^{2-}$. These oxygen species are more or less electrophilic (O_2^-, O_2^{2-}, O^-), nucleophilic (O^{2-}) in the nomenclature proposed by Bielanski & Haber [10]. There have been extensive and controversial discussions about oxygen species in terms of their nature and location. Activation of molecular oxygen takes place by its reduction at the oxide surface which should present four free electrons per O_2 molecule, whereas the highest oxidation state cations cannot activate oxygen. In other words, structural defects, encompassing lower oxidation state cations, have to be present.

Another general feature of such reactions in heterogeneous catalysis, which has appeared over the years, is that the oxide surface could be considered as living, as in a breathing motion, to allow the Mars and van Krevelen (MvK) mechanism to take place (vide infra). Such a property was clearly shown for iron phosphate catalysts used for isobutyric acid ODH to methacrylic acid (vide infra), implying that the oxide structure should be strong but mobile enough to allow the redox mechanism to occur, without structural collapse such as $2PO_4^{3-} \leftrightarrow P_2O_7^{4-} + O^{2-}$ as schematised in Figure 1.

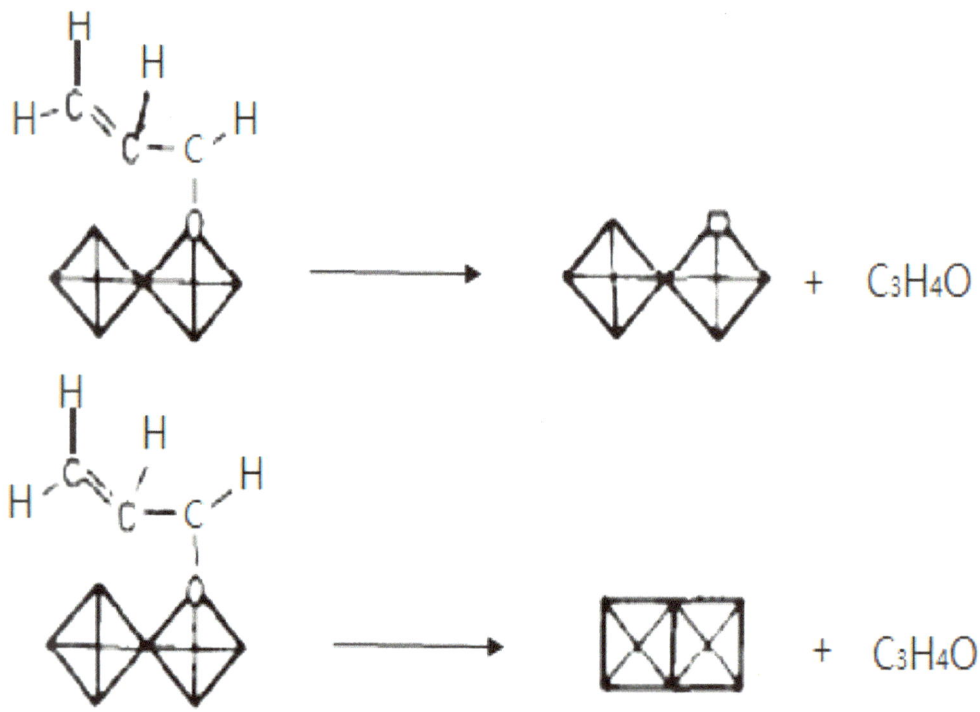

Figure 1: Schematic representation of the mechanism of O insertion into a hydrocarbon molecule (left) and its consequences in creation of a reversible O vacancy in a metal oxide (right) $P_2O_7^{4-} + O^{2-} \leftrightarrow 2\ PO_4^{3-}$. Reproduced from [11]. Copyright 2015, Academic Press.

It was also shown that such a property may suggest the existence of amorphous surface layers over a well crystallised underlying bulk phase, for example, for V_2O_5/TiO_2 catalysts as observed by HR-TEM analysis [12], and for VPO catalysts in *n*-butane oxidation to maleic anhydride [13].

Table 1: Main selective oxidation reactions in heterogeneous catalysis from [14]. Process: NI = not industrialised yet, I: industrialised; P: pilot, R: research

Reaction Type	Reactants	Products	Nb of e^-	Catalysts	Yield mol %	State
oxidative coupling	CH_4	ethane + ethylene	2 4	Li_2O/MgO	25	NI
oxidative dehydrogenation	ethane	ethylene	2	Pt, mixed MoVTeNb-oxides	60 75	NI NI
	propane	propene	2	VMgO	40	NI
	n-butane	butene, butadiene	2 4	metal molybdates	38	NI
	ethylbenzene	styrene	2	$FeO\text{-}AlPO_4$	70	P
	methanol	formaldehyde	4	$FeMO_4$	81	I
	isobutyric acid	methacrylic acid	2	$FePO_4$	75	NI
oxychloration	ethane, Cl_2	vinyl chloride	4	AgMnCoO	15	NI
	ethene, Cl_2		2	CuPdCl	90	I
hydroxylation	benzene	phenol	2	Ti or Fe zeolite	-	I
selective oxidation	methane	$CO + H_2$	2	Pt or Ni	90	R
		formaldehyde	4	MoSnPO	16	R
	ethene	ethyloxide	2	Ag/Al_2O_3	8	I
		acetaldehyde	2	$V_2O_5 + PdCl_2$	50	I
		acetic acid	4	MoVNbO + Pd/Al_2O_3		
	ethane	acetic acid	6	MoVNb-O	10	P
	propene	acrolein	4	BiFeCoMoO	86	I
	propane	acrylic acid	6	MoVNb(Te,Sb)O	8	NI
	n-butane	maleic anhydride	14	$(VO)_2P_2O_7$	70	I
	i-butane	methacrylic acid	8	HPA, oxides	11	R
	i-butene	methacrolein	6	SnSbO	65	NI
	o-xylene	phthalic anhydride	12	V_2O_5/TiO_2	82	I
	acrolein	acrylic acid	2	VMoWO	85	I
	methacrolein	methacrylic acid	2	-	95	I
	t-butylic alcohol	methacrolein	4	BiMoFeCoO	88	I
ammoxidation	Propene + NH_3	acrylonitrile	6	MoBiFeCoNiO	80	I
	Propane + NH_3		8	VSbO, MoVO	30	I

Another important aspect to consider is the number of electrons necessary for a given reaction such as $CH_3\text{–}CH_2\text{–}CH_3 + O^{2-} \rightarrow CH_3\text{–}CH{=}CH_2 + H_2O + 2e^-$ or dissociation of molecular oxygen O_2 into $2O^{2-}$ requiring $4e^-$ and their transfer ability within the material. In Table 1, some important partial oxidation reactions are reported as well as their state with respect to actual developments and number of electrons involved in the reaction. It then follows that electron mobility on the surface and even bulk catalyst layers may play an important role in determining catalytic properties. This feature could be determined via electrical conductivity measurements as it was observed in many cases (vide infra Section 4.2).

GENERAL REACTION MECHANISMS

Many partial oxidation reaction mechanisms have been proposed in the literature [2,15], since the pioneering industrial work from Sohio scientists for propene oxidation/ammoxidation to acrolein/acrylonitrile on bismuth molybdate-based multicomponent catalysts ($Bi_2MO_3O_{12}$/Co, $FeMoO_4$) in the 1960s [2]. Propene chemisorption was suggested to occur on Mo centres, and the rate determining step was suggested to be the α-H atom abstraction by Bi-O leading to a radical-like α-allyl Mo complex. O or N insertion centres consist of coordinated unsaturated $O{=}Mo^{VI}{=}O$ or $HN{=}Mo^{VI}{=}NH$ sites, respectively. Reversible formation of α-O or N-allyl molybdenum complex occurs prior to subsequent (*i.e.*, 2nd and 3rd) allylic H abstractions. The presence of Bi was suggested to facilitate the second H-abstraction, which has a lower activation energy E_a for oxidation than for ammoxidation.

For the partial oxidation reaction, the main reaction mechanism is the Mars and van Krevelen (MvK) one [16], schematised below. This mechanism concerns cations which can easily change their oxidation state such as Ce, Co, Cr, Cu, Fe, Mo, Ni, ,Ti, V, W, *etc*.

$$2[\mathrm{CatO}] + \mathrm{R-CH} \rightarrow 2[\mathrm{Cat}] + \mathrm{R-C-O} + \mathrm{H_2O}$$

$$2[\mathrm{Cat}] + \mathrm{O_2\,(gas)} \rightarrow 2[\mathrm{CatO}]$$

In this scheme, [CatO] corresponds to the oxidised surface and [Cat] its reduced state; r_{red} is the rate of catalyst reduction by a reactant and r_{ox} the rate of its re-oxidation by co-fed oxygen, R–CH and R–C–O the reactant and the product. The kinetic equation contains the concentration of reduced (θ) and oxidised (1-θ) sites of the catalyst. At the steady state, $r_{red} = r_{ox}$ or $k_{red} \cdot p_{HC}(1-\theta) = k_{ox} \cdot p_{O2}\theta$ with p_{HC}, p_{O2} partial pressure of HC and O_2 and k_{red}, k_{ox} rate constants of reduction of catalyst (1st step) and r_{ox} rate of oxidation by O_2 (2nd step). Their relative rate values are important for the selectivity and the reaction necessitates lattice oxygen anions, which are incorporated into the reactant molecule, whereas the corresponding vacancy created is replenished by gaseous oxygen in the re-oxidation step. If $k_{red} \cdot p_{HC} >> k_{ox} \cdot p_{O2}$, reoxidation of the surface is the rate-determining step while if $k_{red} \cdot p_{HC} << k_{ox} \cdot p_{O2}$, reduction of the surface is the rate-determining step. Thus, reoxidisability and reducibility of the catalyst will lead to different kinetic expressions. Lattice oxygen is introduced into the substrate or into H_2O for ODH reaction. This process involves two active sites: an active cationic site and a site active for oxygen reduction. The process starts with the abstraction of a proton in α of the double bond by surface O^{2-} and the sharing of electron between C=C and the empty d° orbital, by a two-electron transfer that reduces the transition metal cations. Reaction mechanisms in partial oxidation reactions may also include homogeneous phase reaction (radical-type mechanism), in addition to the usual heterogeneous one, even at relatively low temperatures such as below 300 °C while it occurs usually at high temperatures (600 °C and above) [17].

Note that on Me_xO_y oxide when Me^{n+} is a transition metal cation in d^0 electron configuration and containing an oxygen vacancy, olefins ($C_3^=$, $C_4^=$, *etc.*) and aromatics (benzene, xylenes, *etc.*) are efficiently activated, the cation acting as an electron acceptor [18] (see Figure 2a). For V-based catalysts, it is well accepted that the V^{5+} ion activates an alkane by H abstraction first and that a second H abstraction gives the allylic radical before insertion of oxygen. Catalytic activity was shown to be proportional to the total V concentration on the surface [19], while the higher selectivity was assigned to VO_4 monomeric species or low-size oligomeric $(VO_x)_n$ species. The MvK mechanism has also been found to be quite valid for alkane partial oxidation and ODH reactions. However, in some cases, power law model, rake model, Langmuir-Hinshelwood model (LH) (uniform surface with one type of site), Eley Rideal (ER) model were also considered with more or less success. A steady-state adsorption model (SSAM), which can be regarded as a surface variant of the MvK model, was developed by Downie *et al.* [20] and applied also to describe kinetics of the vapor-phase oxidation of *o*-xylene over the vanadium oxide catalyst.

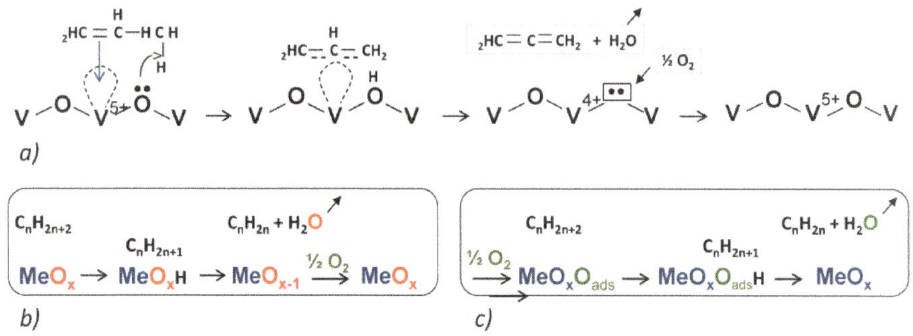

Figure 2: Simplified schemes of: (**a**) propene adsorption O–V–O and formation of a π-allyl complex intermediate; (**b**) case of an alkane; (**c**) effect of adsorbed oxygen.

STRUCTURAL ASPECTS OF M_XO_Y OXIDES AND DESCRIPTION OF ACTIVE SITES IN SOME PARTIAL OXIDATION REACTIONS

It follows, from the general features described above, that the structural properties of M_xO_y oxides confer specific selectivity in oxidation reactions that depends on redox couples $M^{n+}/M^{(n-p)+}$, length/strength/energy of M–O bonds, anionic defects (vacancies or anion excess, in particular O^{2-}) and/or cationic (vacancies or cation excess) [21]. Surface and bulk mobility of oxygen species is also an important characteristic of the catalyst. We have seen above that partial oxidation catalysts generally contain a transition metal cation such as Ti, V, Cr, Mo, W, highly charged, small and polarising. In vanadyl (VO^{2+}), molybdenyl (MoO_2^{2+}) oxo-cations, multiple bonds are formed between the cation and oxygen. V^{5+} cations (ionic radius $r_i = 0.054$ nm) float in an octahedral environment of six O^{2-} ($r_i = 0.140$ nm) and form V=O bond to be more stable. The V...O bond in *trans* V=O is longer, and its binding energy is weaker. Thus, this oxygen may easily be withdrawn, and the V coordination becomes five. Another striking feature is that octahedral environments of VO_6 in V_2O_5 and in $V_2O_4$are very close. Compared to V^{5+}, O=V^{4+} and V^{4+}...O bonds are longer (from 1.58 to 1.64 nm) and shorter (from 2.78 to 2.70 nm), and octahedron height changes very little. Therefore, at first approximation, one may expect an easy e^- transfer: $V^{5+} \leftrightarrow V^{4+}$ or $Mo^{6+} \leftrightarrow Mo^{5+}$ couples (r_i = 0.059 nm et 0.061 nm for Mo^{6+} and Mo^{5+}) and thus improved catalytic properties. As a matter of fact, a catalyst will be more active as its structural changes are small, and one may expect such reactions to be structure sensitive as described by Boudart [22] for metallic catalysts. Structure sensitivity of metal oxides for oxidation reactions was first demonstrated by Volta [23] for propene partial oxidation. Volta has used a novel method to prepare MoO_3 crystals with specific orientations. It was observed that propene was oxidised almost exclusively to acrolein on the (100) lateral face and CO_2 on the apical (10-1), (101) and basal (010) faces. For the oxidation of isobutene and but-1-ene to 1-3-butadiene, the selectivities observed were completely different, namely (100) for methacrolein CH_2=C(CH_3)–CHO and CO_x from isobutene and (010) for 1–3 butadiene from but-1-ene, showing that the geometry of the reacting molecule and atomic arrangements at the oxide surface are important in partial oxidation reactions [24]. This is schematised in Figure 3and was published by Tatibouët *et al.* [25] for butene oxidation.

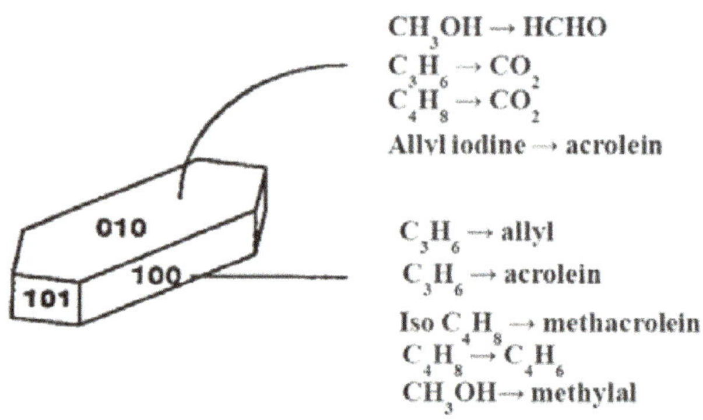

Figure 3: Structural sensitivity of selective oxidation reactions of olefins or methanol on single crystals of MoO_3 as plates (mainly (010) faces). Assignment of reactions to surfaces as indicated came from comparison of many sample shapes. Reproduced from [26]. Copyright 2015, Elsevier.

Structure sensitivity was also observed by Germain *et al.* [27] for the oxidation of methanol to formaldehyde and ethanol to acetaldehyde over MoO_3 catalyst. However, in this case, the basal (010) face was observed to be selective for aldehyde formation. A review article [26] has summarized these findings. Later on, many selective oxidation reactions have been shown to be structure sensitive [28] such as n-butane to maleic anhydride on the (100) face of $(VO)_2P_2O_7$ [29] or propane to acrylic acid on terminating (001) plane *vs.* lateral faces of needle shaped M1 phase (MoVTeNb-O) [30,31,32], *etc.* This structure sensitivity feature holds particularly true when the hydrocarbon molecule is large and the number of electrons involved in the reaction is high (see 4th column in Table 1), which implies that active sites for such reactions are quite complex.

In 1925, Taylor [33] proposed that "a chemical reaction is not catalysed over the entire solid surface of the catalyst but only at certain 'active sites' or centres". He has also proposed that "chemisorption is an activated process, and may occur slowly". Moreover, Taylor suggested that chemically active sites are "composed of an atom or an ensemble of atoms and could be sparse on the surface of a catalyst". It follows that a catalyst surface can be poisoned by only a few molecules. Since that time, many descriptions of active sites have been proposed, in particular for metal oxides, as noted in [34].

For single oxides, e.g., MgO, surface sites could be Mg^{2+} or O^{2-} ions with different coordination numbers depending on the defectuous surface structure as schematised in Figure 4.

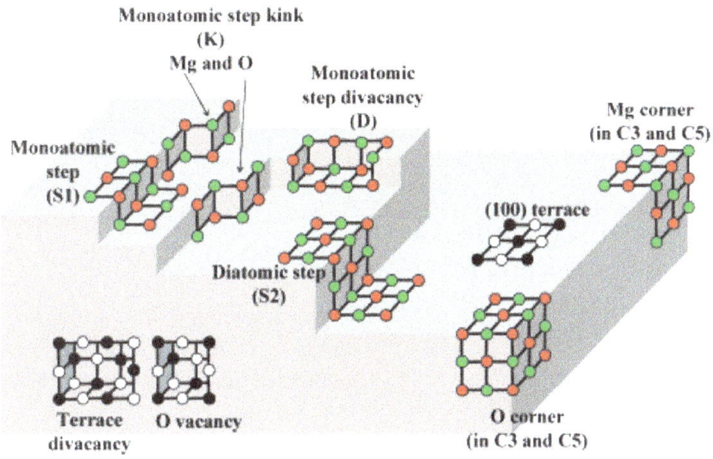

Figure 4: Schematical representation of MgO surface. Reproduced from [35] Copyright 2016, ACS.

For large organic molecules and complex metal oxides used industrially, the active sites are usually ensembles of atoms in order to fulfill all requirements, to activate the reactants, to provide the number of electrons necessary for the reaction and thus to exhibit multifunctionality properties. In selective oxidation reactions, active sites have been shown to be isolated from each other to avoid overoxidation [2], which was assigned to quickly to lattice oxygen ions' mobility. For instance, in the case of iron hydroxyphosphates, such as $Fe_2^{3+}Fe^{2+}(P_2O_7)_2$ (+ H_2O & 400 °C) ↔ $Fe_2^{3+}Fe^{2+}(PO_3OH)_4$ (+O_2) → $Fe_{2+x}^{3+}Fe_{1-x}^{2+}(PO_3OH)_{4-x}(PO_4)_x$, and used for oxidative dehydrogenation of isobutyric acid to methacrylic acid, the most active and selective sites have been described [36,37] as ensembles of FeO_6 octahedra trimers in $\alpha Fe_3P_2O_7$, where electron transfer during the redox process occurs between Fe^{2+} and Fe^{3+} in $\alpha Fe_3P_2O_7$, while in $\beta Fe_3P_2O_7$ no electron exchange was possible as calculated by extended Hückel theory [38] and shown in Figure 5, and, also, no catalytic property was observed, which certainly corresponds to low redox efficiency in the MvK mechanism.

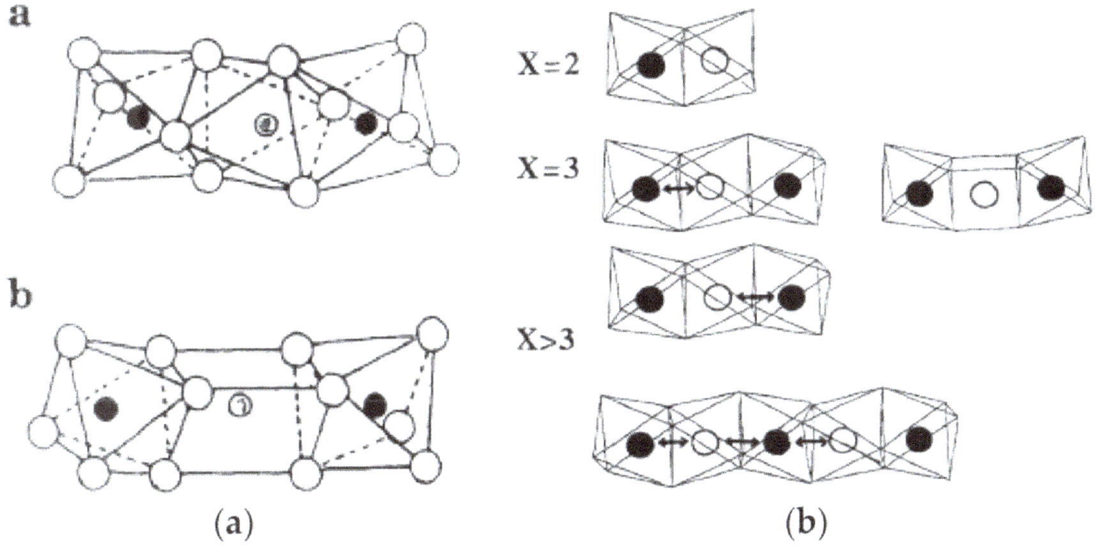

Figure 5: Arrangements of FeO_6 octhedra in trimeric clusters isolated one from the others by P_2O_7 groups: (a) αP_2O_7; (b) $\beta Fe_3P_2O_7$ grey circles: Fe^{2+}, filled: Fe^{3+}. Electron exchange between iron cations in FeO_6 octahedra clusters calculated by extended Hückel molecular orbital theory. Reproduced from [36]. copyright 2016, Elsevier BV.

In fact, isobutyric acid (IBA) ODH to methacrylic acid and its esterification to methyl methacrylate (MMA) using FePO or (Cu)MoPV (heteropolyacid-type) or iron or nickel oxide based catalysts have been industrialized by different companies. There are some alternate methods to producing methyl methacrylate. Among them is the oxidative dehydrogenation of methyl isobutyrate, (MIB) which is formed by the carbonylation of propene in the presence of methanol [39].

In MIB ODH reactions, co-feeding water showed an increase in conversion of MIB over the catalysts. Along with higher conversion of MIB, the selectivity towards MMA was lower than in the case without co-feeding of water catalysts, due to the formation of the hydrolysis products of MMA and IBA. The mechanisms of formation of the products from MIB over the iron phosphate catalysts involves the Fe^{2+}/Fe^{3+} redox couple, and the mechanism follows a similar pathway of Mars and van Krevelen mechanism.

Another example of active site description corresponds to $(VO)_2P_2O_7$, designated as VPO catalyst, used industrially for butane oxidation to maleic anhydride. The active sites have been described as ensembles of four dimers of octahedral VO_6entities isolated by P_2O_7 groups as schematised in Figure 6 [40].

Many other examples of active sites described as isolated atoms or ensembles of atoms can be found in the literature. Let us take some of them. Panov *et al.*, in their study of benzene direct oxidation to phenol by N_2O on Fe-MFI catalysts, have shown that active sites are FeO dimers [41] inside the pores of the zeolite. Janas *et al.* [42] have studied Co(II) ions in SiBEA zeolite. At low Co content (0.8 wt. %), cobalt is present as lattice tetrahedral Co(II) species. For much higher Co content (11 wt. %), mainly extra-lattice octahedral Co(II) species and/or cobalt oxides are present as shown by DR UV-vis spectroscopy. It was then concluded that zeolite with isolated tetrahedral Co(II) species are active in selective catalytic reduction of NO by ethanol with selectivity towards N_2 exceeding 85% for NO conversion from 30% to 70%. At variance, when extra-lattice octahedral Co(II) species and/ or cobalt oxides are present, the full oxidation of ethanol and NO by O_2 to CO_2 and NO_2, respectively, were observed to be the main reaction pathways. Isolated Co(II) cations in lattice position were thus suggested to be active sites for the reaction. Another example follows the pioneered works on metallic gold particles for CO oxidation at room temperature by Haruta [43], who has shown that small particles, in a given range of nm sizes, are active while bulk metal is inactive. This work has been extended to Co_3O_4 nanorods exhibiting (100) faces, for CO oxidation at low temperature [44] was also related to structure sensitivity face, although CO and O_2 are rather small molecules.

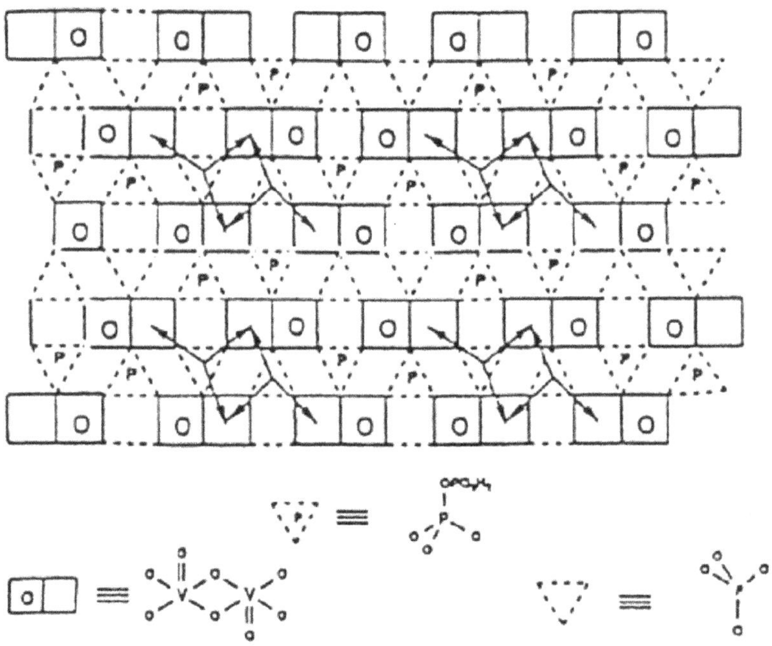

Figure 6: Schematic representation of a surface structure of a type of polytype of $(VO)_2P_2O_7$. The arrows represent the facile pathways for surface oxygen mobility. This illustrates the site isolation principle by surface P_2O_7 groups which constitute a barrier to oxygen diffusion. Reproduced from [8] Copyright 2016, Elsevier BV.

Such an ensemble of atoms as active sites has also been proposed for enzymatic oxidation/hydroxylation reactions using cytochrome P-450 enzyme [45]. The major challenge in mimicking cytochrome-P450 enzymes is the development of catalysts able to induce insertion of one O atom from O_2 on the substrate and a $2e^-$ reduction of the other O atom to water. In nature, this is carried out by NAD(P)H reductase protein. To overcome this problem, addition of several reductants such as zinc dust or aldehydes to the metalloporphyrins' oxidation catalytic systems or use of high temperatures and pressures have been reported. For instance, porphyrin-like complexes, such as *N*-bridged di-iron oxo phthalocyanine associating cytochrome P-450 and a monooxygenase enzyme and using H_2O_2 [46] has been shown to oxidise methane [47] into formic acid and secondarily to formaldehyde and methanol in specific conditions (40 °C, 3.2 MPa, CH_3CN as solvent), while monomeric Fe complexes were found inactive. However, transformation of metalloporphyrins as valuable oxidation catalysts for large scale purposes is still a challenging issue, mainly due to porphyrin stability, cost and reutilization. This could depend on the development of efficient procedures for the immobilization of porphyrins into/onto solid supports such as mesoporous aluminosilicates, polymers' double layered hydroxides, which unfortunately generally require the use of polluting oxidants and O_2 with co-reductants.

Let us take another and more recent example. In the past 15 years, a tremendous amount of work has been performed and published on a complex mixed oxide based on MoVTe(Sb) Nb-O, with V/Mo, Nb/Mo and Te(Sb)/Mo ratios variable but preferentially close to 0.3, 0.1 and 0.2, respectively, able to (amm)oxidise directly propane to (acrylonitrile) acrylic acid in the presence of water vapour [48]. In the absence of water vapour, mainly propylene is formed. The highest acrylonitrile yield was found for $V_{0.3}Te_{0.23}Nb_{0.12}MoO_x$ supported on silica after calcination under nitrogen at 620 °C [49]. In fact, this catalyst family was discovered by researchers of Union Carbide, who published their work in 1978 [50] on ethane oxidation to ethylene and acetic acid. This material was shown to be Mo-based bronzes, the preparation and activation in air or N_2 of which was observed to greatly influence their structure and catalytic properties [51]. The best catalysts for (amm) oxidation were further observed to contain a phase designated M1 phase, often associated with another phase designated M2. It was suggested [52] that the reaction needs either four monomeric or two dimeric units of redox storage to deal with one molecule of oxygen, according to:

$$4V^{4+}-OH + OH^- \rightarrow 4V^{5+}=O + 4e^- + 4H_2O$$

$$O_2 + 2H_2O + 4e^- \rightarrow 4OH^-$$

$$4V^{4+}-OH + O_2 \rightarrow 4V^{5+}=O + 2H_2O$$

$$O=V-O-V=O + C_3H_6 \rightarrow O=V-O-V-OC_3H_5 + H^+ + e^-$$

$$O=V-O-V=OC_3H_5 + H^+ + e^- \rightarrow HO-V-O-V-OC_3H_5$$

$$HO-V-O-V-OC_3H_5 + 3OH^- \rightarrow HO-V-O-V-OC_3H_4O + 2H_2O + 3e^-$$

This oxidation leads to acrolein being still bounded to the active site, as the electrons released from the organic substrate do not reduce the V centre. Active sites were suggested [53] to be isolated ensembles of atoms as schematised in Figure 7.

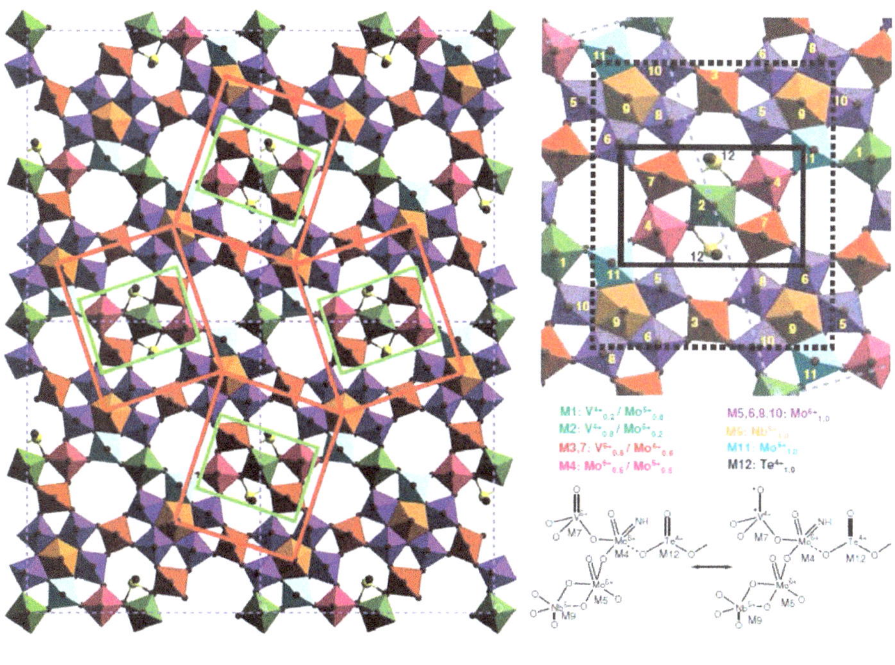

Figure 7: Active site (left) in $Mo_{7.8}V_{1.2}NbTe_{0.94}O_{28.9}$ (M1): 2 × 2 unit cell structure model of M1 in (001) projection showing four isolated catalytically active centres. The site sites are isolated from each other by four Nb-bipyramids that are surrounded by five MO_n-octahedra. ChemDraw depiction (right). Adapted from figures 15 & 16 in [8], copyright 2016, Elsevier BV.

In a study of V_2O_5/TiO_2 catalyst studied by the Eurocat oxide group, it was found that: [54]

- Monomeric VO_4^{3-} species exhibit an acid character and gives propene in isopropanol conversion, used as a test reaction, and total oxidation for *o*-xylene oxidation (coverage <0.43% in V_2O_5) as expected for acid sites;
- Polyvanadate species exhibit redox properties and give acetone in isopropanol conversion and phtalic anhydride in *o*-xylene conversion;
- V_2O_5 crystallites give low activity in *o*-xylene conversion and high selectivity in CO_2.

This shows that the size of the active sites is quite important for catalytic partial oxidation reactions depending also on the compound size and the multifunctionality necessary for the reaction. Such concept has been largely discussed in the literature for many years and monomeric isolated species or low-size oligomeric vanadium oxide species were suggested for light alkane oxidation and ammoxidation reactions (vide infra) as suggested in [3] (Figure 24-3, p. 779).

MAIN FACTORS ACTING IN PARTIAL OXIDATION REACTIONS

General Aspects of the Acid-Base Properties of Metal Oxides

Metal oxides are composed of redox metal cation and lattice oxygen anions which are Lewis acid and basic sites, respectively. The acid-base characteristics of the oxide have three major effects, namely on reactant molecule activation, on the rates of competitive pathways of transformation and on the rate of adsorption and desorption of reactants and products.

It is known [55,56] that during a reaction, an acid surface would favour desorption of an acid intermediate while a basic surface would favour desorption of a basic intermediate, which then would avoid further over-oxidation of the adsorbed intermediate to CO_2. Thus, acidic products, such as carboxylic acids, should favourably be formed on acid surfaces while basic products, such as olefins in ODH reactions, should favourably be formed on basic surfaces. This is why the presence and the strength of Lewis acid sites (cations) and of basic sites (O^{2-} or OH^-) are important and should be characterised carefully. For such a characterisation [57], one usually uses FTIR or microcalorimetry of adsorbed probe molecules (basic NH_3, pyridine or acid CO_2, SO_2, respectively) or test reactions such as isopropanol conversion which gives propene by dehydration on acid sites or acetone by dehydrogenation on basic sites in the absence of air in the feed or on redox sites in the presence of air [58]. Another test reaction has been proposed for acid base characterisation, namely 2-methyl but-3-en-1-yn-2-ol (MBOH) conversion, which follows three different pathways depending on the acid-base properties of the catalyst, as schematised in Figure 8.

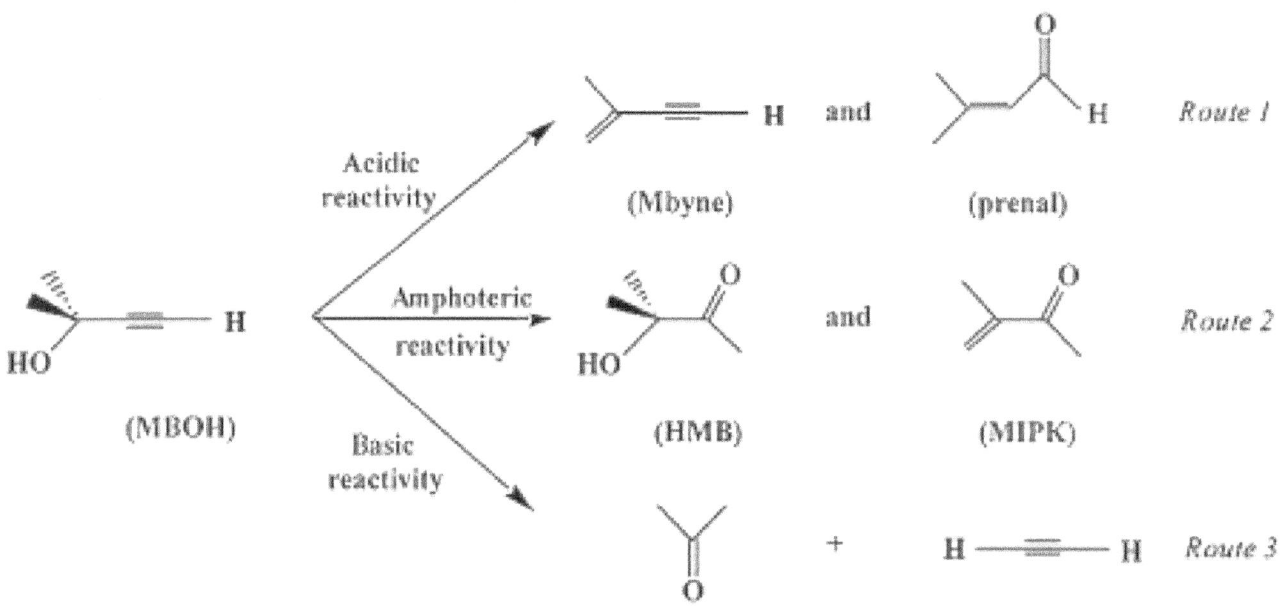

Figure 8: Scheme of 2-methyl but-3-en-1-yn-2-ol (MBOH) reaction pathways for an acid-base catalyst Reproduced from [59] copyright 2016, Taylor & Francis.

An oxygen element has a highly electronegative character and its bond with metals is ionic, thus more basic, or covalent with non-metals thus less basic. The electronegativity of a cation depends on its oxidation state and increases as the oxidation state increases. Oxides of metals at high oxidation state, such as V^{5+} or Mo^{6+}, are characterized by a covalent metal to oxygen bond and behave as acid oxides, whereas the same elements at lower oxidation state have a more ionic character and behave more as basic oxides. Note that metal cations, active for oxidation reactions, have several oxidation states and are able to easily change their oxidation state and, thus, their acid base character under certain catalytic reaction conditions.

The strength and nature of acid/basic sites are important parameters as they are expected to participate in the molecule activation as for the first H abstraction. Let us take as an example the MoV(Nb)Te(Sb)–O based M1 and M2 phases, mentioned above. Their acidity was determined for [60] with pyridine as a probe molecule as shown in Figure 9. Both Brønsted and Lewis acid sites were detected and it was observed that a too strong acidity favours total oxidation, as expected, and, thus, that a moderate acidity is favourable for better selectivity to acrylic acid/acrylonitrile.

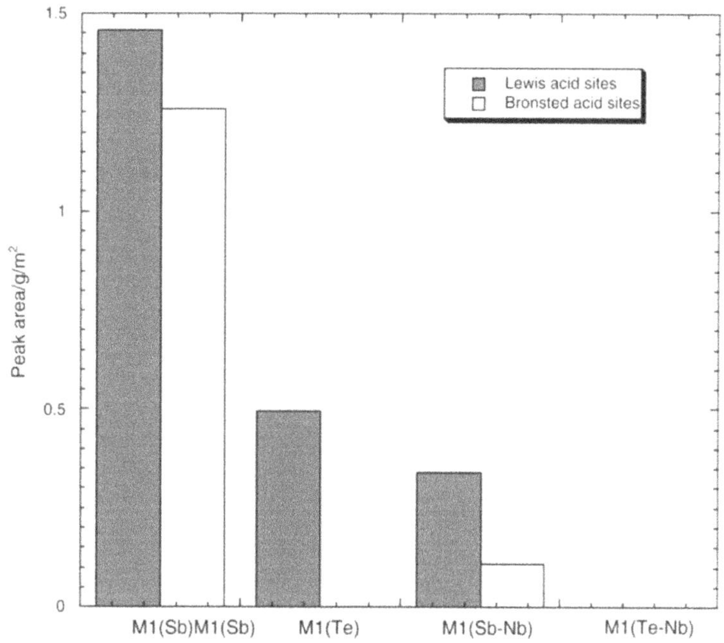

Figure 9: Comparison of the Brønsted and Lewis relative acidity of the M1 phases of MoVTe(Sb)NbO, expressed as area under the pyridine peaks at 1540 and 1450 cm^{-1}normalized by surface unit in the spectra recorded after desorption at 100 °C, From [60], Copyright 2015, Elsevier BV.

In another example in the same family, namely K-free and K-MoVSbO bronzes of formula $(SbO)_2M_{20}O_{56}$, thus of M1-type structure [61], the selectivity to acrylic acid was observed to be higher than for the starting sample while, when P was added to M1 phase [62], acrylic acid selectivity was also improved. It was also observed that Sb samples had medium strength acidity and Te samples even lower strength and that Nb insertion decreases both acidities, while WO_x added to MoVTeNb-O increases propane to acrylic acid by improving propene formation [63]. In a recent study [64] on Bi promoted MoVTeNb-$O_x$catalyst for propane ammoxidation, it was suggested that Bi suppressed acrylonitrile's further transformation and decreases the acidity of the catalyst.

Acidic properties were also studied [65] using NH_3 TPD and acid-base titration with a standardized NH_4OH solution (−log $[NH_3]$ = 6.7) for $Mo_1V_{0.3}Sb_{0.35}Nb_{0.08}O_n$ mixed oxide, calcined under N_2. This compound was composed mainly of $Sb_4Mo_{10}O_{11}$ (isomorphous of M2 phase) and $(V_xNb_yMo_{1-x-y})_5O_{14}$ (M1-type) phases. Low Lewis acidity and no Brønsted acidity were observed for such sample.

Importance of Electron Transfer Properties

In the first step of a redox mechanism surface, oxygen anions O^{2-} are extracted to be inserted in products while cations are reduced according to: $M^{n+} + pe^- \rightarrow M^{(n-p)+}$. For example: $V^{5+}{}_2O_5 + 2e^- \rightarrow V^{4+}{}_2O_4 + \square$ (where □ represents an oxygen vacancy) or $V_2O_5 \rightarrow V_2O_4 + ½\,O_2$ as usually written for an oxide reduction. In the absence of oxygen in the feed, these vacancies are filled via O^{2-} diffusion from the bulk, as this occurs in ionic conductors and the solid is more and more reduced. In general, O_2 is added to the reactants to promote reoxidation as diffusion of O^{2-} ions is too slow. The catalyst surface permits reduction of O_2 according to: $O_2 + 4e^- + 2\square_{surf} \rightarrow 2O^{2-}{}_{surf}$. The electron transfer property of a solid catalyst is thus an important aspect of the Mars and van Krevelen mechanism, which implies high mobility of lattice oxygen anions and of electrons to ensure a fast reaction.

Some typical reactions have already been listed above in Table 1, including the number of e$^-$ involved in them. The majority of selective catalysts are semiconductors due to defects present naturally on their surface. They could be *n*-type (*n*: negative or e$^-$ donors in the valence band) or *p*-type (*p*: positive or e$^-$ hole acceptor in the conduction band). Note that insertion of dopants may allow to modulate semi conductivity. N-type semi conduction is favourable to O_2 activation to $2O^{2-}$, while *p*-type and/or, if the cation is more electropositive, other oxygen species can be formed at the surface according to:

$$O_2 + e^- \rightarrow O_2^-;\ O_2^- + e \rightarrow O_2^{2-};\ O_2^{2-} \rightarrow 2O^-;\ O^- + e^- \rightarrow O^{2-}$$

Such electrophilic species are more stable for electropositive cations such as Mg^{2+}, Ca^{2+}, La^{3+}, Ce^{4+}, *etc.* and would give carbon oxides for the oxidation of an unsaturated hydrocarbon. On the contrary, O_2^- is selective in ethylene epoxidation on Ag^0, and O^- species are very reactive at low temperatures for aldehydes to acids, benzene to phenol on titanosilicate or alkane activation using N_2O instead of O_2.

Under an oxygen atmosphere, the electron trapped by adsorbed O_2 is provided by the valence band of the oxide and the electrical conductivity increases with the consumption of oxygen. Conversely, exposure to a reducing gas creates oxygen vacancies, releases electrons in the valence band and creates new positive holes with a decrease of electrical conductivity. This is why electrical conductivity measurement has been used for characterising redox properties of metal oxides. In Figure 10 and Figure 11, the variations of electrical conductivity are shown respectively *vs.* the local atmosphere indicated in the caption for VMgO catalysts with different V content, used for propane ODH to propene [66] (Figure 10) and in relation to selectivity for ethane to ethene on Nb doped NiO catalysts [67] (Figure 11).

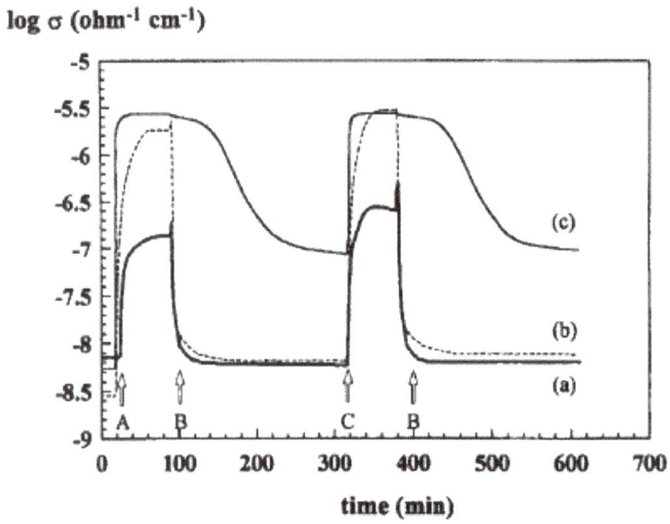

Figure 10: Variations of electrical conductivity during redox cycles [66] Copyright 2015, Elsevier BV: (**A**) at 500 °C for 14V/VMgO, (**B**) at 800 °C for 14 V/MgO and (**C**) for 45 V/MgO. (a) $C_3H_8/O_2 = 1.2/1$; (b) vacuum; and then O_2; (c) C_3.

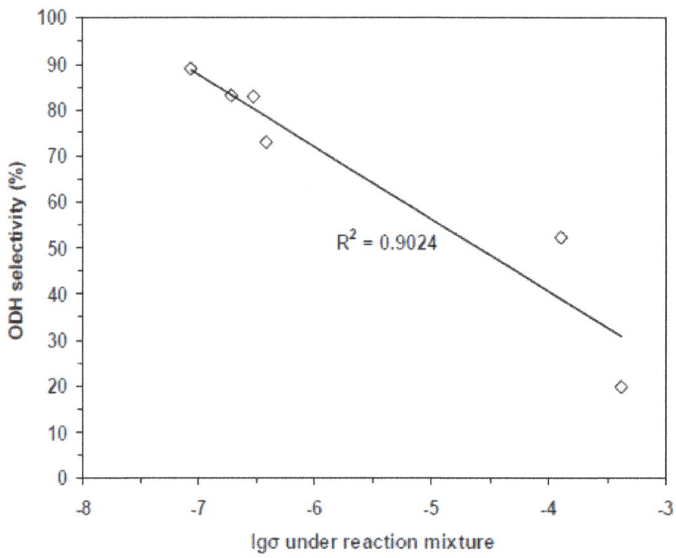

Figure 11. Variation of ethylene selectivity at ethane isoconversion at 350 °C *vs.* electrical conductivity for NiO and Nb-doped NiO catalysts (From [67] copyright 2015, RSC Great Britain).

Synergy or Cooperation Effect in Mixtures of Phases [68,69]

Synergy effects have been very often met in heterogeneous catalysis. They occur when the following conditions are met: (i) bi- or multi-phasic systems, and (ii) catalytic properties are drastically enhanced compared to that of each single component. One of the most striking examples came from the bismuth molybdate-type catalysts developed at SOHIO in the 1960s for propene oxidation/ammoxidation in acrolein/acrylonitrile [2]. It was further shown that the different phases α, β and γ are active for propene oxidation to acrolein, the α-phase being the best [70]. Moreover, industrial catalysts, designated as multicomponent catalysts, contain also other molybdates, in particular $Co(Fe)MoO_4$. Using electrical conductivity measurements and Mössbauer spectroscopy [71], it was shown that the synergy effect was in fact due to the enhanced electrical conductivity of the Co(Fe) molybdate support due to the presence of both Fe^{II} and Fe^{III} in a Co^{II} molybdate which facilitates the redox Mars and van Krevelen mechanism [72] (Figure 12). It was then suggested that intimate contact between the active Bi molybdate phase and its support $Co(Fe)MoO_4$, as illustrated in Figure 13, allows an improved redox mechanism by an easy electron transfer $Fe^{2+} \leftrightarrow e^- + Fe^{3+}$. Such synergy effect holds true for many other systems, like, for instance, V_2O_5/TiO_2-anatase or TiO_2-B for o-xylene to phthalic anhydride [73,74], or ammoxidation of toluene [75]. It was observed for many mixed metal oxides (Ti, V, Mo, Nb, Te, Fe, Co, Bi, Sb, P), whose structures allow for local elastic constraints by distortion and connection of MO_n coordinated polyhedra. The effect was interpreted as being related to phase cooperation between different phases due to a close crystallographic adaptation of lattices at their mutual interface as schematized in Figure 13, and is designated as coherent interfaces [76].

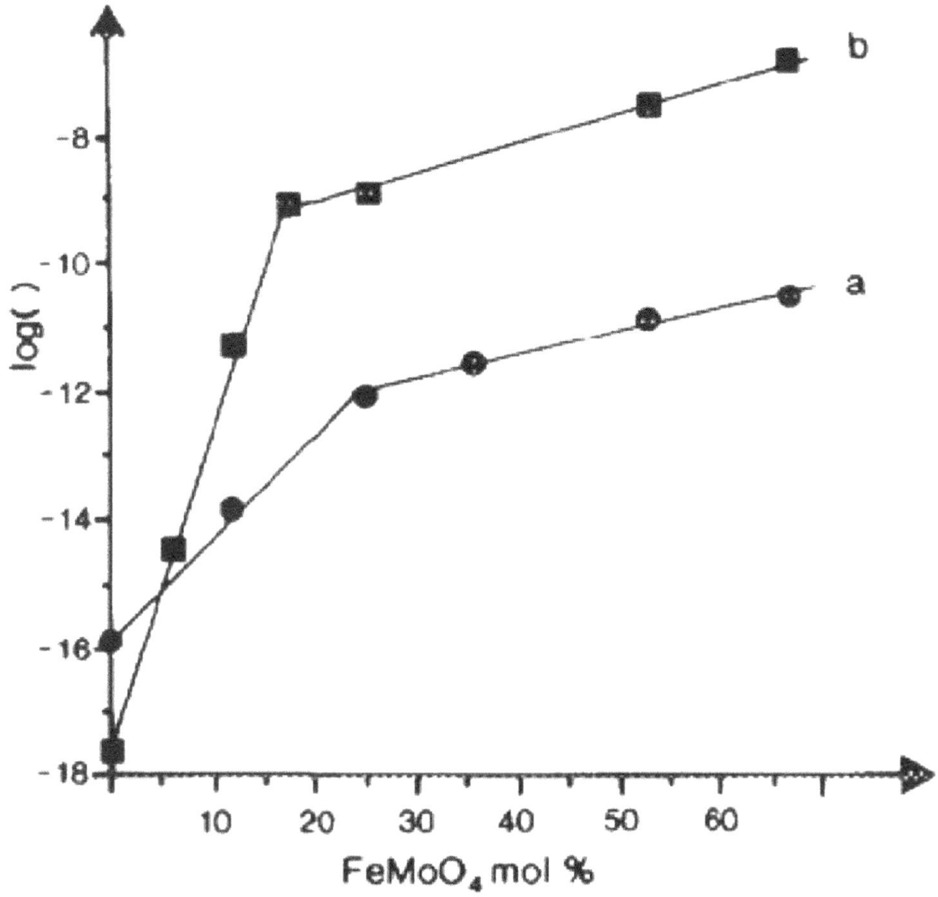

Figure 12: (a) Log rate of acrolein formation in propene oxidation on $Bi_2Mo_3O_{12}$:$Fe_xCo_{1-x}MoO_4$ (0.043:1) at 380 °C; (b) log σ of $Fe_xCo_{1-x}MoO_4$ solid solution*vs.* its iron molybdate content. Reproduced from [72] Copyright 2015, Elsevier BV.

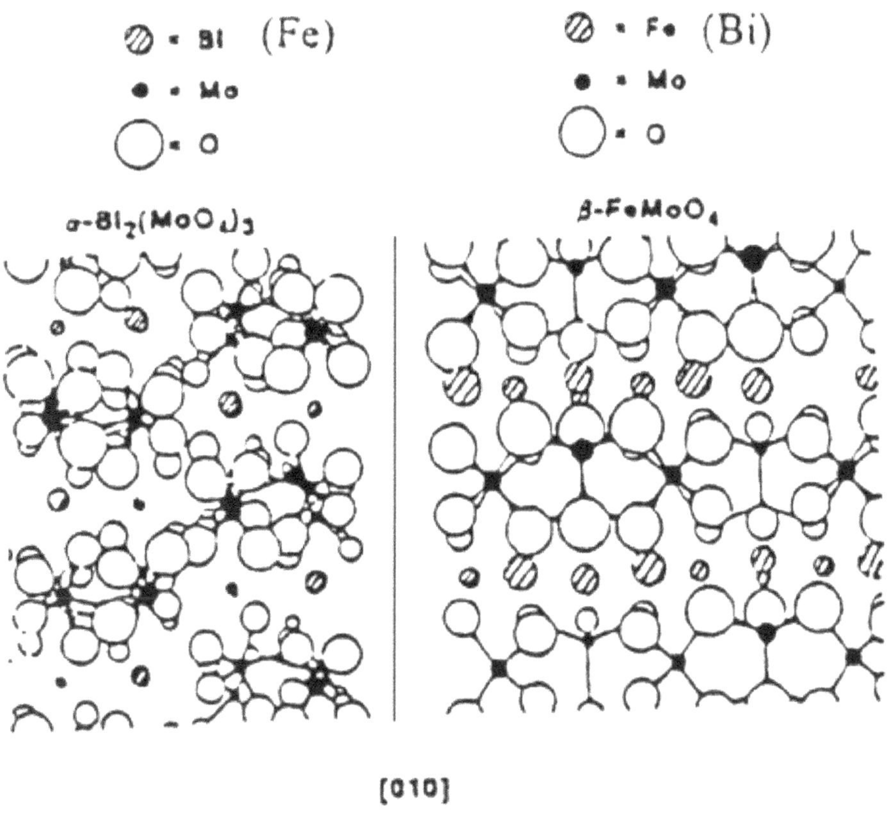

Figure 13: Epitaxial match of (010) face of $Bi_2Mo_3O_{12}$ and β-$FeMoO_4$. (From [77] copyright 2016, Elsevier BV.).

Similar cooperation effects have been observed for multicomponent antimoniates (MCA) such as $FeSb_{7.7}O_x$ or $W_{0.25}TeFe_{10}Sb_{25}O_x$ or KCsNiMgFeMnBiCrMo-O with phase cooperation $FeSbO_4 + Sb_2O_{4(5)}$ or $FeSbO_4+Sb_2O_{4(5)} + FeMoO_y$, for VPO catalysts [78] and for $Mo_{7.8}V_{1.2}NbTe_{0.94}O_{28}$(M1) + $Mo_{4.67}V_{1.33}Te_{1.82}O_{19.82}$ (M2) [79]. Delmon *et al.* [80] have given an explanation for this synergy effect, what they designated as "remote control effect", which corresponds to "a multiplying phenomenon, *i.e.*, that a spillover species creates and/or regenerates catalytic sites which work many times". In such a mechanism, one phase, designated as donor phase, dissociates oxygen to form a surface mobile species, which spills over to the other phase, which is the acceptor active phase. In a multicomponent catalyst system, "donor" and "acceptor" sites are present as it was shown for metal oxides [79] or sulfides [81] able to provide oxygen/sulfide ions that spill over from one oxide/sulfide to the other.

Recently, a new way of preparing bismuth molybdates by mild hydrothermal synthesis and new mixed Bi, Ce molybdates were found to be quite active for propylene oxidation [82] and ammoxidation [83], respectively, even more than the classical α-$Bi_2Mo_3O_{12}$ phase. The catalyst $Bi_xCe_{2-x}Mo_3O_{12}$, e.g., $BiCe_{1.5}Mo_3O_{12}$ of solid solution nature has a scheelite structure, *i.e.*, ABO_4, where A^{2+} is a large cation 8-coordinated to oxygen and B^{6+} a small cation tetrahedrally coordinated to oxygen. Substitution of a divalent cation by a trivalent one such as Bi results in a defect scheelite $A_2\square(BO_4)_3$ where $\square$ indicates a vacancy. The active site has been described as Bi cation surrounded by two Ce cations, which readily undergo $Ce^{3+} \leftrightarrow Ce^{4+}$redox phenomenon, and are located next to a vacant cation site, thus able to accommodate a lone pair of electrons.

CHANGES IN CATALYST SURFACE STRUCTURE AND CHEMICAL COMPOSITION DURING CATALYTIC REACTION

This is a general aspect of dynamic structure "which is constantly under going changes in its chemical composition and geometrical structure during reaction" and which has been known for a very long time for many reactions in heterogeneous catalysis, as reminded by Tim from BASF in his introductory lecture to 8th ICC in Berlin (p. 21 in [84]). For instance, in oxidation catalysis, it is well known that VPO catalyst necessitates many hours on stream in n-butane oxidation to maleic anhydride before optimal performance is reached, whereas excess of P at the surface *vs.* P:V = 1:1 stoichiometry was detected by LEIS and XPS techniques. This phenomenon corresponds to a steady state of the catalyst surface in redox conditions, with often demonstrates some elemental migration toward or away from the surface. This may characterise elemental segregation or the formation of a sublayer of different composition than the bulk as shown by Millet *et al.* [85] for MoVTeNb-O catalyst used for propane oxidation to acrylic acid or ammoxidation to acrylonitrile. These authors have observed [86] that the surface composition of the M1 phase differs from the bulk which implies that the catalytically active sites are no part of the M1 crystal structure but part of all terminating planes. It follows that acrylic acid formation may be related to surface depletion in Mo^{6+} and enrichment in V^{5+} sites. In the presence of steam in the feed, the active sites for acrylic acid formation consist of V^{5+} oxo-species in close vicinity to Te^{4+} sites in a Te/V ratio of 1.4. The active sites formed under propane oxidation conditions are embedded in a thin layer enriched in V, Te, and Nb on the surface of the M1 phase. It follows that structure sensitivity properties as described above have to be considered with caution.

In a study using microcalorimetry, Schlögl *et al.* have shown that on activated catalyst (calcined) the adsorption energy of propene and propane are approximately constant *vs.* coverage (57 $kJ{\cdot}mol^{-1}$) but after reaction the energy decreases and is heterogeneous in strength *vs.* coverage. The first part is thus characteristic of "single site heterogeneous catalyst" (SSHC site as defined by Thomas [87]), *i.e.*, sites of similar strength and structure and thus of similar selectivity and activity. However, under reaction conditions, the surface is modified and the sites are heterogeneous in strength (not satisfying the SSHC principle anymore) and thus heterogeneous in their activity and/or selectivity. Moreover, *in-situ* XPS analysis has shown that depletion of Mo and that V enrichment did occur as discussed in the previous paragraph. This is shown in Figure 14.

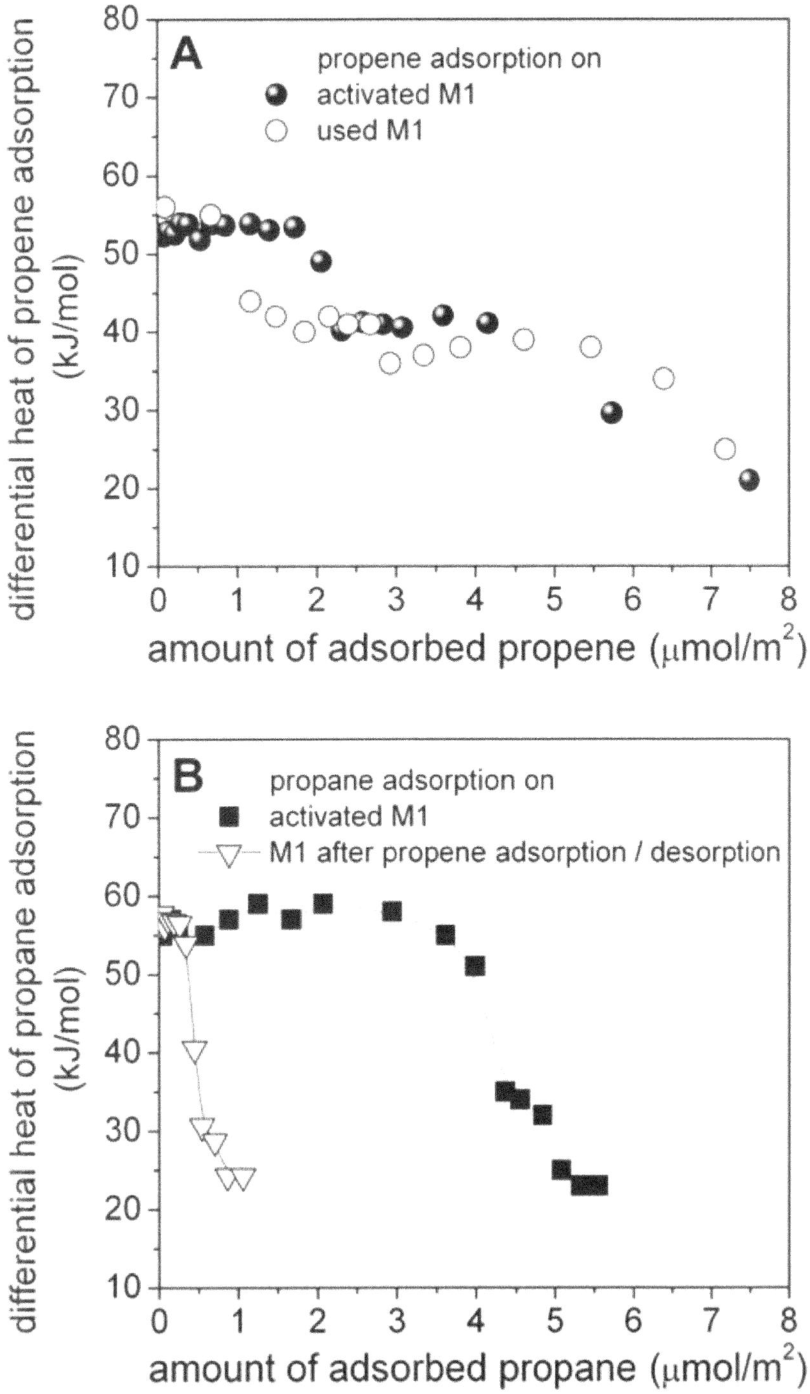

Figure 14: (A) Heat of propene adsorption on M1 MoVTeNb oxide before (filled circles) and after propane oxidation (open circles; reaction conditions of propane oxidation: 400 °C, $C_3/O_2/H_2O/N_2 = 3/6/40/51$). (B) Heat of propane adsorption on activated (fresh) M1 MoVTeNb oxide, (filled squares) and after adsorption and subsequent desorption of propene at 40 °C (open triangles). The differential heat of adsorption is plotted as a function of coverage normalized to the specific surface area (from [88] copyright 2015, Elsevier BV).

This phenomenon indicates that under steady state conditions, as usually met for partial oxidation reactions, which necessitate hours on stream before steady state is reached, surface active sites are different from the starting surface and are the best sites for the reaction in terms of selectivity [66]. It also shows that caution has to be taken when catalyst characterisation is done on the starting material and that *in-situ* characterisation, *i.e.*, under real working conditions and after reaction steady state has been reached, is necessary to better characterise the catalyst surface and the active catalytic sites.

DEHYDROGENATION, OXIDATIVE DEHYDROGENATION AND SELECTIVE OXIDATION OF LIGHT PARAFFINS

Low molecular weight alkenes, such as ethene, propene, and butenes, can be formed via non-oxidative dehydrogenation of the corresponding alkane. However, direct dehydrogenation (DH) reactions are endothermic, thermodynamically limited reactions and lead to the formation of coke deposits and of lower molecular weight alkanes, which decrease alkene yields but are commercially well established and still in use.

There are various processes for light alkanes' DH such as Catofin (Lummus), Olefex (UOP), Linde-BASF, Snamprogetti-Yarsintez, and Star (Philipps petroleum) with chromia-alumina and fixed bed, supported Pt in moving bed, chromia-alumina and fixed bed, chromia-alumina and fluid bed and supported Pt and moving bed, respectively. However, endothermicity, on one side—which thus necessitates heat supply and high coke formation, due to thermal cracking, as well as frequent regenerations—and, on the other side, thermodynamic olefin yield limitation make the process unappealing and open the way to oxidative dehydrogenation (ODH) processes [89].

Oxidative dehydrogenation (ODH) of light alkanes provides another way to obtain alkenes, as the reaction is exothermic and avoids the thermodynamic constraints of direct DH approaches by forming water as a by-product. In addition, carbon deposition during ODH is eliminated, leading to stable catalytic activity. Compared to the conventional steam cracking process of alkanes' dehydrogenation to olefins and catalytic direct DH processes, ODH can allow for a less expensive process, lowering greenhouse gas emissions and saving energy. However, the yield of alkenes obtained by ODH on most catalysts is limited by alkene oxidation to CO and CO_2. In fact, at present, no catalysts have been found to be active and selective enough to convert light alkanes to olefins by ODH to permit an industrial process to be acceptable.

Ethane, propane, butane or pentane ODH reactions are governed by the C–H bond strengths, which equal 376 kJ·mol^{-1} for C–C, 420 for C–H primary, 401 C–H secondary, 390 for C–H tertiary, 361 for C–H allyl and 445 for C–H vinyl. For instance, as ethane contains only primary C–H bonds and its dehydrogenated product ethylene contains only vinylic C–H bonds, its activation requires the highest temperature *vs.* C_3, C_4 or C_5, but the reaction is the most selective in terms of the formation of alkene. In Figure 15, we reported literature data for such ODH reactions, where it clearly appears that selectivity decreases with an increase in conversion, due to higher oxidability of the olefins *vs.* the starting alkane. This is why no new process for the production of light olefins from their corresponding alkane has been commercialised [90].

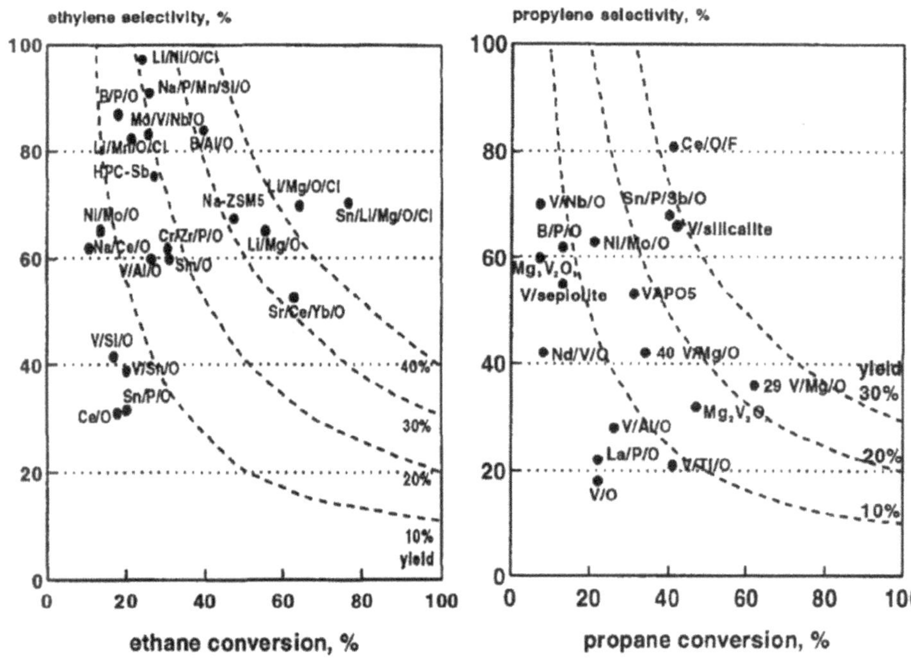

Figure 15: Best yields in olefin selectivity *vs.* conversion for ethane and propane ODH. Reproduced from [6]. Copyright 2015, Elsevier BV.

Case of Methane

Because of its natural high abundance on earth, methane oxidative transformation to C_2-hydrocarbons, methanol or formaldehyde has led to intense research efforts worldwide [91] by partial oxidation (POM) to methanol or formaldehyde (MoO_x or VO_x species) or oxidative coupling (OCM) to ethane and ethylene or by halogenation and oxihalogenation, but no industrialisation processes have been developed yet. An alternative process is the transformation of methane to syngas, a mildly exothermal reaction with $H_2/CO = 2$ as an alternative to the well-known steam reforming process. Another possibility is to transform methane to hydrogen and benzene and secondarily naphthalene, anthracene and higher aromatics on Mo/MFI or Zn-MFI catalysts at very high temperatures (>700 °C) but yield in aromatics has been very low and deactivation has been very high due to the large amount of aromatics formed [92].

Case of Ethane

Ethene is one of the most important building blocks in the chemical industry and is used to synthesize polymers. It was surprisingly shown that NiO presents high catalytic activity for ethane conversion and ethylene selectivity [93] and that the method of preparation of the catalysts governs its catalytic activity [94,95]. NiO prepared by the sol-gel method [67] starts converting ethane at ~275 °C, and the conversion increases up to 350 °C (23% conversion and 47% selectivity in ethylene), while with NiO prepared by the precipitation method [67], the yield decreases to ~8% and to 3% with the NiO prepared by a microemulsion method, *i.e.*, both conversion and selectivity in the catalytic oxidation of ethane on NiO catalyst depend on the way in which the catalyst is prepared. When Nb was deposited on NiO, the catalytic activity for ethane conversion and ethylene selectivity were improved. This phenomenon was assigned to electron transfer enhancement played by Nb, leading to easier activation of the C-H bond of ethane (see above Section 4.2 & Figure 11 from [67]). Other promoters such as Zr, which acts as a Lewis acid, but also Mg, Li, Ga, Ti, Ta,[96] or alumina have also been proposed for NiO catalyst to increase ethylene yield.

However, ethane oxidation was also found to give acetic acid in addition to ethylene. The main processes for synthesis of acetic acid include methanol carbonylation on Rh or Ir-based catalysts (originally by Monsanto, now also by Celanes, BP-Cativa, *etc.*), liquid phase oxidation of acetaldehyde, n-butane and naphta and, more recently, ethylene oxidation in gas phase (Showa Denko), butene liquid phase oxidation (Wacker) and gas phase ethane oxidation. The later reaction $C_2H_6 + (3/2)\ O_2 \rightarrow CH_3COOH + H_2O$ is only viable if ethane is easily available on the industrial platform and cheap. Different classes of catalysts have been proposed but the highest performing one is a mixed $(VNbMo)_5O_{14}$ based oxide described by Thorsteinson *et al.* [50] with a d-spacing close to 0.40 nm The best composition was claimed to be $Mo_{0.73}V_{0.18}Nb_{0.09}O_x$ with 10% ethane conversion at 286 °C and almost total selectivity to ethylene. At 200 °C, catalysts based on $(MoVNb)_5O_{14}$microdomains in MoO_3 were found to give 45% selectivity both in acetic acid and ethylene, at 25% ethane conversion. Even better yields in ethylene (75%) have been obtained at 400 °C over the same type of material further on [97]. It was suggested that ethane first adsorbs on Mo^{6+} or V^{5+} to form an ethoxide species which transforms to ethylene by β-elimination, but also possibly oxidises to acetaldehyde by α-elimination, transformed to surface acetate and then to acetic acid by its hydrolysis. MoVNb-O catalyst family thus appears as the most promising possibility at present for ethane to ethylene and acetic acid, the problem being that both products are formed, and limiting acetic acid formation, if desired, is difficult. However, when adding Pd, the selectivity to acid acetic may reach 80% [98,99], and even more when adding water in the feed [100]. The same effect holds true for propane oxidative dehydrogenation by Pt-$H_3PMo_{12}O_{40}$ [101] with O_2 and H_2 and for propane ammoxidation to acrylonitrile. High transformation to acetic acid was claimed by Saudi Basic on MoVNbP-O; MoVLaPd-O; MoVNbPd-O under a pressure of 1 to 2 MPa. The technology developed by Saudi Basic is now commercial [102], using the catalyst presenting multifunctionality properties. Ethylene as well as acetic acid are rather stable compounds and do not transform to CO_2 in the catalytic conditions chosen, with unconverted ethane and ethylene being recycled. Other systems have also been proposed such as MoVAl-O with Fe, Ga, Ti starting from the Anderson-type polyoxometallates, supported Keggin-type MoVP-O polyoxometallates and titania supported vanadium oxide or V(Mo)P-O. A recent review paper [103] summarizes the present situation.

Case of Propane

A large number of catalysts, mostly V based, have been investigated for propane ODH but propene yield varies in the 8% to 20% range. The upper yield was attained with VMgO, $Mg_2V_2O_7$ or $MgV_xSb_yO_z$, or with orthovanadate on hydrotalcite [104], while V_2O_5 supported on Al_2O_3, Bi_2O_3, La_2O_3, Nb_2O_5, SiO_2, Sm_2O_3 or TiO_2 gives inferior performance. Co and Ni molybdates were also investigated with different dopants such as Fe, Mg, Ni or P and gave yields near 15% with high productivity and no production of undesirable oxidised products. The major problem is that, in addition to alkane activation, such metal oxide catalysts may possess also O– or NH– insertion properties and thus may give undesired oxygenates or nitriles in addition to the olefin. The use of Pt based catalyst and of short contact times may be a solution to get mainly the olefin, which is known to be a reaction intermediate. At variance, one may add a catalyst known to transform the olefin to unsaturated carboxylic acids or nitriles for complete transformation of alkanes to carboxylic acids or nitriles, but both operating temperature and chemical compatibility of both catalysts need to be optimised.

For propane ammoxidation, it has been shown that the mixed oxides' surfaces are nitrided [19]. For instance, $AlO_3(NH)VO_{5.6}$ was suggested to be an active site, while insertion of nitrogen in the bulk was observed to enhance electrical conductivity and, thus, electron transfer ability in the redox mechanism.

Another aspect in ODH reaction is the nature and surface density of surface oxygen species. For instance, in the case of propane ODH, both propylene and ethylene can be formed depending on the reaction mechanism by dehydrogenation or cracking. Comparing non reducible (rare earth/alkaline earth oxides) and reducible (V based oxides) catalysts, it was shown [105] that the ratio propylene over ethylene depends on the strength and mainly the density of oxidising active sites (DOS), with high DOS favouring propylene. The oxidising sites are those active for H-atom abstraction, usually V^{5+}=O and or bridged V^{5+}–O–M (M = metal cation in the next cationic position, *i.e.*, V or another metal atom from the structure or support).

The best results obtained for ethane and propane partial oxidation or ammoxidation are given in Table 2.

Table 2: Summary of the best results in the literature for partial oxidation of C_2–C_3alkanes from [3] p. 772 and references herein

Catalyst	T °C	Conversion %	Selectivity %	Yield %
		Ethane → ethylene		
MoVTeNbO	400	85	88	75
NiNbO	400	51	90	47
VCoAPO-18	600	57	60	34
		Ethane → acetonitrile		
$NbSbO/Al_2O_3$	540	40	50	20
Co-BEA	476	47	57	26
		Ethane → acetic acid		
MoVSbNbReCaO	277	14	78	11
$MoVNbPdO_x$	300	5	82	4.1
		Propane → propylene		
VMgO	540	62	38	24
V-silicalite	550	30	70	21
		Propane → acrylonitrile		
MoVTeNbO	420	86	72	62
$VSbWO_x/SiO_2$-Al_2O_3	500	67	60	40
		Propane → acrylic acid		
MoVTeNbO	432	69	67	46
MoVTeNbO A relevant	380	80	61	49

Case of Butane

Here, ODH reaction for *n*-and *iso*-butanes to butene and isobutene on catalysts of the same type as described for ethane or propane such as VMgO or metal molybdates is examined. Another example corresponds to the catalytic conversion of *n*-butane to butenes by oxidative dehydrogenation (ODH). It is well known that supported vanadium pentoxide is a promising catalyst for the ODH of *n*-butane. MgO supported vanadium was reported as a very selective catalyst in propane and *n*-butane ODH, while V/γ-Al_2O_3 was observed to exhibit a good selectivity to ethene for ethane ODH but a poor selectivity in the ODH of *n*-butane. The acid-base character of the support was suggested to lead to this different behaviour. On MgO, a support with basic properties, the interaction between vanadium species and the support leads to the formation of vanadate compounds. In the case of more acidic supports, such as SiO_2 or Al_2O_3, a weak interaction is expected leading to less dispersed vanadium species on the surface favouring the formation of V_2O_5 crystallites.

However, the most important process concerns butane partial oxidation on $(VO)_2P_2O_7$ catalyst, designated as VPO, and which substituted the old benzene oxidation process in the 1980s. A description of active sites as ensembles of four dimers of octahedral VO_6 entities isolated by P_2O_7 groups (Section 3 and Figure 6) and surface composition during catalytic reaction (Section 5) has already been given above. In such a reaction, maleic anhydride is relatively stable to further oxidation which permits high yields. This is also why, in a moving bed technology, developed in Asturias, Spain by Du Pont Co, the catalyst, encapsulated by porous silica to limit attrition of catalyst grains, that maleic anhydride was further transformed to less stable tetrahydrofurane (THF) by reduction on Pd based catalyst (see below Section 8).

ROLE OF OTHER OXIDANTS

Instead of molecular oxygen as the oxidant, carbon dioxide can also be used as a non-traditional oxygen source or oxidant [106], for instance, according to:

$$CH_3\text{-}CH_3 + CO_2 \rightarrow CH_2 = CH_2 + CO + H_2O$$

However, in addition to ethylene, CO and water, propane and methane can be formed as a function of the catalysts used. The use of CO_2 has the advantage of producing CO and thus recovering some of the fuel capacity of the disappearing hydrogen while O_2 produces water. In addition, utilisation of CO_2 should be an alternative to their valorisation. CO_2 is less aggressive than oxygen, and it is less likely to favour secondary reactions by reacting with the ethylene produced by dehydrogenation. The amount of oxygen in the ODH feed with O_2 is limited by the danger of explosion. The use of CO_2 as an oxidant can overcome the high energy consumption for the endothermic pyrolysis and prevents deep oxidation, but unfortunately, oxidative dehydrogenation of ethane with CO_2 remains an endothermic reaction and requires addition of heat for its activation [102,107]. However, the low chemical reactivity of CO_2 leads to a difficult activation, which usually results in higher reaction temperatures. This drawback makes the use of CO_2 as a selective oxidant well suited for the activation of ethane, which also requires high temperatures. However, the first coupling of a dehydrogenation reaction with a hydrogen-consuming reaction, known as the Deacon Process [108], was performed in 1868 and can be viewed as an oxidative dehydrogenation of HCl ($2HCl + 1/2O_2 \rightarrow Cl_2 + H_2O$). The use of CO_2 to improve catalytic conversion and selectivity in methane oxidative coupling and ethane ODH was reported in 1993 [109] on Li^+/MgO catalyst. Several metal oxides have been used as catalysts [110,111], and the most effective catalysts with high ethane conversion (>60%) and ethylene selectivity (>90%) contained Cr or Ce elements.

Other oxidants instead of molecular oxygen and CO_2 have been studied, including sulfur oxide, N_2O, H_2O_2, organic peroxides, hypochlorite, bromine, iodine and their compounds. The advantage of using these oxidants is that high selectivity for dehydrogenation could be obtained. The corrosive nature of the halogen and sulfur oxide gases and the potential environmental concern over their use for N_2O or their synthesis for H_2O_2 have deferred commercialisation of these processes.

GENERAL CONCLUSIONS AND PERSPECTIVES

In this review, we have limited ourselves to studying heterogeneous partial oxidation reactions on metallic mixed oxides, and we have presented the main features and requirements for better activity and overall better selectivity.

The majority of the processes described are rather old, namely more than half a century old, although many of them have been greatly improved over the years either by adjusting their preparation and activation procedures or by chemical engineering in reactors and process improvements. For environmental reasons, some processes had to be abandoned due to the need to use environmentally friendly reactants or solvents such as sulfuric acid, cyanic acid, *etc*. Although many challenges have been solved, more or less completely, improvements are still possible. One of the most important challenges remains alkane upgrading to olefins [112] or carboxylic acids/aldehydes, in particular methane. The most researched possibility is to discover single phase structures, containing the desired catalytic functionalities, without the need to use two separate catalysts in two separate reactors. There is much room to further optimise the structure of known catalysts. Reactions such as simultaneous dehydration and selective oxidation or ammoxidation of biomass derived intermediates such as glycerin, lactic acid or 3-hydroxypropionic acid to acrylic acid or acrylonitrile, respectively, are important perspectives to follow up.

It is clear that partial oxidation solid catalysts are very complex as they must have many functionalities and behave most often as "living" or "breathing" materials at their surface, leading to element migration toward or from the surface to the bulk during catalyst activation or under reaction conditions up until reaching steady state. This migration phenomenon was observed using surface techniques such as XPS and LEIS, showing that surface structure varied from the bulk structure and was even observed to be amorphous using XRD (epitaxial close contact being maintained between surface and bulk layers). It has been shown that oxidation reactions are structure sensitive, *i.e.*, necessitate ensembles of atoms in specific structural arrangements as active sites, isolated from each other to avoid over-oxidation to CO_2. Acid-base and overall redox and electron conductivity properties are quite important to activate reactant molecules and to allow the redox mechanism to occur. Note that a too strong acidity could be deleterious and is often neutralized with low amounts of a base such as alkaline ions. Redox, a phenomenon which is crucial to permitting the Mars and van Krevelen mechanisms to function, is an important property which could be enhanced by higher electrical conductivity of the surface or by using semi-conducting materials or semi-conducting supports. Such oxidation reactions obey the new approach of nanocatalysts, with control of the size, shape and chemical composition of the catalysts as highlighted recently in a review article devoted to selectivity in heterogeneous catalysis [113].

In the present stage, the most important trends concern oxidation of ethane to ethylene or acetic acid and propane (amm)oxidation on MoVTe(Sb)Nb-O catalysts (Mo_5O_{14} based catalysts with Te, Sb or Nb incorporated into the structure, leading to M1 and M2 phases). A desirable aim/challenge is to substitute Te, which is easily reducible, volatile and poisonous, by any other similarly selective element such as Bi or Sb. Note that Bi has already been tried but without success, while Sb has interesting properties.

As we have described above, catalytic cracking of alkanes gives many by-products and is not necessarily optimised. In the case of propane dehydrogenation, propene is formed as the main product. The drawback is that the dehydrogenation equilibrium favours propene only at high temperatures or low pressure, increasing the overall cost of the process. The need for cryogenic separation of the unconverted propane and produced propene also increases the process costs. Improvements of the dehydrogenation process to make it commercially more attractive can be done by shifting the equilibrium by removal of one of the reaction products. This can be done either physically by means of a membrane, or chemically by *in-situ*catalytic oxidation using for instance a transition metal oxide. This approach has the advantage of an exothermic oxidation which facilitates the dehydrogenation. However, mixing H_2, O_2 and hydrocarbons at high temperatures creates a highly dangerous mixture and a high level of risk of explosion, which can be reduced by using oxygen-selective membranes.

Another approach is based on the reactors used for the reaction chosen. Usually, fixed beds are preferred as thermal control is easier, but considering the size of the tubes, secondary reactions may occur if the first intermediate product (e.g., the olefin) is easily converted. Moreover, safety reasons may impose a low reactant over oxygen ratio (to be out of explosion limits). This is why fluidized or moving bed reactors have been developed, as it was the case by DuPont Co in Asturias, Spain, for butane to maleic anhydride on VPO catalysts [114]. Further developments of chemical engineering and reactors remain an attractive possibility for the future, to circumvent problems in selective oxidation reactions [115]. A possibility resides in using membrane reactors to control the relative feed rate of oxygen and of the organic compound, mainly with the aim of increasing organic compound concentration (limited by explosion limits in fixed be reactors) and, thus, providing higher production rates [116]. For instance, a VPO catalyst was deposited as a thin layer on mesoporous MFI membrane [117] or packed in a tube of porous alumina. Metal oxide nanoparticles entrapped in porous materials may well be interesting for selective oxidation reactions, such as alkane upgrading ones, but up until now, and to our knowledge, no success was mentioned. A conventional fixed bed reactor could also be used with a porous metallic membrane immersed in a gas-solid fluid bed reactor [118].

REFERENCES

1. Duprez, D.; Cavani, F. *Handbook of Advanced Methods and Processes in Oxidation Catalysis*; Imperial College Press: London, UK, 2014.
2. Callahan, J.L.; Grasselli, R.K. A selectivity factor in vapor-phase hydrocarbon oxidation catalysis. *AIChE J.* 1963, *9*, 755–760.
3. Ivars, F.; López Nieto, J.M. Light alkanes oxidation: Targets and current challenges. In *Handbook of Advanced Methods and Processes in Oxidation Catalysis*; Duprez, D., Cavani, F., Eds.; Imperial College Press: London, UK, 2014; pp. 767–834.
4. Corma, A.; Garcia, H. Lewis acids as catalysts in oxidation reactions: From homogeneous to heterogeneous systems. *Chem. Rev.* 2002, *102*, 3837–3892.
5. Liu, Y.Y.; van der Voort, P. Metal-organic frameworks as selective or chiral oxidation catalysts. *Catal. Rev. Sci. Eng.*2014, *56*, 1–56.

6. Cavani, F.; Trifiro, F. The oxidative dehydrogenation of ethane and propane as an alternative way for the production of light olefins. *Catal. Today* 1995, *24*, 307–313.
7. Védrine, J.C.; Millet, J.M.M. Heteropolyoxometallates catalysts for partial oxidation. In *Metal Oxide Catalysis*; Jackson, S.D., Hargreaves, J., Eds.; Wiley-VCH Verlag: Weinheim, Germany, 2009; Volume 2, pp. 561–594.
8. Grasselli, R.K. Site isolation and phase cooperation: Two important concepts in selective oxidation catalysis: A retrospective. *Catal. Today* 2014, *238*, 10–27.
9. Che, M.; Tench, A.J. Characterisation and reactivity of mononuclear oxygen species on oxide surfaces. *Adv. Catal.* 1982, *31*, 77–133.
10. Bielanski, A.; Haber, J. Oxygen in oxidation on transition metal oxides. *Catal. Rev. Sci. Eng.* 1979, *79*, 1–31.
11. Haber, J.; Janas, J.; Schiavello, M.; Tilley, R.J.D. Tungsten oxides as catalysts in selective oxidation. *J. Catal.* 1983, *82*, 395–403.
12. Bond, G.C.; Védrine, J.C. General Conclusions. *Catal. Today* 1994, *20*, 171–178.
13. Hutchings, G.J. Vanadium phosphate: A new look at the active components of catalysts for the oxidation of butane to maleic anhydride. *J. Mater. Chem.* 2004, *14*, 3385–3395.
14. Bordes-Richard, E.; Védrine, J.C. Catalyse sélective redox. *Tech. Ingénieur* 2013, *1215*, 1–30.
15. Grasselli, R.K.; Burrington, J.D. Selective oxidation and ammoxidation of propylene by heterogeneous catalysis. *Adv. Catal.* 1980, *30*, 133–163.
16. Mars, P.; van Krevelen, D.W. Oxidations carried out by means of vanadium oxide catalysts. *Chem. Eng. Sci.* 1954, *3*, 41–49.
17. Vislovskiy, V.P.; Suleimanov, T.E.; Sinev, M.Y.; Tulenin, Yu.P.; Margolis, L.Y.; Cortés Corberán, V. On the role of heterogeneous and homogeneous processes in oxidative dehydrogenation of C_3-C_4 alkanes. *Catal. Today* 2000, *61*, 287–293.
18. Germain, J.E. *Catalytic Conversion of Hydrocarbons*; Academic Press: New York, NY, USA, 1969.
19. Bilde, J.; Janke, C.; Lorentz, C.; Delichère, P.; Popescu, I.; Marcu, I.C.; Loridant, S.; Brückner, A.; Millet, J.M.M. Molecular level insights into the structure of active sites of VAlO mixed oxides in propane ammoxidation. *J. Phys. Chem. C* 2013, *117*, 22926–22938.
20. Jussola, J.A.; Mann, R.F.; Downie, J. The kinetics of the vapor-phase oxidation of *o*-xylene over a vanadium oxide catalyst. *J. Catal.* 1970, *17*, 103–106.
21. Bordes, E. The role of structural chemistry of selective catalysts in heterogeneous mild oxidation of hydrocarbons. *C. R. Acad. Sci. Ser. IIc: Chem.* 2000, *2*, 725–733.
22. Boudart, M. Effect of surface structure on catalytic activity. In Proceedings of the 6th International Congress on Catalysis, London, UK, 12–16 July 1976; Bond, G.C., Wells, P.B., Tompkins, F.C., Eds.; The Chemical Society: London, UK, 1977; pp. 1–9.
23. Volta, J.C.; Desquesnes, W.; Moraweck, B.; Coudurier, G. A new method to obtain supported oriented oxides: MoO_3graphite catalyst for propylene oxidation to acrolein. *Kinet. Catal. Lett.* 1979, *12*, 241–245.
24. Volta, J.C.; Tatibouët, J.M.; Phichitkul, C.; Germain, J.E. Structural-sensitive catalytic oxidation on MoO_3 catalysts. In Proceedings of the 8th International Congress on Catalysis, Berlin, Germany, 2–6 July 1984; Verlag Chemie: Frankfurt am Main, Germany, 1984; Volume 4, pp. 451–458.
25. Tatibouët, J.M.; Phichitkul, C.; Germain, J.E. Structure-sensitive catalytic oxidation of butenes on molybdenum trioxide crystallites. *J. Catal.* 1986, *99*, 231–238.
26. Volta, J.C.; Portefaix, J.L. Structure sensitivity of mild oxidation reactions on oxide catalysts: A review. *Appl. Catal.* 1985, *18*, 1–32.
27. Tatibouët, J.M.; Germain, J.E.; Volta, J.C. Structure-sensitive catalytic oxidation: Alcohols on graphite-supported molybdenum trioxide. *J. Catal.* 1983, *82*, 240–248.
28. Haber, J. The concept of structure sensitivity in catalysis by oxides. *Stud. Surf. Sci. Catal.* 1989, *48*, 447–467.
29. Ziolkowski, J.; Bordes, E.; Courtine, P. Oxidation of butane and butene on the (100) face of $(VO)_2P_2O_7$: A dynamic view in terms of crystallochemical model of active sites. *J. Catal.* 1990, *122*, 126–150.
30. Ueda, W.; Oshihara, K. Selective oxidation of light alkanes over hydrothermally synthesized Mo–V–M–O (M = Al, Ga, Bi, Sb, and Te) oxide catalysts. *Appl. Catal. A Gen.* 2000, *200*, 135–143.
31. Guliants, V.V.; Brongersma, H.H.; Knoester, A.; Gaffney, A.M.; Han, S. Surface active sites present in the orthorhombic M1 phases: Low energy ion scattering study of methanol and allyl alcohol chemisorption over Mo–V–Te–Nb–O and Mo–V–O catalysts. *Top. Catal.* 2006, *38*, 41–50.
32. Celaya Sanfiz, A.; Hansen, T.W.; Sakthivel, A.; Trunschke, A.; Schlögl, R.; Knoester, A.; Brongersma, H.H.; Looi, M.H.; Hamid, S.B.A. How important is the (001) plane of M1 for selective oxidation of propane to acrylic acid? *J. Catal.* 2008,*258*, 35–43.
33. Taylor, H.S. A theory of the catalytic surface. *Proc. R. Soc. Lond.* 1925, *A108*, 105–111.
34. Védrine, J.C. Revisiting active sites in heterogeneous catalysis: Their structure and their dynamic behaviour. *Appl. Catal. A Gen.* 2014, *474*, 40–50.
35. Chizallet, C.; Costentin, G.; Che, M.; Delbecq, F.; Sautet, P. Revisiting acido-basicity of the MgO surface by periodic density functional theory calculations: role of surface topology and ion coordination on water dissociation. *J. Phys. Chem. B* 2006, *110*, 15878–15886.
36. Védrine, J.C.; Coudurier, G.; Millet, J.M.M. Molecular design of active sites in partial oxidation reactions on metallic oxides. *Catal. Today* 1997, *33*, 3–13.
37. Millet, J.M.M. FePO catalysts for the selective oxidative dehydrogenation of isobutyric acid into methacrylic acid.*Catal. Rev. Sci. Eng.* 1998, *40*, 1–38.

38. Robert, V.; Borshch, S.A.; Bigot, B. Molecular orbital model of electron delocalization in mixed-valence trimeric clusters. *J. Phys. Chem.* 1996, *100*, 580–584.

39. Haynes, A.; Maitlis, P.M.; Morris, G.E.; Sunley, G.J.; Adams, H.; Badger, P.W.; Bowers, C.M.; Cook, D.B.; Elliott, P.I.P.; Ghaffar, T.; *et al.* Promotion of Iridium-Catalyzed Methanol Carbonylation: Mechanistic Studies of the Cativa Process.*J. Am. Chem. Soc.* 2004, *126*, 2847–2861.

40. Agashar, P.A.; de Caul, L.; Grasselli, R.K. A molecular level mechanism of *n*-butane oxidation to maleic anhydride over vanadyl pyrophosphate. *Catal. Lett.* 1994, *23*, 339–346.

41. Panov, G.I.; Kharitonov, A.S.; Sobolev, V.I. Oxidative hydroxylation using dinitrogen monoxide: A possible route for organic synthesis over zeolites. *Appl. Catal. A* 1993, *98*, 1–20.

42. Janas, J.; Shishido, T.; Che, M.; Dzwigaj, S. Role of tetrahedral Co(II) sites of CoSiBEA zeolite in the selective catalytic reduction of NO: XRD, UV-vis, XAS and catalysis study. *Appl. Catal. B* 2009, *89*, 196–203.

43. Haruta, M.; Kobayashi, T.; Sano, H.; Yamada, N. Novel gold catalysts for the oxidation of carbon monoxide at a temperature far below 0 °C. *Chem. Lett.* 1987, *16*, 405–408.

44. Haruta, M.; Shen, W. Low-temperature oxidation of CO catalysed by Co_3O_4 nanorods. *Nature* 2009, *458*, 746–749.

45. Ogliaro, F.; Harris, N.; Cohen, S.; Filatov, M.; de Visser, S.P.; Saik, S. A model "rebound" mechanism of hydroxylation by cytochrome P450: Stepwise and effectively concerned pathways and their reactivity patterns. *J. Am. Chem. Soc.*2000, *122*, 8977–8989.

46. Afanasiev, P.; Kudrik, E.V.; Millet, J.M.M.; Bouchu, D.; Sorokin, A. High-valent di iron species generated from N-bridged diiron phthalocyanine and H_2O_2. *J. Chem. Soc. Dalton Trans.* 2011, *40*, 701–710.

47. Kudrik, E.V.; Afanasiev, P.; Alvarez, L.X.; Dubourdeaux, P.; Clémancey, M.; Latour, J.M.; Blondin, G.; Bouchu, D.; Albrieux, F.; Nefedov, S.E.; *et al.* An N-bridged high-valent diiron-oxo species on a porphyrin platform that can oxidize methane. *Nat. Chem.* 2012, *4*, 1024–1029.

48. Ushikubo, T.; Oshima, K.; Kayou, A.; Vaarkamp, M.; Hatano, M. Ammoxidation of propane over catalysts comprising mixed oxides of Mo and V. *J. Catal.* 1997, *169*, 394–396.

49. Oshama, K.; Kayo, A.; Umezawa, T.K.; Kiyono, K.; Sawaki, I. Synthesis of MoVTe(Sb)-O composite oxide catalyst via reduction of polyoxometalates in an aqueous medium. EP 529 853, 1992.

50. Thorsteinson, E.M.; Wilson, T.P.; Young, F.G.; Kasai, P.H. The oxidative dehydrogenation of ethane over catalysts containing mixed oxides of molybdenum and vanadium. *J. Catal.* 1978, *52*, 116–132.

51. Botella, P.E.; García-González, E.; López Nieto, J.M.; González-Calbet, J.M. MoVTeNbO multifunctional catalysts: Correlation between constituent crystalline phases and catalytic performance. *Solid State Sci.* 2005, *7*, 507–519.

52. Schlögl, R. Active sites for propane oxidation: Some generic considerations. *Top. Catal.* 2011, *54*, 627–638.

53. Grasselli, R.K.; Burrington, J.D.; Buttrey, D.J.; de Santo, P., Jr.; Lugmair, C.G.; Volpe, A.F., Jr.; Weingand, T. Multifunctionality of active centers in (amm)oxidation catalysts: From Bi–Mo–O_x to Mo–V–Nb–(Te, Sb)–O_x. *Top. Catal.*2003, *23*, 5–22.

54. Védrine, J.C. Eurocat oxide V_2O_5/TiO_2. *Catal Today* 1994, *20*, 1–178.

55. Védrine, J.C. The role of redox, acid-base and collective properties and of crystalline state of heterogeneous catalysts in the selective oxidation of hydrocarbons. *Top. Catal.* 2002, *21*, 97–106.

56. Tielens, F.; Dzwigaj, S. Probing acid-base sites in vanadium redox zeolites by DFT calculation and compared with FTIR results. *Catal. Today* 2010, *152*, 66–69.

57. Védrine, J.C. Acid-base characterization of heterogeneous catalysts: An up-to-date overview. *Res. Chem. Intermed.*2015.

58. Noller, H.; Tomke, K. Transition states of catalytic dehydration and dehydrogenation of alcohols. *J. Mol. Catal.* 1979, *6*, 375–392.

59. Lauron-Pernot, H. Evaluation of surface acido-basic properties of inorganic-based solids by model alcohol catalytic reaction networks. *Catal. Rev. Sci. Eng.* 2006, *348*, 315–361.

60. Baca, M.; Pigamo, A.; Dubois, J.L.; Millet, J.M.M. Fourier transform infrared spectroscopic study of surface acidity by pyridine adsorption on the M1 active phase of the MoVTe(Sb)NbO catalysts used in propane oxidation. *Catal. Commun.* 2005, *6*, 215–220.

61. Ivars-Barceló, F.; Millet, J.M.M.; Blasco, T.; Concepción, P.; Valente, J.S.; López Nieto, J.M. Understanding effects of activation-treatments in K-free and K-MoVSbO bronze catalysts for propane partial oxidation. *Catal. Today* 2014, *238*, 41–48.

62. Grasselli, R.K.; Lugmair, C.G.; Volpe, A.E.; Andersson, A.; Burrington, J.D. Enhancement of acrylic acid yields in propane and propylene oxidation by selective P doping of MoV(Nb)TeO-based M1 and M2 catalysts. *Catal. Today*2010, *157*, 33–38.

63. Jo, B.Y.; Kum, S.S.; Moon, S.H. Performance of WO_x-added Mo–V–Te–Nb–O catalysts in the partial oxidation of propane to acrylic acid. *Appl. Catal. A* 2010, *378*, 76–82.

64. Andrushkevich, T.V.; Popova, G.Y.; Chesalov, Y.A.; Ischenko, E.V.; Khramov, M.I.; Kaichev, V.V. Propane ammoxidation on Bi promoted $MoVTeNbO_x$ oxide catalysts: Effect of mixture composition. *Appl. Catal. A* 2015, *506*, 109–117.

65. Novakova, E.N.; Védrine, J.C. Propane selective oxidation to propene and oxygenates on metal oxides. In *Metal Oxides: Chemistry and Applications*; Fierro, J.L.G., Ed.; CRC: Boca Raton, FL, USA; Taylor & Francis: Baton Rouge, LA, USA, 2006; pp. 444–446.

66. Pantazidis, A.; Burrows, A.; Kiely, C.J.; Mirodatos, C. Direct evidence of active surface reconstruction during

oxidative dehydrogenation of propane over VMgO catalyst. *J. Catal.* 1998, *177*, 325–334.

67. Popescu, I.; Skoufa, Z.; Heracleous, E.; Lemonidou, A.; Marcu, I.C. A study of electrical conductivity measurements of the semi conductive and redox properties of Nb-doped NiO catalysts in correlation with the oxidative dehydrogenation of ethane. *Phys. Chem. Chem. Phys.* 2015, *17*, 8138–8147.

68. Courtine, P.; Bordes, E. Synergistic effects in multicomponent catalysts for selective oxidation. *Stud. Surf. Sci. Catal.*1997, *110*, 177–184.

69. Bordes, E. Synergistic effects in selective oxidation catalysis: Does phase cooperation result in site isolation? *Top. Catal.*2001, *15*, 131–137.

70. Carson, D.; Coudurier, G.; Védrine, J.C.; Laarif, A.; Theobald, F. Synergy effects in the catalytic properties of Bi molybdate. *J. Chem. Soc. Faraday Trans. I* 1983, *79*, 1921–1929.

71. Ponceblanc, H.; Millet, J.M.M.; Coudurier, G.; Herrmann, J.M.; Védrine, J.C. Study of multiphasic molybdates based catalysts. Part I—Electrical conductivity study of valence states and solubility limits in mixed iron and cobalt molybdates. *J. Catal.* 1993, *142*, 373–380.

72. Millet, J.M.M.; Ponceblanc, H.; Coudurier, G.; Herrmann, J.M.; Védrine, J.C. Study of multiphasic molybdates based catalysts. Part II—Synergy effect between bismuth and mixed iron and cobalt molybdates in mild oxidation of propene. *J. Catal.* 1993, *142*, 381–391.

73. Wainwright, M.S.; Foster, N.R. Catalysts, Kinetics and Reactor Design in Phthalic Anhydride Synthesis. *Catal. Rev. Sci. Eng.* 1979, *19*, 211–292.

74. Papachryssanthou, J.; Bordes, E.; Véjux, A.; Courtine, P.; Marchand, R.; Tournoux, M. TiO_2(B), a new support for V_2O_5in the oxidation of O-xylene. *Catal. Today* 1987, *1*, 219–227.

75. Sanati, M.; Wallenberg, R.; Andersson, A.; Jasen, S.; To, Y. Vanadia catalysts on anatase, rutile, and TiO_2(B) for the ammoxidation of toluene: An ESR and high-resolution electron microscopy characterization. *J. Catal.* 1991, *132*, 128–144.

76. Véjux, A.; Courtine, P. Interfacial reactions between V_2O_5 and TiO_2 (anatase): Role of the structural properties. *J. Solid State Chem.* 1978, *23*, 93–103.

77. Grasselli, R.K. Advances and future trends in selective oxidation and ammoxidation catalysis. *Catal. Today* 1999, *49*, 141–154.

78. Volta, J.C. Vanadium phosphorus oxides, a reference catalyst for mild oxidation of light alkanes. *C.R. Acad. Sci. IIc: Chem.* 2000, *3*, 717–723.

79. Korovchenko, P.; Shiju, N.R.; Dozier, A.K.; Graham, U.M.; Guerrero-Pérez, M.O.; Guliants, V.V. M1 to M2 phase transformation and phase cooperation in bulk mixed metal Mo–V–M–O (M = Te,Nb) catalysts for selective ammoxidation of propane. *Top. Catal.* 2007, *50*, 43–51.

80. Weng, L.T.; Delmon, B. Phase cooperation and remote control effects in selective oxidation catalysts. *Appl. Catal. A*1992, *81*, 141–213.

81. Karroua, M.; Grange, P.; Delmon, B. Existence of synergy between "CoMoS" and Co_9S_8: New proof of remote control in hydrodesulfurization. *Appl. Catal.* 1989, *50*, L5–L10.

82. Schuh, K.; Kleist, W.; Høj, M.; Trouillet, V.; Beato, P.; Degn Jensen, A.; Grunwaldt, J.D. Bismuth molybdate catalysts prepared by mild hydrothermal synthesis: Influence of pH on the selective oxidation of propylene. *Catalysts* 2015, *5*, 1554–1573.

83. Brazdil, J.F.; Toft, M.A.; Lin, S.Y.; McKenna, S.T.; Zajac, G.; Kaduk, J.; Golab, T. Characterization of bismuth-cerium-molybdate selective propylene ammoxidation catalysts. *Appl. Catal. A* 2015, *495*, 115–123.

84. Timm, B. The ammonia synthesis and heterogeneous catalysis. A historical review. In Proceedings of the 8th International Congress on Catalysis, Berlin, Germany; 1984; Volume 1, pp. 7–30.

85. Mehlomakulu, B.; Nguyen, T.T.N.; Delichère, P.; van Teen, E.; Millet, J.M.M. Identification of the active species in oxidation reactions on mixed oxide catalysts: Supra-surface or bulk-surface species. *J. Catal.* 2012, *289*, 1–10.

86. Schlögl, R. Concepts in selective oxidation of small alkane molecules. In *Modern Heterogeneous Oxidation Catalysis*; Mizuno, N., Ed.; Wiley-VCH: Weinheim, Germany, 2009; pp. 1–42.

87. Thomas, J.M. The societal significance of catalysis and the growing practical importance of single-site heterogeneous catalysts. *Proc. Roy. Soc. A* 2012, *468*, 1884–1903.

88. Hävecker, M.; Wrabetz, S.; Kröhnert, J.; Csepei, L.I.; D'Alnoncourt, R.N.; Kolen'ko, Y.V.; Girgsdies, F.; Schlögl, R.; Trunschke, A. Surface chemistry of phase-pure M1 MoVTeNb oxide during operation in selective oxidation of propane to acrylic acid. *J. Catal.* 2012, *285*, 48–60.

89. Kung, H. Oxidative dehydrogenation of light C_2 to C_4 alkanes. *Adv. Catal.* 1994, *40*, 1–38.

90. Blasco, T.; López-Nieto, J.M. Oxidative dehydrogenation of short chain alkanes on supported vanadium oxide catalysts. *Appl. Catal. A* 1997, *157*, 117–142.

91. Tabaka, K.; Teng, Y.; Takemoto, T.; Suzuki, E.; Bañares, M.A.; Peña, M.A.; Fierro, J.L.G. Activation of methane by oxygen and nitrogen oxides. *Catal. Rev.* 2002, *44*, 1–58.

92. Wang, L.H.; Tao, L.X.; Xie, M.S.; Xu, G.F.; Huang, J.H.; Xu, Y. Dehydrogenation and aromatization of methane under non-oxidizing conditions. *Catal. Lett.* 1993, *21*, 35–41.

93. Heracleous, E.; Lemonidou, A. Ni-Nb-O mixed oxides as highly active and selective catalysts for ethane production via ethane oxidative dehydrogenation. Part I: Characterization and catalytic performance. *J. Catal.* 2006, *237*, 162–174.

94. Wu, Y.; Gao, J.; He, Y.; Wu, T. Preparation and characterization of Ni–Zr–O nanoparticles and its catalytic behavior for ethane oxidative dehydrogenation. *Appl. Surf. Sci.* 2012, *258*, 4922–4928.

95. Wu, Y.; He, Y.; Wu, T.; Weng, W.; Wan, H. Effect of synthesis method on the physical and catalytic property of nanosized NiO. *Mater. Lett.* 2007, *61*, 2679–2682.

96. Popescu, I.; Heracleous, E.; Skoufa, Z.; Lemonidou, A.; Marcu, I.C. Study of electrical conductivity measurements of semiconductive and redox properties of M-doped NiO (M = Li, Mg, Al, Ga, Ti, Nb) catalysts for the oxidative dehydrogenation of ethane. *Phys. Chem. Chem. Phys.* 2014, *16*, 4962–4970.

97. López-Nieto, J.M.; Botella, P.; Vázquez, M.L.; Dejoz, A. The selective oxidative dehydrogenation of ethane over hydrothermally synthesised MoVTeNb catalysts. *Chem. Commun.* 2002, 1906–1907.

98. Soliman, M.; Al-Zeghayer, Y.; Al-Awadi, A.S.; Al-Mayman, S. Economics of acetic acid production by partial oxidation of ethane. *APCBEE Procedia* 2012, *3*, 200–208.

99. Linke, D.; Wolf, D.; Zey, B.S.; Baerns, M.; Timpe, O.; Schlögl, R.; Zeyβ, S.; Dingerdissen, U. Catalytic partial oxidation of ethane to acetic acid over $Mo_1V_{0.25}Nb_{0.12}Pd_{0.0005}O_x$, I: Catalyst performance and reaction mechanism. *J. Catal.*2002, *205*, 16–31.

100. Rahman, F.; Loughlin, K.F.; Al-Saleh, M.A.; Saeed, M.R.; Tukur, N.M.; Hossain, M.M.; Karim, K.; Mamedov, A. Kinetics and mechanism of partial oxidation of ethane to ethylene and acetic acid over MoV type catalysts. *Appl. Catal. A* 2010, *375*, 17–25.

101. Kutnetsova, N.I.; Popova, G.Y.; Kuznetsova, L.I.; Zaikovskii, V.I.; Koscheev, S.V.; Andrushkevich, T.V.; Lisitsyn, A.S.; Likholobov, V.A.; Han, S. Improving the performance of $Pt-H_3PMo_{12}O_{40}$ catalysts in the selective dehydrogenation of propane with O_2 and H_2. *Catal. Today* 2015, *245*, 179–185.

102. Mamedov, E.; Karim, K. Selective oxidation at SABIC: Innovative catalysts and technologies. In *Handbook of Advanced Methods and Processes in Oxidation Catalysis*; Duprez, D., Cavani, F., Eds.; Imperial College Press: London, UK, 2014; pp. 291–301.

103. Gärtner, C.A.; van Veen, A.C.; Lercher, J.A. Oxidative dehydrogenation of ethane: Common principles and mechanistic aspects. *ChemCatChem* 2013, *5*, 3196–3217.

104. Valverde, J.A.; Echavarria, A.; Eon, J.G.; Faro, A.C.; Palacio, L.A. V-Mg-Al catalyst from hydrotalcite for oxidative dehydrogenation of propane. *React. Kinet. Mech. Catal.* 2014, *11*, 679–696.

105. Kondratenko, E.V.; Sinev, M.Y. Effect of nature and surface density of oxygen species on product distribution in the oxidative dehydrogenation of propane over oxide catalysts. *Appl. Catal. A* 2007, *325*, 353–361.

106. Valenzuela, R.X.; Bueno, G.; Cortés Corberán, V.; Xu, Y.; Chen, C.G. Selective oxidehydrogenation of ethane with CO_2over CeO_2-based catalysts. *Catal. Today* 2000, *61*, 43–48.

107. Cortés Corberán, V. Novel approaches for the improvement of selectivity in the oxidative activation of light alkanes.*Catal. Today* 2005, *99*, 33–41.

108. Deacon, H. Improvement in the manufacture of chlorine. US Patent 85370A, 29 December 1868.

109. Wang, D.J.; Xu, M.T.; Shi, C.L.; Lunsford, J.H. Effect of carbon dioxide on the selectivities obtained during the partial oxidation of methane and ethane over Li^+/MgO catalysts. *Catal. Lett.* 1993, *18*, 323–328.

110. Wang, S.; Murata, K.; Hayakawa, T.; Hamakawa, S.; Suzuki, K. Dehydrogenation of ethane with carbon dioxide over supported chromium oxide catalysts. *Appl. Catal A* 2000, *196*, 1–8.

111. Xu, L.Y.; Liu, J.X.; Yang, H.; Xu, Y.; Wang, Q.X.; Lin, L.W. Regeneration behaviors of Fe/Si-2 and Fe–Mn/Si-2 catalysts for C_2H_6 dehydrogenation with CO_2 to C_2H_4. *Catal. Lett.* 1999, *62*, 185–189.

112. Cavani, F.; Ballarini, N.; Cericola, A. Oxidative dehydrogenation of ethane and propane: How far from commercial implementation? *Catal. Today* 2007, *127*, 113–131.

113. Somorjai, G.A.; Park, J.Y. Molecular factor of catalytic selectivity. *Angew. Chem. Int. Ed.* 2008, *47*, 9212–9228.

114. Contractor, R.M.; Garnett, D.I.; Horowitz, H.S.; Bergna, H.E.; Patience, G.S. A new commercial scale process for n-butane oxidation to maleic anhydride using a circulating fluidized bed reactor. *Stud. Surf. Sci. Catal.* 1994, *82*, 233–242.

115. Groppi, G.; Beretta, A.; Troconi, E. Structured catalytic reactors for selective oxidations. In *Handbook of Advanced Methods and Processes in Oxidation Catalysis*; Duprez, D., Cavani, F., Eds.; Imperial College Press: London, UK, 2014; pp. 943–997.

116. Menéndez, M. Membrane reactors as tools for improved catalytic oxidation process. In *Handbook of Advanced Methods and Processes in Oxidation Catalysis*; Duprez, D., Cavani, F., Eds.; Imperial College Press: London, UK, 2014; pp. 921–942.

117. Mallada, R.; Menéndez, M.; Santamaria, J. Use of membrane reactors for the oxidation of butane to maleic anhydride under high butane concentration. *Catal. Today* 2000, *56*, 191–197.

118. Bordes-Richard, E.; Shekari, A.; Patience, G.S. Vanadium phosphorus oxide catalyst for n-butane selective oxidation: From catalyst synthesis to the industrial process. In *Handbook of Advanced Methods and Processes in Oxidation Catalysis*; Duprez, D., Cavani, F., Eds.; Imperial College Press: London, UK, 2014; pp. 549–585.

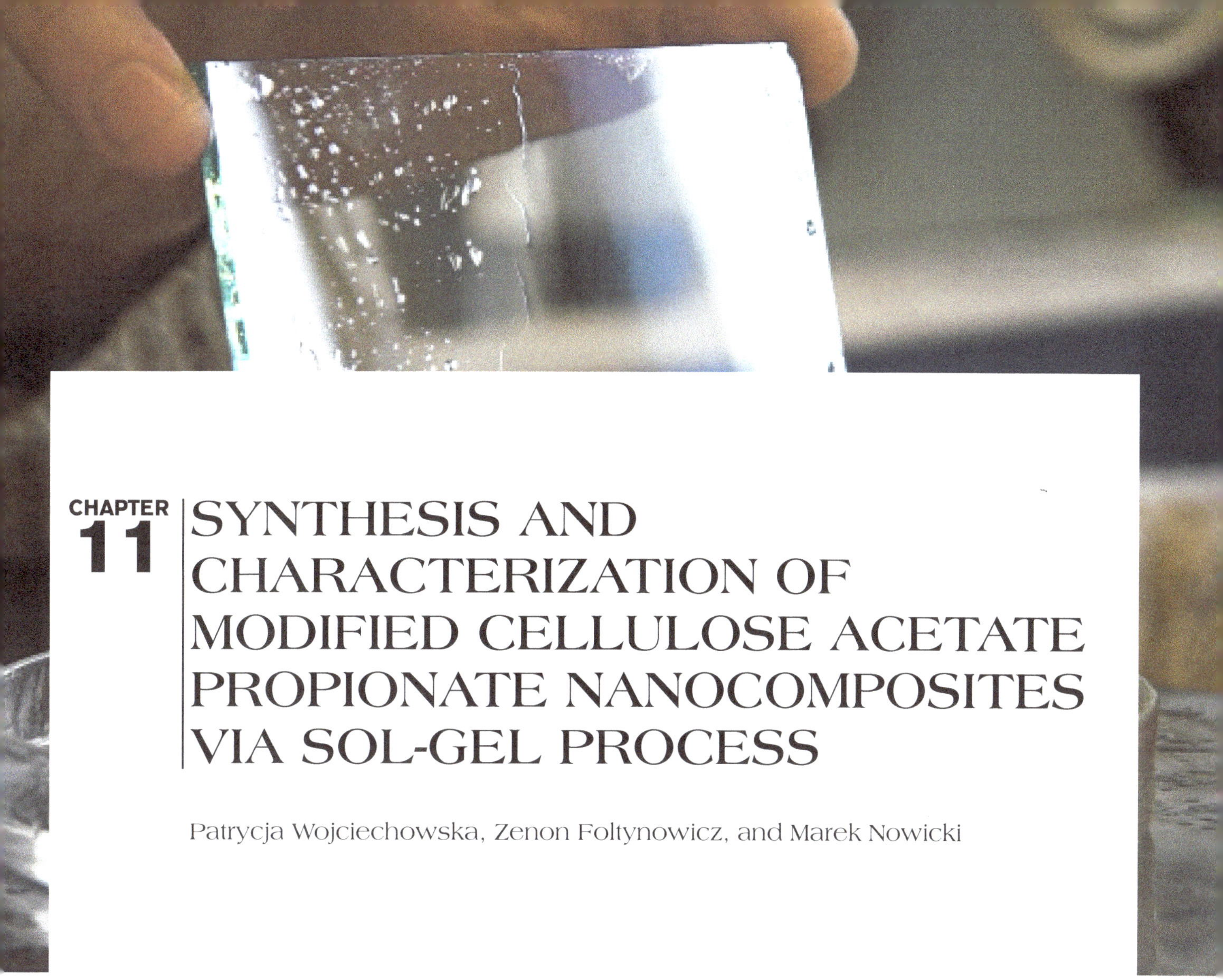

CHAPTER 11 SYNTHESIS AND CHARACTERIZATION OF MODIFIED CELLULOSE ACETATE PROPIONATE NANOCOMPOSITES VIA SOL-GEL PROCESS

Patrycja Wojciechowska, Zenon Foltynowicz, and Marek Nowicki

INTRODUCTION

Organic-inorganic polymer hybrids are rapidly expanding area of research. Such materials offer the opportunity to combine the desirable properties of organic polymers (toughness, elasticity, and formability) with those of inorganic solids (hardness, chemical resistance, and strength). Sol-gel techniques generating nanosized silica particles within a polymer matrix have been widely used for the preparation of organic-inorganic hybrid nanocomposites. The advantage of sol-gel route is the ability to prepare organic-inorganic nanostructured materials at ambient temperature and the possibility to control morphology of the growing inorganic phase by the subtle control of various reaction conditions. Sol-gel technique involves hydrolysis of the precursor (metal alkoxide) followed by condensation reactions of the resulting hydroxyl groups. Hydrolysis of metal alkoxide and condensation of the hydroxyl groups and residual alkoxyl groups occur simultaneously. Once the hydrolysis reaction has been initiated, condensation takes place parallel to hydrolysis and occurs even if not all of the alkoxyl groups are hydrolyzed. Considering the nature of the interface between the organic and inorganic phases, hybrid materials can be categorized into two different classes. The first class corresponds to not covalently bound networks of inorganic and organic phases. These hybrids show weak interactions between the polymer matrix and inorganic phase, such as van der Waals, hydrogen bonding, or weak electrostatic interactions, and can be prepared by physical mixing of an organic polymer with a metal alkoxide. In the second class organic and inorganic phases are linked through strong chemical bonds (covalent or ionic). The use of a suitable coupling agent permits an interconnected network to be obtained. In the second class of the hybrid materials one of the most often employed approaches is to use precursors containing organic functional groups capable of polymerization as well as groups which will form the inorganic network [1–3].

Biobased polymers from renewable resources are attracting interest as a substitute for petrochemical based polymers, predominantly due to environmental concerns (growing problem of plastic waste disposal, increased price of crude oil, and global warming) and secondly because of the realization that our petroleum resources are finite. Cellulose esters such as cellulose acetate (CA), cellulose acetate propionate (CAP), and cellulose acetate butyrate (CAB) are thermoplastic materials produced through the esterification of the most abundant natural polymer on earth: cellulose. Cellulose plastics are interesting materials for nanocomposites preparation due to their good mechanical properties and processability.

Citation: Patrycja Wojciechowska, Zenon Foltynowicz, and Marek Nowicki, "Synthesis and Characterization of Modified Cellulose Acetate Propionate Nanocomposites via Sol-Gel Process," Journal of Spectroscopy, vol. 2013, Article ID 616159, 8 pages, 2013. doi:10.1155/2013/616159

Numerous polymer nanocomposites based on cellulose derivatives have been developed in the past few years [4–10]. In our previous work we reported the preparation of organic-inorganic hybrids of the first class (not covalently bound networks of inorganic and organic phases) based on cellulose acetate butyrate (CAB) [11]. In this paper, novel modified cellulose acetate propionate/silica (MCAP/SiO_2) nanocomposites were synthesized by sol-gel process applying TEOS as a precursor and ICPTES as a couple agent resulting in the formation of a covalent bound between inorganic phase and polymer matrix. The chemical structure and morphological properties of hybrid nanocomposites are discussed.

EXPERIMENTAL

Materials and Reagents

Cellulose acetate propionate (CAP, M_n=75000), 3-isocyanatopropyltriethoxysilane (ICPTES, 95%), dibutyltin dilaurate (95%), and tetraethoxysilane (TEOS, 98%) were purchased from Sigma-Aldrich. Dioctyl phthalate (DOP), acetone, tetrahydrofuran (THF), diethyl ether, and hydrochloric acid (HCl, 35%–38%) were purchased from POCH S.A and used as received.

Modification of Cellulose Acetate Propionate

Cellulose acetate propionate (CAP) was modified through the reaction of 3-isocyanatopropyltriethoxysilane (ICPTES) couple agent with free hydroxyl on the CAP backbone (Scheme 1). The procedure was modeled on the ones reported by Jo and Park [12], and Messori et al. [13]. Functionalization of CAP was performed at a three-necked round-bottom flask equipped with a magnetic stir bar, thermometer, and an argon inlet. In this flask 13.0 g of CAP was dissolved in 300 mL of THF and stirred until a homogenous solution was obtained. To this mixture 1.24 g of ICPTES in a 10 : 1 polymer/silane molar ratio and 0.085 g of dibutyltin dilaurate in a 35 : 1 polymer/catalyst molar ratio were added. For this study one substituent was grafted per ten CAP repeat units (assuming CAP monomeric molecular weight of 260). The solution was kept under stirring for several hours at 60°C until the reaction was complete. The reaction progress was monitored by infrared spectroscopy of the polymer solution at various reaction times. When the reaction was complete, modified polymer (MCAP) was purified by precipitation in diethyl ether and afterwards dried for 24 hours and kept under argon for the preparation of MCAP nanocomposites.

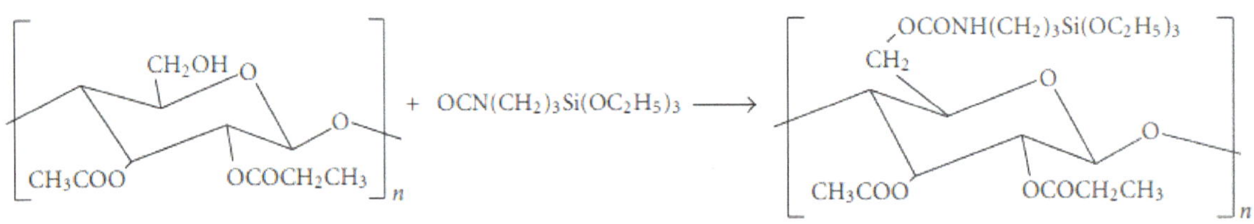

Scheme 1: Chemical modification of CAP.

Preparation of MCAP Nanocomposites

A detailed procedure of organic-inorganic hybrids preparation was described in patent [14]. Sample compositions and characteristics are listed in Table 1. Due to the high brittleness of neat CAP, in order to avoid cracking of the samples, nanocomposites were prepared with 35% of plasticizer (DOP). A typical preparation of organic-inorganic hybrid was as follows: polymer was placed in a polyethylene beaker and dissolved in acetone. Then DOP and TEOS were added and stirred vigorously until a homogenous solution was obtained. To this mixture catalytic amount of HCl (0.1 M) was added to initiate the sol-gel process and mixed until it appeared clear and homogenous. The solution was cast in an evaporating PTFE dish and after the solvent removal dried in a vacuum drier at 40°C for 12 hours to ensure the complete solvent evaporation. For comparison, a reference sample of neat CAP containing 35% of plasticizer was also prepared. Figure 1 shows photographs of unmodified CAP and MCAP nanocomposites with thickness in the range of 200–215 µm. The films' thicknesses were measured using a Sylvac s229 digital indicator equipped with a 10 mm diameter flat point, with a resolution of 0.001 mm.

Table 1: Sample compositions and characteristics.

Sample	Polymer : TEOS (wt. ratio)	TGA SiO_2 wt%	Nature of the sample
CAP	—	—	transparent
MCAP	—	0.6*	translucent
MCAP6.25	93.75/6.25	3	translucent
MCAP12.5	87.5/12.5	4.8	translucent

*SiO_2 from ICPTES.

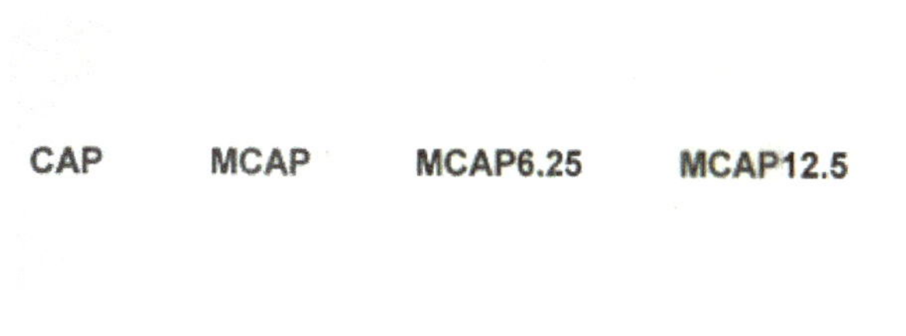

Figure 1: Photographs of the samples: CAP, MCAP, MCAP6.25, and MCAP12.5.

Sample Characterization

IR spectra of the samples were measured with a double-beam spectrophotometer Specord M80 (Carl Zeiss Jena) over a range of 400–4000 cm^{-1}. The ^{29}Si solid-state NMR was performed using a Bruker DSX 300 Avance spectrometer. The spectra of the samples were obtained using cross-polarization and magic-angle spinning (CP/MAS). The spinning rate used was 8 kHz. Standard Bruker 4 mm CPMAS probe was used. Scanning electron microscopy (SEM) analysis was carried out with a Zeiss Evo 40 instrument. Prior to the examination samples were freeze-fractured and gold-coated. SEM images were taken at the magnification 5000 and 10000. AFM imaging was conducted using easyScan 2 atomic force microscopy (AFM) from Nanosurf. Images were recorded in the tapping mode using PPP-NCLR silicon probes (NanoAndMore GmbH). AFM images were collected using Nanosurf Easyscan software and analyzed with Nanotec WSxM [15] computer program. Prior to the AFM analysis samples were twice purified in absolute ethanol using ultrasonic cleaner and mounted onto stainless steel disk with epoxy resin glue. All experiments were carried out at room temperature and ambient humidity. Height, amplitude, and phase images were collected at various scan sizes; maximum x- and y-axis range was 10 micrometers. Thermogravimetric analysis (TGA) was carried out using a Shimadzu TGA-50 thermal analyzer. The sample was heated from room temperature to 900°C at 10°C/min under the air atmosphere.

RESULTS AND DISCUSSION

Modification of Cellulose Acetate Propionate

Figure 2 depicts the IR spectra recorded at different reaction times during the synthesis of modified cellulose acetate propionate (MCAP). The progress of the reaction was followed by the decreasing till disappearing of the absorption band related to isocyanate groups at 2272 cm^{-1} [13]. Under experimental conditions used in this study the modification of CAP was complete within 6 hours. Modified polymer was purified by precipitation in diethyl ether in order to remove the catalyst, soluble in this solvent.

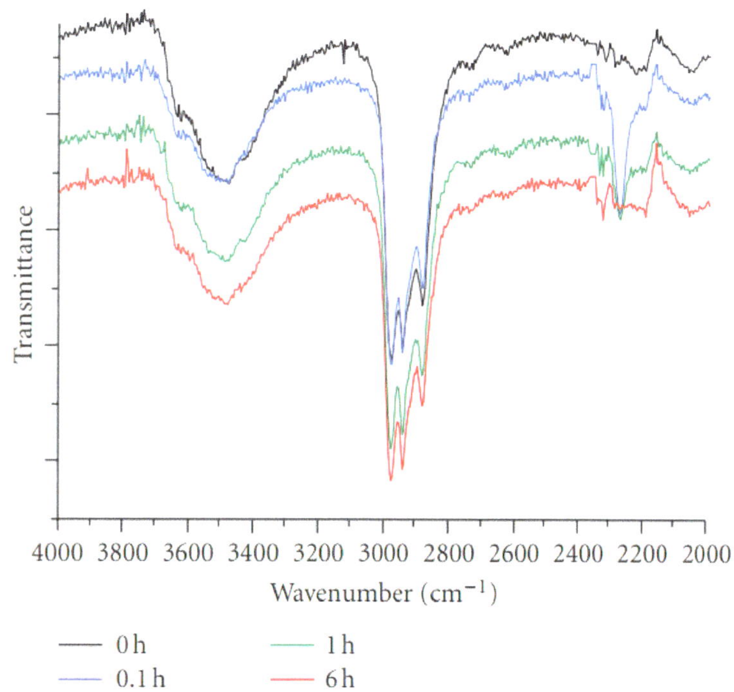

Figure 2: IR spectra of the polymer solution during reaction of free hydroxyls on the polymer with ICPTES for different reaction times, showing the disappearance of isocyanate linkage.

Chemical Structure of Organic-Inorganic Hybrid Nanocomposites

IR spectroscopy and ^{29}Si NMR were used to study the chemical structure of the synthesized nanocomposites. Figure 3 shows IR spectra of neat polymer (CAP) and nanocomposite (MCAP12.5). The bands around 1000–1100 cm^{-1}, 800 cm^{-1}, and 460 cm^{-1} that appeared in the nanocomposite spectrum indicate that silica is formed via the sol-gel reaction. The peak at 1000–1100 cm^{-1} can be assigned to asymmetric stretching vibration of the Si–O–Si bonds of the inorganic component. The band at 800 cm^{-1} is due to symmetric stretching vibration of the Si–O–Si atoms. The presence of the band at 460 cm^{-1} reveals the existence of Si–O–Si bending mode.

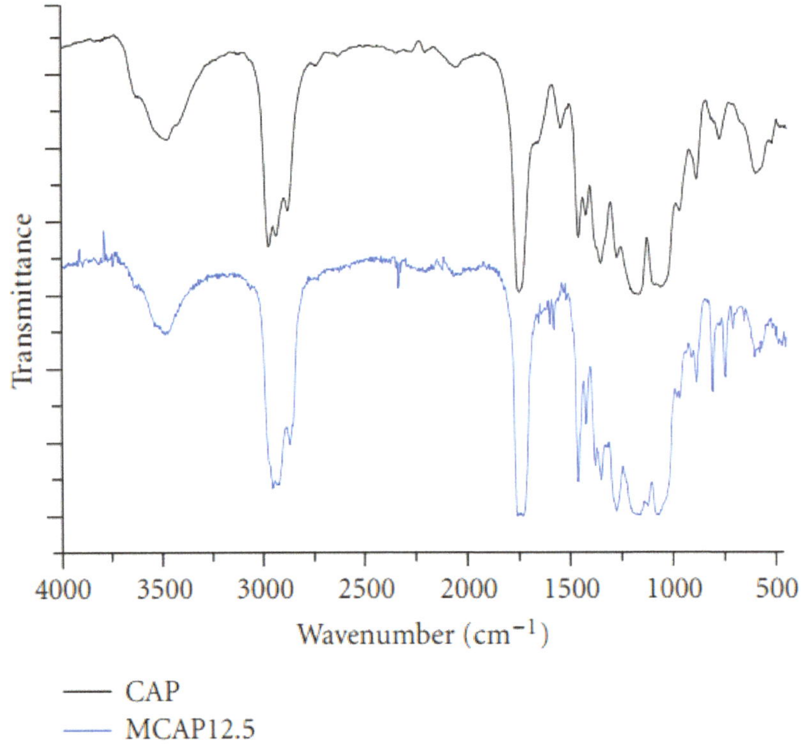

Figure 3: IR spectra of CAP, MCAP12.5.

The ^{29}Si solid state NMR spectrum of the MCAP12.5 nanocomposite is shown in Figure 4. The MCAP12.5 sample was prepared by the grafting of ICPTES onto CAP polymer (resulting in T-type structures) followed by the addition of 12.5 wt% TEOS and sol-gel process (resulting in Q-type structures). Observed in the spectrum T^3 and T^2 structures are related to trisubstituted siloxane bonds and disubstituted siloxane bonds, respectively. The chemical shifts are −64 and −58 ppm, respectively, and agree closely to the literature values [16]. The Q-type structures observed in the spectrum are related to the in situ formed silica network. According to the literature data very high Q^4 and Q^3 structures content but low population of Q^2 states suggest a highly intraconnected or cross-linked silicate nanophase. Figure 4 showed very high Q^3 content (the band at −102 ppm), revealing that this is the major structure forming silica network. The Q^3 structure is assigned to silicon atom connected to the three silicon and one hydrogen atoms or an alkyl group in the second coordination sphere. The peak at −110 ppm (Q^4 structure) indicates fully condensed silica. The high intensity of the Q^3 band (suggesting the existence of –OH or –OR groups) and also Q^2 structures observed in the spectra (at −90 ppm) indicate that the condensation reaction was incomplete. ^{29}Si NMR spectrum of the MCAP nanocomposite (Figure 5) revealed peaks T^2 and T^3 observed at −59 and −71 ppm, respectively. The chemical structure of MCAP and MCAP12.5 nanocomposite is depicted in Schemes 2 and 3, respectively.

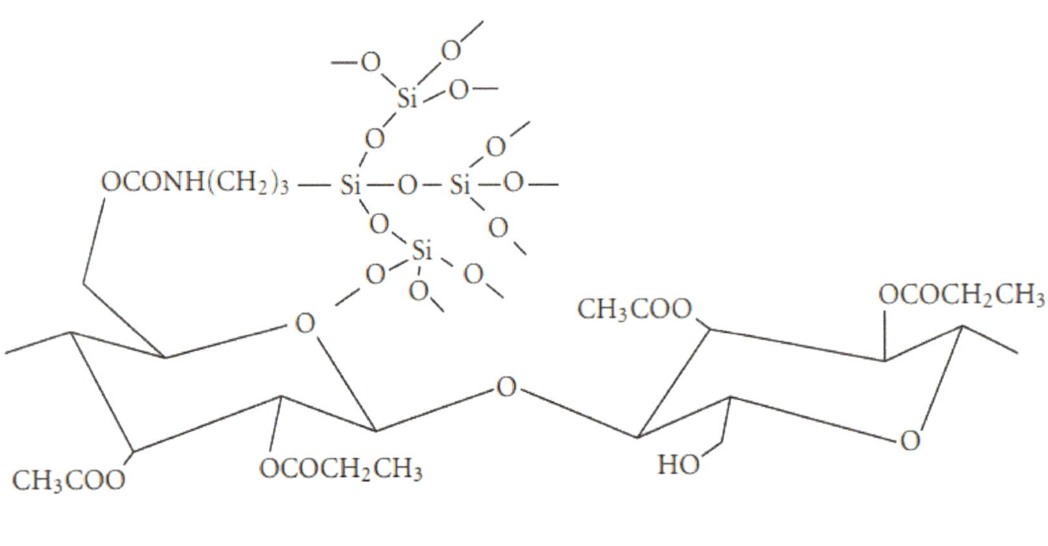

Scheme 2: Chemical structure of MCAP.

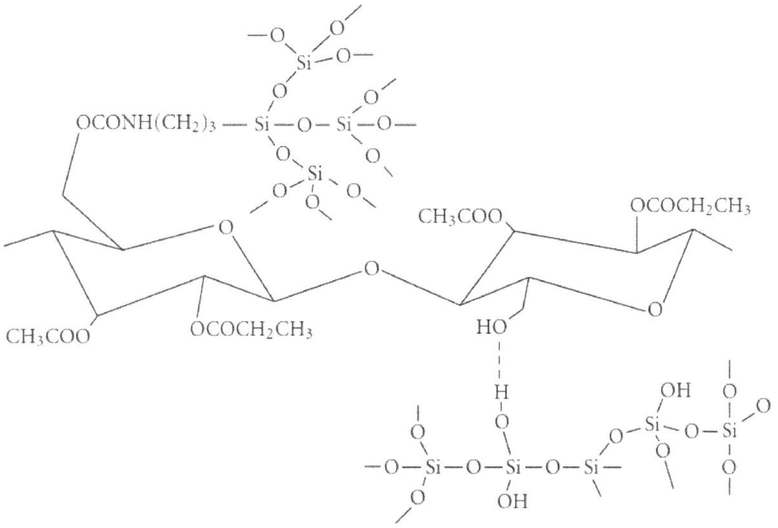

Scheme 3: Chemical structure of MCAP12.5.

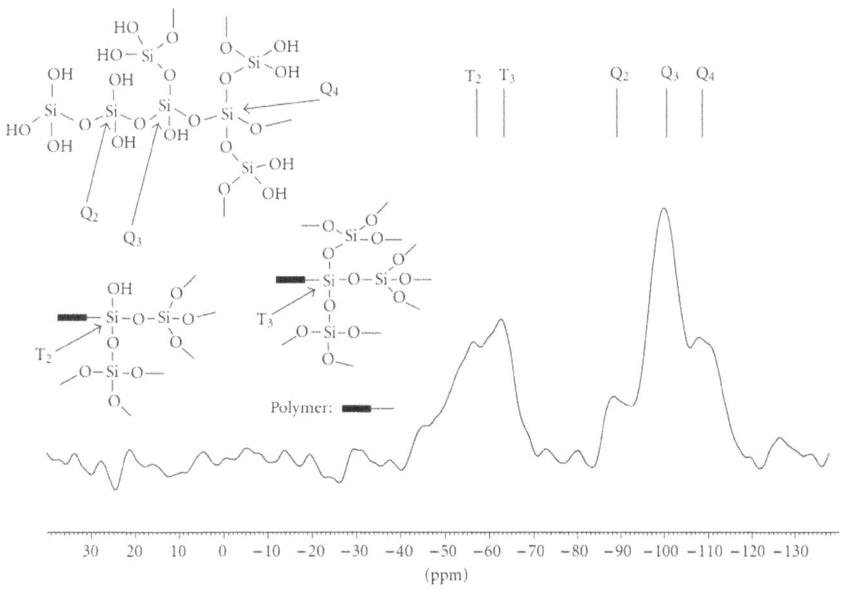

Figure 4: MAS ^{29}Si NMR spectrum of MCAP12.5.

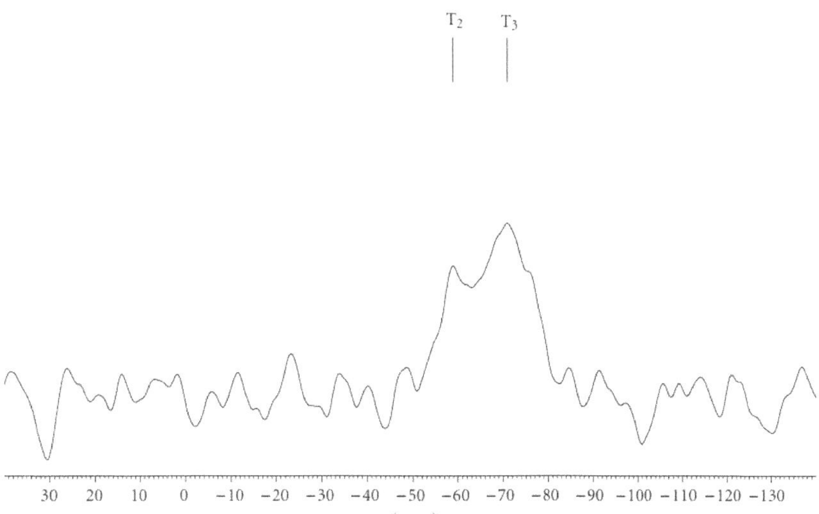

Figure 5: MAS ^{29}Si NMR spectrum of MCAP.

Morphological Investigation

For surface analysis intermittent-contact atomic force microscopy (IC-AFM) was applied. This technique is commonly used for topographical and compositional mapping due to the high spatial and the minimal lateral interaction [17, 18]. This operation mode is based on cantilever vibrations near the sample surface. In this mode tip strikes the surface with sufficiently large amplitudes to eliminate sticking of the tip to the surface. Due to vibrations, tip-sample surface interaction forces are time dependent, and they contain information on the elastic, viscoelastic, and adhesive properties of the sample. In the IC-AFM mode material composition mapping is performed by recording the shift between the cantilever excitation signal and photodetector signal (compare signals in places A and B on Figure 6) [19]. This shift is known as a "phase signal" and depends on the energy dissipation during the tip-sample interactions [20]. Image build from "phase signal" data is usually called "material contrast" image.

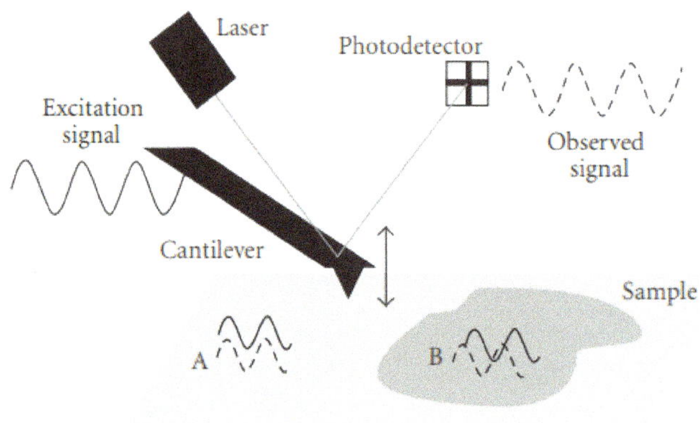

Figure 6: Idea of construction "material contrast" image using IC-AFM (places A and B have highly different mechanical properties).

Phase imaging is commonly used to show material contrast in polymer blends and composites. This method is very precise and allows detecting structures in composite materials in a nanometer range [21–24].

The presence of two mechanically different phases is clearly seen in the material contrast images. Varied colors displayed on this image show different local mechanical properties (adhesion, Young's modulus, and hardness). This allowed to distinguish modified regions of the polymer matrix, new-created by the addition of the inorganic component. Figure 7 shows a comparison of the morphology of unmodified CAP with MCAP nanocomposite. Material contrast image of unmodified polymer (CAP) is homogenous (Figure 7(b)) and opposite to nanocomposites based on MCAP does not contain mechanical different areas (Figure 7(d)). AFM image of MCAP nanocomposite showed that inorganic component modified the mechanical properties of some regions of the polymer matrix. Inorganic phase-rich regions are siloxane network formed during sol-gel process (Schemes 2 and 3).

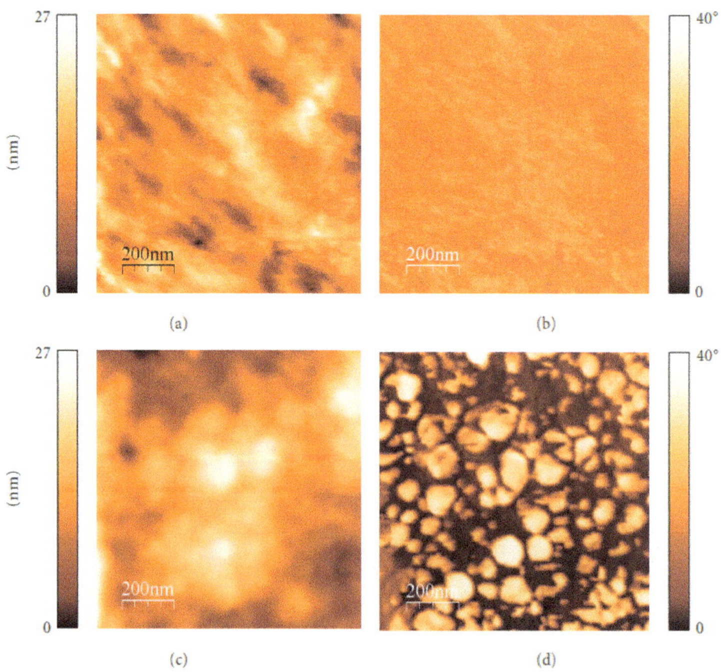

Figure 7: The comparison of AFM images of CAP ((a) topography, and (b) material contrast) and MCAP nanocomposite ((c) topography, and (d) material contrast).

Figures 8(a), 8(b), and 8(c) show the comparison of MCAP, MCAP6.25, and MCAP12.5 samples, respectively. For the MCAP a region with modified mechanical properties by inorganic phase is about 45% of the analyzed area (Figure 8(a)). For MCAP6.25 we noted similar structures with diameters between 50 to 150 nm, constituting about 53% of surface. For MCAP12.5 modified regions are much bigger (ca. 100–300 nm in length) and frequently connected, covering about 72% of analyzed surface.

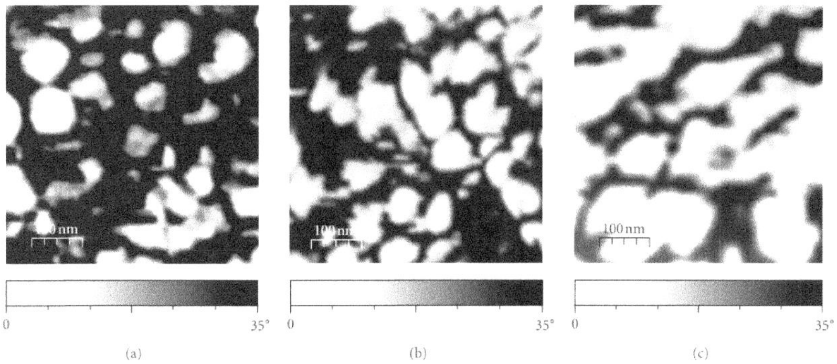

Figure 8: Mechanical contrast images of MCAP (a), MCAP6.25 (b), and MCAP12.5 (c).

Material contrast images showed clear the presence of inorganic phase-rich regions within polymer matrix, varying with increasing amount of TEOS. Topography images of the samples CAP, MCAP, MCAP6.25, and MCAP12.5 showed RMS roughness: 2.8 nm, 2.8 nm, 3.4 nm, and 9.5 nm, respectively. The same value of RMS for CAP and for MCAP indicates that chemical modification of cellulose acetate propionate with ICPTES does not impact on surface roughness. On the contrary, the size of inorganic phase-rich regions formed in the sol-gel reaction from varied amount of TEOS considerably changed the RMS roughness of the nanocomposites in comparison with unmodified CAP.

The representative SEM images of unmodified CAP and nanocomposites (Figure 9) revealed that chemical modification of CAP changed the morphological structure of the polymer. The cross-section SEM images of the freeze-fractured samples revealed that the morphology of unmodified CAP (Figure 9(a)) looks more brittle than of MCAP (Figure 9(b)), MCAP6.25 (Figure 9(c)), and MCAP12.5 (Figure 9(d)) nanocomposites. SEM images of nanocomposites prepared from MCAP by the addition of 6.25 and 12.5 TEOS (MCAP6.25 and MCAP12.5, resp.) showed cavities (ca. 1.5–4.5 μm in length) formed by evaporation of ethanol evolved during the sol-gel process. Moreover, microscale aggregates of silica particles were not observed, indicating good compatibility of the polymer matrix and covalently bound inorganic phase.

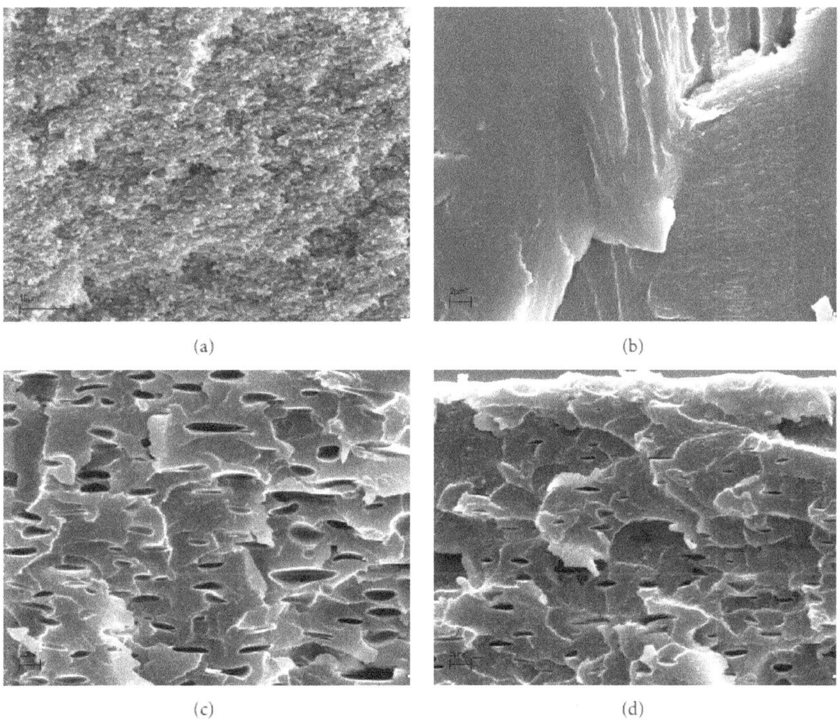

Figure 9: SEM images: (a) CAP (magnification 5000, scale bar: 10 μm), (b) MCAP nanocomposite (magnification 10000, scale bar: 2 μm), (c) MCAP6.25 nanocomposite (magnification 10000, scale bar: 2 μm), and (d) MCAP12.5 nanocomposite (magnification 10000, scale bar: 2 μm).

CONCLUSIONS

New hybrid materials with inorganic phase dispersed at the nanometer scale were successfully prepared using modified cellulose acetate propionate via sol-gel process. IR and solid state ^{29}Si NMR spectroscopic studies indicated formation of Si–O–Si links within these materials and showed that inorganic phase is bounded to the CAP matrix through covalent bounds. SEM data revealed different morphological structures of unmodified polymer and MCAP/SiO_2 nanocomposites. AFM studies confirmed growth of inorganic phase network within hybrid materials. AFM data showed the dependence between the amount of TEOS used for nanocomposites preparation and inorganic phase-rich regions size, dispersion, and RMS roughness of the samples.

REFERENCES

1. G. Kickelbick, "Concepts for the incorporation of inorganic building blocks into organic polymers on a nanoscale," Progress in Polymer Science, vol. 28, no. 1, pp. 83–114, 2003.
2. S. K. Young, G. C. Gemeinhardt, J. W. Sherman et al., "Covalent and non-covalently coupled polyester-inorganic composite materials," Polymer, vol. 43, no. 23, pp. 6101–6114, 2002. ·
3. V. Bounor-Legaré, C. Angelloz, P. Blanc, P. Cassagnau, and A. Michel, "A new route for organic-inorganic hybrid material synthesis through reactive processing without solvent," Polymer, vol. 45, no. 5, pp. 1485–1493, 2004.
4. M. Yoshioka, K. Takabe, J. Sugiyama, and Y. Nishio, "Newly developed nanocomposites from cellulose acetate/layered silicate/poly(ε-caprolactone): synthesis and morphological characterization," Journal of Wood Science, vol. 52, no. 2, pp. 121–127, 2006.
5. R. J. B. Pinto, P. A. A. P. Marques, A. M. Barros-Timmons, T. Trindade, and C. P. Neto, "Novel SiO_2/ cellulose nanocomposites obtained by in situ synthesis and via polyelectrolytes assembly,"Composites Science and Technology, vol. 68, no. 3-4, pp. 1088–1093, 2008.
6. N. A. Zakharov, Z. A. Ezhova, E. M. Koval, V. T. Kalinnikov, and A. E. Chalykh, "Hydroxyapatite-carboxymethyl cellulose nanocomposite biomaterial," Inorganic Materials, vol. 41, no. 5, pp. 509–515, 2005.
7. K. Pirkkalainen, K. Leppänen, U. Vainio et al., "Nanocomposites of magnetic cobalt nanoparticles and cellulose," European Physical Journal D, vol. 49, no. 3, pp. 333–342, 2008.
8. S. Yano, K. Iwata, and K. Kurita, "Physical properties and structure of organic-inorganic hybrid materials produced by sol-gel process," Materials Science and Engineering C, vol. 6, no. 2-3, pp. 75–90, 1998.
9. F. J. Rodriguez, A. Coloma, M. J. Galotto, A. Guarda, and J. E. Bruna, "Effect of organoclay content and molecular weight on cellulose acetate nanocomposites properties," Polymer Degradation and Stability, vol. 97, no. 10, pp. 1996–2001, 2012.
10. J. A. de Lima, C. A. Pinotti, M. I. Felisberti, and M. C. Goncalves, "Blends and clay nanocomposites of cellulose acetate and poly(epichlorohydrin)," Composites Part B, vol. 43, no. 5, pp. 2375–2381, 2012.
11. P. Wojciechowska and Z. Foltynowicz, "Synthesis of organic-inorganic hybrids based on cellulose acetate butyrate," Polimery, vol. 54, no. 11-12, pp. 845–848, 2009.
12. S. Jo and K. Park, "Surface modification using silanated poly(ethylene glycol)s," Biomaterials, vol. 21, no. 6, pp. 605–616, 2000.
13. M. Messori, M. Toselli, F. Pilati et al., "Flame retarding poly(methyl methacrylate) with nanostructured organic-inorganic hybrids coatings," Polymer, vol. 44, no. 16, pp. 4463–4470, 2003.·
14. Polish Patent Office 209829, 2011.
15. I. Horcas, R. Fernandez, J. M. Gomez-Rodriguez, J. Colchero, J. Gomez-Herrero, and A. M. Baro, "WSXM: a software for scanning probe microscopy and a tool for nanotechnology," Review of Scientific Instruments, vol. 78, no. 1, Article ID 013705, 2007.
16. S. K. Young, W. L. Jarret, and K. A. Mauritz, "Nafion/ORMOSIL nanocomposites via polymer-in situ sol-gel reactions. 1. Probe of ORMOSIL phase nanostructures by ^{29}Si solid-state NMR spectroscopy,"Polymer, vol. 43, no. 8, pp. 2311–2320, 2002.
17. Q. Zhong, D. Inniss, K. Kjoller, and V. B. Elings, "Fractured polymer/silica fiber surface studied by tapping mode atomic force microscopy," Surface Science, vol. 290, no. 1-2, pp. L688–L692, 1993.
18. D. Klinov and S. Magonov, "True molecular resolution in tapping-mode atomic force microscopy with high-resolution probes," Applied Physics Letters, vol. 84, no. 14, p. 2697, 2004.
19. R. Garcia, R. Magerle, and R. Perez, "Nanoscale compositional mapping with gentle forces," Nature Materials, vol. 6, pp. 405–411, 2007.
20. J. P. Cleveland, B. Anczykowski, A. E. Schmid, and V. B. Elings, "Energy dissipation in tapping-mode atomic force microscopy," Applied Physics Letters, vol. 72, no. 20, pp. 2613–2615, 1998. ·
21. P. Achalla, J. McCormick, T. Hodge et al., "Characterization of elastomeric blends by atomic force microscopy," Journal of Polymer Science, Part B, vol. 44, no. 3, pp. 492–503, 2006.·
22. R. Y. F. Liu, Y. Jin, A. Hiltner, and E. Baer, "Probing nanoscale polymer interactions by forced-assembly,"Macromolecular Rapid Communications, vol. 24, no. 16, pp. 943–948, 2003.
23. M. S. Sanchez, J. M. Mateo, F. J. R. Colomer, and J. L. G. Ribelles, "Nanoindentation and tapping mode AFM study of phase separation in poly(ethyl acrylate-co-hydroxyethyl methacrylate) copolymer networks," European Polymer Journal, vol. 42, no. 6, pp. 1378–1383, 2006.
24. S. N. Magonov, J. Cleveland, V. Elings, D. Denley, and M. H. Whangbo, "Tapping-mode atomic force microscopy study of the near-surface composition of a styrene-butadiene-styrene triblock copolymer film," Surface Science, vol. 389, no. 1–3, pp. 201–211, 1997.

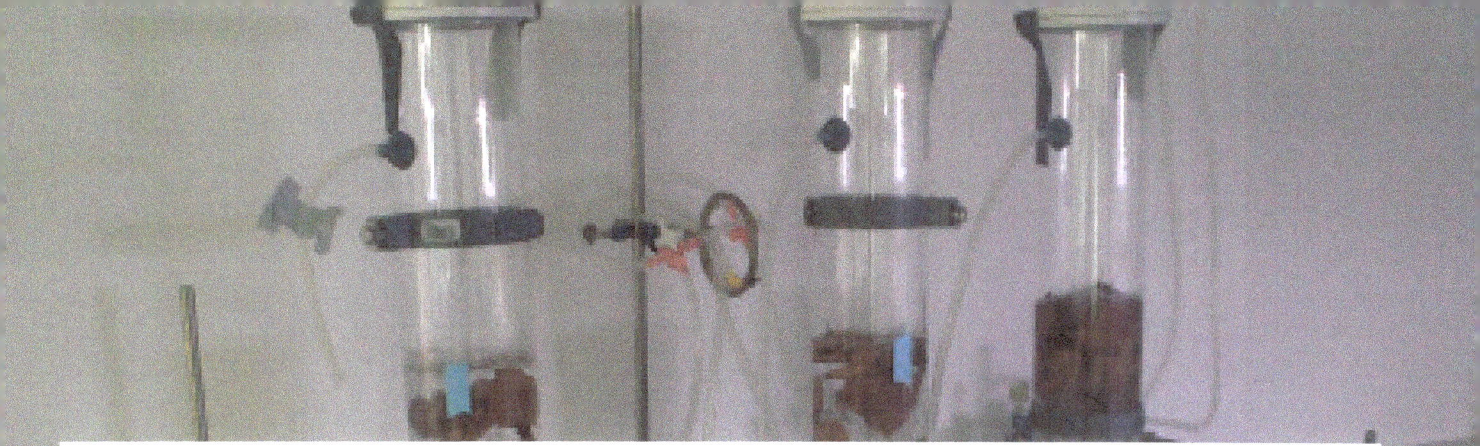

CHAPTER 12

DEGRADATION OF TEXTILE DYES USING IMMOBILIZED LIGNIN PEROXIDASE-LIKE METALLOPORPHINES UNDER MILD EXPERIMENTAL CONDITIONS

Paolo Zucca, Antonio Rescigno, Manuela Pintus, Andrea C Rinaldi and Enrico Sanjust

Keywords: Biomimetic, Dye, Lignin-peroxidase, Porphine, Porphyrin

BACKGROUND

Textile dyes are a broad and heterogeneous class of molecules used in many technological fields (i.e., textile and paper production, food technology, and hair coloring) [1]. Many chemical classes of synthetic dyes are frequently employed on an industrial scale; azo dyes, triarylmethanes, phenothiazine, and anthraquinones are among the most widespread [2, 3].

Because the goal of the dyeing process is color durability, textile dyes of all chemical classes are commonly highly resistant to both chemical and physical degradation. Accordingly, textile dyes represent a serious environmental concern because of their chemical inertness, high annual production (over 10,000 tons per year [1]), toxicity/carcinogenicity [4, 5], and the large amount of dyes released in industrial wastewaters (up to 50% of total process intake [6]).

The European Directive 2002/61/EC forbids the use of some products (in particular some azo dyes). However, these restrictions only marginally solve the environmental problem because of the large volume of other dyes discharged every year.

Several physical, chemical and enzymatic approaches have been proposed to remove dyes from industrial wastewaters [7, 8, 9, 10, 11, 12, 13, 14, 15, 16] that require long reaction times, extreme operational conditions, and high costs [1]. Accordingly, the development of a clean and environmentally friendly process for removing dyes is a challenging issue, far from being solved.

Redox-active metalloporphines are efficient catalysts; in the presence of a proper oxidant [17, 18, 19, 20, 21], they are able to oxidize many substrates, including several textile dyes, under very mild conditions. We previously studied the ability of an immobilized manganese porphine to oxidize alizarin red S (ARS), an anthraquinone dye) [22], and phenosafranine (PNS), a phenylphenazinium dye [23], elucidating several aspects of the catalytic mechanism.

In this study, we compare the catalysts IPS/MnTSPP and PP-PVA/FeTFPP (Figure 1), each containing a metalloporphine (Mn- and Fe-, respectively) immobilized onto a solid support, thus emulating ligninolytic peroxidases [24]. We investigated the ability of these two catalysts to oxidize different chemical classes of textile dyes in the presence of hydrogen peroxide under mild conditions. We also report the results of these catalysts on ARS and PNS to allow a comprehensive comparison. All the dyes included in the study are reported in Table 1.

Citation: Paolo Zucca, Antonio Rescigno, Manuela Pintus, Andrea C Rinaldi and Enrico Sanjust, Degradation of textile dyes using immobilized lignin peroxidase-like metalloporphines under mild experimental conditions, DOI: 10.1186/1752-153X-6-161

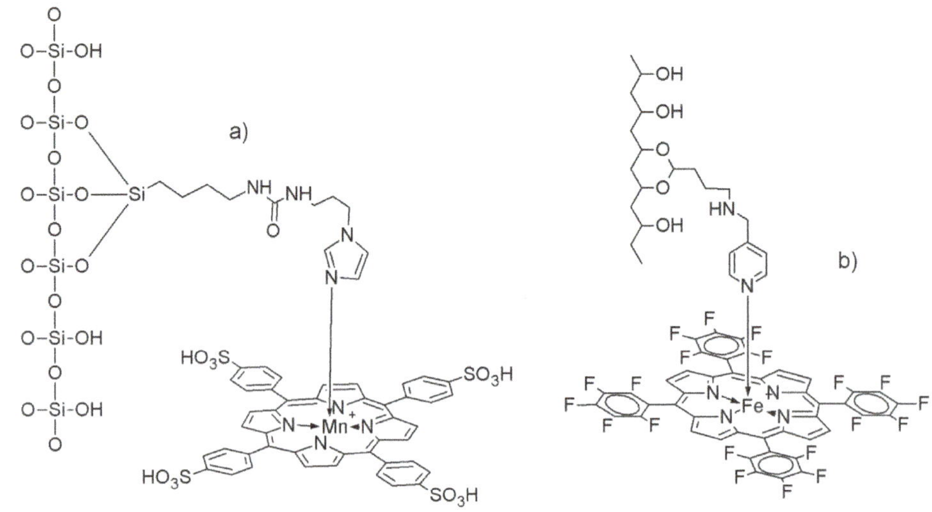

Figure 1: Proposed structure of the catalysts IPS/MnTSPP (a) and PP-PVA/FeTFPP (b). For more details,.

Table 1: The six textile dyes included in this study

Dye		**MW (g/mol)**	**λ_{max}(nm)**	**ε (mol * L^{-1} * cm^{-1})**
Alizarin Red S (ARS)	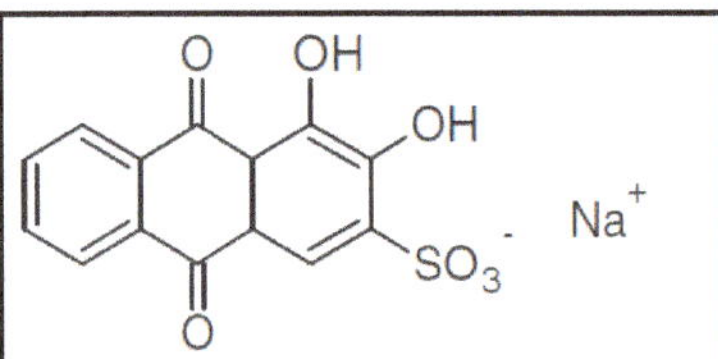	342.26	520	7200
Phenosafranine (PNS)	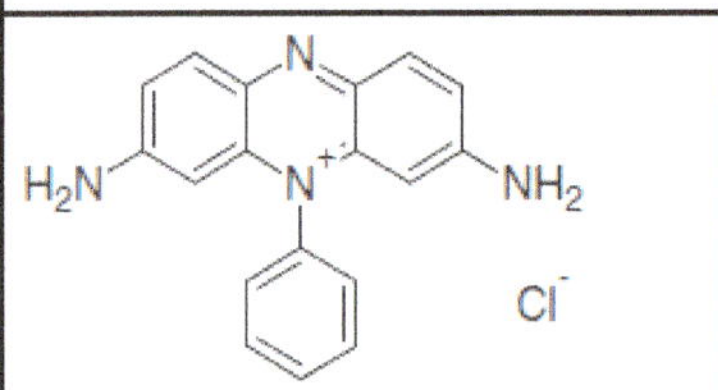	322.80	520	32000
Xylenol Orange (XO)	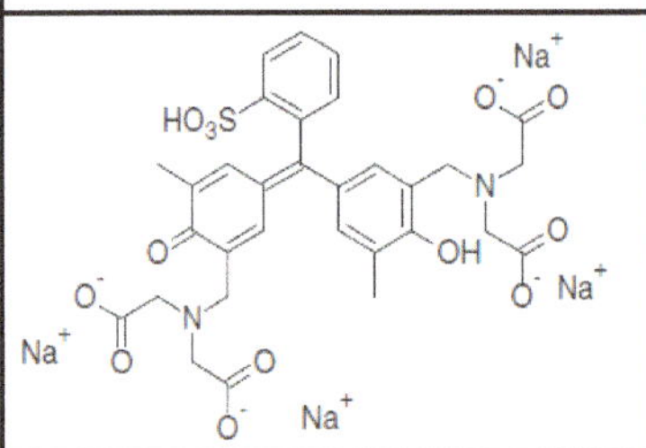	782.56	435	10500
Methylene Blue (MB)	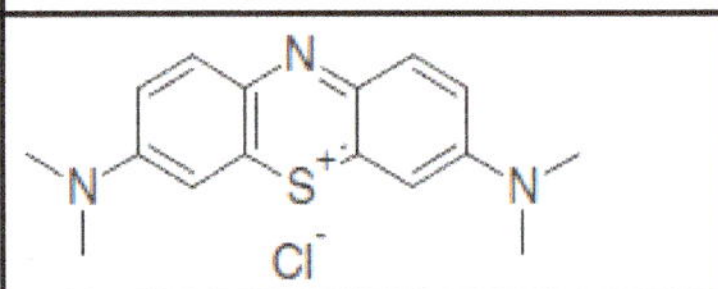	319.86	660	79500
Methyl Green (MG)	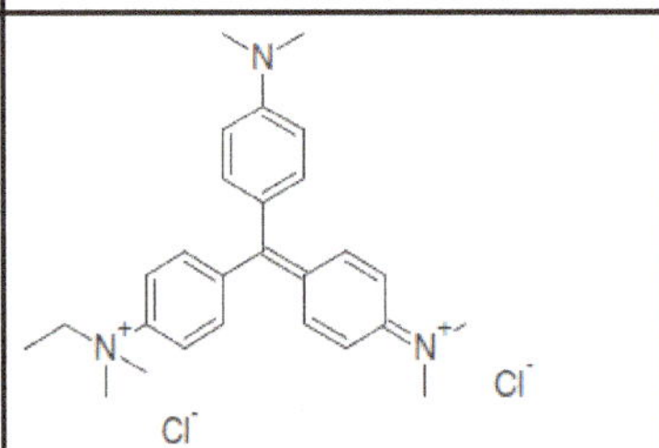	472.51	632	55700
Methyl Orange (MO)	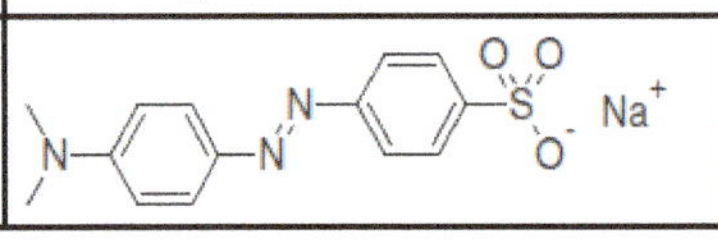	327.34	464	24800

RESULTS AND DISCUSSION

Comparison of IPS/MnTSPP and PP-PVA/FeTFPP

We previously developed two biomimetic catalysts, IPS/MnTSPP and PP-PVA/FeTFPP, which are able to oxidize several lignin-model compounds. IPS/MnTSPP strictly emulates a lignin peroxidase (LiP) active site, having an imidazole-grafted support that coordinates a Mn(III)-porphine [24]. This catalyst showed a LiP-like activity, whereas, unlike an enzyme such as manganese peroxidase (MnP), it was not able to use the Mn^{2+} ion as a redox mediator. Conversely, PP-PVA/FeTFPP was only "bioinspired" (i.e., only inspired by but not strictly emulating, a natural enzyme active site) because the metalloporphine was coordinated by a pyridine-grafted support [25] rather than imidazole, as in ligninolytic peroxidases. This catalyst showed both LiP-like and MnP-like catalytic activities.

These two catalysts were tested for their ability to decolorize several textile dyes in the presence and absence of Mn^{2+} using hydrogen peroxide as the oxidant in accordance with green chemistry principles [26]. The investigation employed very mild conditions (neutral pH and room temperature, using no organic solvents, and the most environmental-friendly oxidant) to serve as the basis for an environmentally sustainable textile wastewater treatment process.

The results of these tests are reported in Table 2. Both biomimetic catalysts easily degraded all tested dyes, showing wide substrate specificity. As previously reported [24], the catalytic behavior of IPS/MnTSPP was not affected by the presence of Mn(II), showing similar bleaching percentages in both the presence and absence of this ion. All six dyes were rapidly bleached under the conditions studied; more than with a removal in all cases higher than 50% of each dye (up to >90% for PNS, MG, and MO) was removed in 1 h. In contrast, PP-PVA/FeTFPP was almost inactive in the absence of a suitable redox mediator. The conversion of all six dyes increased in the presence of Mn^{2+}, leading to values similar on average to those observed for IPS/MnTSPP.

Table 2: Removal of the studied dyes in the presence of hydrogen peroxide and either IPS/MnTSPP or PP-PVA/FeTFPP, in both the presence and absence of Mn(II) as the redox mediator

R	Catalyst	Dyes removal % (1 h)					
		ARS	PNS	XO	MB	MG	MO
No	IPS/MnTSPP	62±1	92±2	42±4	96±4	82±5	87±2
	PP-PVA/FeTFPP	n.d.	n.d.	n.d.	n.d.	n.d.	n.d.
Yes	IPS/MnTSPP	55±1	88±2	44±2	91±3	77±4	85±4
	PP-PVA/FeTFPP	48±3	58±2	55±5	37±1	48±2	91±2

The reaction mixtures contained 5 mg of catalyst in 1 mL of 25 mM buffer solution containing 8.8 mM hydrogen peroxide and a proper concentration of the selected dyes (0.29 mM ARS, 0.31 mM PNS, 2 mM XO, 0.15 mM MB, 1.5 mM MG, and 1.25 mM MO). Quantification of the removed dye was obtained photometrically, using the λ_{max} and molar extinction coefficients reported in Table 1. n.d.: not detectable. ($n=4$).

Considering that textile wastewaters do not usually contain manganous ion and that it would therefore be necessary to add Mn^{2+}, itself a heavy metal pollutant, to the reaction, IPS/MnTSPP was considered the most promising catalyst for future industrial applications Hence, the rest of study focused on IPS/MnTSPP.

The time dependence of dye bleaching by IPS/MnTSPP is shown in Figure 2. Note that most of the bleaching occurs in the first hour of the reaction, consistent with industrial timing needs.

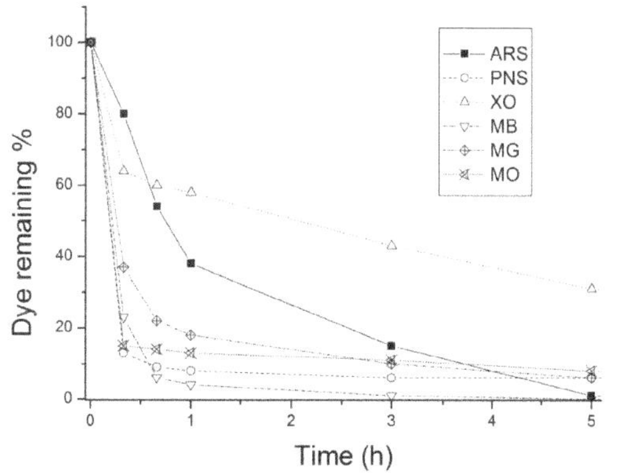

Figure 2: Time dependence of dye removal in the presence of IPS/MnTSPP and hydrogen peroxide. The reaction mixtures contained 5 mg of catalyst in 1 mL of 25 mM buffer solution containing 8.8 mM hydrogen peroxide and a proper concentration of the selected dyes (0.29 mM ARS, 0.31 mM PNS, 2 mM XO, 0.15 mM MB, 1.5 mM MG, and 1.25 mM MO). Quantification of the dyes was obtained photometrically, using the λ_{max} and molar extinction coefficients reported in Table 1.

Characterization of bleaching by IPS/MnTSPP and H_2O_2

IPS/MnTSPP catalysis was strongly pH-dependent; an approximately neutral pH was optimal for all dyes tested except XO and PNS, which had optimal bleaching at pH 6 and 8, respectively (data not shown). However, XO and PNS were sufficiently bleached at pH 7 that IPS/MnTSPP remains a viable catalyst for environmentally friendly wastewater treatment.

An exhaustive kinetic modeling of this catalysis is beyond the scope of our work. However, the decolorization of all six dyes apparently followed saturation Michaelis-Menten-like kinetics, allowing the calculation of catalytic parameters (Table 3). Reported Michaelis constants (K_M) showed that IPS/MnTSPP had significantly higher affinity for cationic dyes than anionic dyes (lower K_M for PNS and MB, higher K_M for ARS and XO). This lower affinity led to the lowest bleaching rate for XO and ARS (compare Figure 2). Notably, in all cases, K_M values were in the millimolar range and did not suffer from substrate inhibition. This can be a crucial issue for potential industrial applications because some bleaching approaches (i.e., enzymatic, *vide ultra*) are not able to work at these concentrations.

Table 3: Kinetic parameters measured for the bleaching of each dye by IPS/MnTSPP and hydrogen peroxide

Sub-strate	Kinetic parameter	Dye					
		ARS	PNS	XO	MB	MG	MO
Reducing substrate	K_M(mM)	2.11±0.74	0.32±0.04	1.33±0.38	0.11±0.02	0.72±0.12	0.63±0.05
	k_{cat} (min^{-1})	0.14±0.03	2.2±0.1	98±14	19.0±1.8	108±9	67±4
	k_{cat}/K_M(mM^{-1} min^{-1})	0.066±0.037	7.0±1.2	74±31	166±52	150±37	105±16
H_2O_2	K_M(mM)	3.21±0.12	3.0±0.4	7.9±1.9	2.4±0.5	0.43±0.08	7.4±1.5
	k_{cat} (min^{-1})	0.13±0.01	1.9±0.1	78±8.9	15.2±1.3	159±12	71±8
	k_{cat}/K_M(mM^{-1} min^{-1})	0.040±0.004	0.63±0.13	9.9±3.6	6.4±2.1	369±100	9.6±3.2

Data for ARS and PNS were previously reported and are added here to allow a full comparison [22, 23] (n=5).

To test the hypothesis of a one-electron abstraction mechanism as the first step of the bleaching process, the influence of several well-known redox mediators was also studied: 4-hydroxy-2,2,6,6-tetramethylpiperidine-1-oxyl (OH- TEMPO), *N*-hydroxyphthalimide (NPH), *N*-hydroxysuccinimide (NHS), and *N*-hydroxybenzotriazole (NHT). The results are summarized in Table 4. Only in the case of MG was an increase (three-fold) of the catalytic activity observed; in all the other cases, the redox-mediators mainly decreased IPS/MnTSPP catalysis (this quenching could be due to the lower reactivity of the oxygen-centered radical form of these mediators). Even in the best case, however, because the increase in catalytic activity was moderate, the industrial benefit of these redox mediators may not justify their environmental impact.

Table 4: Effect of well-known redox-mediators on the bleaching of the six tested dyes by IPS/MnTSPP

Redox mediator	% Catalytic activity					
	ARS	PNS	XO	MB	MG	MO
OH-TEMPO	98	45	277	56	336	16
NHT	142	46	162	38	299	121
NHS	141	89	146	90	313	122
NPH	103	32	74	56	340	57

Results are expressed by referring to a 100% reference activity in the absence of any mediator. Data for ARS and PNS were previously reported and are added here to allow a full comparison [22, 23].

The catalyst was also investigated for its reuse potential. The results are reported in Table 5. On average, a rapid decrease of catalytic activity was observed in the first 2–3 cycles. Subsequently, activity was stable at approximately 25-30% of maximum for at least 8–10 cycles. A potential explanation for this pattern is the possibility that a fraction of some dyes remains tightly adsorbed on the catalyst even when the supernatant is completely decolorized. This adsorbed dye would likely reduce the catalytic performance by steric hindrance and/or electrostatic repulsion of incoming substrate molecules. Nevertheless, the catalyst keeps a significant part of its catalytic activity during multi-cycle employment. This is an important quality for large-scale applications.

Table 5: Multi-cycle activity of IPS/MnTSPP

Catalytic cycle	**% Catalytic activity**					
	ARS	**PNS**	**XO**	**MB**	**MG**	**MO**
1	100	100	100	100	100	100
2	68	73	100	99	98	44
3	49	70	93	73	97	37
4	33	61	88	49	97	34
5	30	51	47	37	97	28
6	28	49	41	34	96	28
7	28	44	31	34	96	28
8	26	41	30	26	94	27
9	25	39	29	25	94	20
10	25	35	27	24	90	19

Catalytic activity of the IPS/MnTSPP adduct upon multicyclic use during the decolorization of the studied dyes (*n*=3). Data for ARS and PNS were previously reported and are added here to allow a full comparison [22, 23].

Significant removal of total organic carbon (TOC) was also observed during the bleaching by H_2O_2/IPS/MnTSPP, as shown in Figure 3. In the case of ARS, PNS, and MB, over 50% of TOC removal was observed in 24 hours, whereas lower decreases were recorded for XO and MG. Conversely, in the case of MO no TOC modification occurred during the catalysis. These results suggest that a decarboxylative degradation occurred for some of the dyes, whose TOC values decreased significantly during the bleaching experiments.

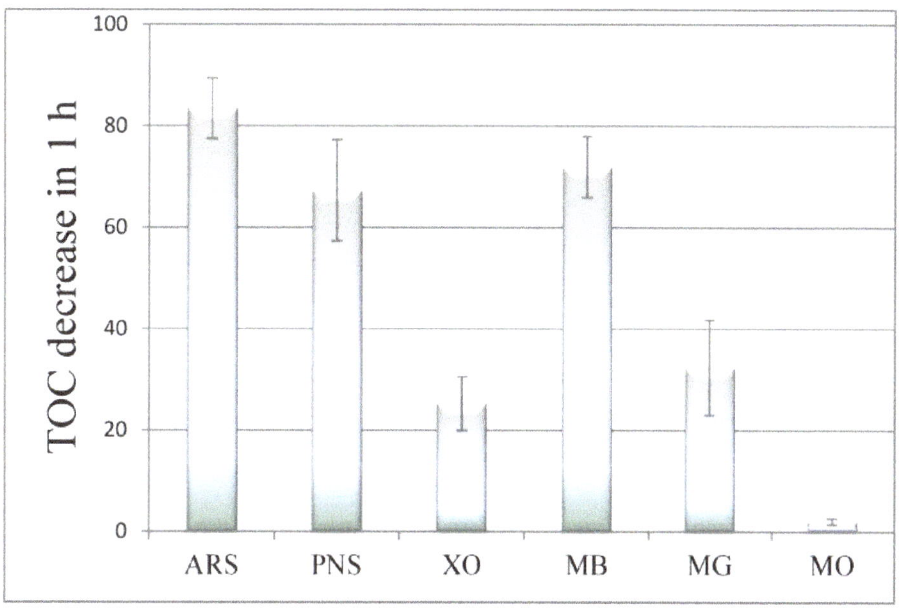

Figure 3: Total organic carbon (TOC) decrease during the bleaching of the textile dyes by IPS/MnTSPP (*n*=3).

IPS/MnTSPP was also tested for its ability to decolorize a complex mixture of all the six dyes, emulating realistic textile plant waste. The six dyes were each present in the saturating concentrations used above, leading to high impacting TOC (approximately 1580 mg/L). The changes in the UV–vis spectra (Figure 4) show a significant decolorization, although with a slower kinetics (approximately 5 hours for a 50% decrease of the main visible peak). However, the feasibility of using this treatment in an industrial process was confirmed by a significant TOC decrease (approximately 20%) in the same time frame.

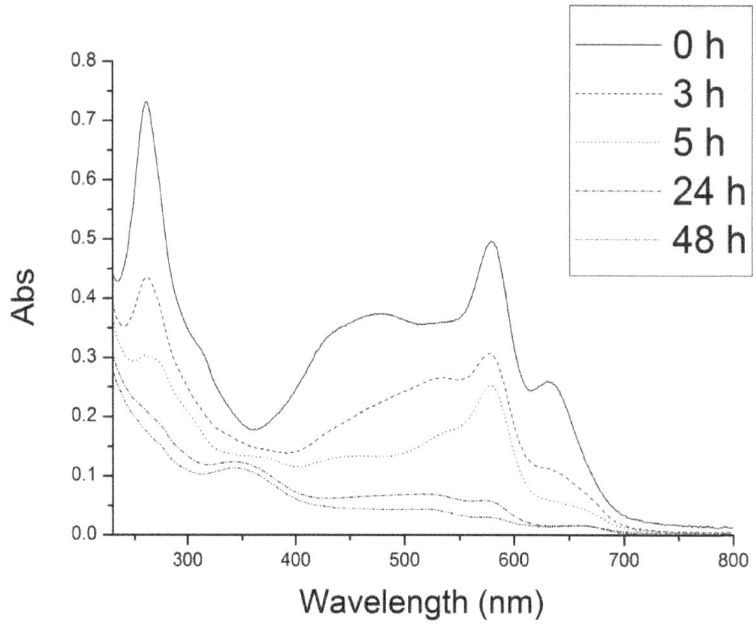

Figure 4: Changes in UV–vis spectra during the bleaching of a mix of the six studied dyes by IPS/MnTSPP.

Comparison with Enzymatic Bleaching

A full comparison with the enzymatic bleaching of the dyes was also performed. The ligninolytic enzymes tested were laccase (LC), lignin peroxidase, and the non-ligninolytic enzyme horseradish peroxidase (HRP). In addition, Mn(III) was also used as oxidizing agent, mimicking manganese peroxidase action [27]. These biological catalysts suffered from significant substrate inhibition, preventing their use at the dye concentrations used for IPS/MnTSPP. Usually, concentrations of dyes at least of one order of magnitude lower were necessary. These enzymes also suffer from H_2O_2 inhibition, needing lower concentrations of oxidizing substrate. Accordingly, under the same conditions, it is not surprising that enzymatic bleaching efficiency was significantly lower than IPS/MnTSPP bleaching efficiency (<1% dye removal in all cases). Enzymatic bleaching activity, under the conditions studied, is reported in Table 6, allowing a qualitative analysis of substrate specificity only.

Table 6: Enzymatic bleaching of the dyes

Dye	LC	LC + H_2O_2	HRP	LiP	Mn^{3+}
ARS	+	++	+/–	+/–	+
PNS	-	-	-	+/–	-
XO	+	+	+	-	+
MB	-	+/–	+/–	+/–	-
MG	+	+	++	+/–	++
MO	+	+/–	+	+/–	+

In the conditions tested, ligninolytic enzymes present a narrower substrate specificity towards the dyes included in this study than IPS/MnTSPP. Legend: LC laccase, HRP horseradish peroxidase, LiP lignin peroxidase, ++ good substrate (complete bleaching<1 h), + substrate, +/– weak substrate (bleaching<10% in 24 h), - no substrate.

None of the examined enzymes were able to bleach all the studied dyes. For instance, PNS is resistant to LC, HRP, and Mn^3+, whereas only a partial degradation with concomitant irreversible enzyme inactivation is observed with LiP. MB underwent only partial oxidation with the enzymatic catalysts.

IPS/MnTSPP is able to effectively bleach all the dyes studied, as described above. This feature, combined with its ability to oxidize relatively high concentration of dye (in the millimolar range), and with its partial multi-cycling, may make IPS/MnTSPP catalysis a feasible alternative for industrial treatment of textile wastewaters.

EXPERIMENTAL SECTION

Chemicals and instrumentation

All reagents were the best commercial grade available and were used without further purification. In particular, silica gel 100 came from Fluka (cat. No. 60746), and MnTSPP and FeTFPP were purchased from Sigma-Aldrich (cat. No. 441813 and 252913, respectively), PVA from Aldrich (cat. No. 363138), glutaraldehyde from Fluka (cat. no. 49629), horseradish peroxidase (E.C. 1.11.1.7) and lignin peroxidase (E.C. 1.11.1.14) were from Sigma-Aldrich (cat No. P-6782 and 42603, respectively). Laccase from *Pleurotus sajor-caju* was purified through two chromatographic steps (details in [28, 29]).

Spectrophotometric measurements were carried out with an UltroSpec 2100pro (Amersham Bioscience). Concentrations of the substrates were also measured by UV-HPLC and GC-MS, using previously described methods [30, 31].

Preparation of IPS/MnTSPP

The complex IPS/MnTSPP was synthesized as previously described [24]. Briefly, 10 g plain silica gel was functionalized by overnight reaction at 80°C with 10 mmol 3-(1-imidazolyl) propylcarbamoyl-3'-aminopropyl-triethoxysilane solubilized in 10 mL dioxane, in the presence of 10 μL *N*-methylmorpholine acting as the catalyst. The activated silica (IPS) was washed consecutively with 0.5 M HCl, H_2O, 0.1 M NaOH and H_2O again. The wet silica was then carefully dried overnight in a vacuum oven.

The bound MnTSPP was quantified by the difference in absorption at 400 nm (the λ_{max} of the MnTSPP Soret band), as described in [24].

Preparation of PP-PVA/FeTFPP

The complex PP-PVA/FeTFPP was synthesized as previously described [25]. Briefly, 1 g of aminopropyl crosslinked PVA (prepared as in [32]) was suspended in an excess of 0.05 M sodium acetate buffer pH 5, and treated with 0.1 mL 4-pyridinecarboxaldehyde and 0.5 g sodium cyanoborohydride. After 24 h of incubation at 25°C, the support was exhaustively washed with water, and dried at 50°C, to obtain PP-PVA. One-gram aliquots of PP-PVA were treated with 20 mg of FeTFPP, solubilized in 10 mL DMSO. The slurries were kept stirring in the dark for 24 h, washed exhaustively with DMSO, and carefully dried at 50°C.

Bound metalloporphine was quantified by difference through spectrophotometric measurement at 411 nm (ε_{411} = 115,000 M^{-1} cm^{-1} in DMSO) [25].

Activity Measurements

Routine measurements of the catalytic activity of immobilized metalloporphines were performed through spectrophotometric assays. The suspension contained 5 mg of catalyst in 1 mL of 25 mM buffer solution containing 8.8 mM hydrogen peroxide and a proper concentration of the selected dye. The routine analyses were performed using the following concentrations of dyes (chosen to be only slightly higher than respective K_M, to better observe influences on catalytic activity): 0.29 mM ARS, 0.31 mM PNS, 2 mM XO, 0.15 mM MB, 1.5 mM MG, and 1.25 mM MO. During the kinetic experiments, wide ranges of substrate concentrations were studied (0.01-2 mM for the dyes, 0.1-40 mM for hydrogen peroxide).

The reaction mixture was then stirred at 25°C in the dark, and the reaction was monitored photometrically using the λ_{max} and molar extinction coefficients, reported in Table 1.

The actual hydrogen peroxide concentration was determined by $KMnO_4$ titration.

When the reaction pH was not 7, the pH was corrected to 7 with 0.1 mL of potassium phosphate buffer 1 M (pH 7) before an absorbance measurement. UV–vis spectra in the range 230–700 nm were also recorded. Different McIlvaine buffers were used in the pH range 3–8.

When used, redox mediators (4-hydroxy-2,2,6,6-tetramethylpiperidine-1-oxyl OH- TEMPO; *N*-hydroxyphthalimide NPH; *N*-hydroxysuccinimide NHS; and *N*-hydroxybenzotriazole NHT) were added to the reaction medium at a final concentration of 1 mM. For the MnP-like assays, all the experiments were repeated as above in the presence of 1 mM $MnSO_4$ and 50 mM malonic acid. During multi-cyclic runs, the catalysts were regenerated between consecutive cycles by repeated washings with water and 2-propanol and by drying at 50°C.

Total organic carbon determination was performed by automatic analyzer TOC 5000A (Shimadzu), equipped with Pt-Al_2O_3 oxidizing catalyst. The volume of injection was 25 μL. The calibration curve was obtained using a range of standard solutions of potassium phthalate (0–200 ppm).

Michaelis-Menten kinetic parameters were calculated by using R 2.5.1 software (*R Foundation for Statistical Computing*, Vienna).

Enzymatic Assays

When horseradish peroxidase (HRP) was used, up to 20 E.U. were present in a final volume of 1 mL of 25 mM buffer (pH 4, 5, 6, or 7) with 8.8 mM H_2O_2. Due to significant substrate inhibition, lower dyes concentrations were used: 119 μM ARS; 31 μM PNS; 200 μM XO; 15 μM MB; 150 μM MG; 65 μM MO.

The same conditions were adopted for the experiments with laccase (23 E.U. added); every experiment was performed in the presence and absence of H_2O_2.

In the case of lignin peroxidase (LiP), 0.2 E.U. were present in the final volume of 1 mL, with 0.176 mM H_2O_2, due to strong substrate inhibition.

Moreover, to mimic manganese peroxidase (MnP) action, other experiments were conducted using Mn(III) as the putative oxidizing agent. One mM manganese triacetate was dissolved in 50 mM sodium malonate buffer, pH 6.5 [27], and the final mixture contained this solution along with dyes at the same concentrations, as mentioned above.

CONCLUSION

IPS/MnTSPP has proved able to efficiently degrade all tested dyes, regardless of their chemical class, under very mild conditions: standard pressure, room temperature and neutral pH, using no organic solvents, and the most environmental-friendly oxidant. The yields and time requirements of the process were quite promising for a potential industrial application. Furthermore, a multi-cycle capability was described, even if only 30% of catalytic activity is retained for 8–10 cycles.

In comparison to enzymatic systems, the biomimetic catalyst was more efficient and had wider substrate specificity. Therefore, after proper optimization of the process, this catalyst is worth consideration as a possible medium/large scale application in textile wastewater treatment.

ACKNOWLEDGMENTS

This research was partly funded by Fondazione Banco di Sardegna (Project Code 861/2012) and Regione Autonoma Sardegna (PO Sardegna FSE 2007–2013, L.R.7/2007, Bando Giovani Ricercatori, Project CRP1_27, and PO Sardegna FSE 2007–2013, L.R.7/2007, Ricerca di Base Bando 2008, Project CRP1_408).

AUTHORS' CONTRIBUTIONS

PZ performed the main part of the experiments and drafted the manuscript. MP performed TOC and enzymatic assays. ACR, and AR helped analyzing the data and draft the manuscript. ES conceived the study, coordinated the study, and worked on data analysis and interpretation. All the authors read and approved the final manuscript.

REFERENCES

1. Forgacs E, Cserhati T, Oros G: Removal of synthetic dyes from wastewaters: a review. Environ Int. 2004, 30: 953-971. 10.1016/j.envint.2004.02.001.
2. Azizi A, Moghaddam MRA, Arami M: Application of wood waste for removal of reactive blue 19 from aqueous solutions: optimization through response surface methodology. Environ Eng Manag J. 2012, 11: 795-804.
3. Brînzilă CI, Ciobanu R, Brillas E: Effect of experimental parameters on crystal violet mineralization by electro-fenton processes. Environ Eng Manag J. 2012, 11: 517-523.
4. Tsuda S, Murakami M, Matsusaka N, Kano K, Taniguchi K, Sasaki YF: DNA Damage induced by red food dyes orally administered to pregnant and male mice. Toxicol Sci. 2001, 61: 92-99. 10.1093/toxsci/61.1.92.
5. Walthall WK, Stark JD: The acute and chronic toxicity of two xanthene dyes, fluorescein sodium salt and phloxine B, to Daphnia pulex. Environ Pollut. 1999, 104: 207-215. 10.1016/S0269-7491(98)00189-4.
6. Pierce J: Colour in textile effluents - the origins of the problem. J Soc Dye Colour. 1994, 110: 131-133.
7. Colar LA, Ilinoiu EC, Manea F, Pode R, Lazau C: Photocatalytic activity of a silver doped TIO 2 modified zeolite in the degradation of reactive yellow 125 azo dye. Environ Eng Manag J. 2012, 11: 61-68.
8. Zucca P, Rescigno A, Olianas A, Maccioni S, Sollai F, Sanjust E: Induction, purification, and characterization of a laccase isozyme from Pleurotus sajor-caju and the potential in decolorization of textile dyes. J Mol Catal B Enzym. 2011, 68: 216-222. 10.1016/j.molcatb.2010.11.008.
9. Pakshirajan K, Jaiswal S, Das RK: Biodecolourization of azo dyes using phanerochaete chrysosporium: effect of culture conditions and enzyme activities. J Sci Ind Res. 2011, 70: 987-991.
10. Córdoba A, Magario I, Ferreira ML: Experimental design and MM2-PM6 molecular modelling of hematin as a peroxidase-like catalyst in Alizarin Red S degradation. J Mol Catal A Chem. 2012, 355: 44-60.
11. Zhu L, Ghosh T, Park CY, Meng ZD, Oh WC: Enhanced sonocatalytic degradation of rhodamine B by graphene-TiO 2 composites synthesized by an ultrasonic-assisted method. Chin J Catal. 2012, 33: 1276-1283. 10.1016/S1872-2067(11)60430-0.
12. Oh WC, Chen M, Cho K, Kim C, Meng Z, Zhu L: Synthesis of graphene-CdSe composite by a simple hydrothermal method and its photocatalytic degradation of organic dyes. Chin J Catal. 2011, 32: 1577-1583. 10.1016/S1872-2067(10)60264-1.
13. Asgher M, Kamal S, Iqbal HMN: Improvement of catalytic efficiency, thermo-stability and Dye decolorization capability of pleurotus ostreatus IBL-02 laccase by hydrophobic Sol Gel entrapment. Chem Centr J. 2012, 110:
14. Stengl V, Matys Grygar T, Henych J, Kormunda M: Hydrogen peroxide route to Sn-doped titania photocatalysts. Chem Centr J. 2012, 113:
15. Velmurugan R, Sreedhar B, Swaminathan M: Nanostructured AgBr loaded TiO 2: an efficient sunlight active photocatalyst for degradation of reactive Red 120. Chem Centr J. 2011, 5:
16. Chiavola A: Textiles. Water Environ Res. 2012, 84: 1511-1532. 10.2175/106143012X13407275695391.
17. Zucca P, Rescigno A, Sanjust E: Ligninolytic peroxidase-like activity of a synthetic metalloporphine immobilized onto mercapto-grafted crosslinked PVA inspired by the active site of cytochrome P450. Chin J Catal. 2011, 32: 1663-1666. 10.1016/S1872-2067(10)60277-X.
18. Faria AL, Mac Leod TOC, Barros VP, Assis MD: Hydrocarbon oxidation catalyzed by iron and manganese porphyrins anchored on aminofunctionalized supports. J Brazil Chem Soc. 2009, 20: 895-906. 10.1590/S0103-50532009000500014.
19. Dos Santos JS, Faria AL, Da Silva Amorin PM, La Luna FM, Caiado KL, Silva DOC E, Sartoratto PPC, Assis MD: Iron(III) porphyrin covalently supported onto magnetic amino-functionalized nanospheres as catalyst for hydrocarbon and herbicide oxidations. J Brazil Chem Soc. 2012, 23: 1411-1420. 10.1590/S0103-50532012005000005.
20. Díaz-Díaz G, Blanco-López MC, Lobo-Castañón MJ, Miranda-Ordieres AJ, Tuñón-Blanco P: Hemo-acrylic polymers as catalyst for the oxidative dehalogenation of 2,4,6-trichlorophenol. Chloroperoxidase's mimic imprinting effects. J Mol Catal A Chem. 2012, 353–354: 117-121.

21. Tian P, Gao B, Chen Y: Synchronous synthesis and immobilization of pyridyl porphyrins on crosslinked polystyrene microspheres and catalytic oxidation performance of immobilized cationic Co-porphyrins. Chin J Catal. 2011, 32: 483-489.

22. Zucca P, Vinci C, Sollai F, Rescigno A, Sanjust E: Degradation of Alizarin Red S under mild experimental conditions by immobilized 5,10,15,20-tetrakis(4-sulfonatophenyl)porphine-Mn(III) as a biomimetic peroxidase-like catalyst. J Mol Catal A Chem. 2008, 288: 97-102. 10.1016/j.molcata.2008.04.001.

23. Zucca P, Vinci C, Rescigno A, Dumitriu E, Sanjust E: Is the bleaching of phenosafranine by hydrogen peroxide oxidation catalyzed by silica-supported 5,10,15,20-tetrakis-(sulfonatophenyl)porphine-Mn(III) really biomimetic?. J Mol Catal A Chem. 2010, 321: 27-33. 10.1016/j.molcata.2010.01.015.

24. Zucca P, Mocci G, Rescigno A, Sanjust E: 5,10,15,20-Tetrakis(4-sulfonato-phenyl)porphine-Mn(III) immobilized on imidazole activacted silica as a novel lignin-peroxidase-like biomimetic catalyst. J Mol Catal A Chem. 2007, 278: 220-227. 10.1016/j.molcata.2007.09.017.

25. Zucca P, Sollai F, Garau A, Rescigno A, Sanjust E: Fe(III)-5,10,15,20-Tetrakis(pentafluorophenyl)porphine supported on pyridyl-functionalized, crosslinked poly(vinylalcohol) as a biomimetic versatile-peroxidase-like catalyst. J Mol Catal A Chem. 2009, 306: 89-96. 10.1016/j.molcata.2009.02.029.

26. Cojocariu AM, Mutin PH, Dumitriu E, Vioux A, Fajula F, Hulea V: Removal of dimethylsulfoxide from wastewater using mild oxidation with H2O2 over Ti-based catalysts. Chemosphere. 2009, 77: 1065-1068. 10.1016/j.chemosphere.2009.08.059.

27. Wariishi H, Valli K, Gold MH: Manganese(II) oxidation by manganese peroxidase from the basidiomycete phanerochaete-chrysosporium - kinetic mechanism and role of chelators. J Biol Chem. 1992, 267: 23688-23695.

28. Rescigno A, Zucca P, Flurkey A, Inlow J, Flurkey WH: Identification and discrimination between some contaminant enzyme activities in commercial preparations of mushroom tyrosinase. Enzyme Microb Technol. 2007, 41: 620-627. 10.1016/j.enzmictec.2007.05.009.

29. Sollai F, Zucca P, Sanjust E, Steri D, Rescigno A: Umbelliferone and esculetin: inhibitors or substrates for polyphenol oxidases?. Biol Pharm Bull. 2008, 31: 2187-2193. 10.1248/bpb.31.2187.

30. Zucca P, Littarru M, Rescigno A, Sanjust E: Cofactor recycling for selective enzymatic biotrasformation of cinnamaldehyde to cinnamyl alcohol. Biosci Biotechnol Biochem. 2009, 73: 1224-1226. 10.1271/bbb.90025.

31. Rescigno A, Casañola-Martin GM, Sanjust E, Zucca P, Marrero-Ponce Y: Vanilloid derivatives as tyrosinase inhibitors driven by virtual screening-based QSAR models. Drug Test Anal. 2011, 3: 176-181. 10.1002/dta.187.

32. Sanjust E, Cocco D, Curreli N, Rescigno A, Sollai F, Bannister JV: Flavin-grafted poly(vinyl alcohol): preparation and properties. J Appl Polym Sci. 2002, 85: 2471-2477. 10.1002/app.10877.

CHAPTER 13

PHOTO-ELECTROCHEMICAL TREATMENT OF REACTIVE DYES IN WASTEWATER AND REUSE OF THE EFFLUENT: METHOD OPTIMIZATION

Mireia Sala, Víctor López-Grimau and Carmen Gutiérrez-Bouzán

Keywords: Electrochemical, UV irradiation, indirect-oxidation, effluents reuse, reactive dyes, mono-functional, bi-functional, tri-functional

INTRODUCTION

The textile industry produces large volumes of wastewater in its dyeing and finishing processes. These effluents have as common characteristic their high coloration. This work is focused on reactive dyes, mainly used in the cotton industry. They were selected because of their high consumption, which implies an environmental and aesthetic problem caused both in dyeing and soaping effluents. The conventional biological plant treatments are not effective in color removal because reactive dyes have aromatic rings in their large molecules that provide them chemical stability and resistance to the microbiological attack [1]. Although the effluent color regulations are very variable depending on the country, the biological treatments are insufficient to remove color and to accomplish with current regulations. Consequently, the application of tertiary treatments is required.

Several methods are used for the removal of organic dyes from wastewaters. The current physic-chemical methods, based on the separation between dye and effluents, produce a residue which requires an additional treatment to be destroyed. In addition, absorbent materials (such as active carbon, silica gel or alumina) require regeneration after several treatments [2]. The use of low cost adsorbents and natural adsorbents obtained from agricultural waste can avoid this expensive regeneration process [3,4,5]. The filtration and membranes methods need thorough chemical cleaning treatments to achieve a good performance [6,7]. Chemical oxidation methods are rather expensive and involve some operational difficulties [8,9]. Enzymatic decomposition requires further investigation in order to know which enzymatic process takes place [10]. Moreover, temperature and pressure have to be controlled to avoid enzyme denaturalization. For these reasons, the electrochemical methods are nowadays the subject of a wide range of investigations at laboratory and pilot plant scale. The advantage of the electrochemical techniques is that the electron is a clean reagent. They also have good versatility and high energy efficiency. They are safe and easy for automation, because it is possible to operate at smooth conditions [11]. On the other hand, one of the main disadvantages of electrochemical treatments is their high energy cost which can be reduced by indirect electrochemical treatments in the presence of chloride.

As some industrial wastewaters contain large amounts of chloride, the indirect electrochemical oxidation method with

Citation: Mireia Sala, Víctor López-Grimau and Carmen Gutiérrez-Bouzán, Photo-Electrochemical Treatment of Reactive Dyes in Wastewater and Reuse of the Effluent: Method Optimization, doi:10.3390/ma7117349

active chlorine is the most suitable technique to treat this kind of effluent. The addition of chemical products in the electrolysis is not required because chloride is used as an electrolyte. In contrast, the combination of electrochemistry and chloride can produce haloforms such as chloroform, although it is not an inconvenience if the treated water is degraded lately in a biological plant to accomplish its mineralization. In fact, it has been verified that the concentration of haloforms is very low and they do not show any toxic effect on the plant microorganisms [12]. Moreover, the combination of UV exposure with the electrochemical treatment (photo-electrochemical methods) achieves the decomposition of the small amount of haloforms generated during the electrochemical treatment [13].

The electrochemistry method using chlorine as indirect oxidant has been noted to be effective in several kind of dyes, such as azo dyes [14], acid dyes [15] or disperse dyes [16] and combined with photo-electrochemistry has also provided good results for phtalocyanine dyes degradation [17], but in this case, the metal ions liberated (*i.e.*, copper) have to be removed.

The indirect electro-oxidation occurs when strong oxidants are generated *in situ* during the electrolysis and react with the organic pollutants such as dyestuffs, resulting in their total or partial degradation.

In the electro-oxidation with active chlorine [18,19] (which is a major oxidizing agent), the free chlorine gaseous and/or the generated chlorine-oxygen species such as hypochlorous acid (HClO) or hypochlorite ions (ClO^-) depending on the pH, oxidize the organic matter present in the effluents, according to the reactions (1)–(3):

$$2Cl^- \rightarrow Cl_{2(aq)} + 2e^- \quad (1)$$

$$Cl_{2(aq)} + H_2O \rightarrow ClO^- + Cl^- + H^+ \quad (2)$$

$$\text{Dye} + ClO^- \rightarrow \text{intermediate compounds} \rightarrow CO_2 + H_2O + Cl^- \quad (3)$$

Moreover, the current policies concerning water and energy consumption are conducive to recycling and reuse treatments. In this sense, recent studies demonstrate the possibility of reusing these discolored effluents for new dyeing processes. The reuse of 70% of discolored dyebaths, after electrochemical treatment assisted by UV irradiation, provides in most of cases, low color differences ($DE_{CMC\,(2:1)} \leq 1$, which is the limit for acceptance in the textile industry) with respect to the original dyeing with decalcified tap water.

Most electrochemical decolorization studies are focused on reactive dyes. They represent about 20%–30% of the total market [20], because of their washing fastness and brilliant color. Their structure contains one or several *reactive groups*(which react with the fiber) and a *chromophore group* (which gives the color). The most used chromophore group is the "azo" (R–N=N–R'), followed by the anthraquinone group [21]. Azo dyes constitute more than half of worldwide production [22], approximately 65% [23,24,25].

In the case of triazine reactive dyes, there is a dyeing reaction with the fiber that occurs by a nucleophilic displacement of the substituent atom from the reactive group of the dye (*i.e.*, chlorine) to the hydroxyl group from the cellulose in alkaline medium, according to the reaction (4) [26].

$$\text{dye-X} + \text{HO-cellulose} \rightarrow \text{dye-O-cellulose} + \text{HX} \quad (4)$$

Simultaneously, there is a competitive reaction between dye and water which produces dye hydrolysis (reaction (5)). The hydrolyzed dyes cannot react with the fiber, and this is the major cause of the effluent color.

$$\text{dye-X} + H_2O \rightarrow \text{dye–OH} + \text{HX} \quad (5)$$

Initially, reactive dyes contained only one functional group in their molecule. However, since the 1970s there has been an increase of the number of functional groups. Thus, reactive dyes can be classified as mono, bi, or tri-functional depending on the number of groups able to react with the fiber. The dye exhaustion is related to this factor. The exhaustion value for mono, bi, tri-reactive dyes goes from 60% to 90%. As it is expected, the bi- and tri-reactive dyes react faster with the fiber and consequently, they achieve higher dyeing exhaustion (which implies less dyes hydrolysis).

This work was structured in two parts. In the first one, the study was carried out with the bi-functional dye Procion Navy HEXL (referred to as PN): different intensities were applied to a synthetic dyeing effluent, defining their kinetic rate order. Then the optimization of the photo-electrochemical method was carried out with the appropriate intensity.

The second part of this paper is based on the application of the optimized photo-electrochemical method to eight additional reactive dyes (mono-, bi- and tri-functional), with the aim to obtain their kinetic rates. Subsequently, the reuse of the treated effluents was studied. In all cases, 70% of the dyeing wastewater was treated and reused, as 30% remains on the fabric. Finally, the influence of the total organic matter content on the dyeing obtained with the treated effluent was studied.

Consequently, the main goal of this study is to optimize the photo-electrochemical method for the treatment of dyeing effluents and their reuse in a new dyeing process, with the aim to reduce the consumption of water and the discharge of salts.

RESULTS AND DISCUSSION

Electrochemical Treatment: Effect of the Intensity

The electrochemical kinetic rate degradation of PN dye was studied at three intensities (2 A, 5 A and 10 A). Results are plotted in Figure 1.

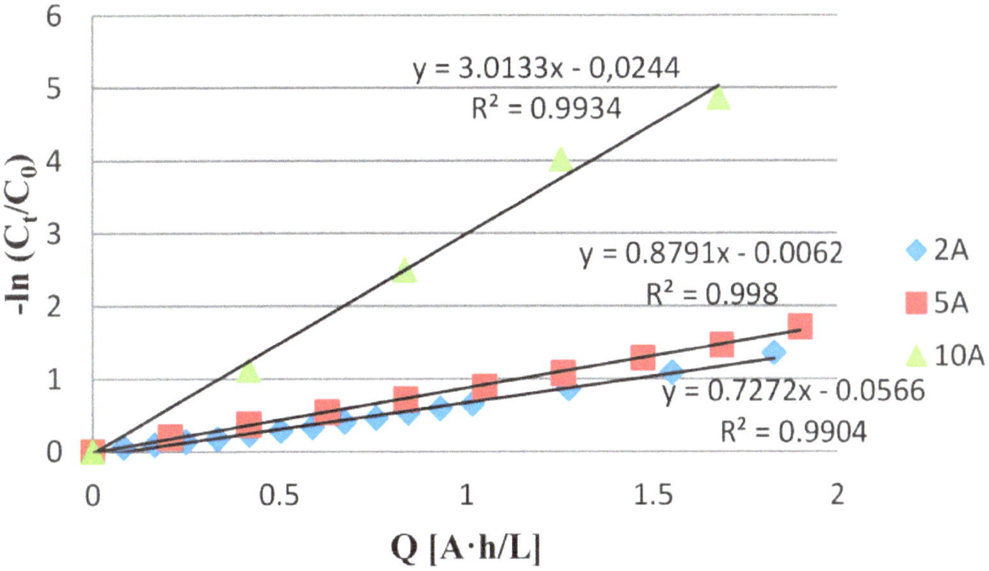

(a)

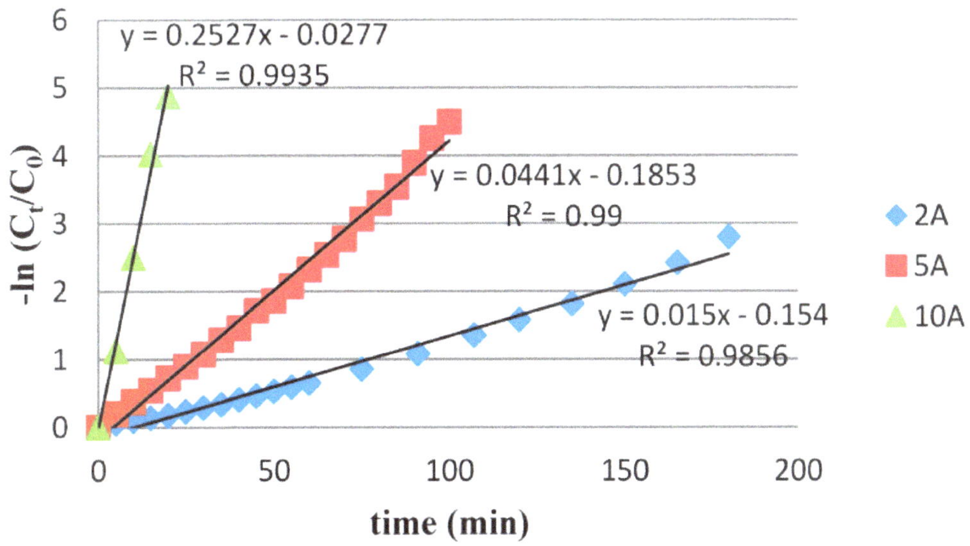

(b)

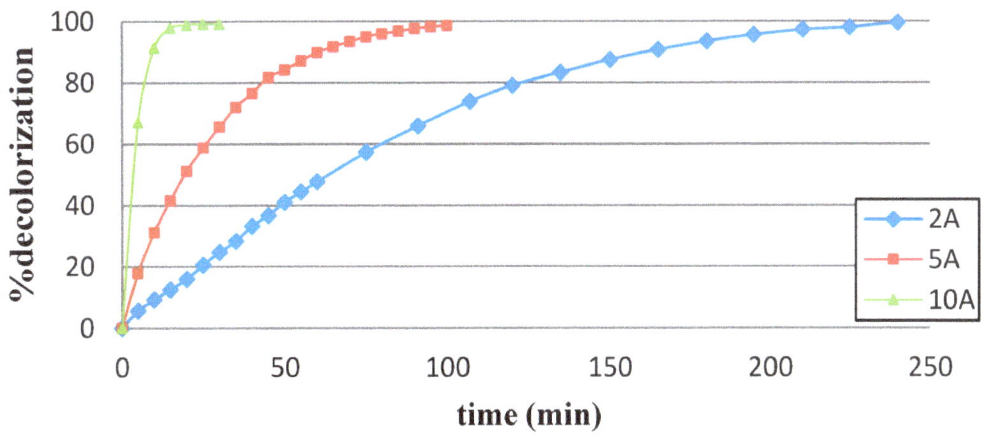

(c)

Figure 1: Procion Navy HEXL (PN) Decolorization with electrochemical treatments at 2 A, 5 A and 10 A. (**a**) PN kinetic rate in time; (**b**) PN kinetic rate in applied charge; (**c**) Evolution of PN decolorization.

The representation of $-\ln(C_t/C_0)$ *vs.* time (Figure 1a) showed that in all the studied cases, the dye degradation follows a first-order reaction kinetics (see Section 3.5.2, Equations (7) and (8)). As expected, the treatment time required to obtain a specific degradation was shorter when the intensity applied was higher. In order to know the efficiency of the treatments, $-\ln(C_t/C_0)$ *vs.* the specific charge applied (A·h/L), which is a normalized parameter, were plotted in the Figure 1b. The results showed that the slopes at 2 A and 5 A were of the same order (0.73 and 0.88 respectively). In this case, the time required to achieve a fixed decolorization had an inverse linear relationship with the intensity (at constant volume). On the other hand, when the intensity is increased to 10 A, the slope becomes much higher. The time necessary to obtain the fixed decolorization is much lower than expected. Probably this higher current value promotes the generation of additional oxidant species produced from chloride oxidation (such as Cl_2, OCl^- and chlorine radicals) or produced from water oxidation, such hydroxyl radical (•OH), atomic oxygen (•O), hydrogen peroxide or ozone [27].

Decolorization values of 99% can be reached with all the studied intensities (Figure 1c). The treatments at 2 A and 5 A requires a similar power consumption (29.70 W·h and 28 W·h respectively), whereas at 10 A, the consumption is reduced to 15.17 W·h to achieve the same decolorization rate. The power consumption was calculated from the corresponding voltage value for each intensity value (5.4 V, 5.2 V and 9.1 V at 2 A, 5 A and 10 A respectively). The obtained power values were multiplied by the time of treatment in hours.

Consequently, it can be stated that the decolorization treatment performed at 10 A is more efficient according to the kinetic rate, the specific charge applied and power consumption results. Therefore, the subsequent studies were carried out at this intensity.

Optimization of the Photo-Electrochemical Treatment for PN at 10 A

When an electrochemical treatment is applied to a solution which contains chloride and organic matter, small amounts of halogenated volatile compounds can be generated, mainly haloforms. In order to solve this problem, UV radiation is applied. Several studies were carried out by combining the electrochemical treatment with UV radiation. Table 1 shows the results corresponding to the final decolorization achieved and the decolorization kinetic rate, for electrochemical treatment (trial A) and the different combination of electrochemical and UV treatment studied on the PN effluent (trials B–E).

The results shown in Table 1 indicate that the electrochemical treatment was the faster technique to achieve 99% of decolorization because its kinetic rate was higher. When UV radiation was applied simultaneously with the electrochemical treatment (UVEC), 99% decolorization was also achieved but the kinetic rate decreased due to the UV light degradation of the indirect oxidant compounds, generated during the electrolysis ($HClO/ClO^-$). On the other hand, it can be appreciated that the UV radiation after the electrochemical treatment enhances the dyes decolorization.

Taking into account the local regulations on discharge limits, the full decolorization is generally not required. In fact, it is commonly established that no color must be appreciated when the effluent is diluted 1/10–1/30 (this value change depending on the Country and even of the different regions). Then, the more appropriate selection is an initial electrochemical treatment followed by the UV irradiation.

Among the combinations of these techniques, 5 min of electrochemical treatment provided decolorization yields around 57%. If the treatment was continued with the combined treatment UVEC, 99% of decolorization would also be achieved, whereas 66% is obtained with the UV radiation alone. For that reason, the authors decided to follow up the electrochemical treatment during 10 min (80% of decolorization) with further UV radiation to achieve 86% decolorization.

Table 1. Characterization of the electrochemical and photo-electrochemical treatments in kinetic rate, total organic carbon (TOC) removal and decolorization, applied to PN at 10 A

Trial	Treatments	Decolorization (%)		Kinetic rate (min^{-1})		R^2	
		Treat 1	Treat 1 + 2	K_1	K_2	K_1	K_2
A	EC	99	-	0.2729	-	0.99	-
B	UVEC	99	-	0.1332	-	0.99	-
C	(1) 5 min EC	55	-	0.2715	-	0.99	-
	(2) UVEC		99		0.1598		0.99
D	(1) 5min EC	59	-	0.2738	-	0.99	-
	(2) UV		66		0.0014		0.97
E	(1) 10 min EC	80	-	0.2722	-	0.99	-
	(2) UV		86		0.0037		0.98

In this section, the kinetic rates expressed in charge or in time units will have the same evolution because all the experiments were run at the same intensity (10 A). Consequently, they were only reported in time units (min^{-1}).

As can be seen in Figure 2, the decolorization value was stabilized after a certain treatment time. From that moment, the decolorization value is not enhanced when the treatment time is increased.

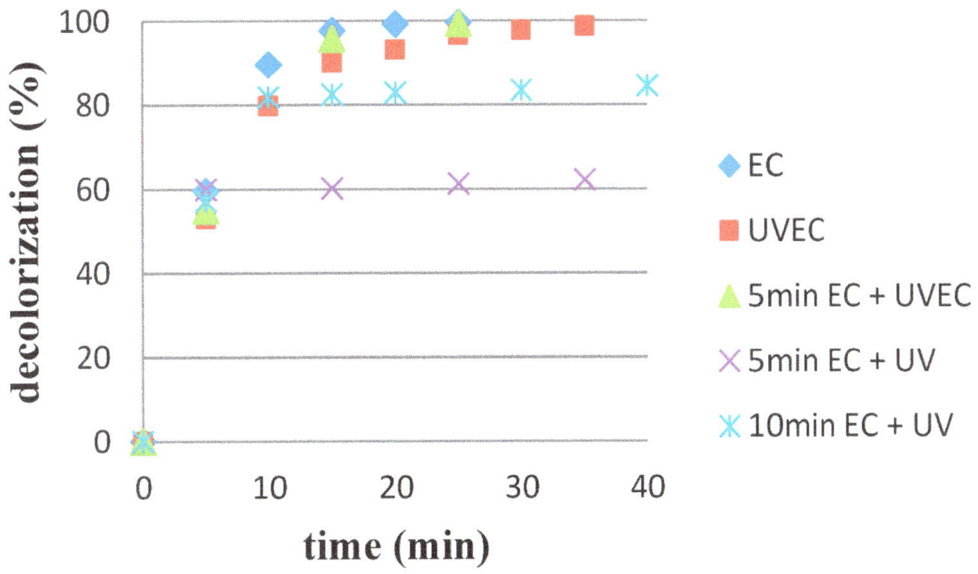

Figure 2: PN effluent decolorization evolution with different combination of electrochemical (10 A) and photo-electrochemical treatments.

The samples for haloforms detection were collected when the higher decolorization value was achieved. Chloroform was the unique halogenated compound detected by the GCMS analysis. The concentration of this compound generated in the different treatments is plotted in Figure 3.

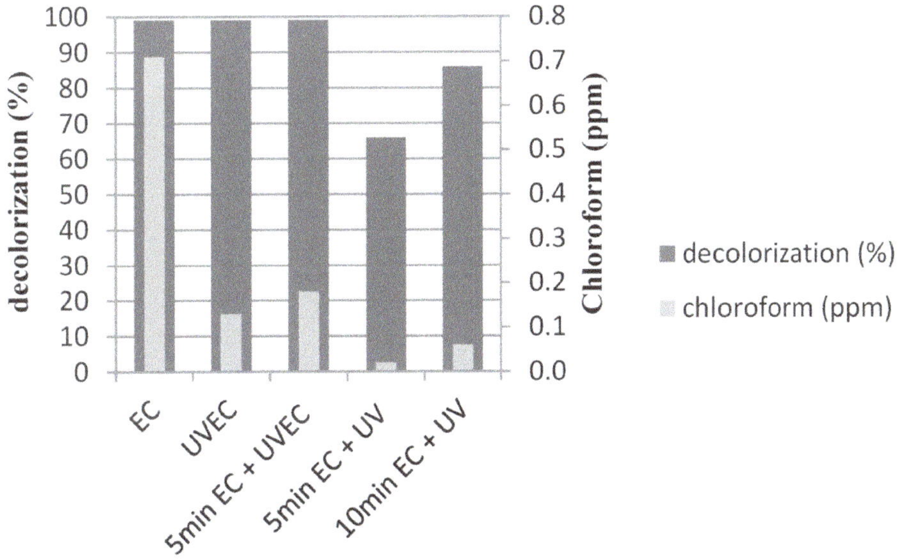

Figure 3: Final decolorization of PN effluent and chloroform generated *versus* the different electrochemical (10 A) and photo-electrochemical treatments.

According to Figure 3, it can be seen that the electrochemical treatment alone is able to achieve 99% decolorization, but at the same time, 0.7 ppm of chloroform are generated. A clear reduction of this concentration is achieved with the UV radiation treatments, both applied simultaneously during the electrochemical treatment or after it, as a final treatment. UV treatment is able to reduce chloroform concentration below 0.2 ppm.

Taking into account all the results obtained, including decolorization and chloroform removal, we consider that the best method is the trial E, where the electrochemical treatment was applied until 80% of decolorization, with posterior UV radiation (10 min EC + UV). This procedure achieved 86% effluent decolorization (which is already an acceptable value for the subsequent discharge). Under these conditions, the final chloroform concentration was lower than 0.1 ppm and the final dye concentration was lower than 0.5 ppm, which is acceptable for the reuse step. In addition, in the case of effluent discharge, this concentration is not appreciated by the human eye after a 1/10 dilution which corresponds to the mixture of the dyeing bath with the rest of effluents (rinsing and washing baths).

Application of the Optimized Photo-Electrochemical Treatment to Nine Reactive Dyes

The photo-electrochemical treatment optimized in the previous steps was applied to nine reactive dyes (described in Section 3.1). Solutions were prepared as indicated in Section 3.2 to simulate a dyeing effluent. In Table 2 are summarized the kinetic rates, total organic carbon (TOC) removal and the chloroform concentration for each dyeing effluent.

Table 2: Optimized electrochemical and photo-electrochemical treatments applied to 9 reactive dyes: kinetic rate, TOC removal and chloroform generation

Dye	Kinetic rate (min^{-1})	R^2	TOC removal (%)	Chloroform (ppm) [EC]	Chloroform (ppm) [EC + UV]
PMX2R	0.1795	0.9919	12	0.57	0.05
PC	0.1162	0.9998	6.81	1.66	0.14
PY	0.0833	0.9963	11.86	0.65	0.05
PB	0.2622	0.9907	8.33	0.09	0.01
PN	0.2686	0.9954	9.78	0.70	0.06
RB5	0.2471	0.9927	4.31	0.78	0.07
CD	0.1301	0.9985	7.84	0.45	0.04
CR	0.1593	0.9975	9.38	0.57	0.05
CA	0.1295	0.9973	7.04	0.33	0.03

No tendency is detected with respect to the number of functional groups in the dye molecule. In all cases, dyes degradation follows a first order kinetic rate. In addition, in all the studied cases, 86% decolorization was achieved. The TOC removal values are poor, in a range from 4% to 11% depending on the dye. However, recent studies [8] with more concentrated effluents showed that the TOC removal is dependent on dye concentration. 50%–80% TOC removal can be achieved when the dye concentration is 1 g/L or when the treatment time is increased. The authors consider that the organic matter removal is not the goal of this work and consequently, an extended electrochemical treatment is not necessary. There are more efficient methods for dye mineralization, from the economic point of view, such as biological plants. Thus, the aim of this study is the decolorization of the effluent and its further reuse or discharge to a biological plant (where the total mineralization is carried out). In Section 2.4, the influence of TOC removal in the effluent reuse for a new dyeing process is studied. In addition, the chloroform concentration was analyzed after the EC and the EC + UV treatment. As expected, the combination of both techniques (electrochemistry and UV irradiation) reduces the amount of chloroform. The chloroform concentrations detected after the EC + UV treatment were lower than 0.1 ppm, except for PC, for which the value was 0.14 ppm. On average, 90% of chloroform reduction is obtained in the EC + UV treatment with respect to EC treatment.

Reuse of the Decolorized Effluents

The decolorized effluents were reused in a new dyeing process. The color differences (calculated as detailed in Section 3.6) were evaluated with the parameter $DE_{CMC\,(2:1)}$ with respect to a reference dyeing. In all the studied cases, these values were lower than 1, which is the acceptance limit for the textile industry (Table 3). This implies that all the new dyeings are within the acceptance range. Consequently, the dyes mineralization is not necessary for the effluent reuse because a shorter decolorization treatment provide satisfactory results.

The reuse procedure enables saving 70% of the dyeing water and leads a considerable reduction of effluents salinity. Hence, the photo-electrochemical technique described can be considered an ecofriendly method for the treatment and reuse of dyeing effluents.

Table 3: Color differences: fabrics dyed with the treated and reused effluent *vs.* a reference dyeing

Dye	DL_{CMC}	DC_{CMC}	DH_{CMC}	$DE_{CMC\,(2:1)}$
PMX2R	0.36	−0.26	0.10	0.46
PC	0.06	−0.14	−0.07	0.16
PN	−0.43	0.13	0.22	0.50
PY	−0.42	−0.21	−0.33	0.57
PB	0.04	−0.23	0.04	0.24
RB5	−0.05	0.36	0.03	0.36
CD	−0.51	−0.20	0.11	0.56
CR	0.04	0.05	0.09	0.11
CA	0.01	0.04	−0.01	0.05

In order to evaluate the influence of the residual organic matter in the dyeing color differences, longer electrochemical treatments were applied to the PN effluent. Higher TOC degradations were obtained, and the decolorized effluents were reused in new dyeing processes. The dyeings performed with these reused baths showed the color differences (*versus* a reference) listed in Table 4.

Table 4: Influence of the TOC removal in the PN dyeing result. Color differences of fabrics dyed with the treated and reused effluent *vs.* the reference

TOC removal (%)	$DE_{CMC\,(2:1)}$	DL_{CMC}	DC_{CMC}	DH_{CMC}
4.74	0.6	−0.42	0.09	0.42
5.97	0.59	0.38	0.40	0.22
7.21	0.48	0.02	0.34	0.34
11.91	0.37	0.14	0.23	0.24

This table and Figure 4 showed a clear relationship between color differences and the organic matter removal. The higher the TOC removal, the lower the color differences. However, it must be underlined that in all the studied cases, the color differences are into the limits of acceptance for the industry. For this reason, it is not necessary to increase the TOC removal rate as an extended electrochemical treatment would imply higher power consumption.

In addition, it can be stated that all dyes behave in a similar way. Then, the number of functional groups is not an influent parameter in the reuse process.

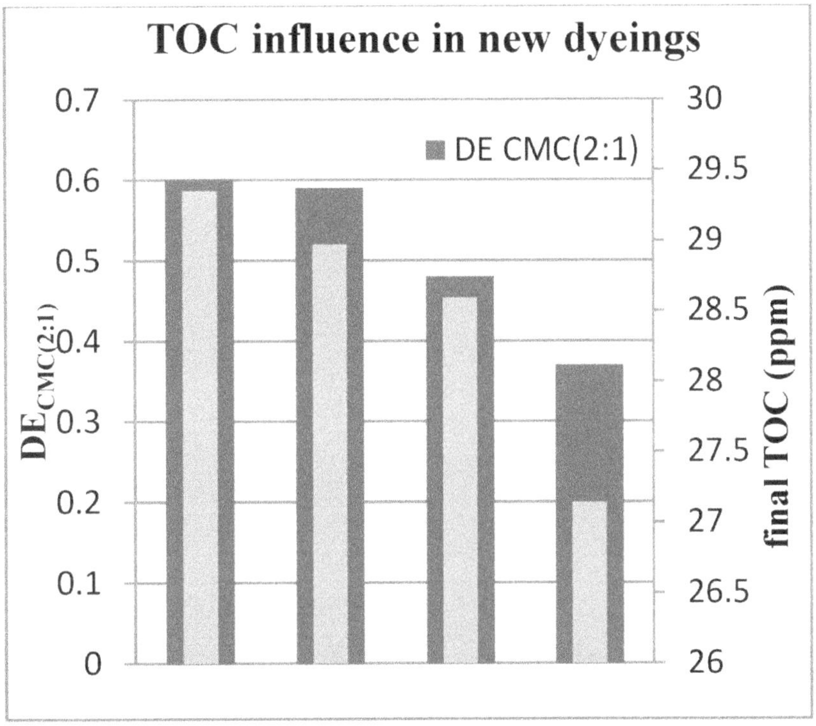

Figure 4: Influence of TOC removal in the reuse of treated effluent for PN: color differences ($DE_{CMC\,(2:1)}$) of dyeings performed with different TOC content effluents.

EXPERIMENTAL PROCEDURES

Dyes and Reagents

Nine reactive dyes, including monofunctional, bifunctional and trifunctional structures, were kindly provided by DyStar. Their name and functional groups are listed in Table 5.

Table 5: Description of selected Reactive Dyes

Abbr.	Commercial name	C.I. name	Chromophore	Reactive groups	Type of Reactive group	λ_{max} (nm)
PN	Procion Navy H-EXL	Not registered	Disazo	2	monochlorotriazine	606
RB5	Remazol Black 133B	Reactive Black 5	Disazo	2	sulphatoethylsulfone	598
PMX2R	Procion Orange MX-2R	Reactive Orange 4	Monoazo	1	dichlorotriazine	489
PY	Procion Yellow H-EXL	Reactive Yellow 138:1	Disazo	2	monochlorotriazine	416
PC	Procion Crimson H-EXL	Reactive Red 231	Disazo	2	monochlorotriazine	545
PB	Procion Blue H-EXL	Not registered	Not known	2	Not known	624
CD	Cibacron Deepnight S-R	Not registered	Not known	3	Not known	583
CR	Cibacron Ruby S-3B	Not registered	Not known	3	Not known	543
CA	Cibacron Yellow S-3R	Not registered	Not known	3	Not known	417

Figure 5 shows their chemical structure. The formula corresponding to PB, CD, CR, CA have not already been published.

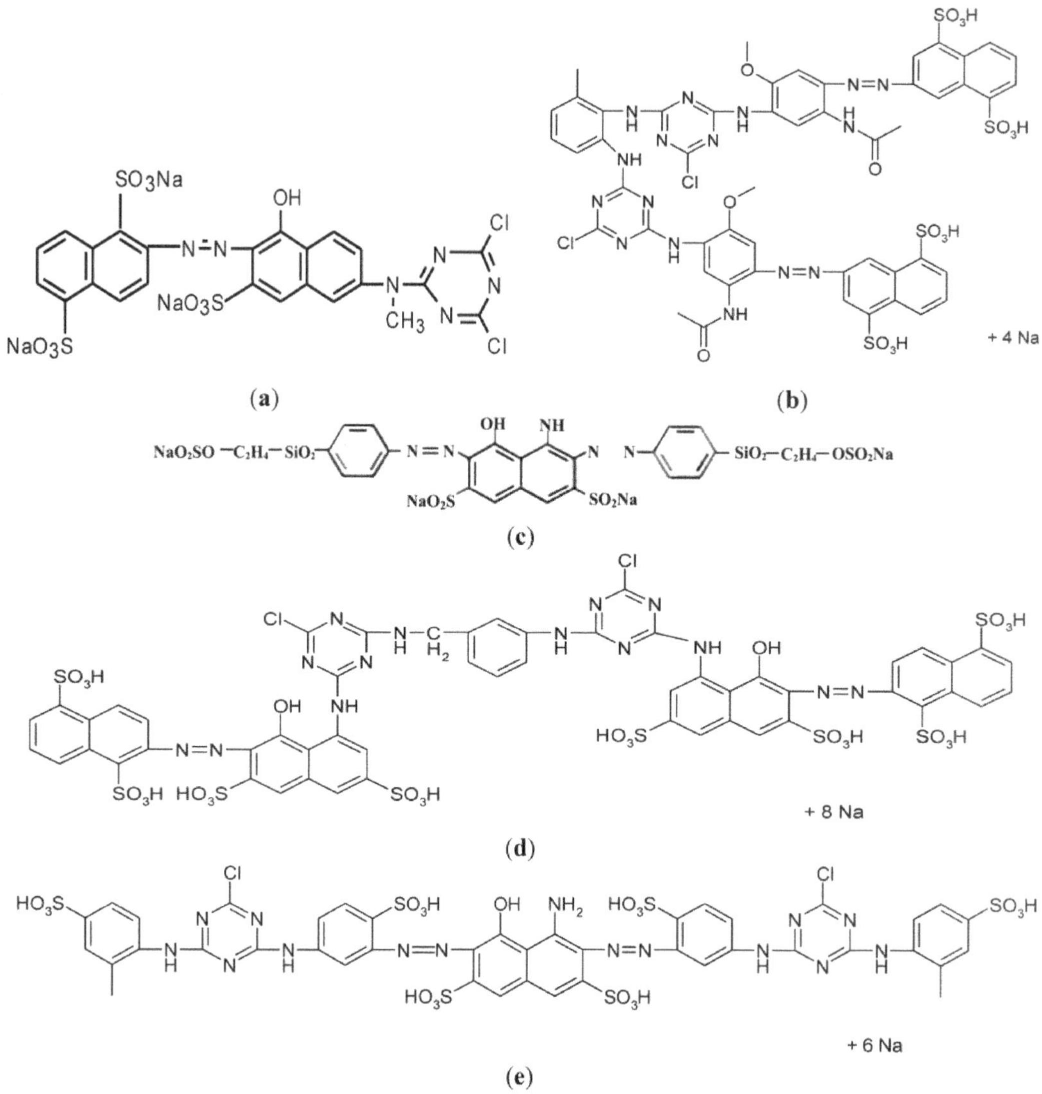

Figure 5: Chemical structure of (**a**) PMX2R; (**b**) PY; (**c**) RB5; (**d**) PC; (**e**) PN.

Dyeing was performed on 100% cotton fabrics, kindly provided by TIPSA.

The chemical products used in the effluents and the dyebath preparation (NaOH, Na_2SO_4, and NaCl) were of analysis quality supplied by Merck and Fluka (Madrid, Spain). The Internal Standard for the gas chromatography was p-Bromofluorobenzene chromatography grade provided by AccuStandard (Barcelona, Spain).

Effluent Preparation

Dyebath effluents were prepared in the laboratory to simulate an industrial effluent after the dyeing process with reactive dyes. During the dyeing process, the dye unfixed to the fiber remains hydrolyzed in the bath solution as the reactive groups of the dye molecule were replaced by OH groups. The effluents containing 0.1 g/L of hydrolyzed reactive dyes were prepared previously to the electrochemical treatment. The hydrolysis was carried out by heating at 80 °C for 2 h at pH 12 with 0.01 M·NaOH solution (about 10% of the total dyebath volume).

As indicated, chloride plays an important role in dyes indirect oxidation. For this reason, sodium chloride (0.3 g/L) was added to simulate the concentration of tap water (industrial dyeings are performed with decalcified water). The final conductivity of the simulated effluent was adjusted to 35 mS/cm with sodium sulfate in order to reach the industrial effluents conductivity.

Electrochemical Treatment

The synthetic effluents were treated in an undividable electrolytic cell, showed in Figure 6.

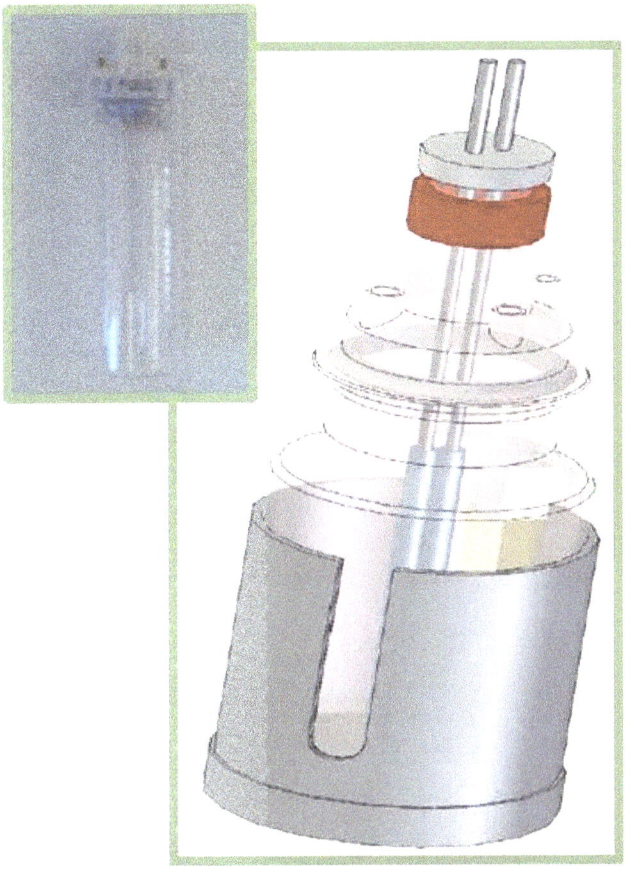

Figure 6: Electrolytic cell with the UV source.

In all cases, the treated volume was 2 L and studies were carried out under galvanostatic conditions, with a power supply Grelco GVD310 0–30 Vcc/0–10 A (Cornellà de Llobregat, Barcelona, Spain) at intensities 2 A, 5 A and 10 A. The cathode was made of stainless steel and the anode was a DSA (Dimensionally Stable Anode) made of Ti/Pt, with 59.32 cm^2 of active surface.

The UV radiation was performed with a Philips TUV lamp PL-S UV-C (Sanchinarro, Madrid, Spain) at 254 nm and 9 watts.

Effluent Reuse

Dyeing tests were performed with a TICOLOR instrument under the following conditions: 10 g of cotton fabric, dyestuff concentration 3% o.w.f (on the weight of fiber), liquor ratio 1/10 and 35 g/L of Na_2SO_4. NaOH was used as dyeing alkali (until pH 12).

According to the information supplied by DyStar [28], the dyeing method "all in" was selected. The dye and the entire electrolyte were included in the initial dyeing bath and the alkali was added when the dyeing process was already started. The dyeing started at 50 °C for 15 min, then the temperature was raised to 80 °C at a gradient of 1.4 °C/min. The alkali was added after 30 min at 80 °C and the dyeing lasted for 60 min more.

All the experiments were run in triplicate.

After the dyeing process, a washing process takes place where the cotton dyed fabrics were washed to eliminate the dye not fixed to the fabric. This process consists on nine successive washes at different temperatures. The first ones (1st–3rd washes) were at 50 °C with tap water during 10 min, then a soaping step (4th wash) took place with 2 g/L of COTEMOLL TLTR at 95 °C during 15 min. Then, a tap water step (5th wash) and another soap step (6th wash) were carried out with the same conditions described above. Finally, another three water washes at 50 °C were carried out during 10 min (7th–9th washes). All steps were performed at liquor ratio 1:10.

Analyses and Instruments

Spectroscopic Analysis

The decolorization process was studied by spectroscopy analysis, where the initial dyes absorbance (Abs_0) was compared with the absorbance of the samples collected during the treatment (Abs_t). The absorbance was measured at the visible maximum dye absorption wavelength (606 nm for PN, 598 nm for RB5, 489 nm PMX2R, 416 nm for PY, 545 nm for PC, 624 nm for PB, 583 nm for CD, 543 nm for CR and 417 nm for CA). Samples were collected each 5min during the electrochemical treatment, and the decolorization was reported in % (Equation (6)).

$$D(\%) = \frac{(Abs_0 - Abs_t) \times 100}{Abs_0} \quad (6)$$

Absorbance measurements were carried out with a UV-Vis spectrophotometer (Shimadzu UV-2401 PC, Kyoto, Japan).

KINETIC EVOLUTION

In accordance with previous studies [13], the dyes degradation with the electrochemical treatment follows a first-order reaction. The decolorization rate constants (K_d) were calculated from the slope of semilogarithmic absorbance values rate*versus* exposition time (t) or charge (Q), in accordance with the kinetic Equations (7)–(9):

$$-\ln\frac{Abs_t}{Abs_0} = K_d \times t \quad (7)$$

$$-\ln\frac{Abs_t}{Abs_0} = K_{d'} \times Q \quad (8)$$

where:

$$Q\left(\frac{A \cdot h}{L}\right) = \left(\frac{I \times (t_t - t_{t-1})}{vol}\right) \quad (9)$$

TOC Removal

The Total Organic Carbon was determined with a Shimadzu TOC5050A instrument. The TOC degradation was reported in % (see Equation (10)).

$$TOC\ removal(\%) = \frac{(TOC_0 - TOC_t) \times 100}{TOC_0} \quad (10)$$

GCMS Analysis

In order to evaluate the possible generation of volatile halogenated compounds, a GCMS method was used for the identification of 23 halogenated compounds listed by the Environmental Protection Agency (EPA-624 method) [29]. An

Internal Standard (IS), p-bromofluorobenzene, was used for the quantification of all these compounds.

The gas chromatography analyses were carried out in a Shimadzu QP 2010 (GCMS) system with a Mass Spectrum detector. The sample injection was carried out by a headspace technique, where 15 mL of the sample was heated to 80 °C for 45 min in a 20 mL vial. The vial was sealed with silicone/PFTE septum and 1 mL of gas was injected to the GCMS. The injection temperature was 200 °C. The chromatography program started at 35 °C during 10 min, the gradient rate was increased at a rate of 4 °C/min until it reached 150 °C, which is the final temperature (10 min). The column selected was TRB-624 (length 30 m, internal diameter 0.25 mm and 1.4 µm packing). The carrier gas was Helium at a 0.95 mL/min column flow rate.

The identification of halogenated compounds was performed using Nist 147, Nist 27 and Wiley 229 as reference libraries. Compounds, detection limits and the retention times are listed in Table S1 of Supporting information.

The quantitative analysis is carried out by using the Equation (11):

$$\text{Concentration}_c = \frac{\text{Area}_c \times \text{Concentration}_{IS}}{\text{Area}_{IS} \times R_f} \quad (11)$$

where R_f is the response factor, which is specific for each compound. It corresponds to the rate between the chromatographic response of the halogenated compound detected and the Internal Standard (1-bromo, 3-fluorobenzene).

Color Evaluation

The final dyed fabric color was measured by a spectrophotometer Macbeth Color Eye 7000, with illuminate D65 and 10° of standard observer. The instrument evaluates the chromatic coordinates of each dyed fabric. These coordinates are defined by three parameters (Lightness DL_{cmc}; Chrome DC_{cmc}, and Hue DH_{cmc}) which compromise the color difference between the standard and the reused effluent dyeing. The color difference with the formula $DE_{CMC\,(2:1)}$ is represented in the Equation (12) according to UNE-EN ISO 105-J03: 1997 [30].

$$DE_{CMC\,(2:1)} = [(DL^*/2S_L)^2 + (DC^*_{ab}/S_c)^2 + (DH^*_{ab}/S_H)^2]^{1/2} \quad (12)$$

$S_L = 0.040975L^*_R/(1 + 0.01765L^*_R)$ if $L^*_R \geq 16$;
Or $S_L = 0.511$ if $L^*_R < 16$;
$S_c = [0.0638\, C^*_{ab,R}/(1 + 0.0131C^*_{ab,R})] + 0.638$;
$S_H = (FT + 1 - F)\, S_c$;

and:

$F = \{(C^*_{ab,R})^4/[(C^*_{ab,R})^4 + 1900]\}^{1/2}$;
$T = 0.36 + |0.4\cos(35 + h_{ab,R})|$ if $h_{ab,R} \geq 345°$ or $h_{ab,R} \leq 164°$;
Or $T = 0.56 + |0.2\cos(168 + h_{ab,R})|$ if $164° < h_{ab,R} < 345°$.

In general, the acceptance limit for color differences in the textile industry is one unit ($DE_{CMC\,(2:1)} \leq 1$). This criterion is widely used in dyeing quality control to compare the color differences between two fabric samples.

CONCLUSIONS

The electrochemical treatment achieved 99% of dye decolorization at all the studied intensities, although it was more efficient at the higher current intensity, 10 A. When this technique was applied with simultaneous UV irradiation, the decolorization kinetic rates decreased, whereas dye decolorization was enhanced by the application of UV after the electrochemical treatment. In any case, the UV irradiation reduced the chloroform content.

The most efficient combination to treat reactive dye effluents is the electrochemical treatment up to 80% decolorization followed by UV irradiation. In this way, the chloroform concentration was lower than 0.1 ppm except for PMX-2R. No influence of the number of functional groups in the dye molecule was evidenced.

It was also demonstrated that the complete dye mineralization was not necessary to reuse the treated effluent in new dyeing processes. Decolorized effluents with only 4% TOC removal provided color differences into the acceptable range ($DE_{CMC\,(2:1)} < 1$), which resulted in reducing significantly effluent salinity and saving 70% water.

ACKNOWLEDGMENTS

This work was supported by the Spanish Ministry of Science and Innovation (Project CTM 2012-31461).

AUTHOR CONTRIBUTIONS

Carmen Gutiérrez-Bouzán planed the experimental part. Mireia Sala and Victor López carried out the experimental work. All the authors participated in analysis of results. The manuscript was written by Mireia Sala and Carmen Gutiérrez-Bouzán.

REFERENCES

1. Dos Santos, A.B.; Cervantes, F.J.; van Lier, J.B. Review on current technologies for decolorisation of textile wastewaters: Prespectives for anaerobic biotechnology. *Bioresour. Technol.* 2007, *98*, 2369–2385.
2. Robinson, T.; McMullan, G.; Marchant, R.; Nigman, P. Remediation of dyes in textile effluent: A critical review on current treatment technologies with a proposed alternative. *Bioresour. Technol.* 2001, *77*, 247–255.
3. Kyzas, G.Z. A decolorization technique with spent "Greek Coffee" grounds as zero-cost adsorbents for industrial textile wastewaters. *Materials* 2012, *5*, 2069–2087.
4. Kyzas, G.Z.; Fu, J.; Matis, K.A. The change from past to future for adsorbent materials in treatment of dyeing wastewaters. *Materials* 2013, *6*, 5131–5158.
5. Kyzas, G.Z.; Kostoglou, M. Green adsorbents for wastewaters: A critical review. *Materials* 2014, *7*, 333–364.
6. Sun, S.P.; Hatton, T.A.; Chan, S.Y.; Chung, T.S. Novel thin-film composite nanofiltration hollow fiber membranes with double repulsion for effective removal of emerging organic matters from water. *J. Membr. Sci.* 2012, *401–402*, 152–162.
7. Ong, Y.K.; Li, F.Y.; Sun, S.P.; Zhao, B.W.; Liang, C.Z.; Chung, T.S. Nanofiltration hollow fiber membranes for textile wastewater treatment: Lab-scale and pilot-scale studies. *Chem. Eng. Sci.* 2014, *114*, 51–57.

8. Gutierrez, M.C.; Crespi, M. A review of electrochemical treatments for color elimination. *J. Soc. Dyes Colour.* 1999, *115*, 342–345.

9. Naim, M.M.; el Abd, Y.M. Removal and recovery of dyestuff from dyeing wastewaters. *Separ. Purif. Meth.* 2002, *31*, 171–228.

10. Forgacs, E.; Cserhati, T.; Oros, G. Removal of synthetic dyes wastewaters: A review. *Environ. Int.* 2004, *330*, 953–971.

11. Martinez-Huitle, C.A.; Ferro, S. Electrochemical oxidation of organic pollutants for the wastewater treatment: Direct and indirect processes. *Chem. Soc. Rev.* 2006, *35*, 1324–1340.

12. Vilaseca, M.; Gutierrez, M.C.; Lopez-Grimau, V.; Lopez-Mesas, M.; Crespi, M. Biologial treatment of a textile effluent after electrochemical oxidation of reactive dyes. *Water Environ. Res.* 2009, *82*, 176–182.

13. Riera-Torres, M.; Gutierrez, M.C. Colour removal of three reactive dyes by UV light exposure after electrochemical treatment. *Chem. Eng. J.* 2010, *156*, 114–120.

14. Lopez-Grimau, V.; Gutierrez, M.C. Decolorisation of simulated reactive dyebath effluents by electrochemical oxidation assisted by UV light. *Chemosphere* 2006, *62*, 106–112.

15. Oliveira, F.H.; Osugi, M.E.; Paschoal, F.M.M.; Profeti, D.; Olivi, P.; Zanoni, M.V.B. Electrochemical oxidation of an acid dye by active chlorine generated using Ti/$Sn_{(1-x)}Ir_xO_2$ electrodes. *J. Appl. Electrochem.* 2007, *37*, 583–592.

16. Osugi, M.E.; Rajeshwar, K.; Ferraz, E.R.A.; Oliveira, D.P.; Araujo, A.R.; Zanoni, M.V.B. Comparison of oxidation efficiency of disperse dyes by chemical and photoelectrocatalytic chlorination and removal of mutagenic activity.*Electrochim. Acta* 2009, *54*, 2086–2093.

17. Osugi, M.E.; Umbuzeiro, G.A.; de Castro, F.J.V.; Zanoni, M.V.B. Photoelectrocatalytic oxidation of remazol turquoise blue and toxicological assessment of its products. *J. Hazard. Mater.* 2006, *137*, 871–877.

18. Panizza, M.; Barbucci, A.; Ricotti, R.; Cerisola, G. Electrochemical degradation of methylene blue. *Separ. Purif. Technol.*2007, *54*, 328–387.

19. Daneshvar, N.; Aber, S.; Vatanpour, V.; Rasoulifard, M.H. Electro-Fenton treatment of dye solution containing Orange II: Influence of operational parameters. *J. Electroanal. Chem.* 2008, *615*, 165–174.

20. Carneiro, P.A.; Osugi, M.E.; Fugivara, C.S.; Boralle, N.; Furlan, M.; Zanoni, M.V.B. Evaluation of different electrochemical methods on the oxidation and degradation of Reactive Blue 4 in aqueous solution. *Chemosphere* 2005,*59*, 431–439.

21. Lee, Y.H.; Pavlostathis, S.G. Decolorization and toxicity of reactive anthraquinone textile dyes under methanogenic conditions. *Water Res.* 2004, *38*, 1838–1852.

22. Oliveira, G.A.R.; Ferraz, E.R.A.; Chequer, F.M.D.; Grando, M.D.; Angeli, J.P.F.; Tsuboy, M.S.; Marcarini, J.C.; Mantovani, M.S.; Osugi, M.E.; Lizier, T.M.; *et al.* Chlorination treatment of aqueous samples reduces, but not eliminate, the mutagenic effect of the azo dyes Disperse Red 1, Disperse Red 13 and Disperse Orange 1. *Mutat. Res.*2010.

23. Carneiro, P.A.; Umbuzeiro, G.A.; Oliveira, D.P.; Zanoni, M.V.B. Assessment of wáter contamination caused by mutagenic textile effluent/dyehouse effluent bearing disperse dyes. *J. Hazard. Mater.* 2010, *174*, 694–699.

24. Guaratini, C.C.I.; Zanoni, M.V.B. Corantes texteix. *Quim. Nova* 2000, *23*, 71–78.

25. Zanoni, M.V.B.; Fogg, A.G.; Barek, J.; Zima, J. Electrochemical investigations of reactive dyes; cathodic stripping voltammetric determination of anthraquinone-based chlorotriazine dyes at a hanging mercury drop electrode. *Anal. Chim. Acta* 1997, *349*, 101–109.

26. Zollinger, H. *Color Chemistry—Synthesis, Propierties and Applications of Organic Dyes and Pigments*; V.C.H Publishers: New York, NY, USA, 1991.

27. Martínez-Huitle, C.A.; Brillas, E. Electrochemical alternatives for drinking water disinfection. *Angew. Chem. Int. Ed.*2008, *47*, 1998–2005.

28. DyStar Inc. *Work Procedures*. In *Industrial Washing Procedure, Excel Washing*; DyStar, Inc.: Frankfurt, Germany, 2004.

29. U.S. Environment Protection Agency. Methods for Organic Chemical Analysis of Municipal and Industrial Wastewater. Method 624—Purgeables. Available online: http://water.epa.gov/scitech/methods/cwa/organics/upload/2007_07_10_methods_method_organics_624.pdf.

30. AENOR: UNE-EN ISO105-J03. In *Ensayos de Solidez del Color: Cálculo de Las Diferencias de Color*; Asociación Española de Normalización y Certificación: Madrid, Spanish, 1997. (In Spanish)

CHAPTER 14

HIGHLY ISOTACTIC POLY(N-BUTENYL-CARBAZOLE): SYNTHESIS, CHARACTERIZATION, AND OPTICAL PROPERTIES

Antonio Botta, Vincenzo Venditto, Alfredo Rubino, and Stefania Pragliola

INTRODUCTION

The recent growing interest in carbazole containing polymers is due to the discovery of polymeric light emitting diodes [1] and organic photorefractive materials [2], but also to their potential application as components in photovoltaic devices [3]. The first and the most widely studied organic polymeric photoconductor is the poly(N-vinylcarbazole) (PVK) [4–6]. This polymer was promptly used by IBM company to develop the PVK-TNF (2,4,7-trinitrofluorenone) system as organic photoconductor for Copier I series [7]. Although PVK-TNF system can be considered nowadays completely obsolete, due to the relatively low photosensitivity, high toxicity of TNF, and poor mechanical strength of the PVK films, the interest in carbazole containing polymers remains high. To find new application fields of PVK derivatives, attempts have been made to modify PVK structure. The aim was to provide the rational design and optimisation of high performance organic photoactive polymeric materials. Polymers with pendant or in chain isolated carbazolyl groups were synthetized by using different polymerization techniques [8–11]. In particular, in literature, a detailed investigation of photophysical properties of polymers with different pendant carbazolyl groups and the comparison with PVK has been reported [9–13]. It was found that the nature and length of spacer linking carbazole moiety to polymer backbone, as well as the stereoregularity of polymer microstructure, modify the photophysical properties and photoconductive behaviour of carbazole containing polymers [10].

It is well known that a precise stereocontrol of microstructures for polymers obtained by vinyl monomers can be achieved by coordination Ziegler-Natta polymerization [14]. In particular, the development of homogeneous catalytic systems based on metallocenes and postmetallocenes allowed in the latest two decades reaching an important goal in the realm of specialty polymers [15–18]. Recently, we reported the polymerization of a vinyl monomer with a pendant carbazole group (9-(pent-4-en-1-yl)- 9H-carbazole). Isotactic poly(N-pentenyl-carbazole) (*i*-PPK) was obtained with high yield by using the isospecific catalytic system rac-[(CH_3)2Si(indenyl)$_2$]$ZrCl_2$/MAO [19, 20]. For the first time, isotactic microstructure of PPK was unambiguously determined by 13C and 1 H NMR analysis [19, 20]. *i*- PPK optical properties were also investigated and an electroluminescent behaviour of polymer was observed. Innovative organic light emitting diodes (OLEDs) were achieved using this new material as emitting layer, showing white light emission and promising performances [19, 20]. In this paper we present the synthesis of isotactic

Citation: Antonio Botta, Vincenzo Venditto, Alfredo Rubino, and Stefania Pragliola, "Highly Isotactic Poly(N-butenyl-carbazole): Synthesis, Characterization, and Optical Properties," Journal of Chemistry, vol. 2016, Article ID 1656459, 8 pages, 2016. doi:10.1155/2016/1656459

poly(Nbutenyl-carbazole) (*i*-PBK), a polymer homologue to *i*-PPK but having the linker connecting the carbazole to the main chain one carbon atom shorter. *i*-PBK was obtained by using the homogeneous isospecific Ziegler-Natta catalytic system rac-$[(CH_3)_2Si(indenyl)_2]ZrCl_2$/methylallumoxane (MAO). The achieved polymer was structurally analyzed by 13C NMR, FTIR, and X-ray analysis; moreover, thermal analysis, UV-Vis, and photoluminescence analysis were also carried out. In addition, the infrared and fluorescence behaviors of *i*-PBK were compared with those of the largely investigated PVK [4–6, 10] and of the recently achieved *i*-PPK [19, 20]. On the basis of this comparison, the influence of the flexible alkylene spacer, absent in PVK and longer by one carbon atom in *i*-PPK, on the structural and optical properties of this class of polymers, was also determined

EXPERIMENTAL

All manipulations involving air and moisture sensitive compounds were carried out under an atmosphere of dried and purified nitrogen with standard vacuum-line, Schlenk, or glovebox techniques.

MATERIALS

All reagents and solvents were purchased from SigmaAldrich s.r.l. (Milan, Italy). Toluene was refluxed over sodium for 48 h and distilled before use, methylalumoxane was used as a solid after distillation of solvent (toluene), and other reagents were used without further purification. The catalytic precursor rac-dimethylsilylbis(1-indenyl)zirconium dichloride was purchased from Strem Chemicals, Inc. Poly(9- vinylcarbazole) was purchased from Sigma-Aldrich s.r.l. (Milan, Italy) with an average M_n of 25,000–50,000.

Synthesis of 9-(But-3-en-1-yl)-9H-carbazole.

In a 100 mL round bottom flask, to a suspension of carbazole (0,0598 mol) and 4-bromo-1-butene (0,598 mol), benzyltriethylammonium bromide (TBEA) (0,00179 mol) and 35 mL of 16 M NaOH solution were added. The mixture was left stirring for 24 h at 60° C. Afterwards the mixture was poured into water and extracted twice with ethyl acetate. The organic layer was dried over anhydrous $MgSO_4$ and filtered and the solvent removed by evaporation to give the crude product. The resulting residue was purified on a silica gel column chromatography in hexane/ethyl acetate (8 : 2) to give a white solid (87% yield). 1 H NMR (300 MHz, $CDCl_3$) δ: 2.61–2.68 (2H, m, CH_2), 4.40 (2H, t, J = 7.2 Hz, NCH_2), 5.06–5.12 (2H, dd, J = 10.2, 17.1 Hz, CH2=CH), 5.82–5.93 (1H, m, CH), 7.25 (2H, d, J = 7.5 Hz, Ar), 7.45 (2H, t, J=8 Hz, Ar), 7.50 (2H, t, J=8 Hz, Ar), 8.12, (2H, d, J = 7.8 Hz Ar); 13C NMR (75 MHz, CDCl3) δ: 140.49, 135.11, 125.83, 123.12, 120.58, 119.07, 117.52, 108.88, 42.92, 33.42 [21].

Synthesis of Isotactic PBK (*i*-PBK)

Isotactic PBK was prepared by introducing 7 mL of dry toluene, 4.52×10^{-3} mol (1 g) of monomer, and 8.27 × 10–3 mol (480 mg) of MAO into a 50 mL round bottom flask equipped with magnetic stirrer. Polymerization was started by introducing 1.43×10^{-5} mol (6.4 mg) of rac-dimethylsilylbis(1-indenyl) zirconium dichloride. Polymerization was stopped, after 72 hours, by introducing a few milliliters of ethanol. Then, the polymer was coagulated in an excess of acidified ethanol (solution of HCl in ethanol 0.1–0.2 M), washed several times with fresh ethanol, and dried in vacuo at room temperature (yield 0.575 g). The polymer as obtained from the synthesis and was extracted in boiling acetone and two fractions, an insoluble (0.375 g) and a soluble (0.200 g) acetone fractions, were separated. The acetone insoluble fraction was soluble in toluene, chlorobenzene and tetrachloroetane.

CHARACTERIZATIONS

NMR Analysis

NMR spectra of monomer were recorded on an AX 300 Bruker spectrometer operating at 298 K. The sample was prepared by dissolving 5 mg of monomer in 0.5 mL of deuterated chloroform ($CDCl_3$). Tetramethylsilane was used as internal chemical shift reference. NMR spectra of polymer were recorded on an AX 300 Bruker spectrometer operating at 75 MHz and 373 K. The samples were prepared by dissolving 40 mg of polymer in 0.5 mL of 1,1,2,2-Tetrachloroethane-d_2 (TCDE). Hexamethyldisiloxane was used as internal chemical shift reference.

Gel Permeation Chromatography Measurements

The molecular weight distribution and polydispersity index were estimated by GPC at 140° C using 1,2,4-trichlorobenzene as solvent and narrow molecular weight distribution polystyrene (PS) standard samples as reference. The measurements were performed on PLGPC210 with four PL-Gel Mixed A Columns, a refractive detector, and a DM400 data manager (Viscotek, Malvern Instruments Ltd., Worcestershire, UK).

Calorimetric Measurements. Calorimetric measure

ments were carried out on a DSC Q20 apparatus manufactured by TA Instruments, in flowing N_2. Polymer samples of 5–10 mg were placed in aluminum pans and heated/cooled at a rate of 10° C/min. Measurements were done in the range 30–280° C.

X-Ray Diffraction Patterns

X-ray diffraction patterns of polymer powder sample were obtained by an automatic Bruker D8 Advance diffractometer, in reflection, by using the nickel filtered Cu-Kα radiation (1.5418A).

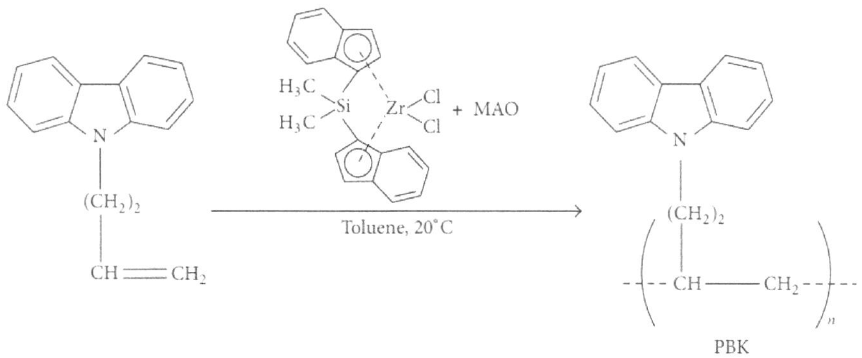

Scheme 1

Fourier Transform Infrared Measurements.

Infrared spectra were obtained at a resolution of 2.0 cm^{-1} with a Tensor 27 Bruker spectrometer equipped with deuterated triglycine sulfate (DTGS) detector and a Ge/KBr beam splitter. The frequency scale was internally calibrated to 0.01 cm−1 using a He-Ne laser. Thirty-two scans were signal averaged to reduce the noise. Before infrared analysis, samples were prepared on KBr pellets, at room temperature. The pellets were then dried in vacuum at room temperature.

Optical Characterization

Thin polymer films have been prepared by spin coating on quartz slides substrate. The film thickness and roughness were measured by KLA Tencor P- 10 surface profiler. UV-Vis measurement was performed by a Varian Cary 50 spectrophotometer and photoluminescence recorded by a Varian Cary Eclipse spectrophotometer.

RESULTS AND DISCUSSION

Synthesis and ^{13}C-NMR Characterization

Monomer, 9- (but-3-en-1-yl)-9H-carbazole, was synthesized following a synthetic procedure accurately described in the experimental part. Polymerization of monomer was performed at 20° C in toluene solution by using the rac-$[(CH_3)_2$Si(indenyl)2]ZrCl2/ methylallumoxane (MAO) catalytic system (see Scheme 1). The catalytic precursor rac-$[(CH_3)_2$Si(indenyl)$_2$]$ZrCl_2$, activated by MAO, is well known to promote the isospecific polymerization of α-olefins [15, 22, 23]. Recently, its high performances in the isospecific 9-(pent-4-en-1-yl)-9H-carbazole polymerization was also reported [19, 20]. The obtained polymer (PBK) was fractionated by exhaustive extraction with boiling acetone and two different fractions, an acetone insoluble (65% of whole starting material) and a soluble (35% of whole starting material), were obtained. Both PBK fractions, analyzed by GPC, showed monomodal curves and narrow molecular weight distributions. The molecular weight of the acetone insoluble fraction (Mn = 3000 Dalton, PDI = 1.6) becomes higher than the acetone soluble fraction (Mn = 1922 Dalton, PDI = 1.10). The microstructure of PBK fractions was fully determined by 13C NMR analysis, utilizing the carbon attributions in similar environments, reported in literature [19, 20, 24– 27]. In Figure 1(a), the ^{13}C NMR spectrum of the acetone insoluble PBK fraction sample is reported. As expected, this spectrum is fully consistent with both regio- and stereoregular polymer microstructure and, in particular, from the chemical shifts related to carbons 1 and 2, an isotactic microstructure of polymer (*i*-PBK) is assumed [24–26].

The acetone soluble PBK fraction ^{13}C NMR spectrum (Figure 1(b)) is instead indicative of a less stereoregular polymer microstructure. Regioerrors are also recognized in this fraction.

X-Ray, DSC and Infrared Characterization

In Figure 2(a), the X-ray diffraction pattern of the as polymerized *i*-PBK is reported. The X-ray pattern shows a small but well-defined reflection at $2\theta \approx 7°$ and other less defined reflections in the 2θ range of 10° –15° overlapped to a large amorphous halo centered at $2\theta \approx 20°$. Such X-ray pattern indicates that a limited amount of crystalline phase is present in the as polymerized sample.

X-ray spectra of acetone soluble and insoluble fractions of *i*-PBK are shown in Figures 2(b) and 2(c), respectively. The pattern of the acetone-soluble fraction, Figure 2(b), presents only a broad amorphous halo with two well defined humps at $2\theta \approx 12°$ and 20° , while the pattern of the acetone-insoluble fraction, Figure 2(c), shows many welldefined crystal reflections. The results of WAXD analysis of the amorphous and crystalline fractions of the *i*-PBK are in agreement with the NMR analysis, which shows that the acetone-insoluble fraction (crystalline fraction) is more stereoregular of the acetone-soluble fraction (amorphous fraction). It is worth noting that the two humps in the X-ray pattern of the amorphous fraction correspond to those observed in the spectrum of the as polymerized sample, and the few indistinct crystal reflections observed in the X-ray pattern of the as polymerized sample correspond to peaks of the crystalline fraction pattern.

In Figure 3, the second DSC heating run of the unfractionated polymer (Figure 3(a)) and that of its amorphous fraction (Figure 3(b)) are compared. The observed difference in glass transition (T_g) value can be rationalized on the basis of the fact that the molecular weight of the amorphous fraction, evaluated by GPC analysis, is smaller ($M_n \approx 2000$ Da) compared to that of the unfractionated polymer ($M_n \approx 3000$ Da).

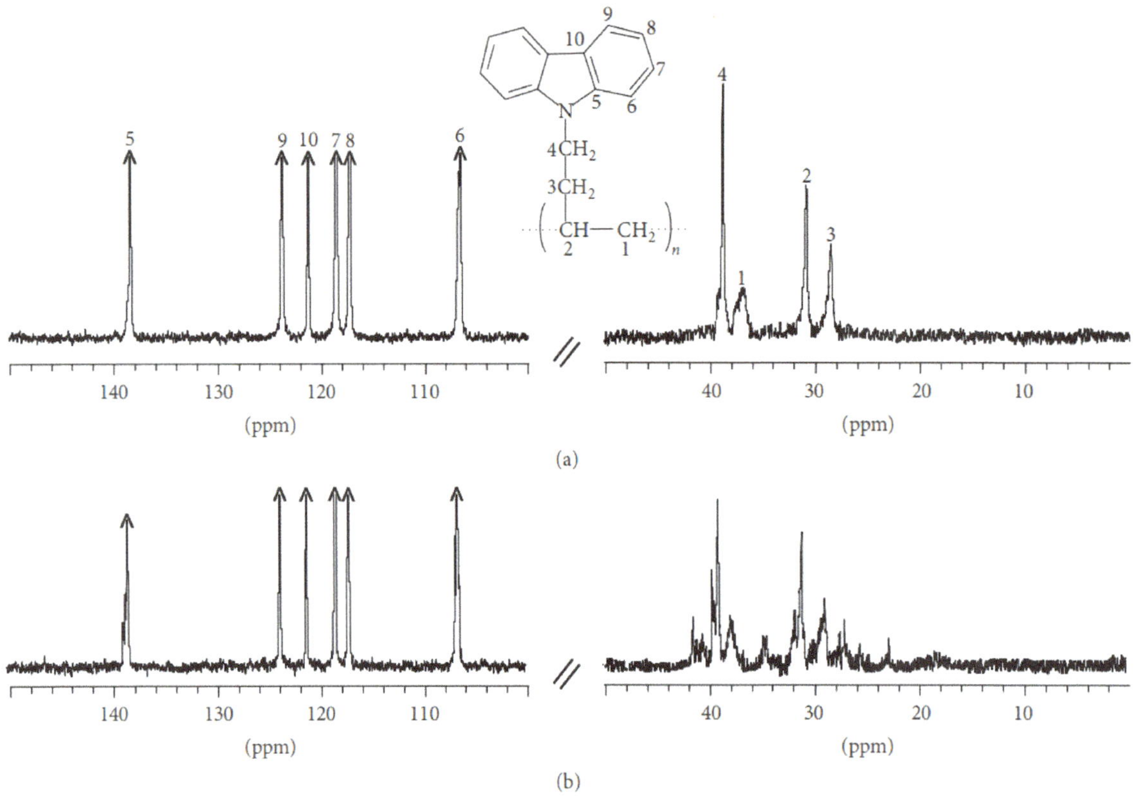

Figure 1: ^{13}C NMR spectra of PBK fractions, (a) acetone insoluble, (b) acetone soluble (solvent: tetrachlorodideteuroethane; scale: hexamethyldisiloxane).

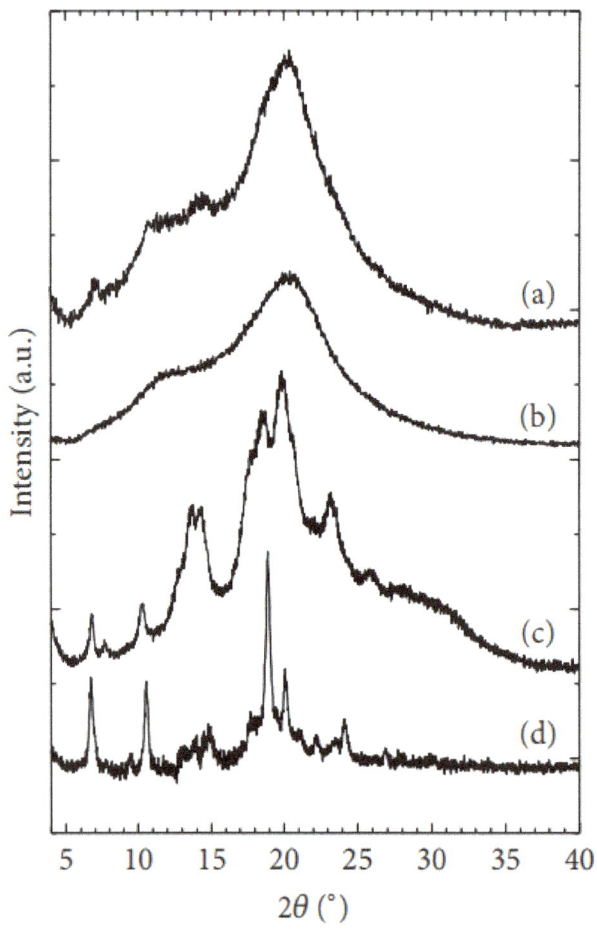

Figure 2: X-ray diffraction patterns of (a) as polymerized sample, (b) acetone-soluble fraction, (c) acetone-insoluble fraction, and (d) crystalline fraction after annealing at $T = 200°$ C for 20 min.

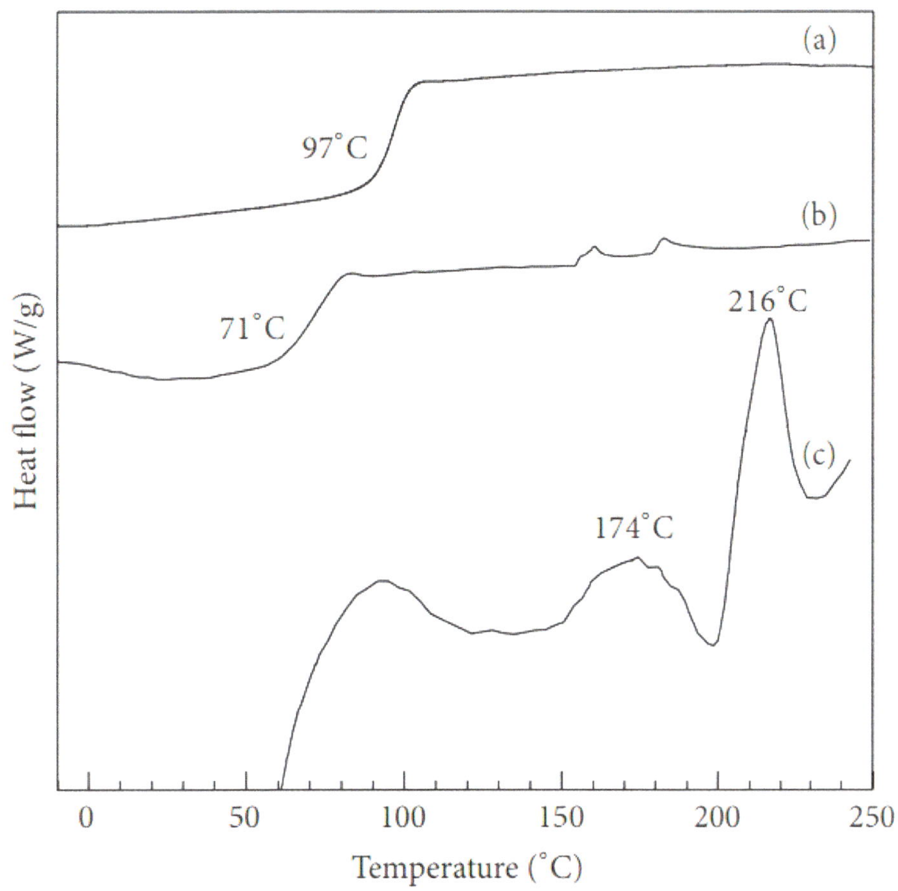

Figure 3: Second heating of DSC runs of (a) as polymerized (unfractionated) sample, (b) acetone-soluble fraction, and (c) first heating of DSC run of acetone-insoluble fraction.

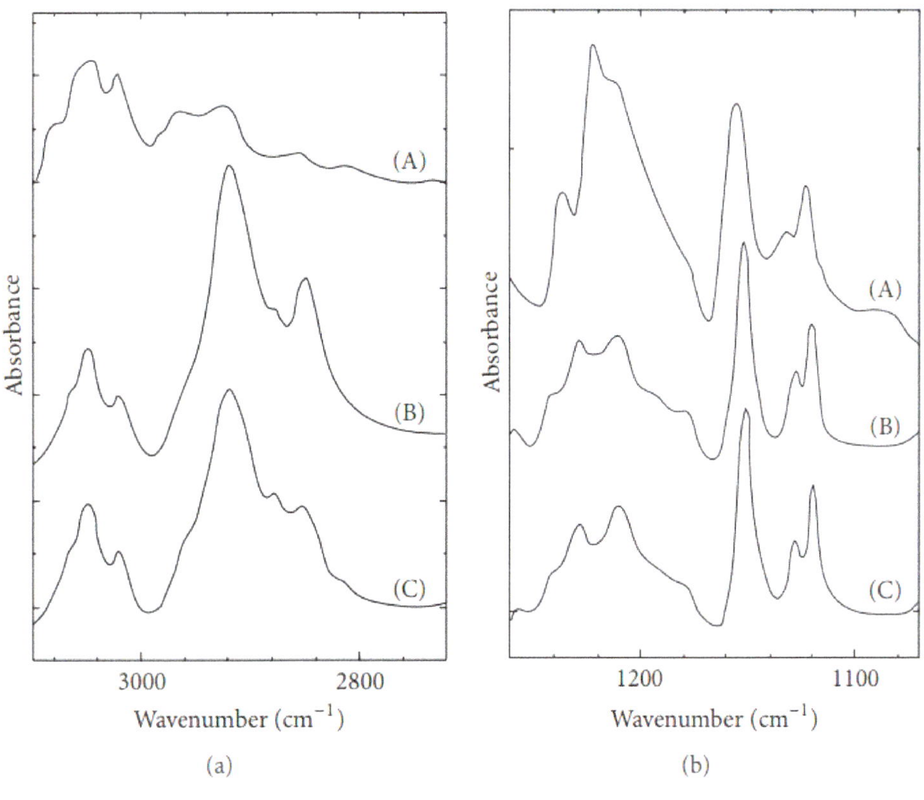

Figure 4: FTIR spectra of PVK (line (A)); crystalline (line (B)); and amorphous (line (C)) fractions of *i*-PBK.

The first DSC heating run of the polymer crystalline fraction, is also shown in Figure 3. Two separate and welldefined endothermic peaks, at $T \approx 174°$ C and $T = 216°$ C, associable with distinct crystalline phases, are observed. Between the two endothermic peaks a weak exothermic peak at $T \approx 200°$ C, is also present.This peak can be tentatively interpreted as a crystallization phenomenon associated with the solid-solid transition between the two crystal forms which melt at $T \approx 174$ and 216° C, respectively. In Figure 2(d), the X-ray pattern of the *i*-PBK crystalline fraction after annealing at $T = 200°$ C for 20 minutes is shown. Only some of narrow peaks of the pattern (c) of Figure 2 are still present, although slightly shifted. On the contrary, many reflections of the two patterns do not match each other. These differences can be rationalized by assuming that the two patterns correspond to two different crystalline phases, which are probably very similar

Both of these two crystalline forms cannot be obtained by crystallization from the melt, so that in the second DSC heating run (not showed in Figure 3) of the *i*-PBK crystalline fraction, only a glass transition at $T = 97°$ C is observed. This T_g value, which is very similar to that of the as polymerized *i*-PBK sample (see Figure 3(a)), appears to be intermediate between that reported for PVK ($T_g = 211°$ C) (obtained from the second DSC heating run of poly(9- vinylcarbazole) purchased from Sigma-Aldrich s.r.l. (Milan, Italy) with an average *Mn* of 25,000–50,000) and *i*-PPK ($T_g = 90°$ C) [20]. The observed differences between T_g of these polymers accord well with the widely accepted hypothesis that rigidity of polymer chains, on which Tg value depends, is affected by the side branches length of polymer chain. In detail, as for the PVK, the carbazole units directly connected to the main chain make the polymer chain extremely rigid [28], while, for PBK and PPK, the alkylene linkers, which connect the carbazole unit with the main chain spacing the polymer chains each other, increases the free volume and reduces T_g. The different T_g values of *i*-PBK and *i*-PPK, on the basis of this hypothesis, can be justified by the different length of the alkylene linker which is 2 and 3 C atoms for *i*-PBK and *i*-PPK, respectively.

As for the infrared characterization of the *i*-PBK, particularly informative is the comparison with the PVK. The FTIR spectra of amorphous samples of PVK and *i*-PBK are very similar to each other. The most important difference is in the region of C-H stretching (Figure 4(a)) where the intensity ratio of the aliphatic to aromatic C-H stretching (at 2848– 2918 cm−1 and 3021–3048 cm^{-1}, resp.) increases in the *i*-PBK spectra, due to the $(CH_2)_2$ alkylene linker connecting the carbazole moiety to the main chain. The same phenomenon has already been reported for *i*- PPK [20], a polymer homologue to i-PBK but having the linker connecting the carbazole to the main chain one carbon atom longer. In addition to the C-H stretching region, the region in which there are more differences between the PVK infrared spectrum and those of the amorphous and crystalline fractions of *i*-PBK is in the wavelength range 1070–1260 cm−1 (Figure 4(b)). In Figure 4(b), the absorption bands of carbazole ring vibrational modes, at 1121, 1132, and 1151 cm−1 as valued in the spectrum of PVK, assigned to C-H in-plane bending and ring deformation [29], remain substantially unchanged in the *i*-PBK spectra, although slightly shifted.

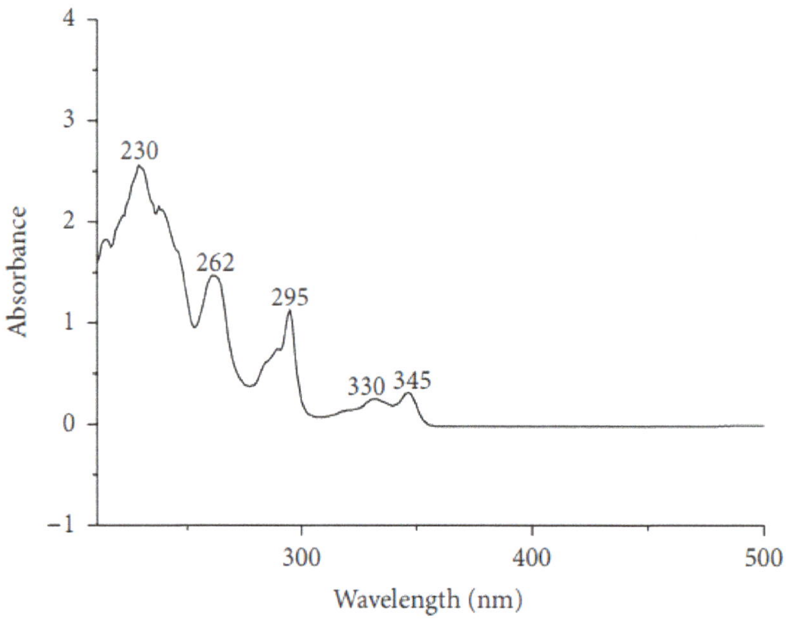

Figure 5: Absorption spectrum of *i*-PBK film.

On the contrary, the PVK absorption bands at 1224 and 1236 cm^{-1}, assigned, respectively, to Ccarbazole-N and Caliphatic-N stretching [29–31], change position and intensity in the spectra of *i*-PBK fractions, and they are blue shifted at 1128 and 1241 cm^{-1}, respectively. In addition a new band at 1180 cm^{-1} appears in both amorphous and crystalline *i*-PBK fractions. These differences are probably due to the influence that the alkylene linker has on *i*-PBK vibrational modes and in particular on vibrational modes involving the nitrogen of the carbazole group. Finally, it should be noted that additional bands at 1195 and 1224 cm^{-1} are present only in the spectrum of *i*-PBK crystalline fraction. These bands are possibly associated with the vibrational modes influenced by polymer chains packing in the *i*-PBK crystalline structure.

Optical Characterization

The good solubility of *i*-PBK in chlorobenzene allowed the preparation of uniform and homogenous films of 100 nm thick. Figure 5 shows the absorption spectrum of a *i*-PBK film sample. It presents several peaks maxima at around 230, 262, 295, 330, and 345 nm, where the two longest wavelengths are due to the carbazole group [32]. *i*-PBK fluorescence spectrum of the acetone insoluble fraction film sample measured under an excitation wavelength of 298 nm at room temperature is reported in Figure 6 (bottom curve). In Figure 6, for comparison, the fluorescence spectra of an *i*-PPK film sample [19, 20] (medium curve) and a commercial PVK sample (trade sample purchased from Sigma Aldrich Company presenting a partially isotactic microstructure) (curve on the top) are also shown. It is worth nothing that both *i*-PBK and *i*-PPK fluorescence spectra present very similar curve shape. The *i*-PBK spectrum exhibits two emission bands in the high energy region presenting peaks at 357 nm and 374 nm, respectively, which correspond to those from isolate monomer units [33, 34]. The same bands are also present in the *i*-PPK spectrum, but not in the PVK spectrum.This result can be related to the presence of the flexible spacer (-$(CH_2)_3$- for PPK and -$(CH_2)_2$- for PBK), absent in PVK, between the main chain and the carbazole pendant. It was found that other vinyl polymers with pendant carbazole groups connected to the skeleton chain by a flexible spacer like -O-(CH2)2- (poly[2-N-carbazolyl]ethyl vinyl ether) also show the typical monomer emission bands in their fluorescence film spectra [35, 36].

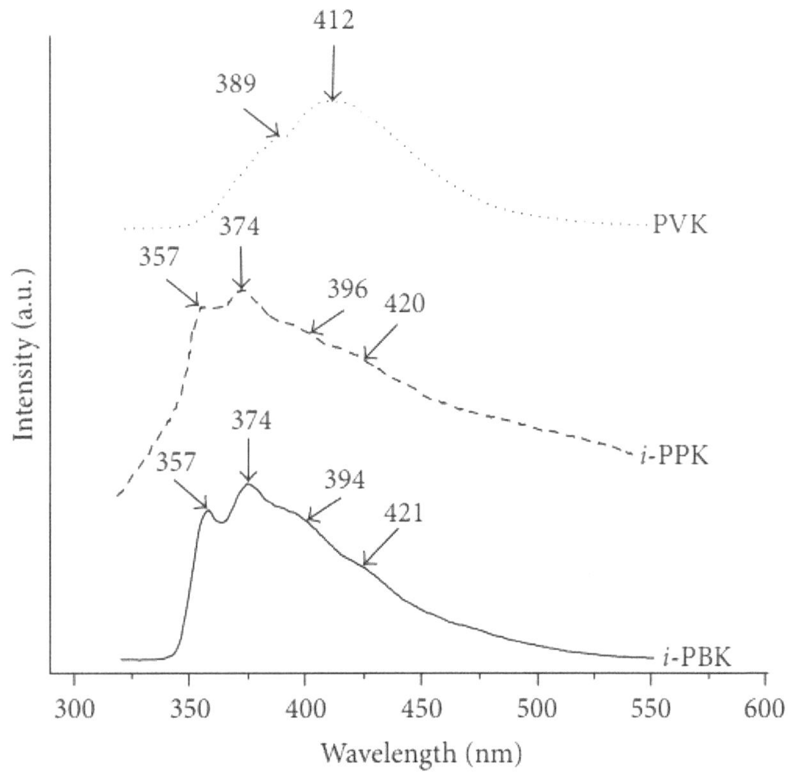

Figure 6: Fluorescence spectra of *i*-PBK, *i*-PPK, and PVK films.

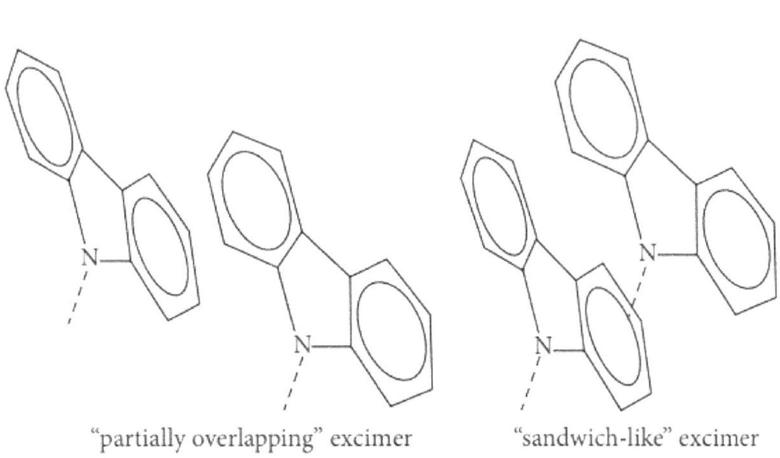

Scheme 2

As in *i*-PPK sample, other two strong, broad, and partially overlapping fluorescence bands, centred at 394 and 421 nm, respectively, are also detected in the *i*-PBK spectrum. As already reported in our previous papers on *i*-PPK [19, 20], these bands can be attributed to two different excimers formed by close carbazole groups. In particular, the band at 421 nm was attributed to a fully overlapping conformation of two carbazole groups (sandwich-like excimer fluorescence), while the band at 394 nm was attributed from a partially overlapping conformation of two carbazole groups (second excimer fluorescence) (see Scheme 2)

The formation of these two excimers was previously detected in PVK films and was also described in literature [35, 37–41]. Nevertheless, it is worth noting that the fluorescence due to "sandwich-like" excimers that is also common to all vinyl aromatic polymers is always observed in PVK. Instead, the fluorescence due to the high energy "partially overlapping" excimers, which is peculiar of vinyl polymers with pendant carbazole groups [39], is only associated with PVK syndiotactic sequences [39]. It was also reported that the polymer conformation allowing the formation of the high energy "partially overlapping" excimer state has to exist prior to initial excitation step [37, 41]. Partially isotactic PVK commercial sample, used for comparison in this work, shows an emission spectrum where the fluorescence generated by the "sandwich-like" excimer state is predominant. Instead, *i*-PBK spectrum, as in the case of *i*-PPK, presents, in addition to the characteristic monomer emission bands, both the fluorescence bands attributable to two different excimers, "sandwich-like" and "partially overlapping," having comparable intensity

CONCLUSIONS

Isotactic poly(N-butenyl-carbazole) (*i*-PBK) was synthesized by using the homogeneous isospecific Ziegler-Natta catalytic system rac-$[(CH_3)_2Si(indenyl)_2]ZrCl_2$/methylallumoxane (MAO) and 9-(but-3-en-1-yl)-9H-carbazole monomer. The microstructures of PBK fractions were fully determined by 13C NMR analysis and, for the first time, the isotactic microstructure of PBK (*i*-PBK) was unambiguously determined. WAXD analysis showed that *i*-PBK insoluble acetone fraction is crystalline, while *i*-PBK soluble acetone fraction is amorphous. DSC analysis of *i*-PBK crystalline fraction showed two separate and well-defined endothermic peaks, at $T \approx 174$∘ C and $T = 216°$ C, associable with two distinct crystalline phases. The solid-solid transition between these two crystal forms was also observed as an exothermic peak at 200∘ C. *i*-PBK T_g value (96.8∘ C) became intermediate between that previously reported for PVK ($_{Tg}$ = 211∘ C) and *i*-PPK (T_g = 90∘ C). As for the infrared analysis, *i*-PBK showed FTIR spectra similar to that of PVK as regards the bands related to the carbazole group, but it presented a number of differences tentatively associated to the $(CH_2)_2$ alkylene group connecting the carbazole group to the main chain. In fact, this group, distancing the main chains, increases the flexibility of polymer chains and modifies the intra- and interchain interactions and therefore the normal vibrational modes of polymer. To optical analysis, *i*-PBK (crystalline fraction), similarly to *i*-PPK, showed a fluorescence film spectrum presenting both bands relative to isolate monomer units and other two, broad and partially overlapp ing bands due to different excimer states, "sandwich-like" and "partially overlapping," respectively. It is worth noting that OLEDs showing white light emission, although of low efficiency, were obtained by using *i*-PPK as emitting layer. The similar optical behavior of *i*-PBK with respect to that of *i*-PPK makes this material very promising for optical applications and deserving greater attention. The development of OLED devices based on *i*-PBK is in progress.

ACKNOWLEDGMENTS

The authors thank Dr. Patrizia Oliva and Dr. Ivano Immediata for technical assistance. Financial support of the Relight project within PON 2007–2013 program by the Ministero dell'Universita e della Ricerca (MIUR) is gratefully acknowl-` edged.

REFERENCES

1. J. H. Burroughes, D. D. C. Bradley, A. R. Brown et al., "Light-emitting diodes based on conjugated polymers," Nature, vol. 347, no. 6293, pp. 539–541, 1990.
2. K. Meerholz, B. L. Volodin, Sandalphon, B. Kippelen, and N. Peyghambarian, "A photorefractive polymer with high optical gain and diffraction efficiency near 100%," Nature, vol. 371, no. 6497, pp. 497–500, 1994.
3. G. Wang, S. Qian, J. Xu et al., "Enhanced photovoltaic response of PVK/C60 composite films," Physica B: Condensed Matter, vol. 279, no. 1–3, pp. 116–119, 2000.
4. H. Hoegl, "On photoelectric effects in polymers and their sensitization by dopants," Journal of Physical Chemistry, vol. 69, no. 3, pp. 755–766, 1965.
5. W. Reppe and E. Keyssner, "N-Vinyl Compounds," DRP 618 120, 1935.
6. W. Reppe, E. Keyssner, and E. Dorrer, "Polymerized vinyl compounds," DRP 664 231, 1938.
7. M. D. Schattuck and U. Vahtra, "Photoconductive compositions," U.S. Patent, 3,484,327, 1969.
8. B. Wang, F. Shen, P. Lu et al., "New ladder-type conjugated polymer containing carbazole and fluorene units in backbone: synthesis, optical, and electrochemistry properties," Journal of Polymer Science, Part A: Polymer Chemistry, vol. 46, no. 9, pp. 3120–3127, 2008.
9. N. Blouin, W. Mitchell, C. Wang, and S. Tierney, "Polymers of 8,9-dihydrobenzo[def]carbazole and their use as organic semiconductors," U.S. Patent 20130001476 A1, 2013.
10. J. V. Grazulevicius, P. Strohriegl, J. Pielichowski, and K. Pielichowski, "Carbazole-containing polymers: synthesis, properties and applications," Progress in Polymer Science, vol. 28, no. 9, pp. 1297–1353, 2003. ·
11. Y.-S. Cho, S.-W. Kim, C.-S. Ihn, and J.-S. Lee, "Anionic polymerization of 4-(9-carbazolyl) methylstyrene,"Polymer, vol. 42, no. 18, pp. 7611–7616, 2001.
12. R. Stanionyte, G. Buika, J. V. Grazulevicius, and S. Grigalevicius, "Synthesis and properties of oxetane monomers and oligomers with electro-active pendent groups," Polymer International, vol. 57, no. 9, pp. 1036–1041, 2008.

13. J. V. Grazulevicius, I. Soutar, and L. Swanson, "Photophysics of carbazole-containing systems. 3 Fluorescence of carbazole-containing oligoethers in dilute solution," Macromolecules, vol. 31, no. 15, pp. 4820–4827, 1998.

14. E. J. Arlman and P. Cossee, "Ziegler-Natta catalysis III. Stereospecific polymerization of propene with the catalyst system $TiCl_3AlEt_3$," Journal of Catalysis, vol. 3, no. 1, pp. 99–104, 1964.

15. L. Resconi, L. Cavallo, A. Fait, and F. Piemontesi, "Selectivity in propene polymerization with metallocene catalysts," Chemical Reviews, vol. 100, no. 4, pp. 1253–1345, 2000.

16. P. Longo, E. Siani, S. Pragliola, and G. Monaco, "Copolymerization of ethene and propene in the presence of Cs symmetric group 4 metallocenes and methylaluminoxane," Journal of Polymer Science Part A: Polymer Chemistry, vol. 40, no. 19, pp. 3249–3255, 2002.

17. A. C. Boccia, C. Costabile, S. Pragliola, and P. Longo, "Selective dimerization of γ-branched α-olefins in the presence of C2v group-4 metallocene-based catalysts," Macromolecular Chemistry and Physics, vol. 205, no. 10, pp. 1320–1326, 2004.

18. L. Annunziata, S. Pragliola, D. Pappalardo, C. Tedesco, and C. Pellecchia, "New (anilidomethyl)pyridine titanium(IV) and zirconium(IV) catalyst precursors for the highly chemo-and stereoselective cis-1,4-polymerization of 1,3-butadiene," Macromolecules, vol. 44, no. 7, pp. 1934–1941, 2011. ·

19. A. Botta, S. Pragliola, V. Venditto et al., "Highly Isotactic Poly(N-pentenyl-carbazole): a challenging polymer for optoelectronic applications," in Proceedings of the Times of Polymers (TOP) and Composites, vol. 1599 of AIP Conference Proceedings, pp. 493–497, Ischia, Italy, June 2014.

20. A. Botta, S. Pragliola, V. Venditto et al., "Synthesis, characterization, and use as emissive layer of white organic light emitting diodes of the highly isotactic poly(N-pentenyl-carbazole)," Polymer Composites, vol. 36, no. 6, pp. 1110–1117, 2015.

21. X. Huang, S. Zhong, X. Yan, X. Ke, N. Srisanit, and M. R. Wang, "The synthesis and nonlinear optical property of carbazole-azo binary compounds," Synthetic Metals, vol. 140, no. 1, pp. 79–86, 2004.

22. S. Pragliola, E. Forlenza, and P. Longo, "C_2-symmetric zirconocenes in the polymerization of conjugated diolefins," Macromolecular Rapid Communications, vol. 22, no. 10, pp. 783–786, 2001.

23. S. Pragliola, C. Costabile, and V. Venditto, "Ethylene/1,3-butadiene cyclocopolymerization catalyzed by zirconocene systems," European Polymer Journal, vol. 58, pp. 157–163, 2014.

24. T. Asakura, M. Depura, and Y. Nishiyama, "Carbon-13 NMR spectral assignment of five polyolefins determined from the chemical shift calculation and the polymerization mechanism," Macromolecules, vol. 24, no. 9, pp. 2334–2340, 1991.

25. T. Asanuma, Y. Nishimori, M. Ito, N. Uchikawa, and T. Shiomura, "Preparation of syndiotactic polyolefins by using metallocene catalysts," Polymer Bulletin, vol. 25, no. 5, pp. 567–570, 1991.

26. K.-I. Okamoto, M. Yamada, A. Itaya, T. Kimura, and S. Kusabayashi, "Polymerization of N-vinylcarbazole, N-vinyl-5H-benzo[b]carbazole, and N-vinyl-7H-benzo[c]carbazole," Macromolecules, vol. 9, no. 4, pp. 645–649, 1976.

27. G. Barrera Galland, L. Forgiarini Da Silva, and A. Nicolini, "Tacticity of poly-α-olefins from poly-1-hexene to poly-1-octadecene," Journal of Polymer Science Part A: Polymer Chemistry, vol. 43, no. 20, pp. 4744–4753, 2005.

28. A. Kimura, S. Yoshimoto, Y. Akana et al., "Crystallinity of poly-N-vinylcarbazole," Journal of Polymer Science Part A-2: Polymer Physics, vol. 8, no. 4, pp. 643–648.

29. P. Bertoncello, A. Notargiacomo, and C. Nicolini, "Synthesis, fabrication and characterization of poly[3- 3′(vinylcarbazole)] (PVK) Langmuir-Schaefer films," Polymer, vol. 45, no. 5, pp. 1659–1664, 2004.

30. B. Mailhot-Jensen, S. Robu, A. Rivaton et al., "Carbazole containing copolymers: synthesis, characterization, and applications in reversible holographic recording," International Journal of Photoenergy, vol. 2010, Article ID 945242, 11 pages, 2010.

31. P. X. Thinh, C. Basavaraja, W. J. Kim, and D. S. Huh, "Structure characterization and DC conductivity of honeycomb patterned poly(N-vinylcarbazole)/ multiwalled carbon nanotube composite film via pretreatment at high temperature," Polymer Composites, vol. 32, no. 11, pp. 1772–1780, 2011.

32. X. Gong, S.-H. Lim, J. C. Ostrowski, D. Moses, C. J. Bardeen, and G. C. Bazan, "Phosphorescence from iridium complexes doped into polymer blends," Journal of Applied Physics, vol. 95, no. 3, pp. 948–953, 2004.

33. I. B. Berlman, Handbook of Fluorescence Spectra of Aromatic Molecules, Academic Press, New York, NY, USA, 1971.

34. P. C. Johnson and H. W. Offen, "Excimer fluorescence of poly(N-vinylcarbazole)," The Journal of Chemical Physics, vol. 55, no. 6, pp. 2945–2949, 1971.

35. A. Itaya, K. Okamoto, and S. Kusabayashi, "Emission spectra of the vinyl polymers with pendant carbazolyl groups," Bulletin of the Chemical Society of Japan, vol. 49, no. 8, pp. 2082–2088, 1976.

36. A. Itaya, K. Okamoto, and S. Kusabayashi, "Singlet excitation energy transfer in the vinyl polymers with pendant carbazolyl groups," Bulletin of the Chemical Society of Japan, vol. 50, no. 1, pp. 22–26, 1977.

37. G. E. Johnson, "Emission properties of vinylcarbazole polymers," The Journal of Chemical Physics, vol. 62, no. 12, pp. 4697–4709, 1975.

38. K. P. Ghiggino, R. D. Wright, and D. Phillips, "Excimer kinetics in poly(n-vinyl carbazole)," European Polymer Journal, vol. 14, no. 8, pp. 567–569, 1978.

39. C. E. Hoyle, T. L. Nemzek, A. Mar, and J. E. Guillet, "Time resolved fluorescence studies of poly(N-vinylcarbazole),

poly(1-vinylnaphthalene), and 1,3-bis(N-carbazolyl) propane," Macromolecules, vol. 11, no. 2, pp. 429–431, 1978.

40. F. Evers, K. Kobs, R. Memming, and D. R. Terrell, "Intramolecular excimer emission of poly(N-vinylcarbazole) and rac- and meso-2,4-di-N-carbazolylpentane. Model substances for its syndiotactic and isotactic dyads," Journal of the American Chemical Society, vol. 105, no. 19, pp. 5988–5995, 1983. ·

41. A. Rivaton, B. Mailhot, G. Derderian, P. O. Bussiere, and J.-L. Gardette, "Investigation of the photophysical processes and photochemical reactions involved in PVK films irradiated at $\lambda > 300$ nm,"Macromolecules, vol. 36, no. 15, pp. 5815–5824, 2003.

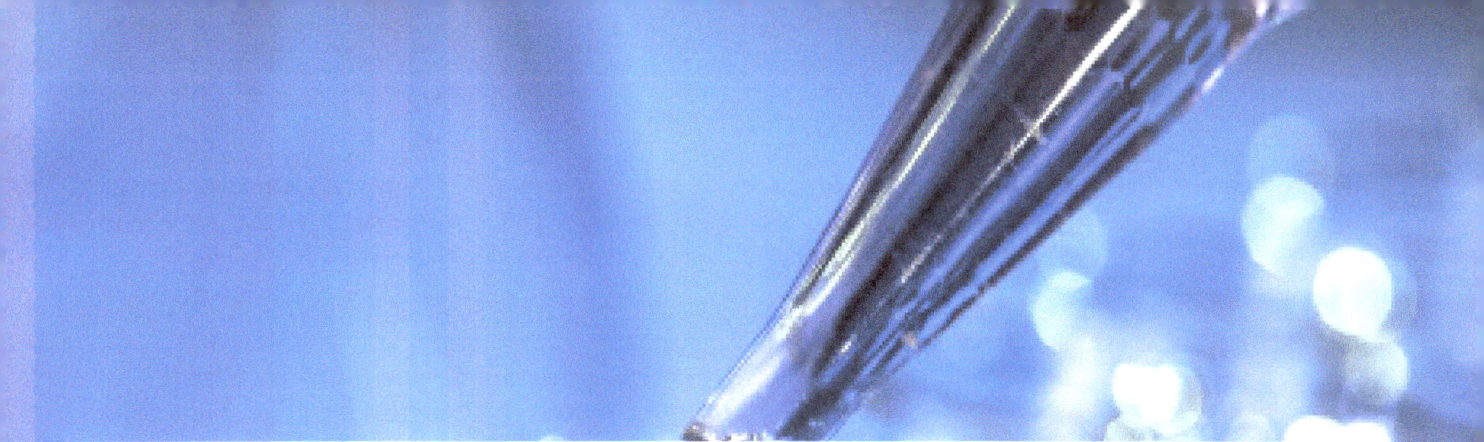

CHAPTER 15 RECENT DEVELOPMENTS IN CHEMICAL SYNTHESIS WITH BIOCATALYSTS IN IONIC LIQUIDS

Mahesh K. Potdar, Geoffrey F. Kelso, Lachlan Schwarz, Chunfang Zhang and Milton T. W. Hearn

Keywords: Enzyme catalysis, ionic liquids, bioseparation, whole cell biocatalysis, process intensification

INTRODUCTION

As a consequence of rapid developments in many fields of synthetic organic chemistry, researchers from both academia and industry now have to pay much greater attention to the detrimental effects of chemicals and chemistry processes on the environment. Against the backdrop of climate change and increasing public concern about global sustainability, the increasing needs for environmental protection and sustainable development are forcing chemists to rethink the way in which chemistry is conducted, resulting in new synthetic strategies that are more atom efficient, achieve better mass intensification factors, generate less waste and utilize less hazardous conditions through the deployment of the twelve principles of Green Chemistry [1].

Various approaches can be practiced in green organic synthesis, including ways to enhance atom utilization, replace stoichiometric reagents with catalysts, use benign solvents or solvent-free processes, design and generate safer chemical products that are biodegradable, and reduce or totally eliminate waste products. A variety of next-generation solvents, including liquid polymers, fluorous solvents, supercritical fluids and ionic liquids have been explored as greener alternatives to those conventional organic solvents that exhibit undesirable flammable, explosive or toxic properties.

Ionic liquids, as the name suggests, are liquids comprised of ions. In practical terms, it is desirable for an ionic liquid to exist in the liquid state at temperatures below 100 °C and, preferentially, at ambient temperature. Although ionic liquids have been known for at least 100 years, their application in various chemistry processes has only been studied extensively over the last three decades [2]. Research into ionic liquids has ranged from their application as solvents in electrochemistry [3], organic and inorganic transformations [2,4], biotechnology processes [5], as solar cell electrolytes [6], and as supported phases in organic chemistry [7,8]. For organic synthesis with transition metal catalysis, ionic liquids were initially employed as biphasic systems [9]. Since then, their application has expanded to include the development of task-specific ionic liquids, wherein the constituent ions of the ionic liquid are designed to fulfill the requirements of a specific reaction [10,11] or alternatively a biotransformation [12,13]. In addition, substrates have been covalently attached to an ionic liquid prior to the reaction, after which the desired product is isolated and the ionic liquid recycled [14,15].

Citation: Mahesh K. Potdar, Geoffrey F. Kelso, Lachlan Schwarz, Chunfang Zhang and Milton T. W. Hearn, Recent Developments in Chemical Synthesis with Biocatalysts in Ionic Liquids, doi:10.3390/molecules200916788

Ionic liquids have long been considered as green alternatives to organic solvents, although extensive debate and some controversy have surrounded their "so called" green properties. For example, although a specific ionic liquid may have negligible vapor pressure and low flammability, the anions and/or the cations, which comprise the ionic liquid, may individually, both, or when in combination, be toxic. Therefore, like any other solvent, all the potential detrimental properties, e.g., toxicity and low biodegradability, of a specific ionic liquid needs to be assessed before it can be regarded as a green alternative to conventional organic solvents. In addition to these considerations, the production and life-cycle of an ionic liquid also needs to be evaluated in order to fully determine its "greenness" compared to conventional organic solvents. A recent study by Deetlefs and Seddon on the "greenness" of various ionic liquids made by laboratory-scale synthetic methods found that there is considerable room for improvement in terms of E-factor, energy consumption and purification [16]. Hence, for future industrial processes it will be desirable from the outset to design the ionic liquid to be easily and cleanly synthesized, to have *inter alia* the requisite green properties of low vapor pressure, low flammability and low toxicity, and possess practical working attributes such as low viscosity and easy recyclability, by taking advantage of the enormous combinatorial and structural diversity of ionic liquids in terms of their anion/cation composition [17,18].

Enzymes are nature's catalysts and have been employed for millennia in the manufacture of food and beverage products. With the advent of genetic engineering and advanced biotechnology processes, their use has greatly increased over several decades in industry for the synthesis of important pharmaceuticals, agrochemicals and fine chemicals [19,20]. Due to their excellent catalytic, stereo-selective and chemoselective properties, low toxicity, high biodegradability and efficiency under mild reaction conditions, enzymes are regarded as excellent green catalysts. Whilst some enzymes need to be associated with biological membranes to elicit function, many enzymes exist in nature in an aqueous environment. Many enzymes have the specific ability to perform well in organic solvents. For example, in his pioneering work, Klibanov showed that certain enzymes are stable and can catalyze reactions in various organic solvents like hexane, toluene, acetonitrile or tetrahydrofuran [21,22].

This versatile aspect of enzymes has prompted chemists to explore their utility in ionic liquids as the combination of biocatalysts and ionic liquids have the potential to provide a more sustainable approach for organic transformations to be carried out, particularly when the substrates have poor solubility in water. Interest in the use of ionic liquids dates back to 1984, when Magnuson *et al.* [23] investigated the effect of ethylammonium nitrate on the activity and stability of the enzyme alkaline phosphatase. Subsequently, considerable research effort has been committed, particularly since 2000, to evaluate the impact of the physicochemical properties and the design of ionic liquids suitable for the stabilization or activation of enzymes in several different formats, such as immobilization onto solid carriers, as sol-gel encapsulation, as solid phase complexes or following chemical modification [24,25,26]. It is widely appreciated that many ionic liquids exhibit considerable toxicity and poor biodegradability, which preclude their use in many cases as "green" solvents, whilst the effects of ionic liquids on the functional status and solubility of enzymes are difficult to accurately predict with the current state of knowledge.

Nevertheless, it is well known that the effect of ions from low to relatively high concentrations, e.g., in excess of 4 M for ammonium sulphate, on enzyme activity and stability usually follows the Hofmeister series, whereby strongly hydrated ions (called kosmotropes) increase the water structure and thus the conformational stability of proteins, whilst weakly hydrated ions (called chaotropes) generally operate in the opposite way, e.g., destabilize protein conformation. However, as a generalization, the behavior of the anions of more hydrophilic ionic liquids tends to correlate well with the Hofmeister series. For example, Yang [27] and Zhao [28] demonstrated that an ionic liquid composed of a chaotropic cation and a kosmotropic anion can favorably influence an enzyme to retain good activity and stability. When both the anion and cation are chaotropic, then loss of activity and stability tend to follow. The more hydrophilic ionic liquids that combine powerful hydrogen bonding capacity tend to act as water-mimicking liquids with the ability to dissolve an enzyme for monophasic catalysis while retaining a high level of enzyme activity [29,30].

The first examples of the use of biocatalysts in the presence of an ionic liquid were reported by Erbeldinger *et al.* [31] and Lau *et al.* [32] in 2000. Since then, a large amount of research has been performed in this field as illustrated by Figure 1, which summarizes the number of publications per year from 2000. The number of issued or pending patent applications mirrors a similar trend. The importance of this rapidly growing field of research has been documented by various authors, including the different research groups of Sheldon *et al.*, Kragl *et al.* and Goto *et al.*, highlighting practical developments, as well as providing compilations of data for different ionic liquids/ enzyme combinations [13,24,25,26,33,34,35,36,37,38,39,40,41]. Collectively, these works have summarized different facets of the use of biocatalysts in ionic liquids, wherein key issues related to the effects of ionic liquids on the structure, activity and stability of enzymes have been explored and the impact of the water content in the ionic liquid assessed, together with several biocatalytic reactions employing a range of different enzymes and ionic liquid combinations for the design of reaction systems, including biocatalyst recovery, product isolation and choice of biphasic systems.

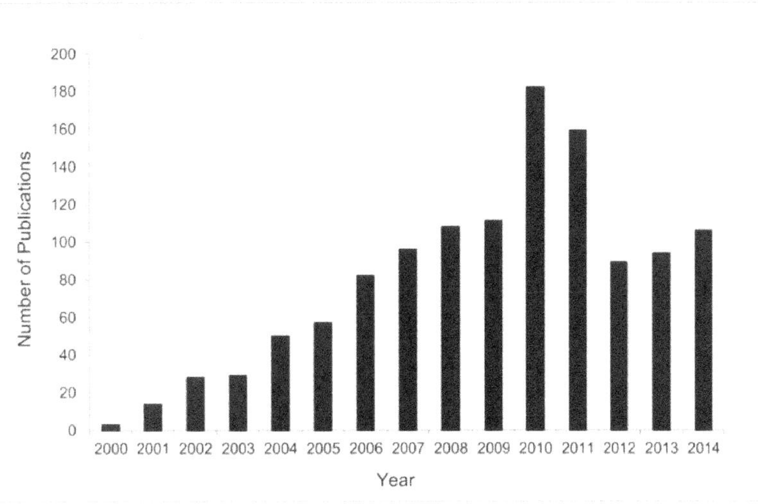

Figure 1: Year-wise reports published in the area of enzyme-based catalytic reactions in ionic liquids.

In this current article, an overview of innovations that have been developed in this field over the last decade is presented with a focus on the synthetic chemistry outcomes and benefits derived from carry out biocatalytic reactions in suitable ionic liquids. Key aspects behind different chemical synthetic processes that are destined upon scale-up to truly be more sustainable are described. Emphasis has thus been mainly placed on the synthetic processes, including approaches whereby the product can be efficiently separated and the ionic liquid and biocatalyst can be recycled or employed in continuous processes, rather than on the physicochemical properties of the ionic liquids *per se*, which have been eloquently summarized in earlier reviews, (e.g., [2,5,13,16,24,25,26,33,34,35,36,37,38,39,40]). Importantly, such uses of biocatalysts in appropriate ionic liquids offers important advantages for process intensification and enhancement of synthetic productivity for the manufacture of chiral molecules. An overview of the application of a special class of biodegradable ion-based liquids, known as deep eutectic solvents, in biocatalysis, and recent adaptations of the use of ionic liquids to whole-cell biocatalysis, which from a green chemical perspective have potential for new applications in chemical manufacturing, medicine and the environmental sciences are also provided.

CHEMICAL SYNTHESIS WITH ENZYMES IN IONIC LIQUIDS

Ionic Liquids as Solvents for Enzyme Catalysis

The first definitive example of enzyme-catalyzed reactions in a pure ionic liquid was reported by Sheldon's group. Thus, Lau *et al.* [32] reported lipase-catalyzed transesterification, amidation and epoxidation reactions in ionic liquids (Scheme 1). Reaction rates for the immobilized *Candida antarctica* lipase B (Novozym® 435)-catalyzed transesterification in anhydrous [bmim][PF_6] were found to be comparable, faster or slower than those in organic solvents depending on the type of ester and alcohol employed, with good to excellent conversions (56%–81%) obtained. The Novozym® 435-catalysed condensation of ammonia with octanoic acid in [bmim][BF_4] proceeded in quantitative conversion after 4 days. Epoxidation of cyclohexene by peroctanoic acid was also investigated using the Novozym® 435 to catalyze the production of the peroxy-acid *in situ* from octanoic acid and 60% aqueous H_2O_2 in [bmim][BF_4], documenting an example of a safer reaction process that did not require the use of a flammable solvent or the handling of a hazardous peroxy-acid. A yield of 83% was achieved after 24 h, comparable to that obtained when the same reaction was carried out in acetonitrile (93%). Overall, this pioneering study highlighted the potential of low vapor pressure, non-flammable ionic liquids as greener solvent alternatives to organic solvents for enzyme-catalyzed production of important industrial chemicals.

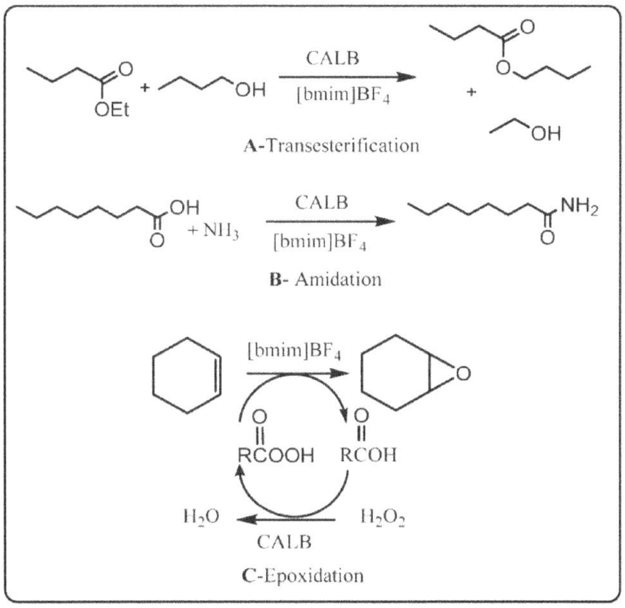

Scheme 1: *Candida antarctica* lipase (CALB)-catalyzed transesterification (**A**); amidation (**B**) and epoxidation (**C**) reactions in ionic liquids [32].

The important roles that enzymes can play as catalysts are particularly prominent in enantioselective transformations. Pioneering examples of enantioselective reactions using enzymes in ionic liquids were demonstrated independently in 2001 by Schöfer *et al.* [40] and Itoh *et al.* [41]. For example, Itoh *et al.* [41] carried out the enantioselective acylation of allylic alcohols in an ionic liquid solvent. The group successfully demonstrated that the lipase *Candida antarctica* lipase B

(Novozym® 435) was anchored to the imidazolium based ionic liquid solvent. These experiments documented the power of a recyclable system of enzyme and ionic liquid for enantioselective transformation (Scheme 2). Overall, these demonstrations of the potential of lipase-catalyzed reactions in ionic liquids [32,40,41] represent landmark studies, since lipases are now the most widely used enzymes in organic synthesis.

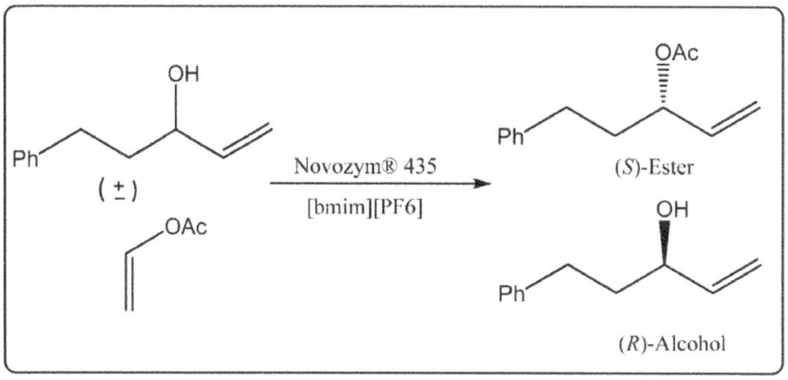

Scheme 2: *Candida antarctica* lipase (CALB)-catalyzed enantioselective transesterification of allylic alcohols in ionic liquid [41].

Subsequent to this pioneering study of lipase-catalyzed biocatalysis in ionic liquids, several other different reactions encompassing a wider array of substrates have been reported [13,42,43]. For example, Mohile *et al.* have demonstrated the use of ionic liquid as a co-solvent with aqueous buffer for the *Candida rugosa* lipase-catalyzed enantioselective hydrolysis of racemic butyl 2-(4-chlorophenoxy)propionate (Scheme 3) [44].

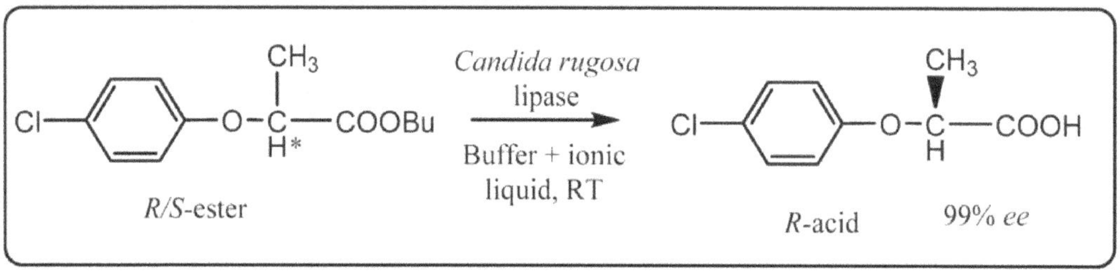

Scheme 3: Enantioselective hydrolysis of racemic butyl-2-(4-chlorophenoxy)propionate in an aq. buffer/ionic liquid system [44].

In biphasic systems employing the hydrophobic [bmim][PF_6] or 1-hexyl-3-methyl-imidazolium [hmim][BF_4] ionic liquids, reaction times increased by about 10-fold compared to when an aqueous buffer alone was employed, but a remarkably large enhancement in enantioselectivity up to 99% compared to 47%, respectively, was observed. Similar results have been obtained with a monophasic system when the more hydrophilic [bmim][BF_4] ionic liquid was used. In the case of the biphasic systems, the ionic liquids were able to be recycled with only marginally reduced conversion rates after four cycles and with no loss of enantioselectivity. In subsequent work [bmim][PF_6] was assessed as a solvent for the resolution of racemic alcohols by enantioselective acylation with succinic anhydride using *Pseudomonas cepacia* lipase supported on celite (lipase PS-C) as a catalyst (Scheme 4) [45].

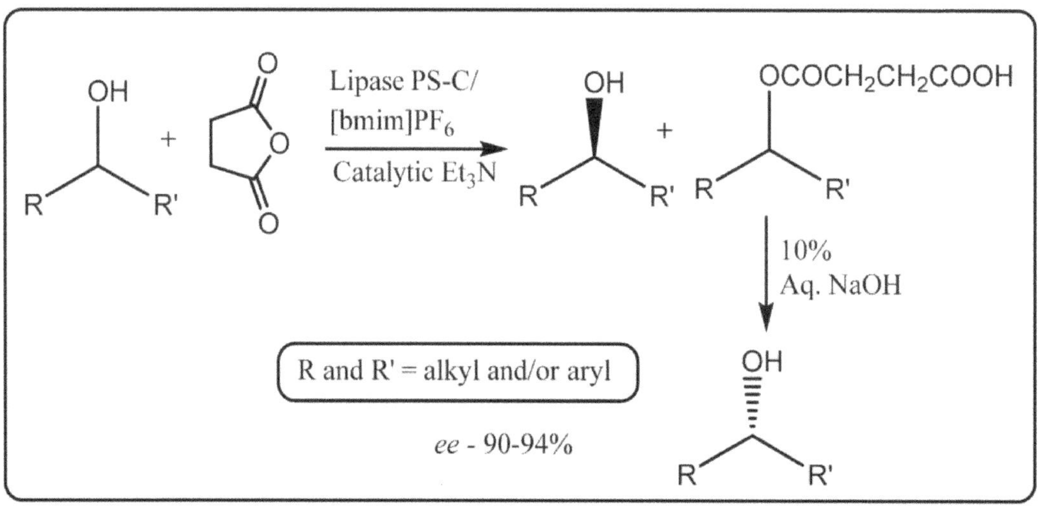

Scheme 4: Lipase PS-C catalyzed resolution of racemic alcohols by enantioselective acylation with succinic anhydride in [bmim][PF_6] [45].

A distinct advantage of using succinic anhydride as the acylating agent lies in the easy separation of the resultant ester by extraction from the ionic liquid into a mildly basic aqueous solution, followed by hydrolysis with a strong base to give back the enantiomerically enriched alcohol. The ionic liquid [bmim][PF_6] was found to be a suitable recyclable solvent for lipase PS-C-catalyzed enantioselective acylation, in up to 94% enantioselectivity. Including a catalytic amount of triethylamine as an additive enhanced the rate of the reaction by about 1.5-fold.

As an example of pharmaceutical synthesis, Lourenco *et al.* [46] developed an efficient lipase-mediated resolution of the HIV protease inhibitor Indinavir precursor (±)-*cis*-benzyl *N*-(1-hydroxyindan-2-yl) carbamate in [aliq][NCN_2] ionic liquid (Scheme 5). Novozym® 435 was again used to selectively acylate the 1*S*,2*R*-*cis*-benzyl enantiomer with vinyl acetate in up to 97% enantioselectivity. The catalyst was removed by filtration and the product and unreacted 1*R*,2*S*-enantiomer were then isolated from the non-volatile ionic liquid by sublimation, and then separated by column chromatography. This approach allowed reuse of the [aliq][NCN_2] and enzyme with comparable yield and enantioselectivity.

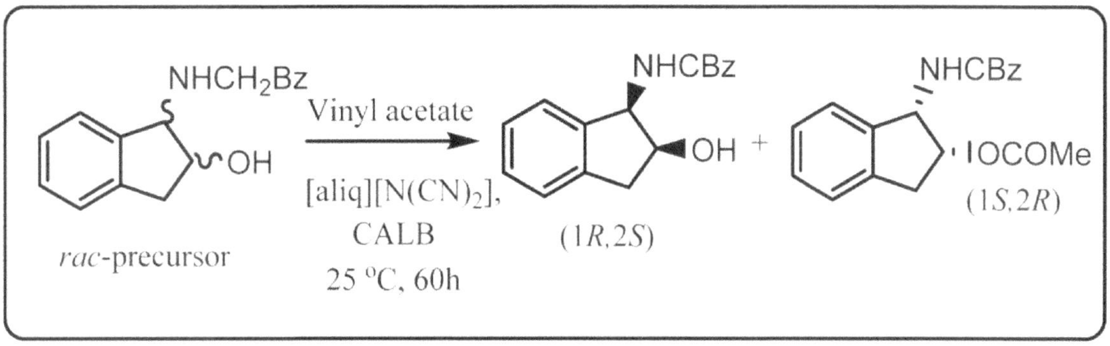

Scheme 5: Resolution of *cis*-benzyl *N*-(1-hydroxyindan-2-yl)carbamate using CALB (Novozym 435®) in ionic liquid [46].

Mai *et al.* have demonstrated the enzymatic synthesis of sugar fatty acids in ionic liquids [47]. Sugar fatty acid esters are environment friendly bio-surfactants known for their non-toxic, non-ionic, and high biodegradability. Transesterification of vinyl laurate with glucose was catalyzed by Novozym® 435 in a mixture of [Bmim][TfO]/[Bmim][Tf_2N] ionic liquid (1:1 *v*/*v*) in 96% conversion (Scheme 6). The mixture of [Bmim][TfO]/[Bmim][Tf_2N] permitted good solubility of glucose (12 g/L) and enzyme stability critical for its reuse. The reaction was scaled up to 2.5 L with comparable conversion efficiency. After 10 cycles, the activity of Novozym® 435 was 75% of the initial activity.

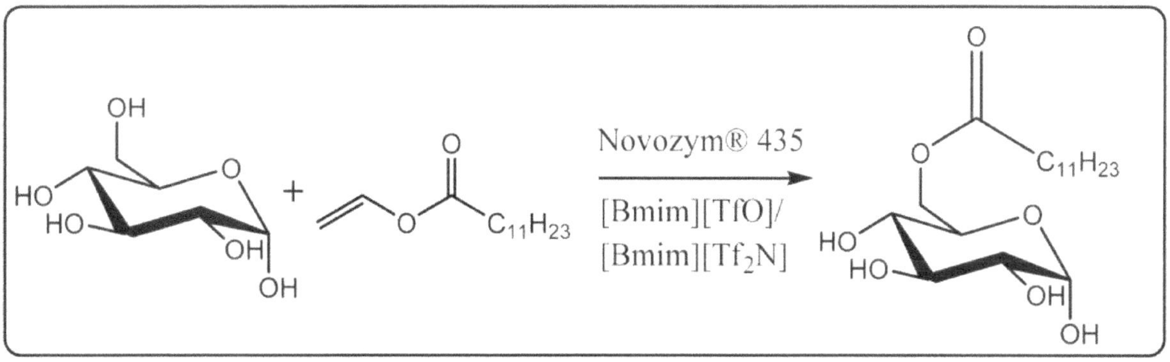

Scheme 6: Transesterification of vinyl laurate with glucose in ionic liquids catalyzed by Novozym® 435 [47].

Developments in Sustainable Biocatalysis Employing Ionic Liquids

Ionic Liquid-Coated Enzymes as Heterogeneous Catalysts

Initial applications of ionic liquids in biocatalysis concentrated on their use as a replacement for conventional solvents, with subsequent studies yielding greater understanding of how they interact with enzymes. An example of such an application was reported by Lee *et al.* [48], in which an ionic liquid-coated enzyme (ILCE) was prepared as a heterogeneous catalyst for use in the enantioselective transesterification of vinyl acetate with racemic alcohols in toluene. In this work, a lipase from *Pseudomonas cepacia* (PS) was first dissolved in the ionic liquid 1-(3′-phenylpropyl)-3-methylimid-azolium hexafluorophosphate [ppmim][PF_6], which has a melting point of 53 °C. Upon cooling the solution to room temperature, a solid ILCE formed, which was then used as a heterogeneous catalyst for the conversion of the substrate to product in toluene. Compared to the native enzyme in toluene, the ILCE enhanced the enantioselectivity of the products by 1.5–2 folds. Furthermore, the ILCE was easily removed from the reaction mixture by filtration and reused as an ionic liquid coated biocatalyst for up to 5 times with comparable activity.

Similarly, Itoh *et al.* [49] have prepared a 1-butyl-2,3-dimethylimidazolium [bdmim][cetyl-PEG10-sulfate] coated lipase by extracting a commercial lipase immobilized onto a ceramic matrix (lipase PS-C) into aqueous buffer containing the ionic liquid followed by freeze-drying. The ILCE was used as a heterogeneous catalyst for transesterification of vinyl acetate with racemic alcohols in diisopropyl ether (Scheme 7) and gave rate enhancements of up to 1000-fold compared to lipase PS-C while maintaining excellent enantioselectivity (>99%). Reusing the ILCE up to five times gave comparable product yield and enantioselectivity. To demonstrate the versatility of the protocol, similar activation effects were also demonstrated for [bdmim][cetyl-PEG10-sulfate]-coated *Candida Rugosa* lipase.

Electron microscopy of the ILCE showed it possessed a higher surface area than the commercial lipase PS-C immobilized on ceramic, whilst MALDI-TOF mass spectrometry experiments indicated the enzyme in the ILCE forms a discrete non-covalent complex with the ionic liquids. Together, these results suggested that the enhanced activity of the ILCE may be due to easier access of the substrate to the more porous ILCE and favorable enzyme flexibility and conformational changes imparted by its non-covalent interactions with the ionic liquid.

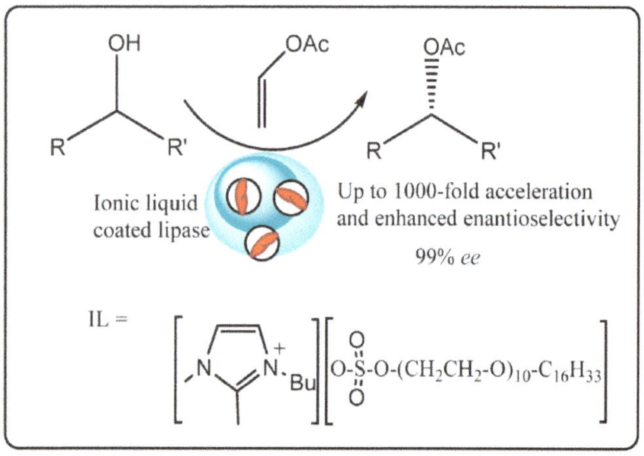

Scheme 7: ILCE-catalyzed enantioselective transesterification of vinyl acetate with racemic alcohols [48].

Encapsulation of enzymes in polymers prepared from ionic liquid monomers has also been investigated [50]. A useful feature in designing ionic liquids is that pendant functional groups can be incorporated into the cation or anion for polymerization. Thus, 1-vinyl-3-ethylimidazolium bromide was polymerized with the cross-linker *N,N'*-methylenebis(acrylamide) in the presence of native horse-radish peroxidase (HRP) or HRP modified with a polyethylene comb polymer (PM_{13}-HRP) to protect it from conformational deactivation by the ionic liquid bromide anion (Figure 2). This method led to the formation of enzyme-encapsulated polymer micro particles, which functioned as heterogeneous catalysts for the oxidation of guaiacol by H_2O_2 in water. The activity of the encapsulated PM_{13}-HRP was found to be 3 times higher than that of encapsulated native HRP and 2 times higher than PM_{13}-HRP encapsulated in non-ionic polyacrylamide micro particles. Furthermore, the encapsulated PM_{13}-HRP was easily recycled by centrifugation, with reduction in activity occurring only at the first recycling stage, which was attributed to loss of PM_{13}-HRP loosely immobilized on the surface of the micro particles.

Cross-linked enzyme aggregates (CLEAs) of mung bean epoxide hydrolases (*mb*EH) has been used to catalyze the asymmetric hydrolysis of styrene oxide in a water-ionic liquid biphasic system (Scheme 8) [51]. Styrene oxide remains dispersed in the ionic liquid [bmim][PF_6] and the hydrolysis product, (*R*)-phenylethanediol ((*R*)-PED), was obtained in the aqueous phase, facilitating its easy separation from starting material and avoiding non-enzymatic hydrolysis. On a 500 mL preparative scale the yield of (*R*)-PED was 49% in 94.6% enantioselectivity.

Rehmann and co-workers have recently demonstrated that a variety of ionic liquids can stabilize the activity of the laccase from *Trametes versicolour* [52]. Laccase, an oxidative mediator system, is important owing to its potential for numerous green oxidation processes but its activity is inhibited by oxidation mediators like TEMPO (2,2,6,6-tetramethyl-1-piperidinyl)oxidanyl, ABTS (2,2′-azino-bis-(3-ethylbenzo-thiazoline-6-sulfonic acid) or 4-hydroxybenzyl alcohol. Such inactivation was avoided by using a biphasic ionic liquid-water system, wherein the mediator predominantly partitions into the ionic liquid, restricting its contact with the laccase, which predominantly resided in the aqueous phase.

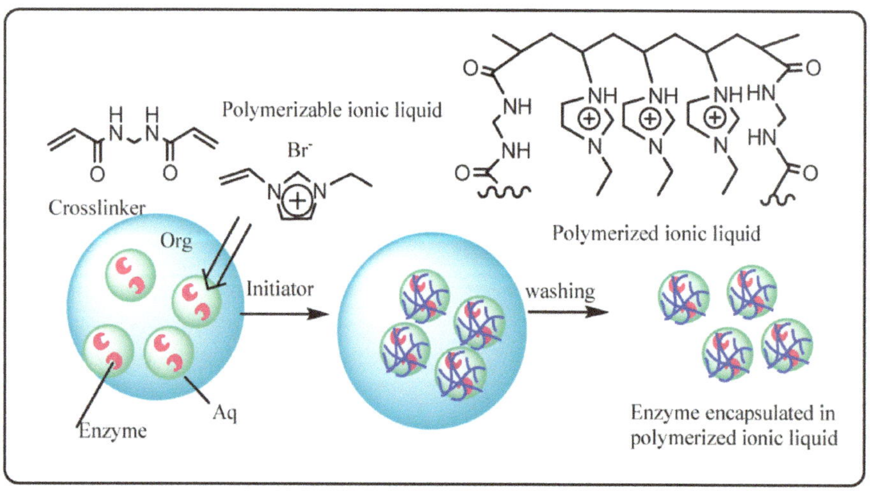

Figure 2: Enzyme encapsulation in an ionic liquid-based polymer [50].

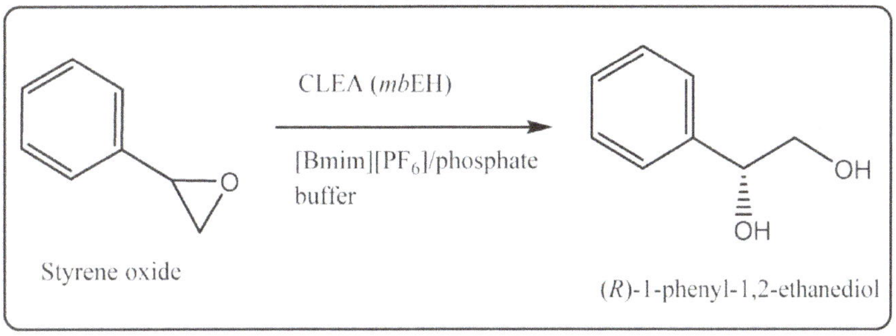

Scheme 8: Cross linked enzyme aggregate of *mb*EH catalyzed asymmetric hydrolysis of styrene oxide [51].

Role of Enzymes and Ionic Liquid System in Biomass Conversion and Biofuel Production

To address the expected increase in demand for energy and transport fuel from a rapidly increasing global population and developing economies, the sustainable conversion of renewable resources to liquid transport fuel and commodity chemicals has become an important area of research. Lignocellulosic (woody) biomass is the key sustainable biomass resource for transformation into biofuels and bio-based chemical products. However, the chemical composition of lignocellulosic biomass differs significantly from that of fossil fuels and efficient catalytic processes need to be developed in order to convert it to fuel precursors and platform chemicals, and the use of ionic liquids and enzymes in addressing this challenge is an active area of research.

Over the past decade, there has been increased interest in using cellulose from renewable biomass resources as a more sustainable alternative raw material to petrochemical feedstock for producing chemical products. A barrier to chemically transform cellulose to other products is its poor solubility in many different solvents. Although ionic liquids containing anions, which are strong hydrogen-bond acceptors, e.g., halides, have been found to dissolve cellulose, these types of ionic liquids tend to denature enzymes, hampering biocatalytic processing of the dissolved cellulose [53,54]. The recent observations of Zhao *et al.* [55] with polyethylene glycol-based ionic liquids that are able to dissolve carbohydrates and cellulose, and retain enzyme activity has provided one option to overcome these limitations. Moreover, these systems are also relevant to the transesterification of methylmethacrylate with glucose or cellulose using Novozym® 435 with the biocatalytic reaction proceeded in up to 89% conversion and 66% yield [55].

Currently, research efforts are also being focused on the use of ionic liquid and deep eutectic solvents for lignocellulosic pre-treatment [56,57] in conjunction with enzymatic hydrolysis of cellulose to fermentable sugars [58,59]. Lignocellulosic biomass is not amenable to facile microbial or enzymatic industrial biotransformation, which limits its economic conversion to fuel precursors and platform chemicals by a biotechnological process. Moniruzzaman and Ono have investigated ways to overcome this barrier by developing an enzymatic biomass process employing an ionic liquid pre-treatment to enable efficient access for enzyme laccase to reactive sites [60]. This enabled an efficient delignification in an ionic liquid/aqueous buffer system, giving cellulose fibers in improved yields and shorter reaction times (Figure 3).

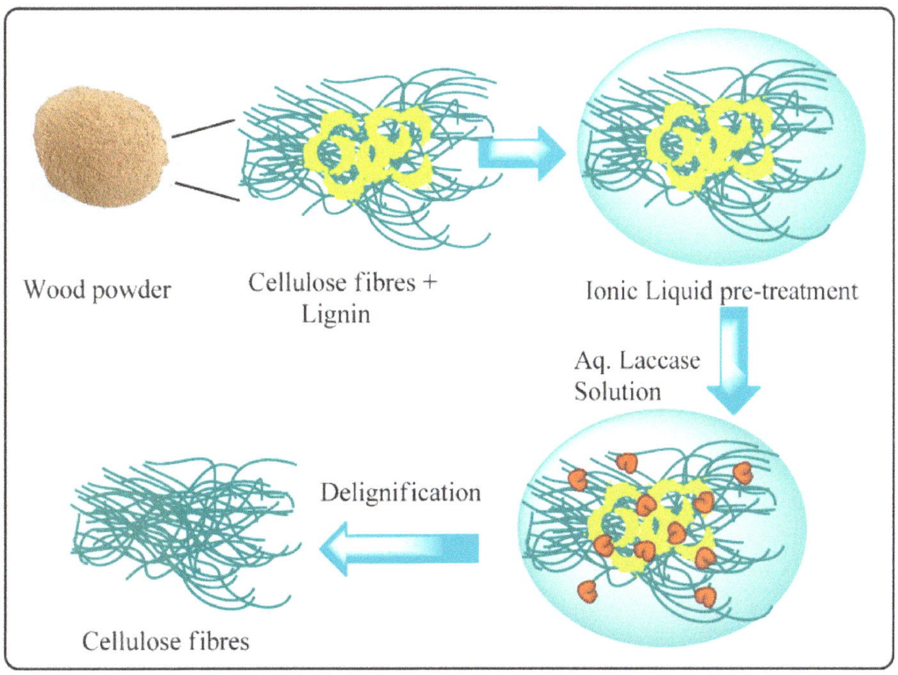

Figure 3: Delignification of woody biomass using ionic liquid and enzymes [60].

Biodiesel is a transport fuel that consists largely of fatty acid methyl esters (FAMEs) prepared by transesterification of natural triglycerides in plant oils or animal fats with methanol. These fuels have generated interest as renewable, low-carbon impact alternatives to fossil fuels. Current biodiesel production uses homogenous alkaline catalysts for transesterification and although this process is efficient, it complicates downstream product separation and increases energy costs [61]. Biocatalytic production of biodiesel in ionic liquids merits consideration as a green and potentially environmentally friendlier alternative to eliminate issues associated with product separation, homogenous catalyst recycling and energy consumption. The first demonstration of the potential of biocatalytic biodiesel production in ionic liquids was reported by Ha *et al.* [62], who used the Novozym® 435 lipase to catalyze the transesterification of soya bean oil with methanol. 1-Ethyl-3-methylimidazolium trifluoromethanesulfonate ([Emim][TfO]) was found to be the best solvent, giving an 80% yield of fatty acid methyl esters in 12 h compared to 65% in *tert*-butanol and 10% under solvent-free reaction conditions. An additional advantage of the ionic liquid system was phase separation of the biodiesel as it was formed. Gamba *et al.* subsequently developed an efficient biphasic system for biodiesel production comprised of a [bmim]bis(trifluoro-methylsulfonyl)-imide ([NTf$_2$]) ionic liquid phase containing methanol and lipase, and a soya bean oil phase (Figure 4) [63].

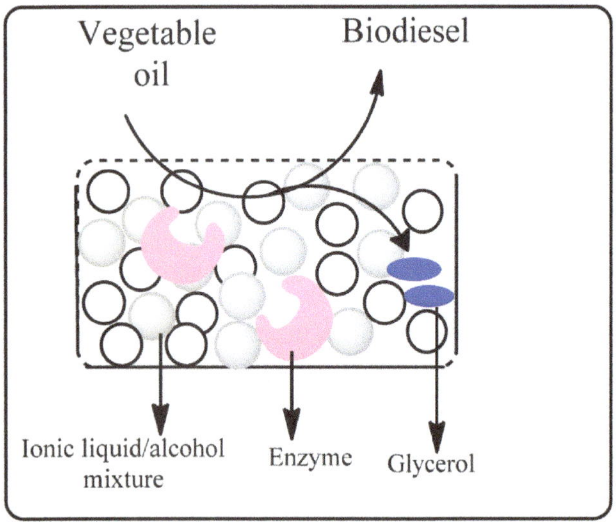

Figure 4: Lipase-catalyzed production of biodiesel in a biphasic ionic liquid-soybean oil system with concomitant separation of the glycerol by-product [63].

This transesterification proceeded at room temperature to give biodiesel in 95% yield in 24 h and was enhanced by the addition of water. Furthermore, the ionic liquid phase served as an extracting agent for removing glycerol from the reaction mixture and the biodiesel was separated by simple decantation. The ionic liquid phase containing the enzyme was recycled four times without significant loss of catalytic activity. After these four cycles the enzyme was filtered from the ionic liquid and the glycerol by-product extracted with water in 98% purity.

Further, Zhao *et al.* have reported the synthesis and application of novel triethyleneglycol-functionalised ionic liquids (Figure 5) in biodiesel production [64]. These ionic liquids dissolved both triglycerides and lipases, and a high level of catalytic activity was observed with different commercially available lipases; quantitative conversions of Miglyol® oil 812 prepared from coconut and palm kernel oils was achieved at 96 h in a 70/30 (*v*/*v*) ionic liquid/MeOH mixture. Overall, these preliminary studies on the transesterification of vegetable oils in ionic liquids indicate they are promising solvent systems for industrial biocatalytic biodiesel production.

The enzyme catalysed esterification of soybean oil in [Emim][PF$_6$] with microwave heating has been demonstrated by Yu*et al.* to give faster conversion with a 1.8-7.8-fold increase in enzyme activity compared to *t*-butanol and a solvent-free system using conventional heating [65]. This is the first example of using synergistic effect of microwave irradiation and ionic liquids for enzyme catalyzed biofuel production. The study also demonstrated recycling of the ionic liquid/enzyme phase for five cycles with only a slight decrease (8%) in enzyme activity.

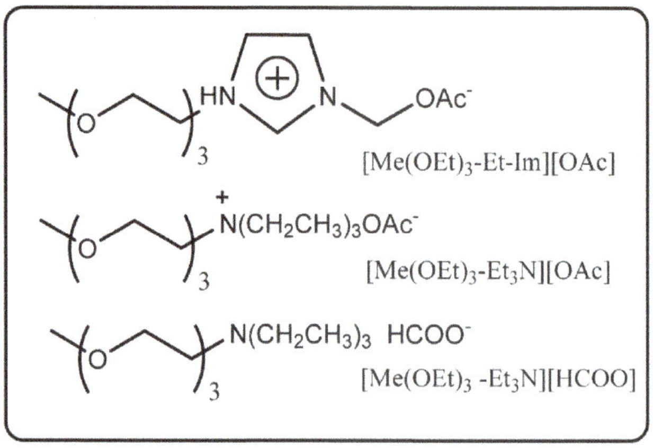

Figure 5: Novel triethyleneglycol-functionalized ionic liquids for biodiesel synthesis [64].

Zhang *et al.* have demonstrated the activity of *Penicillium expansum* lipase (PEL) in ionic liquids and reported the application of this system for the synthesis of corn oil-based biodiesel [66]. The ionic liquid [Bmim][PF$_6$] was found to be an optimal solvent for the transesterification reaction, giving a higher conversion (86%) compared to *t*-butanol (52%), and PEL was found to be more tolerant to the reaction conditions compared to other lipase enzymes routinely used. Furthermore, the biphasic ionic liquid-FAME reaction system facilitated separation of the biodiesel and recycling of the catalyst. Lai *et al.*, from the same group, have reported the use of PEL for the transesterification of triglycerides from the microalgae *Chlorella vulgaris*, the first example of ionic liquid mediated biocatalytic conversion of microalgal oil to biodiesel [67].

Ionic Liquid-Based Supported Liquid Membranes in Separation Processes

Selectively permeable membranes that separate bulk solvent phases have potential application in separation processes. For example, kinetic resolution employing enzyme catalysis followed by product separation using selectively permeable membranes allows the conversion and separation of the desired product in a single continuous process which is attractive from an environmental and industrial perspective. Supported liquid membranes (SLMs) have been identified as suitable materials for such continuous separation processes. These materials are usually comprised of a solvent immobilized in the porous structure of polymeric or ceramic membrane [68]. The SLM serves to separate a feedstock solvent and a receiving solvent with the immobilized solvent mediating solute transport across the membrane. A critical feature of SLMs in continuous extraction processes is negligible loss of the immobilized solvent to either of the solvent phases and to the atmosphere.

The potential of using SLMs incorporating ionic liquids as the immobilized solvent for transport of organic molecules was first demonstrated by Branco *et al.*, who employed [bmim][PF_6] in a polyvinyl-idene fluoride membrane owing to its poor solubility in the feedstock and receiving solvents, and negligible vapor pressure [69]. This system was able to selectively transport diisopropylamine from a mixture of amines in the feed solvent to the receiving solvent.

Miyako *et al.* have subsequently demonstrated a lipase-facilitated transport of organic acids across SLMs employing ionic liquids as transport solvents (Scheme 9) [70]. Based on this technology, organic acids were esterified by *Candida rugosa* lipase (CRL) in the feedstock phase with the resulting ester partitioning into the ionic liquid phase in the SLM and then diffusing into a receiving phase that contained a porcine pancreatic lipase (PPL) to hydrolyze the ester back to the initial organic acid. Accumulation of organic acids in the receiving phase was not observed in the absence of CRL due to their poor solubility in the ionic liquid used to construct the SLM. The use of ionic liquids as transport solvents also enabled long-term stability of the SLM compared to conventional organic solvent.

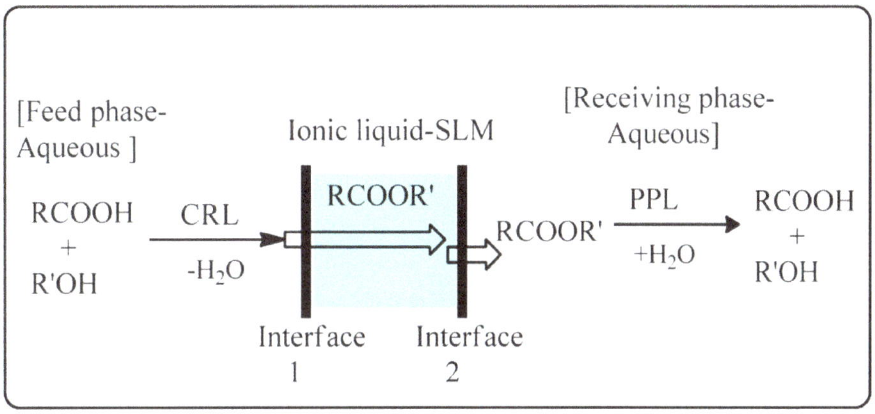

Scheme 9: Lipase-facilitated transport of carboxylic acids across an ionic liquid-based supported liquid membrane (SLM) [70].

This methodology was further extended to the resolution of (*S*)-ibuprofen through selective CRL-catalyzed esterification of the (*S*)-enantiomer in the feedstock, diffusion of the ester through the SLM ionic liquid and regeneration of (*S*)-ibuprofen in the receiving phase by PPL (Scheme 10) [71]. This process provided (*S*)-ibuprofen in up to 75% enantioselectivity.

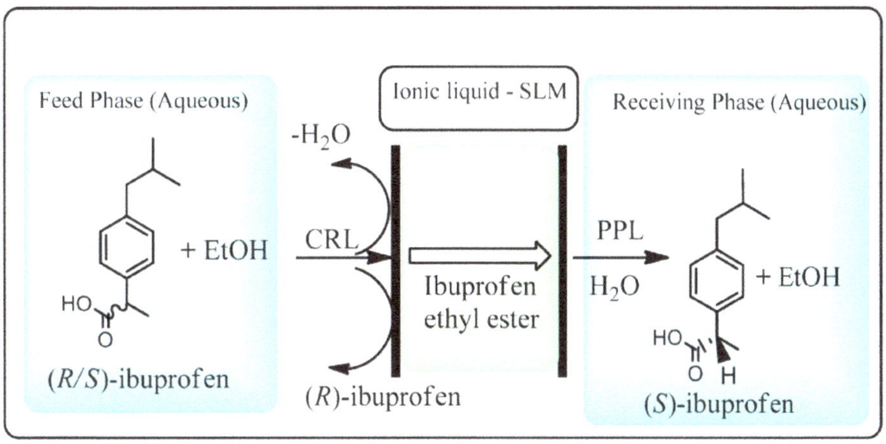

Scheme 10: Lipase-facilitated enantioselective transport of (*S*)-ibuprofen across an ionic liquid based supported liquid membrane (SLM) [71].

An ionic liquid-based SLM system for the kinetic resolution of racemic 2-pentanol and separation of the unreacted enantiomer has been developed by Hernández-Fernández *et al.* [72]. (*S*)-2-Pentanol is an interesting chiral intermediate in the synthesis of drug candidates for Alzheimer's disease [73]. Resolution and separation of (*S*)-2-pentanol from a racemic mixture was achieved by Novozym® 435 lipase-catalyzed enantioselective transesterification of vinyl esters resulting in (*R*)-ester and (*S*)-alcohol (Scheme 11). In the SLM system, the alcohols readily diffused across the membrane whilst the diffusion rate of the esters was relatively low. Thus, enantioselective transesterification resulted in (*S*)-2-pentanol selectively accumulating in the receiving phase, enabling its separation from the (*R*)-enantiomer. Of six ionic liquids tested, [bmim][BF_4] was found to be the most suitable as a transport solvent, giving 60–70-fold enrichment of the (*S*)-2-pentanol enantiomer in the receiving phase.

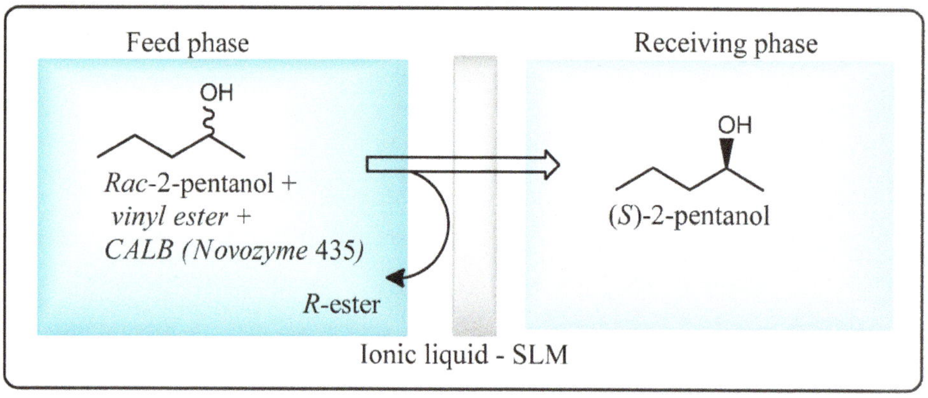

Scheme 11: Lipase-facilitated resolution of *rac*-2-pentanol using an ionic liquid-based supported liquid membrane (SLM) [72].

This methodology was further extended to the separation of (*S*)-phenylethanol, a commercially important chiral building block in the fine chemical and pharmaceutical industries [74]. The kinetic resolution of *rac*-1-phenylethanol was achieved by transesterification of a vinyl ester with the (*R*)-enantiomer, catalysed by Novozym® 435 lipase. (*S*)-1-Phenylethanol and the (*R*)-vinyl ester derivative were separated in the reactor using a [bmim][BF_4]-based SLM to give the (*S*)-enantiomer in 99% enantioselectivity.

Biocatalysis in Biphasic Ionic Liquid-Supercritical Carbon Dioxide Systems

Some industrially important enzymes have been found to retain activity in a variety of non-aqueous solvents, including supercritical carbon dioxide ($ScCO_2$) with several types of enzyme catalyzed reactions documented in this alternative solvent [75,76]. $ScCO_2$ is an attractive industrial solvent owing to its non-flammable properties and low toxicity, and is relatively easily produced in terms of pressure and temperature compared to other supercritical fluids. A number of reports have demonstrated the value of combining ionic liquids and $ScCO_2$ for green biocatalytic processes [42,77]. Because $ScCO_2$is immiscible with several different ionic liquids, it can be used to deliver substrates to, and extract products from, the ionic liquid phase. This particular use of $ScCO_2$ makes it an attractive candidate for continuous flow and batch industrial processes for reactions involving the use of enzymes in ionic liquids. A critical feature of such systems is efficient mass transfer of substrates and products between $ScCO_2$ and ionic liquids, which is dependent on ionic liquid hydrophobicity and viscosity.

The group of Iborra have carried out detailed work on the use of biphasic systems consisting of ionic liquid and $ScCO_2$phases for combined biocatalysis and separation. Thus, Lozano *et al.* developed [78] a continuous process for CALB-catalyzed synthesis of short chain esters by transesterific-ation of vinyl esters with alcohols in an ionic liquid/Sc-CO_2biphasic system with $ScCO_2$ acting as a substrate feedstock solvent and product extraction solvent (Scheme 12). Synthetic efficiency was greater in the biphasic system employing butyltetramethyl ammonium [btma][NTf_2] compared to 3-cyanopropyl-trimethylammonium [NTf_2] which was attributed to better mass transfer of substrates between $ScCO_2$ and the more hydrophobic [btma][NTf_2] ionic liquid.

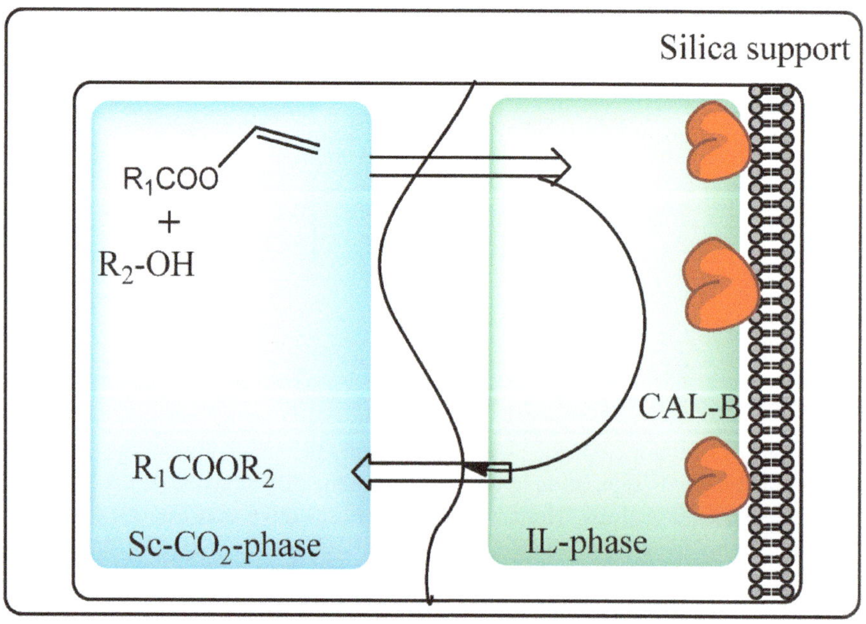

Scheme 12: Transesterification of vinyl esters in a biphasic ionic liquid-$ScCO_2$ system [78].

Immobilization of ionic liquids onto an insoluble solid support is an attractive strategy for continuous biocatalytic processes employing a mobile $ScCO_2$ phase since this approach minimizes the amount of ionic liquid used and enables easy product separation and catalyst recycling [79,80,81]. Lozano *et al* have synthesized immobilized ionic liquid phases by grafting or copolymerizing imidazolium-based ionic liquids to generate macroporous polymer monoliths with imidazolium loading ranging from 40% to 55% *w/w* [82]. The enzyme CALB was non-covalently adsorbed onto polymer monoliths from an aqueous solution. The systems were then used as heterogeneous catalysts for the continuous flow transesterification synthesis of citronellyl propionate from citronellol and vinyl propionate in $ScCO_2$ (Figure 6). Productivity was highest when a styrene/divinylbenzene/imidazolium-based monolith was used compared to a 2-hydroxyethyl/ethylene glycol dimethyl acrylate/imidazolium-based monolith which was attributed to better mass transfer properties due to its greater hydrophobicity and porosity. Yields of up to 93% were achieved with no enzyme leaching observed.

The same group later reported studies on the immobilization of CALB by adsorption onto silica gel supports containing covalently grafted [btma] or trioctyl-methyl ammonium [toma][NTf_2] ionic liquids [83]. The immobilized CALB was then assessed for its ability to kinetically resolve *rac*-1-phenylethanol by enantioselective transesterification of vinyl propionate in a continuous flow reactor using $ScCO_2$ as the mobile phase. A yield of up to 96% and an enantioselectivity of >99% was achieved for the *trans*-esterification with the (*R*)-enantiomer. The productivity of this reaction system was found to be up to six times greater when using $ScCO_2$ compared to hexane, attributed to better ability of $ScCO_2$ to transport solutes through the grafted ionic liquid phase.

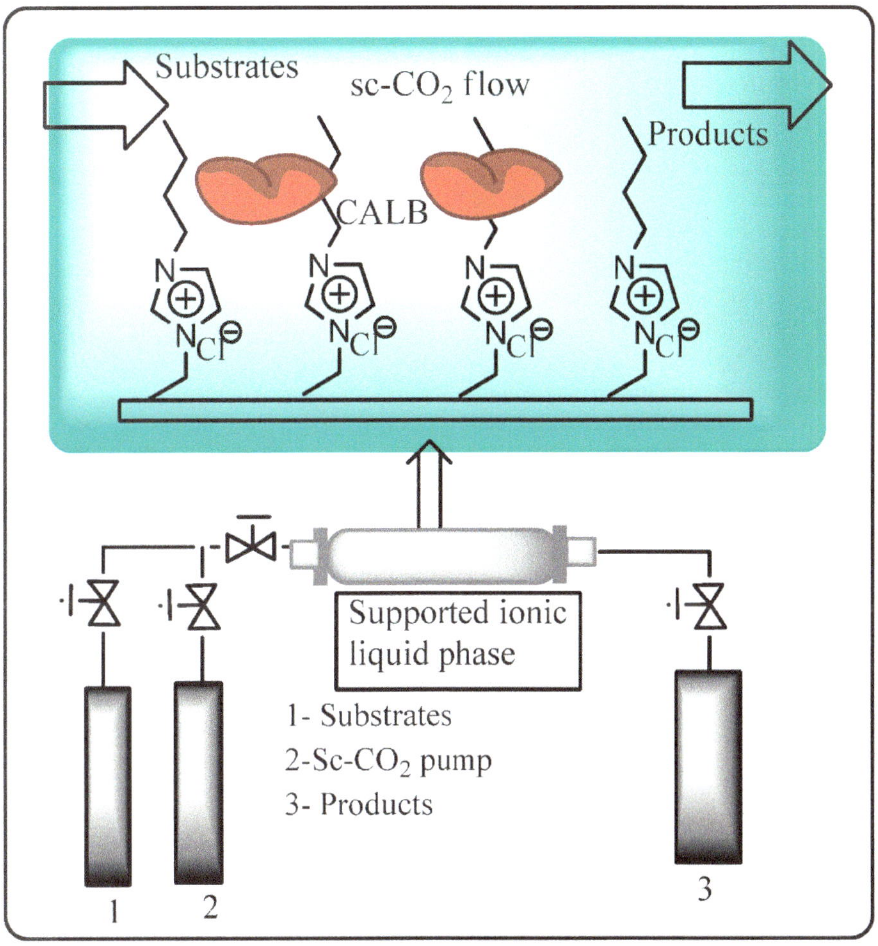

Figure 6: Continuous flow reactor setup employing CALB supported on an immobilized ionic liquid phase and sc-CO_2 as a mobile phase [82].

In a seminal study, Lozano *et al.* [84,85] developed a continuous dynamic kinetic resolution of *rac*-1-phenylethanol employing a mixture of Novozym® 435 lipase and a heterogeneous acidic zeolite in an ionic liquid-$ScCO_2$ biphasic system. The role of the lipase was to catalyze transesterification of vinyl propionate with (*R*)-1-phenylethanol while the acidic zeolite was included to catalyze the isomerization of (*S*)-1-phenylethanol to the (*R*)-enantiomer. Both catalysts were first coated with ionic liquids by adsorption to stabilize the enzyme from the strongly acidic sites of the zeolite and the mobile $ScCO_2$phase. Overall, this catalytic system provided (*R*)-phenylethylpropionate in up to 98% yield and 97% enantioselectivity (Scheme 13) without any loss of catalyst activity after 14 days of operation. These findings clearly demonstrate the industrial potential of a combination of heterogeneous biological and chemical catalysis employing ionic liquids and $ScCO_2$ in continuous flow process for green production of commercially important enantiomers.

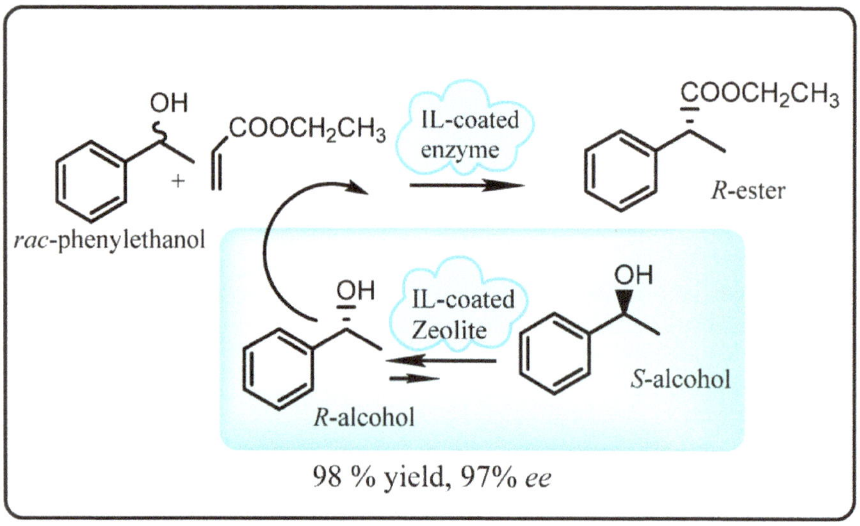

Scheme 13: Dynamic kinetic resolution of *R*-phenylethylpropionate with ionic liquid-coated enzyme and zeolite [85].

Use of Biodegradable Ionic-Based Deep Eutectic Solvents for Biocatalysis

Although ionic liquids are often claimed to be new generation green solvents due to their low flammability and vapor pressure, their component ions may be toxic, slow to biodegrade and may accumulate in marine organisms. Hence, the presence of ionic liquids in industrial waste streams may result in an environmental hazard. In light of this potential problem, the design and use of biodegradable ionic liquids that will not persist in the environment is a matter of importance. The building blocks used to synthesize such a class of ionic liquids are generally those regarded as being biocompatible, e.g., amines, amino acids, amino alcohols, essential oils, carbohydrates and carboxylic acids [86]. An added advantage is that these building blocks are obtained from renewable resources, which further enhances the sustainability profile of the ionic liquid.

An interesting class of designed biodegradable ionic liquids are physical mixtures of biological salts, such as choline chloride and uncharged biological hydrogen bond donors, e.g., urea or glycerol, which exist as deep eutectic solvents (DESs). Such eutectic solvents exist as a physical mixture with a specific component composition that imparts a melting temperature much lower than that of any of the individual components. Compared to earlier generation ionic liquids, ionic-based DESs are relatively easy to prepare in accord with green chemistry principles since they simply comprised of mixtures of compounds in defined ratios and hence do not require any purification [43].

Currently, there are only a few examples of enzyme catalysis in a DES compared to earlier generations of ionic liquids. The first report of enzymes used economically with a DES was a lipase-catalyzed transesterification of ethyl valerate with 1-butanol [87]. The specific activity of immobilized CALB (iCALB) was found to be 2–7 times higher in a choline chloride:glycerol (ChCl:Gly) DES compared to earlier generation ionic liquids [bmim][Tf_2N] or [bmim][BF_4], with conversion of 96% achieved. Remarkably, transesterification with the glycerol component of the DES was extremely low (<0.5%). iCALB-catalysed aminolysis of ethyl valerate with 1-butylamine also proceeded efficiently in ChCl:Gly with >90% conversion after 1 h. ChCl:Gly was also found to be a suitable co-solvent with aqueous buffers; a significant 20-fold enhancement in the rate of epoxide hydrolase-catalyzed hydrolysis of styrene oxide was observed in 25% *v*/*v*ChCl:Gly/buffer compared to buffer alone.

The biocatalytic oxidation of cellulose-derived carbohydrates has potential application in bioelectrochemical synthesis and energy production in biofuel cells. Recently, the potential of a biocompatible hydrated choline dihydrogen phosphate (choline dhp) ionic liquid containing 35 wt % water as a solvent for the cellobiose dehydrogenase (CDH)-catalyzed oxidation of cellobiose to cellobionolactone (Scheme 14) was demonstrated [88]. Inter-electron transfer to a synthetic dye acceptor, DCIP, was observed, indicating the system may also have use in bioelectrochemical synthesis. Since CDH is known to be able to transfer electrons to electrodes, the CDH-DES system may have application in ionic liquid/carbohydrate-based bioenergy cells.

Recently, Zhao *et al* have employed DES-made from choline acetate and glycerol in the enzymatic preparation of biodiesel from Miglyol® oil 812 [89]. A eutectic mixture of choline acetate/glycerol (1:1.5) had a lower viscosity than choline chloride/urea eutectic mixtures, and was found to support CALB enzyme activity. Transesterification reactions performed in DES and 20% (*v*/*v*) MeOH and 1% water using Novozym® 435 gave a faster reaction rate compared to that in organic solvents or hydrophobic ionic liquids, and afforded 97% conversion. The biphasic separation of the resultant biodiesel allowed for its easy separation from the DES.

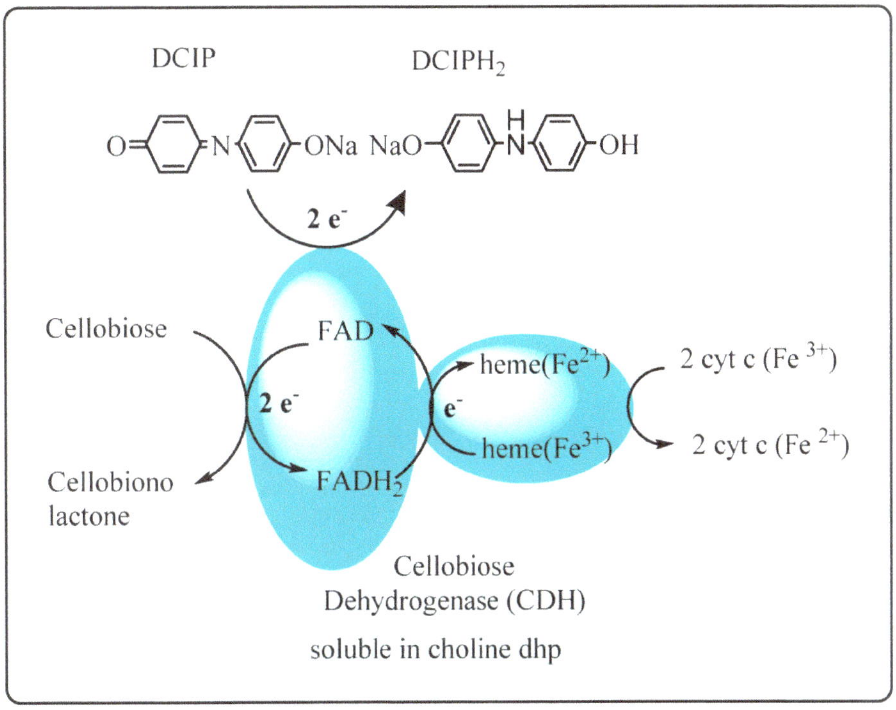

Scheme 14: CDH-catalyzed oxidation of cellobiose by inter and intra-electron transfer pathways [88].

Application of Ionic Liquids in Whole-Cell Biocatalysis: A Greener Perspective

Biocatalysis in organic synthesis employing recombinant enzymes requires an expensive set of enzyme purification unit operations, which themselves generate undesirable waste streams. Using whole cells to catalyze chemical transformations avoids this problem. However, technological challenges with whole-cell biocatalysis often include poor substrate and product solubility in the aqueous buffer/medium critical for cell survival. Furthermore, product accumulation in the cell culture medium can lead to cell toxicity. To overcome these problems, multiphase solvent systems have been developed using immiscible organic solvents, which act as substrate reservoirs and product extraction agents. However, the limitations of these systems are the flammability/toxicity of these solvents, and their deleterious effects on cell viability due to their destructive effects on cell membrane structure [77]. Hydrophobic ionic liquids are an attractive alternative to overcome these problems. Compared to conventional organic solvents, the ionic liquids [bmim][PF_6], [bmim][NTF_2] and methytrioctyl-ammonium [NTf_2] have been found not to affect cell membrane integrity of *Lactobacillus kefir*, *Escherichia coli* and*Saccharomyces cerevisiae* in biphasic systems, thus offering the advantage of preserving cell function and allowing intracellular co-factor regeneration in organic synthesis.

The first example of whole-cell biocatalysis employing a water-ionic liquid biphasic system was reported by Cull *et al.*using *Rhodococcus* R312 [90]. These cells contain nitrile hydratase which catalyzes the hydration of the nitrile group to an amide, a process of synthetic interest for producing fine chemicals and pharmaceutical intermediates. Toluene could be replaced by [bmim][PF_6] as a reservoir for the substrate 1,3-dicyanobenzene and as an extraction solvent for the product 3-cyano-benzamide in this biphasic system (Scheme 15). It was observed that the initial rate of 3-cyanobenz-amide production was lower in the [bmim][PF_6]-water biphasic system compared to the toluene-water system but that the overall yield was slightly higher. Since the specific activity of nitrile hydratase activity was found to be higher in the ionic liquid-water system and that the ionic liquid did not reduce cell viability, it was concluded the lower rate of product formation was due to reduced mass transfer of substrate from the more viscous ionic liquid phase. An advantage of [bmim][PF_6] in downstream processing was reduced cell aggregation at the phase interface, which made product recovery and material recycling easier.

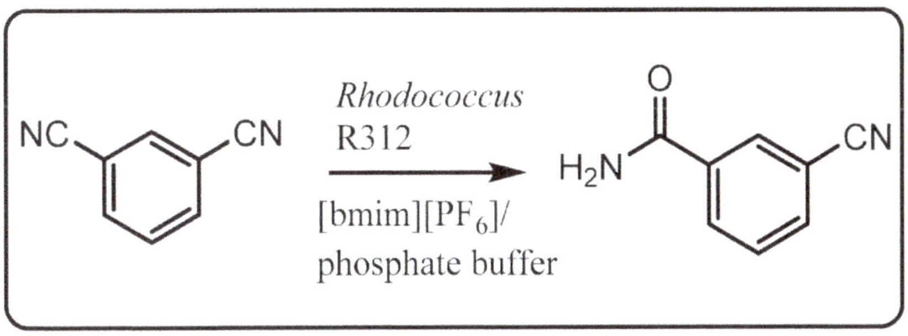

Scheme 15. Hydration of 1,3-dicyanobenzene to 3-cyanobenzamide using *Rhodococcus*R312 in biphasic system [90].

Asymmetric reduction of prochiral ketones to chiral alcohols is an important reaction in the development of chiral pharmaceuticals. Howarth *et al.* have reported whole-cell biocatalysis for asymmetric reduction of prochiral ketones in a biphasic ionic liquid-water system [91]. In this work, yeast was immobilized by encapsulation in low-cost calcium alginate beads and the reduction of several different prochiral ketones performed in a biphasic system of water and [bmim][PF_6] containing the substrate (Scheme 16). Yields ranged from poor to good depending on the chosen substrate, with some excellent enantioselectivities (>95%) obtained. The asymmetric reduction of prochiral ketones in ionic liquid-water biphasic systems has since been further developed [92,93].

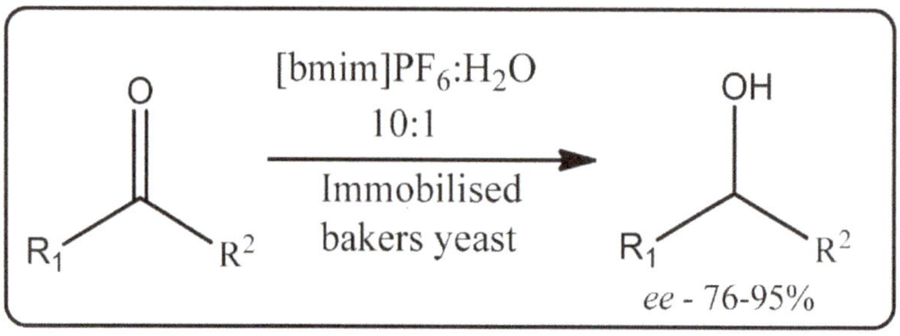

Scheme 16: Immobilized bakers yeast-catalyzed reduction of ketones in a biphasic ionic liquid-water system [91].

Asymmetric reduction of 4-chloro-acetophenone to (*R*)-1-(4-chlorophenyl)ethanol with *Lactobacillus kefir* was carried out as a representative example of a whole-cell-catalyzed multiphase process for the synthesis of fine chemicals (Scheme 17) [92]. In this biphasic system, the ionic liquids employed did not disrupt the necessary cellular cofactor generation systems required for performing the reaction, thus eliminating the need to add these expensive cofactors externally, and provided the product in 92.8% yield with 99.7% enantioselectivity. The potential industrial application of this system was demonstrated by scaling up the reaction volume nearly a hundred fold with the same yield and product purity. Furthermore, no emulsion formed when mixing the two phases, enabling their easy separation for subsequent product recovery and solvent recycling. Similar biphasic ionic liquid-water systems have been used for the *Saccharomyces cerevisiae*asymmetric reduction of 4-chloro acetoacetate to (*S*)-4-chloro-3-hydroxy-butanoate, an intermediate in the synthesis of cholesterol-reducing statin inhibitors, in good yield (80%) and enantioselectivity (84%). In another example of asymmetric reduction, Xu *et al.* [93] used immobilized *Acetobacter* sp. CCTCC M209061 cells in a deep eutectic comprised of chloline chloride and urea as a biocatalytic system to reduce 3-chloro-propiophenone to (*S*)-3chloro-1-phenylpropanol in 82% yield and >99% enantioselectivity. The reaction could be scaled up to 500 mL with comparable yield and enantioselectivity.

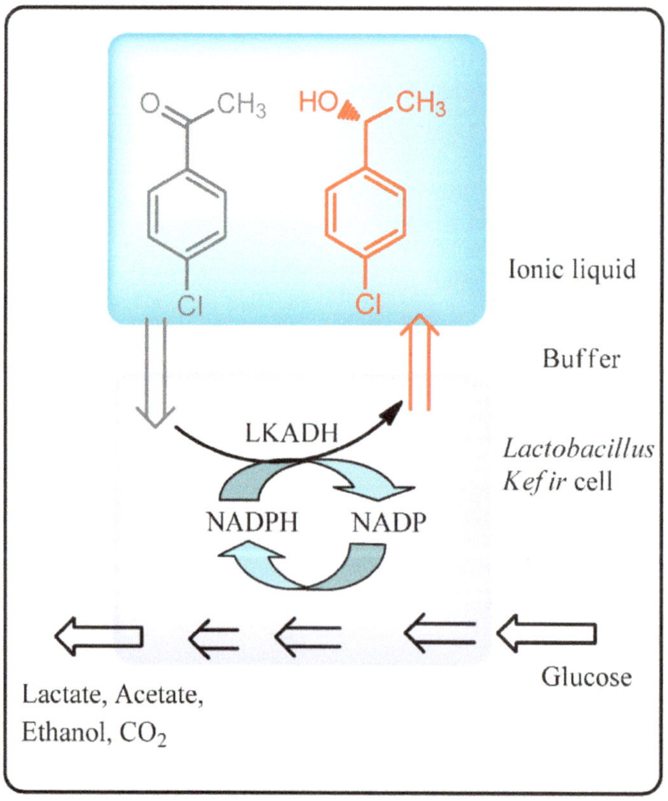

Scheme 17: Asymmetric reduction of 4-chloroacetophenone by *Lactobacillus kefir* in a biphasic ionic liquid-water system [92].

In addition to being employed in whole-cell biphasic systems for the production of fine chemicals or pharmaceutical intermediates, ionic liquids have also been assessed as post-fermentation extraction solvents of amino acids and antibiotics [90,93,94,95,96]. The partition coefficients of erythromycin in biphasic water-butyl acetate and water-[bmim][PF_6] were, for example, found by Pfruender *et al.* to be comparable at physiological pH, with about a 10-fold accumulation into the non-aqueous phase [93]. At pH 10 or above, partitioning of erythromycin into [bmim][PF_6] decreased to 2-fold while accumulation into butyl acetate increased. These results suggest a potential use of [bmim][PF_6] as a non-flammable extraction solvent for erythromycin at physiological pH followed by back extraction with an alkaline aqueous solution to enable recycling of the ionic liquid. Similarly, a pH-dependent partitioning of the antibiotics amoxicillin and ampicllin between water and ionic liquids was observed by Soto *et al.*; the antibiotics accumulated 3–5-fold into ionic liquid at pH 8 while the reverse occurred at pH 4 [96]. Many different amino acids are prepared by cell fermentation but their isolation by liquid-liquid extraction technologies can be problematic due to their hydrophilic properties. Smirnova *et al.* assessed the ability of a solution of the crown ether dicyclohexano-18-crown-6 in [bmim][PF_6] to extract amino acids from aqueous solution [95]. In this system, the crown ether served as a complexing agent with the ammonium group at acidic pH to enhance the solubility of the amino acid in the ionic liquid. Excellent recoveries (>90%) of hydrophilic and hydrophobic amino acids were obtained from acidic aqueous solutions and fermentation broths. Furthermore, the amino acids were efficiently back-extracted from the ionic-liquid using an alkaline aqueous solution.

Although enzyme-based production of biodiesel is an attractive alternative to homogenous alkali catalysis, a whole-cell process for producing biodiesel would remove the need for an enzyme purification step. To this end, Arai *et al.*demonstrated the transesterification of soybean oil with methanol using fungi in an ionic liquid-oil biphasic system with [emim][BF_4] and [bmim][BF_4] [97]. These ionic liquids were chosen to act as a reservoir for methanol and the transesterification by-product glycerol, both of which can have detrimental effects on whole cell catalysts, while allowing phase separation of the biodiesel for easy product separation. Four types of whole-cell biocatalysts supported on polyurethane biomass support particles were assessed in this study; wild type *Rhizopus oryzae* producing triacylglycerol lipase (w-ROL) and genetically engineered *Aspergillus oryzae* expressing either *Fusarium heterosporum* lipase (r-FHL), *Candida antarctica* lipase B (r-CALB) or *A. oryzae* mono- and diacylglycerol lipase (r-mdlB). Supported w-ROL was found to have the highest activity achieving 60% FAME content. Its use alone did not result in complete FAME production because of the inherent positional specificity of its triacylglycerol lipase. However, a higher FAME content of 90% was achieved using a mixture of supported w-ROL and r-mdlB. A difficulty associated with the long-term stability of w-ROL in the presence of ionic liquid was overcome by treating it with glutaraldehyde to cross-link its endogenous lipases. This greatly stabilized w-ROL and enabled its recycling. Overall, this research demonstrated the potential application of ionic liquids and whole-cell biocatalysis in developing a more sustainable and environmentally friendly route toward biodiesel production.

CONCLUSIONS

Ionic liquids as bulk solvent phases offer new opportunities as recyclable alternatives to conventional organic solvents for biocatalytic production of commercially important chemicals including asymmetric synthesis, with enhancement in certain cases of the rate of product formation and enantioselectivity. The application of benign ionic liquids that can both dissolve cellulose and maintain enzyme activity bodes well for future developments in utilizing cellulose as raw chemical resource. Similar strategies can be forecasted as apparent from our recent work with biphasic biocatalytic conversion of xylan, the second most abundant biomaterial found in nature after cellulose, using recombinantly expressed, thermostable xylanases, genetically engineered to have high activity in a variety of non-polar ionic liquids, leading to the production of a range of chemical intermediates that can be directly recovered from the aqueous phase in continuous processes [98]. Similar approaches and processes are equally germane in applications with biphasic ionic liquid-vegetable oil mixtures and show great potential as efficient reaction systems for producing biodiesel using enzyme or whole-cell-catalyzed transesterification. In addition to their use as bulk solvents, ionic liquids have been applied as enzyme coatings for the development of recyclable heterogeneous biocatalysts, as immobilized supports for enzymes in biphasic systems employing $ScCO_2$ for continuous reaction systems and as components of supported liquid membranes for continuous separation of reactants and products in enzyme catalyzed reactions. These are exciting developments as the amounts of ionic liquid required in these systems are much less than those for bulk solvent applications, which will reduce any unfavorable environmental impacts of synthesizing or using ionic liquids. The inherent biphasic nature of these systems also enables easier recycling of the ionic liquid component. The CDH-catalyzed oxidation of cellobiose in a biodegradable ionic-based DES demonstrates the potential of employing ionic liquids as components of bioelectrochemical energy cells. Overall, these developments highlight the potential of ionic liquid-based biocatalytic processes in designing more benign sustainable chemical processes for enhancing the quality of life of future generations.

REFERENCES

1. Anastas, P.T.; Kirchhoff, M.M. Origins, current status, and future challenges of green chemistry. *Acc. Chem. Res.* 2002,*35*, 686–694.
2. Welton, T. Room-temperature ionic liquids. Solvents for synthesis and catalysis. *Chem. Rev.* 1999, *99*, 2071–2083.
3. MacFarlane, D.R.; Forsyth, M.; Howlett, P.C.; Pringle, J.M.; Sun, J.; Annat, G.; Neil, W.; Izgorodina, E.I. Ionic liquids in electrochemical devices and processes: Managing interfacial electrochemistry. *Acc. Chem. Res.* 2007, *40*, 1165–1173.
4. Dupont, J.; de Souza, R.F.; Suarez, P.A.Z. Ionic liquid (molten salt) phase organometallic catalysis. *Chem. Rev.* 2002,*102*, 3667–3691.
5. Roosen, C.; Muller, P.; Greiner, L. Ionic liquids in biotechnology: Applications and perspectives for biotransformations. *Appl. Microbiol. Biotechnol.* 2008, *81*, 607–614.

6. Wang, Z.S.; Koumura, N.; Cui, Y.; Miyashita, M.; Mori, S.; Hara, K. Exploitation of ionic liquid electrolyte for dye-sensitized solar cells by molecular modification of organic-dye sensitizers. *Chem. Mater.* 2009, *21*, 2810–2816.
7. Riisager, A.; Fehrmann, R.; Haumann, M.; Wasserscheid, P. Supported ionic liquid phase (silp) catalysis: An innovative concept for homogeneous catalysis in continuous fixed-bed reactors. *Eur. J. Inorg. Chem.* 2006, *2006*, 695–706.
8. Gruttadauria, M.; Riela, S.; Aprile, C.; Lo Meo, P.; D'Anna, F.; Noto, R. Supported ionic liquids. New recyclable materials for the l-proline-catalyzed aldol reaction. *Adv. Synth. Catal.* 2006, *348*, 82–92.
9. Wasserscheid, P.; Keim, W. Ionic liquids—New "solutions" for transition metal catalysis. *Angew. Chem. Int. Ed.* 2000,*39*, 3773–3789.
10. Fei, Z.F.; Geldbach, T.J.; Zhao, D.B.; Dyson, P.J. From dysfunction to bis-function: On the design and applications of functionalised ionic liquids. *Chem. -Eur. J.* 2006, *12*, 2123–2130.
11. Lee, S.G. Functionalized imidazolium salts for task-specific ionic liquids and their applications. *Chem. Commun.* 2006, 1049–1063.
12. De Maria, P.D. "Nonsolvent" applications of ionic liquids in biotransformations and organocatalysis. *Angew. Chem. Int. Ed.* 2008, *47*, 6960–6968.
13. Van Rantwijk, F.; Sheldon, R.A. Biocatalysis in ionic liquids. *Chem. Rev.* 2007, *107*, 2757–2785.
14. Miao, W.S.; Chan, T.H. Ionic-liquid-supported peptide synthesis demonstrated by the synthesis of leu^5-enkephalin. *J. Org. Chem.* 2005, *70*, 3251–3255.
15. Fraga-Dubreuil, J.; Bazureau, J.P. Grafted ionic liquid-phase-supported synthesis of small organic molecules. *Tetrahedron Lett.* 2001, *42*, 6097–6100.
16. Deetlefs, M.; Seddon, K.R. Assessing the greenness of some typical laboratory ionic liquid preparations. *Green Chem.*2010, *12*, 17–30.
17. Harjani, J.R.; Singer, R.D.; Garciac, M.T.; Scammells, P.J. Biodegradable pyridinium ionic liquids: Design, synthesis and evaluation. *Green Chem.* 2009, *11*, 83–90.
18. Harjani, J.R.; Farrell, J.; Garcia, M.T.; Singer, R.D.; Scammells, P.J. Further investigation of the biodegradability of imidazolium ionic liquids. *Green Chem.* 2009, *11*, 821–829.
19. Tao, J.H.; Zhao, L.S.; Ran, N.Q. Recent advances in developing chemoenzymatic processes for active pharmaceutical ingredients. *Org. Process Res. Dev.* 2007, *11*, 259–267.
20. Straathof, A.J.J.; Panke, S.; Schmid, A. The production of fine chemicals by biotransformations. *Curr. Opin. Biotechnol.*2002, *13*, 548–556.
21. Klibanov, A.M. Improving enzymes by using them in organic solvents. *Nature* 2001, *409*, 241–246.
22. Klibanov, A.M. Enzymatic catalysis in anhydrous organic-solvents. *Trends Biochem. Sci.* 1989, *14*, 141–144.
23. Magnuson, D.K.; Bodley, J.W.; Evans, D.F. The activity and stability of alkaline phosphatase in solutions of water and the fused salt ethylammonium nitrate. *J. Sol. Chem.* 1984, *13*, 583–587.
24. Moniruzzaman, M.; Nakashima, K.; Kamiya, N.; Goto, M. Recent advances of enzymatic reactions in ionic liquids.*Biochem. Eng. J.* 2010, *48*, 295–314.
25. Zhao, H. Methods for stabilizing and activating enzymes in ionic liquids. *J. Chem. Technol. Biotechnol.* 2010, *85*, 891–907.
26. Moniruzzaman, M.; Kamiya, N.; Goto, M. Activation and stabilization of enzymes in ionic liquids. *Org. Biomol. Chem.*2010, *8*, 2887–2899.
27. Yang, Z. Hofmeister effects: An explanation for the impact of ionic liquids on biocatalysis. *J. Biotechnol.* 2009, *144*, 12–22.
28. Zhao, H. Effect of ions and other compatible solutes on enzyme activity, and its implication for biocatalysis using ionic liquids. *J. Mol. Catal. B Enzym.* 2005, *37*, 16–25.
29. De Gonzalo, G.; Lavandera, I.; Durchschein, K.; Wurm, D.; Faber, K.; Kroutil, W. Asymmetric biocatalytic reduction of ketones using hydroxy-functionalised water-miscible ionic liquids as solvents. *Tetrahedron Asymmetry* 2007, *18*, 2541–2546.
30. Lau, R.M.; Sorgedrager, M.J.; Carrea, G.; van Rantwijk, F.; Secundo, F.; Sheldon, R.A. Dissolution of candida antarctica lipase b in ionic liquids: Effects on structure and activity. *Green Chem.* 2004, *6*, 483–487.
31. Erbeldinger, M.; Mesiano, A.J.; Russell, A.J. Enzymatic catalysis of formation of z-aspartame in ionic liquid—An alternative to enzymatic catalysis in organic solvents. *Biotechnol. Prog.* 2000, *16*, 1129–1131.
32. Lau, R.M.; van Rantwijk, F.; Seddon, K.R.; Sheldon, R.A. Lipase-catalyzed reactions in ionic liquids. *Org. Lett.* 2000, *2*, 4189–4191.
33. Harjani, J.R.; Naik, P.U.; Nara, S.J.; Salunkhe, M.M. Enzyme mediated reactions in ionic liquids. *Curr. Org. Synth.*2007, *4*, 354–369.
34. Yang, Z.; Pan, W.B. Ionic liquids: Green solvents for nonaqueous biocatalysis. *Enzym. Microb. Technol.* 2005, *37*, 19–28.
35. Sureshkumar, M.; Lee, C.K. Biocatalytic reactions in hydrophobic ionic liquids. *J. Mol. Catal. B Enzym.* 2009, *60*, 1–12.
36. Sheldon, R.A. Biocatalysis in ionic liquids. *RSC Catal. Ser.* 2014, *15*, 20–43.
37. Lozano, P.; Bernal, J.M.; Garcia-Verdugo, E.; Vaultier, M.; Luis, S.V. *Biocatalysis in Ionic Liquids*; CRC Press: Boca Raton, FL, USA, 2015; pp. 31–66.
38. Stein, F.; Kragl, U. *Biocatalytic Reactions in Ionic Liquids*; John Wiley & Sons, Inc.: Hoboken, NJ, USA, 2014; pp. 193–216.
39. Klembt, S.; Dreyer, S.; Eckstein, M.; Kragl, U. Biocatalytic reactions in ionic liquids. In *Ionic Liquids in Synthesis*, 2nd ed.; Wasserscheid, P., Welton, T., Eds.; Wiley-VCH Verlags GmbH & Co KGaA: Weinheim, Germany, 2008; Volume 2, pp. 641–661.

40. Schöfer, S.H.; Kaftzik, N.; Wasserscheid, P.; Kragl, U. Enzyme catalysis in ionic liquids: Lipase catalysed kinetic resolution of 1-phenylethanol with improved enantioselectivity. *Chem. Commun.* 2001, 425–426.

41. Itoh, T.; Akasaki, E.; Kudo, K.; Shirakami, S. Lipase-catalyzed enantioselective acylation in the ionic liquid solvent system: Reaction of enzyme anchored to the solvent. *Chem. Lett.* 2001, *30*, 262–263.

42. Cantone, S.; Hanefeld, U.; Basso, A. Biocatalysis in non-conventional media-ionic liquids, supercritical fluids and the gas. *Green Chem.* 2007, *9*, 954–971.

43. Gorke, J.; Srienc, F.; Kazlauskas, R. Toward advanced ionic liquids. Polar, enzyme-friendly solvents for biocatalysis.*Biotechnol. Bioprocess Eng.* 2010, *15*, 40–53.

44. Mohile, S.S.; Potdar, M.K.; Harjani, J.R.; Nara, S.J.; Salunkhe, M.M. Ionic liquids: Efficient additives for candida rugosa lipase-catalyzed enantioselective hydrolysis of butyl 2-(4-chlorophenoxy)propionate. *J. Mol. Catal. B Enzym.* 2004, *30*, 185–188.

45. Rasalkar, M.S.; Potdar, M.K.; Salunkhe, M.M. Pseudomonas cepacia lipase-catalysed resolution of racemic alcohols in ionic liquid using succinic anhydride: Role of triethylamine in enhancement of catalytic activity. *J. Mol. Catal. B Enzym.* 2004, *27*, 267–270.

46. Lourenco, N.M.T.; Barreiros, S.; Afonso, C.A.M. Enzymatic resolution of indinavir precursor in ionic liquids with reuse of biocatalyst and media by product sublimation. *Green Chem.* 2007, *9*, 734–736.

47. Mai, N.L.; Ahn, K.; Bae, S.W.; Shin, D.W.; Morya, V.K.; Koo, Y.-M. Ionic liquids as novel solvents for the synthesis of sugar fatty acid ester. *Biotechnol. J.* 2014, *9*, 1565–1572.

48. Lee, J.K.; Kim, M.-J. Ionic liquid-coated enzyme for biocatalysis in organic solvent. *J. Org. Chem.* 2002, *67*, 6845–6847.

49. Itoh, T.; Matsushita, Y.; Abe, Y.; Han, S.-H.; Wada, S.; Hayase, S.; Kawatsura, M.; Takai, S.; Morimoto, M.; Hirose, Y. Increased enantioselectivity and remarkable acceleration of lipase-catalyzed transesterification by using an imidazolium peg-alkyl sulfate ionic liquid. *Chem. Eur. J.* 2006, *12*, 9228–9237.

50. Nakashima, K.; Kamiya, N.; Koda, D.; Maruyama, T.; Goto, M. Enzyme encapsulation in microparticles composed of polymerized ionic liquids for highly active and reusable biocatalysts. *Org. Biomol. Chem.* 2009, *7*, 2353–2358.

51. Yu, C.-Y.; Wei, P.; Li, X.-F.; Zong, M.-H.; Lou, W.-Y. Using ionic liquid in a biphasic system to improve asymmetric hydrolysis of styrene oxide catalyzed by cross-linked enzyme aggregates (cleas) of mung bean epoxide hydrolases.*Ind. Eng. Chem. Res.* 2014, *53*, 7923–7930.

52. Rehmann, L.; Ivanova, E.; Gunaratne, H.Q.N.; Seddon, K.R.; Stephens, G. Enhanced laccase stability through mediator partitioning into hydrophobic ionic liquids. *Green Chem.* 2014, *16*, 1462–1469.

53. Turner, M.B.; Spear, S.K.; Huddleston, J.G.; Holbrey, J.D.; Rogers, R.D. Ionic liquid salt-induced inactivation and unfolding of cellulase from trichoderma reesei. *Green Chem.* 2003, *5*, 443–447.

54. Toral, A.R.; de los Rios, A.P.; Hernandez, F.J.; Janssen, M.H.A.; Schoevaart, R.; van Rantwijk, F.; Sheldon, R.A. Cross-linked candida antarctica lipase b is active in denaturing ionic liquids. *Enzym. Microb. Technol.* 2007, *40*, 1095–1099.

55. Zhao, H.; Baker, G.A.; Song, Z.; Olubajo, O.; Crittle, T.; Peters, D. Designing enzyme-compatible ionic liquids that can dissolve carbohydrates. *Green Chem.* 2008, *10*, 696–705.

56. Vancov, T.; Alston, A.-S.; Brown, T.; McIntosh, S. Use of ionic liquids in converting lignocellulosic material to biofuels.*Renew. Energy* 2012, *45*, 1–6.

57. Long, J.; Li, X.; Guo, B.; Wang, F.; Yu, Y.; Wang, L. Simultaneous delignification and selective catalytic transformation of agricultural lignocellulose in cooperative ionic liquid pairs. *Green Chem.* 2012, *14*, 1935–1941.

58. Shi, J.; Gladden, J.M.; Sathitsuksanoh, N.; Kambam, P.; Sandoval, L.; Mitra, D.; Zhang, S.; George, A.; Singer, S.W.; Simmons, B.A.; *et al.* One-pot ionic liquid pretreatment and saccharification of switchgrass. *Green Chem.* 2013, *15*, 2579–2589.

59. Sun, N.; Parthasarathi, R.; Socha, A.M.; Shi, J.; Zhang, S.; Stavila, V.; Sale, K.L.; Simmons, B.A.; Singh, S. Understanding pretreatment efficacy of four cholinium and imidazolium ionic liquids by chemistry and computation.*Green Chem.* 2014, *16*, 2546–2557.

60. Moniruzzaman, M.; Ono, T. Ionic liquid assisted enzymatic delignification of wood biomass: A new green' and efficient approach for isolating of cellulose fibers. *Biochem. Eng. J.* 2012, *60*, 156–160.

61. Melero, J.A.; Iglesias, J.; Morales, G. Heterogeneous acid catalysts for biodiesel production: Current status and future challenges. *Green Chem.* 2009, *11*, 1285–1308.

62. Ha, S.H.; Lan, M.N.; Lee, S.H.; Hwang, S.M.; Koo, Y.-M. Lipase-catalyzed biodiesel production from soybean oil in ionic liquids. *Enzym. Microb. Technol.* 2007, *41*, 480–483.

63. Gamba, M.; Lapis, A.A.M.; Dupont, J. Supported ionic liquid enzymatic catalysis for the production of biodiesel. *Adv. Synth. Catal.* 2008, *350*, 160–164.

64. Zhao, H.; Song, Z.; Olubajo, O.; Cowins Janet, V. New ether-functionalized ionic liquids for lipase-catalyzed synthesis of biodiesel. *Appl. Biochem. Biotechnol.* 2010, *162*, 13–23.

65. Yu, D.; Wang, C.; Yin, Y.; Zhang, A.; Gao, G.; Fang, X. A synergistic effect of microwave irradiation and ionic liquids on enzyme-catalyzed biodiesel production. *Green Chem.* 2011, *13*, 1869–1875.

66. Zhang, K.P.; Lai, J.Q.; Huang, Z.L.; Yang, Z. Penicillium expansum lipase-catalyzed production of biodiesel in ionic liquids. *Bioresour. Technol.* 2011, *102*, 2767–2772.

67. Lai, J.-Q.; Hu, Z.-L.; Wang, P.-W.; Yang, Z. Enzymatic production of microalgal biodiesel in ionic liquid [bmim][pf6].*Fuel* 2012, *95*, 329–333.

68. Pellegrino, J.J.; Noble, R.D. Enhanced transport and liquid membranes in bioseparations. *Trends Biotechnol.* 1990, *8*, 216–224.

69. Branco, L.C.; Crespo, J.G.; Afonso, C.A.M. Highly selective transport of organic compounds by using supported liquid membranes based on ionic liquids. *Angew. Chem. Int. Ed.* 2002, *114*, 2895–2897.

70. Miyako, E.; Maruyama, T.; Kamiya, N.; Goto, M. Use of ionic liquids in a lipase-facilitated supported liquid membrane. *Biotechnol. Lett.* 2003, *25*, 805–808.

71. Miyako, E.; Maruyama, T.; Kamiya, N.; Goto, M. Enzyme-facilitated enantioselective transport of (*S*)-ibuprofen through a supported liquid membrane based on ionic liquids. *Chem. Commun.* 2003, 2926–2927.

72. Hernandez-Fernandez, F.J.; de los Rios, A.P.; Tomas-Alonso, F.; Gomez, D.; Villora, G. On the development of an integrated membrane process with ionic liquids for the kinetic resolution of rac-2-pentanol. *J. Membr. Sci.* 2008, *314*, 238–246.

73. Audia, J.E.; Britton, T.C.; Droste, J.J.; Folmer, B.K.; Huffman, G.W.; John, V.; Latimer, L.H.; Mabry, T.E.; Nissen, J.S. Preparation of *N*-(phenylacetyl)di- and Tripeptide Derivatives for Inhibiting β-Amyloid Peptide Release. WO9822494, 28 May 1998.

74. Hernandez-Fernandez, F.J.; de los Rios, A.P.; Tomas-Alonso, F.; Gomez, D.; Villora, G. Kinetic resolution of 1-phenylethanol integrated with separation of substrates and products by a supported ionic liquid membrane. *J. Chem. Technol. Biotechnol.* 2009, *84*, 337–342.

75. Kamat, S.V.; Beckman, E.J.; Russell, A.J. Enzyme-activity in supercritical fluids. *Crit. Rev. Biotechnol.* 1995, *15*, 41–71.

76. Mesiano, A.J.; Beckman, E.J.; Russell, A.J. Supercritical biocatalysis. *Chem. Rev.* 1999, *99*, 623–633.

77. Lozano, P. Enzymes in neoteric solvents: From one-phase to multiphase systems. *Green Chem.* 2010, *12*, 555–569.

78. Lozano, P.; de Diego, T.; Gmouh, S.; Vaultier, M.; Iborra, J.L. Criteria to design green enzymatic processes in ionic liquid/supercritical carbon dioxide systems. *Biotechnol. Prog.* 2004, *20*, 661–669.

79. Fonseca, G.S.; Scholten, J.D.; Dupont, J. Iridium nanoparticles prepared in ionic liquids: An efficient catalytic system for the hydrogenation of ketones. *Synlett* 2004, 1525–1528.

80. Valkenberg, M.H.; de Castro, C.; Holderich, W.F. Immobilisation of ionic liquids on solid supports. *Green Chem.* 2002,*4*, 88–93.

81. Mehnert, C.P. Supported ionic liquid phases. *Chem. -Eur. J.* 2004, *11*, 50–56.

82. Lozano, P.; Garcia-Verdugo, E.; Piamtongkam, R.; Karbass, N.; De Diego, T.; Burguete, M.I.; Luis, S.V.; Iborra, J.L. Bioreactors based on monolith-supported ionic liquid phase for enzyme catalysis in supercritical carbon dioxide. *Adv. Synth. Catal.* 2007, *349*, 1077–1084.

83. Lozano, P.; De Diego, T.; Sauer, T.; Vaultier, M.; Gmouh, S.; Iborra, J.L. On the importance of the supporting material for activity of immobilized candida antarctica lipase b in ionic liquid/hexane and ionic liquid/supercritical carbon dioxide biphasic media. *J. Supercrit. Fluids* 2007, *40*, 93–100.

84. Lozano, P.; de Diego, T.; Mira, C.; Montague, K.; Vaultier, M.; Iborra, J.L. Long term continuous chemoenzymatic dynamic kinetic resolution of rac-1-phenylethanol using ionic liquids and supercritical carbon dioxide. *Green Chem.*2009, *11*, 538–542.

85. Lozano, P.; de Diego, T.; Vaultier, M.; Iborra, J.L. Dynamic kinetic resolution of sec-alcohols in ionic liquids/supercritical carbon dioxide biphasic systems. *Int. J. Chem. React. Eng.* 2009, *7*.

86. Imperato, G.; Konig, B.; Chiappe, C. Ionic green solvents from renewable resources. *Eur. J. Org. Chem.* 2007, *2007*, 1049–1058.

87. Gorke, J.T.; Srienc, F.; Kazlauskas, R.J. Hydrolase-catalyzed biotransformations in deep eutectic solvents. *Chem. Commun.* 2008, 1235–1237.

88. Fujita, K.; Nakamura, N.; Igarashi, K.; Samejima, M.; Ohno, H. Biocatalytic oxidation of cellobiose in an hydrated ionic liquid. *Green Chem.* 2009, *11*, 351–354.

89. Zhao, H.; Baker, G.A.; Holmes, S. New eutectic ionic liquids for lipase activation and enzymatic preparation of biodiesel. *Org. Biomol. Chem.* 2011, *9*, 1908–1916.

90. Cull, S.G.; Holbrey, J.D.; Vargas-Mora, V.; Seddon, K.R.; Lye, G.J. Room-temperature ionic liquids as replacements for organic solvents in multiphase bioprocess operations. *Biotechnol. Bioeng.* 2000, *69*, 227–233.

91. Howarth, J.; James, P.; Dai, J. Immobilized baker's yeast reduction of ketones in an ionic liquid, [bmim]PF6 and water mix. *Tetrahedron Lett.* 2001, *42*, 7517–7519.

92. Pfruender, H.; Midjojo, M.; Kragl, U.; Weuster-Botz, D. Efficient whole-cell biotransformation in a biphasic ionic liquid/water system. *Angew. Chem. Int. Ed.* 2004, *43*, 4529–4531.

93. Xu, P.; Xu, Y.; Li, X.-F.; Yi Zhao, B.-Y.; Zong, M.-H.; Lou, W.-Y. Enhancing Asymmetric Reduction of 3-Chloropropiophenone with Immobilized Acetobacter sp. CCTCC M209061 Cells by Using Deep Eutectic Solvents as Cosolvents. *ACS Sustain. Chem. Eng.* 2015, *3*, 718–724.

94. Pfruender, H.; Jones, R.; Weuster-Botz, D. Water immiscible ionic liquids as solvents for whole cell biocatalysis. *J. Biotechnol.* 2006, *124*, 182–190.

95. Smirnova, S.V.; Torocheshnikova, I.I.; Formanovsky, A.A.; Pletnev, I.V. Solvent extraction of amino acids into a room temperature ionic liquid with dicyclohexano-18-crown-6. *Anal. Bioanal. Chem.* 2004, *378*, 1369–1375.

96. Soto, A.; Arce, A.; Khoshkbarchi, M.K. Partitioning of antibiotics in a two-liquid phase system formed by water and a room temperature ionic liquid. *Sep. Purif. Technol.* 2005, *44*, 242–246.

97. Arai, S.; Nakashima, K.; Tanino, T.; Ogino, C.; Kondo, A.; Fukuda, H. Production of biodiesel fuel from soybean oil catalyzed by fungus whole-cell biocatalysts in ionic liquids. *Enzym. Microb. Technol.* 2010, *46*, 51–55.

98. Lim, A.; Zhang, C.; Oktavianawati, I.; Hearn, M.T.W. Continuous enzymatic conversion of xylan with product recovery by liquid-liquid two-phase extraction. Clayton, Victoria, Australia, Unpublished work . 2015.

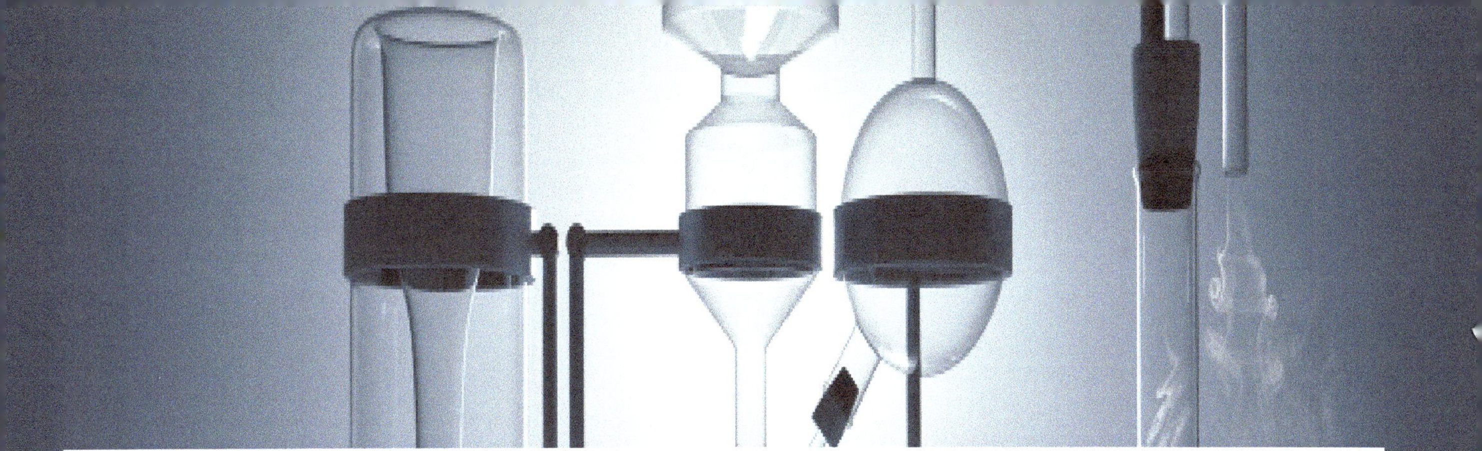

CHAPTER 16

FLOW MICROREACTOR SYNTHESIS IN ORGANO-FLUORINE CHEMISTRY

Hideki Amii, Aiichiro Nagaki and Jun-ichi Yoshida

Keywords: Defluorination, fluorine, fluorination, flow microreactor, organo-fluorine, perfluoroalkylation

REVIEW

Fluorine is a key element in the development of materials and biologically active agents. The introduction of fluorine atoms into organic molecules often exerts influences upon physical, chemical, and biological properties[1-8]. Although the fluorine atom is larger than the hydrogen atom, the replacement of hydrogen by fluorine leads to the smallest steric perturbation available to us. In pharmaceuticals, fluorine is often introduced to increase lipophilicity, bioavailability, and metabolic stability; these unique properties are otherwise difficult to obtain. At present, nearly 15% of medicines and 20% of agrochemicals on the market contain fluorine atoms [6,9,10].

Despite increased demands in wide fields of science, synthesis of fluoro-organic compounds is still often faced with problems such as the difficulty in handling of fluorinating reagents and in controlling of chemical reactions. Furthermore, low stability of fluorine-containing intermediates and low selectivity (chemo-, regio-, and/or stereo-) of the reactions have disturbed the progress of synthesis of fluorochemicals. New methodologies that can cure the weak points in synthesizing fluoro-organic compounds would open up a new perspective for fluorine chemistry.

Meanwhile, flow microreactor synthesis, the use of microfluidic devices, has emerged as a new method for producing chemical substances with high efficiency [11-13]. Now, the introduction of continuous-flow synthesis technique to laboratory synthesis represents a highly useful and increasingly popular method in organic chemistry[14-17]. Flow microreactor systems serve as an effective method for precise control of chemical reactions. Not only operational safety, but also extremely fast mixing, efficient heat transfer, and precise residence time control by virtue of the characteristic features of microstructures are responsible for their effectiveness. In particular, brilliant works of the generation and reactions of reactive intermediates that cannot be done in batch reactors have been developed by means of 'space integration methodology', where a sequence of reactions is conducted in one flow by adding the reagents at different places in continuous-flow synthesis [18-25].

This review deals with the successful examples on synthesis and reactions of fluorine-containing molecules by the use of flow microreactor systems. Useful applications using flow microreactor technology to overcome long-standing problems in synthetic organofluorine chemistry are showcased.

Citation: Hideki Amii, Aiichiro Nagaki and Jun-ichi Yoshida, Flow microreactor synthesis in organo-fluorine chemistry, doi:10.3762/bjoc.9.314

REACTIONS USING HAZARDOUS FLUORINATING REAGENTS

Direct fluorination employing elemental fluorine (F_2) is one of the most straightforward methods to make fluorine-containing molecules with high atom economy. However, handling F_2 needs special care and there are a lot of practical difficulties associated with direct fluorination. Elemental fluorine is a poisonous pale yellow gas at room temperature (bp −188 °C). Reactions of organic substances with elemental fluorine are typically strongly exothermic and extremely fast, frequently explosive. So it is difficult to control direct fluorination reactions with F_2gas in conventional batch reactors. In 1999, the pioneering work using flow microreactor technology for direct fluorination was reported by Chambers and Spink (Scheme 1) [26]. Utilizing the flow microreactor system, selective fluorination reactions were accomplished. For instance, fluorination of β-dicarbonyl compounds proceeds with high efficiency, whereas the use of conventional macrobatch systems suffered from low conversion[27,28]. Since then, direct fluorination processes using flow microreactor systems have been studied extensively[29-36]. The reactions of arenes with F_2 gas cause not only substitution (aromatic fluorination) but also side reactions such as fluorine-addition, polymerization and so on due to the high reactivity of F_2 gas [37-40]. Direct fluorination of aromatics in microreactor system took place cleanly [36].

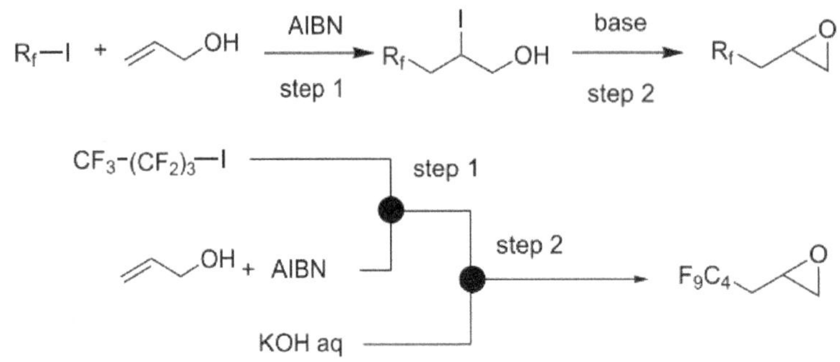

	step 1	step 2
in flow microreactor system	20 min, 97% yield	17 min, 97% yield
in macrobatch system	8 h, 97% yield	2 h, 84% yield

Scheme 1: Direct fluorination using microreactor systems.

As another splendid example for safe handling of hazardous fluorinating reagents, the flow microreactor synthesis using diethylaminosulfur trifluoride (Et_2NSF_3, abbreviated as DAST) was demonstrated. DAST is a commercially available liquid reagent that is utilized for the conversion of alcohols and carbonyl compounds into the corresponding fluoro derivatives. In spite of the effectiveness and convenience for nucleophilic fluorination, DAST is volatile and quite moisture sensitive, and readily undergoes disproportionation to SF_4 and explosive $(Et_2N)_2SF_2$when heating at over 90 °C [41-43]. Therefore, special care is required in handling a large quantity of DAST concerning the danger of the explosion under heating conditions. To solve these technical issues, the use of DAST in a continuous-flow reactor employing inert plastic flow tubes provided flexibility, scale-up production, and improved safety of operation. Using flow chemistry, the conversion of alcohols to fluorides was achieved by Seeberger and Ley, independently (Scheme 2) [44-47]. Usually, *gem*-difluorination of ketones by DAST is known to be less efficient because it needs heating and/or long reaction time to complete the fluorination [48,49]. Nevertheless by the use of a flow microreactor device, isatin underwent *gem*-difluorination to give the corresponding difluoride in 73% yield (residence time in the reactor: within 1 h).

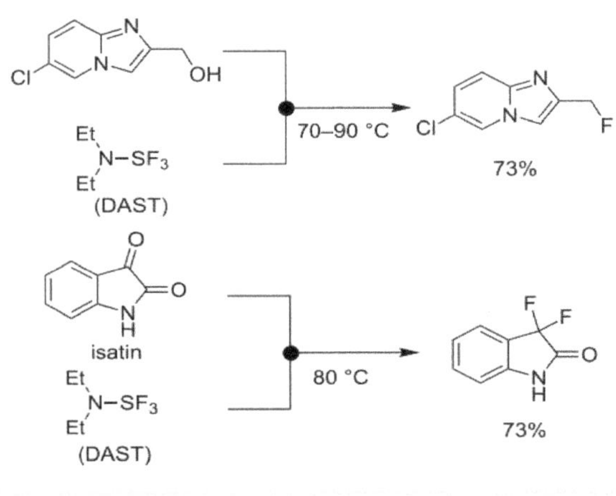

Scheme 2: Use of DAST in continuous-flow reactors.

REACTIONS THAT CAN BE WELL-CONTROLLED AND ACCELERATED USING FLOW MICROREACTOR SYSTEMS

The success of continuous flow chemistry in organic synthesis has enlarged rapidly last decade. There have been numerous examples using flow microreactors with improvement of chemical conversions and selectivities compared to conventional batch equipments. There are several beneficial effects using of flow microreactor systems; for instance, extremely fast mixing ability attributed to shortened diffusion path length, and highly efficient heat transfer ability based on large surface-area-to-volume ratio. In some cases, product selectively can be enhanced dramatically by the use of continuous flow microreactor devices. Daikin Industries, Ltd. developed the flow microreactor synthesis of fluorinated epoxides. In the first step (radical addition of perfluoroalkyl iodides to unsaturated alcohols), there is a problem concerning the violent exothermic reaction induced by decomposition of AIBN [50]. Furthermore, in the second step (intramolecular nucleophilic substitution of β-iodoalcohols to afford epoxides), efficient mixing technique is required for biphasic aqueous–organic systems. To resolve all the troublesome issues, flow microreactor provided improved reaction control over traditional batch reactors; high-yield synthesis of fluorinated epoxides was achieved (Scheme 3).

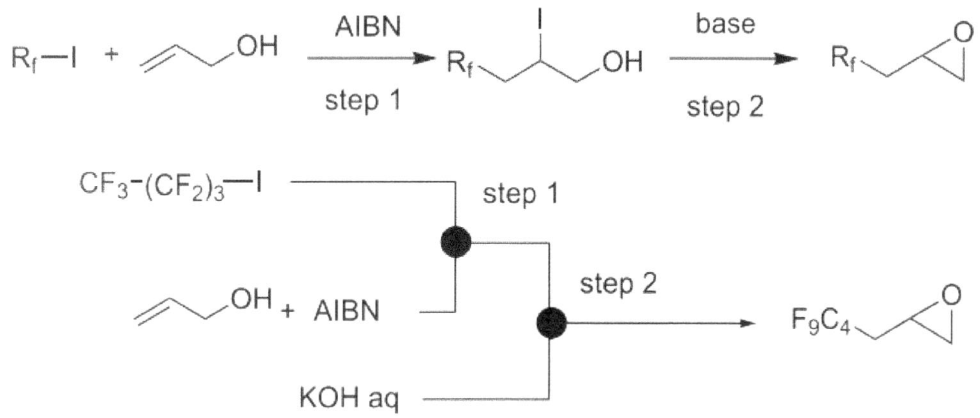

	step 1	step 2
in flow microreactor system	20 min, 97% yield	17 min, 97% yield
in macrobatch system	8 h, 97% yield	2 h, 84% yield

Scheme 3: Flow microreactor synthesis of fluorinated epoxides.

Kitazume et al. demonstrated the benefit of flow microreactors for a highly stereoselective synthesis of difluoromethylated alkenes [51]. They succeeded in perfect control of rapid base-catalyzed isomerization to give (*E*)-ethyl 3-difluoromethylpropenates (Scheme 4). Due to the benefit of temperature control in flow microreactor, further isomerization affording **B** was restricted.

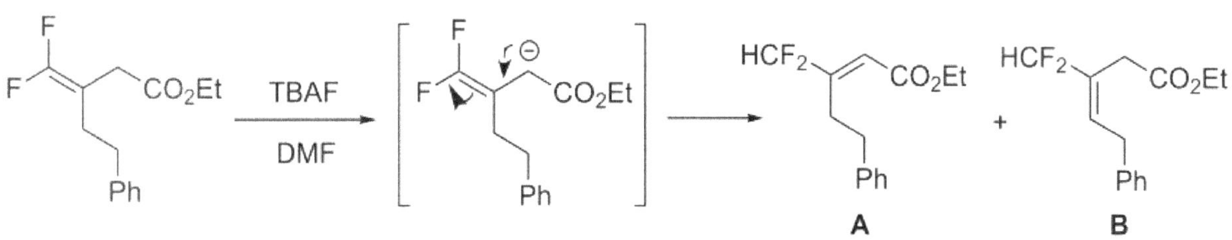

flow microreactor	98%	**A/B** =>99/1
in a flask	63%	**A/B** = 80/20

Scheme 4: Highly controlled isomerization of *gem*-difluoroalkenes.

The Pd-catalyzed fluorination of aryl triflates with fluoride was developed by Buchwald and co-workers in 2009[52]. Both electron-rich and -poor aryl triflates underwent Pd-catalyzed aromatic fluorination. As a fluoride source, they used CsF, which is an expensive reagent. They found that the use of a large excesses of CsF shortened reaction times for catalytic aromatic fluorination. Furthermore, efficient mixing becomes difficult when dealing with large quantities of insoluble CsF. They invented a packed-bed flow reactor, which allowed for easy handling of large quantities of insoluble CsF with efficient mixing (Scheme 5) [53].

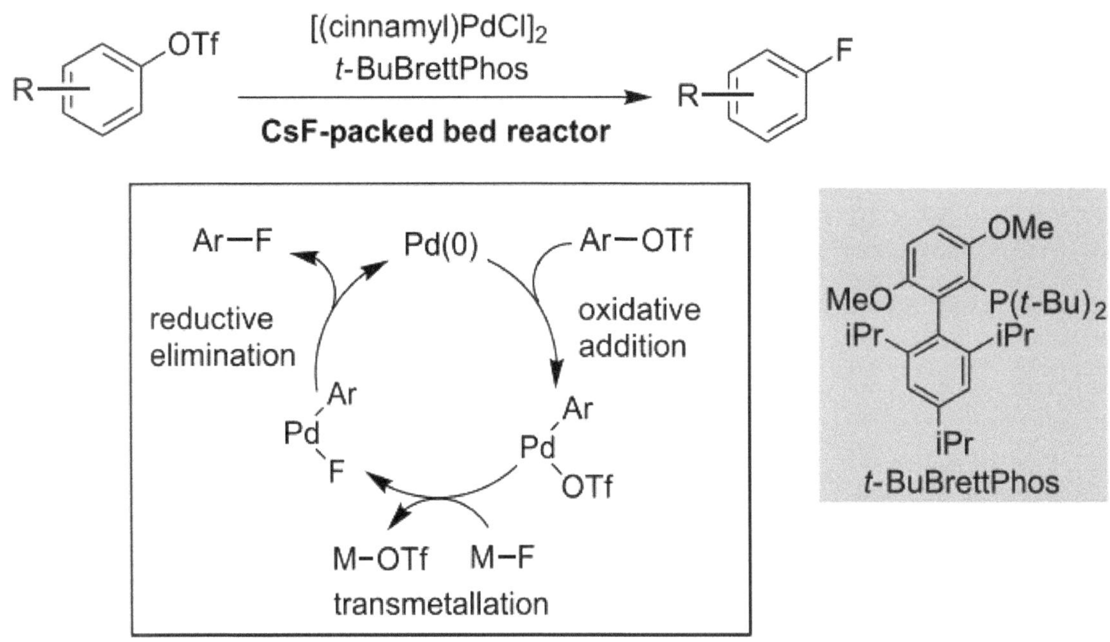

Scheme 5: Flow system for catalytic aromatic fluorination.

As an alternative method for aromatic fluorination, Yoshida and Nagaki reported the reactions of aryllithiums with electrophilic fluorinating agents such as NFSI and *N*-fluorosultam (Scheme 6) [54]. The present flow microreactor method showed a high level of functional group compatibility; a wide repertoire of aryl fluorides possessing electron-withdrawing, electron-donating, and sterically hindered functional groups were obtained in good yields. This is an important feature of 'late-stage fluorination methodology' for efficient drug screening.

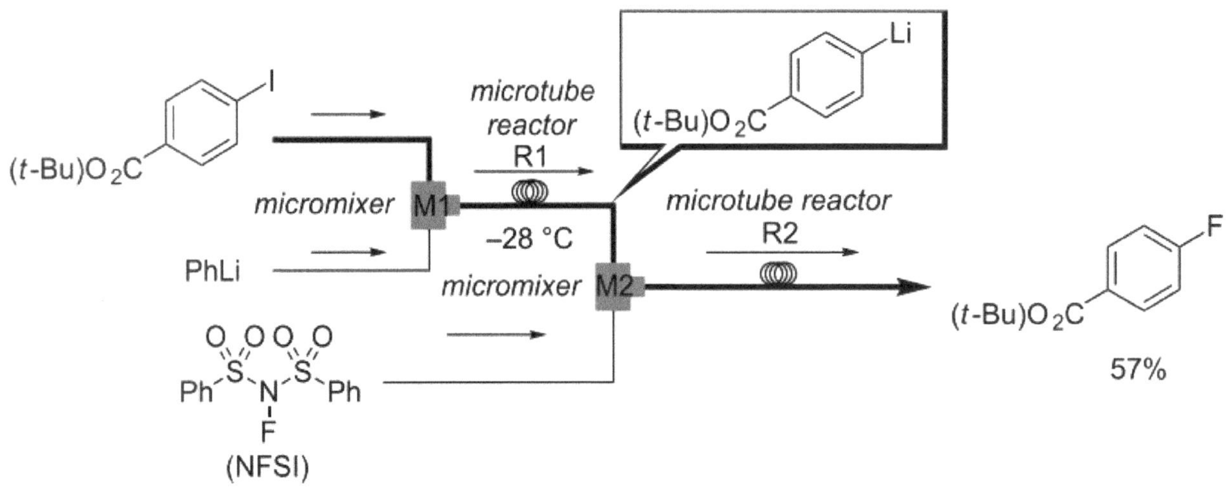

Scheme 6: Continuous-flow reactor for electrophilic aromatic fluorination.

[^{18}F] is a fluorine radioisotope which acts as an important source of positrons. In the medical imaging modality, [^{18}F] FDG is radiopharmaceutically used for positron emission tomography (PET). However, due to the short radioactivity elimination half-life of [^{18}F] ($t_{1/2}$ = 110 min), protocols for rapid and selective synthesis of [^{18}F]-labeled compounds have been required. In addition, on-site and on-demand chemical production of PET agents with automation and ease of in-line purification is suitable for hospital use. To date, continuous flow microreactor technology has shown potential for synthesis of [^{18}F]-radiolabeled molecular imaging probes such as [^{18}F]FDG, [^{18}F]fallypride, [^{18}F]annexin, and so on, which were made by nucleophilic substitution reactions using [^{18}F]-fluoride ion (Scheme 7) [55-67].

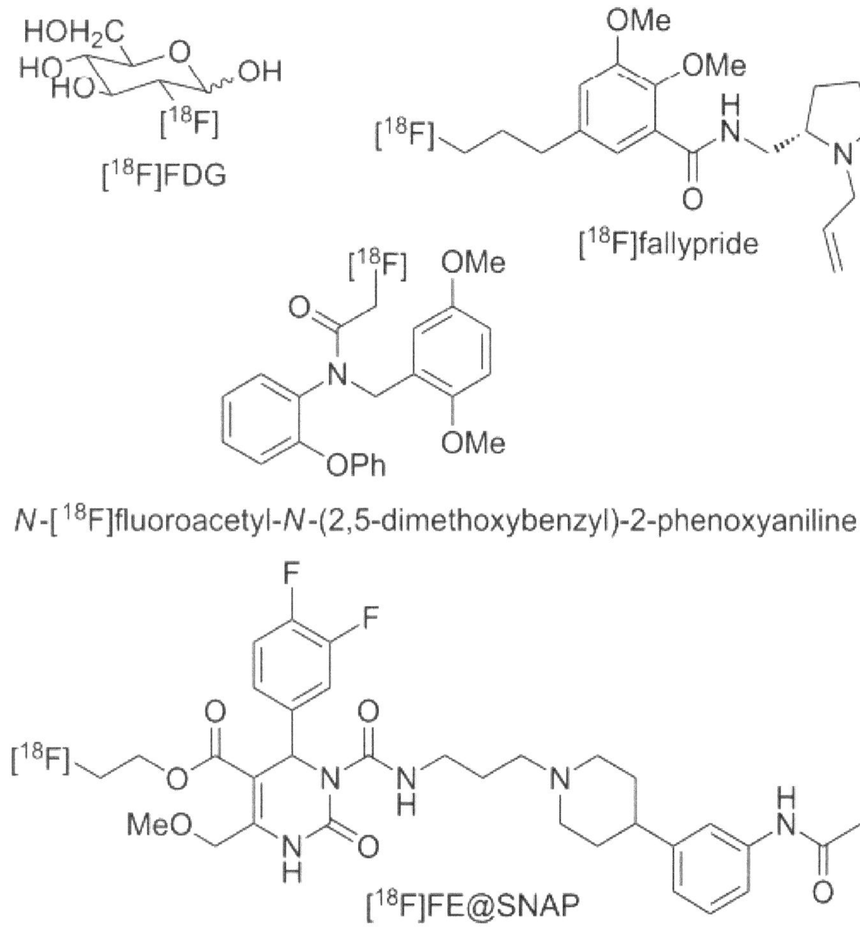

Scheme 7: Examples of [^{18}F]-radiolabeled molecular imaging probes.

For not only fluorination/fluoroalkylation, but also for defluorination, flow microreactor synthesis is quite effective. Defluorination involving a nucleophilic addition–elimination mechanism is a fundamental reaction in organic synthesis. Acid fluorides are known to be more stable to hydrolysis than acid chlorides under acidic or neutral conditions; on the other hand, they react with nucleophiles vigorously under basic conditions. Amino acid fluorides serve as one of the most efficient reagents for peptide bond formation. Seeberger and co-workers demonstrated the use of a silicon-based microreactor for the effective synthesis of peptides (Scheme 8) [68]. The condensation reaction completed in only 3 min at 90 °C to afford the dipeptide in high yield. Under the conditions at higher temperatures and/or leaving in prolonged contact with the reagents, decreasing the chemical yield of the desired dipeptide and increasing the amount of the tripeptide were observed. Using a flow microreactor system, precise residence-time control avoided the formation of undesired tripeptide byproduct.

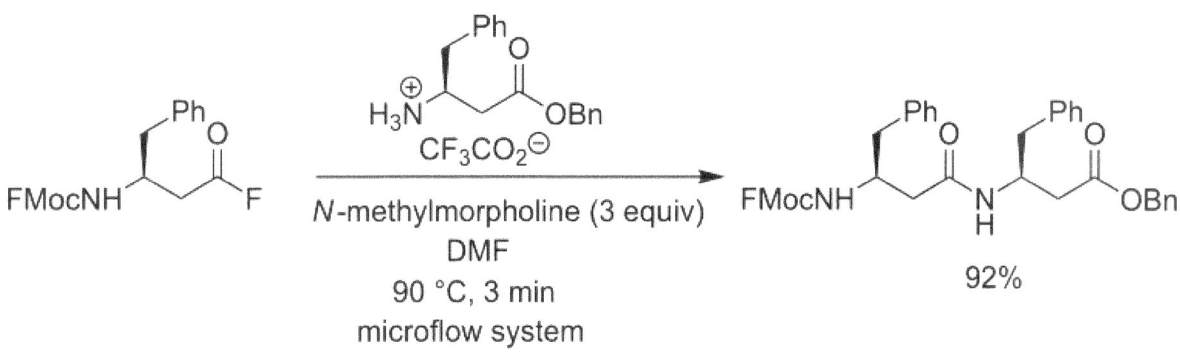

Scheme 8: Flow microreactor synthesis of dipeptides.

Nucleophilic aromatic substitution (S_NAr) chemistry contributes to creating useful materials. In 2005, Comer and Organ reported S_NAr reactions of 2-fluoronitrobenzene using a flow microreactor system with microwave irradiation (Scheme 9) [69]. Toward making compound-libraries, Schwalbe explored a flow microreactor system for sequential transformation towards fluoroquinolone antibiotics such as ciprofloxacin via both inter- and intramolecular S_NAr reactions (Scheme 10) [70]. Starting from the acylation reaction of (*N*-dimethylamino)acrylate with 2,4,5-trifluorobenzoic acid chloride, followed by two kinds of library diversification involving 1) Michael addition of a set of primary amines and the subsequent nucleophilic ring closure to give the difluoroquinolone system, 2) intermolecular nucleophilic aromatic substitution with various amines and hydrolysis of ester moieties to afford a number of ciprofloxacin analogues in good overall yields with high purity.

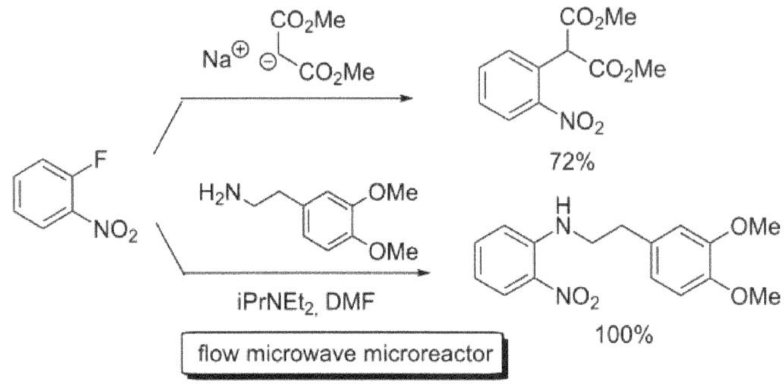

Scheme 9: Flow synthesis involving S_NAr reactions.

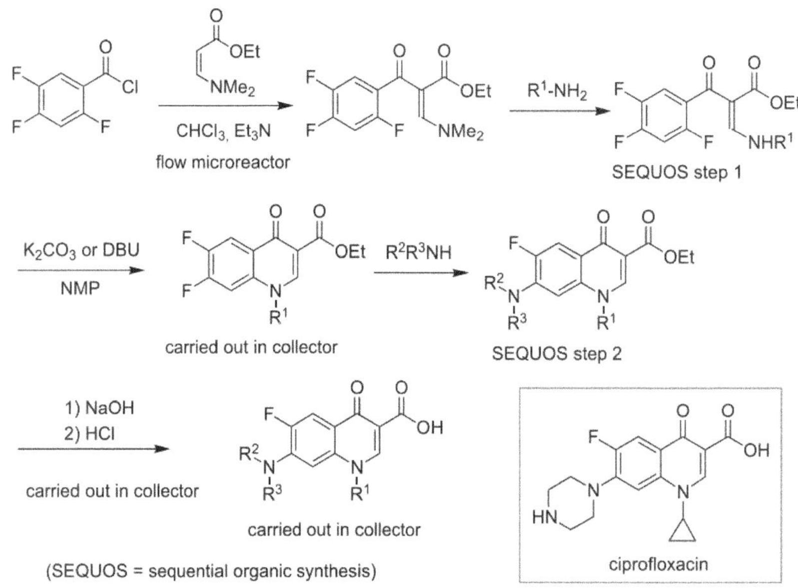

Scheme 10: Flow synthesis of fluoroquinolone antibiotics.

REACTIONS IN WHICH A REACTIVE INTERMEDIATE EASILY DECOMPOSES IN BATCH REACTORS

As aforementioned, the benefits of flow microreactor synthesis are 1) accurate control of reaction time, temperature, and pressure, 2) efficient mixing, 3) improved product selectivity (yield and purity), 4) rapid removal of heat of reaction for increased process safety, 5) ease of scale-up from lab to plant scale, and so on.

The reaction times are governed by the residence times (concerning path lengths and flow rates) inside flow microreactor which allow the generation and reactions of reactive intermediates before decomposition. By virtue of these characteristic features, the residence time control in flow microreactors makes the generation of highly reactive species possible, including their reactions on a preparative scale within a second or less.

Pentafluorophenylmagnesium bromide (PFPMgBr) is industrially produced by means of the halogen–metal exchange reaction of ethylmagnesium bromide and bromopentafluorobenzene [71]. However, this process is a highly exothermic reaction, and therefore, the use of a slow addition technique is required to avoid rapid increase in temperature. The flow reactor systems with micromixer and microheat exchangers show efficacy of forming PFPMgBr (Scheme 11) [72]. In the well-controlled formation of PFPMgBr and the subsequent protonation, continuous running of the pilot plant-scale flow microreactor gave pentafluorobenzene (PFB) in 92% yield (14.7 kg).

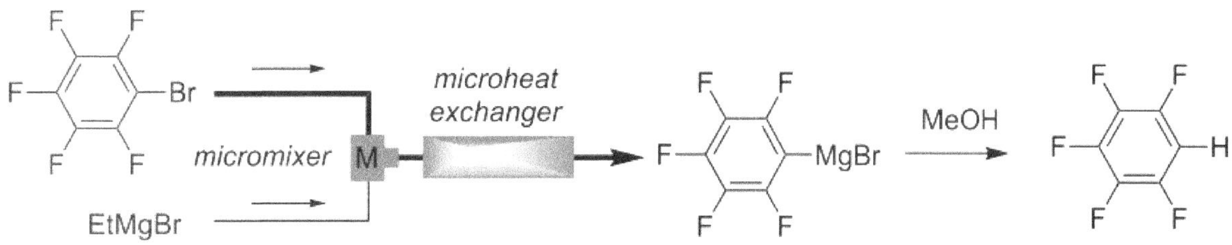

Scheme 11: Highly controlled formation of PFPMgBr.

Molecular switches based on photochromic diarylethenes are one of the most promising electronic materials [73]. In particular, a number of 1,2-bis(heteroaryl)-substituted perfluorocyclopentenes have been synthesized for use as photosensitive components of photochromic recording media for optical memory. For the synthesis of 1,2-diarylperfluorocyclopentenes, octafluorocyclopentene undergoes very quick reaction with nucleophiles because of the extremely high electrophilicity of the double bond [74]. The nucleophilic substitutions take place in a stepwise manner via an addition–elimination pathway. However, it is difficult to synthesize unsymmetrical diarylethenes with conventional macro batch systems because of contamination of the symmetrical diarylethenes (two identical nucleophiles are incorporated at the same time). Integrated flow microreactor synthesis of photochromic diarylethenes was found to be effective; the generation of heteroaryllithiums and the subsequent nucleophilic addition/elimination with octafluorocyclopentene were successfully achieved (Scheme 12) [75]. As a significant progress, the selective synthesis of unsymmetrical diarylethenes was accomplished by stepwise introduction of different heteroaryl nucleophiles into octafluorocyclopentene with highly controlled manner [76].

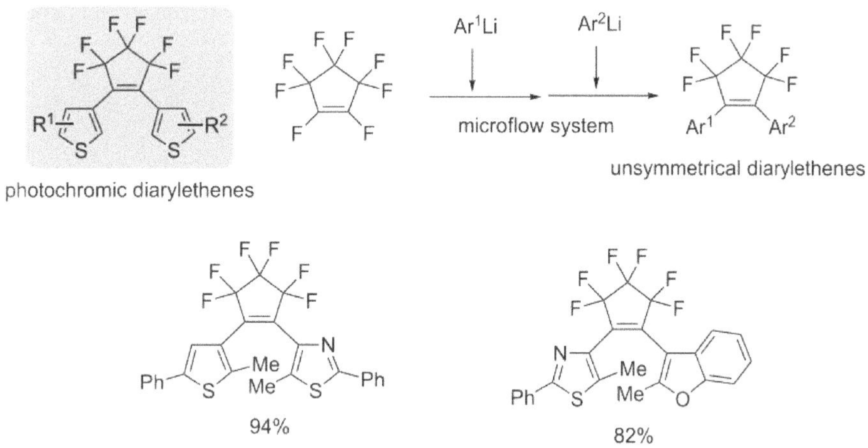

Scheme 12: Selective flow synthesis of photochromic diarylethenes.

The installation of perfluoroalkyl moieties into organic molecules can substantially changes the electronic properties of the parent organic compounds, and improves lipophilicity leading to enhanced solubility in fatty tissue and more efficient transport in the body. Nucleophilic introduction of perfluoroalkyl (R_F) groups are one of the most effective and general methods for the synthesis of perfluoroalkylated compounds [77]. Perfluoroalkyllithiums, which are usually prepared by halogen–lithium exchange reactions of perfluoroalkyl halides with alkyllithiums are useful for nucleophilic perfluoroalkylation [78], however they readily undergo β-elimination to form perfluoroalkenes [79]. Flow microreactor systems were proved to be effective for generation and reactions of R_FLi avoiding the β-elimination [80]. In the case of benzaldehyde as an electrophile, the reactions could be conducted at 0 °C, although much lower temperatures such as −78 °C are required to avoid the decomposition of perfluoroalkyllithium intermediates in batch processes (Scheme 13). However, the use of highly reactive electrophiles such as trimethylsilyl triflate and chlorotributylstannane gave rise to very low yields of the desired products, because the reactions of MeLi with such electrophiles are faster than those with perfluoroalkyl halides. In such cases, the generation of R_FLi and the reactions with electrophiles should be separated to give the desired R_FSiMe_3 and R_FSnBu_3 in high yields. To solve the issue, the use of the integrated flow microreactor system was effective; well-controlled generation of highly reactive R_FLi intermediates in the absence of active electrophiles and transfer of R_FLi into the next step were accomplished by taking short residence times in the reactors before they decomposed (Scheme 14).

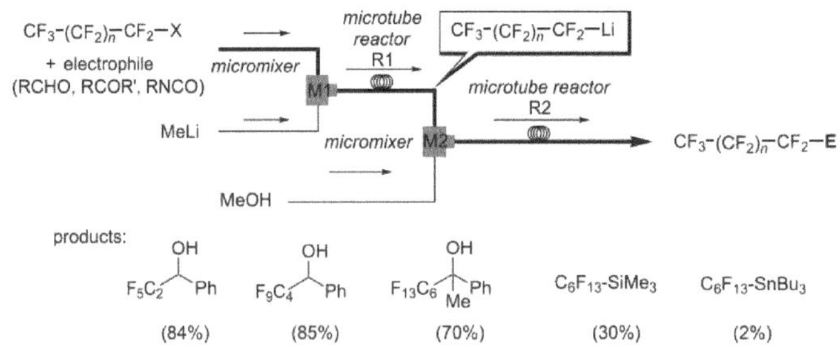

Scheme 13: Flow microreactor system for perfluoroalkylation by generation of perfluoroalkyllithiums in the presence of electrophiles.

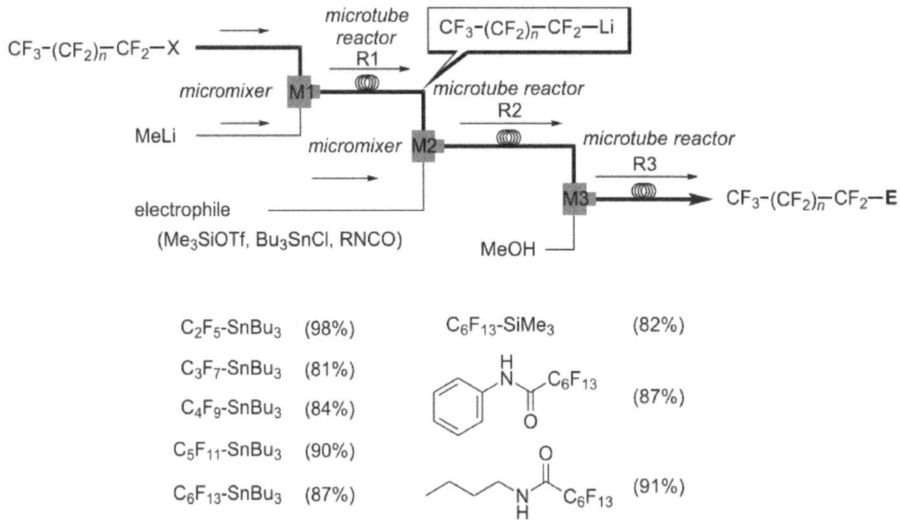

Scheme 14: Integrated flow microreactor system for perfluoroalkylation by generation of perfluoroalkyllithiums in the absence of electrophiles.

CONCLUSION

The illustrious achievements have been made in developing organic reactions by the use of flow microreactor devices. Fluorinated organic compounds have been gaining a significant importance in wide field of science and technology, and there have been the associated growing need for new and more efficient synthetic methods. As highlighted here, each of these elegant flow microreactor syntheses surely supplies the innovative methods for the synthesis of organo-fluorine compounds. For the future, continuous flow microreactor chemistry will make a breakthrough in developing new chemical production with highly efficiency and unexploited fluorinated intermediates by the ingenious strategy, and will afford fluorine-containing useful materials.

ACKNOWLEDGEMENTS

The financial support of the Ministry of Education, Culture, Sports, Science and Technology of Japan (Grant-in-Aid for Scientific Research on Innovative Areas "Organic Synthesis Based on Reaction Integration") is gratefully acknowledged.

REFERENCES

1. Banks, R. E.; Smart, B. E.; Tatlow, J. C. Organofluorine Chemistry: Principles and Commercial Applications; Plenum Press: New York, 2000.
2. Hiyama, T.; Kanie, K.; Kusumoto, T.; Morizawa, Y.; Shimizu, M. Organofluorine Compounds: Chemistry and Application; Springer-Verlag: Berlin, 2000. doi:10.1007/978-3-662-04164-2
3. Kirsch, P. Modern Fluoroorganic Chemistry; Wiley-VCH: Weinheim, 2004. doi:10.1002/352760393X
4. Chambers, R. D. Fluorine in Organic Chemistry; Blackwell: Oxford, 2004. doi:10.1002/9781444305371
5. Uneyama, K. Organofluorine Chemistry; Blackwell: Oxford, 2006. doi:10.1002/9780470988589
6. Bégué, J.-P.; Bonnet-Delpon, D. Bioorganic and Medicinal Chemistry of Fluorine; John Wiley & Sons, Inc.: Hoboken, NJ, 2008. doi:10.1002/9780470281895
7. Return to citation in text: [1] [2]
8. Ojima, I. Fluorine in Medicinal Chemistry and Chemical Biology; Wiley-Blackwell: Chichester, West Sussex, 2009.
9. Gouverneur, V.; Müller, K., Eds. Fluorine in Pharmaceutical and Medicinal Chemistry: From Biophysical Aspects to Clinical Applications; World Scientific Publishing Company: London, 2012. doi:10.1142/p746
10. Thayer, A. M. Chem. Eng. News 2006, 84, 15–24.
11. Hagmann, W. K. J. Med. Chem. 2008, 51, 4359–4369. doi:10.1021/jm800219f

12. Yoshida, J. Flash Chemistry: Fast Organic Synthesis in Microsystems; Wiley: Chichester, West Sussex, 2008.
13. Wiles, C.; Watts, P. Micro Reaction Technology in Organic Synthesis; CRC Press: Boca Raton, FL, 2011.
14. Wirth, T., Ed. Microreactors in Organic Chemistry and Catalysis; Wiley-VCH: Weinheim, 2013. doi:10.1002/9783527659722
15. Yoshida, J.; Nagaki, A.; Yamada, D. Drug Discovery Today 2013, 10, e53–e59. doi:10.1016/j.ddtec.2012.10.013
16. Yoshida, J. Chem. Rec. 2010, 10, 332–341. doi:10.1002/tcr.201000020
17. McQuade, D. T.; Seeberger, P. H. J. Org. Chem. 2013, 78, 6384–6389. doi:10.1021/jo400583m
18. Schwalbe, T.; Simoms, K. Chim. Oggi 2006, 24, 56–61.
19. Suga, S.; Yamada, D.; Yoshida, J. Chem. Lett. 2010, 39, 404–406. doi:10.1246/cl.2010.404
20. Yoshida, J.; Saito, K.; Nokami, T.; Nagaki, A. Synlett 2011, 1189–1194. doi:10.1055/s-0030-1259946
21. Nagaki, A.; Kim, H.; Usutani, H.; Matsuo, C.; Yoshida, J. Org. Biomol. Chem. 2010, 8, 1212–1217. doi:10.1039/b919325c
22. Tomida, Y.; Nagaki, A.; Yoshida, J. J. Am. Chem. Soc. 2011, 133, 3744–3747. doi:10.1021/ja110898s
23. Nagaki, A.; Kim, H.; Yoshida, J. Angew. Chem., Int. Ed. 2008, 47, 7833–7836. doi:10.1002/anie.200803205
24. Usutani, H.; Tomida, Y.; Nagaki, A.; Okamoto, H.; Nokami, T.; Yoshida, J. J. Am. Chem. Soc. 2007, 129, 3046–3047. doi:10.1021/ja068330s
25. Kim, H.; Nagaki, A.; Yoshida, J. Nat. Commun. 2011, 2, No. 264. doi:10.1038/ncomms1264
26. Saito, K.; Ueoka, K.; Matsumoto, K.; Suga, S.; Nokami, T.; Yoshida, J. Angew. Chem., Int. Ed. 2011, 50, 5153–5156. doi:10.1002/anie.201100854
27. Chambers, R. D.; Spink, R. C. H. Chem. Commun. 1999, 883–884. doi:10.1039/a901473j
28. Chambers, R. D.; Greenhall, M. P.; Hutchinson, J. J. Chem. Soc., Chem. Commun. 1995, 21–22. doi:10.1039/C39950000021
29. Chambers, R. D.; Greenhall, M. P.; Hutchinson, J. Tetrahedron 1996, 52, 1–8. doi:10.1016/0040-4020(95)00883-A
30. Chambers, R. D.; Sandford, G. Chim. Oggi 2004, 22, 6–8.
31. Jähnisch, K.; Baerns, M.; Hessel, V.; Ehrfeld, W.; Haverkamp, V.; Löwe, H.; Wille, C.; Guber, A. J. Fluorine Chem. 2000, 105, 117–128. doi:10.1016/S0022-1139(00)00300-6
32. Chambers, R. D.; Holling, D.; Spink, R. C. H.; Sandford, G. Lab Chip 2001, 1, 132–137. doi:10.1039/b108841f
33. Löb, P.; Löwe, H.; Hessel, V. J. Fluorine Chem. 2004, 125, 1677–1694. doi:10.1016/j.jfluchem.2004.09.006
34. Chambers, R. D.; Fox, M. F.; Sandford, G. Lab Chip 2005, 5, 1132–1139. doi:10.1039/b504675k
35. Sato, T. Preparation of fluorinated organic compounds by direct fluorination. Jpn. Kokai Tokkyo Koho JP 2006-001881, Jan 5, 2006.
36. de Mas, N.; Günther, A.; Schmidt, M. A.; Jensen, K. F. Ind. Eng. Chem. Res. 2009, 48, 1428–1434. doi:10.1021/ie801232d
37. Chambers, R. D.; Fox, M. A.; Sandford, G.; Trmcic, J.; Goeta, A. J. Fluorine Chem. 2007, 128, 29–33. doi:10.1016/j.jfluchem.2006.09.010
38. Return to citation in text: [1] [2]
39. Hessel, V.; Angeli, P.; Gavriilidis, A.; Löwe, H. Ind. Eng. Chem. Res. 2005, 44, 9750–9769. doi:10.1021/ie0503139
40. And references therein.
41. Grakauskas, V. J. Org. Chem. 1969, 34, 2835–2839. doi:10.1021/jo01262a009
42. Cacace, F.; Giacomello, P.; Wolf, A. P. J. Am. Chem. Soc. 1980, 102, 3511–3515. doi:10.1021/ja00530a034
43. Purrington, S. T.; Woodard, D. L. J. Org. Chem. 1991, 56, 142–145. doi:10.1021/jo00001a029
44. Messina, P. A.; Mange, K. C.; Middleton, W. J. J. Fluorine Chem. 1989, 42, 137–143. doi:10.1016/S0022-1139(00)83974-3
45. von Allenstein, E.; Schrempf, G. Z. Anorg. Allg. Chem. 1981, 474, 7–17.
46. Lal, G. S.; Pez, G. P.; Pesaresi, R. J.; Prozonic, F. M.; Cheng, H. J. Org. Chem. 1999, 64, 7048–7054. doi:10.1021/jo990566+
47. Gustafsson, T.; Gilmour, R.; Seeberger, P. H. Chem. Commun. 2008, 3022–3024. doi:10.1039/b803695k
48. Baumann, M.; Baxendale, I. R.; Ley, S. V. Synlett 2008, 2111–2114. doi:10.1055/s-2008-1078026
49. Baumann, M.; Baxendale, I. R.; Martin, L. J.; Ley, S. V. Tetrahedron 2009, 65, 6611–6625. doi:10.1016/j.tet.2009.05.083
50. Carter, C. F.; Lange, H.; Ley, S. V.; Baxendale, I. R.; Wittkamp, B.; Goode, J. G.; Gaunt, N. L. Org. Process Res. Dev. 2010, 14, 393–404. doi:10.1021/op900305v
51. Torres, J. C.; Garden, S. J.; Pinto, A. C.; da Silva, F. S. Q.; Boechat, N. Tetrahedron 1999, 55, 1881–1892. doi:10.1016/S0040-4020(98)01229-0
52. Li, G.; van der Donk, W. A. Org. Lett. 2007, 9, 41–44. doi:10.1021/ol062401a
53. Taguchi, M.; Nakaya, H.; Ichihara, K.; Senba, Y.; Hiraga, Y. Manufacture of fluorine-containing epoxides from fluoroalkyl iodides and unsaturated alcohols. Jpn. Kokai Tokkyo Koho JP 2009-67687, April 2, 2009.
54. Kitazume, T.; Kawai, K.; Nihei, T.; Miyake, N. J. Fluorine Chem. 2005, 126, 59–62. doi:10.1016/j.jfluchem.2004.10.008
55. Watson, D. A.; Su, M.; Teverovskiy, G.; Zhang, Y.; Garcia-Fortanet, J.; Kinzel, T.; Buchwald, S. L. Science 2009, 325, 1661–1664. doi:10.1126/science.1178239
56. Noël, T.; Maimone, T. J.; Buchwald, S. L. Angew. Chem., Int. Ed. 2011, 50, 8900–8903. doi:10.1002/anie.201104652

57. Nagaki, A.; Uesugi, Y.; Kim, H.; Yoshida, J. Chem.–Asian J. 2013, 8, 705–708. doi:10.1002/asia.201201191

58. Lee, C.-C.; Sui, G.; Elizarov, A.; Shu, C. J.; Shin, Y.-S.; Dooley, A. N.; Huang, J.; Daridon, A.; Wyatt, P.; Stout, D.; Kolb, H. C.; Witte, O. N.; Satyamurthy, N.; Heath, J. R.; Phelps, M. E.; Quake, S. R.; Tseng, H.-R. Science 2005, 310, 1793–1796. doi:10.1126/science.1118919

59. Elizarov, A. M.; van Dam, R. M.; Shin, Y. S.; Kolb, H. C.; Padgett, H. C.; Stout, D.; Shu, J.; Huang, J.; Daridon, A.; Heath, J. R. J. Nucl. Med. 2010, 51, 282–287. doi:10.2967/jnumed.109.065946

60. Gillies, J. M.; Prenant, C.; Chimon, G. N.; Smethurst, G. J.; Dekker, B. A.; Zweit, J. Appl. Radiat. Isot. 2006, 64, 333–336. doi:10.1016/j.apradiso.2005.08.009

61. Steel, C. J.; O'Brien, A. T.; Luthra, S. K.; Brady, F. J. Labelled Compd. Radiopharm. 2007, 50, 308–311. doi:10.1002/jlcr.1259

62. Gillies, J. M.; Prenant, C.; Chimon, G. N.; Smethurst, G. J.; Perrie, W.; Hamblett, I.; Dekker, B.; Zweit, J. Appl. Radiat. Isot. 2006, 64, 325–332. doi:10.1016/j.apradiso.2005.08.007

63. Lu, S.; Giamis, A. M.; Pike, V. W. Curr. Radiopharm. 2009, 2, 49–55. doi:10.2174/1874471010902010049

64. Briard, E.; Zoghbi, S. S.; Siméon, F. G.; Imaizumi, M.; Gourley, J. P.; Shetty, H. U.; Lu, S.; Fujita, M.; Innis, R. B.; Pike, V. W. J. Med. Chem. 2009, 52, 688–699. doi:10.1021/jm8011855

65. Lu, S.-Y.; Watts, P.; Chin, F. T.; Hong, J.; Musachio, J. L.; Briard, E.; Pike, V. W. Lab Chip 2004, 4, 523–525. doi:10.1039/b407938h

66. Pascali, G.; Mazzone, G.; Saccomanni, G.; Manera, C.; Salvadori, P. A. Nucl. Med. Biol. 2010, 37, 547–555. doi:10.1016/j.nucmedbio.2010.03.006

67. Pascali, G.; Nannavecchia, G.; Pitzianti, S.; Salvadori, P. A. Nucl. Med. Biol. 2011, 38, 637–644. doi:10.1016/j.nucmedbio.2011.01.005

68. Chun, J.-H.; Lu, S.; Lee, Y.-S.; Pike, V. W. J. Org. Chem. 2010, 75, 3332–3338. doi:10.1021/jo100361d

69. Philippe, C.; Ungersboeck, J.; Schirmer, E.; Zdravkovic, M.; Nics, L.; Zeilinger, M.; Shanab, K.; Lanzenberger, R.; Karanikas, G.; Spreitzer, H.; Viernstein, H.; Mitterhauser, M.; Wadsak, W. Bioorg. Med. Chem. 2012, 20, 5936–5940. doi:10.1016/j.bmc.2012.07.051

70. Keng, P. Y.; Chen, S.; Ding, H.; Sadeghi, S.; Shah, G. J.; Dooraghi, A.; Phelps, M. E.; Satyamurthy, N.; Chatziioannou, A. F.; Kim, C.-J.; van Dam, R. M. Proc. Natl. Acad. Sci. U. S. A. 2012, 109, 690–695. doi:10.1073/pnas.1117566109

71. Flögel, O.; Codée, J. D. C.; Seebach, D.; Seeberger, P. H. Angew. Chem., Int. Ed. 2006, 45, 7000–7003. doi:10.1002/anie.200602167

72. Comer, E.; Organ, M. G. J. Am. Chem. Soc. 2005, 127, 8160–8167. doi:10.1021/ja0512069

73. Schwalbe, T.; Kadzimirsz, D.; Jas, G. QSAR Comb. Sci. 2005, 24, 758–768. doi:10.1002/qsar.200540005

74. Respess, W. L.; Tamborski, C. J. Organomet. Chem. 1968, 11, 619–622. doi:10.1016/0022-328X(68)80092-0

75. Wakami, H.; Yoshida, J. Org. Process Res. Dev. 2005, 9, 787–791. doi:10.1021/op0501500

76. Irie, M. Chem. Rev. 2000, 100, 1685–1716. doi:10.1021/cr980069d

77. Amii, H.; Uneyama, K. Chem. Rev. 2009, 109, 2119–2183. doi:10.1021/cr800388c

78. Ushiogi, Y.; Hase, T.; Iinuma, Y.; Takata, A.; Yoshida, J. Chem. Commun. 2007, 2947–2949. doi:10.1039/b702277h

79. Asai, T.; Takata, A.; Nagaki, A.; Yoshida, J. ChemSusChem 2012, 5, 339–350. doi:10.1002/cssc.201100376

80. Schwalbe, T.; Autze, V.; Hohmann, M.; Stirner, W. Org. Process Res. Dev. 2004, 8, 440–454. doi:10.1021/op049970n

81. See for a Grignard exchange reaction of C2F5I with MeMgCl followed by reaction with benzophenone in a microflow system.

82. Uno, H.; Suzuki, H. Synlett 1993, 91–96. doi:10.1055/s-1993-22361

83. Uneyama, K.; Katagiri, T.; Amii, H. Acc. Chem. Res. 2008, 41, 817–829. doi:10.1021/ar7002573

84. Nagaki, A.; Tokuoka, S.; Yamada, S.; Tomida, Y.; Oshiro, K.; Amii, H.; Yoshida, J. Org. Biomol. Chem. 2011, 9, 7559–7563. doi:10.1039/c1ob06350b

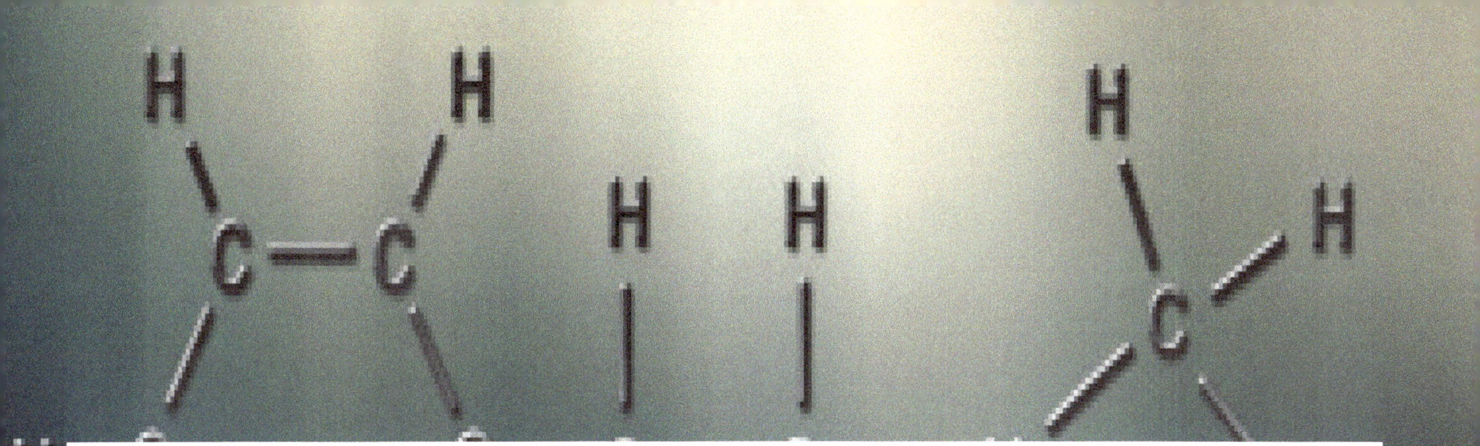

CHAPTER 17

SYNTHESIS AND REACTIONS OF FIVE-MEMBERED HETEROCYCLES USING PHASE TRANSFER CATALYST (PTC) TECHNIQUES

Ahmed M. El-Sayed, Omyma A. Abd Allah, Ahmed M. M. El-Saghier, and Shaaban K. Mohamed

INTRODUCTION

Organic synthesis is an essential way to get chemical products having practical applications such as pharmaceuticals, plant protection agents, dyes, photographic sensitizers, and monomers. Transformations of starting materials into desired final products usually require number of chemical operations in which additional reagents, catalysts, and solvents are employed. Thus, during any synthetic method, besides the desired products, many waste materials are produced because transformations of reactants into products are neither quantitative nor selective processes. This waste could be regenerated, destroyed, or disposed. This will lead to consuming much energy and creating heavy burden on the environment. Therefore it is of a great importance to develop and use synthetic methodologies that minimize or eliminate such problems. One of the most common and efficient methodologies that fulfill this requirement is employing phase-transfer catalyses techniques [1–4]. The most significant advantages of use of PTCs in industrial applications are [5]

(1)elimination of organic solvents,(2)elimination of dangerous, inconvenient, and expensive reagents such as NaH and $NaNH_2$,(3)increasing the reactivity and selectivity of the active species,(4)improving the yield and purity of products to the optimum records,(5)simplifying the whole synthetic process and making it safer and objective,(6)reducing industrial wastes and overall costs and saving energy which gives a positive impact on economic and environmental interests,(7) accelerating and performing mimic reactions in an efficient mode.

Fundamentals of Phase Transfer Catalysis (PTC)

First phase transfer catalysis was discovered by Jarrouse and Hebd in 1951 when they observed that the quaternary ammonium salt and benzyltriethylammonium chloride accelerated two-phase reaction of benzyl chloride with cyclohexanol (Figure 1; (1)) and the two-phase alkylation reaction of phenylacetonitrile with benzyl chloride or ethyl chloride [6] (Figure 1; (2)).

Citation: Ahmed M. El-Sayed, Omyma A. Abd Allah, Ahmed M. M. El-Saghier, and Shaaban K. Mohamed, "Synthesis and Reactions of Five-Membered Heterocycles Using Phase Transfer Catalyst (PTC) Techniques," Journal of Chemistry, vol. 2014, Article ID 163074, 47 pages, 2014. doi:10.1155/2014/163074

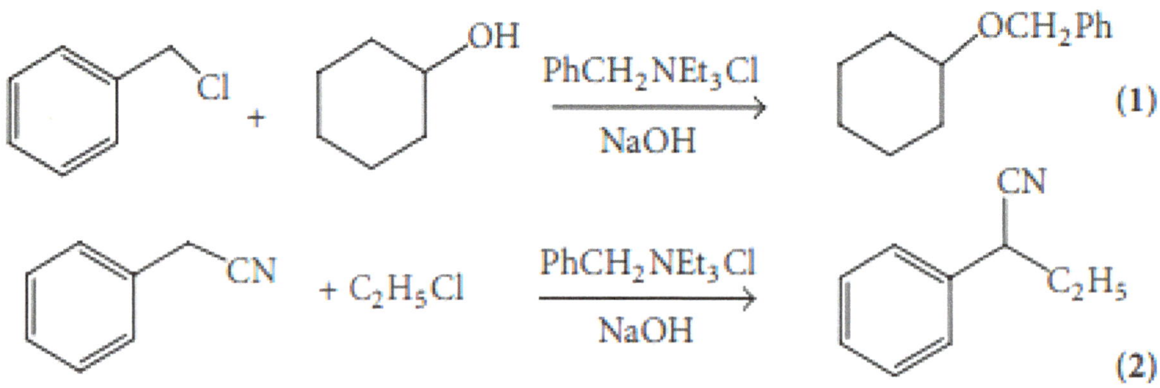

Figure 1

In addition, numerous publications and patents [7–12] have been reported during the period of 1950–1965. PTCs techniques have been further developed by Makosza et al. for the purpose of obtaining more efficient and pure yield [13–16].

Mechanism of Phase Transfer Catalysis

All phase-transfer catalyzed reactions involve at least two steps:(1)transfer of one reagent from its ground phase into the second phase as an intermediate,(2)reaction of the transferred reagent with the nontransferred reagent, for example, the alkylation of phenylacetonitrile with alkyl halide using aqueous NaOH as a base and tetrabutylammonium halide (QX) as a catalyst can be formulated as shown in Figure 2.

$$\underset{\text{org.}}{PhCH_2CN} + NaOH \rightleftharpoons \underset{\text{int}}{Ph\overset{-}{C}HCN\,\overset{+}{Na}} + \underset{\text{aq.}}{H_2O}$$

$$Ph\overset{-}{C}HCN\,\underset{\text{int}}{\overset{+}{Na}} + \underset{\text{org.}}{\overset{+\,-}{QX}} \rightleftharpoons \underset{\text{org.}}{Ph\overset{-}{C}HCN\,\overset{+}{Q}} + \underset{\text{aq.}}{NaX}$$

$$\underset{\text{org.}}{Ph\overset{-}{C}HCN\,\overset{+}{Q}} + \underset{\text{org.}}{R\text{-}X} \longrightarrow \underset{\text{org.}}{Ph(R)CHCN} + \underset{\text{org.}}{\overset{+\,-}{QX}}$$

Figure 2

The concept of phase-transfer catalysis is not limited to anion transfer but is much more general, so that, in principle, one could also transfer cations, free radicals, or whole molecule. Phase transfer catalysis is classified as liquid-liquid, liquid-solid, liquid-gas, solid-gas, or solid-solid systems.

APPLICATION OF PHASE TRANSFER CATALYSIS IN SYNTHESIS OF FIVE-MEMBERED RING OF HETEROCYCLIC COMPOUNDS

Synthesis of Five-Membered Ring Heterocycles Containing One Heteroatom

Treatment of anilinomethylenemalononitrile 1 with ethyl chloroacetate, chloroaceta-mide, phenacyl chloride, or diethyl bromomalonate in 1:1:1 molar ratio (dioxan/K_2CO_3/tetrabutylammonium bromide (TBAB)) afforded the corresponding substituted pyrroles 2a–c and 3, respectively [17] (Figure 3).

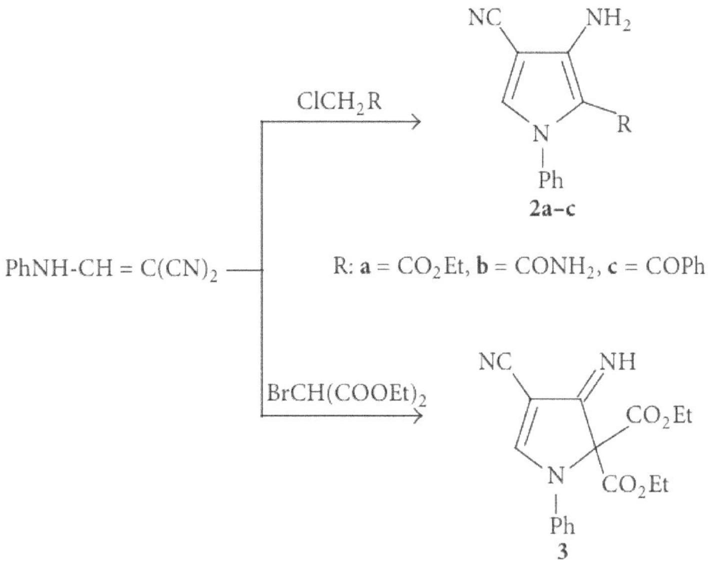

Figure 3

The catalytic cycle which explains the reaction pathway can be simplified as in Scheme 1.

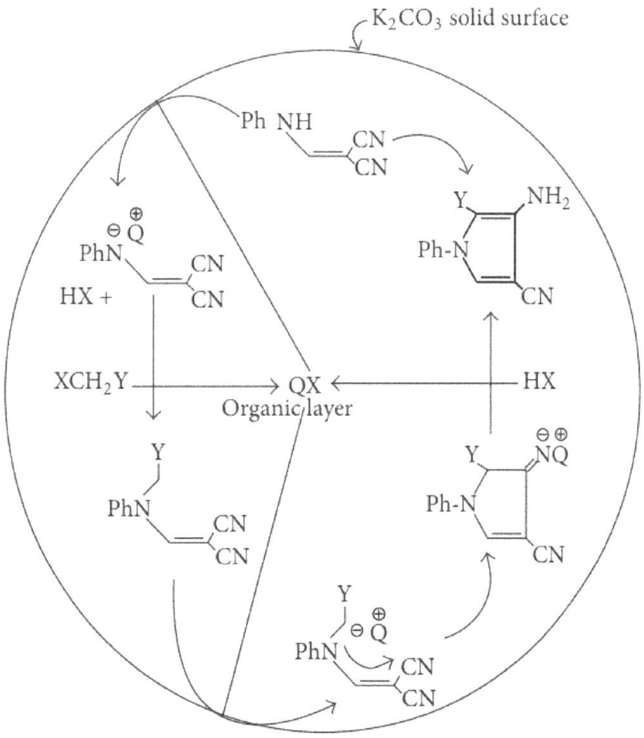

Scheme 1

A simple and convenient synthesis of 2-amino-1-aryl-5-oxo-4,5-dihydro-1H-pyrrole-3-carbonitriles 6 was achieved by reaction of N-aryl-2-chloroacetamide 4 and malononitrile 5 under solid-liquid PTC condition (CH_3CN/K_2CO_3/18-crown-6) at room temperature [18] (**Figure** 4).

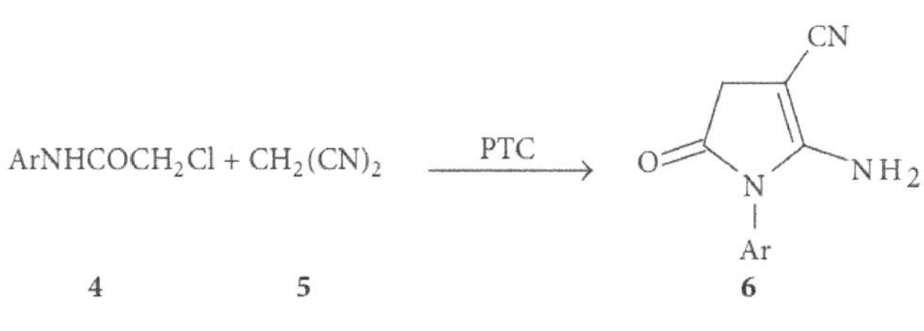

Figure 4

3-Amino-4-ethoxycarbonyl-1-phenyl-5-thioxo-2,5-dihydro-2-pyrrolidene malononitrile 9a was prepared via reaction of malononitrile 5, phenyl isothiocyanate 7, and ethoxymethylenemalononitrile 8a in 1 : 1 : 1 molar ratio under the PTC mixture of dioxan/K_2CO_3/(TBAB). Replacing of ethyl ethoxymethylenecyanoacetate 8bwith 8a gave ethyl[3-amino-4-ethoxy-carbonyl-1-phenyl-5-thioxo-2,5-dihydro-2-pyrrolidene]cyanoacetate 9band 10 under the same experimental conditions [19] (Figure 5).

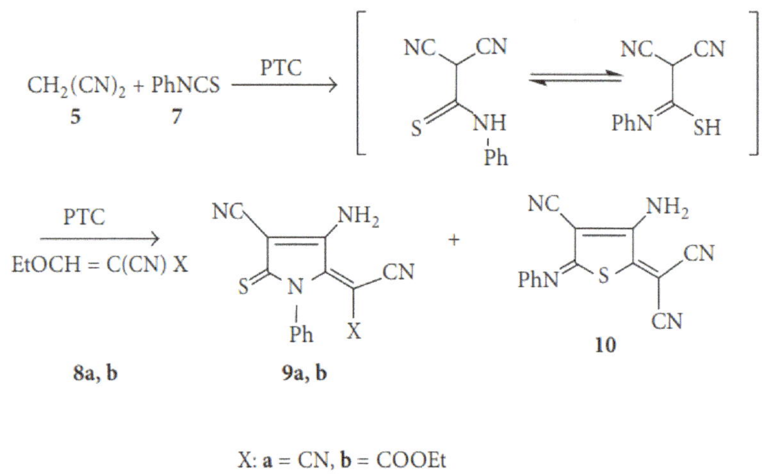

Figure 5

Moreover, the reaction of ethyl cyanoacetate 11 with phenyl isothiocyante 7 and ethoxymethylenemalononitrile8a or ethyl ethoxymethylenecyanoacetate 8b in 1 : 1 : 1 molar ratio under same PTC experimental conditions afforded the corresponding pyrrole derivatives 12a, b, and 13 [19] (Figure 6).

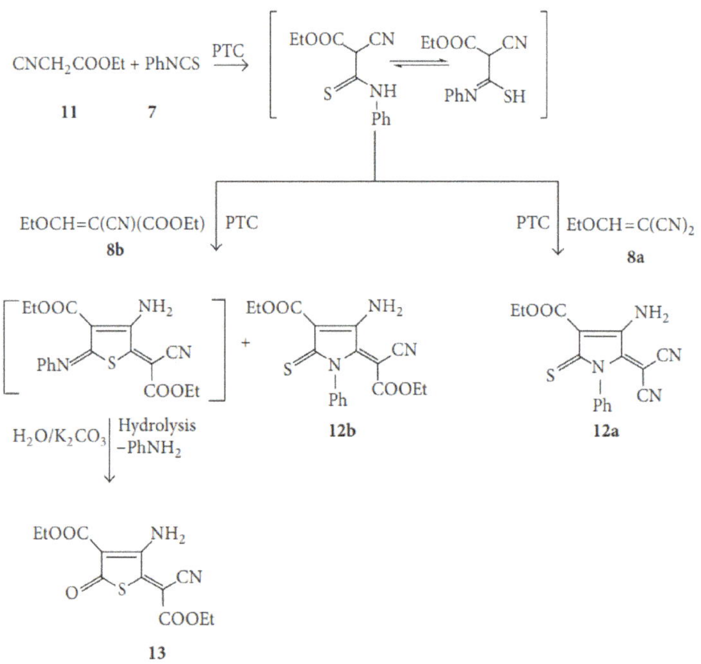

Figure 6

Reaction of methyl 2,5-dibromopentanoate 14 with trichloroacetamide 15 under solid-liquid technique (CH_3CN/K_2CO_3/TEBACl) yielded methyl N-trichloroacetyl-2-pyrrolidine carboxylate 16 [20] (Figure 7).

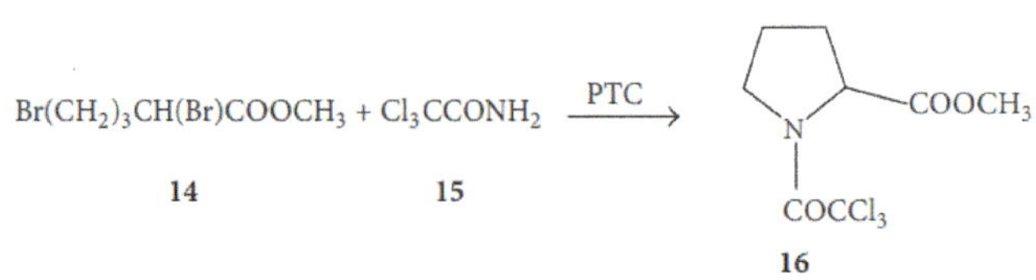

Figure 7

Substituted pyrrolidines 19 were synthesized via the 1,3-dipolar cycloaddition reaction between imino esters 17(derived from alanine and glycine with alkanes) and alkyl acrylate 18 in a mixture of THF or toluene, KOH, and TBACl [21] (Figure 8).

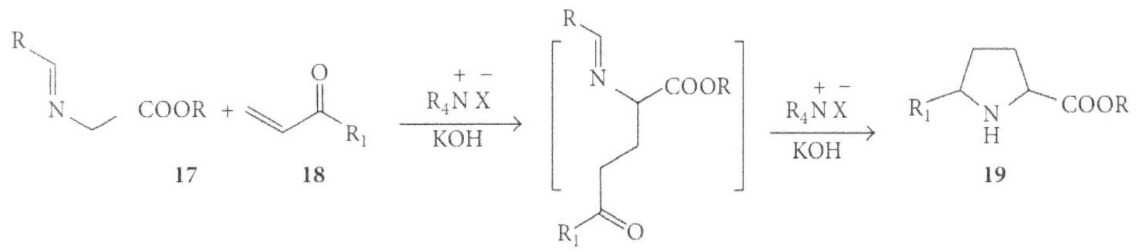

Figure 8

1,4-Di(malononitrilemethyleneamino)benzene 20 was allowed to react with ethyl chloroacetate, chloroacetamide, phenacyl chloride, or diethyl bromomalonate in 1 : 2 molar ratios under solid-liquid PTC technique to give the corresponding 1,4-bi(1-pyrrolyl) benzene derivatives 21a–c and 22 [17] (Figure 9).

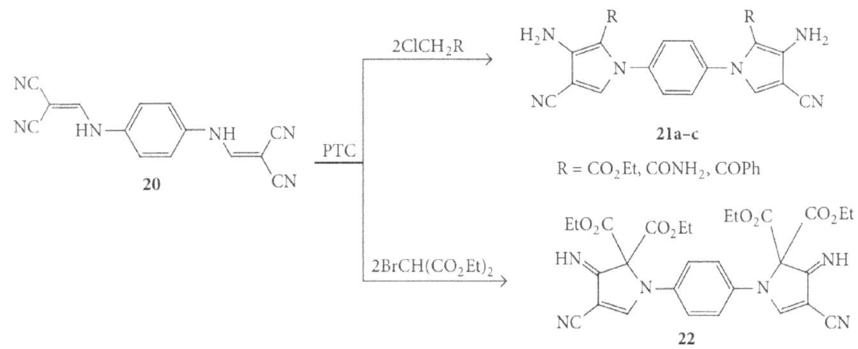

Figure 9

Pyrrolidine derivatives 24 were obtained with thiophenes 25 from the reaction of diethyl cyanomalonate 23with phenyl isothiocyanate 7 and chloroacetonitrile or chloroacetamide using the PTC of (DMF/K_2CO_3/TBAB) [22] (Figure 10).

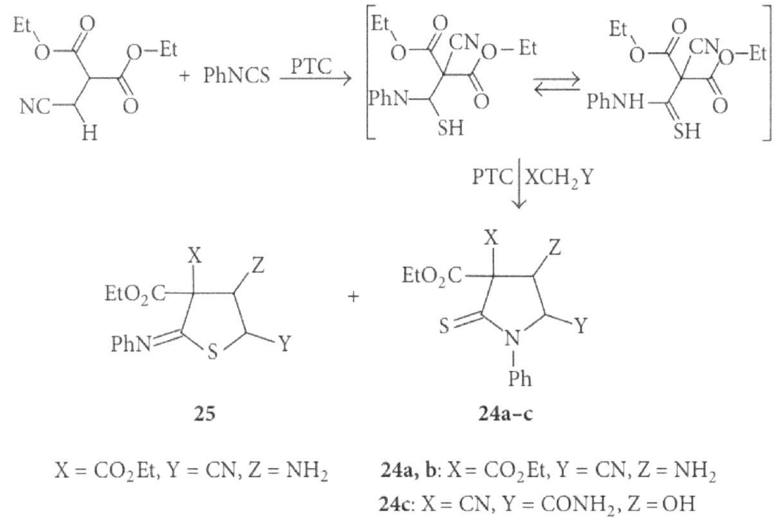

Figure 10

One-pot reaction of malononitrile 5, carbon disulphide, and chloroacetamide in 1 : 1 : 1 molar ratio under solid-liquid PT catalysis (benzene/K_2CO_3/TBAB) has furnished 3-amino-2-carboxyamido-4-cyano-5-mercaptothiophene 26 [23] (Figure 11). Also, treatment of malononitrile 5 with CS_2 and ethyl chloroacetate under the same experimental conditions gave the corresponding thiophene derivative 27 [23] (Figure 11).

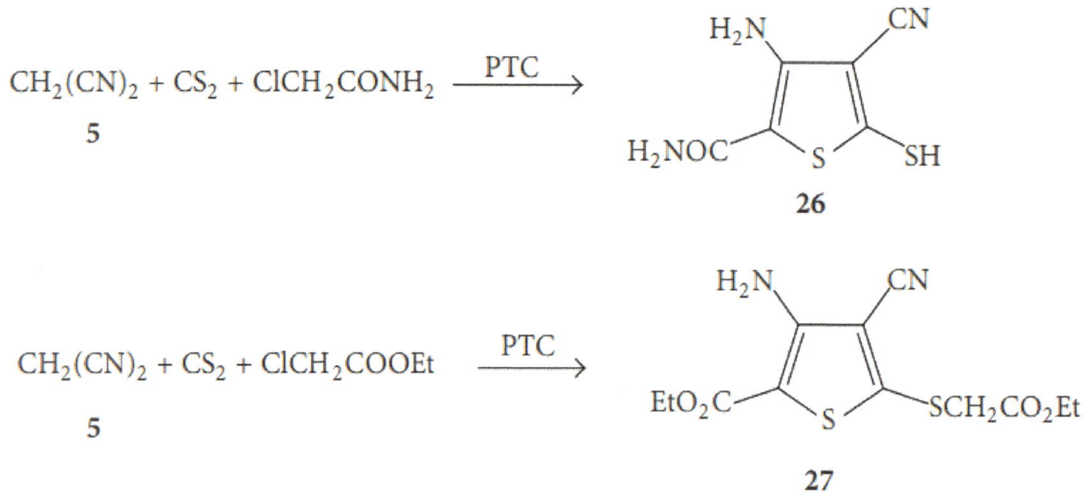

Figure 11

Different thiophene compounds (24 and 28–30)prepared from reaction of diethyl cyanomalonate 23 with carbon disulphide or phenyl isothiocyanate 7 along with different active halocompounds under solid-liquid phase condition (dioxan/K_2CO_3/TBAB) have been reported by Abd Allah and El-Sayed [22] (Figure 12).

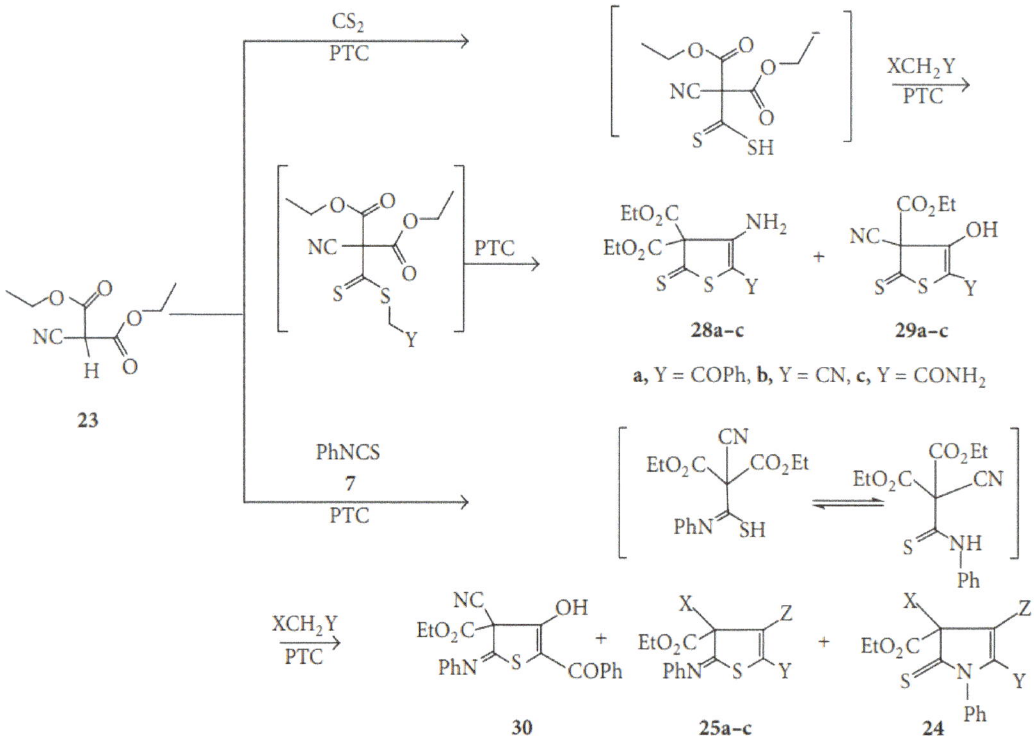

24a, b, 25a-b; X = $CO_2C_2H_5$, Y = CN, Z = NH_2

24c, 25c; X = $CO_2C_2H_5$, Y = $CONH_2$, Z = NH_2

Figure 12

2-Thiophenylidenes 31–33 were obtained through the reaction of either malononitrile 5 or ethyl cyanoacetate11 with carbon disulphide along with ethoxymethylenemalononitrile 8a or ethyl ethoxymethylenecyanoacetate8b using liquid-solid technique (dioxan/K_2CO_3/TBAB) [19] (Figure 13).

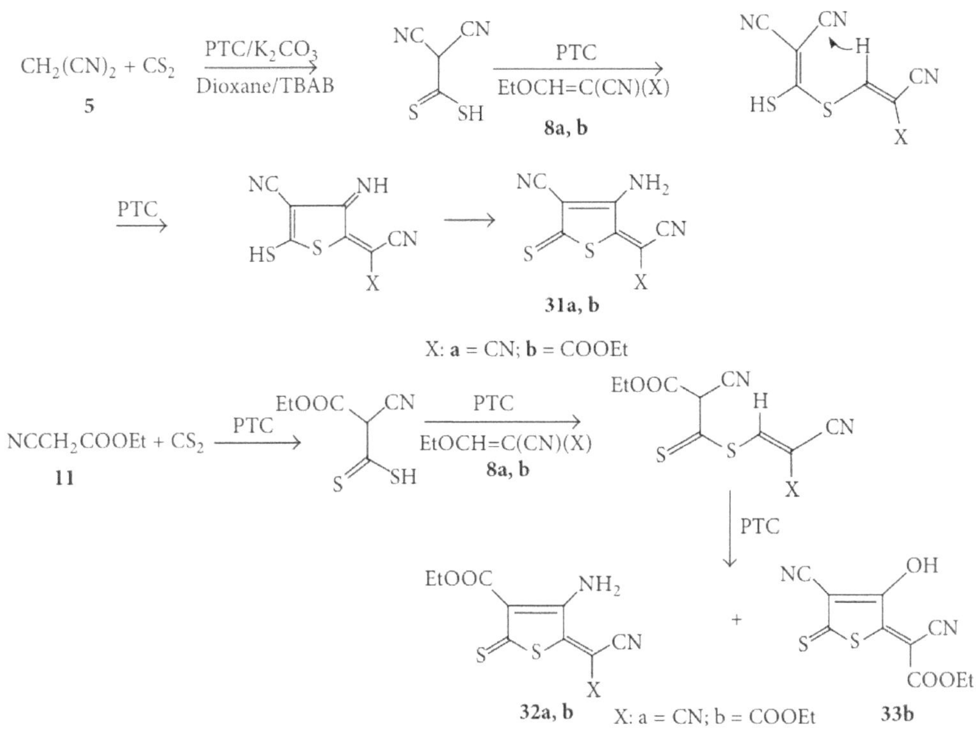

Figure 13

In this reaction, the yield of products was found to be a temperature dependent. Thus, at low temperature (40°C) it affords the aminothiophene derivative 32a in high yield while at high temperature hydroxythiophene32b or 33b was the solely product but in low yield.

The reaction of pyridosultam 34 with 3-chloropropyl thiocyanate under the experimental condition (toluene/aq. NaOH/TBAB) yielded the corresponding dihydrothiophene derivative 35 [24] (Figure 14).

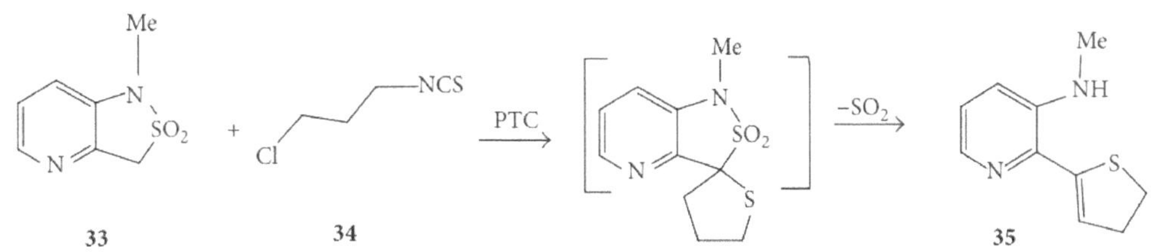

Figure 14

4-Amino-5-cyano-2-hydroxy-3-furancarboxylic acid 37 and compound 38 were synthesized by treating 1-phenyl-3,5-pyrazolinedione 36 with bromomalononitrile under solid-liquid condition (dioxan/K_2CO_3/TBAB) (Figure 15) [25].

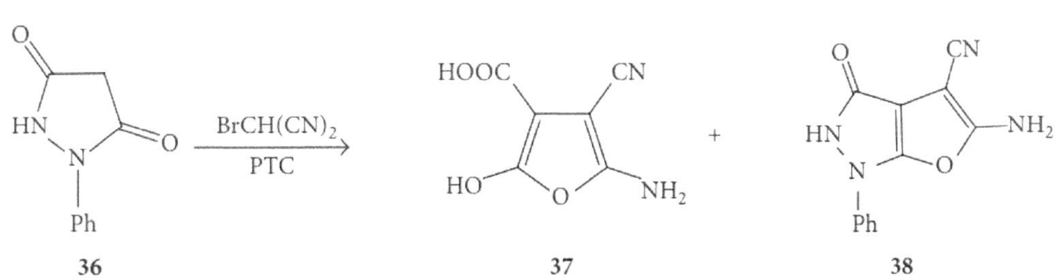

Figure 15

The reaction pathway for the formation of unexpected compound 37 was assumed to proceed via catalytic hydrolysis of compound 38 into the furan derivative 37 and phenyl hydrazine [25] (Figure 16).

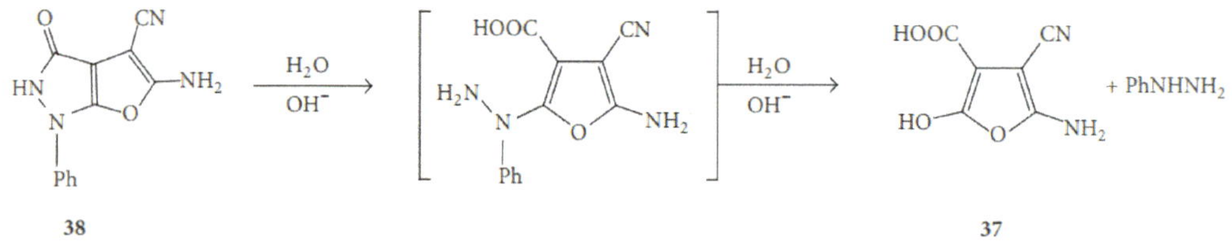

Figure 16

Condensation of salicylaldehyde 39 and its derivatives with various esters of chloroacetic acids 40 in the presence of TBAB led to the synthesis of benzo[b]furans 41 under solventless and microwave irradiation technique [26] (Figure 17).

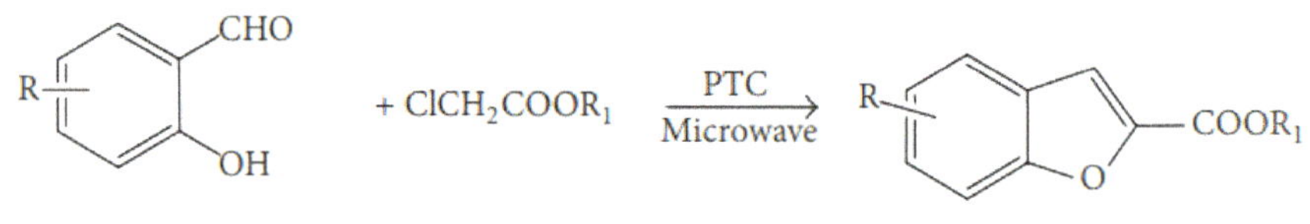

39 40 41

Figure 17

Liquid-liquid PTC reaction of 4-chlorobutyronitrile 42 with nonenolizable aldehydes 43 via sequence of an addition cyclization reaction gave tetrahydrofuran-3-carbonitrile derivative 44 [27] (Figure 18).

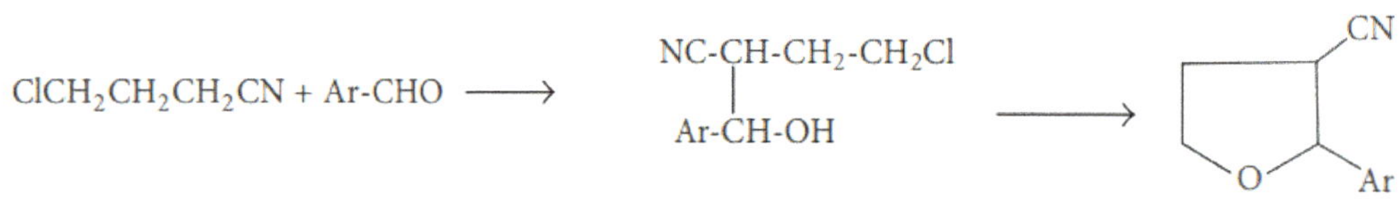

42 43 44

Figure 18

Fused bis-aroylbenzodifurans 46 have been synthesized by condensing 2,4-diacetyl resorcinol 45 with various p-substituted phenacyl bromides [28] (Figure 19).

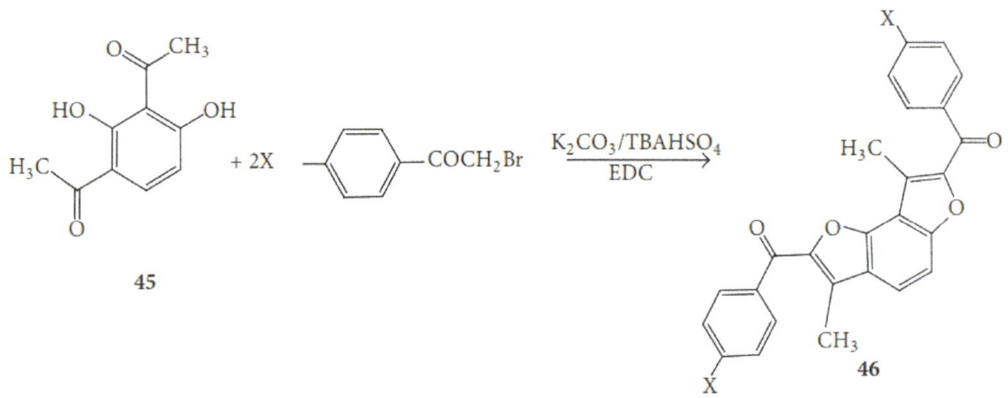

X = H, Cl, NO_2, CH_3, OCH_3, OCH_3, Br

Figure 19

Methyl or ethyl 3-amino-4-arylthiophenes-2-carboxylates 48a–f were synthesized by Thorpe reaction through the treatment of 3-hydroxy-2-arylacrylonitriles 47a–f and methyl or ethyl thioglycolates with hydrochloric acid by using different PTC conditions (solid-liquid or liquid-liquid). The solid-liquid PTC conditions using 18-crown-6 along with potassium hydroxide as a catalyst are the method with excellent yields. The reactions were carried out in acetonitrile at RT [29] (Figure 20).

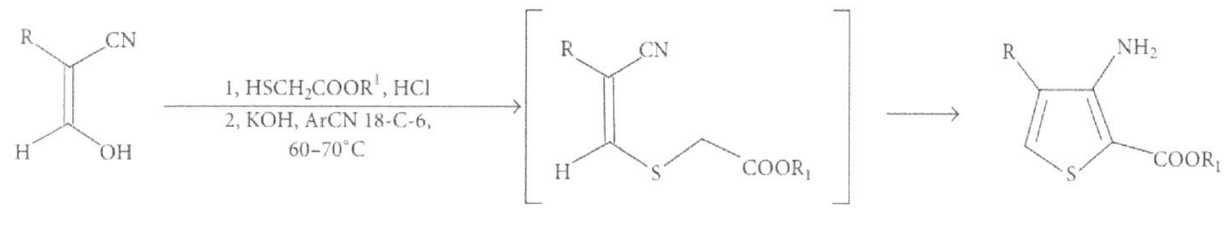

47a–f

48a–f

R	R^1
a, ph	COOMe
b, 4-O-CH_3	COOMe
c, benzo[b]furyl-2	COOMe
d, thinyl-2	COOMe
e, dimethyl 2,5-thinyl	COOMe
f, C_6H_5	COOEt

Figure 20

In Japp-Klingemann reaction the indole derivatives 52 have been prepared as indicated in Figure 21. The addition of a suitable PTC catalyst to the reaction could significantly improve the reaction process. Firstly, aryl amines 49 were diazotized and reacted under alkaline conditions ethyl 2-[2-(1,3-dioxo-1,3-dihydro-2H-isoindol-2-yl)ethyl]-3-oxobutanoate 50 using PTC-promoted Japp-Klingemann reaction to form the ring-opened aryl hydrazones 51. These aryl hydrazones were cyclized and converted into the corresponding indole derivatives 52 by adding hydrochloric acid in ethanol [30] (Figure 21).

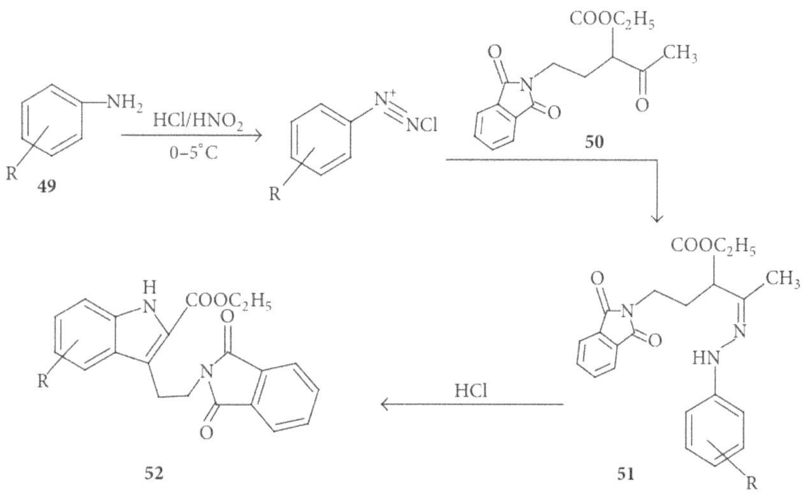

Figure 21

Synthesis of dimethyl phosphinyl substituted tetra-hydropyrroles 55 via 1,3-cycloaddition reaction of the azomethine 53 to electron deficient alkenes such as ,-unsaturated ketones (e.g., benzylideneacetophenone54), esters, and nitriles took place in high yield and low diastereoselectivity in phase-transfer catalysis conditions (solid potassium carbonate, triethylbenzylammonium chloride (TEBA)) in acetonitrile as a solvent [31] (Figure 22).

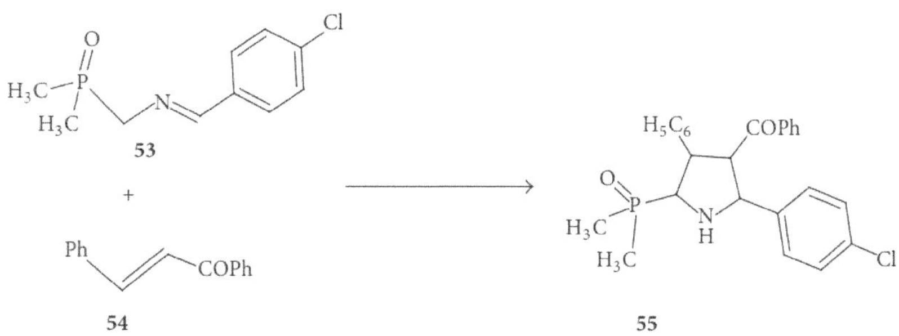

Figure 22

When Schiff base 53 and ethyl cinnamate 56 reacted under different phase-transfer catalysis conditions (10 equivalents of aqueous NaOH (50%)/TEBA/DMSO), the ester of pyrrolidinecarboxylic acid 57 has not been formed; instead the acid itself 58 was obtained as a mixture of two diastereoisomers [31] (Figure 23).

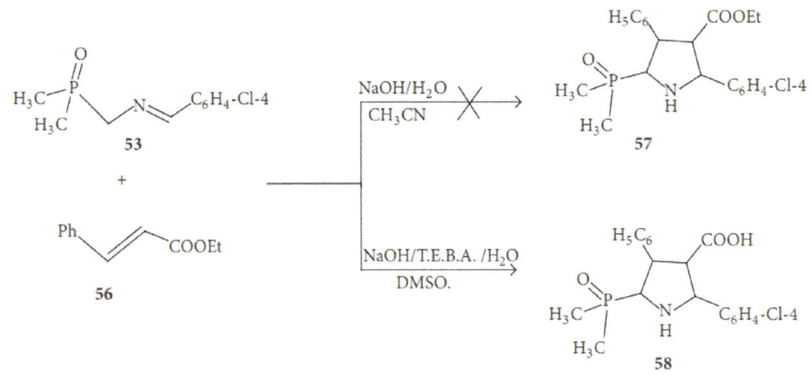

Figure 23

Synthesis of Five-Membered Ring Heterocycles Containing Two Heteroatoms

(4Z)-4-(4-amino-1,3-dithiol-2-ylidene)-5-methyl-2-phenyl-2,4-dihydro-3H-pyrazol-3-one derivative 60 was the product of a reaction of 3-methyl-1-phenyl-2-pyrazoline-5-one 59 with CS_2 and chloroacetonitrile in equimolar ratio under solid-liquid technique (benzene/K_2CO_3/TBAB) [23] (Figure 24).

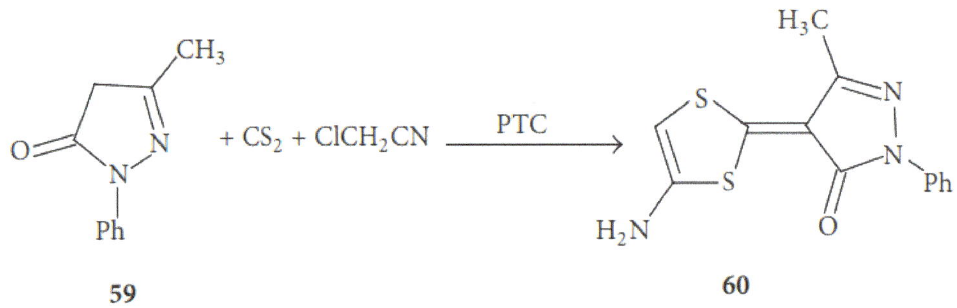

Figure 24

Similarly, 1-phenyl-3,5-pyrazolidinedione 36 was treated with CS_2 and ethyl chloroacetate under the same condition (benzene/K_2CO_3/TBAB) to give the corresponding (4Z)-4-(4-oxo-1,3-dithiolan-2-ylidene)-1-phenylpyrazolidine-3,5-dione 61 [25] (Figure 25).

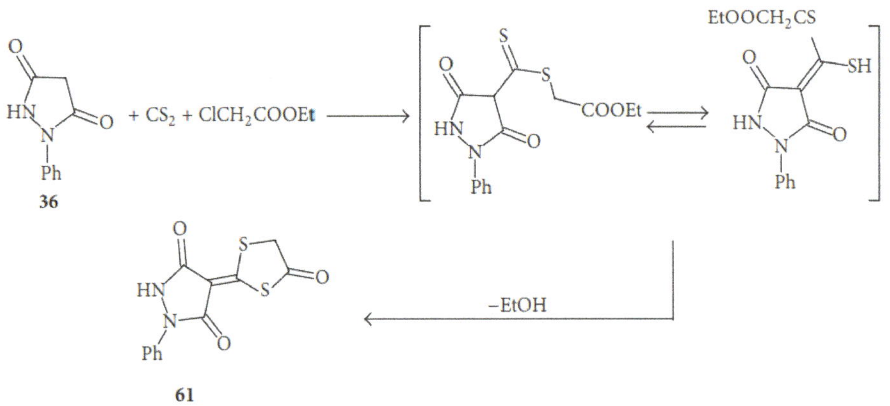

Figure 25

The addition reaction of ethyl 3,4-diamino-5-cyanothieno[2,3-b]thiophene-2-carboxylate 62 to carbon disulfide, followed by cyclization reaction through the treatment with chloroacetonitrile or ethyl chloroacetate under PTC conditions (K_2CO_3, TBAB, dioxane), furnished 3,4-bis(4-amino-2-thioxo-1,3-thiazol-3(2H)-yl)-5-propanoylthieno[2,3-b]thiophene-2-carbonitrile 63 and 3,4-bis(4-oxo-2-thioxo-1,3-thiazolidin-3-yl)-5-propanoylthieno[2,3-b]thiophene-2-carbonitrile 64, in satisfactory yields [32] (Figure 26).

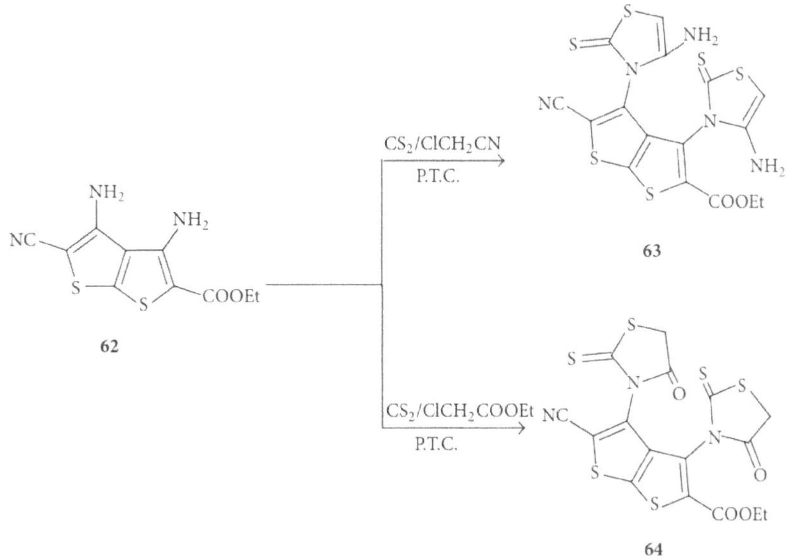

Figure 26

Moreover, the addition reaction of compound 62 to carbon disulfide followed by cycloalkylation reaction with 1,2-dibromoethane, at 1 : 1 : 1 molar ratio under same PTC conditions, afforded the corresponding bis-1,3-dithiolan-2-ylidene 65 [32] (Figure 27).

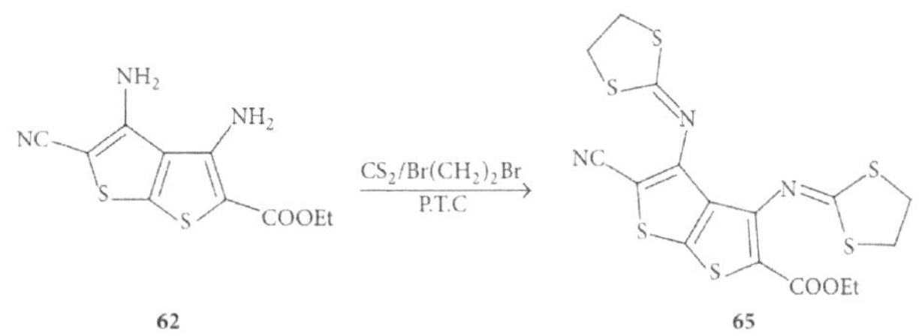

Figure 27

Treatment of triazepene compound 66 with dibromides using liquid-liquid condition (benzene/NaOH/TBAB) furnished pyrazole derivatives 67a, b [33] (Figure 28).

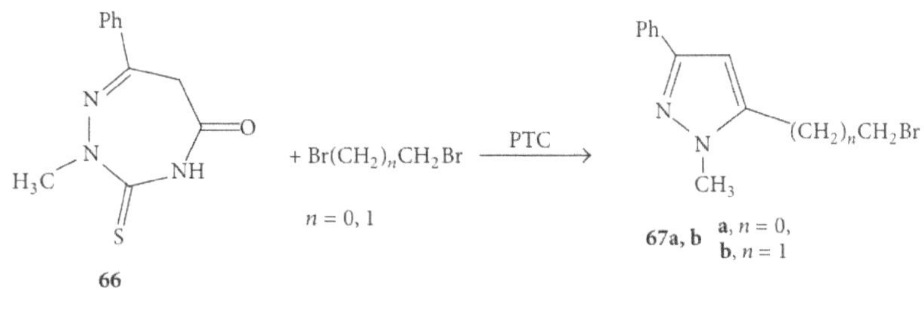

Figure 28

1,3-Dipolar cycloaddition of various substituted nitrilimines 68 to the appropriate alkenyl dipolarophiles 69 in aqueous media and in presence of a surfactant afforded a number of 1-aryl-5-substituted-4,5-dihydropyrazoles70 [34] (**Figure** 29).

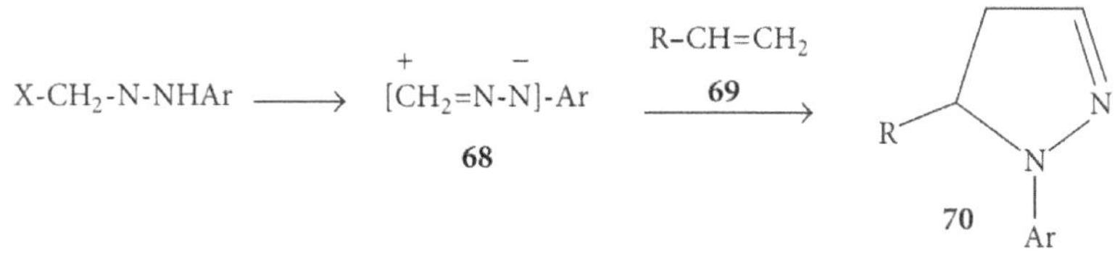

Figure 29

Preparation of 5-substituted oxazoles 74 or imidazoles 75 was achieved by reaction of p-tolylsulphonylmethyl isocyanide 71 with aldehydes 72 $R_1CH=NR_2$ or 73 (R_1, R_2 = substituted phenyl, heteroaryl, and alkyl) under liquid-liquid condition [35, 36] (**Figure** 30).

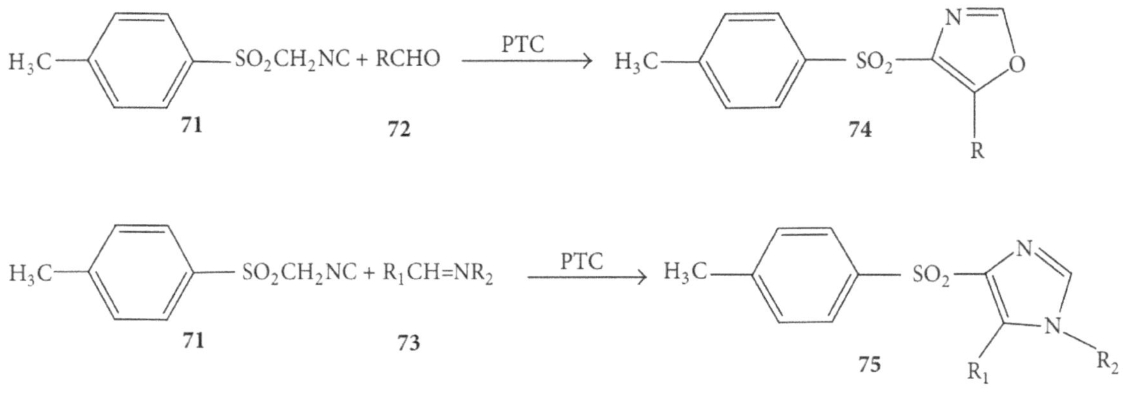

Figure 30

The interaction of 3-chloro-3-methyl-1-butyne 76 with 4,6-dimethyl-5-arylazo-2-thiopyrimidine 77 under liquid-liquid phase (benzene/aq. KOH/BTEACl) afforded 4,6-dimethyl-5-[2-(2-methylprop-1-enyl)-1H-benzimidazol-1-yl]pyrimidinyl-(5H)-thiones 78 [37] (**Figure** 31).

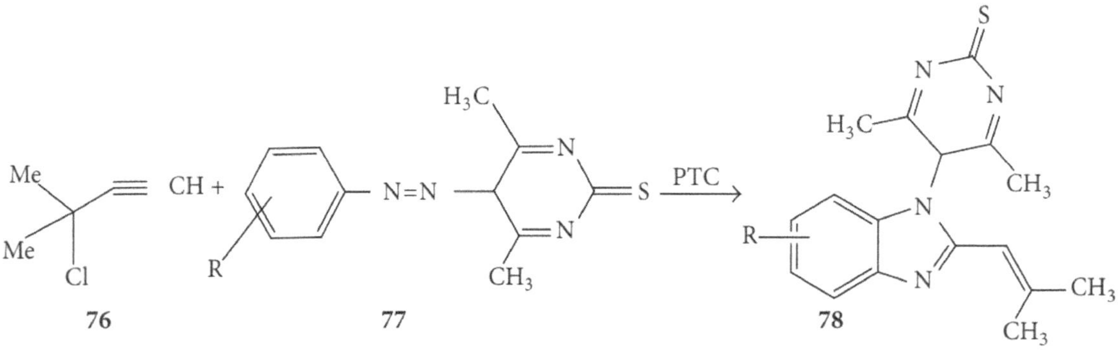

Figure 31

Reaction of 5-phenylmethylene thiohydantoin 79 with 1,2-dibromoethane 80 under liquid-liquid conditions (benzene/aq. NaOH/TBAB) afforded 2,3-dihydro-6-phenylmethylene-5-oxo-imidazo[2,3-b]thiazole 81 [38] (**Figure** 32).

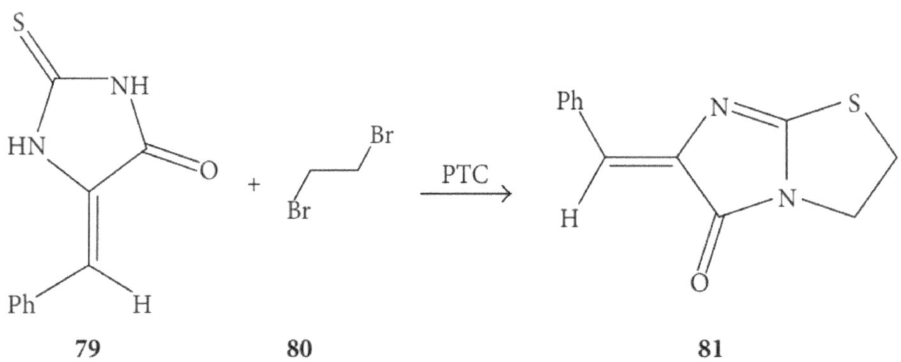

Figure 32

A variety of 4-thiazolidinone derivatives 83a–g were successfully synthesized via in situ formation of ketene-N,S-acetals 82a–g which in turn was reacted with ethyl chloroethyl acetate, chloroacetamide, or chloroacetyl chloride followed by ring closure to afford the desired 4-thiazolidinones 83a–g [39] (**Figure** 33).

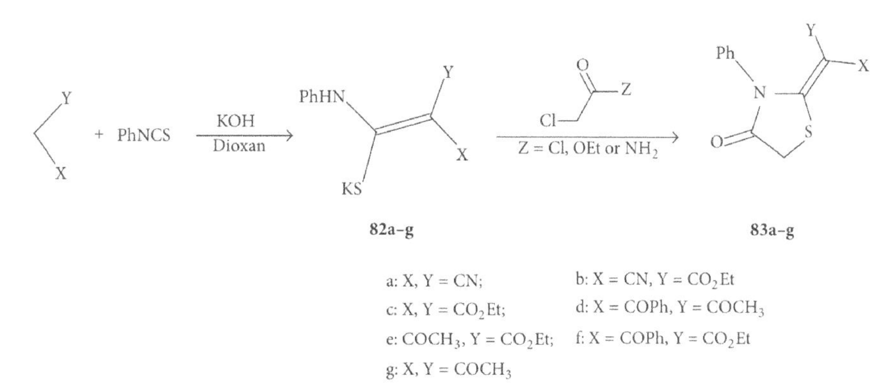

Figure 33

One of the medicinal applications for PTC techniques is synthesis of sibenadet hydrochloride 87 which is a potent drug used for treatment of chronic obstructive pulmonary disease. This bioactive molecule was synthesized by O-alkylation of phenylethanol 84 with the alkyl bromide 85 under PTC condition to form the alkylated product 86 in 97% yield. Reaction of 86 with benzothiazole derivative led to formation of the desired product 86 [40] (**Figure** 34).

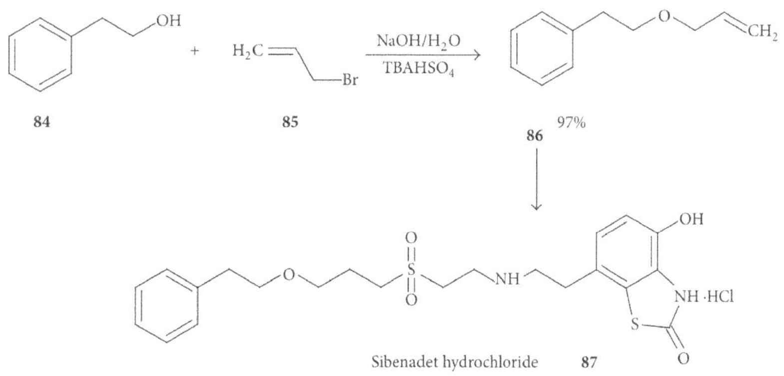

Figure 34

Treatment of an equimolar amount of 2-mercaptoquinazolin-4(3H)-one 88 and dihalocompounds such as 1,2-dibromoethane and chloroacetyl chloride underwent S- and N-cyclo-alkylation by using PTC conditions, giving 2,3-di-hydro-5H-[1,3] thiazolo[2,3-b]quinazolin-5-one 89 and 5H-[1,3]-thiazolo-[2,3-b]quinazoline-3,5(2H)-dione 90, respectively [41] (**Figure** 35).

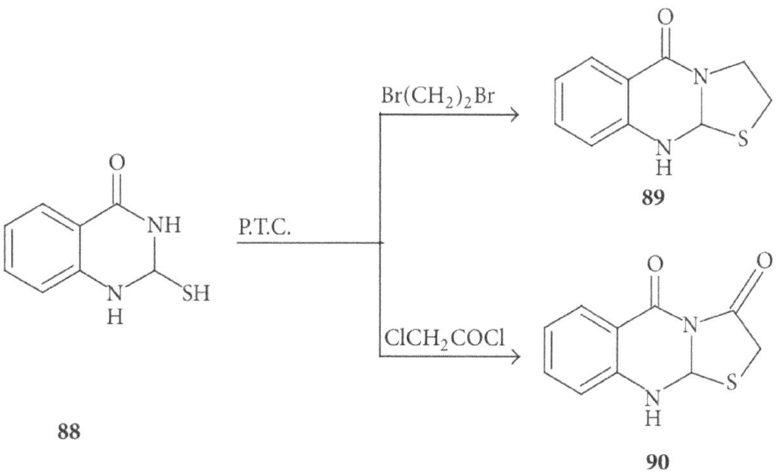

Figure 35

The PTC reaction of 3-phenyl-2-thiohydantoin 91 with dihalocompounds, namely, ethylene dibromide or 1,3-dibromopropane, with CS_2, yielded 1,3-dithioxane derivative 92a or 5-(1,3-dithian-2-ylidene)-3-phenyl-2-thioxoimidazolidin-4-one derivative 92b, respectively [42] (**Figure** 36).

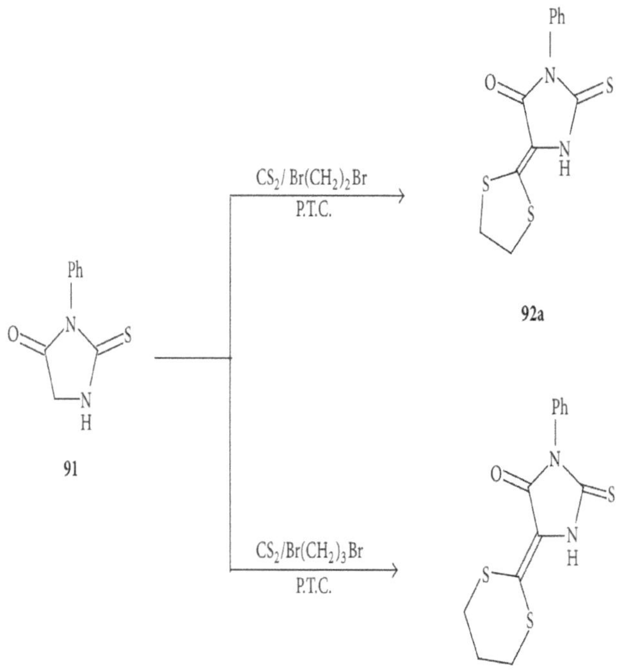

Figure 36

Treatment of the thiourea 93 and urea 94 derivatives with ethyl bromoacetate under PTC condition using TBAB as a catalyst and benzene/anhydrous K_2CO_3 as liquid-solid phases gave imidazolidinediones 95a, b via ring closure pathway [43] (**Figure** 37).

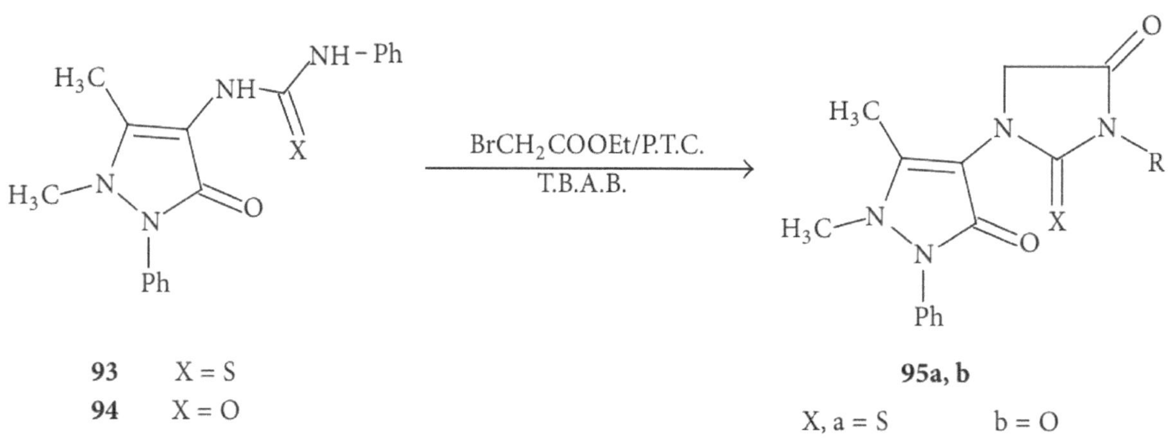

Figure 37

Synthesis of Five-Membered Ring Heterocycles Containing Three Heteroatoms

Alkylation of (2Z,5Z)-5-benzylidene-2-(hydroxyimino)imidazolidin-4-one 96 with methylene bromide in benzene/aq. NaOH in presence of TBAB as a catalyst gave 2-phenyl-2,3-dihydro-6-phenylmethylene-5-oxo-imidazo[2,1-c]-1,2,4-triazole 97. On the other hand, alkylation of the oxime derivative of 5-phenylmethylene thiohydantoin 98 with methylene bromide gave 3-(1H)-6-phenylmethylene-5-oxo-imidazo[2,1-c]-1,2,4-oxadiazoles 99a, b [38] (**Figure** 38).

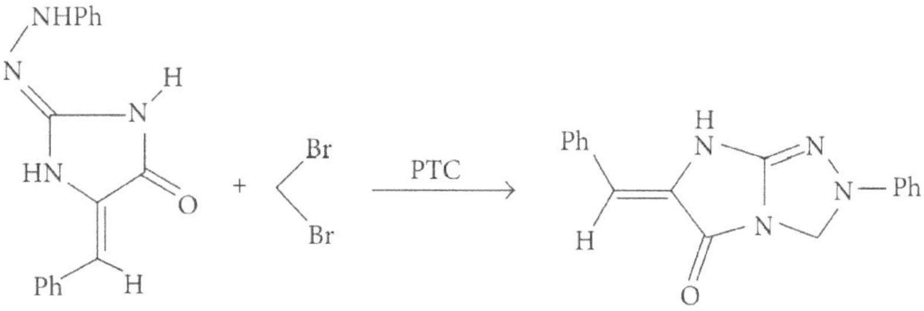

96 97

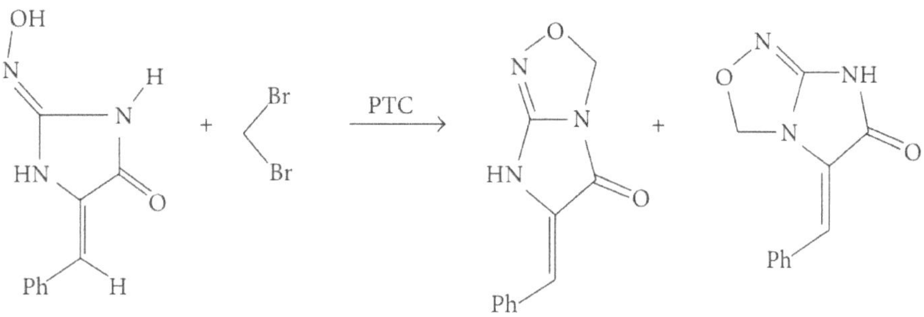

98 99a 99b

Figure 38

Anastrozole 101 which acts as selective aromatase inhibitor and is employed effectively in treatment of advanced breast cancer in postmenopausal women has been synthesized in good yield by methylation reaction of 3,5-bis(cyanomethyl)toluene 100, using methyl chloride and 50% aq. NaOH/TEBA as a PTC condition [40] (**Figure** 39).

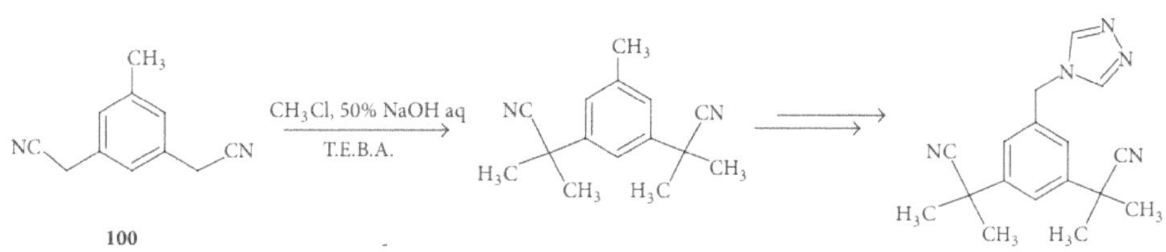

100 101 Anastrozole

Figure 39

Synthesis of Five-Membered Ring Heterocycles Containing Four Heteroatoms

5-Alkyl and 5-arylthiotetrazoles 104 were synthesized in good yield from reaction of alkyl(aryl)thiocyanates102 with azide 103 under PTC conditions [44] (**Figure** 40).

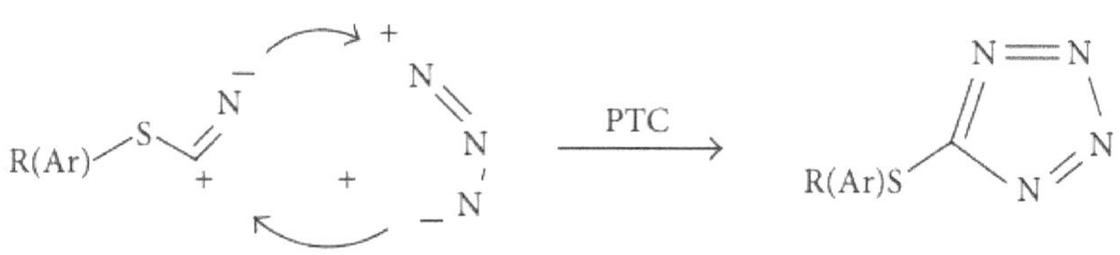

102 103 104

Figure 40

Functionally substituted tetrazoles have been synthesized from the corresponding N,N′,N″-triarylbenzene-1,3,5-tricarboxamides 105 via sequential transformation of these compounds into imidoyl chlorides and treatment of the latter with sodium azide under conditions of phase-transfer catalysis. As a result, a number of heterocyclic structures 106a–106e containing three tetrazole rings have been isolated [45] (**Figure** 41).

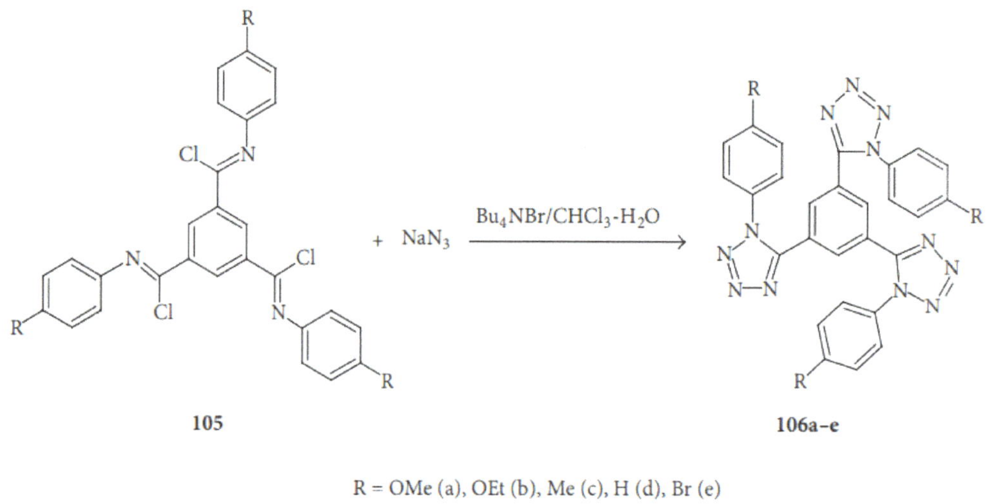

Figure 41

SYNTHESIS OF FUSED FIVE-MEMBERED RING HETEROCYCLES

Functionally substituted thieno(2,3-b)thiophenes 107a, b were synthesized in a one-pot reaction of malononitrile 5, CS_2, and -halocarbethoxy or -halonitrile electrophiles in 1 : 1 : 2 molar ratios using (benzene/K_2CO_3/TBAB) as a PTC condition [23] (**Figure** 42).

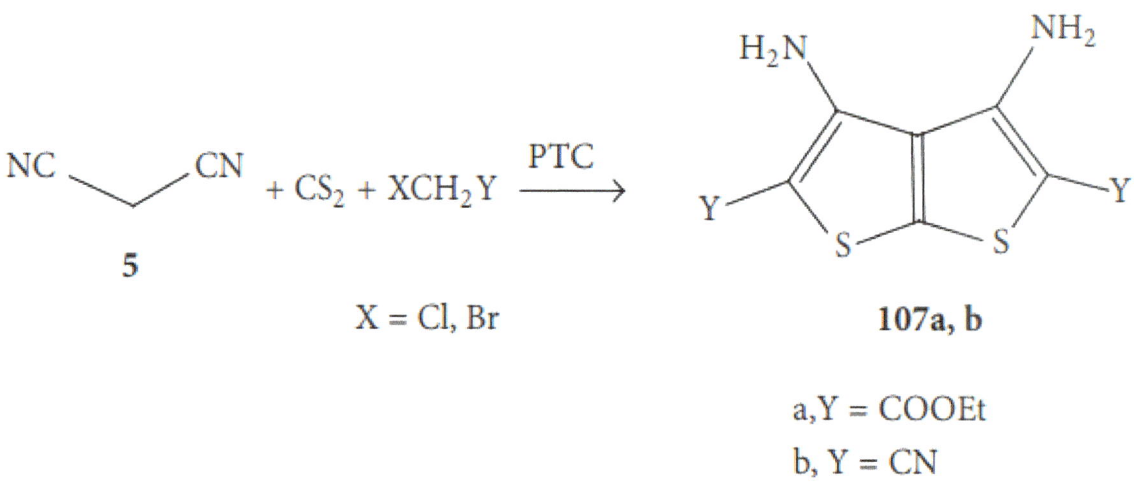

Figure 42

Another different thienothiophene 109 was synthesized via the reaction of acetylacetone 108 with CS_2 and ethyl chloroacetate in 1 : 1 : 2 molar ratio under the same experimental conditions [46] (**Figure** 43).

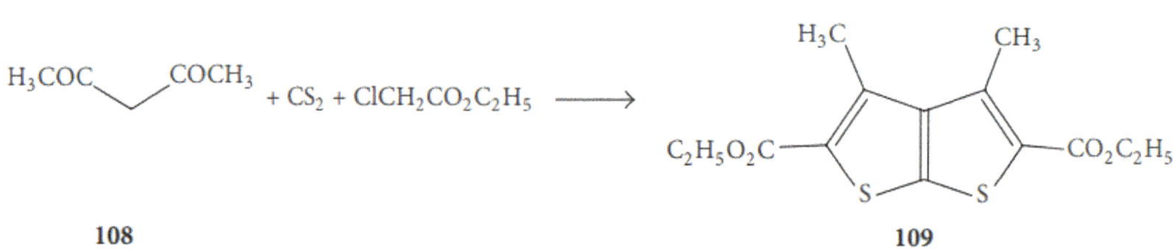

Figure 43

Thieno(2,3-b)pyrrole derivative 110 was obtained by treating malononitrile 5 with phenyl isothiocyante 7 and ethyl chloroacetate in 1 : 1 : 2 molar ratio using (benzene/K_2CO_3/TBAB) as a phase transfer catalyst [23] (**Figure** 44).

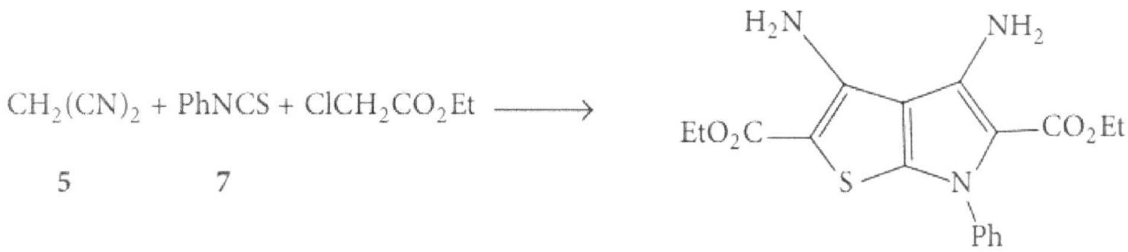

5 7 110

Figure 44

Likewise, under the same conditions, 3-methyl-1-phenyl-2-pyrazoline-5-one 63 was subjected to react with CS_2 and chloroacetonitrile in 1 : 1 : 1 molar ratio where 6-cyano-4,6-dihydro-3-methyl-1-phenylthieno(3,4-c)pyrazol-4-thione 111 was obtained in good yield [23] (Figure 45).

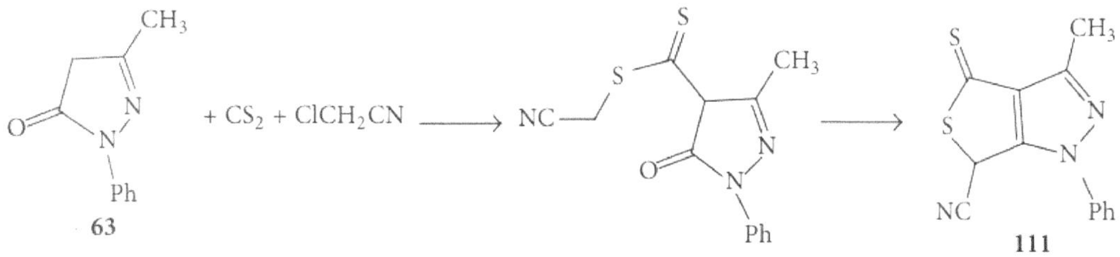

63 111

Figure 45

1-Phenyl-3,5-pyrazolinedione 36 was allowed to react with CS_2 along with chloroacetonitrile or ethyl chloroacetate under the same conditions to give the corresponding thioxo-thienopyrazolone derivatives 112[25] (Figure 46).

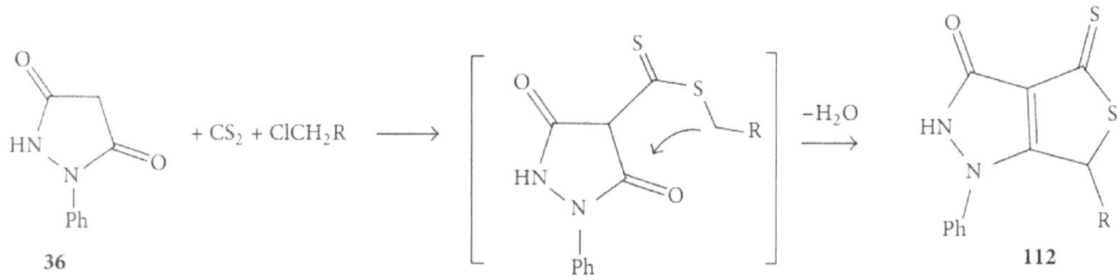

36 112

R = CN, COOEt

Figure 46

Reaction of 4-hydroxydithiocoumarin 113 with allyl bromide 114 under liquid-liquid technique (CH_2Cl_2/aq. NaOH/TBAB or TBACl) gave 2-methyl-2,3-dihydro-4H-thieno[2,3-b]benzothiopyran-4-one 115 [47] (Figure47).

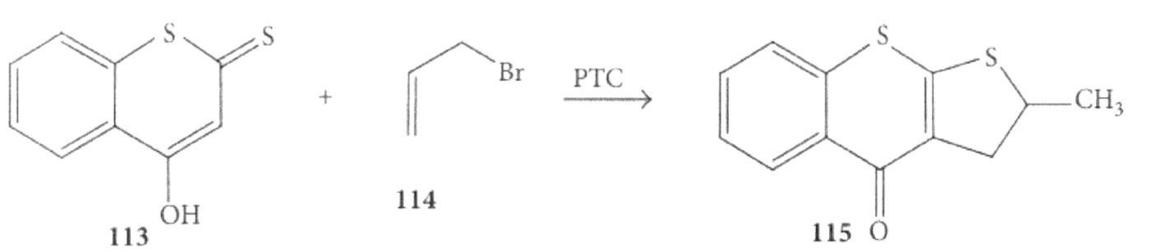

113 114 115

Figure 47

Treating of pyridazinium ylides 116 with N-phenylmaleimide, maleic and fumaric esters, resulted in the cycloadduct products 117–120 with high stereospecificity in the presence of KF and trioctylmethylammonium chloride or without solvent in the presence of aliquat 336 as phase transfer catalyst [48] (**Figure** 48).

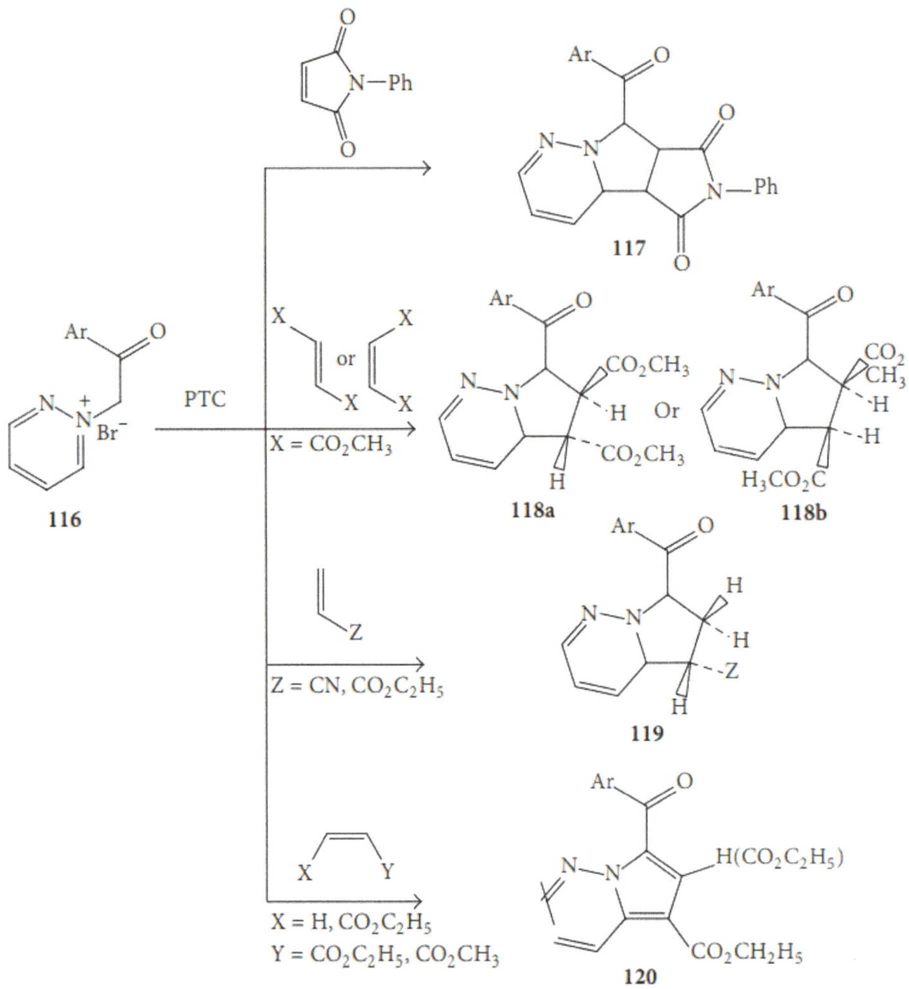

Figure 48

Under solid-liquid technique (dioxan/K_2CO_3/TBAB), bis-thieno(1,8)naphthyridines 122 were prepared by reaction of 4,5-diamino-3,6-dicyano-(1,8)naphthyridine-2,7-dithiol 121 with ethyl chloroacetate, chloroacetonitrile, phenacyl bromide, or chloroacetamide in 1:2 molar ratio or from reaction of 4,5-diamino-2,7-dichloro-3,6-dicyano(1,8)naphthyridine 123 with two equivalents of ethyl mercaptoacetate [49] (Figure49).

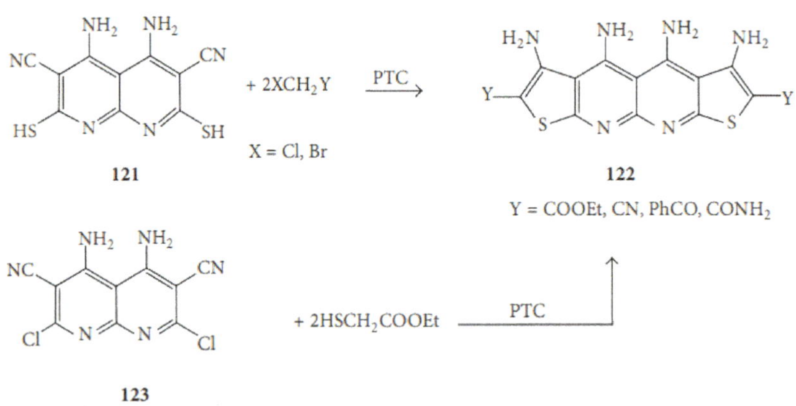

Figure 49

Similarly 4-amino-3,5-dicyano-2-mercapto-6-methylthiopyridine 124 was reacted with ethyl chloroacetate, chloroacetonitrile, phenacyl bromide, or chloroacetamide in 1:1 molar ratio using the same PTC condition to give the corresponding thieno(2,3-b) pyridines 125 or pyrrolo(2,3-d)thieno(2,3-b)pyridines 126 [49] (Figure50).

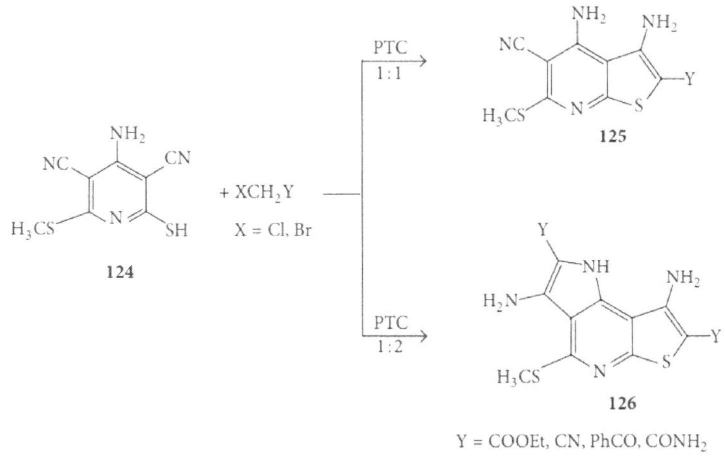

Figure 50

Using solid-liquid technique (dioxan/K_2CO_3/TBAB), 2-ylidenemalononitrile and ethyl 2-ylidenecyanoacetate of both 2,3-dihydrobenzothiazole 127 and 2,3-dihydrobenzoxazole 128 were allowed to react with chloroacetonitrile, ethyl chloroacetate, or phenacyl bromide in equimolar ratio which afforded the corresponding substituted pyrrolo[2,1-b]benzothiazoles 129 or substituted pyrrolo[2,1-b]benzoxazoles 130, respectively [42]. Similarly, (3-amino-2-(1,3-dihydro-2H-benzimidazol-2-ylidene) propanenitrile malononitrile131a or ethyl 3-amino-2-(1,3-dihydro-2H-benzimidazol-2-ylidene)propanoate 131b were treated under the same PTC conditions, where the corresponding pyrrolo[2,3-b]pyrrolo[1,2-f]benzoimidazoles 132 were obtained [50] (**Figure** 51).

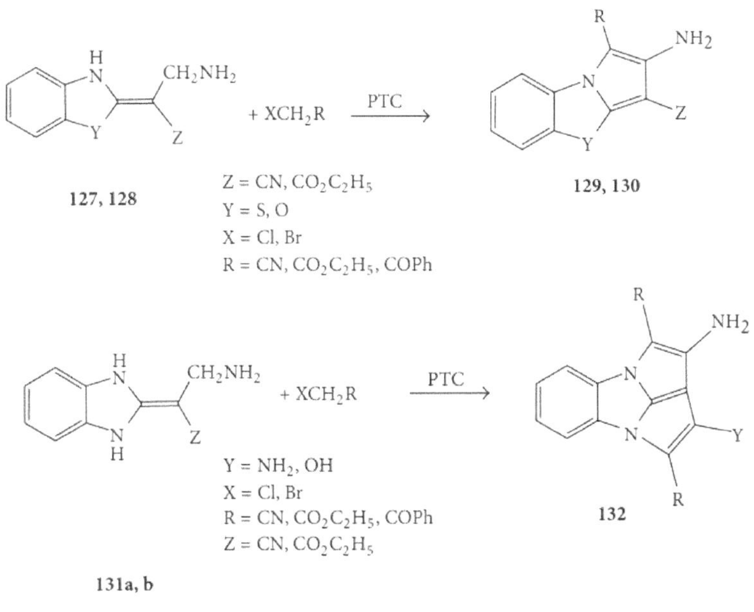

Figure 51

Reaction of 2,7-dichloro-1,8-naphthyridine derivative 133 with ethyl glycinate hydrochloride in 1 : 1 molar ratio under solid-liquid technique (dioxan/K_2CO_3/TBAB) afforded the corresponding bis-pyrrolo(1,8)naphthyridine derivative 134 [49] (**Figure** 52).

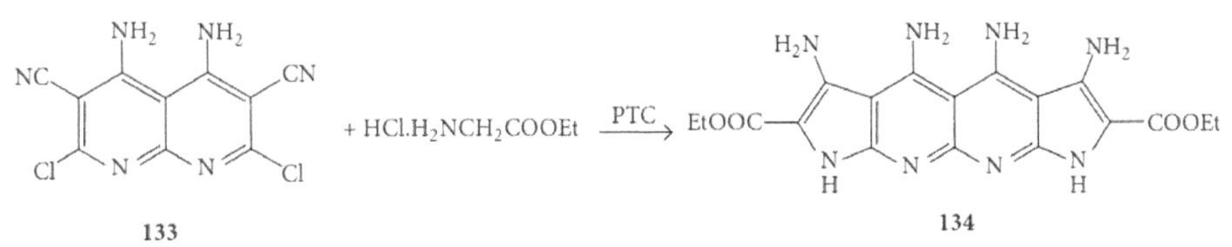

Figure 52

Thieno(2,3-b) thiopyran derivatives 136 were obtained from reaction of the thiopyran derivative 135 with ethyl chloroacetate in 1 : 1 molar ratio (dioxan/K_2CO_3/TBAB), while on carrying out the same reaction under the same conditions but in 1 : 2 molar ratio gave the corresponding thienopyrrolothiopyran derivatives 137 [51] (**Figure** 53).

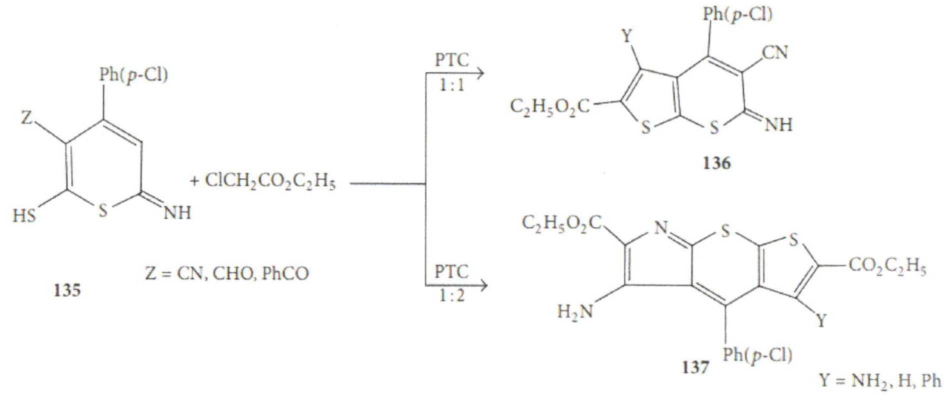

Figure 53

Treatment of 3-amino-1-phenyl-2-pyrazoline-5-one 138 with CS_2 along with chloroacetonitrile, ethyl chloroacetate, or phenacyl bromide in 1:1:1 and 1:1:2 molar ratios using [dioxan/K_2CO_3/TBAB] as a PTC gave the corresponding thienopyrazoles 139–143, 145 and respectively, 144, 146 and 147, fused thiazines [52] (**Figure** 54).

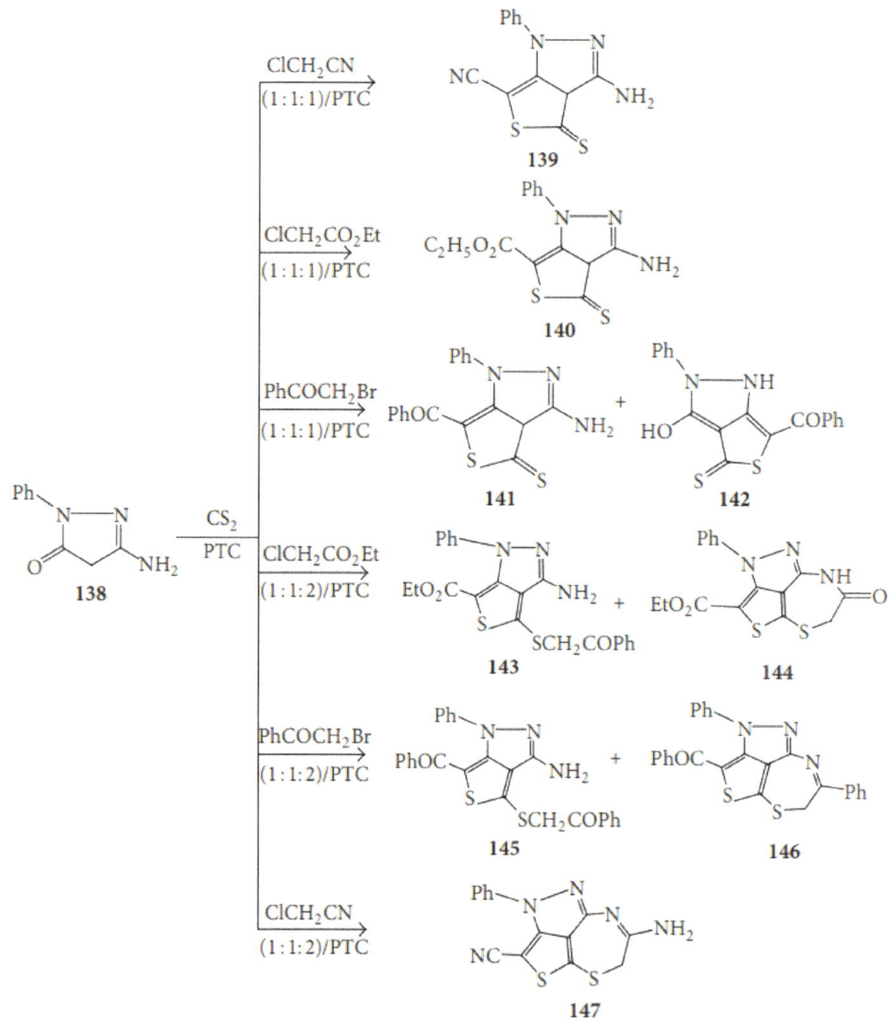

Figure 54

Thieno(3,4-b)pyrrole derivative 148 was obtained by reacting ethyl thiophene-2-ylidenecyanoacetate derivative31b with phenacyl bromide in a heterogeneous mixture of (dioxan/K_2CO_3/TBAB) as a PTC [19] (**Figure** 55).

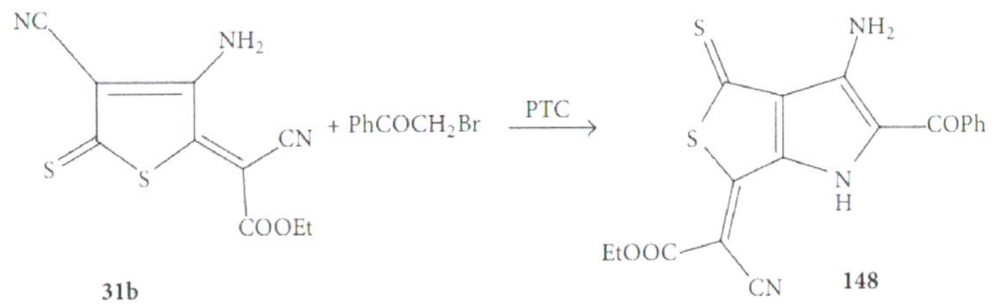

Figure 55

In the same manner, pyridin-2-ylidenemalononitriles 149a, b were reacted with phenacyl bromide under the same reaction condition and yielded the corresponding thieno(2,3-b)pyridine derivatives 150a, b [18] (Figure56).

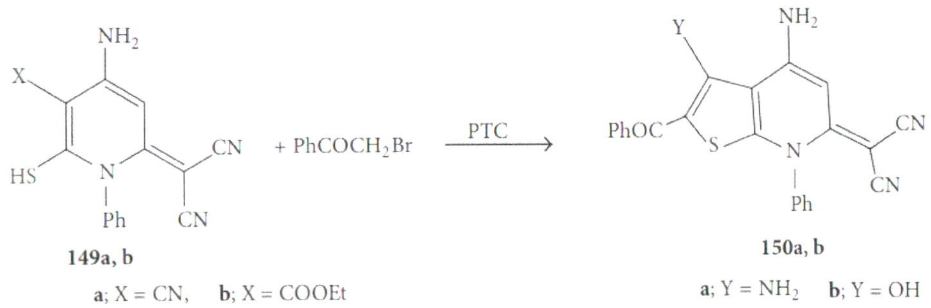

Figure 56

2,6-Dibenzoyl-5-methyl-3-(substituted styryl)benzo[1,2-b:5,4-b]difurans 152 were synthesized from reaction of cinnamoylbenzofurans 151 with phenacyl bromide under liquid-liquid technique (benzene/aq. K_2CO_3/$TBAHSO_4$) [53] (**Figure** 57).

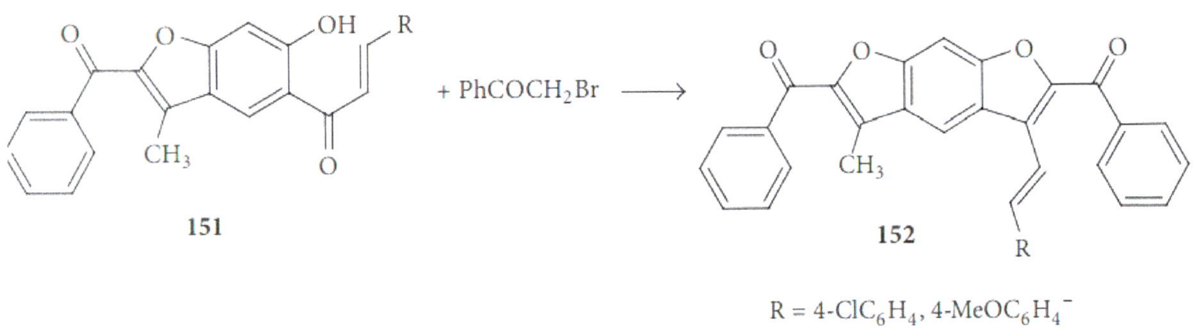

Figure 57

Similarly, 2,6-dibenzoyl-3,5-distyrylbenzo(1,2-b:5,4′-b)difurans 154 were prepared by condensing substituted dichalcones 153 with phenacyl bromide under the same preceding PTC conditions [54] (**Figure** 58).

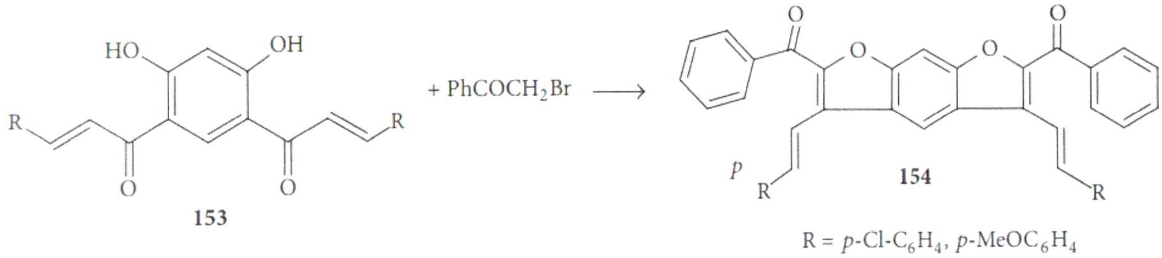

Figure 58

Also, under the same PTC conditions, dibenzoylbenzodifurans 156 were obtained from the reaction of substituted resorcinols 155 with phenacyl bromide [55] (**Figure** 59).

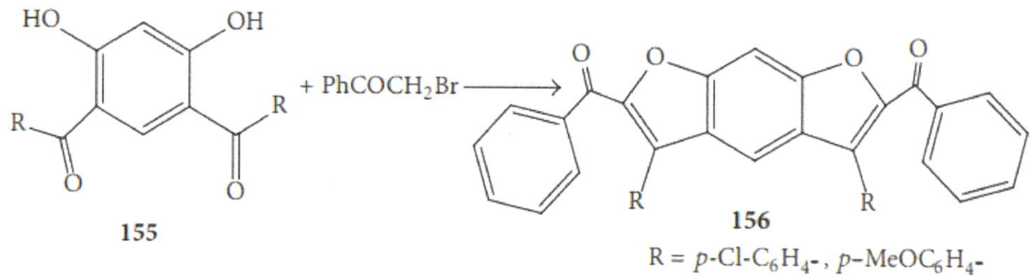

Figure 59

Moreover, 2,6-diaroyl/naphthoyl-3,5-dialkyl/phenylbenzo[1,2-b:5,4-b′]difurans 158 were synthesized by condensing 2,4-diacyl/diaroylresorcinols 157 with various p-substituted -bromo ketones [56] (**Figure** 60).

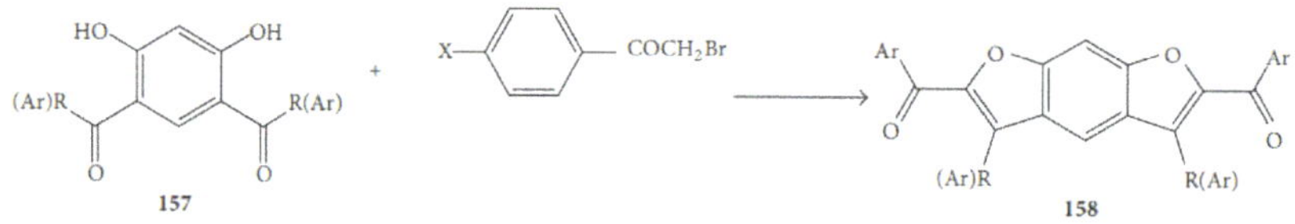

Figure 60

Treatment of 3-methyl-1-phenyl-2-pyrazolin-5-one 63 with bromomalononitrile in 1 : 1 molar ratio yielded 5-amino-4-cyano-3-methyl-1-phenylfuro(2,3-c)pyrazoline 159 [23] (**Figure** 61).

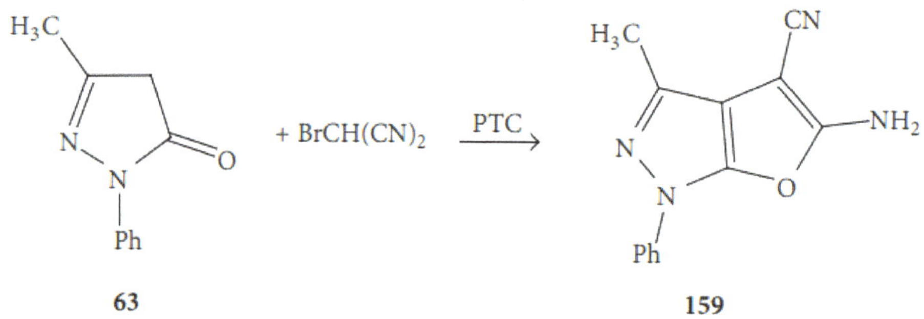

Figure 61

Furo(1,5)benzodiazepin derivative 161 was obtained by reaction of 1,4-dimethyl-3H-(1,5)benzodiazepin-2-one160 with chloroacetonitrile under solid-liquid condition (dioxan/K_2CO_3/TBAB) [57] (**Figure** 62).

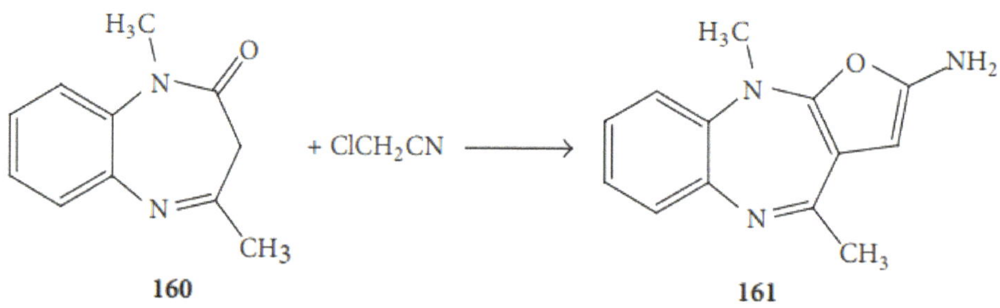

Figure 62

By reaction of 4-methyl-1,3-dihydro-2H-1-benzazepin-2-one 162 with chloroacetonitrile or phenacyl bromide under the solid-liquid phase transfer catalyst (dioxan/K_2CO_3/TBAB), the corresponding pyrrolo[1,2-a](1,5)benzodiazepine 163 was obtained [58] (**Figure** 63).

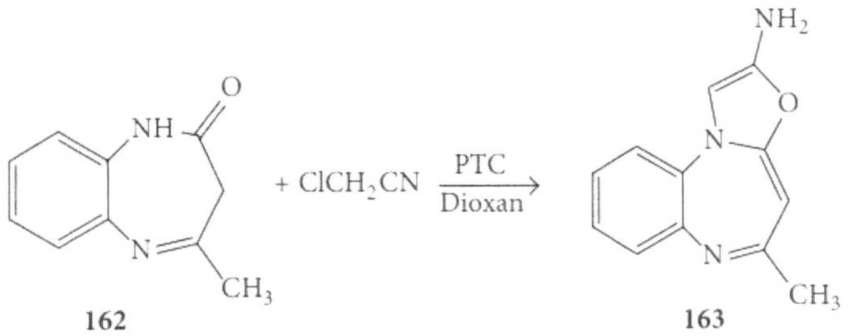

Figure 63

Synthesis of 2,6-di(1-acetyl-2-oxopropylidene)dithiolo[4,5-b:4′,5′-e]-4,8-benzoquinone 166 was achieved in one-pot reaction under solid-liquid technique (benzene/K_2CO_3/TBAB) starting from acetylacetone, CS_2 to give 165 which in turn reacted with tetrabromo p-benzo-quinone 164 in 2 : 1 molar ratio [59] (**Figure** 64).

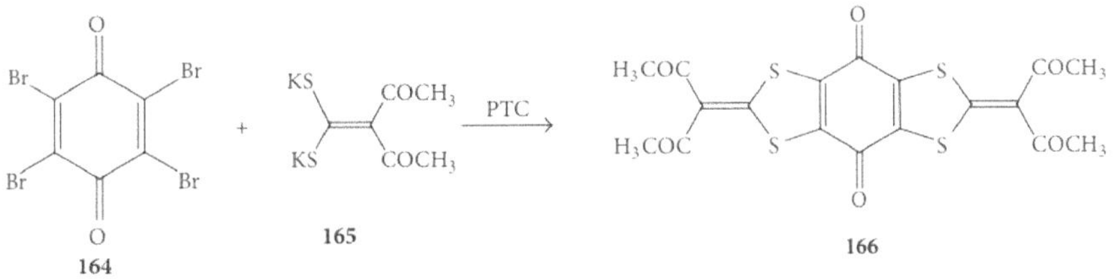

Figure 64

Treatment of 4-arylidene-1-phenyl-3,5-pyrazolinediones 167 with S,S-acetal derivatives using solid-liquid technique (dioxan/K_2CO_3/TBAB) yielded the corresponding pyrazolino-(1,3)-dithiolane derivatives 168 [25] (**Figure** 65).

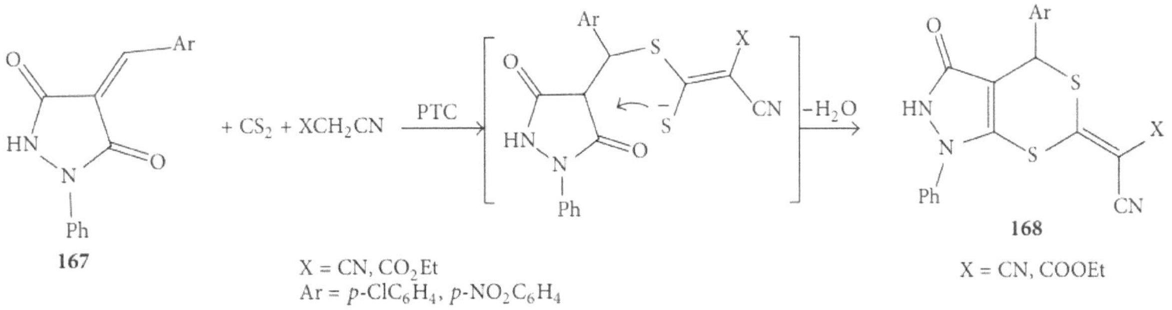

Figure 65

Some condensed heterocyclic systems 171 were obtained by reacting 3-phenyl-4-amino-s-triazole-5-thiol 169as a dianionic compound containing N and S poles with some di- and tetrahaloderivatives 170 as well as -haloketones and -halonitrile derivatives under solid-liquid technique [60] (**Figure** 66).

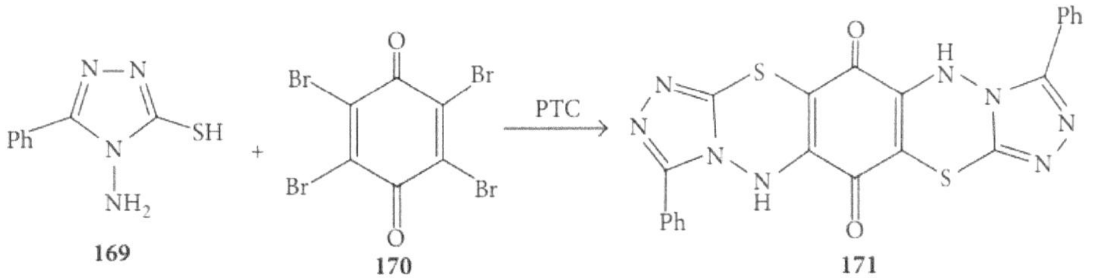

Figure 66

2-Aminoprop-1-ene-1,1,3-tricarbonitrile 172 was reacted with CS_2 or PhNCS along with 3-methyl-1-phenyl-2-pyrazoline-5-one 63 or 2-iminothiazolidin-4-one 173 (dioxan/K_2CO_3/TBAB) to give the corresponding fused pyrazoles 174 and thiazoles 175, respectively [61] (**Figure** 67).

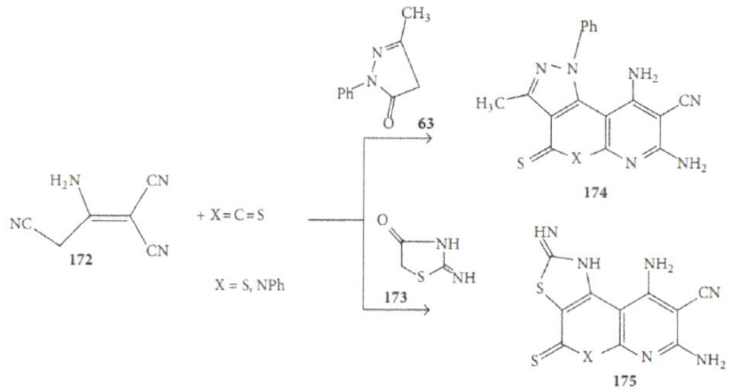

Figure 67

The reaction of 1,3-dihydro-4-methyl(1,5)benzodiazepin-2-one 176 with chloroacetonitrile using solid-liquid technique (dioxan/K_2CO_3/TBAB) gave the corresponding oxazolo-(1,5)benzodiazepin derivative 177 [57] (**Figure** 68).

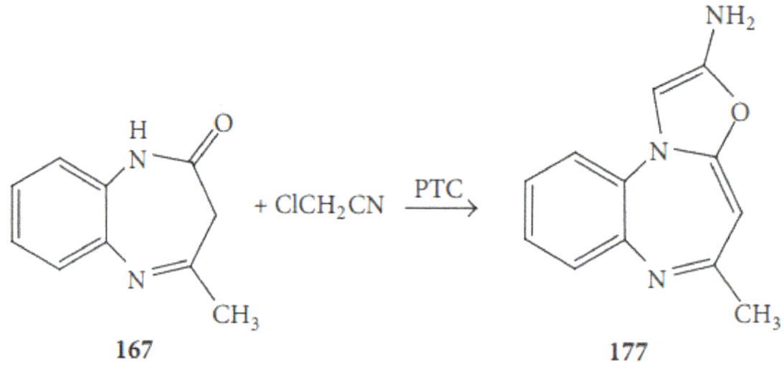

Figure 68

When 7-(methoxyimino)-4-methyl-2H-chromene-2,8-(7H)-dione 178 was treated with crotyltriphenylphosphonium chloride (CTPPCl) 179 (CH_2Cl_2/Li(OH)/CTPPCl), the corresponding chromeno-oxazolone 180 was obtained in which one of the reactants (CTPPCl) acts also as a catalyst [62] (**Figure** 69).

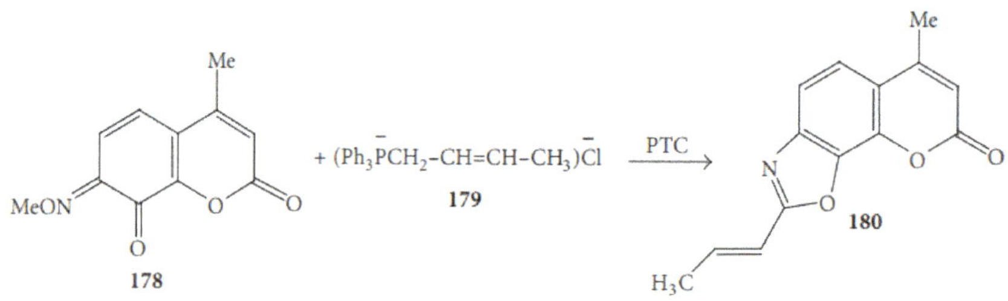

Figure 69

The triazepine 181 was treated with 1,2-dibromoethane under liquid-liquid technique (benzene/aq. NaOH/BTEACl) to afford the corresponding 6-methyl-8-phenyl-2,3-dihydro[1,3]thiazolo[3,2-d][1,2,4]triazepine-5(6H)-thione 182 [33] (**Figure** 70).

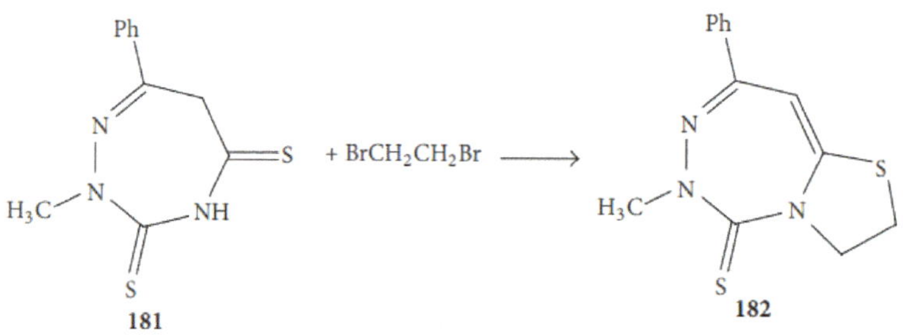

Figure 70

Synthesis of Spiro Five-Membered Ring Heterocycles

Synthesis of spirorhodanine heterocycles 184–187 was achieved by treating 5,5-dibromo-3-phenyl-2-thioxo-1,3-thiazolidin-4-one 183 with different bidentates or with mixture of CS_2 and some active methylene compounds under solid-liquid condition (dioxan/K_2CO_3/TBAB) [63] (**Figure** 71).

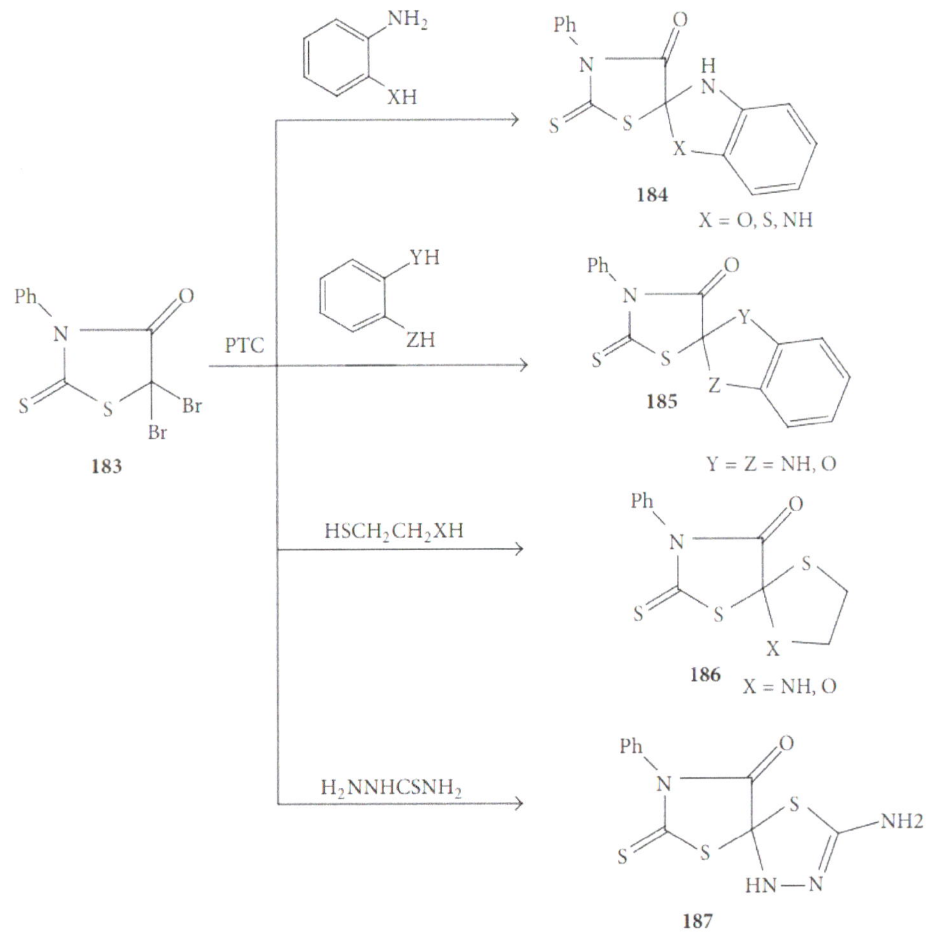

Figure 71

Reaction of 4-ethoxymethylene-1-phenyl-3,5-pyrazolidinedione 188 with CS_2 and malononitrile 5 or ethyl cyanoacetate 11 using solid-liquid technique (dioxan/K_2CO_3/TBAB) yielded the corresponding spiro-1,3-dithiolane derivatives 189 [25] (**Figure** 72).

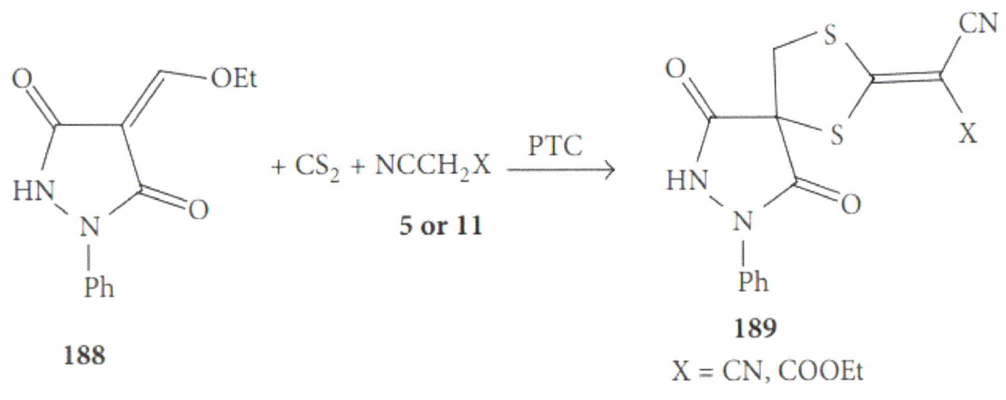

Figure 72

1-Benzoyl-3,3-dibromo-4-phenyl-(1,5)benzodiazepin-2-one 190 was reacted with different dinucleophiles under PTC condition (dioxan/K_2CO_3/TBAB) to furnish the corresponding spiroheterocycles attached to benzodiazepine moiety 191–193 [64] (**Figure** 73).

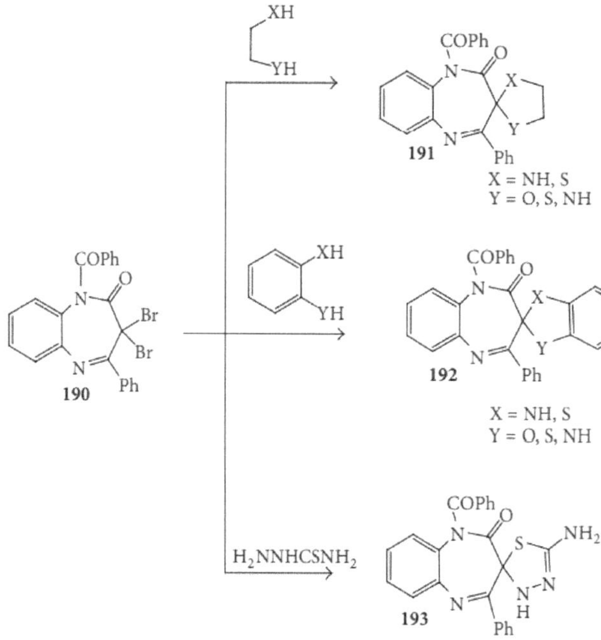

Figure 73

USES OF PHASE TRANSFER CATALYSIS TECHNIQUES IN REACTIONS OF FIVE-MEMBERED HETEROCYCLES

Alkylation and Acylation

N-Alkylation or Acylation

Diez-Barra et al. [65] studied the alkylation of pyrrole 194 by PTC technique in absence of the solvent. The study revealed that the N versus C ratio is not affected by the nature of the catalyst. The nature of the leaving groups have a significant influence where N-alkylation increases in the sequence I < Br < Cl < OTs (**Figure** 74).

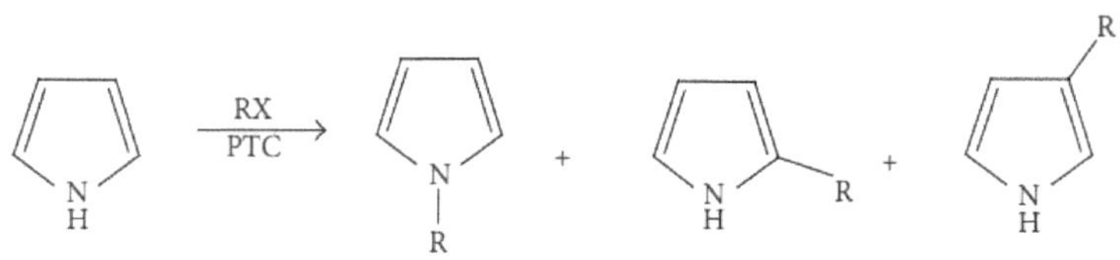

161 162 163 164

Figure 74

N-alkyl pyrroliden-2,5-diones 200 [66, 67] were synthesized via reaction of pyrroliden-2,5-diones with different alkylating reagents under phase transfer catalysis condition (toluene/K_2CO_3/$TBAHSO_4$) according to (**Figure** 75).

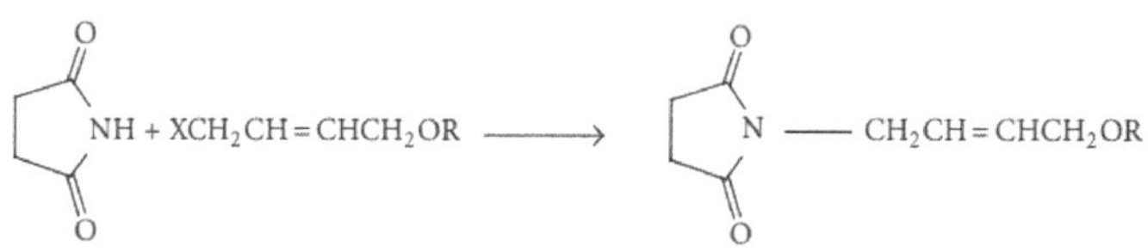

198 199 200

Figure 75

N-allylindoles 203 were easily carried out via N-allylation of the proper indoles 201 with the suitable allyl halides 202. The reaction was accomplished in diethyl ether via a phase transfer process in which 18-crown-6 was employed as the transfer agent and t-BuOK as the base [68] (Figure 76).

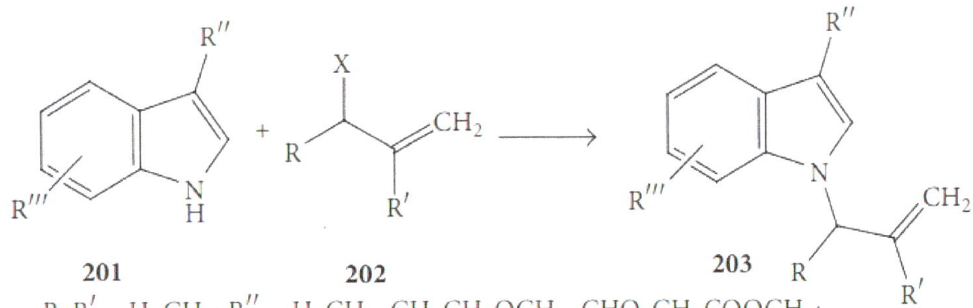

R, R' = H, CH_3; R'' = H, CH_3, $CH_2CH_2OCH_3$, CHO, CH_2COOCH_3;

R''' = H, 5-CH_3, 7-CH_3, X =

Figure 76

2-Substituted 1-allylpyrroles 206a–c were easily prepared from reaction of 2-formylpyrrole or 2-acetylpyrrole204 with either 3-chlorobut-1-ene 205a or 3-chloro-2-methylprop-1-ene 205b, respectively, under phase transfer conditions (toluene/NaOH/TBAHSO$_4$) [69] (**Figure** 77).

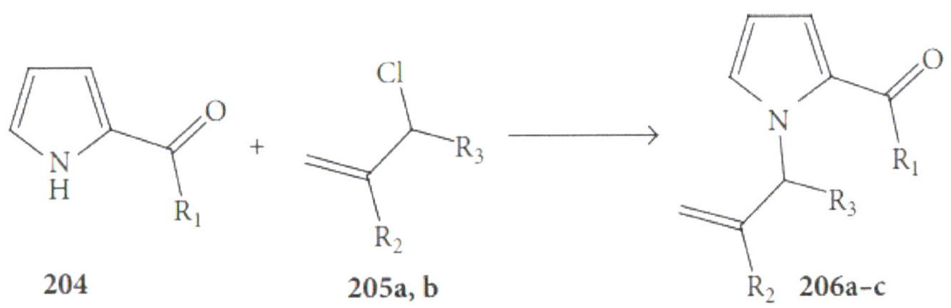

a; R_1 = H, R_2 = CH_3

b, **c**; R_1 = CH_3, R_2 = CH_3, R_3 = H

Figure 77

Treatment of pyrroles 204 with 1,2-dichloroethane in presence of 50% NaOH and TBAB as a catalyst yielded the corresponding N-chloroethyl-pyroles 207 in excellent yield which was transformed into 1,2-divinylpyrroles208 via its reaction with sodamide using solid-liquid technique (THF/NaNH$_2$/CH$_3$Ph$_3$Br) [70] (**Figure** 78).

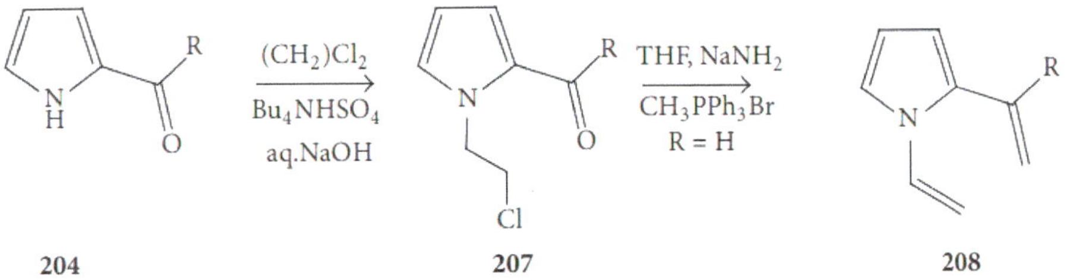

(a) R = H
(b) R = CH_3

Figure 78

Synthesis of a series of N-alkylpyrrolidino[59]fullerenes was achieved via the combination of PTC without solvent and microwave irradiation technique. Thus, 2-phenylpyrrolidinofullerene 209 was treated with benzyl,p-nitrobenzyl, p-methyloxy-carbonylbenzyl, n-octyl, or allyl bromides in microwave oven in presence of K_2CO_3 and TBAB to yield the corresponding N-alkylpyrrolidino[59]fullerenes 210 [71] (**Figure** 79).

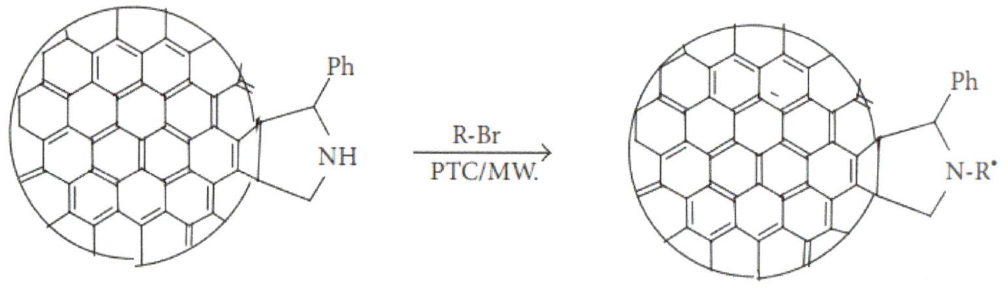

209 210

Figure 79

1,3-Bis[2-(aryl)indol-1-yl]propanes 213 and 1,3-bis[3-(aryl)-5-(aryl)pyrazol-1-yl]-propanes 214 were prepared from reaction of appropriate 2-arylindoles 211 and 4,5-dihydropyrazoles 212, respectively, with 1,3-dibromopropane, via liquid-liquid technique using TBAHS as a catalyst, benzene as the organic phase, and 50% KOH as an aq. phase [72] (**Figure** 80).

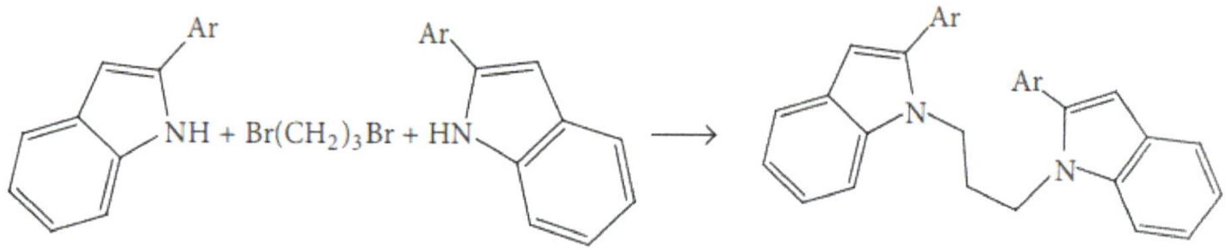

211 213

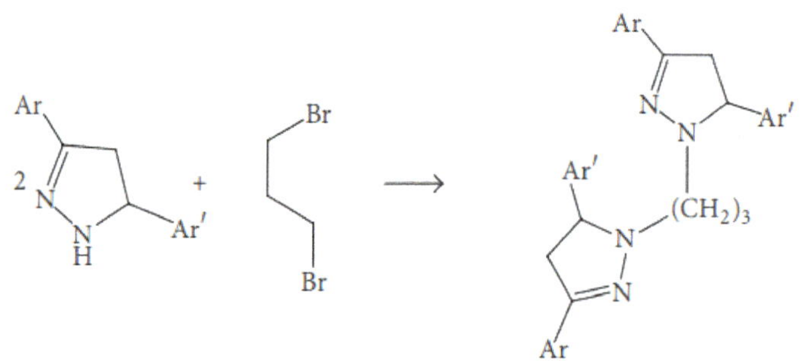

212 214

Figure 80

Indole-2-acetonitrile 215 was treated with $(CH_3)_2CO_3$ in a mixture of [DMF/TBAB/K_2CO_3] to afford 1-methylindole-2-acetonitrile 216a and 2-(1-methylindole-3-yl)propionitrile 216b [73] (**Figure** 81).

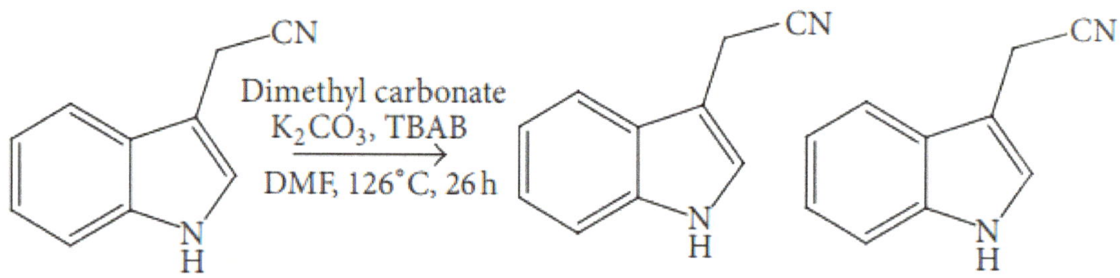

215 216a 216b

Figure 81

Alkylation of indole 217, 2,3-dimethylindole 218 with chlorodifluromethane under liquid-liquid technique (CH_2Cl_2/aq. NaOH/BzTEACl) afforded the corresponding 1-alkylated derivatives 219 and 220, respectively [74] (**Figure** 82).

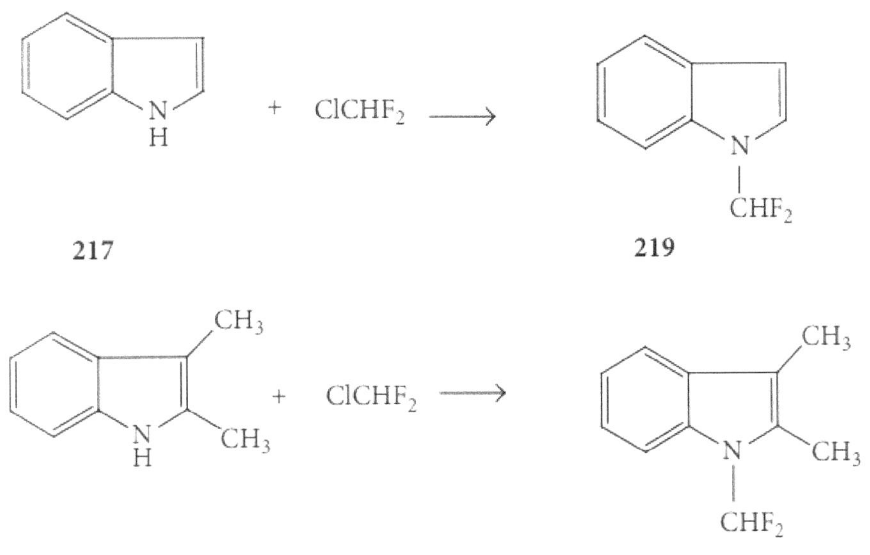

217 **219**

218 **220**

Figure 82

Barraja et al. heated 2-substituted-3-bromoindoles 221 with excess of different nucleophiles, solid KOH, and dibenzo-18-crown as a phase transfer catalyst to afford the corresponding 3-substituted indoles 222 in a satisfactory increase of yield [75] (**Figure** 83).

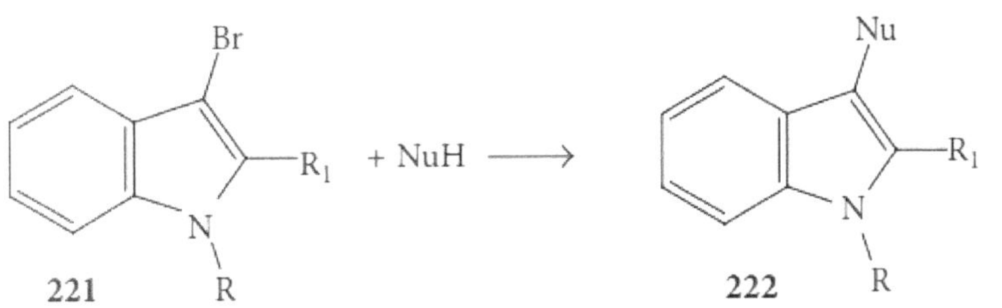

221 **222**

$R = H{\cdot}CH_3$; R_1 = Ph, 2-$NO_2C_6H_5$, 2-$NH_2C_6H_5$, CO_2Et

Nu = SAr, O-Ar, NR(Ar, cylic).

Figure 83

Carbazole 223 was alkylated with 1,6-dibromohexane in 1 : 1 molar ratio to give N-alkyl derivative 224 in a mixture of (benzene/NaOH/TBAB) [76] (**Figure** 84).

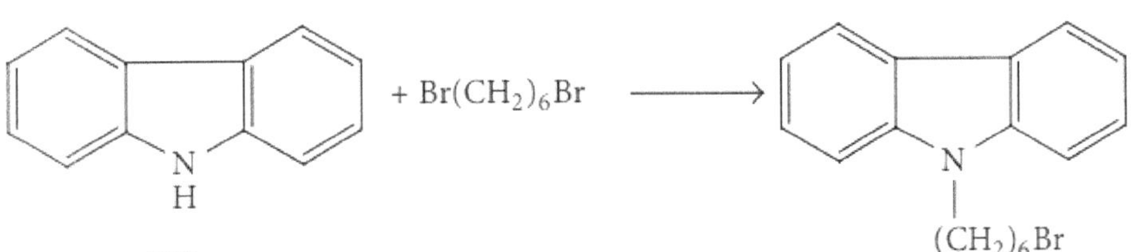

223 **224**

Figure 84

Alkylation of pyrazole 225 with cyclopentyl or cyclohexyl bromides without solvent with PTC system (KOH/TBAB) afforded the corresponding 1-cyclopentyl or 1-cyclohexylpyrazoles 226a, b, respectively. Likewise treatment of pyrazole with 1,2-dibromoethane in 1 : 1 or 2 : 1 molar ratio afforded 1-bromoethylpyrazole or 1,2 bispyrazolylethane 227, 228 [76](**Figure** 85). Also, alkylation reaction of 1H-pyrazole 225 with linear or branched alkyl halide in liquid-liquid PTC system gave 1-alkyl-1H-pyrazole 229[77] (**Figure** 85).

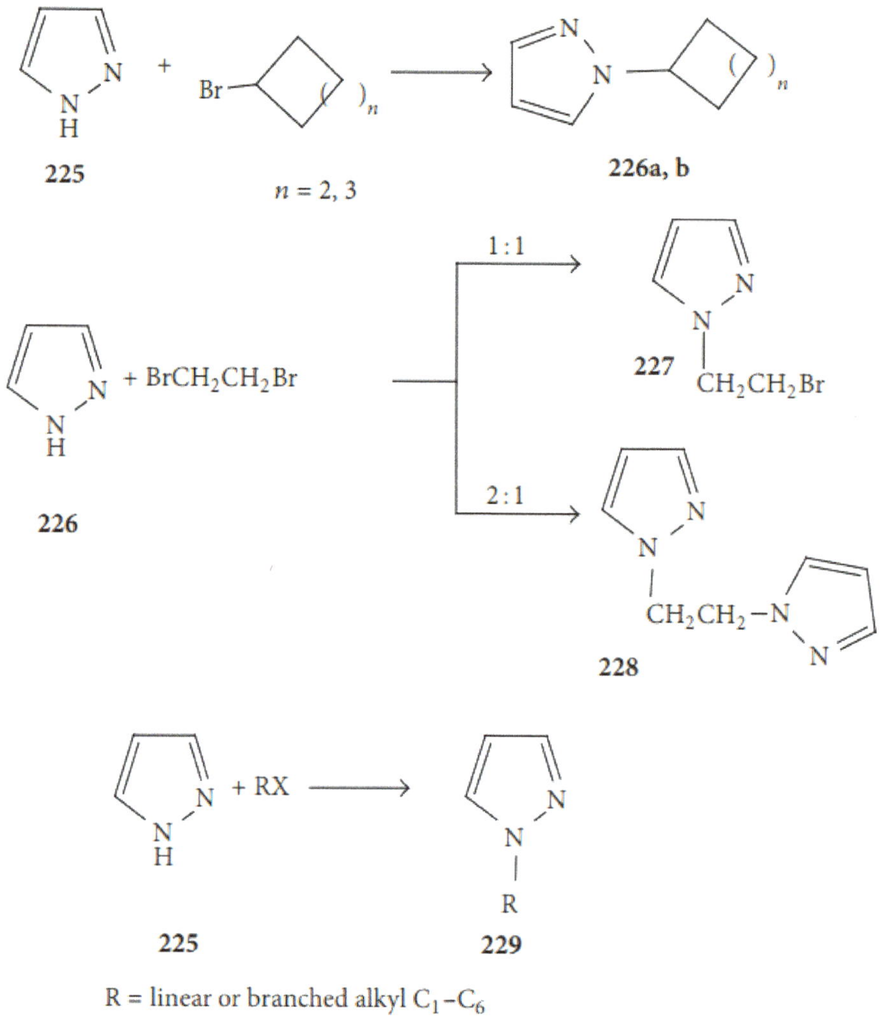

Figure 85

The reaction of 3-tert-butylpyrazole 230 with dibromomethane afforded bis(tert-butylpyrazol-1-yl)methane [CH_2(3-t-BuPz)$_2$] 231. However, the reaction of 3-isopropylpyrazole 232 with dibromomethane under the same condition yielded three isomers, namely, bis(5-isopropylpyrazol-1-yl)methane[CH_2(5-isoPrPz)$_2$] 233, (3-isopropylpyrazol-1-yl-5′-isopropylpyrazol-1′-yl)methane[CH_2(3-iPrPz-5′-isoPrPz)$_2$] 234, and bis(3-isopropylpyrazol-1-yl)methane[CH_2(3-iPrPz)$_2$] 235 [78] (**Figure** 86).

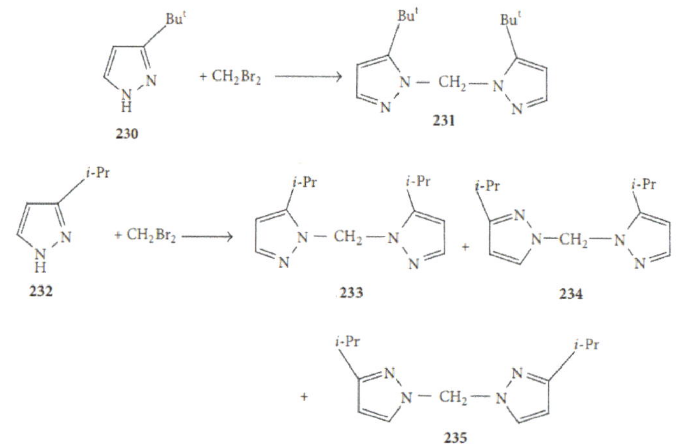

Figure 86

N-Substituted-3,5-diarylpyrazolines 237 and 238 were prepared via liquid-liquid system in ($CHCl_3$ or benzene/aq. KOH/TBAB) using alkyl and allyl halides, respectively, as alkylating agents for pyrazoles 236 [79] (**Figure** 87).

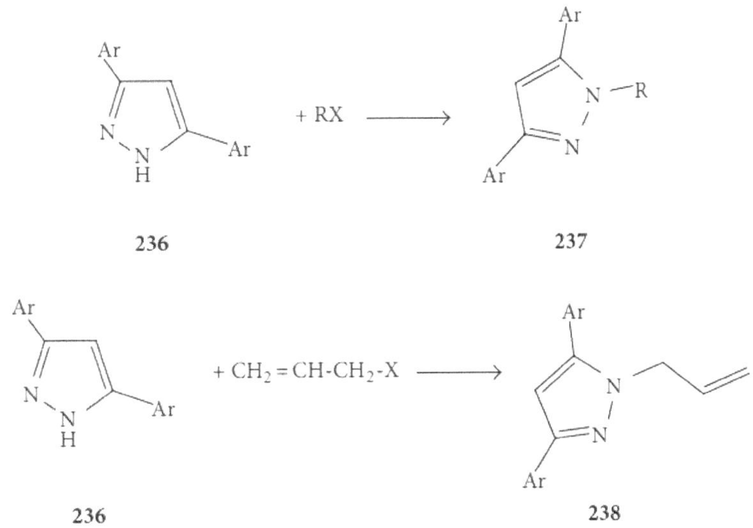

Figure 87

N-Alkylation reactions of 4,9-dihydro-9-methyl-4,10,10-trioxo-1(2H)-pyrazolo[3,4-c][2,1]-benzothiazepine239 with dimethyl sulphate, diethyl sulphate, benzyl bromide, benzyl chloride, phenacyl bromide, phenacyl chloride, or cyclohexyl bromide under the liquid-liquid PTC condition of (toluene/NaOH(25%)/BTEACl, TBAB, or $TBAHSO_4$) produced 1- and 2-alkylated isomers 240a, b. The ratios between these two isomers were calculated [80] (**Figure** 88).

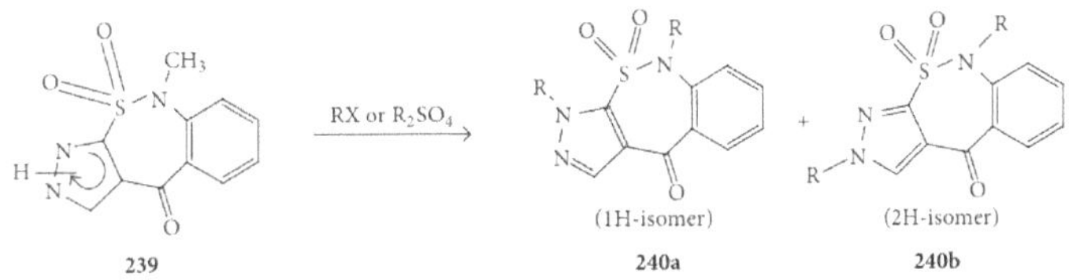

Figure 88

A number of 4-substituted pyrazolo[3,4-d]pyrimidines 241 have been reacted with (2-acetoxyethoxy)methyl bromide using liquid-liquid and solid-liquid techniques. The influences of the solvent and catalyst have been studied [81, 82] (**Figure** 89).

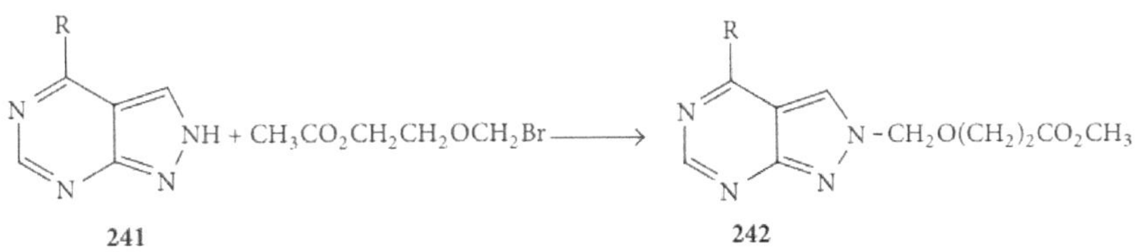

Figure 89

Imidazole 243 was alkylated with different alkyl halides under solid-liquid PTC in absence of solvent where the N-alkyl imidazoles 244a and imidazolium salts 244b were obtained [83] (**Figure** 90).

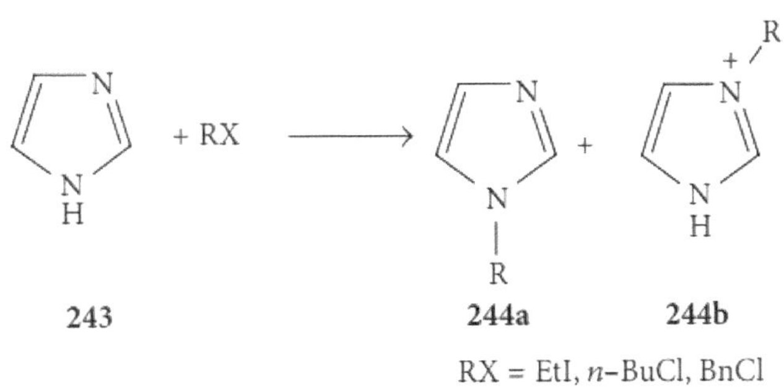

Figure 90

1-Alkyl(C_4–C_{14})imidazoles 245 were obtained from the alkylation of imidazole 243 with the appropriate n-bromoalkane via PTC technique (benzene/KOH/TBAI) [84] (**Figure** 91).

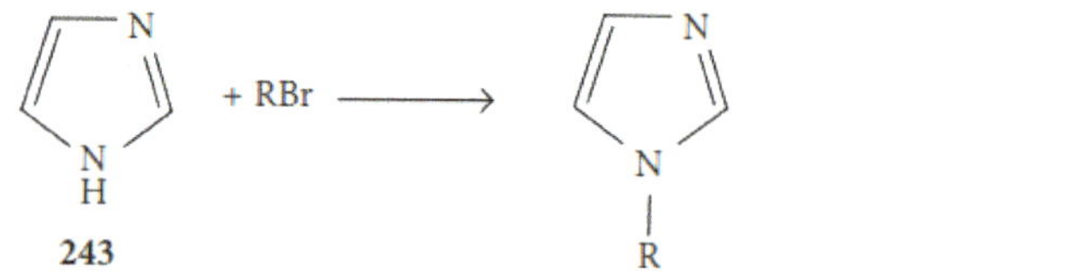

243

245 R = $CH_3(CH_2)_n$, n = 3–13

Figure 91

1-Alkyl-2-methyl-2-imidazolines 247 were obtained in good to excellent yields by alkylation of 2-methyl-2-imidazoline 246 with organic halides in the absence of solvent [85] (**Figure** 92).

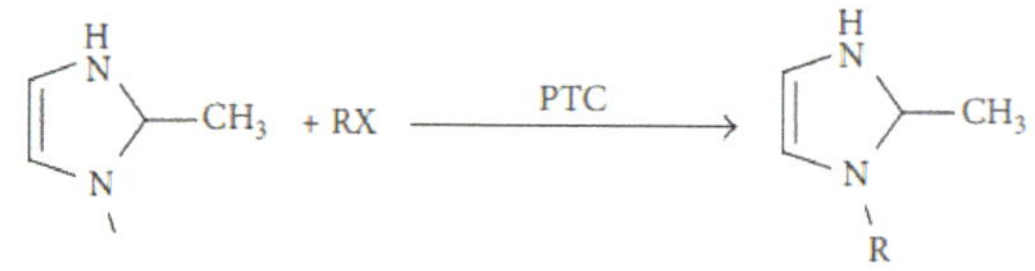

246

247

Figure 92

Using the PTC condition such as (CH_2Cl_2/KOH/TEBA), a mixture (1 : 1) of N-alkylated 5-nitroimidazoles250a, b or 6-nitrobenzimidazoles 251a, b was obtained by the reaction of 5-nitroimidazoles 248 or 6-nitrobenzimidazoles 249 with methyl iodide or benzyl chloride, respectively, in a satisfactory yield [86] (Figure93).

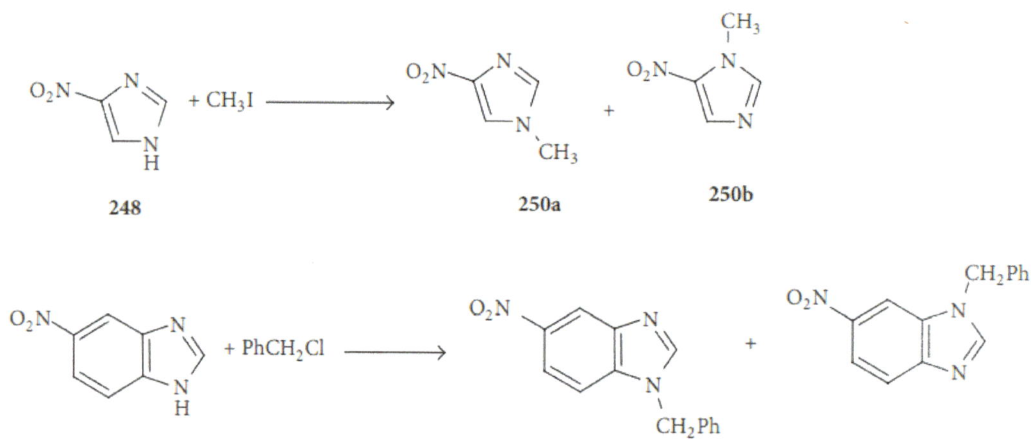

248 250a 250b

249 251a 251b

Figure 93

Different imidazole compounds such as 2-methylimidazole 252a and 2-methyl-4-nitro imidazole 252b were reacted with different alkyl bromides 253 in an alkaline media, at reflux temperature and in the presence of tetraethylammonium iodide (TEAI) or TBAB as PTC and afforded 1-alkyl imidazole derivatives 254 which exhibited a variety of valuable pharmacological properties [87] (**Figure** 94).

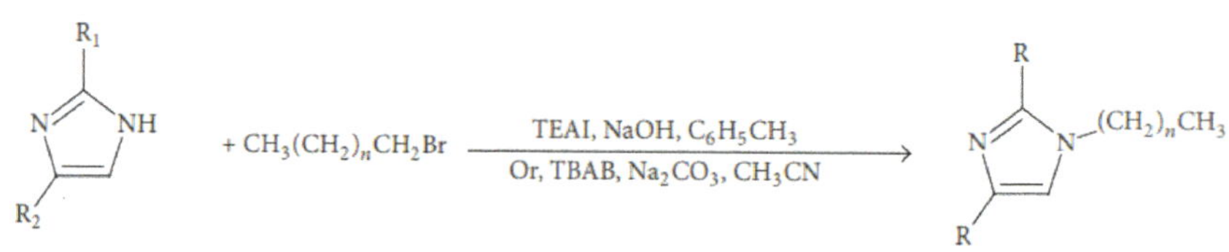

252a, b 253 254

a, R_1 = CH_3, R_2 = H
b, R_1 = CH_3, R_2 = NO_2

Figure 94

N-Alkyl-2-acetylbenzimidazoles 257 have been obtained in over 90% yield by alkylating 2-acetylbenzimidazole256 using the PTC mixture of (CH_3CN/K_2CO_3/TEBACl) at room temperature [88] (**Figure** 95).

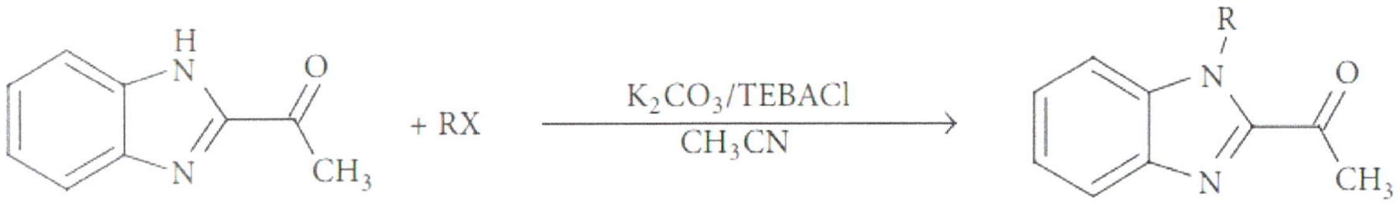

256 **257**

Figure 95

2-Phenylbenzimidazole 258 was alkylated with chlorodifluromethane under liquid-liquid condition (CH_2Cl_2/aq. NaOH/ BzTEACl) to give 1-difluoromethyl-2-phenylbenz-imidazole 259 [75] (Figure 96).

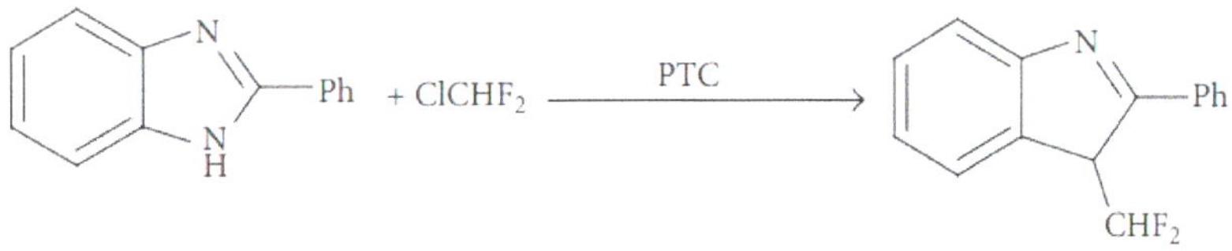

258 **259**

Figure 96

The N-alkylated imidazole derivatives 262 were achieved by treating compounds 260 with different alkyl bromide 261 under PTC condition (TBAB/K_2CO_3/CH_3CN) [89] (**Figure** 97).

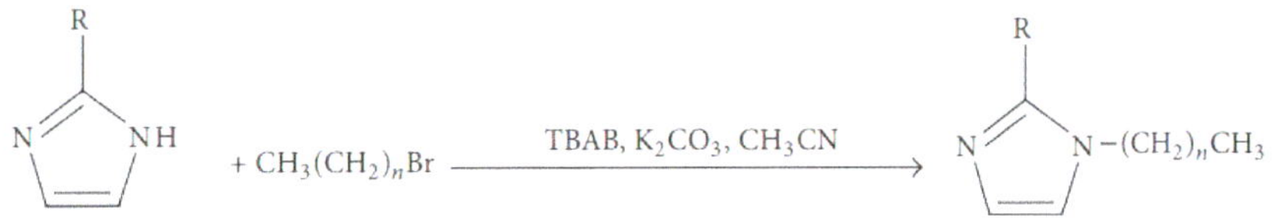

260 **261** **262**

Figure 97

Competitive N versus O benzylation of 5,5-dimethyl-3-isoxazolidinone 263 using (CH_2Cl_2/NaOH) using different PTC catalysts was studied. The ratio of N-/O-alkylations 264a/264b was 2/3 for most cases of catalysts [90] (**Figure** 98).

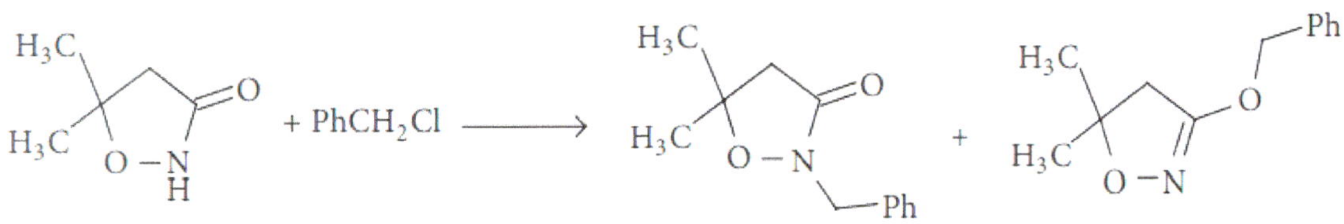

263 **264a** **264b**

Figure 98

Alkylation of 1,2,4-triazole 265 and benzotriazole 267 has been performed either in basic media under solvent free PTC conditions or in absence of base by conventional and microwave heating. Several parameters affecting the selectivity have been studied [91]. Arylation of 1H-1,2,3-benzotriazole 267 with activated aryl halides in a medium of aromatic hydrocarbons under PTC condition using inorganic bases and acetyltrimethylammonium bromide as a phase transfer catalyst was studied. Both N(1)- and N(2)-arylated products 269a and 269b, respectively, have been isolated. Their ratio has depended on the nature of the employed base and the reactivity of arylating agent [92] (**Figure** 99).

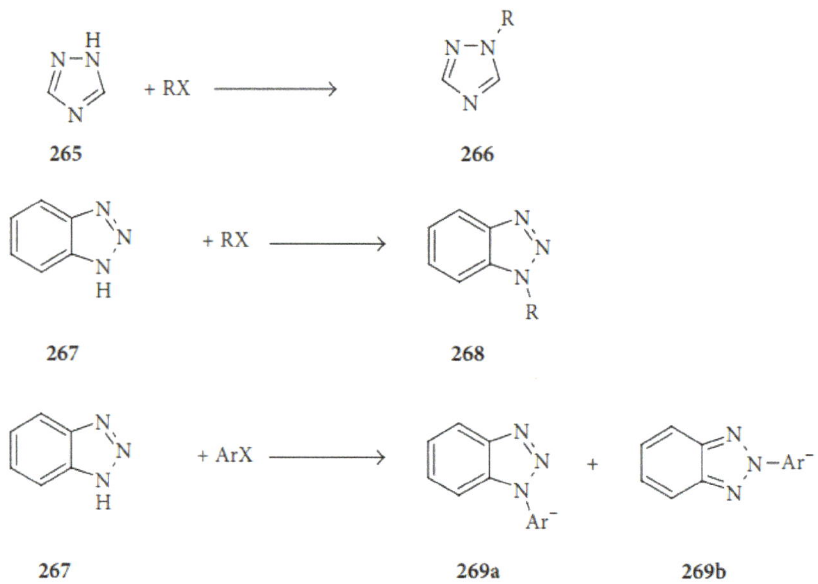

Figure 99

Treatment of benzotriazole 267 with chlorodifluromethane under liquid-liquid conditions (CH_2Cl_2/aq. NaOH/BzTEACl) has afforded 1-(difluoromethyl)-1H-benzotriazole 270 [74] (**Figure** 100).

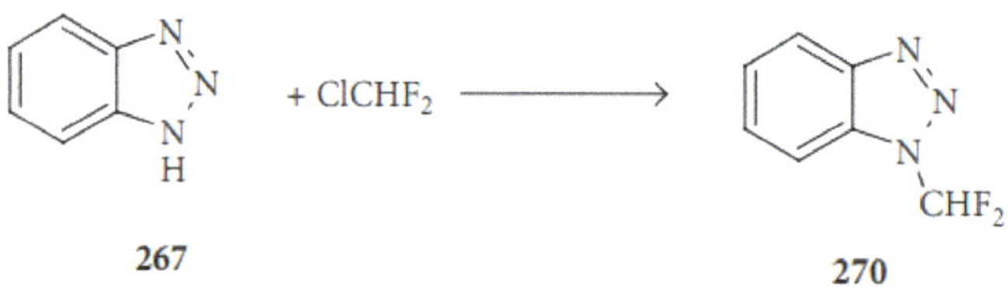

Figure 100

Reaction of 4-substituted-1,2,3-triazol 271 with alkyl bromide 272 under basic condition using PTC Bu_4NBr produced N-substituted 1,2,3-triazole derivatives 273a–c [93] (Figure 101).

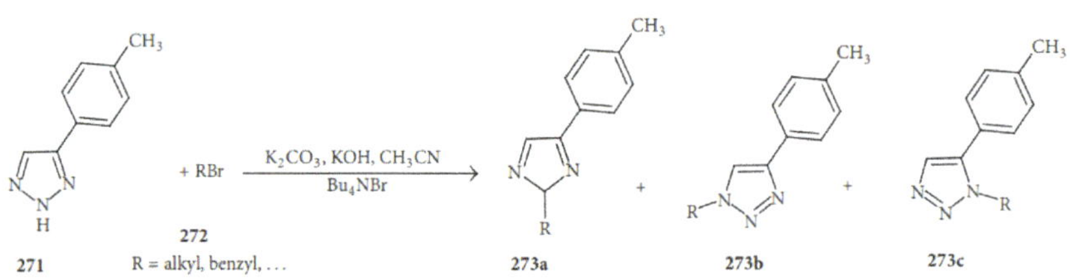

Figure 101

The alkylation of 5-benzyl-1H-tetrazole 274 under liquid- liquid technique such as (CH_2Cl_2/aq. NaOH/BzTEACl) gave a mixture of 5-benzyl-1-difluoromethyl-1,2,3,5-tetrazole 275a in 23% and 4-benzyl-1-difluoromethyl-1,2,3,5-tetrazole 275b in 15% [74] (Figure 102).

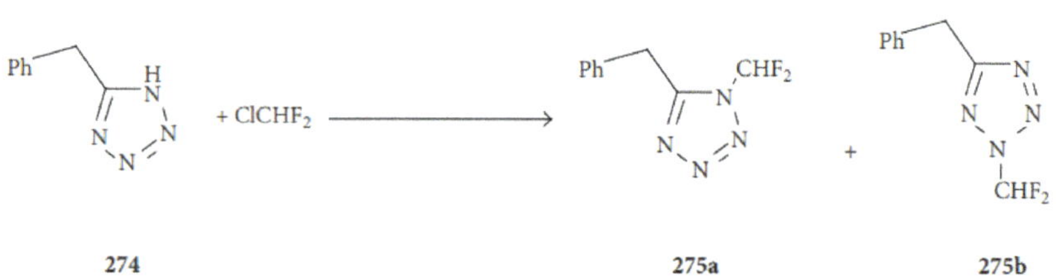

Figure 102

A series of 5-alkylthio-1-aryltetrazoles 277 were prepared by alkylation reaction of the corresponding 1-aryltetrazole-5-thiols 276 with alkyl bromides under solid-liquid PTC technique [94], whereas without solvent and also under PTC technique several N-p-nitrophenylazoles 279–288 were synthesized by direct arylation of the corresponding azolo compound 278 with p-fluoronitrobenzene [95], while the alkylation reaction of 1-aryltetrazol-5-ones 289 with dibromoalkanes under both liquid-liquid and solid-liquid PTC systems provided a convenient route to prepare bis-tetrazolon derivatives 290 and 4-bromoalkyl derivatives 291 [96] (Figure 103).

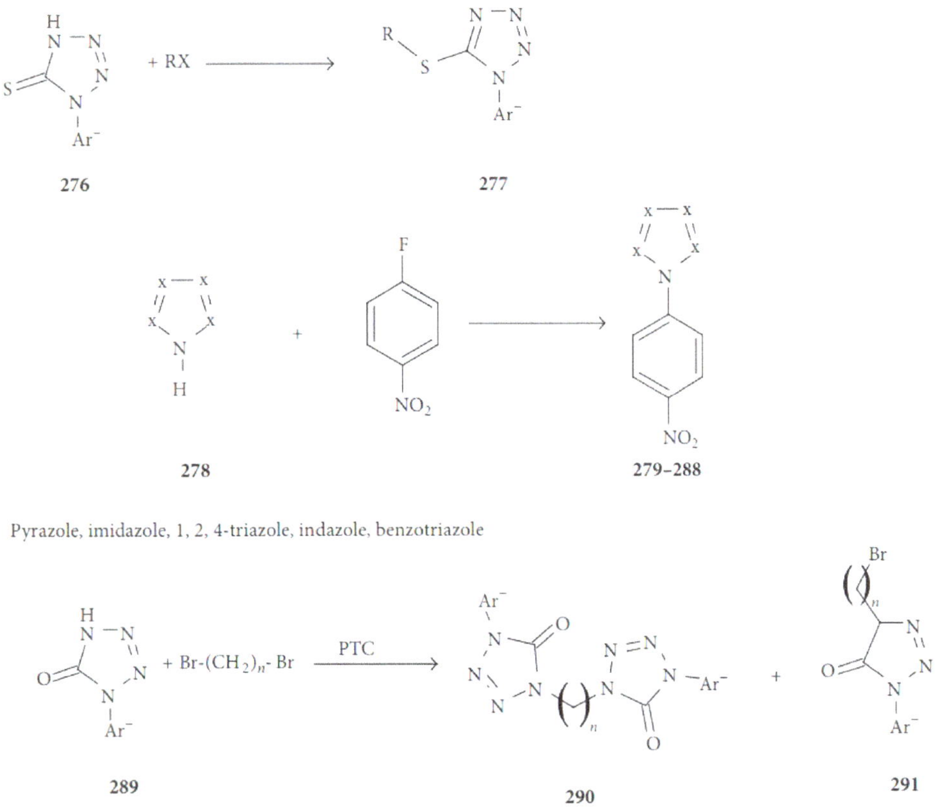

Figure 103

N-Acetylated adenosine 292 was alkylated with benzyl bromide, methyl iodide, or allyl bromide in the presence of tetrabutylammonium bromide as a PTC catalyst in CH_2Cl_2/NaOH to give the corresponding N-alkyl-6-N-acetyl adenosines 293a, b [97] (Figure 104).

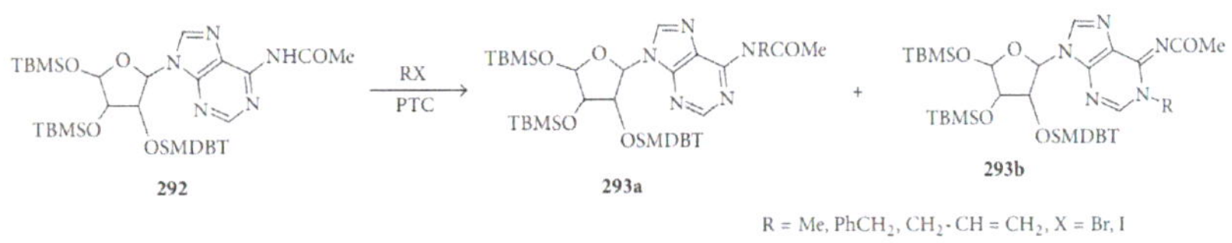

Figure 104

N-alkylation reaction of adenine 294 took place with different alkyl halides under the phase transfer conditions using microwave irradiation assistance [98] (Figure 105).

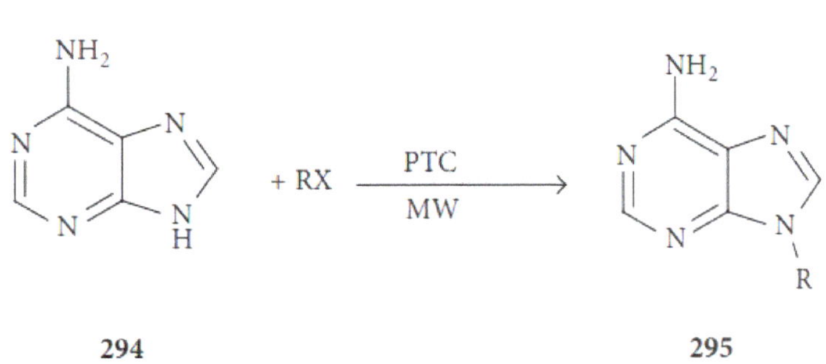

Figure 105

5-Aryloxymethyl-2-[(2-chlorophenyl)oxyacetylamido]-1,3,4-thiadiazoles 280 are synthesized from reaction of 2-chlorophenyloxyacetyl chloride 297 with 5-aryloxymethyl-2-amino-1,3,4-thiadiazoles 296 under liquid-liquid condition using PEG-400 as a catalyst [99] (**Figure** 106).

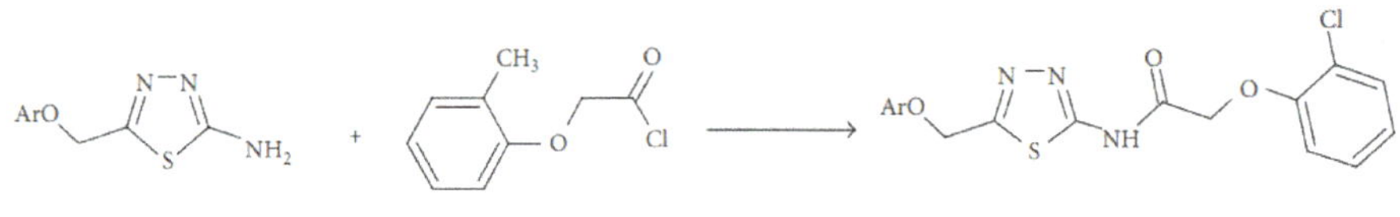

296 297 298

Figure 106

N-Alkylation of 3-phenyl-2-thiohydantoin-5-arylidene derivative 299 has been carried out by using monohalocompounds such as allyl bromide as an alkylating agent. The reaction was proceeded through nucleophilic displacement under PTC conditions of solid-liquid phases such as a mixture of anhydrous potassium carbonate, dioxane, and TBAB as a heterogeneous catalyst. The corresponding alkylated product 300was obtained in a good yield [42] (**Figure** 107).

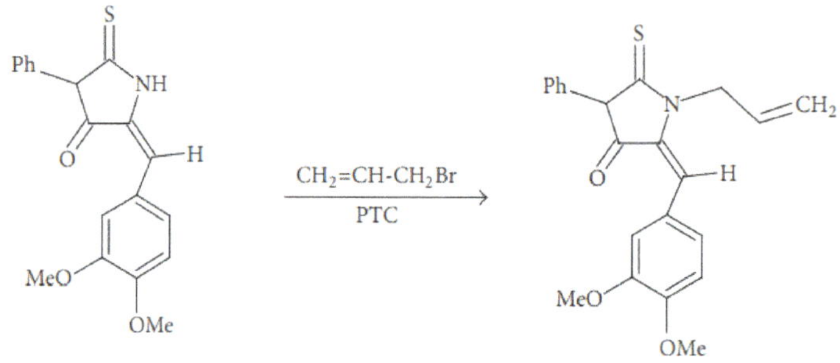

299 300

Figure 107

Alkylation reaction of pyrazoles 301–303 with 2-chloroethylamine under the condition of phase-transfer catalysis was carried out in a liquid-solid system using benzene as an organic solvent and BTEAC as a catalyst. Reaction between equimolar amounts of reactants led to poor yields (20–40%) of the alkylated products 304–306 due to concurrent dehydrochlorination of 2-chloroethylamine. The yield was considerably decreased in going from pyrazole (301) to 3,5 dimethylpyrazole (303) whereas the maximal yield of alkylated products (70–80%) was obtained by using one of the following ratios: pyrazole : NaOH : 2-chloroethylamine 1 : 2 : 2, 3-(5)methylpyrazole : NaOH : 2 chloroethylamine 1 : 3 : 3, or 3,5-dimethylpyrazole : NaOH : 2-chloroethylamine 1 : 5 : 5 [100] (**Figure** 108).

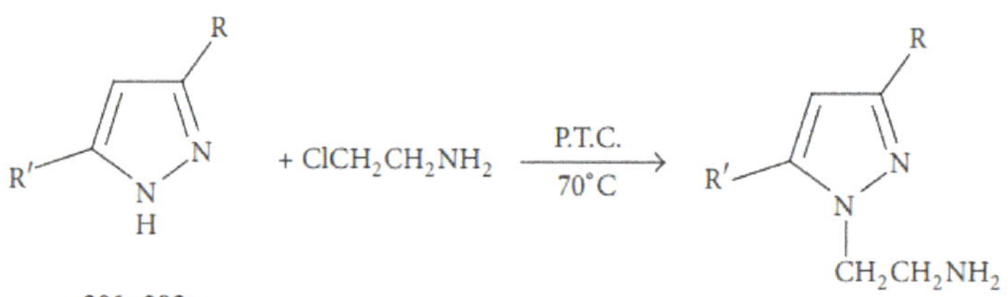

301-303 304-306

301, 304 R = R′ = H; 302, 305 R = CH_3, R′ = H; (a), R′ = CH_3, R = H; (b) 303, 306 R = R′ = CH_3

Figure 108

In the N-alkylation reaction of pyrazole, 3-(5)-methylpyrazole, and 3,5-dimethylpyrazole 301–303 with dichloroethane (DCE), the dehydrochlorination of the obtained 1-(β-chloroethyl)pyrazoles has been carried out to give the corresponding products N-vinylpyrazoles 307–309 (**Figure** 9) in low yield. Attempts to carry out the reaction under standard conditions (water/benzene/NaOH/TEBAC) did not lead to the desired result. The yield of all products was sharply increased when benzene was replaced with an excess of dichloroethane. The reaction investigation showed that the ease of alkylation depends strongly on the basicity of the pyrazole. The introduction of an electron-donating substituent (e.g., Me) into the molecule of pyrazole 301 increases the electron density at the "pyrazole" nitrogen atoms. As a result, deprotonation was hindered and the base was consumed in elimination of dichloroethane. It must also be mentioned that a 5- to 7-folds excess of dichloroethane was necessary to obtain optimal yields on alkylation of compounds 301–303 [101] (**Figure** 109).

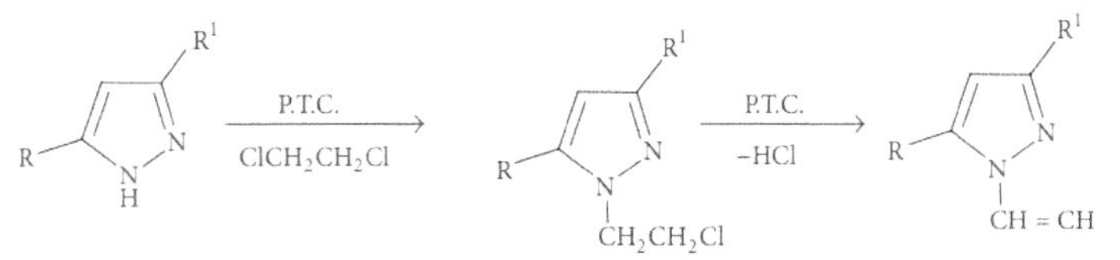

301–303 307–309

301, 304, 307 R= R_1 = H; 302, 305, 308 R = H, R_1 = Me, R = Me, R_1 = H; 303, 306, 309. R, R_1 = Me

Figure 109

O-Alkylation or Acylation

2,5-Didodecyloxymethylfuran 311 was obtained via alkylation of 2,5-dihydroxymethylfuran 310 with dodecyl bromide without solvent using (KOH/aliquat) as a catalyst, while alkylation of furfuryl alcohol 312 with 1,12-dibromododecane under the same PTC conditions produced the corresponding furanic diether 313 [102] (**Figure** 110).

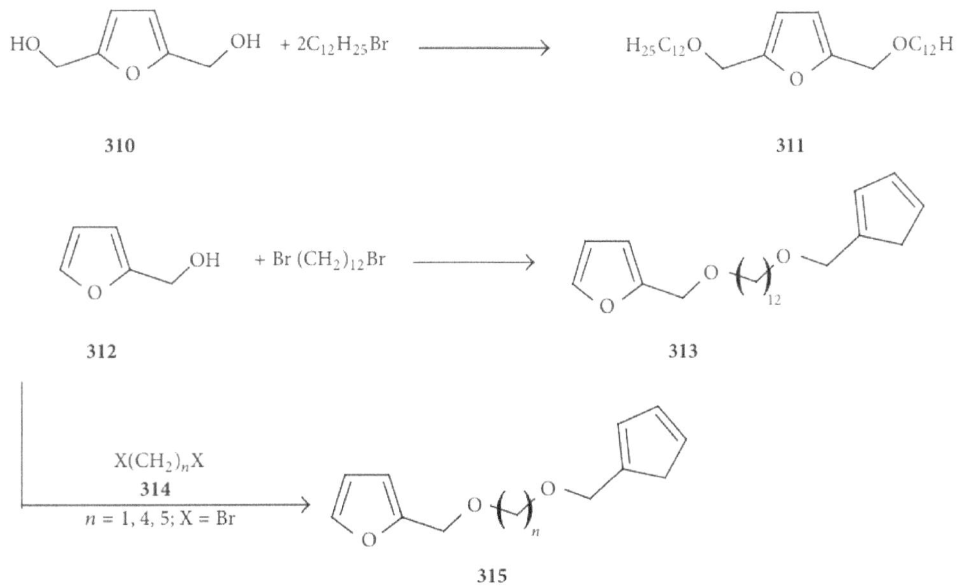

Figure 110

Similarly, 2-furfuryl alcohol 312 was subjected to react with alkyl dihalides 314 using (benzene/KOH/polyethylene glycol (PEG-400)) to yield the corresponding difurfuryl diethers 315 [102] (**Figure** 110).

4-(benzyloxy)-2-[(2E)-but-2-en-1-yloxy]-5-methoxytetrahydrofuran-3-ol 317 was obtained from benzylation of 2-O-methyl-5-O-crotyl-,-D-ribofuranoside 316 using benzyl bromide in aq. solution of NaOH/$CH_2Cl_2$and TBAB as a catalyst [103] (**Figure** 111).

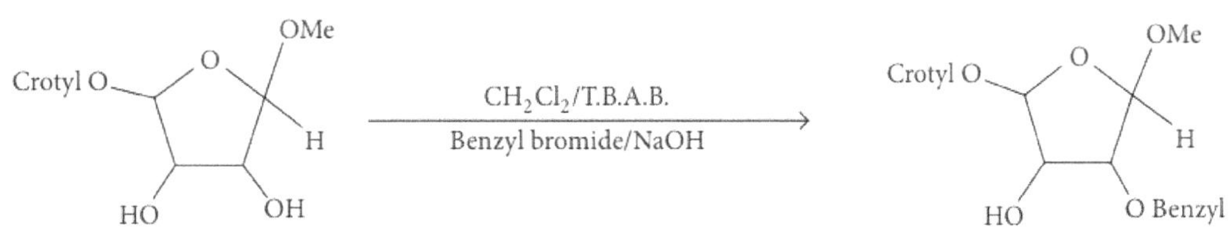

316 317

Figure 111

S-Alkylation

PTC alkylation of arylidene derivative 318 using dihalocompounds such as ethylene dibromide or 1,3-dibromopropane afforded the thioalkylated dimers 319a, b, respectively, via s-alkylation pathway followed by dimerization reaction [42] (**Figure** 112).

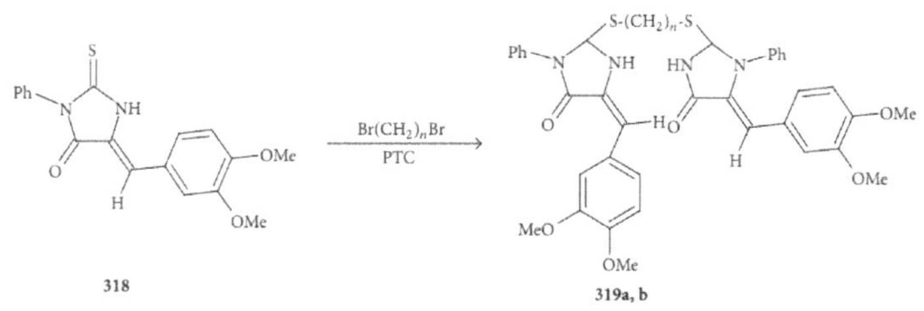

Figure 112

C-Alkylation

Reaction of 2-thiono-N(m-tolyl)thiazolidin-4-one 320 with methyl bromide, ethyl bromide, chloroacetaldehyde, ethyl chloroacetate, diethyl malonate, acetyl chloride, ethyl chloroformate, or p-nitrobenzenediazonium tetrafluoroborate under solid-liquid system and using the heterogeneous mixture of (dioxan/K_2CO_3/TBAB) as a PTC yielded the corresponding 5-substituted thiazolidin-4-ones 321a–h and yielded the corresponding spiro thiazolidin-4-one derivatives 322 when reacted with 1,2-dibromoethane under the same PTC condition [104] (**Figure** 113).

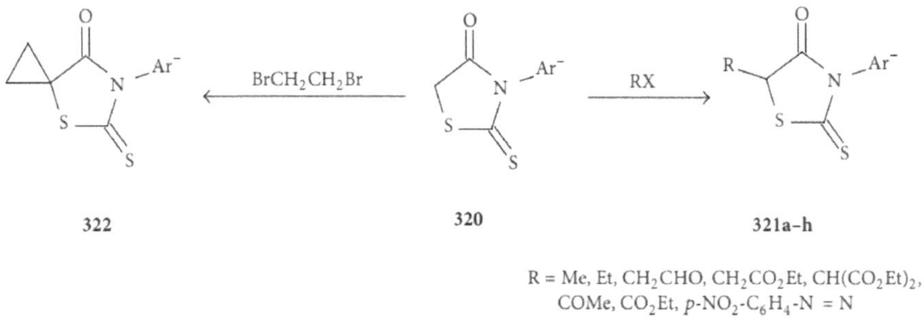

Figure 113

4-Benzyl-2-phenyl-2-oxazoline-4-carboxylic acid tertbutyl ester 324 was prepared via the alkylation of 2-phenyl-2-oxazoline-4-carboxylic acid tert-butyl ester 323 with benzyl bromide using (toluene/KOH(s)/Cinchona alkaloids) as a PTC [105] (**Figure** 114).

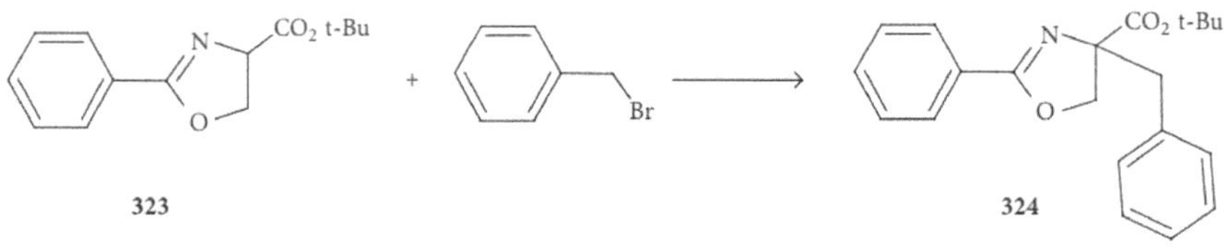

Figure 114

2-Thioxo-3,5,7-trisubstituted-1-(2,3,5-tri-O-benzoyl-β-D-ribofuranosyl)pyrido[2,3-d]-pyrimidin-4(1H)-ones327 have been prepared via phase transfer ribosylation of 2-thioxo-3,5,7-trisubstituted pyrido[2,3-d]pyrimidin-4(1H)-ones 325 with 2,3,5-tri-O-benzoyl-β-D-ribofuranosyl bromide 326 in biphase solvents such as CH_2Cl_2/50% aq. NaOH using TBAB as a catalyst [106] (**Figure** 115).

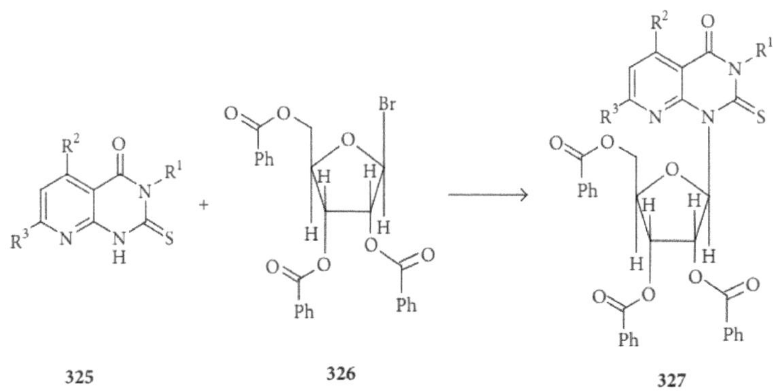

Figure 115

In the presence of dioxane as a solvent under PTC conditions, one-pot reaction of 3-phenyl-2-thiohydantoin 91with CS_2 and halocompounds such as ethyl bromide or benzyl chloride afforded 5-[bis(ethylsulfanyl)methylidene]-2-(ethylsulfanyl)-3-phenyl-3,5-dihydro-4H-imidazol-4-one 328a and 5-[bis(benzylsulfanyl)methylidene]-2-(benzylsulfanyl)-3-phenyl-3,5-dihydro-4H-imidazol-4-one 328b, respectively [42] (**Figure** 116).

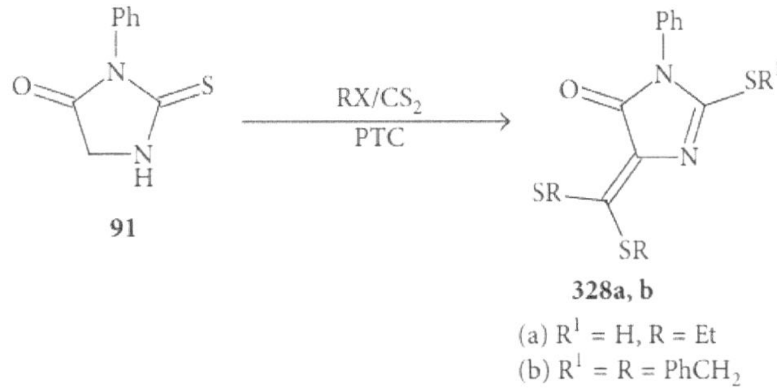

Figure 116

Acylation of imidazolones 318 using chloroacetyl chloride under PTC conditions yielded the corresponding acylated arylidene derivative 329 [42] (**Figure** 117).

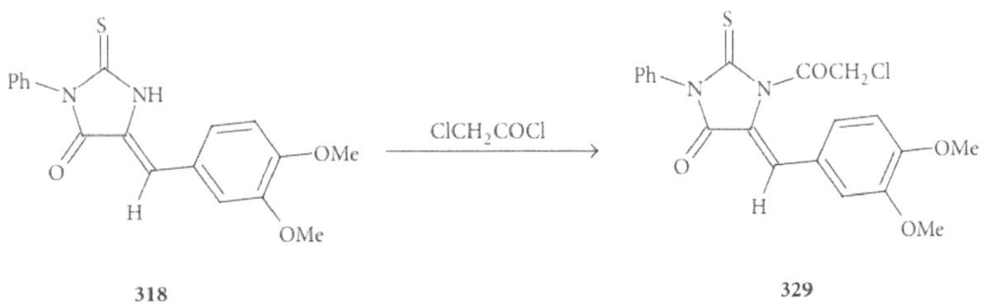

Figure 117

Miscellaneous Alkylation

5-Bromorhodanine derivative 330 was effectively used as an alkylating agent for some mono- or dianionic moieties containing S, N, or O under solid-liquid phase transfer catalysis conditions. Thus, with 4-hydroxy-2-mercaptopyrimidine, where possible S, N, or O sites are available, the reaction afforded the S-substituted product. With 2-aminobenzothiazole or piperazine, the reaction yielded the corresponding N-substituted 331a,b or the disubstituted products 332 [104] (**Figure** 118).

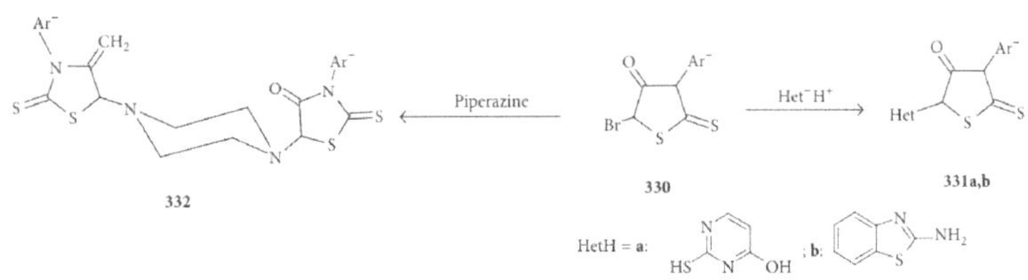

Figure 118

In a plan to study C- versus O-alkylation of the pyrazoline derivatives, 3-methyl-1-phenyl-2-pyrazolin-5-one333 was treated under PTC condition with different bromoorganic compounds such as benzyl bromide, 1,3-dibromopropane, methyl bromoacetate, bromoacetaldehyde, or diethylacetal as alkylating agents either in the absence or the presence of carbon disulphide and afforded numerous of C- or O-alkylated derivatives 334a–e,335a–e [107] (**Figure** 119).

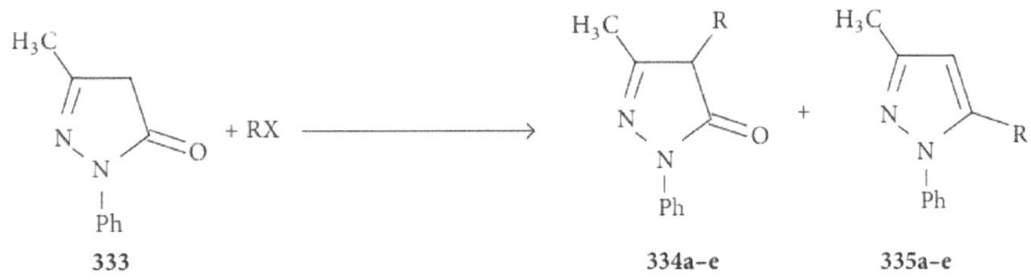

Figure 119

Reaction of 6-hydroxy-3,4-dihydroquinoline 337 and 1-cyclohexyl-5-(4-bromo-butyl)-1,2,3,4-tetrazole 336 in a heterogeneous mixture of toluene/H_2O, K_2CO_3 and TBACl as a catalyst gave 6-[4-(1-cyclohexyl-1,2,3,4-tetrazol-5-yl)butoxyl]-3,4-dihydrocarbostyril 338 in 99% purity [108] (**Figure** 120).

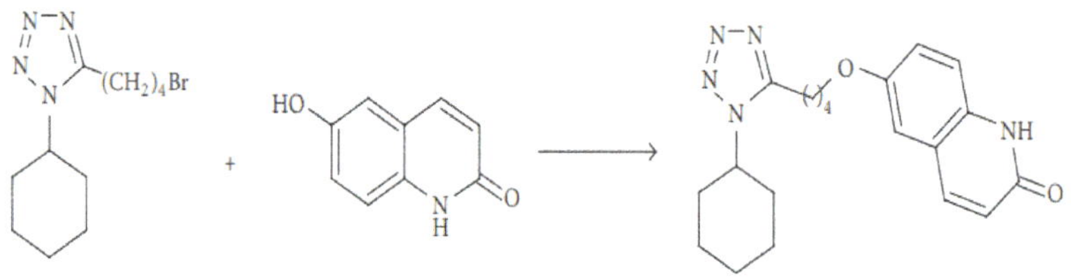

336 337 338

Figure 120

Sulphonylation

4,5-Dihydro-3,5-diaryl-1-(4-fluorophenylsulphonyl)pyrazoles 341 and 2-aryl-1-(4-fluorophenylsulphonyl)indoles 343 were prepared from reaction of 4-fluorophenylsulphonyl chloride 339 with the appropriate 4,5-dihydro-3,5-diarylpyrazoles 340 or 2-arylindoles 342, via solid-liquid phase (THF/KOH/TBAB) [107] (**Figure** 121).

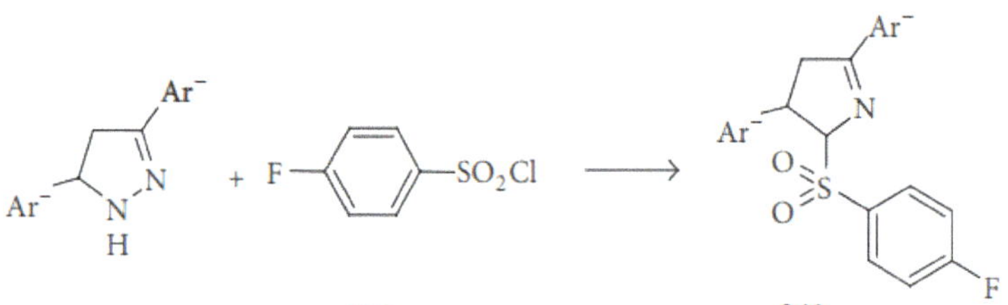

340 **339** **341**

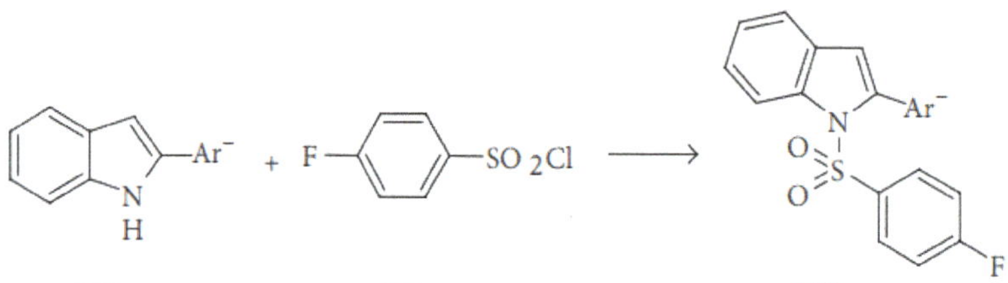

342 **339** **343**

Figure 121

N-Sulphonylation product 346 was prepared through treating 2-chloro-6,6-difluro-1,3-dioxolo[4,5-f]benzimidazole 344 with 3,5-dimethylisoxazole-4-sulphonyl chloride 345 via PTC protocol (toluene/aq. K_2CO_3/TBAB) [109] (Figure 122).

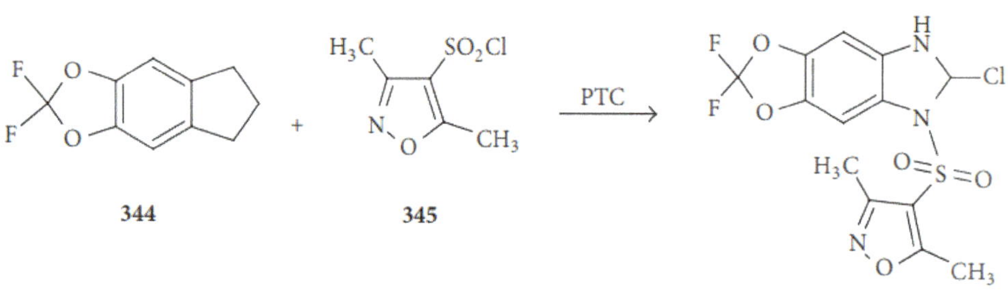

344 **345** **346**

Figure 122

N-Carbonylation

Carbonyldiimidazole 349 was prepared in 87% yield via reaction of imidazole 347 with phosgene 348 under PTC condition (chlorobenzene/NaOH/tributylhexadecylphosphonium bromide (TBHDPB)) [110] (Figure123).

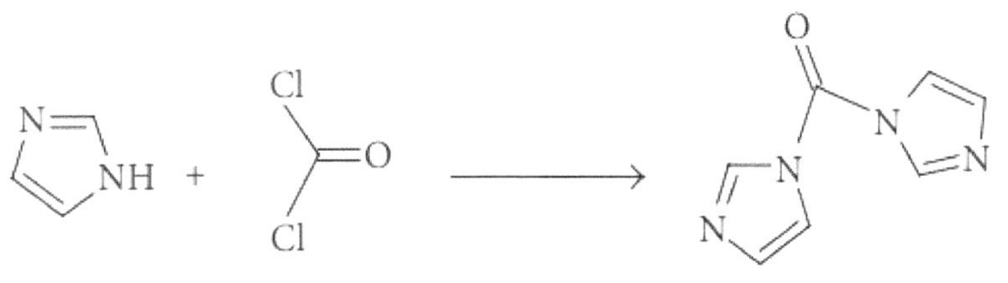

347 348 349

Figure 123

N-Phosphorylation

Reaction of 1,3-thiazolidin-2-one 350 with S-sec-Bu-O-Et chlorophosphorothiolate 351 in the presence of NaOH and N-dodecyl-N-methylephedrinium bromide gave s-sec-Bu O-Et(2-oxo-3-thiazolidinyl) phosphothiolate 352 in 84% yield [111] (**Figure** 124).

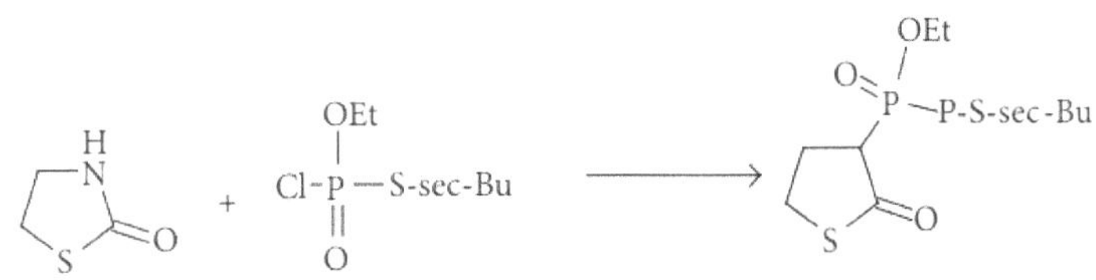

350 351 352

Figure 124

Elimination Reactions

β-Elimination of bromomethyl cyclic ketyl acetals 353 was carried out under solid-liquid phase (THF//Aliquat 336) to prepare the corresponding cyclic ketene acetals 354 [112] (**Figure** 125).

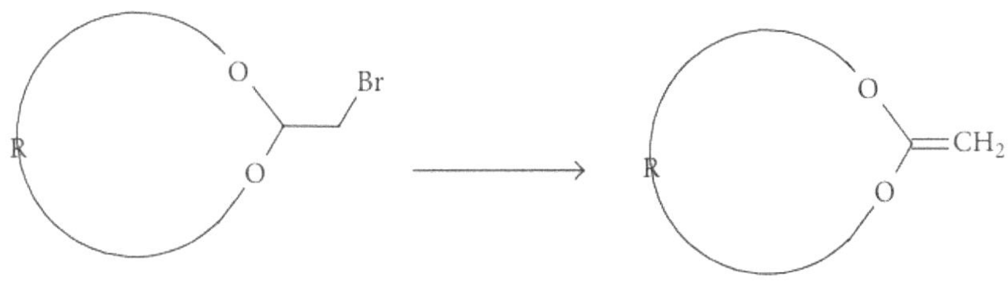

353 354

Figure 125

Reaction of 6-chloro-2-chloromethyl-3-nitroimidazo[1,2-b]pyridazine 355 with 3 equivalents of nitroalkane anion derivative 356 under liquid-liquid PTC protocol (CH_2Cl_2/H_2O/TBA(OH)) gave 3-nitroimidazo[1,2-b]pyridazines 357 bearing trisubstituted ethylenic double bond derivatives in a good yield [113] (**Figure** 126).

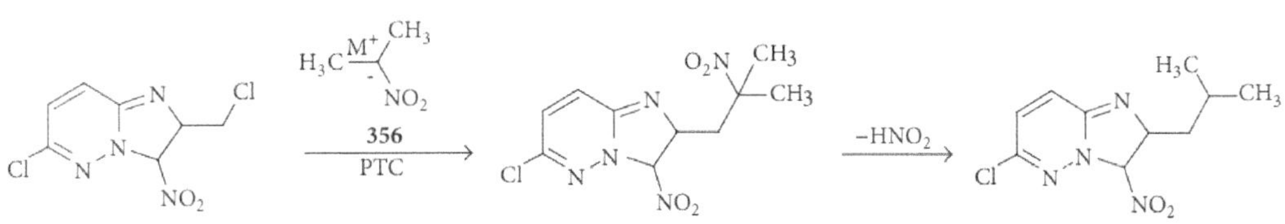

355 357

Figure 126

Vinylation of indole 358 with $ClCH_2CH_2Br$ was achieved under PTC condition (toluene/KOH/18-crown-6) and afforded the corresponding N-vinyl derivative 359. Similarly, under the same reaction conditions N,S-divinyl derivatives 362 and 363 were prepared from 3-mercaptoindole 360 and 2-mercaptobenzimidazole 361, respectively [109] (**Figure** 127).

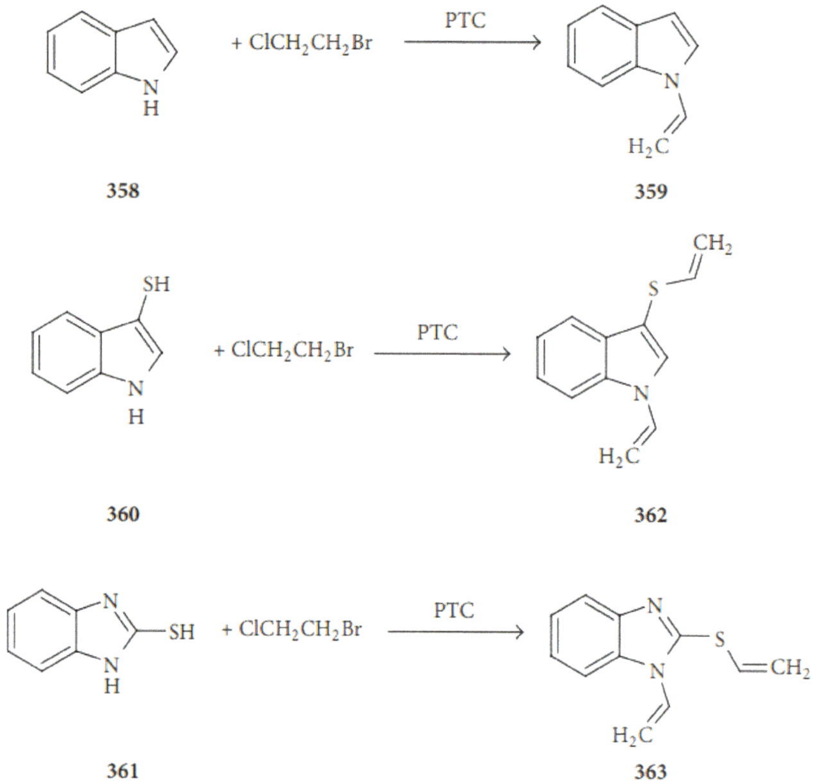

Figure 127

Condensation Reactions

Arylidenethiazolidin-4-ones 385 were synthesized from condensation reaction of 2-thiono-N-(m-tolyl)thiazolidin-4-one 384 with aromatic or heteroaldehydes under solid-liquid PTC technique [104] (Figure128).

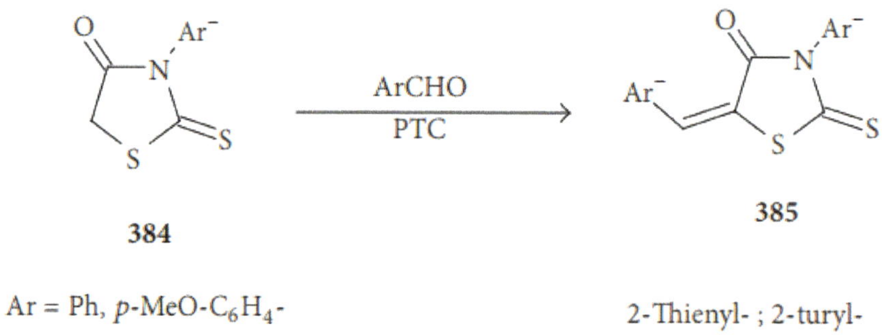

Figure 128

Addition Reactions

Reaction of 2-thioxo-N-(m-tolyl)thiazolidin-4-one 386 with formaldehyde or some arylidenemalononitriles via PTC technique (dioxane/K_2CO_3/TBAB) afforded the corresponding 5-hydroxymethyl derivative 387 or Michael type adduct 388, respectively. While the addition reaction of cyanoacetamide to 5-(p-methoxybenzylidene)-2-thioxo-N-(m-tolyl)thiazolidin-4-one 389 followed by condensation reaction under the same technique gave 7-anisyl-6-cyano-2-thioxo-3-(m-tolyl)thiazolo[4,5-b] pyridin-5-one 390 [104] (Figure129).

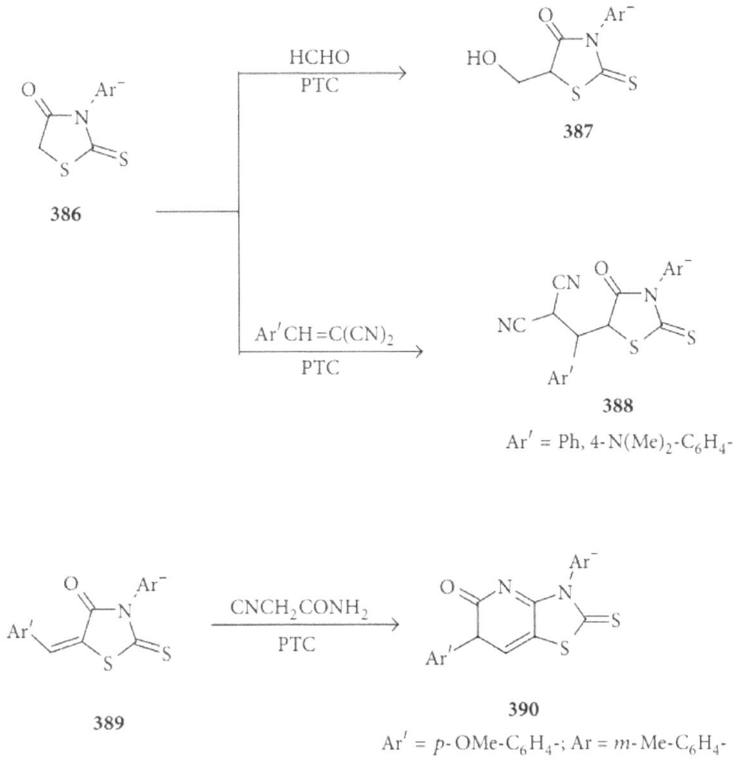

Figure 129

Ring Expansion

Treatment of 5-alkyl(aryl)-3H-pyrrolin-2-ones 391 with dichlorocarbene using ($CHCl_3$/NaOH/BTEAB) at 20–30°C afforded the intermediates 1-alkyl-6,6-dichloro-2-azabicyclo[3.1.0]hexan-3-ones which directly rearranged and dehydrochlorinated to give 6-alkyl(aryl)-1-phenyl-5-oxohydropyridin-2-ones 392 [110] (Figure130).

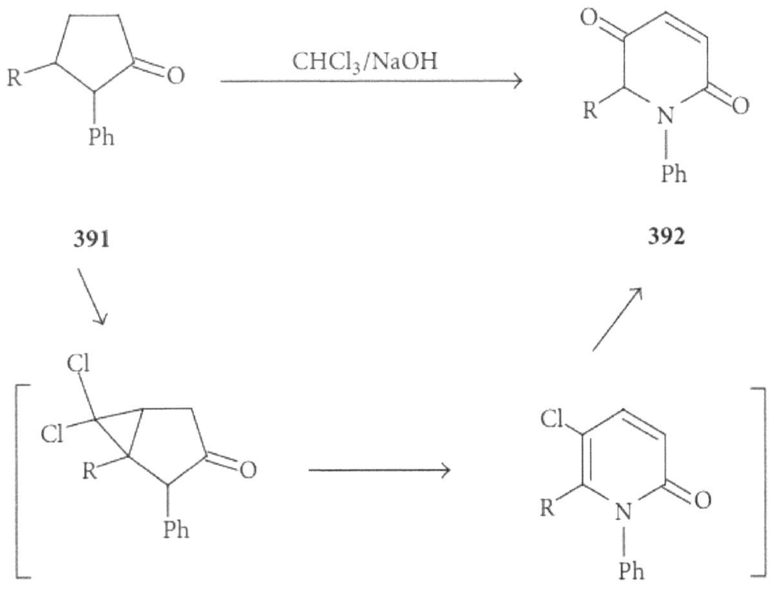

Figure 130

Also, ring transformation of indole derivatives 393 has been induced by dichlorocarbene (aq. KOH/$TBABSO_4H$) to produce 3-haloquinoline 394 in a good yield. Also, under similar conditions, reaction of 3-formyl-2-(3-chloro-4-fluorophenyl) indole 395 with dichlorocarbene gave 3-chloro-4-[2,2-dichloro-3-oxiranyl]-2-(3′-chloro-4′-fluoro-phenyl)quinoline 396. The evolved HCl gas has been eliminated during the reaction [111] (**Figure** 131).

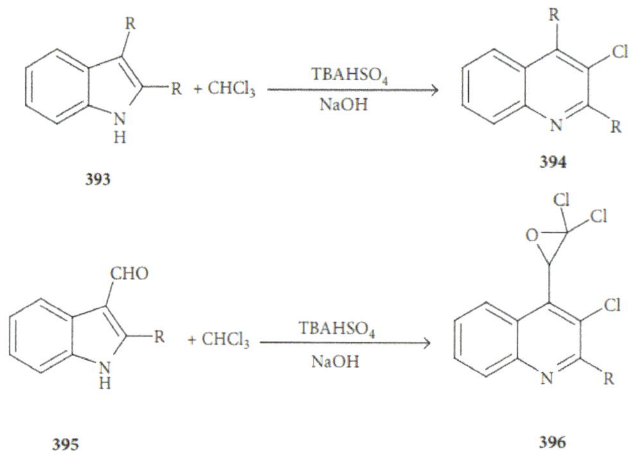

Figure 131

Ring Opening

Treatment of benzosultams 400 with different disulphides under PTC technique ($CH_3CN/K_2CO_3/TBAHSO_4$) gave the corresponding 2-methylamino-benzaldehyde dithioacetals 401 [24] (**Figure** 132).

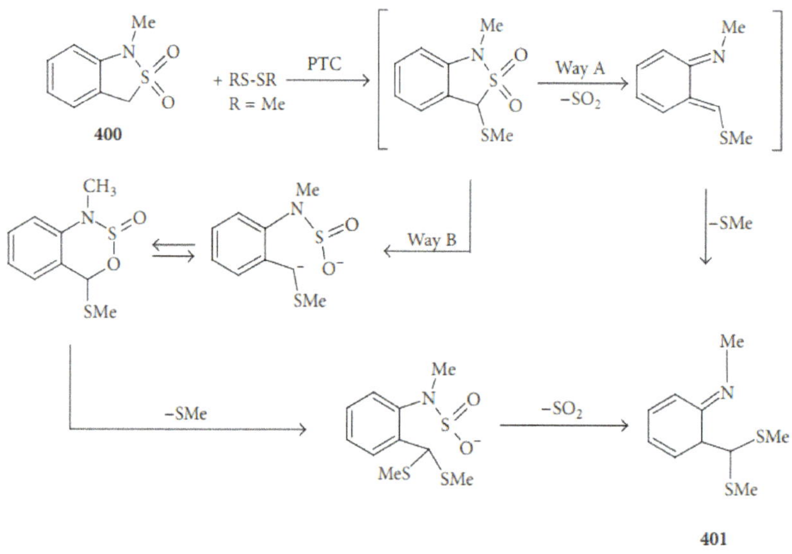

Figure 132:

REFERENCES

1. M. Makosza and M. Fedoryński, "Catalysis in two-phase systems: phase transfer and related phenomena," Advances in Catalysis, vol. 35, pp. 375–422, 1987.
2. E. V. Dehmlow and S. S. Dehmlow, Phase-Transfer Catalysis, Chemie, Weinheim, Germany, 3rd edition, 1993.
3. C. M. Starks, C. L. Liotta, and M. Halpern, Phase Transfer Catalysis: Fundamentals, Applications and Industrial Perspectives, Chapman and Hall, New York, NY, USA, 1994.
4. R. R. Nadendla, Indian Pharmacist, vol. 2, p. 13, 2003.
5. M. Makosza, "Phase-transfer catalysis. A general green methodology in organic synthesis," Pure and Applied Chemistry, vol. 72, no. 7, pp. 1399–1403, 2000.
6. J. Jarrouse and C. R. Hebd, Seances Acad Sci Ser C, vol. 232, p. 1424, 1951.
7. H.RathandU.Einsele,"ÜberdiechemischeModifizierung der Zellulose durch Alkylierung," Melliand Textilber, vol. 40, p. 526, 1959.
8. F. Bayer, German Patent, 959, 497, 1957.
9. N. V. Gavaert, "Photo produce," Belgiam Patent, 602, 793, 1961.
10. M. A. Iskenderov, V. V. Korshak, and S. V. Vinogradova, Vysokomol Soedin, vol. 4, p. 637, 1962.
11. W. S. Port, British Patent, 912, 104, 1962.

12. R. W. Kay, British Patent, 916, 772, 1963.
13. S. Desikan and L. K. Doraiswamy, "Enhanced activity of polymer-supported phase transfer catalysts,"Chemical Engineering Science, vol. 55, no. 24, pp. 6119–6127, 2000.
14. T. Shioiri, "Chiral phase transfer catalysis," in Handbook of Phase-Transfer Catalysis, Y. Sasson and R. Neumann, Eds., chapter 14, Blackie Academic & Professional, London, UK, 1997.
15. M. Makosza, "Two-phase reactions in the chemistry of carbanions and halocarbenes. A useful tool in organic synthesis," Pure and Applied Chemistry, vol. 43, p. 439, 1975.
16. M. Makosza and I. Krylowa, "Remarks on the mechanism of phase-transfer catalyzed carbanion generation in two-phase systems," Tetrahedron, vol. 55, pp. 6395–6402, 1999.
17. H. Abdel-Ghany, A. M. El-Sayed, and A. K. El-Shafei, "Synthesis of pyrrole, pyridinone and pyrimidinone derivatives using PTC conditions," Synthetic Communications, vol. 25, no. 8, pp. 1119–1131, 1995.
18. C. G. Dave and V. A. Parikh, "Heterocyclization using phase transfer catalysis: a simple and convenient synthesis of 2-amino-1-aryl-5-oxo-4,5-dihydro-1H-pyrrole-3-carbonitriles," Synthetic Communications, vol. 31, no. 9, pp. 1301–1306, 2001.
19. A. M. El-Sayed, Trends in Heterocycles, vol. 9, p. 33, 2003.
20. D. Albanese, D. Landini, and M. Penso, "Synthesis of 2-amino acids via selective mono-N-alkylation of trichloroacetamide by 2-bromo carboxylic esters under solid-liquid phase-transfer catalysis conditions,"Journal of Organic Chemistry, vol. 57, no. 5, pp. 1603–1605, 1992.
21. J. Casas, R. Grigg, C. Nájera, and J. M. Sansano, "The effect of phase-transfer catalysis in the 1,3-dipolar cycloaddition reactions of azomethine ylides - Synthesis of substituted prolines using AgOAc and inorganic base in substoichiometric amounts," European Journal of Organic Chemistry, no. 10, pp. 1971–1982, 2001.
22. O. A. Abd Allah and A. M. El-Sayed, "Application of phase transfer catalysis in heterocyclic synthesis: synthesis of some new polyfunctional thiophenes, thiopyrans, pyridines and oxazines," Phosphorus, Sulfur and Silicon and Related Elements, vol. 177, no. 5, pp. 1291–1301, 2002.
23. A. K. El-Shafei, H. Abdel-Ghany, A. A. Sultan, and M. M. El-Saghier, "Synthesis of thieno (2,3-b) thiophenes and related structures," Phosphorus, Sulfur, and Silicon and the Related Elements, vol. 73, pp. 15–25, 1992.
24. H. Modrzejewska and K. Wojciechowski, "Synthesis and reactions of 3-alkylsulfanyl-1,3-dihydro-2,1-benzisothiazole 2,2-dioxides," Tetrahedron, vol. 61, no. 37, pp. 8848–8854, 2005.
25. A. Khodairy, "Synthesis of fused and spiro heterocyclic compounds derived from 3,5-pyrazolidinedione derivatives," Phosphorus, Sulfur and Silicon and Related Elements, vol. 160, pp. 159–180, 2000.
26. D. Bogdal and M. Warzala, "Microwave-assisted preparation of benzo[b]furans under solventless phase-transfer catalytic conditions," Tetrahedron, vol. 56, no. 44, pp. 8769–8773, 2000. ·
27. M. Makosza, J. Przyborowski, R. Klajn, and A. Kwast, "Simple synthesis of 2-substituted tetrahydrofuran-3-carbonitriles," Synlett, no. 12, pp. 1773–1774, 2000.
28. K. S. Krishna Murthy, B. Rajitha, and M. Kanakalingeswara Rao, "Synthesis of biologically active angularly fused bisaroylbenzodifurans by PTC and solvent free microwave irradiation," Indian Journal of Chemistry B, vol. 42, no. 2, pp. 425–428, 2003.
29. R. D. Shah, "Phase transfer catalysis assisted thorpe reaction for the synthesis of 3-aminothiophene-2-carboxylates," E-Journal of Chemistry, vol. 8, no. 1, pp. 368–372, 2011.
30. W. He, B.-L. Zhang, Z.-J. Li, and S.-Y. Zhang, "PTC-promoted Japp-Klingmann reaction for the synthesis of indole derivatives," Synthetic Communications, vol. 35, no. 10, pp. 1359–1368, 2005. ·
31. T. Cholakova, Y. Zagraniarsky, S. Simova, S. Varbanov, and A. Dobrev, "A simple synthesis of dimethylphosphinyl-substituted tetrahydropyrroles," Phosphorus, Sulfur and Silicon and the Related Elements, vol. 180, no. 7, pp. 1721–1728, 2005.
32. A. Khodairy and A. M. El-Saghier, "Thieno[2,3-b] thiophenes: part 7. Some heterocyclization reactions with ethyl 3,4-diamino- 5-cyanothieno[2,3-b]thiophene-2-carboxylate," Acta Chimica Slovenica, vol. 58, no. 2, pp. 360–366, 2011.
33. M. Y. Ait Itto, A. Hasnaoui, A. Riahi, and J.-P. Lavergne, "Regioselective synthesis of new biheterocyclic triazepines," Tetrahedron Letters, vol. 38, no. 12, pp. 2087–2090, 1997.
34. G. Molteni, A. Ponti, and M. Orlandi, "Uncommon aqueous media for nitrilimine cycloadditions. I. Synthetic and mechanistic aspects in the formation of 1-aryl-5-substituted-4,5-dihydropyrazoles," New Journal of Chemistry, vol. 26, no. 10, pp. 1340–1345, 2002.
35. B. W. LeBlanc and B. S. Jursic, "Preparation of 5-alkylthio and 5-arylthiotetrazoles from thiocyanates using phase transfer catalysis," Synthetic Communications, vol. 28, no. 19, pp. 3591–3599, 1998.
36. A. Ogihara, H. Sakai, N. Matsui, and H. Miyazaki, PCT International Applications WO0276,958 (Cl.C/32), 2002.
37. P. Sharma, A. Kumar, and M. Sharma, "Generation of 4,6-dimethyl-5-[2-(2-methylprop-1-enyl)-1H-benzimidazol-1-yl]pyrimidine-2(5H) -thiones under kinetically controlled phase transfer catalysis conditions," Journal of Molecular Catalysis A, vol. 237, no. 1-2, pp. 191–198, 2005. ·
38. A. M. Youssef, Egyptian Journal of Chemistry, vol. 45, p. 777, 2002.
39. M. F. Farhat, A. M. M. El-Saghier, M. A. Makhlouf, K. M. Kreddan, and A. B. Elmezoughi, "Ketene N,S-acetals in heterocyclic synthesis: part 1: synthesis of N-phenyl-2-ylidene and 2,5-diylidene-4-thiazolidinone

derivatives," Journal of Sulfur Chemistry, vol. 28, no. 6, pp. 563–572, 2007.

40. M. Fedoryński, M. Jezierska-Zięba, and B. Kąkol, "Phase transfer catalysis in pharmaceutical industry—where are we?" Acta Poloniae Pharmaceutica, vol. 65, pp. 647–654, 2008.

41. A. K. Khalil, "Phase-transfer catalyzed alkylation and cycloalkylation of 2-mercaptoquinazolin-4(3H)-one," Phosphorus, Sulfur and Silicon and the Related Elements, vol. 180, no. 11, pp. 2533–2541, 2005.

42. M. K. Abou El-Regal, A. A. Abdalha, M. A. El-Kassaby, and A. T. Ali, "Synthesis of new thiohydantoin derivatives under phase transfer catalysis," Phosphorus, Sulfur and Silicon and the Related Elements, vol. 182, no. 4, pp. 845–851, 2007.

43. S. El-Metwally and A. K. Khalil, "Reactions of 1,3-diphenyl-2-pyrazolin-5-one and 4-amino-1,5-dimethyl-2- phenyl-1H-pyrazol-3(2H)-one. Synthesis of some new pyrazoles and pyrazolones," Acta Chimica Slovenica, vol. 57, no. 4, pp. 941–947, 2010.

44. O. Atsushi, S. Hiroshi, M. Nobuo, and M. Hidekazu, PCT International Applications, WO 0276, 958, 2002.

45. M. V. Zatsepina, T. V. Artamonova, and G. I. Koldobskii, "Tetrazoles: LII. Synthesis of functionally substituted tetrazoles from benzene-1,3,5-tricarboxylic acid derivatives," Russian Journal of Organic Chemistry, vol. 44, no. 4, pp. 577–581, 2008.

46. A. M. M. El-Saghier, "Simple one-pot synthesis of thieno[2,3-b]thiophene derivatives under solid-liquid PTC conditions. Useful starting material for the synthesis of biological active compounds," Bulletin of the Chemical Society of Japan, vol. 66, no. 7, pp. 2011–2015, 1993. ·

47. K. C. Majumdar, S. Saha, and A. T. Khan, "Indian Journal of Chemistry B," vol. 33, p. 216, 1994.

48. R. M. Butnariu and I. I. Mangalagiu, "New pyridazine derivatives: Synthesis, chemistry and biological activity," Bioorganic and Medicinal Chemistry, vol. 17, no. 7, pp. 2823–2829, 2009. ·

49. A. M. El-Sayed, "One—pot synthesis of pyridines, thienopyridines, pyrrolothienopyridines and (1,8) naphthyridines under phase-transfer catalysis conditions," Phosphorus, Sulfur and Silicon and Related Elements, vol. 163, pp. 29–40, 2000.

50. A. K. El-Shafei, A. M. Soliman, A. A. Sultan, and A. M. M. El-Saghier, "Synthesis of some new fused heterocycles containing a bridgehead nitrogen atom under ptc conditions—a new application of cyanoketene s,s-acetals," Gazzetta Chimica Italiana, vol. 125, p. 115, 1995.

51. A. Khodairy and A. M. El-Sayed, "Synthetic studies on the synthesis of pyridine, α-pyran, α-thiopyran, and Thienothiopyranopyrrol derivatives using PTC technique," Synthetic Communications, vol. 31, no. 4, pp. 475–486, 2001.

52. G. A. El-Saraf, A. M. El-Sayed, and A. M. M. El-Saghier, "One-pot PTC synthesis of polyfused pyrazoles,"Heteroatom Chemistry, vol. 14, no. 3, pp. 211–217, 2003.

53. P. Sampath Rao, K. Vishnu Vardhan Reddy, and D. Ashok, "An efficient synthesis of dibenzoyl dialkyl benzodifurans under phase transfer catalysis conditions and their antifeedant activity," Indian Journal of Heterocyclic Chemistry, vol. 8, no. 1, pp. 51–54, 1998.

54. P. Sampath Rao, K. Vishnu Vardhan Reddy, and D. Ashok, "Synthesis of 2,6-dibenzoyl-3,5-distyryl benzo (1,2-b: 5,4-b′)-difurans as potential antifeedants," Indian Journal of Heterocyclic Chemistry, vol. 7, no. 4, pp. 293–294, 1998.

55. P. Sampath Rao, K. Vishnu Vardhan Reddy, and D. Ashok, "An efficient synthesis of dibenzoyl dialkyl benzodifurans under phase transfer catalysis conditions and their antifeedant activity," Indian Journal of Heterocyclic Chemistry, vol. 8, no. 1, pp. 51–54, 1998.

56. K. S. Krishna Murthy, B. Rajitha, M. Kanakalingeswara Rao, T. Raja Komuraiah, and S. M. Reddy, "Facile synthesis of biologically active linear bisaroyl benzodifurans by ptc and solvent free microwave irradiation," Heterocyclic Communications, vol. 8, no. 2, pp. 179–186, 2002.

57. A. M. El-Sayed, H. Abdel-Ghany, and A. M. M. El-Saghier, "A novel synthesis of pyrano(2,3-c)-, 1,3-oxazino(2,3 b)-,1,2,4- triazolo(3,4-b)-, oxazolo(2,3-b)-, furano(3,2-c)-, and 3-substituted- (1,5) benzodiazepin-2-ones," Synthetic Communications, vol. 29, no. 20, pp. 3561–3572, 1999.

58. A. Khodairy, H. Abdel-Ghany, A. M. El-Sayed, and H. Salah, "Synthesis of new fused and spiro 1,5-benzodiazepines," Journal of the Chinese Chemical Society, vol. 50, no. 6, pp. 1195–1198, 2003.

59. A.M.Soliman,A.A.Sultan,andA.K.El-Shafei,"Synthesis of some new dispiro[dipyrano-(2,4′:6,4″)bidithiolo(4,5-b:4′,5′-e)-4,8-benzoquinones]," Monatshefte für Chemie Chemical Monthly, vol. 126, no. 5, pp. 615–619, 1995.

60. A. A. Sultan, "Synthesis of some condensed s-triazole heterocycles using phase-transfer catalysis technique," Phosphorus, Sulfur, and Silicon and the Related Elements, vol. 105, pp. 123–127, 1995.

61. A. M. M. El-Saghier, "A novel one-pot synthesis of polyfunctionally substituted thiopyrano(2,3-b)pyridine and pyrido(2,3-b)pyridine derivatives under solid-liquid phase-transfer catalysis,"Phosphorus, Sulfur and Silicon and Related Elements, vol. 177, no. 5, pp. 1213–1221, 2002.

62. C. Bezergiannidou-Balouскts, K. E. Litinas, E. M. Xenikaki, and D. N. Nicolaides, "Reactions of 7-(Methoxyimino)-4-methyl-2H-chromene-2,8(7H)-dione with Phosphorus Ylides. Synthesis of 2-Substituted 6-Methyl-8H-benzopyrano[7,8-d]oxazol-8-ones," Liebigs Annalen Der Chemie, vol. 1993, pp. 1175–1177, 1993.

63. H. A. Ghany, "Synthesis of new spiro heterocyclic compounds derived from rhodanine derivatives,"Phosphorus, Sulfur and Silicon and Related Elements, vol. 122, pp. 173–180, 1997.

64. H. Abdel-Ghany, A. M. El-Sayed, A. Khodairy, and H. Salah, "Synthesis of new 3-substituted and spiro 1,5-benzodiazepin-2-ones under phase-transfer catalysis

conditions," Synthetic Communications, vol. 31, no. 16, pp. 2523–2535, 2001.

65. E. Diez-Barra, A. de la Hoz, A. Loupy, and A. Sanchez-Migallon, "Selective alkylation of pyrrole by phase transfer catalysis in the absence of solvent," Journal of Heterocyclic Chemistry, vol. 31, pp. 1715–1717, 1994.

66. A. Nakmura and K. Takamoto, Japanese Kokai Tokkyo Koho JP, 171,379, (Cl.C07D417/12), 2003.

67. A. Nakamura and K. Takamoto, Japanese Kokai Tokkyo Koho JP, 138, 079, (Cl.C07D209/48), 2002.

68. G. Guazzelli and R. Settambolo, "4-Indolylbutanals from rhodium-catalyzed hydroformylation of allylindoles as precursors of benzofused indolizines," Tetrahedron Letters, vol. 48, no. 34, pp. 6034–6038, 2007.

69. R. Settambolo, S. Miniati, and R. Lazzaroni, "One pot hydroformylation/intramolecular aldol condensation reactions of 1-allyl-2-carbonylpyrroles: a new entry into hydroindolizines synthesis,"Synthetic Communications, vol. 33, no. 17, pp. 2953–2961, 2003.

70. R. Settambolo, M. Mariani, and A. Caiazzo, "Synthesis of 1,2- and 1,3-divinylpyrrole," Journal of Organic Chemistry, vol. 63, no. 26, pp. 10022–10026, 1998.

71. P. De La Cruz, A. De La Hoz, L. M. Font, F. Langa, and M. C. Pérez-Rodríguez, "Solvent-free phase transfer catalysis under microwaves in fullerene chemistry. A convenient preparation of N-alkylpyrrolidino[60] fullerenes," Tetrahedron Letters, vol. 39, no. 33, pp. 6053–6056, 1998.

72. I. Almena, E. Díez-Barra, A. De La Hoz, J. Ruiz, A. Sánchez-Migallón, and J. Elguero, "Alkylation and arylation of pyrazoles under solvent-free conditions: conventional heating versus microwave irradiation,"Journal of Heterocyclic Chemistry, vol. 35, no. 6, pp. 1263–1268, 1998.

73. Inc. Hoffman La-Roch, NJ. Nutely,United States Patent, Patent No. US 6,326,501 B1, 2001.

74. A. Jończyk, E. Nawrot, and M. Kisielewski, "Reactions of some nitrogen heterocycles with chlorodifluoromethane under conditions of phase-transfer catalysis," Journal of Fluorine Chemistry, vol. 126, no. 11-12, pp. 1587–1591, 2005.

75. P. Barraja, P. Diana, A. Carbone, and G. Cirrincione, "Nucleophilic reactions in the indole series: displacement of bromine under phase transfer catalysis," Tetrahedron, vol. 64, no. 51, pp. 11625–11631, 2008.

76. C. Barrett, B. Choudhury, A. Natansohn, and P. Rochon, "Azocarbazole polymethacrylates as single-component electrooptic materials," Macromolecules, vol. 31, no. 15, pp. 4845–4851, 1998.

77. I. Almena, E. Díez-Barra, A. De La Hoz, J. Ruiz, A. Sánchez-Migallón, and J. Elguero, "Alkylation and arylation of pyrazoles under solvent-free conditions: conventional heating versus microwave irradiation,"Journal of Heterocyclic Chemistry, vol. 35, no. 6, pp. 1263–1268, 1998.

78. F. L. Tang, Z. H. Wang, W. L. Jia, Y. M. Xu, and J. T. Wang, "Synthesis of diorganotin derivatives containing asymmetric disubstituted bis(pyrazol-1-yl)methanes: X-ray crystal structures of (3-isopropylpyrazol-1-yl-5′-isopropylpyrazol-1-yl)methane diphenyltin(IV) dibromide and bis(5-isopropylpyrazol-1-yl)methane diphenyltin(IV) dibromide," Polyhedron, vol. 19, pp. 381–387, 2000.

79. V. N. Pathak, C. K. Oza, R. Pathak et al., "Fluorophenylsulphonylation of 4,5-dihydro-3,5-diarylpyrazoles and 2-arylindoles via solid-liquid phase transfer catalysis," Indian Journal of Heterocyclic Chemistry, vol. 7, no. 3, pp. 241–242, 1998.

80. E. Arranz, J. A. Díaz, E. Morante, C. Pérez, and S. Vega, "Preparation and regiochemical assignments of new pyrazolo[3,4-c] [2,1] benzothiazepines," Journal of Heterocyclic Chemistry, vol. 33, no. 1, pp. 151–156, 1996.

81. M. L. Taha, H. B. Lazrek, J. L. Barascut, and J. L. Imbach, "Synthesis of some 4-substituted 1-[(2-hydroxyethoxy) methyl]-pyrazolo[3,4-d]pyrimidines," Bulletin des Sociétés Chimiques Belges, vol. 105, pp. 279–285, 1996.

82. H. B. Lazrek, M. Taourirte, J. L. Barascut, and J. L. Imbach, Bulletin Des Sociétés Chimiques Belges, vol. 105, p. 391, 1996.

83. E. Diez-Barra, A. De La Hoz, A. Sanchez-Migallon, and J. Tejeda, "Alkylation of imidazole by solid-liquid phase transfer catalysis in the absence of solvent," Synthetic Communications, vol. 23, no. 13, pp. 1783–1786, 1993.

84. X. Chen, X. Wang, H. Wang, H. Lian, Y. Pan, and Y. Shi, "N-alkylation of 2-methyl-2-imidazolines by phase transfer catalysis without solvent," Synthetic Communications, vol. 29, no. 17, pp. 3025–3030, 1999.

85. X. Chen, X. Wang, H. Wang, H. Lian, Y. Pan, and Y. Shi, "ChemInform abstract: N-Alkylation of 2-Methyl-2-imidazolines by phase transfer catalysis without solvent," Cheminform, vol. 30, no. 43, 1999.

86. E. V. Dehmlow, R. Richter, and A. B. Zhivich, Journal of Chemical Research S, vol. 504, 1993.

87. S. Khabnadideh, Z. Rezaei, A. Khalafi-Nezhad, R. Bahrinajafi, R. Mohamadi, and A. A. Farrokhroz, "Synthesis of N-alkylated derivatives of imidazole as antibacterial agents," Bioorganic and Medicinal Chemistry Letters, vol. 13, no. 17, pp. 2863–2865, 2003.

88. K. Ramaiah, P. K. Dubey, D. Eswara Rao et al., "Synthesis and spectral properties of a variety of 2-styrylbenzimidazoles," Indian Journal of Chemistry B, vol. 39, no. 12, pp. 904–914, 2000.

89. S. Khabnadideh, Z. Rezaei, A. Khalafi-Nezhad, R. Bahrinajafi, R. Mohamadi, and A. A. Farrokhroz, "Synthesis of N-alkylated derivatives of imidazole as antibacterial agents," Bioorganic and Medicinal Chemistry Letters, vol. 13, no. 17, pp. 2863–2865, 2003.

90. E. V. Dehmlow, J. Bollhofer, and G. Thye, "Journal of Chemical Research S," vol. 113, 2001.

91. A. Abenhaim, "Selective alkylations of 1,2,4-triazole and benzotriazole in the absence of solvent,"Heterocycles, vol. 38, no. 4, p. 793, 1994.

92. I. P. Beletskaya, D. V. Davydov, M. S. Gorovoi, and S. V. Kardashov, "Selective N(1)-arylation of benzotriazole with activated aryl halides under conditions of phase

transfer catalysis," Russian Chemical Bulletin, vol. 48, no. 8, pp. 1533–1536, 1999.

93. C. G. Oliva, N. Jagerovic, P. Goya et al., "N-substituted-1,2,3-triazoles: Synthesis, characterization and evaluation as cannabinoid ligands," Arkivoc, vol. 2010, no. 2, pp. 127–147, 2010. ·

94. K. Waisser, J. Kunes, A. Hrabalek, M. Machacek, and Z. Odlerova, "New groups of potential antituberculotics: 5-Alkylthio-1-aryltetrazoles," Collection of Czechoslovak Chemical Communications, vol. 61, p. 791, 1996.

95. M. L. Cerrada, J. Elguero, J. De La Fuente, C. Pardo, and M. Ramos, "Synthesis of p-nitrophenylazoles by phase transfer catalysis without solvent," Synthetic Communications, vol. 23, no. 14, pp. 1947–1952, 1993.

96. Y. V. Poplavskaya, L. V. Alam, and G. I. Koldobskii, "Tetrazoles: XLI. Alkylation of 1-aryltetrazol-5-ones,"Russian Journal of Organic Chemistry, vol. 36, no. 12, pp. 1793–1799, 2000. ·

97. K. Aritomo, T. Wada, and M. Sekine, "Alkylation of 6-N-acylated adenosine derivatives by the use of phase transfer catalysis," Journal of the Chemical Society, no. 14, pp. 1837–1844, 1995.

98. H. Rodríguez, R. Pérez, M. Suárez, A. Lam, N. Cabrales, and A. Loupy, "Alkylation of some pyrimidine and purine derivatives using microwave-assisted methods," Heterocycles, vol. 55, no. 2, pp. 291–302, 2001.

99. L. Zheng, W. Xi-cun, Y. Jing-ya, and Z. Kexueban, Journal of Northwest Normal University, vol. 37, p. 54, 2001.

100. O. S. Attaryan, A. O. Baltayan, R. E. Sagatelyan, and K. T. Takmazyan, "Synthesis of 1-(2-aminoethyl)pyrazoles under phase-transfer catalysis," Russian Journal of General Chemistry, vol. 78, no. 1, pp. 136–138, 2008.

101. O. S. Attarian, S. G. Matsoyan, and S. S. Martirosyan, "Synthesis of N-vinylpyrazoles," Chemistry of Heterocyclic Compounds, vol. 41, no. 4, pp. 452–455, 2005.

102. M. Majdoub, A. Loupy, A. Petit, and S. Roudesli, "Coupling focused microwaves and solvent-free phase transfer catalysis: application to the synthesis of new furanic diethers," Tetrahedron, vol. 52, no. 2, pp. 617–628, 1996.

103. T. M. Slaghek, A. H. van Oijen, A. A. M. Maas, J. P. Kamerling, and J. F. G. Vliegenthart, "Synthesis of structural elements of the capsular polysaccharides of Streptococcus pneumoniae types 6A and 6B,"Carbohydrate Research, vol. 207, no. 2, pp. 237–248, 1990.

104. A. K. El-Shafei, A. M. El-Sayed, A. A. Sultan, and H. Abdel-Ghany, Gazzetta Chimica Italiana, vol. 120, p. 197, 1990.

105. S.-S. Jew, Y.-J. Lee, J. Lee et al., "Highly enantioselectlve phase-transfer-catalytic alkylation of 2-phenyl-2-oxazoline-4-carboxylic acid tert-butyl ester for the asymmetric synthesis of α-alkyl serines," Angewandte Chemie - International Edition, vol. 43, no. 18, pp. 2382–2385, 2004.

106. T. Tamura and E. Ryukoku, Japanese Kokai Tokkyo Koho JP, 128, 688 (Cl. CO7F 9/6539), 2003.

107. S. A. Shiba, N. S. Harb, M. El-Kassaby, M. A. Hassan, and N. M. Abouelregal, "Facile synthesis of 4-substituted 2-pyrazolin-5-ones under phase transfer catalysis," Phosphorus, Sulfur, and Silicon and the Related Elements, vol. 104, pp. 15–20, 1995.

108. USPTO Applicaton #: #20090176826—Class: 514312 (USPTO).

109. E. Abele, O. Dzenitis, K. Rubina, and E. Lukevics, "Synthesis of N- and S-vinyl derivatives of heteroaromatic compounds using phase-transfer catalysis," Chemistry of Heterocyclic Compounds, vol. 38, no. 6, pp. 682–685, 2002.

110. A. Yu Egorova, V. A. Sedavkina, and Z. Y. Temofeyeva, "Interaction of 1,5-substituted pyrrolin-2-ones with dichlorocarbene under phase transfer catalysis conditions," Molecules, vol. 5, pp. 1082–1084, 2000.

111. C. I. Krishna, J. Renuka, and S. Arora, Journal of the Indian Chemical Society, vol. 70, p. 567, 1993.

112. W. J. Bailey and L.-L. Zhou, "A new elimination with phase-transfer catalysis for cyclic ketene acetals,"Tetrahedron Letters, vol. 32, no. 12, pp. 1539–1540, 1991.

113. T. Terme, C. Galtier, J. Maldonado, M. P. Crozet, A. Gueiffier, and P. Vanelle, "Synthesis of 2-substituted-3-nitroimidazo[1,2-b]pyridazines as potential biologically active agents," Journal of Heterocyclic Chemistry, vol. 39, no. 1, pp. 173–177, 2002.

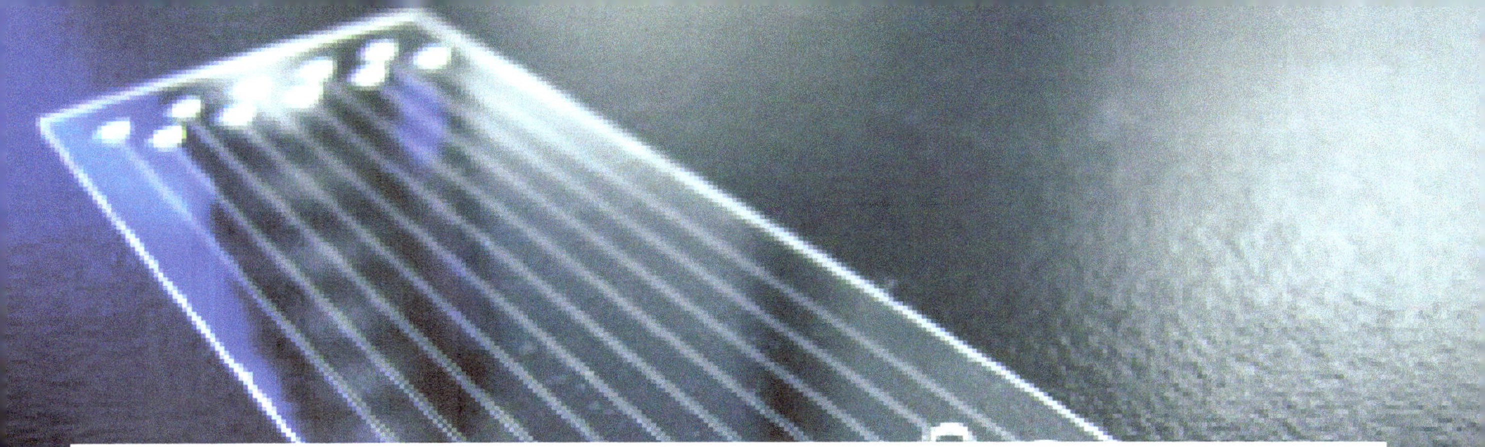

CHAPTER 18

BIO-INSPIRED MICRO-FLUIDIC ANGULAR-RATE SENSOR FOR VESTIBULAR PROSTHESES

Charalambos M. Andreou, Yiannis Pahitas and Julius Georgiou

Keywords: Vestibular prostheses, oscillopsia, inner-ear semi-circular canals, angular-rate sensor, implantable bio-sensors, gyroscope, inertial sensor, motion sensor, MEMS, microfluidics

INTRODUCTION

People suffering from oscillopsia, vertigo and loss of balance due to a vestibular-related disorder could benefit from the development of a vestibular neural prosthesis. The inner ear's vestibular system provides cues about self-motion and balance, body posture and helps stabilize vision during movement. Damage to this system causes failure of vestibulo-ocular and vestibulo-spinal reflexes which results in dizziness, vertigo, nausea, imbalance, blurred vision and instability in locomotion [1]. The restoration of vestibular system can therefore be achieved by bypassing a dysfunctional element in the vestibular pathway and using artificial stimulation [1–7]. There are a number of sites along the vestibular pathway that can be tapped into, like the ampullae, the scarpa's ganglion and individual nerve bundles. A detailed anatomy of the human vestibular system is illustrated in Figure 1. A key component that is required to enable the creation of an ergonomic vestibular prosthesis is that of a suitable Micro-Electro-Mechanical-System (MEMS)-based angular-rate sensor or gyroscope [8,9], since existing gyroscopes are too power-hungry for a totally implantable solution.

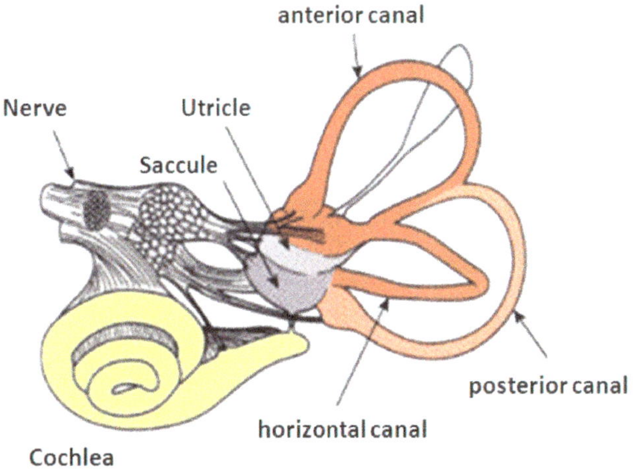

Figure 1: Illustration of the human vestibular system and its components [10].

Citation: Charalambos M. Andreou, Yiannis Pahitas and Julius Georgiou, Bio-Inspired Micro-Fluidic Angular-Rate Sensor for Vestibular Prostheses, doi:10.3390/s140713173.

Low-power, low-cost, small-footprint gyro development has been predominantly driven by the mobile phone and gaming industry. However the power and reliability requirements, in these industries, are far less stringent than those required for implantable devices. In vestibular prototypes, the gyro tends to consume most of the supplied power [11], and as such, the lack of a suitable off-the-shelf gyro is considered one of the stumbling blocks in the development of a fully implantable vestibular prosthesis. State-of-the-art gyros avoid utilizing large masses by vibrating small masses and exploiting the Coriolis effect [12–18]. Vibrating structures need voltages as high as 40V to electro-statically stimulate the vibration [19]. In addition, vibrating these structures at hundreds of kilohertz, for over a decade, is likely to cause fatigue and premature failure of the sensor [20,21].

Although non-vibratory structures have been explored, so far no proof-of-concept gyroscopic device has given promising results [22–25]. Thermal based flow sensing [24,25] is too power hungry and has not been shown to provide the required sensitivity for this application. Various artificial haircell flow sensors [26], have been constructed by growing "cilia" perpendicular to the silicon wafer plane. Using such an approach for sensing motion in a gyroscopic device increases the complexity and cost of fabrication. Furthermore some of these "cilia" that are built with SU-8 epoxy [27] are very likely to detach from the silicon sensing element over time, especially if immersed in a fluid. The surface acoustic wave (SAW) based device, behaves much like a vibrating structure, in that it needs energy to continually keep the atoms in motion via high frequency surface acoustic waves (160 MHz) [28].

In this paper we present a simple but elegant MEMS-Microfluidic solution, which consumes significantly less power than vibration based gyroscopes. The initial concept of this work was previously presented in 2013 [29], with only the simulated results available at that time. The proposed solution mimics the natural mechanism of the vestibular system where a sealed micro-channel is filled with a liquid. Any rotation of the system about the axis of the channel results in a relative motion between the channel itself and the liquid, due to the fluid›s inherent inertia. A flexible object fixed to the channel walls can sense the fluid flow upon deflection, e.g., with a piezo-resistor.

PROPOSED BIO-INSPIRED ANGULAR-RATE SENSOR

The proposed micro-fluidic gyro is based on a relatively large fluidic mass that deflects a sensing element similar in principle to a biological semicircular canal (SCC). The fluid dynamics of a SCC can be mathematically expressed with the "torsion pendulum equation" [30] as:

$$I\ddot{\theta} + B\dot{\theta} + K\theta = -I\alpha \qquad (1)$$

where θ is the mean angular displacement, $\dot{\theta}$ is the mean angular velocity and is $\ddot{\theta}$ the mean angular acceleration of the endolymph fluid. A is the angular acceleration of the head perpendicular to the plane of the canal, I is the inertia, B is the damping coefficient and K is the stiffness of the cupula (in the case of the proposed gyroscope, it's the effective stiffness of the fluid-intercepting cantilevers).

Taking this phenomenon into account, the proposed design integrates flow sensors in the path of the liquid inside the micro-channel. These sensors are cantilevers with embedded piezo-resistors, whose resistance changes according to the deformation caused by pressure applied across the sensing structure.

The hybrid MEMS-microfluidic structure is illustrated in Figure 2 while the physical layout of the design is shown inFigure 3. This structure uses a fluidic mass that is contained in etched glass layers, which are anodically bonded on top and below a bulk-micromachined silicon layer, and whose deflectable structures include buried piezo-resistors for sensing. The deflectable structures consist of four cantilevers (flaps) that intercept the path of the fluid as it flows, serving as the cupula and one piezo-resistor in each cantilever, serving as the hair cells of the SCC.

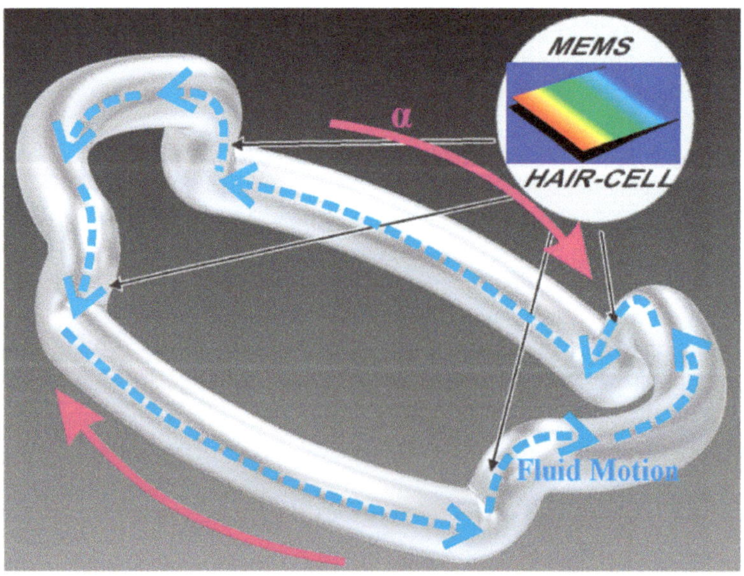

Figure 2: Illustration of the principle behind the sensor, which integrates four cantilevers (MEMS hair cells) at each of the layer transition points, i.e., in the bulk-micromachined silicon layer, positioned at the locations shown by the black arrows.

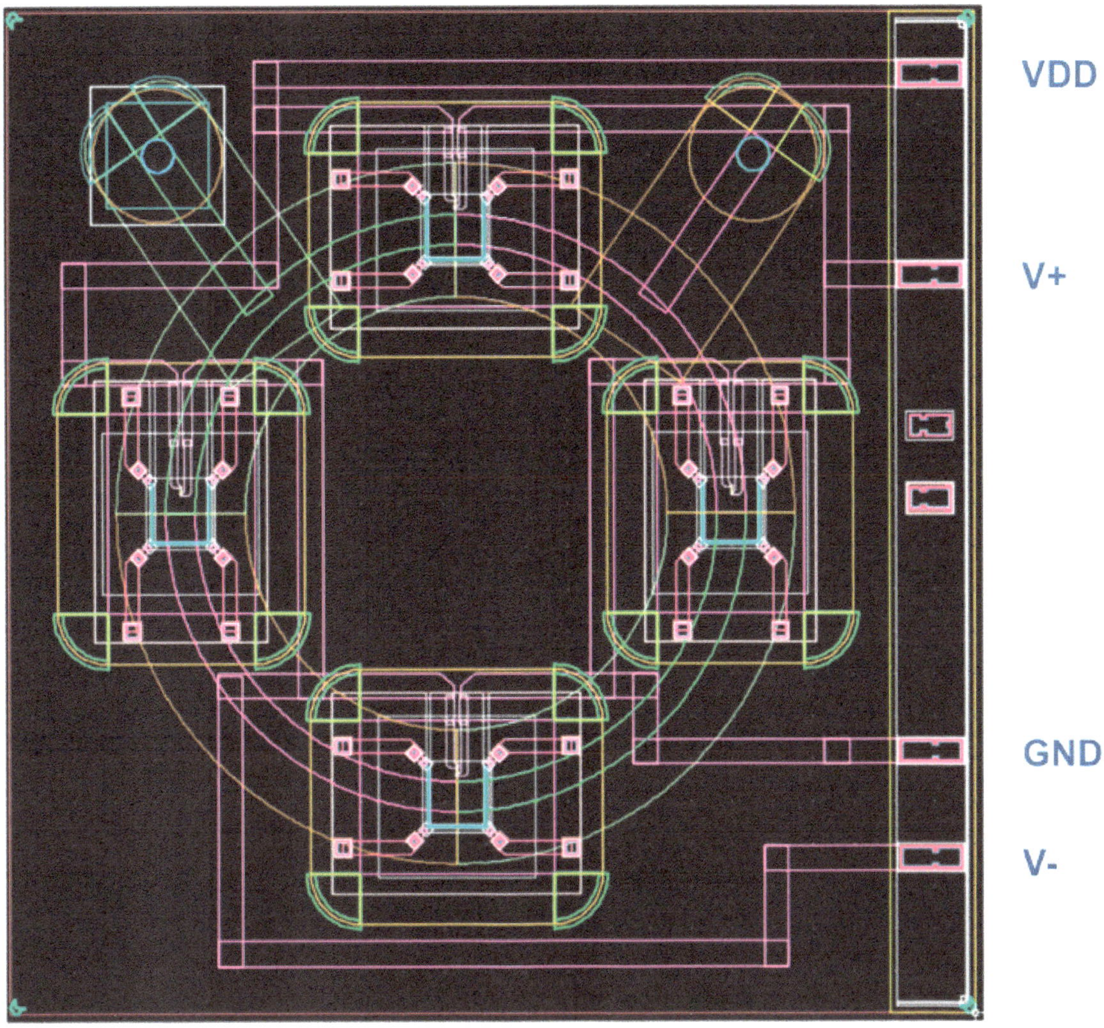

Figure 3: Actual physical layout of the structure consisting of: (1) top-glass canal (green quadrants in upper-left and bottom-right), (2) bottom-glass canal (lilac quadrants in bottom-left and upper right), (3) bulk micromachined silicon layer connecting upper and lower quadrants, intercepted by four flaps (light blue Π-shaped cuts in a 3.1 μm silicon membrane) that are oriented in same direction in all cases and which contain a piezo-resistor centered on each of the membrane›s uncut side, (4) various conductors for powering the Wheatstone Bridge (VDD, GND) and differential outputs (V+, V−).

When angular acceleration is present, the liquid in the quarter-torroidal cavity in the top glass, is deflected through the middle silicon bulk-micromachined layer, where it will deflect a cantilever containing a buried piezo-resistor and progress through the bottom quarter-torroidal cavity etched in the bottom glass, after which it will be deflected back to the top glass cavity, again via the bulk-micromachined layer containing cantilevers and so on. Thus, liquid completes a loop through a distorted torroid, passing through the silicon layer that contains the sensing structures (equivalent to cupula and hair cells).

Angular rotation of the whole device creates an inertial force, which increases the pressure on the sensing cantilever, leading to a detectable deflection that is proportional to the angular acceleration. To cope with the resistors› process variations, the strain gauges are designed in pairs, to form a Wheatstone bridge, which converts the strain induced resistance variations into output voltage variations. A differential voltage read-out circuit then reduces the effects of semiconductor process variations. The sensing piezo-resistive strain gauges are strategically located at regions of maximum stress "σ", which are at the perimeter edges of the cantilever, to give increased electrical sensitivity.

In the sensor of Figures 2 and 3, all four piezo-resistors are placed longitudinally to the stress axis, oriented parallel to the 110 crystallographic direction, in order to maximize the coefficient of piezo-resistivity. During angular motion, two piezo-resistors undergo tensile stress (top to bottom flow), whilst the other two undergo compressive stress (bottom to top flow). These are wired internally, via buried conductors, so as to form a Wheatstone bridge. Thus the two resistive divider legs will determine the output voltage as:

$$\Delta V_{out} \propto \Delta R \propto \Pi_L \sigma_L + \Pi_\tau \sigma_\tau \propto \Pi(P - P_0) \tag{2}$$

where Π is the piezo-resistive coefficient, σ is the mechanical stress, subscripts L and τ denote the longitudinal and transversal coefficients respectively, along the resistor axes, and $(P - P_0)$ is the differential pressure on the sensing element. The thickness of the silicon cantilever (hair-cell) is 3.1 ± 0.3 μm.

The output voltage of the Wheatstone bridge of Figure 4 that was used in the proposed structures of Figures 2 and 3 is described by:

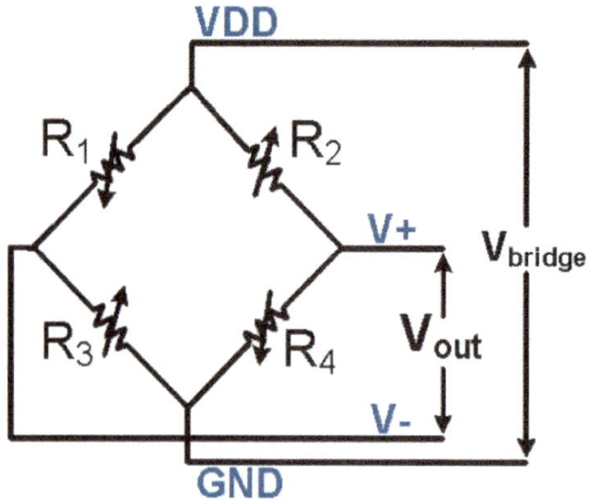

Figure 4: Wheatstone bridge.

$$V_{out} = V_{bridge}\left(\frac{R_3}{R_1 + R_3} - \frac{R_4}{R_2 + R_4}\right) \quad (3)$$

The proposed approach allows an increased inertial (fluidic) mass and thus provides the required sensitivity without utilizing vibrating structures and the Coriolis effect. A key concept of this solution is the decoupling of the structure that holds the (large) mass from the sensing structure; the fluid is supported by the walls of the etched glass, which at the thinnest point is about 200 μm thick, thus providing enough support to the large mass, so as avoid damage if the device is dropped onto a hard surface. The sensing structure thickness is only a few micrometers, but as it is not firmly attached to the mass, it is not susceptible to breakage when dropped.

TESTING PROCEDURE

Given that the MEMS-microfluidic angular rate sensor was a completely novel design, it follows that some challenges were expected to appear whilst attempting to test the functionality of the device.

The working principal of the device lies in the premise that the circular channel is filled with a liquid. For the purposes of these first tests, aimed at a proof-of-concept device, deionized water was chosen for its physical properties—density and viscosity, its chemical proximity to the endolymph and its bio-compatibility/non-toxicity.

Filling the Structures

The proposed structure has two holes that lead from the top external surface to the circular channel. Theoretically, during the filling process this allows for the liquid to enter from one hole, while air is pushed out from the other. Due to the microscopic level of the structure, dislodging the air micro-bubbles was not straight forward. Air micro-bubbles are held in place by forces of adhesion between the liquid around the bubbles and the channel structure as shown in Figure 5. In addition, the cantilevers presented an obstacle to the flow of water and hence the bubbles.

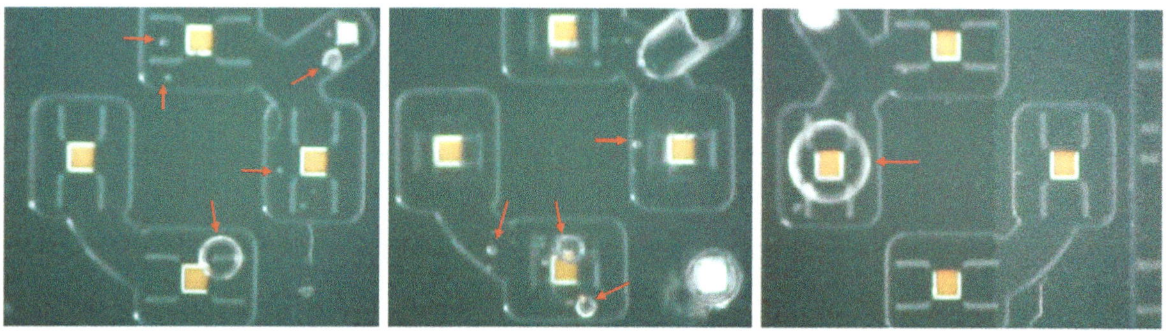

Figure 5: Micro-bubbles that can appear with incorrect filling methods, in the MEMS-microfluidic sensors. The micro-bubbles are indicated with red arrows (photographs).

After testing various filling methods, it was found that the most effective method for filling the micro-fluidic channels, without any trapped micro-bubbles, required the use of a vacuum chamber. This approach for filling the device involves immersing the device in deionized water and placing it in a vacuum chamber. Air is then pumped out of the chamber thereby causing the trapped air in the device to expand and escape. Upon re-pressurization to normal atmospheric pressure, the remaining air is compressed and dissolved in the water drawn into the device. The level of vacuum used was carefully chosen based on the temperature of the chamber to avoid boiling off the water at the lower pressure. The rate of re-pressurization is an important parameter in this process, since if the flow of water is too fast it can damage the MEMS sensing structures. By using this method, the MEMS device was completely filled with deionized water. Furthermore it is fast, easy to set-up and can be scaled up to fill multiple devices simultaneously in a mass production scenario.

Sealing the Structures and Assembling the Test Board

In order to characterize the sensor it has to be ensured that the device remains full with water, without any micro-bubbles for the duration of few weeks. The seal applied for testing purposes was an epoxy resin based UV-A cured adhesive—Panacol Vitralit 1657. Using this method, air only first appeared within the channel a few months after the sensor was sealed. A permanent hermetic seal involves adding an additional glass layer over the holes and using a UV laser to fuse the glasses together, without significantly heating the fluid.

The printed circuit board (PCB) as shown in Figure 6 is designed such that the MEMS device is electrically connected to the PCB through the process of wire bonding. The applied methodology first requires that the die is secured to the PCB with die-attach epoxy (CW200). Next the die is wire-bonded to the board and the wire-bonds are protected with Panacol Vitralit 1671 epoxy. The sensor is subsequently subjected to the above proposed filling and sealing processes.

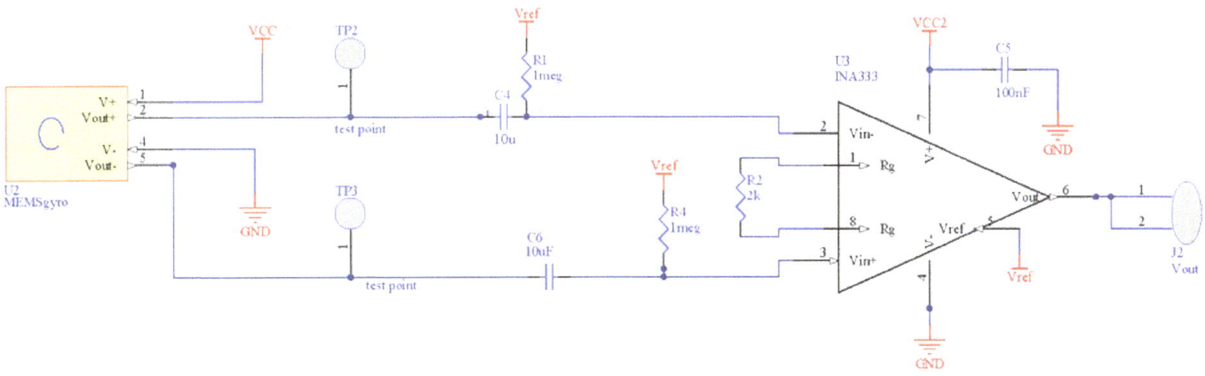

Figure 6: Schematic design of the PCB test board for testing the proposed sensor.

SIMULATED AND MEASURED RESULTS

The simulated frequency response of the sensing elements (cantilevers) of the angular rate sensor, after being designed, was simulated using the Coventor MEMS design software. The displacement versus frequency is shown in Figure 7, where the resonant frequency is at 1.05 MHz, well above the frequencies of interest.

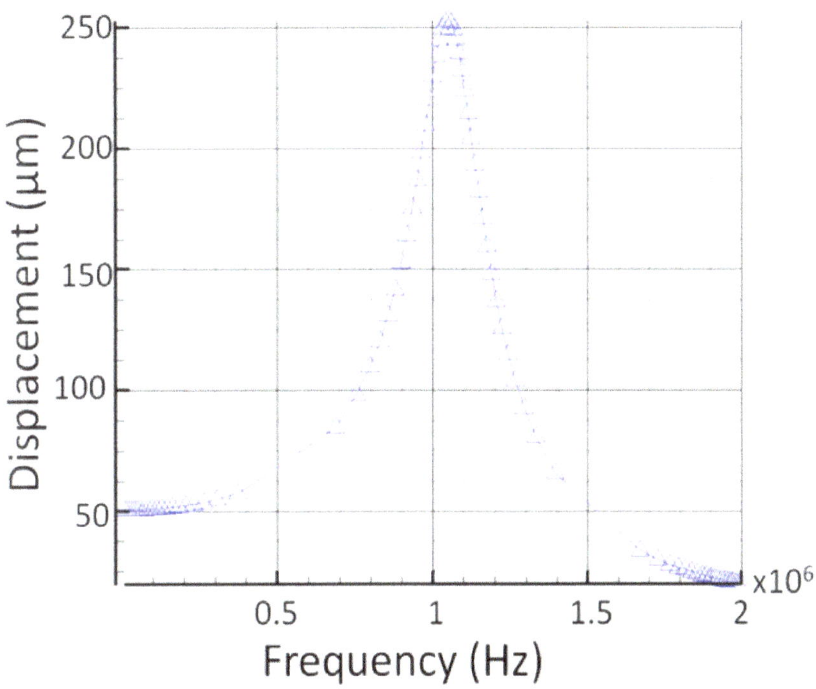

Figure 7: Simulated frequency response of the sensing elements of the proposed sensor.

The angular-rate sensitivity of the sensor can be modified through the selection of the fluidic mass, depending on its density and viscosity. The proposed angular rate sensor was tested after being filled with deionized water and sealed. A PCB integrating the proposed sensor and an instrumentation amplifier was prepared for the measurements. The system was tested using a Tes-3T Motion Dynamic rate table, as shown in Figure 8. The sensor›s Wheatstone bridge output was amplified by an instrumentation amplifier, and routed to the inputs of a National Instruments acquisition card, via the wiring and slip-rings built into the rate table. The motion and the corresponding measurements were controlled through customized Labview scripts that ensured accurate synchronization of the motion and the acquired measurements. The measured results are illustrated in Figure 9a,b, which demonstrate that the sensitivity of the system is comparable to that of humans, whilst keeping the total power consumption down to 300 μW @ 2V, per axis, including the instrumentation amplifier. The power consumption can be attributed to that consumed by the continuous Wheatstone Bridge bias and that of the instrumentation amplifier; in the current proof-of-concept prototype the instrumentation amplifier consumes 100 μW @ 2V, whilst the continuous Wheatstone Bridge bias consumes 200 μW @ 2V. To reduce the power further one could power cycle the Wheatstone Bridge bias to match the sampling rate of the A/D converter. The exact duty cycle would depend on the maximum integrated piezo-resistor value and the maximum capacitance associated with the output node. Furthermore it is possible to reduce the power supply voltage further at the expense of the device sensitivity (see Equation (3)).

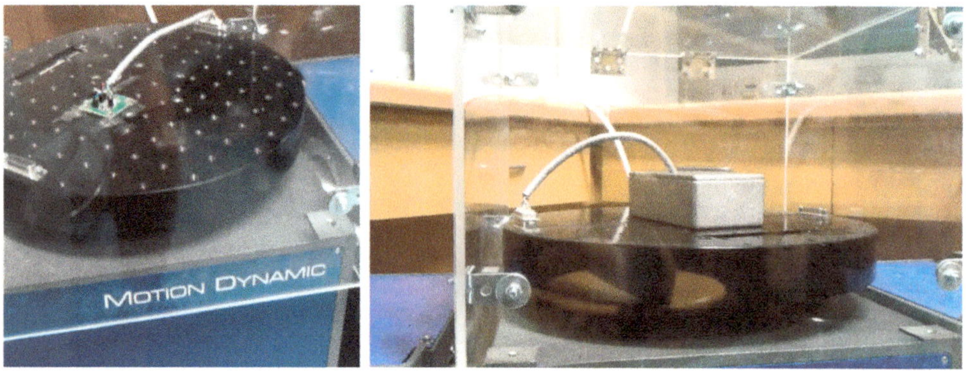

(a) (b)

Figure 8: Test setup for characterization of the proposed angular rate sensor. (**a**) the PCB with the sensor is mounted on top of the rate-table; (**b**) the PCB with the sensor is placed in an aluminum box and mounted on top of the rate-table. The aluminum box ensures electromagnetic shielding of the device under test so as to avoid picking up electrical noise emitted from the rate-table motor.

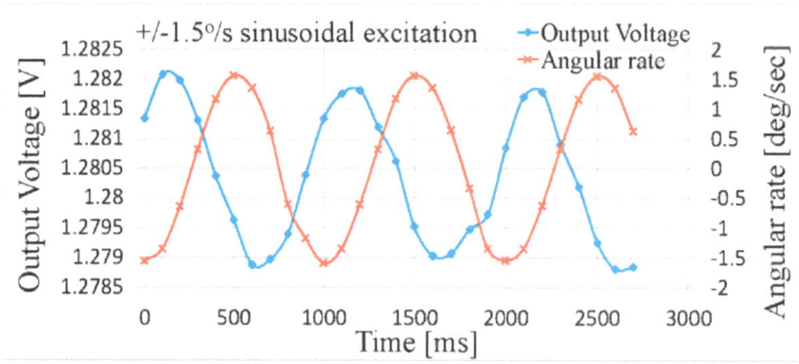

(**a**)

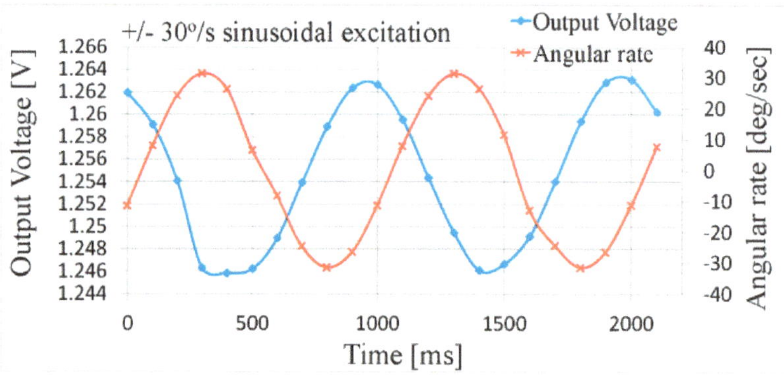

(**b**)

Figure 9: Measured results of the output voltage vs. angular rate. (**a**) Angular rates between 0 °/s and a maximum of ±1.5 °/s (sinusoidally modulated); and (**b**) Angular rates up to ±30 °/s (sinusoidally modulated).

Inherent resistor mismatch between the two legs of the Wheatstone bridge create a small DC component at the output of the sensor. Thus it is necessary to either feed the signal into a high-pass filter (cut-off frequency of 0.1 Hz) prior to being amplified or use an AC coupled amplifier to measure the output of the microfluidic angular rate sensor.

Figure 9a shows the analog output versus time when the sensor is subjected to an angular rate between 0 °/s and a maximum of ±1.5 °/s (sinusoidally modulated). It is clear that the target sensitivity of 1 °/s, which is approximately that of humans, is easily achievable, even without any signal conditioning (e.g., averaging, filtering etc.). Similarly, Figure 9b shows the response of the sensor which is subjected to an angular rate reaching up to ±30 °/s (sinusoidally modulated). For the angular rate of ±1.5 °/s an output voltage of ±1.5 mV is measured, while for ±30 °/s an output voltage of ±8 mV is measured. The approximate 90° lead of the output voltage with respect to the angular rate is because this sensor responds to angular acceleration as described in Equation (1); the angular acceleration is the derivative of the angular rate i.e., the derivative of sinusoidal angular rate input appears as a cosine output voltage. Due to the mechanism used to sense the angular rate the output is not expected to be linear, however this is not a problem for vestibular implants, since the natural system behaves in a similar manner i.e., a form of dynamic range compression takes place. If required, the analog output can be digitalized with a slight increase of the power budget, e.g., 3.1 μW if use MAX1393 ADC at 1.8V with sampling rate of 1 Ksps, totaling approximately 0.3 mW per axis. A 3-axis angular-rate sensor could be realized by mounting three of the proposed sensors perpendicular to each other, within the vestibular implant package, much like the physiological vestibular organ. Alternatively, monolithical integration of these gyros for 3-axis sensing is also possible using the same process, however the sensitivity of other axes would be compromised since the torroidal shape would be greatly deformed and at most two "haircells" could be used.

The natural vestibular system detects angular rates in three perpendicular axes, whilst demonstrating low cross-axis sensitivity. Thus it is important that any rotation that is perpendicular to the sensing axis, of the bioinspired gyroscope presented, should not be detected, i.e., the sensor should have a very low cross-axis sensitivity. In order to verify this feature, the test assembly was designed so as to be able to also mount it perpendicular to the plane of motion of the rate table, and thus be able to test cross axis sensitivity. The measured results of this test are shown in Figure 10, where the sensor was subjected to an amplitude modulated angular rate reaching up to ±3 °/s, for both the sensing axis and a perpendicular axis. It is clearly seen that the cross-axis sensitivity is quite low.

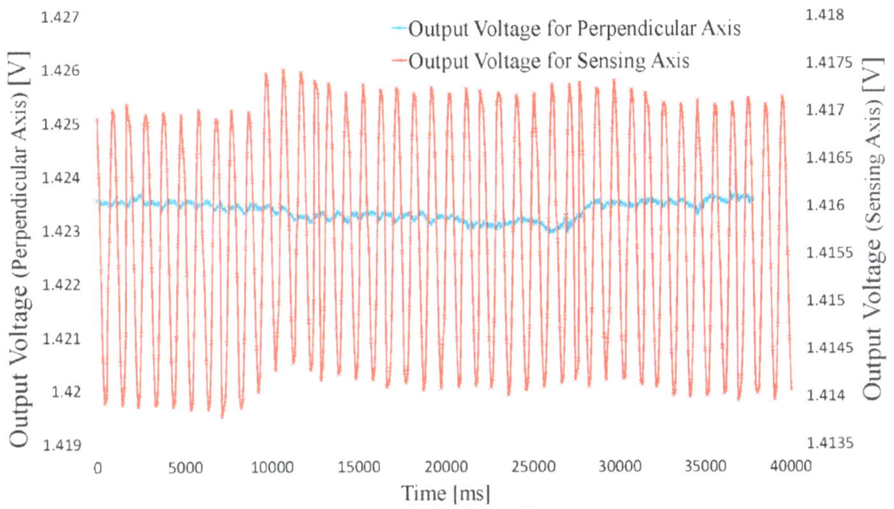

Figure 10: Measured sensor output voltage for angular motion stimuli on the sensing axis and for angular motion stimuli on the cross-axis (the sensor is subjected to the same modulated angular rate of ±3 °/s for both, in-plane and perpendicular motion).

The above tests and measurements demonstrate that this bioinspired, proof-of-concept vestibular gyroscope, fabricated with Sensonor›s standard MEMS technology, has much potential and functions very similarly to the predictions made with back-of-the-envelope calculations [29]. The proposed angular rate sensor has also demonstrated that it can clearly detect motion, even at angular rates below 1 °/s without needing to actively excite the sensing mass, thus providing significant power savings. The MEMS technology used to fabricate the sensor utilizes doped p-type conductors within an n-type substrate for signal routing. The result is that several reverse biased parasitic diodes exist in the structure, which is noisier than if metal routing was used. We believe that with a customized MEMS process the performance of this gyroscope could be improved further, allowing the size to be reduced. Furthermore the measured analog signal was obtained through a relatively noisy connection, provided via the slip rings of the rate table. A summary of the sensor›s specifications is given inTable 1. The total area of the sensor die is 6 mm × 6 mm, as shown in Figure 11a,b. This is approximately the same size as the natural semicircular canals. Similarly the sensor sensitivity is approximately that of the natural vestibular system i.e., 1–2 °/s [31].

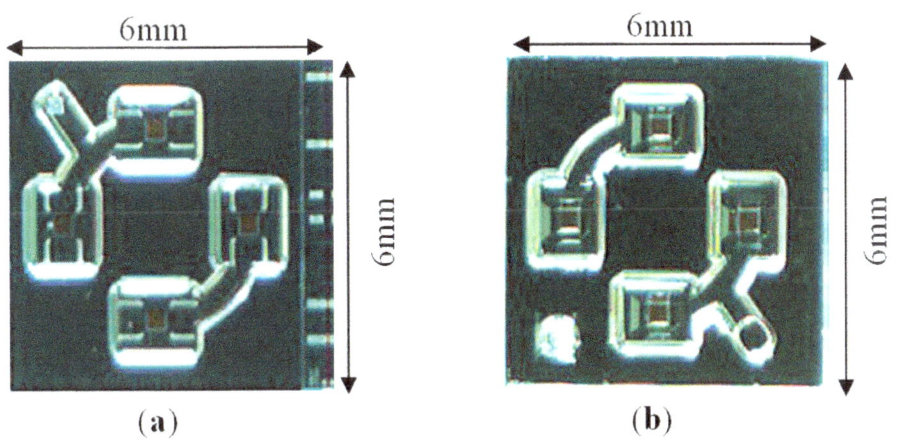

Figure 11: Microphotographs of the gyroscopic sensor structure. (**a**) Top view of the gyro; (**b**) Bottom view of the gyro.

Table 1: Summary of the Specifications of the proposed angular-rate sensor

Parameter	Performance
Sensitivity	< 1 °/s
Supply Voltage	2V
Power Consumption	300 μW
Cantilever Thickness	3.1 ± 0.3 μm
Chip Size (Area)	6 mm × 6 mm
Technology	Sensonor's standard MEMS

CONCLUSIONS

A bio-inspired, proof-of-concept, ultra-low power, MEMS-microfluidic angular-rate sensor, which does not require continually vibrating structures, has been designed, simulated, fabricated and successfully tested. The size of the chip is 6 mm × 6 mm including the pads area and it consumes a mere of 300 μW at 2 V. The proposed gyroscope›s low-power consumption and comparable performance to the vestibular system, which is able to detect angular rates as low as 1 °/s, may enable totally implantable vestibular implants to be developed in the near future. This principle may also be applied to other applications where stringent power constraints prohibit the use of existing gyroscope technology.

ACKNOWLEDGMENTS

This work was partially funded by MED-EL Elektromedizinische Gerate GmbH.

AUTHOR CONTRIBUTIONS

Charalambos M. Andreou designed the sensor, contributed in testing the sensor and prepared this paper. Yiannis Pahitas contributed in testing the sensor. Julius Georgiou conceived the idea, acquired funding and supervised the work presented in this paper.

REFERENCES

1. Della Santina, C.C.; Migliaccio, A.A.; Patel, A.H. A multichannel semicircular canal neural prosthesis using electrical stimulation to restore 3-d vestibular sensation. IEEE Trans. Biomed. Eng. 2007, 54, 1016–1030.
2. Wall, C., III; Merfeld, D.M.; Rauch, S.D.; Black, F.O. Vestibular Prostheses: The Engineering and Biomedical Issues. J. Vestibul. Res. 2003, 12, 95–113.
3. Valentin, N.S.; Hageman, K.N.; Dai, C.; Della Santina, C.C.; Fridman, G.Y. Development of a Multichannel Vestibular Prosthesis Prototype by Modification of a Commercially Available Cochlear Implant. IEEE Trans. Neural Syst. Rehabil. Eng. 2013, 21, 830–839.
4. Gong, W.; Merfeld, D.M. Prototype neural semicircular canal prosthesis using patterned electrical stimulation. Ann. Biomed. Eng. 2000, 28, 572–581.
5. Constandinou, T.G.; Georgiou, J.; Toumazou, C. A partial-current-steering biphasic stimulation driver for vestibular prostheses. IEEE Trans. Biomed. Circuits Syst. 2008, 2, 106–113.
6. Dai, J.; Demosthenous, A.; Perkins, T.; Liu, X.; Donaldson, N. A stimulator ASIC featuring versatile management for vestibular prostheses. IEEE Trans. Biomed. Circuits Syst. 2011, 5, 147–159.
7. Shkel, A.M.; Zeng, F. An Electronic Prosthesis Mimicking the Dynamic Vestibular Function. Audiol. Neurotol. 2006, 11, 113–122.
8. Zeng, H.; Zhao, Y. Sensing movement: Microsensors for body motion measurement. Sensors 2011, 11, 638–660.
9. Shkel, A.M.; Liu, J.; Ikei, C.; Zeng, F. Feasibility Study on a Prototype of Vestibular Implant Using MEMS Gyroscopes. Proceedings of the IEEE Sensors, Orlando, FL, USA, 12–14 June 2002; pp. 1526–1531.
10. Vestibular System. Available online: http://en.wikipedia.org/wiki/Vestibular_system (accessed on 17 July 2014).
11. Chiang, B.; Fridman, G.; Dai, C.; Rahman, M.A.; Della Santina, C.C. Design and performance of a multichannel vestibular prosthesis that restores semicircular canal sensation in rhesus monkey. IEEE Trans. Neural Syst. Rehabil. Eng. 2011, 19, 588–598.
12. Xia, D.; Yu, C.; Kong, L. The Development of Micromachined Gyroscope Structure and Circuitry Technology. Sensors2014, 14, 1394–1473.
13. Wu, L.F.; Peng, Z.; Zhang, F.X. Design and Implementation of a Digital Angular Rate Sensor. Sensors 2010, 10, 9581–9589.
14. Wang, W.; Lv, X.; Sun, F. Design of Micromachined Vibratory Gyroscope with Two Degree-of-Freedom Drive-Mode and Sense-Mode. IEEE Sens. J. 2012, 12, 2460–2464.
15. Wang, W.; Lv, X.Y.; Sun, F. Design of a novel MEMS gyroscope array. Sensors 2013, 13, 1651–1663.
16. Xie, L.; Wu, X.; Li, S.; Wang, H.; Su, J.; Dong, P. A z-axis quartz cross-fork micromachined gyroscope based on shear stress detection. Sensors 2010, 10, 1573–1588.
17. Si, C.; Han, G.; Ning, J.; Yang, F. Bandwidth Optimization Design of a Multi Degree of Freedom MEMS Gyroscope.Sensors 2013, 13, 10550–10560.
18. Feng, L.; Zhao, K.; Sun, Y.; Cui, J.; Cui, F.; Yang, A. A multi-fork z-axis quartz micromachined gyroscope. Sensors 2013,13, 12482–12496.
19. Sharma, A.; Zaman, M.F.; Ayazi, F. A Sub-0.2°/hr

bias drift micromechanical silicon gyroscope with automatic CMOS mode-matching. IEEE J. Solid-State Circuits 2009, 44, 1593–1608.

20. Bagdahn, J.; Schischka, J.; Petzold, M.; Sharpe, W.N., Jr. Fracture toughness and fatigue investigations of polycrystalline silicon. Proceedings of SPIE Reliability, Testing, and Characterization of MEMS/M0EMS, San Francisco, CA, USA, 22 October 2001; 4558, pp. 159–168.
21. Muhlstein, C.L.; Brown, S.B.; Ritchie, R.O. High-cycle fatigue of single-crystal silicon thin films. J. Microelectromech. Syst. 2001, 10, 593–600.
22. Constandinou, T.G.; Georgiou, J.; Andreou, C. An ultra-low-power micro-optoelectromechanical tilt sensor. Proceedings of the IEEE International Symposium on Circuits and Systems, Seattle, WA, USA, 18–21 May 2008; pp. 3158–3161.
23. Bhatti, P.T.; McClain, M.A. Low-power sensing for vestibular prostheses. Proceedings of the IEEE Annual International Conference of the Engineering in Medicine and Biology Society, Boston, MA, USA, 30 August–3 September 2011; pp. 3131–3134.
24. Groenesteijn, J.; Droogendijk, H.; de Boer, M.J.; Sanders, R.G.P.; Wiegerink, R.J.; Krijnen, G.J.M. An angular acceleration sensor inspired by the vestibular system with a fully circular fluid-channel and thermal read-out. Proceedings of the IEEE International Conference on Micro Electro Mechanical Systems, San Francisco, CA, USA, 26–30 January 2014; pp. 696–699.
25. Alrowais, H.; Brand, O.; Bhatti, P.T. Design, simulation and fabrication of thermal angular accelerometers. Proceedings of the Solid-State Sensors, Actuators and Microsystems Workshop, Hilton Head, SC, USA, 8–12 June 2014.
26. Zhou, Z.; Liu, Z. Biomimetic cilia based on MEMS technology. J. Bionic Eng. 2008, 5, 358–365.
27. Chen, N.; Tucker, C.; Engel, J.M.; Yang, Y.; Pandya, S.; Liu, C. Design and characterization of artificial haircell sensor for flow sensing with ultrahigh velocity and angular sensitivity. IEEE J. Microelectromech. Syst. 2007, 16, 999–1014.
28. Wang, W.; Liu, J.L.; Xie, X.; Liu, M.H.; He, S.T. Development of a new surface acoustic wave based gyroscope on a X-112°Y $LiTaO_3$ substrate. Sensors 2011, 11, 10894–10906.
29. Andreou, C.M.; Pahitas, Y.; Pilavaki, E.; Georgiou, J. Bio-mimetic Gyroscopic Sensor for Vestibular Prostheses. Proceedings of the IEEE Biomedical Circuits and Systems Conference, Rotterdam, The Netherlands, 31 October–2 November 2013; pp. 17–20.
30. Van Buskirk, W.C.; Watts, R.G.; Liu, Y.K. The fluid mechanics of the semicircular canals. J. Fluid Mech. 1976, 78, 87–98.
31. Jamali, M.; Mitchell, D.E.; Dale, A.; Carriot, J.; Sadeghi, S.G.; Cullen, K.E. Neuronal detection thresholds during vestibular compensation: Contributions of response variability and sensory substitution. J. Physiol. 2014, 592, 1565–1580.

CHAPTER 19

RICE HUSK ASH TO STABILIZE HEAVY METALS CONTAINED IN MUNICIPAL SOLID WASTE INCINERATION FLY ASH

Laura Benassi, Federica Franchi, Daniele Catina, Flavio Cioffi, Nicola Rodella, Laura Borgese, Michela Pasquali, Laura E. Depero and Elza Bontempi

Keywords: Fly ash, rice husk ash, DIC, heavy metals stabilization, cosmos-rice

INTRODUCTION

Because of poor combustion efficiency and the lack of air pollution control systems, from its early beginning in the 1870s, waste incineration was considered a very large source of pollution [1]. From 1950, the first efforts to limit the pollution of incinerators were imposed by law [2]. However these first limitations did not consider dust emissions, until 1970, when an aggravation of dust emission standards occurred. As a consequence, in the next years, the development of the air pollution control systems started [1].

Today, while modern waste incineration represents a consolidate technology for volume and mass reduction of the waste, combined with efficient energy recovery and systems to reduce and treat emissions, some concerns regarding the production of toxic waste still exist. In particular, fly ash generated by the air pollution control systems are hazardous waste materials for which landfilling is still considered the most appropriate management strategy. Indeed, this ash contains leachable toxic metals, as for example Pb and Zn [3].

Great efforts of researchers are now devoted to developing sustainable stabilization technologies able to reuse this toxic ash, also in view of the European need of new solutions for raw materials substitution that are employed in several applications.

In recent works, we proposed the use of silica extracted from rice husk ash (RHA) for heavy metals stabilization of municipal solid waste incineration (MSWI) fly ash [4]. The new proposed technology was named COSMOS-RICE.

The idea to use this agricultural byproduct is due to its high availability. World rice production in 2011–2012 exceeded 480 million tons [5]. Rice husk constitutes about 20 wt % of the paddy production and it is an abundantly available biomass, commonly employed as a fuel. Its combustion product, rice husk ash (RHA), usually contains more than 60% silica (SiO_2), and 10%–40% carbon and other minor mineral components. Due to the lower density of RHA, and consequently its bulky form, its disposal can become a problem. In some areas, a large amount of RHA is treated as a waste and landfilled, leading to air and water pollution. Airborne RHA particles have been linked to respiratory disease in humans [6]. Because rice husk is abundantly available and it can be converted into RHA after being subjected to an almost zero-cost thermal valorization process, more and more researchers have become interested in how to use this waste as a resource [7].

Citation: Laura Benassi, Federica Franchi, Daniele Catina, Flavio Cioffi, Nicola Rodella, Laura Borgese, Michela Pasquali, Laura E. Depero and Elza Bontempi, Rice Husk Ash to Stabilize Heavy Metals Contained in Municipal Solid Waste Incineration Fly Ash, doi:10.3390/ma8105346.

The stabilization mechanism involved in the COSMOS-RICE technology was extensively discussed [8]: it is attributed to two different reactions, due to the amorphous silica and carbonation. Indeed, the stabilization mechanism is mainly due to the capability of amorphous silica to entrap heavy metals, but this does not only occur during laboratory treatment, it also proceeds over time thanks to carbonation. The proposed method was very recently improved by the direct use of RHA without passing through silica extraction [8]. Since the organic part of rice husk (RH) can easily be removed by thermal decomposition, thermal treatment of RH is an approach to obtain, not only energy, but also silica from rice husk.

Production of biochar from rice husk has also attracted many researchers' attention. This is a versatile material with good adsorption properties. Biochar has a relatively structured carbon matrix with a high degree of porosity and an extensive surface area. It may act as an adsorbent, which is similar in some aspects to activated carbon and it can play an important role in controlling contaminants in the environment [9]. Chemical impregnation with KOH or NaOH of RHA, followed by activation at 650–850 °C, generates a material with extremely high surface area [10]. Even though biochar from rice husk has already been employed for removing heavy metals [11], it was never considered in the COSMOS-RICE technology.

To also obtain biochar from RHA, pyrolysis of rice husk was realized at 500 °C in O_2-limiting conditions.

To evaluate the effect on heavy metals stabilization by a pre-treatment of rice husk, the employment of Instant Pressure Drop (DIC) technology [12] was proposed for the first time.

The DIC technology is based on a thermo-mechanical process that requires some moisture level in the material to be treated. A rapid pressure drop occurs during the material heating, producing bursts of moisture evaporation inside the bulk of material, with the occurrence of a structural damage. It was reported that DIC pre-treatment is able to make a material porous and expanded within short treatment time [13]. This process is based on the use of vacuum and temperature. Moreover, material properties are fundamental for obtaining different final characteristics of the product.

Several research papers have been published in recent years exploring the use of DIC technology, in particular for food treatment [14]. Some applications that have been investigated include material sterilization, removal of certain inhibitors in legumes, structural expansion of fruit and vegetables, drying process acceleration, enhancement of extraction of essential oils, coffee extraction, and preservation of food materials [15,16,17]. DIC technology was also applied on rice preservation [18]. However, it was never employed as a pre-treatment of rice husk.

This paper reports the first work about the application of DIC as a possible new pre-treatment technology for the waste material. The aim of this project is to obtain a safe filler that can be applied in several application, to produce "green composites". This pioneering work explores the viability of applying DIC technology to pre-treat other waste types.

MATERIALS AND METHOD

Pretreatment of Rice Husk Samples by DIC Technology

Rice husk was provided by an Italian factory (Lombardy region).

DIC technology is extensively described in several papers (see the Introduction Section). In the present work, we employed an experimental reactor designed and realized by Contento Trade srl.

The laboratory DIC plant is composed of a 5 L stainless steel reactor, a refrigerated stainless steel vacuum chamber with 500 L of capacity, a vacuum pump, a steam generator and a steam overheater, and is completely controlled by software developed with Labview. It is able to carry out simple or cyclic DIC tests up to a maximum temperature and pressure of 180 °C and 16 bar, respectively.

Figure 1 shows a picture of the instrument.

Figure 1: Picture of the employed experimental DIC instrument.

For the DIC pre-treatment, rice husk samples were enclosed in the stainless steel container at atmospheric pressure and then this was closed. An initial vacuum was performed. Afterwards, saturated steam was injected into the reactor. The abrupt pressure drop towards a vacuum was carried out by an abrupt opening (<0.2 s) of a valve connecting the reactor with the steel vacuum chamber. This abrupt adiabatic pressure drop triggered auto-vaporization of superheated water contained in the material, instantaneous cooling, structure swelling, and even rupture of the structure.

The best texturing process allows obtaining the final product at lower density and with less residual water content. For this aim, several DIC pre-treatments were tested, changing pressure steam, temperature in the chambers, maximum applied pressure, conditioning time and number of repetitions. Based on these preliminary tests, two DIC pre-treatments were identified as the most suitable to obtain a final moisture content of about 10%. After the DIC pre-treatments, samples were thermally annealed in a Nabertherm electric oven at different temperatures for 270 min (see Table 1 at the end of the experimental section).

Pyrolysis of rice husk, to obtain a carbon-rich solid residue, was realized with thermal treatments at 500 °C in sealed sample-holders (*i.e.*, under O_2-limited conditions), inserted in the oven.

Table 1 summarizes the samples description and treatments. Samples that have been pre-treated by DIC were named T1 and T2. Samples that were not pre-treated are named TQ.

Table 1: Samples description and treatments. All the thermal annealing were made for 270 min

Sample ID	DIC Treatment	Annealing Temperature (°C)	O_2-Limited Conditions
TQ 500	NO	500	No
TQ 550	NO	550	No
TQ 600	NO	600	No
TQ 500 B	NO	500	Yes
T1 500	T1	500	No
T1 550	T1	550	No
T1 600	T1	600	No
T1 500 B	T1	500	Yes
T2 500	T2	500	No
T2 550	T2	550	No
T2 600	T2	600	No
T2 500 B	T2	500	Yes

DIC Treatment	Steam Pressure	Max. Temperature	Max. Pressure	Cycle Repetitions
T1	3 bar	150 °C	0.075 br	1
T2	3 bar	150 °C	0.050 bar	3

Determination of Si in Rice Husk Ash Samples

To evaluate the silica content in RHA, a chemical extraction procedure was realized. Fifty grams of 1 N NaOH were added to 0.5 g of RHA in a low density polyethylene container maintained at room temperature for 24 h under constant agitation to dissolve the silica and produce a sodium silicate solution; then, the solutions were filtered and analyzed by ICP-MS to determinate the silica content according to EPA 6010C 2007 method. The extracted silica content resulted 47% (±12%) and 23% (±2%) for samples treated in the open and sealed sample-holders, respectively. These results highlight the lower content of the extracted silica in samples thermally treated in oxygen deficient conditions, in respect to the samples thermally treated in air.

Morphological and Structural Characterization of Samples

Morphologic characteristics of rice husk ash sample were investigated by SEM and XRD. X-Ray Diffraction (XRD) measurements were performed with Panalytical X'Pert Pro diffractometer (University of Brescia, Brescia, Italy) equipped with the X'Celerator detector and Cu anode. Operating values were 40 kV and 40 mA.

The rice husk ashes morphology was investigated with Scanning Electron Microscopy, LEO EVO 40 (University of Brescia, Brescia, Italy) equipped with an energy dispersive microprobe (EDS) quantitative analysis.

Stabilization of Municipal Solid Waste Incineration Fly Ash

Stabilized samples made from MSWI fly ash by means of COSMOS-RICE technology were prepared by adding to the MSWI fly ash (provided by an Italian incinerator), other waste materials: *i.e.*, coal fly ash, rice husk ash (RHA), and flue gas desulfurization (FGD) residues, in accordance to the already established procedure [8]. All the samples were prepared following the same protocol, *i.e.*, by adding to a mixture of the three powders, MSWI fly ash, FGD residues, and coal fly ash (in the weight quantity of 130 g, 10 g, and 15 g, respectively), 1.6 g of RHA (about 1% of the total weight of ash), wetting with 170 g of water, and stirring for 30 min in a laboratory mixer at 60 °C [4]. Generally, the quantity of added silica in the COSMOS technology, ranges between 5% and 10% of the total weight of ash. The lower quantity of RHA used in this work was chosen to be sure that the complete stabilization of heavy metals does not occur. This is necessary to compare the results obtained by using different RHA pre-treatment conditions and to select the best pre-treatment.

After chemical stabilization of heavy metals, the solid inertized material can be washed out to recover almost pure soluble salts [19]. In the present case, salts recovery was not realized.

Leaching Tests

Leaching tests were performed when samples resulted completely dried (30 days after the stabilization treatment).

Leaching tests were carried out according to the CEN normative [20], in order to quantify the leachability of heavy metals in water. Tests were performed mixing the dried and homogenized sample with MilliQ water at a liquid to solid ratio of 10 L/kg. The particle size was below 4 mm (with size reduction) and a temperature in compliance with the Directive (about 20 °C). The contact time of materials and aqueous solution was 2 h. This time was considered sufficient to establish equilibrium in subsequent experiments with MSWI fly ash [21].

Chemical analysis of the leachate solution was done by means of Total Reflection X-Ray Fluorescence technique (TXRF) by the Bruker TXRF system S2 Picofox (University of Brescia, Brescia, Italy) (air cooled, Mo tube, Silicon-Drift Detector), with operating values of 50 kV and 750 μA using an acquisition time of 600 s. TXRF quantitative analysis of the suspended samples was performed by the internal standard procedure. A proper amount of gallium, used as an internal standard element, was added [22]. Every analysis has been replicated three times.

RESULTS AND DISCUSSION

Figure 2 shows X-ray diffraction analysis of TQ rice husk ash treated at different temperatures and oven conditions (seeTable 1). The XRD patterns of samples T1 and T2 (DIC pre-treated samples), thermally treated in the same conditions are very similar to corresponding reported TQ samples, hence, for the purposes of clarity, they were not shown in the Figure. From XRD patterns shown in Figure 2, it appears that all samples are amorphous, with one peak, due to the presence of a crystalline phase.

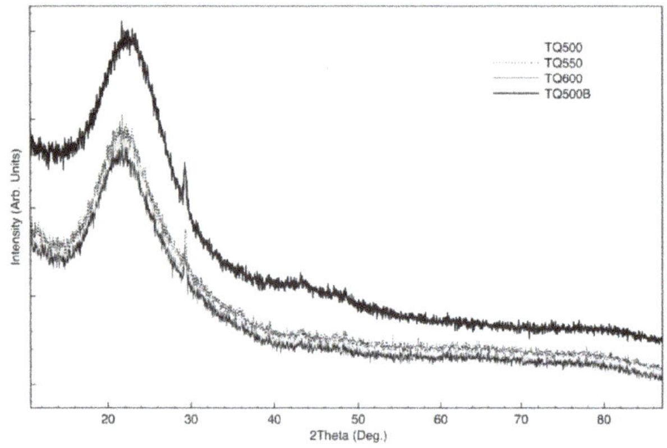

Figure 2: X-ray diffraction analysis of TQ rice husk ashes treated at different temperatures (see Table 1).

Despite the fact that all patterns can essentially be attributed to the presence of amorphous materials, it is evident that basic materials structures are different: indeed, the integrated areas of the XRD patterns are almost the same for samples treated in O2-limited conditions, but they are extremely different from those regarding samples not treated in O2-limited conditions. Moreover, for the last samples, the integrated XRD area is almost the same.

Samples TQ 500, 550 and 600 show a large and intense halo, in the range of 18°–28°. In the literature it was reported that, for similar samples, the typical silica oxide characteristic broad peak is centered at $2\theta = 22.5°$, which can be attributed to the presence of disordered cristobalite [23]. Moreover, the sample TQ 500 B shows a different XRD spectra, with higher intensities, compared to the other shown in the same Figure (the XRD spectra of T1 600 B and T2 600 B are very similar, then are not reported).

The background intensity of sample TQ 500 B, which is extremely high in respect to other samples, can be caused by the amorphous carbon present in the RHA [24].

In particular, for sample TQ 500 B there is a large halo between 16° and 30°), which can be attributed to (002) stacking of the graphitic basal plans in chars [25,26]. Indeed, this halo results at slight higher diffraction angles, in comparison to peak attributed to the silica.

Regarding sample TQ 500 B, another large halo can be found between 38° and 50°. For carbon prepared at different pyrolysis temperatures and for commercial carbon black, X-ray diffraction patterns show two peaks at about 24° and 44°, which were assigned to the planes of graphite (002) and (100) [27].

Therefore, XRD patterns collected on RHA allow concluding that samples annealed in not O2-limited conditions are mainly made of amorphous silica. On the contrary, samples obtained in O2-limited conditions show an important content of biochar. These results are in accord with data about silica content, obtained by NaOH solubilization. Indeed, the content of silica resulted extremely lower for samples treated in the O2-limited conditions, in respect to other samples (see the Experimental Section).

RHA powders were employed for the COSMOS-RICE stabilization tests conducted with MSWI fly ash.

Table 2 reports the elements concentration in the solutions obtained by making the leaching test on MSWI fly ash, before the stabilization treatment. It appears evident that solution contains some metals. In particular, the toxic elements that can be found in high concentration are Pb and Zn. As it be seen in Table 2, the concentrations of Pb and Zn are 60 and 14 mg/L, respectively.

Table 2: Elements concentration and limits of detection (LD) in leaching solutions of MSWI fly ash

Element	Conc. (mg/L)	LD (mg/L)
Cl	9287 ± 928	0.161
K	1385 ± 138	0.092
Ca	6870 ± 687	0.068
Zn	14.3 ± 1.4	0.004
As	1.6 ± 0.2	0.004
Br	118 ± 12	0.004
Rb	4.0 ± 0.4	0.003
Sr	6.2 ± 0.6	0.003
Ba	8.7 ± 0.9	0.033
Pb	60 ± 6	0.004

To verify the stabilization efficacy of the different obtained RHAs, one month after the stabilization procedure (samples were left to dry at room temperature for one month) leaching tests were realized. The data about the concentration of Pb and Zn in the leaching solutions of stabilized samples are shown in Figure 3.

It is very important to highlight that the COSMOS-RICE conditions of synthesis were settled to obtain incomplete stabilization reactions: *i.e.*, the amount of added RHA was about 1% of the total weight of ash and thus, was much lower than that employed in previous works (around 20% in weight of the global fly ash mix) [4,19,28]. This allows finding residual soluble metals in the leaching solutions, to have the possibility to compare the results obtained by using RHA treated in different manners and discriminate the efficacy of different employed RHAs. As a consequence, the samples that have been most effectively stabilized exhibit lower Pb and Zn concentrations in their leaching solutions (see Figure 3). Comparing all the data shown in Figure 3, it results that, despite the low amount of used RHA, the stabilization of MSWI fly ash was obtained: in the leaching solutions, Zn and Pb were reduced at values generally less than one order of magnitude (for example, in sample T2 550, Pb was about 3 mg/L and Zn 1 mg/L).

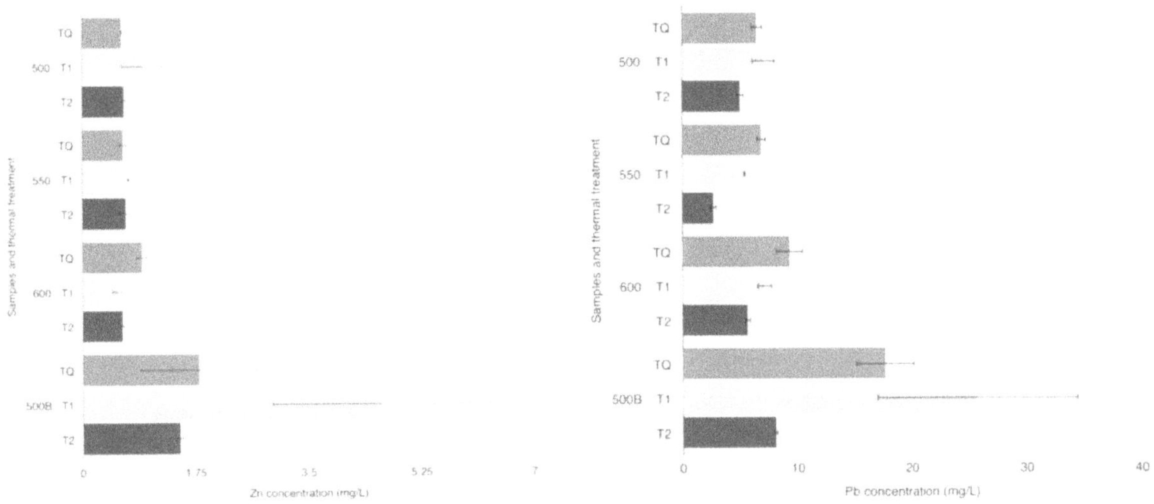

(a) (b)

Figure 3: Concentration of Zn (**a**) and Pb (**b**) in the leaching solutions of the COSMOS-RICE samples obtained by using the whole RHA samples (see Table 1).

In particular, there are two effects to investigate: the presence of biochar in the RHA and the DIC pre-treatments. The first consideration that can be deduced by comparing the data reported in Figure 3 concerns the biochar: RHAs that also contains biochar (see the corresponding stabilized samples, TQ 500 B, T1 500 B, and T2 500 B) are globally less efficient in the heavy metals stabilization because of the higher amount of Pb and Zn found in the leaching solutions. On the basis of these results, it is possible to conclude that the stabilization efficacy of COSMOS-RICE technology is reduced when silica content in RHA is decreased. This is an unexpected result, because there are several examples of use of biochar derived from rice husk as metal absorber reported in the literature [9]. However, the data shown in Figure 3 highlight that the stabilization is better obtained by silica in respect to biochar contained in RHA.

The data in Figure 3 are analyzed in more details below, in order to evaluate the role of the DIC treatment in the stabilization process.

Considering Zn, its concentration is very low (it is generally lower than 1 mg/L), except for samples TQ 500 B, T1 500 B, and T2 500 B, where the RHA do not contain mainly silica but also biochar.

Because of the Zn concentration values are very low, it is not possible to enhance a role of DIC pre-treatment in the stabilization of Zn. However, in the case of samples also containing biochar, it seems that T2 DIC pre-treatment is more efficient in comparison to T1 and not DIC not pre-treated sample (TQ).

Data about Pb concentration in the leaching solutions of stabilized samples are about one order of magnitude higher than Zn. Then, some differences about results of stabilization performed using differently treated RHA are more evident. Considering RHA samples treated in the same thermal conditions, but with different pre-treatment, it appears that leaching solution of samples pre-treated with DIC (T1 and T2) show lower Pb concentration, in respect to TQ (except for samples TQ 500 B, T1 500 B, and T2 500 B). In addition, it clearly appears that T2 pre-treatment results the most efficient in all cases. This is probably also due to the fact that T2 involves three DIC cycle repetitions, and thus it resulted more incisive.

The slight differences among the effectiveness of stabilization procedure for all T1 (and T2) samples can be due to the thermal annealing temperature. Generally, a higher annealing temperature allows obtaining a higher silica content in the RHA, which allows a better heavy metals stabilization [29], in respect to lower annealing temperatures. Indeed, the degradation of organic part present in rice husk occurs starting from 350 °C until about 480 °C [30]. On the other hand, an increase of the annealing temperature can promote the amorphous silica crystallization, with a reduction of the amount of reactive silica. As a consequence, it is extremely difficult, at these temperatures, to have an idea of the predominant factor, related to the amount of amorphous silica present in the RHA.

SEM images of pre-treated rice husk samples and thermally annealed in different conditions are reported in Figure 4.

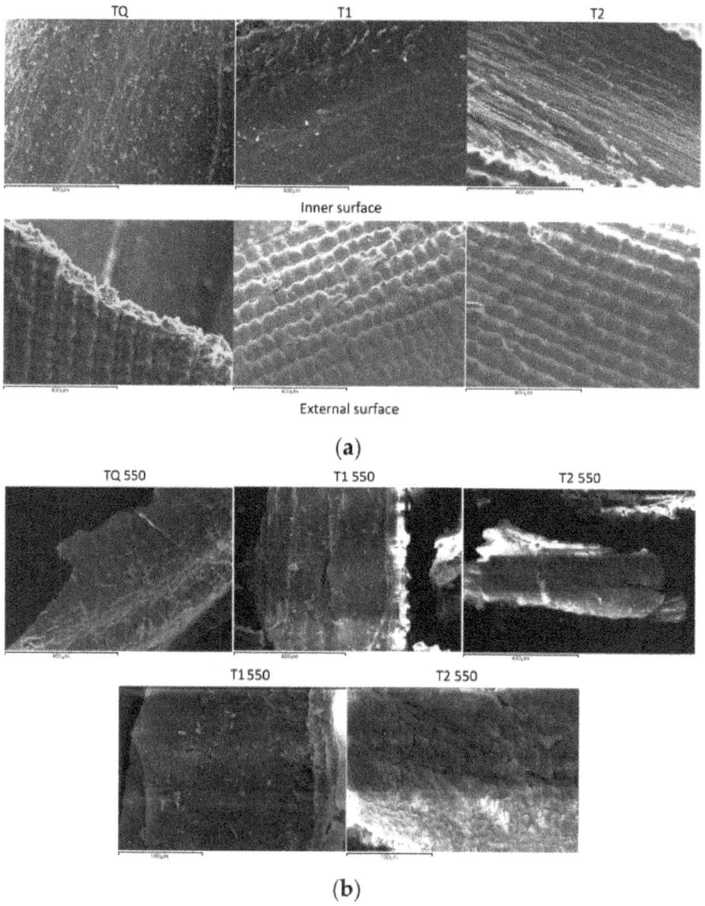

Figure 4: (**a**) SEM images of internal and external surfaces of untreated rice husk (TQ) and pre-treated by T1 and T2 DIC treatments. (**b**) SEM images of internal surfaces of corresponding RHA shown in Figure 4a.

Figure 4a shows the inner and external surfaces of rice husk samples, not pre-treated and pre-treated DIC. As revealed in the SEM image, the external surface of the whole rice husk samples is covered with small irregularly shaped particles that are well aligned. Such domes are shown to be rich in silica. On the contrary, the internal surface of rice husk looks much smoother, and contains a much lower concentration of silicon [31]. External surfaces of all samples do not show any apparent morphological difference. Instead, Figure 4a highlight that internal surface of sample T1, and especially sample T2, seems to be more structured in respect to that of TQ sample, with the appearance of some channels. Since the internal part of rice husk contains less amount of silica, it is possible to conclude that DIC seems to produce some structural changes on the organic part of the sample. Indeed, for organic matter it often produces a structure swelling.

Figure 4b reports the SEM images of corresponding samples of Figure 4a, but after thermal treatment. It was chosen to show the morphology of RHA samples TQ 550, T2 550 and T1 550, to highlight possible morphological differences between samples pre-treated by DIC and not pre-treated (TQ), on the basis of leaching tests results reported in Figure 3. External surface images are not reported because no differences can be highlighted in the SEM images. It appears evident that, comparing the TQ 550 samples with the corresponding two other samples pre-treated by DIC technology, some morphological differences occurred. Sample T1 550 shows a more structured morphology, very similar to that one obtained for sample T2, before making any thermal annealing (see Figure 4a). The appearance of some internal channels occurs that result more evidently in respect to T2 sample. This is probably due to thermal treatment at 550 °C, which produce degradation of organic part of rice husk. Indeed, this occurs starting from 350 °C till about to 480 °C [30], and that may happen in a different manner in respect to TQ sample, because the DIC pre-treatment is done.

Sample T1 and T2 550 have large internal cavities and a more open structure. In order to highlight this strange morphology for these samples, some images were also taken at high magnification (see Figure 4b). These images reveal a great impact of DIC on the microstructure of pre-treated samples. It can be noted that samples T1 and T2 550 show an inner morphology, made by irregular particles with slit-shaped surfaces that provide an internal porous structure producing a higher surface area in respect to TQ sample. In particular, the sample that shows best results in terms of heavy metals stabilization (T2 550) shows particles with dimensions lower in respect to T1 550. This may results in a better RHA degradation and then, in a higher reactivity of RHA.

Figure 5 shows the XRD patterns collected on COSMOS-RICE samples, obtained by using TQ RHA. All COSMOS-RICE samples, obtained by employing the RHA reported in Table 1, show very similar XRD patterns. For this reason, only those of stabilized samples obtained by using TQ RHA are shown.

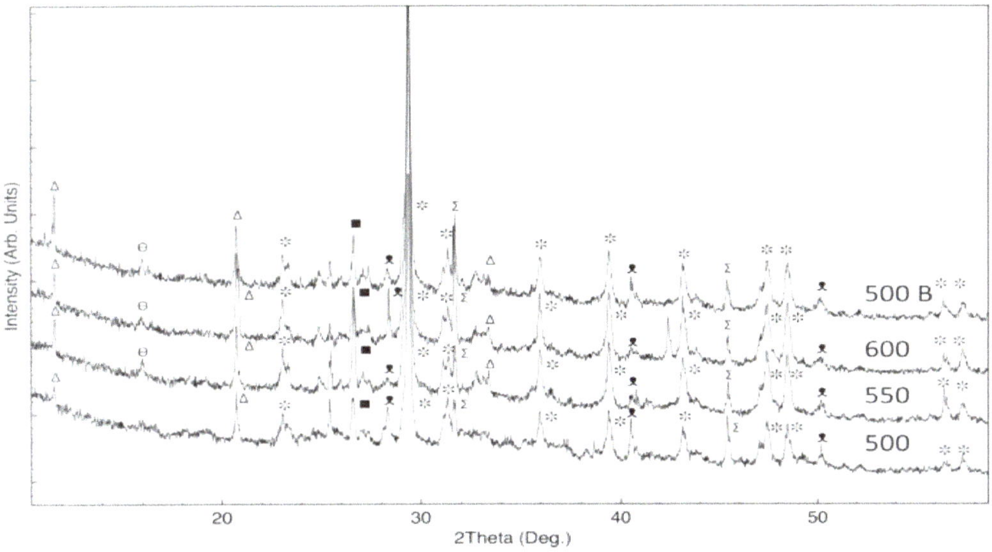

Figure 5: XRD patterns collected on COSMOS-RICE samples, obtained by using TQ RHA (see Table 1). ✻: Calcite ($CaCO_3$); Σ: Halite (NaCl); Δ: Gypsum [$CaSO_4 \cdot 2(H_2O)$]; θ: Thaumasite [$Ca_3(SO_4)[Si(OH)_6](CO_3) \cdot 12(H_2O)$]; ▪: Quartz ($SiO_2$); Potassium chloride (KCl).

It is possible to see that all the samples contain calcite, gypsum, quartz, sylvite, halite, and thaumasite; like the phase mineralogy already reported for samples obtained by COSMOS-RICE technology. These phases are compatible with the employ of this material as a filler, to produce several composites. The stabilization mechanism involving amorphous silica reactions and carbonation has been already reported and discussed [8]. In particular, data shown in Figure 5 allows highlighting that the use of RHA obtained by different DIC pre-treatments does not produce any changes of the crystalline phases in the final stabilized materials. Indeed, DIC may influence the combustion efficiency of rice husk: it was shown that the RHA morphology appears more structured with consequences on the surface area. Therefore, at the same temperature, the combustion of organic part may result more efficient in samples pre-treated by DIC, with an increase of the silica concentration.

Work is in progress to evaluate the capability of DIC pre-treatment to increase the calorific efficiency of rice husk.

Figure 6 show some "green composites" realized by using COSMOS-RICE technology, to obtain a new filler material. The new filler was inserted in polyethylene (30% in weight) and polyethylene foils (5% in weight). In addition, it was also employed to obtain tiles, with different colors (from 30% to 50% in weight). The presence of soluble salts may affect the employment of this material in building application. However, it was already shown that it is possible to wash the filler and recover almost pure soluble salts [19].

Figure 6: Pictures of several "green composites" realized by using the material produced with the COSMOS-RICE technology as a filler. It was inserted in polyethylene (30% in weight) and polyethylene foils (5% in weight). In addition, it was also employed to obtain tiles, with different colors (from 30% to 50% in weight).

CONCLUSIONS

This work investigated the efficiency change of the COSMOS-RICE stabilization procedure, by increasing the biochar quantity contained in the RHA and also by a pre-treatment of rice husk, made by DIC technology. The aim is to produce a filler to be employed to produce new composites.

The increased amount of biochar in the RHA, in respect to the silica content, it is shown to have a negative impact in the stabilization of heavy metals.

On the contrary, the capability of DIC pre-treatment to increase the RHA stabilization of heavy metals was shown. This resulted more evident in the RHA samples containing biochar, which were characterized by high values of Zn and Pb in the leaching solutions.

On the basis of these results, it is possible to conclude that the DIC technology, developed to pre-treat food, can also be applied to pre-treat agricultural waste materials. In addition, the pre-treatment of other waste typologies should be investigated.

ACKNOWLEDGMENTS

The authors acknowledge Eng. Marcello Gelfi for SEM analysis and Eng. Nicolò Bontempi for his support in structural characterization. This work was supported by LIFE+ financial instrument of the European Community (LIFE+ 2011 Project ENV/IT/000256).

AUTHOR CONTRIBUTIONS

Laura Benassi, Federica Franchi and Daniele Catina contributed equally during the samples preparation and characterization analyses by means of TXRF and XRD. Flavio Cioffi provided the starting rice husk and rice husk ash exposed to different treatments. Nicola Rodella contributed with the final XRD data elaboration. Laura E. Depero, Laura Borgese and Michela Pasquali helped with the data elaboration. Elza Bontempi supervised the entire work.

REFERENCES

1. Vehlow, J. Air pollution control systems in WtE units: An overview. *Waste Manag.* 2015, *37*, 58–74.
2. United States Environmental Protection Agency (EPA). Clean Air Act Requirements and History. Available online: http://www.epa.gov/air/caa/requirements.html (accessed on 24 August 2015).
3. Zacco, A.; Borgese, L.; Gianoncelli, A.; Struis, R.P.W.J.; Depero, L.E.; Bontempi, E. Review of fly ash inertisation treatments and re cycling. *Environ. Chem. Lett.* 2014, *12*, 153–175.
4. Bosio, A.; Rodella, N.; Gianoncelli, A.; Zacco, A.; Borgese, L.; Depero, L.E.; Bingham, P.A.; Bontempi, E. A new method to inertize incinerator toxic fly ash with silica from rice husk ash. *Environ. Chem. Lett.* 2013, *11*, 329–333.
5. Muthayya, S.; Sugimoto, J.D.; Montgomery, S.; Maberly, G.F. An overview of global rice production, supply, trade, and consumption. *Ann. N. Y. Acad. Sci.* 2014, *1324*, 7–14.
6. World Health Organization. *IARC Working Group on the Evaluation of Carcinogenic Risks to Humans*; IARC Press: Lyon, France, 1997; Volume 68.
7. Shen, Y.; Zhao, P.; Shao, Q. Porous silica and carbon derived materials from rice husk pyrolysis char. *Microporous Mesoporous Mater.* 2014, *188*, 46–76.
8. Bosio, A.; Zacco, A.; Borgese, L.; Rodella, N.; Colombi, P.; Benassi, L.; Depero, L.E.; Bontempi, E. A sustainable technology for Pb and Zn stabilization based on the use of only waste materials: A green chemistry approach to avoid chemicals and promote CO_2 sequestration. *Chem. Eng. J.* 2014, *253*, 377–384.
9. Beesley, L.; Moreno-Jimenez, E.; Gomez-Eyles, J.L.; Harris, E.; Robinson, B.; Sizmur, T. A review of biochars' potential role in the remediation, revegetation and restoration of contaminated soils. *Environ. Pollut.* 2011, *159*, 3269–3282.
10. Guo, Y.; Yang, S.; Yu, K.; Zhao, J.; Wang, Z.; Xu, H. The preparation and mechanism studies of rice husk based porous carbon. *Mater. Chem. Phys.* 2002, *74*, 320–323.
11. Xu, X.; Cao, X.; Zhao, L. Comparison of rice husk- and dairy manure-derived biochars for simultaneously removing heavy metals from aqueous solutions: Role of mineral components in biochars. *Chemosphere* 2013, *92*, 955–961.
12. Mounir, S.; Allaf, T.; Berka, B.; Hassani, A.; Allaf, K. Instant Controlled Pressure Drop technology: From a new fundamental approach of instantaneous transitory thermodynamics to large industrial applications on high performance–high controlled quality unit operations. *C. R. Chim.* 2014, *17*, 261–267.
13. Maritza, A.M.; Sabah, M.; Anaberta, C.M.; Montejano-Gaitán, J.G.; Allaf, K. Comparative Study of Various Drying Processes at Physical and Chemical Properties of Strawberries (Fragariavarcamarosa). *Procedia Eng.* 2012, *42*, 267–282.
14. Kristiawan, M.; Sobolik, V.; Klìma, L.; Allaf, K. Effect of expansion by instantaneous controlled pressure drop on dielectric properties of fruits and vegetables. *J. Food Eng.* 2011, *102*, 361–368.
15. Albitar, N.; Mounir, S.; Besombes, C.; Allaf, K. Improving the drying of onion using the instant controlled pressure drop technology. *Dry. Technol.* 2011, *29*, 993–1001.
16. Louka, N.; Juhel, F.; Allaf, K. Quality studies on various types of partially dried vegetables texturized by Controlled Sudden Decompression: General patterns for the variation of the expansion ratio. *J. Food Eng.* 2004, *65*, 245–253.
17. Téllez-Pérez, C.; Sobolik, V.; Montejano-Gaitán, J.G.; Abdulla, G.; Allaf, K. Impact of swell-drying process on water activity and drying kinetics of moroccan pepper (capsicum annum). *Dry. Technol.* 2015, *33*, 131–142.
18. Pilatowski, I.; Mounir, S.; Haddad, J.; Cong, D.T.; Allaf, K. The instant controlled pressure drop process as a new post-harvesting treatment of paddy rice: Impacts on drying kinetics and end product attributes. *Food Bioprocess Technol.*2010, *3*, 901–907.

19. Bontempi, E.; Zacco, A.; Borgese, L.; Gianoncelli, A.; Ardesi, R.; Depero, L.E. A new method for municipal solid waste incinerator (MSWI) fly ash inertization, based on colloidal silica. *J. Environ. Monit.* 2010, *12*, 2093–2099.

20. Standard CEN EN 12457-2 Characterisation of waste leaching compliance test for leaching of granular waste materials and sludges part 2: One stage batch test at a liquid to solid ratio of 10 L/kg for materials with particle size below 4 mm (without or with size reduction). European Committee for Standardization: Brussels, Belgium, 2002.

21. Yang, J.; Wang, Q.; Luo, Q.; Wang, Q.; Wu, T. Biosorption behavior of heavy metals in bioleaching process of MSWI fly ash by Aspergillus niger. *Biochem. Eng. J.* 2009, *46*, 294–299.

22. Borgese, L.; Zacco, A.; Bontempi, E.; Colombi, P.; Bertuzzi, R.; Ferretti, E.; Tenini, S.; Depero, L.E. Total reflection of X-Ray fluorescence (TXRF): A mature technique for environmental chemical nanoscale metrology. *Meas. Sci. Technol.*2009, *20*.

23. Liou, T.H. Preparation and characterization of nano-structured silica from rice husk. *Mater. Sci. Eng.* 2004, *364*, 313–323.

24. Guerrero, M.; Pilar Ruiz, M.; Millera, A.; Alzueta, M.U.; Bilbao, R. Characterization of Biomass Chars Formed under Different Devolatilization Conditions: Differences between Rice Husk and Eucalyptus. *Energy Fuel* 2008, *22*, 1275–1284.

25. Kercher, A.; Nagle, D.C. Microstructural evolution during charcoal carbonization by X-ray diffraction analysis. *Carbon*2003, *41*, 15–27.

26. Bourke, J.; Manley-Harris, M.; Fushimi, C.; Dowaki, K.; Nunoura, T.; Antal, M.J. Do all carbonized charcoals have the same chemical structure? 2. A model of the chemical structure of carbonized charcoal. *Ind. Eng. Chem. Res.* 2007, *46*, 5954–5967.

27. Ruland, W.; Smarsly, B. X-ray scattering of non-graphitic carbon: An improved method of evaluation. *J. Appl. Crystallogr.* 2002, *35*, 624–633.

28. Rodella, N.; Bosio, A.; Dalipi, R.; Zacco, A.; Borgese, L.; Depero, L.E.; Bontempi, E. Waste silica sources as heavy metal stabilizers for municipal solid waste incineration fly ash. *Arab. J. Chem* 2015. in press.

29. Yalcin, N.; Sevinc, V. Studies on silica obtained from rice husk. *Ceram. Int.* 2001, *27*, 219–224.

30. Wang, W.; Martin, J.C.; Zhang, N.; Ma, C.; Han, A.; Sun, L. Harvesting silica nanoparticles from rice husks. *J. Nanoparticle Res.* 2011, *13*, 6981–6990.

31. Benassi, L.; Bosio, A.; Dalipi, R.; Borgese, L.; Rodella, N.; Pasquali, M.; Depero, L.E.; Bergese, P.; Bontempi, E. Comparison between rice husk ash grown in different regions for stabilizing fly ash from a solid waste incinerator. *J. Environ. Manag.* 2015, *159*, 128–134.

INDEX